Kinematics of General Spatial Mechanical Systems

Kinematics of General Spatial Mechanical Systems

M. Kemal Ozgoren
Middle East Technical University
Turkey

WILEY

This edition first published 2020

Registered Offices
John Wiley & Sons, Inc., 111 River Street, Hoboken, NJ 07030, USA
John Wiley & Sons Ltd, The Atrium, Southern Gate, Chichester, West Sussex, PO19 8SQ, UK

Editorial Office
The Atrium, Southern Gate, Chichester, West Sussex, PO19 8SQ, UK

For details of our global editorial offices, customer services, and more information about Wiley products visit us at www.wiley.com.

Wiley also publishes its books in a variety of electronic formats and by print-on-demand. Some content that appears in standard print versions of this book may not be available in other formats.

Library of Congress Cataloging-in-Publication Data Applied for

Hardback ISBN: 9781119195733

Cover Design: Wiley
Cover Image: © Phonlamai Photo/Shutterstock

Set in 10/12pt WarnockPro by SPi Global, Chennai, India

Printed and bound by CPI Group (UK) Ltd, Croydon, CR0 4YY

10 9 8 7 6 5 4 3 2 1

To all the readers who may enjoy and make use of this book

Contents

Preface

As implied by the title, *Kinematics of General Spatial Mechanical Systems*, this book is concerned mainly with the kinematic description and analysis of spatial mechanical systems such as *serial manipulators*, *parallel manipulators*, and *spatial mechanisms*. However, a planar mechanical system is also considered occasionally, whenever it is helpful in demonstrating and discussing a kinematic feature more neatly and clearly.

This book may be useful and attractive for a wide spectrum of people interested in the area of robotics and mechanisms ranging from students to specialized scholars. They may keep this book as a handy desktop reference book. Besides, this book may even be adopted as an auxiliary or supplementary textbook in some special courses.

This book places the main emphasis on the *analytical* and *semi-analytical* solution methods for the kinematic problems concerning the systems to be studied. An analytical solution is such that all the unknown variables are obtained with closed-form expressions in terms of the known variables. On the other hand, a semi-analytical solution is such that a large number of the unknown variables are again obtained with closed-form expressions in terms of the known and a small number of judiciously selected special unknown variables. Afterwards, these small number of special unknowns are found by solving an equal number of consistency equations by means of a suitable numerical method. Due to the preference of the analytical and semi-analytical solution methods, the purely numerical solution methods are kept beyond the scope of this book.

One of the major advantages of the analytical and semi-analytical solution methods is that the multiplicities and singularities are readily identified as the by-products of the solution procedure. Owing to the closed-form (or mostly closed-form) expressions, the consequences of the multiplicities and singularities can also be studied easily. This way, the motion planning studies are also facilitated.

On the other hand, the analytical and semi-analytical solution methods necessitate that the orientations and locations of the links (i.e. the rigid body members) of the studied system be expressed separately with all the characteristic details shown explicitly as a preparation for the subsequent symbolic manipulations. Therefore, the compact mathematical formulations that combine the orientations and locations into single algebraic entities (such as homogeneous displacement matrices, screws, dual quaternions, etc.) are not favored in this book. Yet, these compact algebraic entities are actually advantageous in many ways. For example, they are indeed very suitable for writing the equations of the kinematic chains and loops compactly and briefly. They are also suitable for the efficient execution of the computations related to those equations. However, they are not very suitable for the detailed symbolic manipulations that are required for the analytical treatments, which constitute the basis of this book.

The main topics that are covered in this book are indicated and briefly explained below.

The Necessary Mathematics of Spatial Kinematics (Chapters 1, 2, 3, and 4):

The relevant concepts, theorems, and formulas are explained and discussed. Additionally, a simple and neat notation is introduced that clearly distinguishes vectors and their column matrix representations in different reference frames. This notation, together with the accompanying algebraic rules, turns out to be very convenient for the symbolic manipulation of the kinematic relationships. Thus, it facilitates obtaining the analytical and semi-analytical solutions.

Kinematic Constituents of a Mechanical System (Chapters 5 and 6):

The links, the kinematic elements on the links, and the joints (i.e. the kinematic pairs formed by the mating kinematic elements of the connected links) are described mathematically by means of the appropriately defined reference frames, the constant geometric parameters of the links and the joints, and the joint variables that describe the relative positions of the mating kinematic elements with respect to each other.

In this book, the links and the kinematic elements of the joints are assumed to be rigid. In other words, the mechanical systems with flexible links and/or flexural joints are not taken into the scope of this book.

Kinematic Formation and Formulation of a Mechanical System (Chapters 5 and 6):

The necessary equations are written in the position, velocity, and acceleration domains in order to describe the kinematic relationships concerning the open, closed, and hybrid kinematic chains, by which the kinematic constituents are interconnected to form a mechanical system.

Kinematic Treatment of Serial Manipulators (Chapters 7, 8, and 9):

The treatment includes the forward kinematic formulations and the inverse kinematic solutions in the position and velocity domains. The results are extended to the acceleration domain, too. The treatment also includes discussions on the multiplicities in the position domain and the analysis of the position and motion singularities. The singularities are discussed considering their consequences in the task and joint spaces. The singularity analysis suggests certain compatibility conditions on the planned motion of the end-effector. These compatibility conditions, if obeyed, eliminate the necessity of avoiding the singularities. On the contrary, without avoiding the singularities, it becomes possible to execute certain tasks, which could not be executed otherwise.

Kinematic Treatment of Parallel Manipulators (Chapter 10):

The treatment includes the forward and inverse kinematic solutions in the position and velocity domains. The results may be extended, if desired, to the acceleration domain, too. Concerning the multiplicities in the position domain, the treatment includes discussions on the two different sets of posture multiplicities associated with the forward and inverse kinematic solutions. Concerning the singularities, the treatment includes the analysis of four different types of singularities, each of which is designated by one of the following phrases: *position singularity of forward kinematics*, *motion singularity of forward kinematics*, *position singularity of inverse kinematics*, and *motion singularity of inverse kinematics*. The singularity analysis shows that the manipulator becomes uncontrollable through its actuated joints in the position and motion singularities of forward kinematics. Therefore, the singularities of forward kinematics must be avoided. The singularity analysis also shows that the manipulator remains controllable through its actuated joints in the position and motion singularities of inverse kinematics, provided that the desired motion of the end-effector be specified according to certain restrictive compatibility conditions. Therefore, the singularities of inverse kinematics need not be avoided, if the restricted motion of the end-effector is acceptable or desirable for the task to be executed.

Kinematic Treatment of the Mechanisms with Simple Contact Joints (Chapters 5 and 6): In the position domain, the treatment includes the identification of the independent loops, writing the corresponding loop closure equations, and then solving them to obtain the unspecified joint variables as functions of the specified ones. The position domain treatment also includes discussions on the multiple solutions and the position singularities associated with the specified joint variables. In the velocity and acceleration domains, the treatment includes deriving the velocity and acceleration constraint equations and solving them to obtain the unspecified joint variable rates in terms of the specified ones. The velocity and acceleration analyses also include discussions on the motion singularities associated with the specified joint variables. There are several examples of such mechanisms in Chapters 5 and 6.

Kinematic Treatment of the Mechanisms with Rolling Contact Joints (Chapter 6): The rolling contact joints, i.e. the gear and cam joints, need somewhat different treatment as compared with the other simple contact joints. In particular, the mechanisms that contain cam joints with sticking friction, i.e. with the rolling-without-slipping property, happen to be non-holonomic systems that have different degrees of freedom in the position and velocity domains. Therefore, the mechanisms that involve rolling contact joints with the rolling-without-slipping property are treated differently in the order of obtaining the kinematic solutions in the position and velocity domains. For such mechanisms, the kinematic solution is first obtained in the velocity domain and then the corresponding kinematic solution in the position domain is obtained by means of a subsequent numerical integration. The most typical samples of such mechanisms are the mechanisms that involve gear joints, because the gear joints are kinematically equivalent to the cylindrical or conical cam joints with the rolling-without-slipping property. A typical example of such a cam mechanism is presented in Chapter 6. The kinematic solutions of that cam mechanism in the position and velocity domains are obtained both in the cases of rolling with slipping and rolling without slipping.

Readers are encouraged and sincerely welcome to contact me at ozgoren@metu.edu.tr with regard to feedback, suggestions, and questions.

Ankara, March 2019

M. Kemal Ozgoren

Acknowledgments

This book is based on the graduate courses "Advanced Dynamics" and "Principles of Robotics" that I have been teaching for a long time. Therefore, many students of these courses and several colleagues involved with related subjects have so far indicated their valuable opinions and encouraged me to write such a book. I would like to express my thanks to all of them and also to the Mechanical Engineering Department of Middle East Technical University that has provided the excellent academic medium and all the conveniences for me in developing and teaching the mentioned courses.

I would also like to express my thanks to the reviewers for their constructive comments and to all the related personnel of Wiley for their highly appreciated efforts during all the time from the initiation to the finalization of this book.

List of Commonly Used Symbols, Abbreviations, and Acronyms

Symbols Based on the Latin Alphabet

$\vec{a}$	Acceleration of an implied point (in a general use)
$\vec{a}_P$	Acceleration of a point P with respect to an implied reference frame
$\vec{a}_{P/\mathcal{F}_a(Q)}$	Acceleration of a point P with respect to a reference frame $\mathcal{F}_a(Q)$; $\vec{a}_{P/\mathcal{F}_a(Q)} = \vec{a}_{P/Q/\mathcal{F}_a} = D_a^2\vec{r}_{P/Q}$
b_k	Effective length of $\mathcal{L}_k$ between $\mathcal{J}_k$ and $\mathcal{J}_{k+1}$ along $\mathcal{N}_k$
$\mathcal{B}_a$	A rigid body denoted by the index a and represented by the frame $\mathcal{F}_a(O_a)$
$\hat{C}$	Orientation matrix of an implied body or frame (in a general use)
$\hat{C}$	Orientation matrix of the end-effector ($\mathcal{L}_m$) with respect to the base frame; $\hat{C} = \hat{C}_m = \hat{C}^{(0,m)} = [\overline{u}_n\ \overline{u}_s\ \overline{u}_a]$
$\hat{C}^{(a,b)}$	CTM that transforms components from $\mathcal{F}_b$ to $\mathcal{F}_a$; $\overline{r}^{(a)} = \hat{C}^{(a,b)}\overline{r}^{(b)}$
$\hat{C}^{(a,b)}$	Orientation matrix of $\mathcal{F}_b$ with respect to $\mathcal{F}_a$; $\hat{C}^{(a,b)} = \hat{R}_{ab}^{(a)} = \hat{R}_{ab}^{(b)} = \hat{R}_{b/a}^{(a)} = \hat{R}_{b/a}^{(b)}$
$\hat{C}^{(k-1,k)}$	Orientation matrix of $\mathcal{F}_k(O_k)$ with respect to $\mathcal{F}_{k-1}(O_{k-1})$
$\hat{C}_k$	Orientation matrix of $\mathcal{F}_k(O_k)$ with respect to $\mathcal{F}_0(O_0)$; $\hat{C}_k = \hat{C}^{(0,k)}$
$\overline{c}$	A general column matrix; $\overline{c} = c_1\overline{u}_1 + c_2\overline{u}_2 + c_3\overline{u}_3$
$\overline{c}^t$	A general row matrix, i.e. the transpose of $\overline{c}$
$\tilde{c}$	Skew symmetric matrix generated from $\overline{c}$; $\tilde{c} = \mathrm{ssm}(\overline{c})$
D_a	Vector differentiator with respect to $\mathcal{F}_a$; $D_a\vec{r} = (d\vec{r}/dt)\|_{\mathcal{F}_a}$
d_k	Constant offset of $\mathcal{N}_k$ with respect to $\mathcal{N}_{k-1}$ if $\mathcal{J}_k$ is revolute
d_m	Tip point offset; $d_m = RP = O_{m-1}O_m$
E	Elbow point of a manipulator
$e^{\tilde{n}^{(a)}\theta}$	Exponential form of $\hat{R}(\overline{n}^{(a)}, \theta)$
$e^{\tilde{u}_k\theta}$	Exponential form of $\hat{R}_k(\theta) = \hat{R}(\overline{u}_k, \theta)$
$\mathcal{E}_{ab}$	Kinematic element of $\mathcal{B}_a$ that mates with $\mathcal{E}_{ba}$ of $\mathcal{B}_b$
$f_{ab}(x, y, z)$	Surface function that describes the surface $\mathcal{S}_{ab}$ in the link frame $\mathcal{F}_{ab}$
$\mathcal{F}_a$	A reference frame with an orientation index a, whose origin is implied
$\mathcal{F}_a(Q)$	A reference frame $\mathcal{F}_a$ with a specific origin Q
$\mathcal{F}_k(O_k)$	Reference frame attached to $\mathcal{L}_k$
$\mathcal{F}_0(O_0)$	Base frame, i.e. the reference frame attached to the base link $\mathcal{L}_0$

$\vec{g}_{ab}$	Gradient vector of the surface $\mathcal{S}_{ab}$
$\overline{g}_{ab}^{(ab)}$	Column matrix representation of $\vec{g}_{ab}$ in the link frame $\mathcal{F}_{ab}$
$\hat{H}$	Abbreviation for $\hat{H}_m = \hat{H}^{(0,m)}$ of the last link $\mathcal{L}_m$
$\hat{H}^{(k-1,k)}$	HTM that represents the displacement from $\mathcal{F}_{k-1}(O_{k-1})$ to $\mathcal{F}_k(O_k)$
$\hat{H}_{AB}^{(a,b)}$	HTM that transforms coordinates from $\mathcal{F}_b(B)$ to $\mathcal{F}_a(A)$; $\overline{R}_{AP}^{(a)} = \hat{H}_{AB}^{(a,b)}\overline{R}_{BP}^{(b)}$
$\hat{H}_{ab}^{(ab)}$	Hessian matrix of the surface $\mathcal{S}_{ab}$ expressed in the link frame $\mathcal{F}_{ab}$
$\hat{H}_k$	Abbreviation for $\hat{H}^{(0,k)}$ of the link $\mathcal{L}_k$
$\hat{I}$	Identity matrix; $\hat{I} = [\overline{u}_1\ \overline{u}_2\ \overline{u}_3] = \overline{u}_1\overline{u}_1^t + \overline{u}_2\overline{u}_2^t + \overline{u}_3\overline{u}_3^t$
j_k	Number of joints with k degrees of relative freedom
$\hat{J}$	A general Jacobian matrix
$\mathcal{J}_{ab}$	The joint (i.e. kinematic pair) between $\mathcal{B}_a$ and $\mathcal{B}_b$; $\mathcal{J}_{ba} = \mathcal{J}_{ab}$
$\mathcal{J}_k$	Joint between $\mathcal{L}_{k-1}$ and $\mathcal{L}_k$
$\hat{J}_P$	Tip point Jacobian matrix
$\hat{J}_R$	Wrist point Jacobian matrix
L_k	kth leg (or limb) of a parallel manipulator
$\mathcal{L}_a$	A link denoted by the index a
$\mathcal{L}_k$	kth link of a manipulator
$\mathcal{L}_m$	The last link (i.e. end-effector) of a serial manipulator with m links
$\hat{M}$	A general square matrix
m	Number of links and joints of a serial manipulator
$\vec{n}$	A general unit vector
$\overline{n}_k$	Unit column matrix that represents the twisted axis of $\mathcal{J}_k$; $\overline{n}_k = e^{\tilde{u}_1\gamma_k}\overline{u}_3$
n_L	Number of legs (or limbs) of a parallel manipulator
n_{ikl}	Number of independent kinematic loops
n_{kpm}	Number of distinct posture modes of the leg L_k
n_m	Number of moving or movable bodies
n_{pv}	Number of primary variables; $n_{pv} = \mu$
n_{pm}	Number of distinct posture modes of a manipulator
n_{sv}	Number of secondary variables; $n_{sv} = \lambda n_{ikl}$
n_v	Number of variables needed to describe the pose of a mechanical system
$\mathcal{N}_k$	Common normal between the axes of $\mathcal{J}_k$ and $\mathcal{J}_{k+1}$
O	Origin of the base frame $\mathcal{F}_0(O)$; $O = O_0$
O_k	Origin of the reference frame $\mathcal{F}_k(O_k)$
P	An arbitrary point (in a general use)
P	Tip point of a manipulator; $P = O_m$ for a serial manipulator
$\vec{p}$	Tip point position vector with respect to the base frame; $\vec{p} = \vec{r}_P = \vec{r}_{P/O}$
$\overline{p}$	Column matrix representation of $\vec{p}$ in the base frame; $\overline{p} = \overline{p}^{(0)}$
Q_{ab}	Contact point on the surface $\mathcal{S}_{ab}$
$\overline{q}$	Column matrix of the joint variables
q_k	Generalized joint variable of $\mathcal{J}_k$; $q_k = \theta_k$ or $q_k = s_k$
R	Wrist point of a manipulator; $R = O_{m-1}$ for a serial manipulator

$\hat{R}(\overline{n}^{(a)}, \theta)$	Matrix representation of the rotation operator rot($\vec{n}, \theta$) in a frame $\mathcal{F}_a$
$\hat{R}^{(c)}_{ab}$	Matrix representation of the rotation operator rot(a, b) in a frame $\mathcal{F}_c$
$\hat{R}_k(\theta)$	kth basic rotation matrix; $\hat{R}_k(\theta) = \hat{R}(\overline{u}_k, \theta) = e^{\tilde{u}_k\theta}$
$\overline{R}^{(a)}_{AP}$	Augmented position matrix of a point P with respect to a frame $\mathcal{F}_a(A)$
$\vec{r}$	Position vector of an implied point (in a general use)
$\vec{r}$	Wrist point position vector with respect to the base frame; $\vec{r} = \vec{r}_R = \vec{r}_{R/O}$
$\overline{r}$	Column matrix representation of $\vec{r}$ in the base frame; $\overline{r} = \overline{r}^{(0)}$
$\overline{r}^{(a)}$	Column matrix representation of a vector $\vec{r}$ in a frame $\mathcal{F}_a$; $\overline{r}^{(a)} = [\vec{r}]^{(a)}$
$[\vec{r}]^{(a)}$	Column matrix representation of a vector $\vec{r}$ in a frame $\mathcal{F}_a$; $[\vec{r}]^{(a)} = \overline{r}^{(a)}$
$r^{(a)}_k$	kth component of a vector $\vec{r}$ in a frame $\mathcal{F}_a$; $r^{(a)}_k = \vec{r} \cdot \vec{u}^{(a)}_k$
$\vec{r}_{k-1,k}$	Displacement vector from the origin O_{k-1} to the origin O_k
$\overline{r}^{(k-1)}_{k-1,k}$	Column matrix representation of $\vec{r}_{k-1,k}$ in the link frame $\mathcal{F}_{k-1}(O_{k-1})$
$\vec{r}_P$	Position vector of a point P with respect to an implied point
$\vec{r}_{P/Q}$	Position vector of a point P with respect to a point Q; $\vec{r}_{P/Q} = \vec{r}_{QP}$
$\vec{r}_{QP}$	Displacement vector from a point Q to a point P; $\vec{r}_{QP} = \vec{r}_{P/Q}$
s_k	Joint variable (sliding displacement) of the prismatic joint $\mathcal{J}_k$
s_k	Variable offset of $\mathcal{N}_k$ with respect to $\mathcal{N}_{k-1}$ if $\mathcal{J}_k$ is prismatic
S	Shoulder point of a manipulator
$\mathcal{S}_{ab}$	Surface of the kinematic element $\mathcal{E}_{ab}$
$\mathcal{S}_J$	Joint space of a manipulator
$\mathcal{S}_T$	Task space of a manipulator
$\vec{u}_a$	Approach vector of the end-effector; $\overline{u}_a = \overline{u}^{(0)}_a = \hat{C}\overline{u}_3$
$\overline{u}_k$	kth basic column matrix; $\overline{u}_k = \overline{u}^{(a/a)}_k = \overline{u}^{(b/b)}_k = \ldots = \overline{u}^{(z/z)}_k$
$\vec{u}^{(b)}_k$	kth unit basis vector of a frame $\mathcal{F}_b$
$\overline{u}^{(b/a)}_k$	Column matrix representation of $\vec{u}^{(b)}_k$ in a frame $\mathcal{F}_a$; $\overline{u}^{(b/a)}_k = [\vec{u}^{(b)}_k]^{(a)}$
$\vec{u}_n$	Normal vector of the end-effector; $\overline{u}_n = \overline{u}^{(0)}_n = \hat{C}\overline{u}_1$
$\vec{u}_s$	Side vector of the end-effector; $\overline{u}_s = \overline{u}^{(0)}_s = \hat{C}\overline{u}_2$
$\vec{u}^{(k)}_1$	Unit vector along the common normal $\mathcal{N}_k$
$\vec{u}^{(k)}_3$	Unit vector along the axis of $\mathcal{J}_k$
$\mathcal{U}_a$	Basis vector triad of a reference frame $\mathcal{F}_a$; $\mathcal{U}_a = \{\vec{u}^{(a)}_1, \vec{u}^{(a)}_2, \vec{u}^{(a)}_3\}$
$\mathcal{U}_k$	Basis vector triad of the link frame $\mathcal{F}_k(O_k)$; $\mathcal{U}_k = \{\vec{u}^{(k)}_1, \vec{u}^{(k)}_2, \vec{u}^{(k)}_3\}$
$\overline{V}_k$	Tip point velocity influence coefficient due to $\dot{q}_k$; $\overline{V}_k = \partial\overline{v}/\partial\dot{q}_k$
$\vec{v}$	Velocity of an implied point (in a general use)
$\vec{v}$	Tip point velocity with respect to the base frame; $\vec{v} = \vec{v}_P = \vec{v}_{P/\mathcal{F}_o(O)}$
$\overline{v}$	Column matrix representation of $\vec{v}$ in the base frame; $\overline{v} = \overline{v}^{(0)}$
$\vec{v}_P$	Velocity of a point P with respect to an implied reference frame
$\vec{v}_{P/\mathcal{F}_a(Q)}$	Velocity of a point P with respect to a reference frame $\mathcal{F}_a(Q)$; $\vec{v}_{P/\mathcal{F}_a(Q)} = \vec{v}_{P/Q/\mathcal{F}_a} = D_a\vec{r}_{P/Q}$
$\overline{W}_k$	Wrist point velocity influence coefficient due to $\dot{q}_k$; $\overline{W}_k = \partial\overline{w}/\partial\dot{q}_k$
$\vec{w}$	Wrist point velocity with respect to the base frame; $\vec{w} = \vec{v}_R = \vec{v}_{R/\mathcal{F}_o(O)}$
$\overline{w}$	Column matrix representation of $\vec{w}$ in the base frame; $\overline{w} = \overline{w}^{(0)}$

$\vec{w}_{abl}$	Generatrix line vector of a conical surface $\mathcal{S}_{ab}$
$\overline{w}_{abl}^{(ab)}$	Column matrix representation of $\vec{w}_{abl}$ in the link frame $\mathcal{F}_{ab}$
$\overline{x}$	Column matrix of the primary variables
$\overline{y}$	Column matrix of the secondary variables
$\overline{z}$	Column matrix of the primary and secondary variables

Symbols Based on the Greek Alphabet

$\vec{\alpha}$	Angular acceleration of an implied body or frame (in a general use)
$\vec{\alpha}_b$	Angular acceleration of $\mathcal{B}_b$ or $\mathcal{F}_b$ with respect to an implied frame
$\vec{\alpha}_{b/a}$	Angular acceleration of $\mathcal{B}_b$ or $\mathcal{F}_b$ with respect to $\mathcal{F}_a$
β_k	Twist angle of $\mathcal{J}_k$ with respect to $\mathcal{J}_{k-1}$
γ_k	Cumulative twist angle of $\mathcal{J}_k$ with respect to $\mathcal{J}_1$; $\gamma_k = \beta_1 + \beta_2 + \ldots + \beta_k$
$\overline{\gamma}_{ab}^{(ab)}$	Modified gradient of a conical surface expressed in the link frame $\mathcal{F}_{ab}$
δ_{ij}	Kronecker Delta function of the indices i and j
δ_k	Constant rotation angle of $\mathcal{N}_k$ with respect to $\mathcal{N}_{k-1}$ if $\mathcal{J}_k$ is prismatic
ε_{ijk}	Levi-Civita Epsilon function of the indices i, j, and k
$\overline{\eta}$	Column matrix of the end-effector velocity state in the task space $\mathcal{S}_T$
θ_k	Joint variable (angular displacement) of the revolute joint $\mathcal{J}_k$
θ_k	Variable rotation angle of $\mathcal{N}_k$ with respect to $\mathcal{N}_{k-1}$ if $\mathcal{J}_k$ is revolute
λ	DoF of the working or operational space of a mechanical system
λ_{ab}	Gradient ratio between $\vec{g}_{ab}$ and $\vec{g}_{ba}$; $\vec{g}_{ba} = -\lambda_{ab}\vec{g}_{ab}$
μ	Mobility (DoF) of a system in its working or operational space
μ_{ab}	Relative mobility (DoF) of $\mathcal{E}_{ba}$ with respect to $\mathcal{E}_{ab}$; $\mu_{ba} = \mu_{ab}$
$\overline{\rho}$	Column matrix of the end-effector position in the task space $\mathcal{S}_T$
σ	A general sign variable; $\sigma = \pm 1$
σ_k, σ_k'	Sign variables that indicate multiple solutions
σ_{ijk}	Cross product sign variable defined for the indices i, j, and k
$\widehat{\Phi}_k$	Effective orientation matrix of $\mathcal{L}_k$; $\widehat{C}^{(0,k)} = \widehat{C}_k = \widehat{\Phi}_k e^{\tilde{u}_1\gamma_k}$
$\overline{\Omega}_k$	End-effector angular velocity influence coefficient due to $\dot{q}_k$; $\overline{\Omega}_k = \partial\overline{\omega}/\partial\dot{q}_k$
$\vec{\omega}$	Angular velocity of an implied body or frame (in a general use)
$\vec{\omega}$	Angular velocity of the end-effector with respect to the base frame; $\vec{\omega} = \vec{\omega}_m$
$\overline{\omega}$	Column matrix representation of $\vec{\omega}$ in the base frame; $\overline{\omega} = \overline{\omega}^{(0)}$
$\vec{\omega}_b$	Angular velocity of $\mathcal{B}_b$ or $\mathcal{F}_b$ with respect to an implied frame
$\vec{\omega}_{b/a}$	Angular velocity of $\mathcal{B}_b$ or $\mathcal{F}_b$ with respect to $\mathcal{F}_a$
$\vec{\omega}_k$	Angular velocity of $\mathcal{L}_k$ with respect to the base frame: $\vec{\omega}_k = \vec{\omega}_{k/0}$

Abbreviations

$\mathrm{ang}(\vec{p}, \vec{q})$	Angle between the vectors $\vec{p}$ and $\vec{q}$
$\mathrm{atan}_2(y, x)$	Double argument arctangent function; $\theta \equiv \mathrm{atan}_2(r\sin\theta, r\cos\theta)$, $r > 0$
$\mathrm{colm}(\widetilde{c})$	Column matrix generator from a skew symmetric matrix; $\overline{c} = \mathrm{colm}(\widetilde{c})$

$\mathrm{cpm}(\overline{c})$	Cross product matrix generator from a column matrix; $\widetilde{c} = \mathrm{cpm}(\overline{c}) = \mathrm{ssm}(\overline{c})$
$\mathrm{dir}(\vec{v})$	Direction of a vector $\vec{v}$
$\mathrm{mag}(\vec{v})$	Magnitude of a vector $\vec{v}$; $\mathrm{mag}(\vec{v}) = \lvert\vec{v}\rvert$
$\mathrm{rot}(a, b)$	Rotation operator that rotates a frame $\mathcal{F}_a$ into another frame $\mathcal{F}_b$
$\mathrm{rot}(\vec{n}, \theta)$	Rotation operator of an angle θ about an axis parallel to a unit vector $\vec{n}$
$\mathrm{ssm}(\overline{c})$	Skew symmetric matrix generator from a column matrix; $\widetilde{c} = \mathrm{ssm}(\overline{c})$

Acronyms

C	Cylindrical Joint
CTM	Component Transformation Matrix
CPM	Cross Product Matrix
DCM	Direction Cosine Matrix
DoF	Degree of Freedom
D-H	Denavit-Hartenberg
HTM	Homogeneous Transformation Matrix
IFB	Initial Frame Based
IKL	Independent Kinematic Loop
MSFK	Motion Singularity of Forward Kinematics
MSIK	Motion Singularity of Inverse Kinematics
P	Prismatic Joint
PM	Posture Mode
PML	Posture Mode of a Leg
PMCP	Posture Mode Changing Pose
PMCPL	Posture Mode Changing Pose of a Leg
PMFK	Posture Multiplicity of Forward Kinematics
PMIK	Posture Multiplicity of Inverse Kinematics
PSFK	Position Singularity of Forward Kinematics
PSIK	Position Singularity of Inverse Kinematics
R	Revolute Joint
RFB	Rotated Frame Based
S	Spherical Joint
SSM	Skew Symmetric Matrix
TM	Transformation Matrix
U	Universal Joint

About the Companion Website

This book is accompanied by a companion website:

www.wiley.com/go/ozgoren/spatialmechanicalsystems

The website includes:

- A communication medium with the readers
- Solved problems as additional examples
- Unsolved problems as typical exercises

Scan this QR code to visit the companion website.

1

Vectors and Their Matrix Representations in Selected Reference Frames

Synopsis

The main purpose of this chapter is to review the mathematics associated with the vectors and their matrix representations in selected reference frames. This review is expected to be beneficial for the efficient readability of this book. It will also familiarize the reader with the special notation that is used throughout this book. This notation is suitable not only because it can distinguish vectors from their matrix representations, but it can also be used conveniently in both printed texts and handwritten work.

This chapter also explains why and shows how the vectors are treated in this book as mathematical objects that are distinct from the column matrices that represent them in selected reference frames. As the main distinction, the vectors are independent of any reference frame, whereas their matrix representations are necessarily dependent on the selected reference frames. Similarly, a vector equation can be written without indicating any reference frame, whereas the selected reference frame must be indicated for the corresponding matrix equation.

1.1 General Features of Notation

This section gives general information about the special notation that is used throughout the book. This notation is convenient because it can be used not only in printed texts but also in handwritten work. It also has the desirable feature that it can distinguish column matrices from vectors, which are actually different mathematical objects. The main features of the notation are explained below.

A *scalar* is denoted by a plain letter such as s.
A *vector* is denoted by a letter with an *overhead arrow* such as $\vec{v}$.
A *column matrix* is denoted by a letter with an *overhead bar* such as $\bar{c}$.
A *square* or a *rectangular matrix* is denoted by a capital letter with an *overhead circumflex* (a.k.a. *hat*) such as $\hat{M}$.
A *skew symmetric matrix* is denoted by a letter with an *overhead tilde* such as $\tilde{w}$.
The *transpose of a matrix* is denoted by a *superscript t* such as $\bar{c}^t$ and $\hat{M}^t$.

Kinematics of General Spatial Mechanical Systems, First Edition. M. Kemal Ozgoren.

Companion Website: www.wiley.com/go/ozgoren/spatialmechanicalsystems

1.2 Vectors

1.2.1 Definition and Description of a Vector

A vector is defined as an entity that can be described by a *magnitude* and a *direction*. In this book, it is assumed that all the vectors belong to the three-dimensional *Euclidean space*.

The *magnitude* of a vector $\vec{v}$ is denoted as shown below.

$$|\vec{v}| = \text{mag}(\vec{v}) \tag{1.1}$$

A *unit vector* such as $\vec{n}$ is defined so that its magnitude is unity. That is,

$$|\vec{n}| = \text{mag}(\vec{n}) = 1 \tag{1.2}$$

A vector $\vec{v}$ can be expressed as follows by means of a *unit vector* $\vec{n}$, which is introduced to indicate the *direction* of $\vec{v}$.

$$\vec{v} = v\vec{n} \tag{1.3}$$

In Eq. (1.3), v is defined as the *scalar value* of $\vec{v}$ with respect to $\vec{n}$.

Note that the magnitude of a vector is the *absolute value* of its scalar value. That is,

$$|\vec{v}| = |v| \tag{1.4}$$

Note also that the scalar value v can be positive, negative, or zero, but the magnitude $|\vec{v}|$ can only be positive or zero.

The sign variability of the scalar value is demonstrated in the following equation.

$$\vec{v} = v\vec{n} = (-v)(-\vec{n}) \tag{1.5}$$

According to Eq. (1.5), the scalar values of the same vector $\vec{v}$ with respect to $\vec{n}$ and $\vec{n}' = -\vec{n}$ are v and $v' = -v$, respectively.

1.2.2 Equality of Vectors

Two vectors $\vec{v} = v\vec{n}$ and $\vec{w} = w\vec{m}$ are defined to be *equal*, i.e. $\vec{v} = \vec{w}$, if they satisfy the following equations simultaneously, in which $|\vec{n}| = |\vec{m}| = 1$.

$$v = \sigma w \tag{1.6}$$

$$\text{dir}(\vec{n}) = \sigma \text{dir}(\vec{m}) \tag{1.7}$$

In Eqs. (1.6) and (1.7), σ is a sign variable such that

$$\sigma = \pm 1 \tag{1.8}$$

In Eq. (1.7), the notation $\text{dir}(\vec{v})$ indicates the direction of the vector $\vec{v}$. Equation (1.7) implies the following two situations for the unit vectors $\vec{n}$ and $\vec{m}$.

If $\sigma = +1$, $\vec{n}$ and $\vec{m}$ are *codirectional*, i.e. either coincident or parallel with the same direction.

If $\sigma = -1$, $\vec{n}$ and $\vec{m}$ are *opposite*, i.e. either coincident or parallel with opposite directions.

1.2.3 Opposite Vectors

Two vectors $\vec{v} = v\vec{n}$ and $\vec{w} = w\vec{m}$ are defined to be *opposite*, i.e. $\vec{v} = -\vec{w}$, if they are related to each other in either of the following ways.

$$\left.\begin{aligned} v &= -w \text{ but } \mathrm{dir}(\vec{n}) = \mathrm{dir}(\vec{m}) \\ v &= w \text{ but } \mathrm{dir}(\vec{n}) = -\mathrm{dir}(\vec{m}) \end{aligned}\right\} \tag{1.9}$$

1.3 Vector Products

1.3.1 Dot Product

The *dot product* (a.k.a. *scalar product*) of two vectors $\vec{p}$ and $\vec{q}$ is denoted and defined as follows:

$$s = \vec{p} \cdot \vec{q} = |\vec{p}||\vec{q}| \cos\theta_{pq} = |pq| \cos\theta_{pq} \tag{1.10}$$

In Eq. (1.10), θ_{pq} is defined as the angle between the vectors $\vec{p}$ and $\vec{q}$. It is denoted as

$$\theta_{pq} = \sphericalangle(\vec{p}, \vec{q}) = \mathrm{ang}(\vec{p}, \vec{q}) \tag{1.11}$$

Without any significant loss of generality, the range of θ_{pq} may be defined so that $0 \le \theta_{pq} \le \pi$. According to this range definition, it happens that

$$\theta_{qp} = \theta_{pq} \tag{1.12}$$

Besides, $\cos\theta_{pq}$ is not sensitive to the sense of θ_{pq} anyway. Therefore, the order of the vectors in the dot product is immaterial. That is,

$$\vec{q} \cdot \vec{p} = \vec{p} \cdot \vec{q} = s \tag{1.13}$$

If $\vec{q} \perp \vec{p}$, i.e. if $\vec{q}$ is perpendicular (or orthogonal or normal) to $\vec{p}$ so that $\theta_{pq} = \pi/2$, then $s = \vec{p} \cdot \vec{q} = 0$.

If $\vec{q} = \vec{p}$, i.e. if $|q| = |p|$ and $\theta_{pq} = \theta_{pp} = 0$, then $s = \vec{p} \cdot \vec{p} = |p|^2 = p^2$. Hence, the magnitude of a vector $\vec{v}$ can also be expressed as

$$|\vec{v}| = \sqrt{\vec{v} \cdot \vec{v}} \tag{1.14}$$

1.3.2 Cross Product

The *cross product* (a.k.a. *vector product*) of two vectors $\vec{p}$ and $\vec{q}$ is denoted and defined as follows:

$$\vec{r} = \vec{p} \times \vec{q} = \vec{n}_{pq}|\vec{p}||\vec{q}| \sin\theta_{pq} = \vec{n}_{pq}|pq| \sin\theta_{pq} \tag{1.15}$$

In Eq. (1.15), as defined before, θ_{pq} is the angle measured from $\vec{p}$ to $\vec{q}$. As for $\vec{n}_{pq}$, it is defined as a *unit vector*, which is perpendicular to the plane $\mathcal{P}_{pq}$ formed by the vectors $\vec{p}$ and $\vec{q}$.

If $\vec{p}$ and $\vec{q}$ are skew (nonparallel) vectors, then $\mathcal{P}_{pq}$ is formed by imagining that $\vec{p}$ and $\vec{q}$ are translated toward each other until they are connected tail-to-tail. If $\vec{p}$ and $\vec{q}$ are parallel (but not coincident) vectors, then $\mathcal{P}_{pq}$ happens to be the plane that contains them. However, if $\vec{p}$ and $\vec{q}$ are coincident vectors, then $\mathcal{P}_{pq}$ cannot be formed as a definite plane, i.e. it can be any plane that contains them.

The sense of $\vec{n}_{pq}$ is defined conventionally by the *right-hand rule.* This rule is based on the right hand in such a way that $\vec{n}_{pq}$ assumes the orientation of the thumb (directed from root to tip) while the fingers are oriented from $\vec{p}$ to $\vec{q}$.

Since, by definition, $\vec{n}_{pq} \perp \mathcal{P}_{pq}$ and $|\vec{n}_{pq}| = 1$, the following equations can be written for the vectors involved in the cross product.

$$|\vec{r}| = |\vec{p}||\vec{q}||\sin\theta_{pq}| \;\Rightarrow\; |r| = |pq\sin\theta_{pq}| \tag{1.16}$$

$$\vec{r}\cdot\vec{p} = \vec{r}\cdot\vec{q} = 0 \tag{1.17}$$

If $\sin\theta_{pq} = 0$, i.e. if $\text{dir}(\vec{q}) = \text{dir}(\vec{p})$ with $\theta_{pq} = 0$ or $\text{dir}(\vec{q}) = -\text{dir}(\vec{p})$ with $\theta_{pq} = \pi$, then

$$\vec{p}\times\vec{q} = \vec{0} \tag{1.18}$$

If the order of $\vec{p}$ and $\vec{q}$ is reversed, Eq. (1.15) becomes

$$\vec{r}' = \vec{q}\times\vec{p} = \vec{n}_{qp}|\vec{q}||\vec{p}|\sin\theta_{qp} = \vec{n}_{qp}|qp|\sin\theta_{qp} \tag{1.19}$$

According to Eq. (1.12), $\theta_{qp} = \theta_{pq}$. However, according to the right-hand rule,

$$\vec{n}_{qp} = -\vec{n}_{pq} \tag{1.20}$$

Therefore, $\vec{r}' = -\vec{r}$. This verifies the well-known characteristic feature of the cross product that its outcome changes sign when the order of its multiplicands is reversed. That is,

$$\vec{q}\times\vec{p} = -\vec{p}\times\vec{q} \tag{1.21}$$

1.4 Reference Frames

In the three-dimensional Euclidean space, a *reference frame* is defined as an entity that consists of an *origin* and three distinct noncoplanar axes emanating from the origin. The origin is a specified point and the axes have specified orientations. More specifically, the axes of a reference frame are called its *coordinate axes.* For the sake of verbal brevity, a

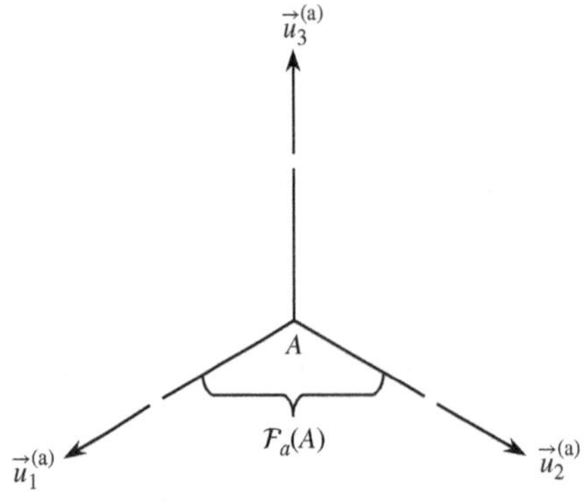

Figure 1.1 A reference frame.

reference frame may sometimes be called simply a *frame*. A reference frame, such as the one shown in Figure 1.1, may be denoted in one of the following ways, which convey different amounts of information about its specific features.

$$\mathcal{F}_a = \mathcal{F}_a(A) = \mathcal{F}_a[A; \vec{u}_1^{(a)}, \vec{u}_2^{(a)}, \vec{u}_3^{(a)}] \tag{1.22}$$

In Eq. (1.22), A is the origin of $\mathcal{F}_a$. The origin of $\mathcal{F}_a$ may also be denoted as O_a. The coordinate axes of $\mathcal{F}_a$ are oriented so that each of them is aligned with one member of the following set of three vectors, which is denoted as $\mathcal{U}_a$ and defined as the *basis vector triad* of $\mathcal{F}_a$.

$$\mathcal{U}_a = \{\vec{u}_k^{(a)} : k = 1, 2, 3\} \tag{1.23}$$

All the reference frames that are used in this book are selected to be *orthonormal, right-handed,* and *equally scaled* on their axes.

A reference frame, say $\mathcal{F}_a$, is defined to be *orthonormal* if its basis vectors are mutually *orthogonal* and each of them is a *unit vector,* i.e. a vector *normalized* to unit magnitude. The orthonormality of $\mathcal{F}_a$ can be expressed by the following set of equations that are obeyed by its basis vectors for all $i \in \{1, 2, 3\}$ and $j \in \{1, 2, 3\}$.

$$\vec{u}_i^{(a)} \cdot \vec{u}_j^{(a)} = \delta_{ij} = \begin{cases} 1 & \text{if } i = j \\ 0 & \text{if } i \neq j \end{cases} \tag{1.24}$$

In Eq. (1.24), δ_{ij} is defined as the *dot product index function,* which is also known as the *Kronecker delta function* of the indices i and j.

A reference frame, say $\mathcal{F}_a$, is defined to be *right-handed* if its basis vectors obey the following set of equations for $i \in \{1, 2, 3\}$, $j \in \{1, 2, 3\}$, and $k \in \{1, 2, 3\}$.

$$\vec{u}_i^{(a)} \times \vec{u}_j^{(a)} = \sum_{k=1}^{3} \varepsilon_{ijk} \vec{u}_k^{(a)} \tag{1.25}$$

In Eq. (1.25), ε_{ijk} is defined as the *cross product index function,* which is also known as the *Levi-Civita epsilon function* of the indices i, j, and k. It is defined as follows:

$$\varepsilon_{ijk} = \begin{cases} +1 & \text{if } ijk = 123, 231, 312 \\ -1 & \text{if } ijk = 321, 132, 213 \\ 0 & \text{otherwise} \end{cases} \tag{1.26}$$

Of course, the cross product formula in Eq. (1.25) produces nonzero results only if the indices i, j, and k are all distinct. Therefore, by allowing the indices i, j, and k to assume only distinct values, i.e. by allowing ijk to be only such that $ijk \in \{123, 231, 312; 321, 132, 213\}$, the considered cross product can also be expressed by the following simpler formula, which does not require a summation operation.

$$\vec{u}_i^{(a)} \times \vec{u}_j^{(a)} = \sigma_{ijk} \vec{u}_k^{(a)} \tag{1.27}$$

In Eq. (1.27), σ_{ijk} is designated as the *cross product sign variable,* which is defined as follows *only* for the *distinct* values of the indices i, j, and k.

$$\sigma_{ijk} = \begin{cases} +1 & \text{if } ijk = 123, 231, 312 \\ -1 & \text{if } ijk = 321, 132, 213 \end{cases} \tag{1.28}$$

1.5 Representation of a Vector in a Selected Reference Frame

A vector $\vec{r}$ can be resolved in a selected reference frame $\mathcal{F}_a$ as shown below.

$$\vec{r} = r_1^{(a)}\vec{u}_1^{(a)} + r_2^{(a)}\vec{u}_2^{(a)} + r_3^{(a)}\vec{u}_3^{(a)} \tag{1.29}$$

Owing to the numerical index notation, Eq. (1.29) can also be written compactly as follows:

$$\vec{r} = \sum_{k=1}^{3} r_k^{(a)}\vec{u}_k^{(a)} \tag{1.30}$$

In Eqs. (1.29) and (1.30), $r_k^{(a)}$ is defined as the kth component of $\vec{r}$ in $\mathcal{F}_a$. It is obtained as

$$r_k^{(a)} = \vec{r}\cdot\vec{u}_k^{(a)} \tag{1.31}$$

The components of $\vec{r}$ can be stacked as follows to form a column matrix $\overline{r}^{(a)}$, which is defined as the *column matrix representation* of the vector $\vec{r}$ in $\mathcal{F}_a$.

$$\overline{r}^{(a)} = \begin{bmatrix} r_1^{(a)} \\ r_2^{(a)} \\ r_3^{(a)} \end{bmatrix} \tag{1.32}$$

In order to show the resolved vector explicitly, $\overline{r}^{(a)}$ may also be denoted as $[\vec{r}]^{(a)}$. That is,

$$\overline{r}^{(a)} = [\vec{r}]^{(a)} \tag{1.33}$$

The basis vector $\vec{u}_k^{(a)}$ of $\mathcal{F}_a$ is represented by the following column matrix in $\mathcal{F}_a$.

$$[\vec{u}_k^{(a)}]^{(a)} = \overline{u}_k^{(a/a)} = \overline{u}_k \tag{1.34}$$

In Eq. (1.34), $\overline{u}_k$ is the kth *basic column matrix*, which is defined as shown below for each $k \in \{1, 2, 3\}$.

$$\overline{u}_1 = \begin{bmatrix} 1 \\ 0 \\ 0 \end{bmatrix},\quad \overline{u}_2 = \begin{bmatrix} 0 \\ 1 \\ 0 \end{bmatrix},\quad \overline{u}_3 = \begin{bmatrix} 0 \\ 0 \\ 1 \end{bmatrix} \tag{1.35}$$

Here, it must be pointed out that, just like a scalar, $\overline{u}_k$ is an entity that is *not* associated with any reference frame. This is because $\overline{u}_k$ represents $\vec{u}_k^{(x)}$ in its own frame $\mathcal{F}_x$, whatever $\mathcal{F}_x$ is. In other words,

$$\overline{u}_k^{(a/a)} = \overline{u}_k^{(b/b)} = \overline{u}_k^{(c/c)} = \cdots = \overline{u}_k^{(z/z)} = \overline{u}_k \tag{1.36}$$

Moreover, the set $\mathcal{U} = \{\overline{u}_1, \overline{u}_2, \overline{u}_3\}$ of the basic column matrices forms the *primary basis* of the space $\mathcal{R}^3$ of the 3×1 column matrices. In other words, any arbitrary column matrix $\overline{c} \in \mathcal{R}^3$ can be expressed as a linear combination of $\overline{u}_1$, $\overline{u}_2$, and $\overline{u}_3$ as shown below.

$$\overline{c} = c_1\overline{u}_1 + c_2\overline{u}_2 + c_3\overline{u}_3 = c_1\begin{bmatrix} 1 \\ 0 \\ 0 \end{bmatrix} + c_2\begin{bmatrix} 0 \\ 1 \\ 0 \end{bmatrix} + c_3\begin{bmatrix} 0 \\ 0 \\ 1 \end{bmatrix} = \begin{bmatrix} c_1 \\ c_2 \\ c_3 \end{bmatrix} \tag{1.37}$$

Therefore, being a column matrix, $\overline{r}^{(a)}$ can also be expressed in terms of the basic column matrices $\overline{u}_1$, $\overline{u}_2$, and $\overline{u}_3$ by writing the following matrix version of Eq. (1.29) in $\mathcal{F}_a$.

$$\overline{r}^{(a)} = [\vec{r}]^{(a)} = [r_1^{(a)}\vec{u}_1^{(a)} + r_2^{(a)}\vec{u}_2^{(a)} + r_3^{(a)}\vec{u}_3^{(a)}]^{(a)} \Rightarrow$$

$$\overline{r}^{(a)} = r_1^{(a)}\overline{u}_1^{(a/a)} + r_2^{(a)}\overline{u}_2^{(a/a)} + r_3^{(a)}\overline{u}_3^{(a/a)} \Rightarrow$$

$$\overline{r}^{(a)} = r_1^{(a)}\overline{u}_1 + r_2^{(a)}\overline{u}_2 + r_3^{(a)}\overline{u}_3 \tag{1.38}$$

For the sake of comparing Eqs. (1.29) and (1.38) from the viewpoint of the *notational logic*, Eq. (1.29) is written again below.

$$\vec{r} = r_1^{(a)}\vec{u}_1^{(a)} + r_2^{(a)}\vec{u}_2^{(a)} + r_3^{(a)}\vec{u}_3^{(a)} \tag{1.39}$$

Here, it is instructive to pay attention to the *interchanged* location of the superscript (a) in Eqs. (1.38) and (1.39). In Eq. (1.39), $\vec{r}$ must not bear (a) because it is a vector that is specified without necessarily knowing anything about the observation frame $\mathcal{F}_a$, whereas $\vec{u}_k^{(a)}$ must necessarily bear (a) because it is one of the basis vectors of $\mathcal{F}_a$. In Eq. (1.38), on the other hand, $\overline{r}^{(a)}$ must necessarily bear (a) because it represents the appearance of $\vec{r}$ as observed in $\mathcal{F}_a$, whereas $\overline{u}_k$ must not bear (a) because it is not tied up to any reference frame as explained above and expressed by Eq. (1.36).

1.6 Matrix Operations Corresponding to Vector Operations

1.6.1 Dot Product

Consider two vectors $\vec{p}$ and $\vec{q}$, which are resolved as follows in a reference frame $\mathcal{F}_a$:

$$\vec{p} = \sum_{i=1}^{3} p_i^{(a)}\vec{u}_i^{(a)} \tag{1.40}$$

$$\vec{q} = \sum_{j=1}^{3} q_j^{(a)}\vec{u}_j^{(a)} \tag{1.41}$$

The dot product of $\vec{p}$ and $\vec{q}$ can be expressed as

$$s = \vec{p} \cdot \vec{q} = \sum_{i=1}^{3}\sum_{j=1}^{3} p_i^{(a)} q_j^{(a)} \vec{u}_i^{(a)} \cdot \vec{u}_j^{(a)} \tag{1.42}$$

On the other hand, according to Eq. (1.24),

$$\vec{u}_i^{(a)} \cdot \vec{u}_j^{(a)} = \delta_{ij} = \begin{cases} 1 \text{ if } i = j \\ 0 \text{ if } i \neq j \end{cases} \tag{1.43}$$

Hence, Eq. (1.42) becomes

$$s = \sum_{i=1}^{3}\sum_{j=1}^{3} p_i^{(a)} q_j^{(a)} \delta_{ij} \tag{1.44}$$

Owing to the definition of δ_{ij}, Eq. (1.44) becomes simplified to

$$s = \sum_{i=1}^{3} p_i^{(a)} q_i^{(a)} = p_1^{(a)} q_1^{(a)} + p_2^{(a)} q_2^{(a)} + p_3^{(a)} q_3^{(a)} \tag{1.45}$$

Equation (1.45) can also be written as follows in terms of $\overline{p}^{(a)}$ and $\overline{q}^{(a)}$, which are the column matrix representations of $\vec{p}$ and $\vec{q}$ in $\mathcal{F}_a$:

$$s = [p_1^{(a)} \;\; p_2^{(a)} \;\; p_3^{(a)}] \begin{bmatrix} q_1^{(a)} \\ q_2^{(a)} \\ q_3^{(a)} \end{bmatrix} = \overline{p}^{(a)t} \overline{q}^{(a)} \Rightarrow$$

$$s = \vec{p} \cdot \vec{q} = \overline{p}^{(a)t} \overline{q}^{(a)} \tag{1.46}$$

Equation (1.46) shows that the dot product of two vectors is equivalent to the *inner product* of their column matrix representations in a reference frame such as $\mathcal{F}_a$.

1.6.2 Cross Product and Skew Symmetric Cross Product Matrices

Consider the same two vectors $\vec{p}$ and $\vec{q}$, which are expressed by Eqs. (1.40) and (1.41) as resolved in the reference frame $\mathcal{F}_a$. Their cross product can be expressed as

$$\vec{r} = \vec{p} \times \vec{q} = \sum_{i=1}^{3} \sum_{j=1}^{3} p_i^{(a)} q_j^{(a)} \vec{u}_i^{(a)} \times \vec{u}_j^{(a)} \tag{1.47}$$

On the other hand, according to Eq. (1.25),

$$\vec{u}_i^{(a)} \times \vec{u}_j^{(a)} = \sum_{k=1}^{3} \varepsilon_{ijk} \vec{u}_k^{(a)} \tag{1.48}$$

Hence, Eq. (1.47) becomes

$$\vec{r} = \sum_{k=1}^{3} r_k^{(a)} \vec{u}_k^{(a)} = \sum_{k=1}^{3} \left[\sum_{i=1}^{3} \sum_{j=1}^{3} \varepsilon_{ijk} p_i^{(a)} q_j^{(a)} \right] \vec{u}_k^{(a)} \tag{1.49}$$

Equation (1.49) implies that

$$r_k^{(a)} = \sum_{i=1}^{3} \sum_{j=1}^{3} \varepsilon_{ijk} p_i^{(a)} q_j^{(a)} \tag{1.50}$$

By using the definition of ε_{ijk} given by Eq. (1.26), Eq. (1.49) can be worked out to what follows:

$$\begin{aligned} \vec{u}_1^{(a)} r_1^{(a)} + \vec{u}_2^{(a)} r_2^{(a)} + \vec{u}_3^{(a)} r_3^{(a)} &= \vec{u}_1^{(a)} [p_2^{(a)} q_3^{(a)} - p_3^{(a)} q_2^{(a)}] + \vec{u}_2^{(a)} [p_3^{(a)} q_1^{(a)} - p_1^{(a)} q_3^{(a)}] \\ &\quad + \vec{u}_3^{(a)} [p_1^{(a)} q_2^{(a)} - p_2^{(a)} q_1^{(a)}] \end{aligned} \tag{1.51}$$

Upon comparing the coefficients of the basis vectors of $\mathcal{F}_a$ on each side of Eq. (1.51), the following column matrix equation can be written.

$$\begin{bmatrix} r_1^{(a)} \\ r_2^{(a)} \\ r_3^{(a)} \end{bmatrix} = \begin{bmatrix} p_2^{(a)} q_3^{(a)} - p_3^{(a)} q_2^{(a)} \\ p_3^{(a)} q_1^{(a)} - p_1^{(a)} q_3^{(a)} \\ p_1^{(a)} q_2^{(a)} - p_2^{(a)} q_1^{(a)} \end{bmatrix} \tag{1.52}$$

Equation (1.52) can be written again as follows by factorizing the column matrix expression on its right side:

$$\begin{bmatrix} r_1^{(a)} \\ r_2^{(a)} \\ r_3^{(a)} \end{bmatrix} = \begin{bmatrix} 0 & -p_3^{(a)} & p_2^{(a)} \\ p_3^{(a)} & 0 & -p_1^{(a)} \\ -p_2^{(a)} & p_1^{(a)} & 0 \end{bmatrix} \begin{bmatrix} q_1^{(a)} \\ q_2^{(a)} \\ q_3^{(a)} \end{bmatrix} \tag{1.53}$$

Furthermore, Eq. (1.53) can be written compactly as

$$\overline{r}^{(a)} = \widetilde{p}^{(a)}\overline{q}^{(a)} \tag{1.54}$$

In Eq. (1.54), $\widetilde{p}^{(a)}$ is defined as the *cross product matrix* (cpm) corresponding to the column matrix $\overline{p}^{(a)}$. When Eqs. (1.53) and (1.54) are compared, it is seen that $\widetilde{p}^{(a)}$ happens to be a *skew symmetric matrix* generated from the column matrix $\overline{p}^{(a)}$.

Considering an arbitrary column matrix $\overline{c}$, the corresponding skew symmetric matrix $\widetilde{c}$ is generated by means of the ssm (skew symmetric matrix) operator as described below.

$$\overline{c} = \begin{bmatrix} c_1 \\ c_2 \\ c_3 \end{bmatrix} \Rightarrow \widetilde{c} = \mathrm{ssm}(\overline{c}) = \begin{bmatrix} 0 & -c_3 & c_2 \\ c_3 & 0 & -c_1 \\ -c_2 & c_1 & 0 \end{bmatrix} \tag{1.55}$$

The inverse of the ssm operator is the colm (column matrix) operator, which is defined so that

$$\overline{c} = \mathrm{ssm}^{-1}(\widetilde{c}) = \mathrm{colm}(\widetilde{c}) = \mathrm{colm}[\mathrm{ssm}(\overline{c})] \tag{1.56}$$

Coming back to the cross product operation, Eqs. (1.47) and (1.54) imply the following mutual correspondence, which shows how the cross product of two vectors can be equivalently expressed by using the matrix representations of the vectors in a reference frame such as $\mathcal{F}_a$.

$$\vec{r} = \vec{p} \times \vec{q} \Leftrightarrow \overline{r}^{(a)} = \widetilde{p}^{(a)}\overline{q}^{(a)} \tag{1.57}$$

1.7 Mathematical Properties of the Skew Symmetric Matrices

The skew symmetric matrices have several mathematical properties that turn out to be quite useful especially in the symbolic matrix manipulations. These properties are shown and explained below by concealing the frame indicating superscripts for the sake of brevity.

∗ Since $\widetilde{p}$ is a skew symmetric matrix,

$$\widetilde{p}^t = -\widetilde{p} \tag{1.58}$$

∗ Since $\vec{p} \times \vec{p} = \vec{0}$,

$$\widetilde{p}\overline{p} = \overline{0} \text{ and } \overline{p}^t\widetilde{p} = \overline{0}^t \tag{1.59}$$

∗ Since $\vec{p} \times \vec{q} = -\vec{q} \times \vec{p}$,

$$\widetilde{p}\overline{q} = -\widetilde{q}\overline{p} \tag{1.60}$$

∗ The product of two skew symmetric matrices can be expanded as follows:

$$\widetilde{p}\widetilde{q} = \overline{q}\overline{p}^t - (\overline{q}^t\overline{p})\hat{I} \tag{1.61}$$

In Eq. (1.61), $\overline{q}^t\overline{p}$ is the *inner product* and $\overline{q}\overline{p}^t$ is the *outer product* of $\overline{q}$ and $\overline{p}$. As for $\widehat{I}$, it is the identity matrix. That is,

$$\widehat{I} = \begin{bmatrix} 1 & 0 & 0 \\ 0 & 1 & 0 \\ 0 & 0 & 1 \end{bmatrix} \tag{1.62}$$

∗ Let $\overline{n}$ be a *unit column matrix* so that $\overline{n}^t\overline{n} = 1$. Then,

$$\widetilde{n}^2 = \overline{n}\overline{n}^t - \widehat{I} \tag{1.63}$$

∗ Equations (1.63) and (1.59) imply that $\widetilde{n} = \mathrm{ssm}(\overline{n})$ is exponentiated as follows:

$$\widetilde{n}^3 = -\widetilde{n},\ \ \widetilde{n}^5 = \widetilde{n},\ \ \widetilde{n}^7 = -\widetilde{n},\ \ \widetilde{n}^9 = \widetilde{n},\ \ \text{etc.} \tag{1.64}$$

$$\widetilde{n}^4 = -\widetilde{n}^2,\ \ \widetilde{n}^6 = \widetilde{n}^2,\ \ \widetilde{n}^8 = -\widetilde{n}^2,\ \ \widetilde{n}^{10} = \widetilde{n}^2,\ \ \text{etc.} \tag{1.65}$$

∗ The ssm operation is applied as shown below on the products $\widetilde{p}\overline{q}$ and $\widehat{R}\overline{p}$.

$$\mathrm{ssm}(\widetilde{p}\overline{q}) = \widetilde{p}\widetilde{q} - \widetilde{q}\widetilde{p} = \overline{q}\overline{p}^t - \overline{p}\overline{q}^t \tag{1.66}$$

$$\mathrm{ssm}(\widehat{R}\overline{p}) = \widehat{R}\widetilde{p}\widehat{R}^t \tag{1.67}$$

Here, it is to be noted that Eq. (1.67) is valid if $\widehat{R}$ is a *rotation matrix*, i.e. an *orthonormal matrix* with $\det(\widehat{R}) = +1$.

∗ It happens that $\widetilde{p}$ is a singular matrix and its rank is two. That is,

$$\det(\widetilde{p}) = 0 \tag{1.68}$$

$$\mathrm{rank}(\widetilde{p}) = 2 \tag{1.69}$$

1.8 Examples Involving Skew Symmetric Matrices

1.8.1 Example 1.1

This example is about the expansion of the following *triple vector product*.

$$\vec{s} = \vec{p} \times (\vec{q} \times \vec{r}) \tag{1.70}$$

In a selected reference frame $\mathcal{F}_a$, the matrix version of Eq. (1.70) can be written as

$$\overline{s} = \widetilde{p}(\widetilde{q}\overline{r}) = \widetilde{p}\widetilde{q}\overline{r} \tag{1.71}$$

In Eq. (1.71), all the matrices are expressed in the same frame $\mathcal{F}_a$. Therefore, the frame indicating superscript (a) is concealed for the sake of brevity. By expanding the product $\widetilde{p}\widetilde{q}$ according to Eq. (1.61)Eq. (1.71) can be written again as follows:

$$\overline{s} = \widetilde{p}\widetilde{q}\overline{r} = [\overline{q}\overline{p}^t - (\overline{q}^t\overline{p})\widehat{I}]\overline{r} = \overline{q}(\overline{p}^t\overline{r}) - (\overline{q}^t\overline{p})\overline{r} \Rightarrow$$

$$\overline{s} = (\overline{p}^t\overline{r})\overline{q} - (\overline{p}^t\overline{q})\overline{r} \tag{1.72}$$

The corresponding vector equation written below turns out to be the required expansion of the triple vector product.

$$\vec{s} = \vec{p} \times (\vec{q} \times \vec{r}) = (\vec{p} \cdot \vec{r})\vec{q} - (\vec{p} \cdot \vec{q})\vec{r} \tag{1.73}$$

1.8.2 Example 1.2

As a typical problem involving the singularity of $\widetilde{p}$, consider the following equation, from which $\overline{q}$ is to be found for given $\overline{p}$ and $\overline{r}$.

$$\widetilde{p}\overline{q} = \overline{r} \tag{1.74}$$

Due to Eqs. (1.68) and (1.69), $\overline{q}$ cannot be found uniquely from Eq. (1.74). However, it can be found with the following expression that contains an arbitrary parameter λ.

$$\overline{q} = \overline{q}^{\circ} + \lambda\overline{p} \ ; \ -\infty < \lambda < \infty \tag{1.75}$$

In Eq. (1.75), $\overline{q}^{\circ}$ is the part of $\overline{q}$ that is orthogonal to $\overline{p}$. So, it can be expressed as

$$\overline{q}^{\circ} = \gamma\widetilde{r}\overline{p} \tag{1.76}$$

The coefficient γ is to be determined so as to satisfy Eq. (1.74). That is,

$$\begin{aligned} &\widetilde{p}(\gamma\widetilde{r}\overline{p} + \lambda\overline{p}) = \overline{r} \Rightarrow \gamma\widetilde{p}\widetilde{r}\overline{p} = \overline{r} \Rightarrow \\ &\gamma[\overline{r}\overline{p}^t - (\overline{r}^t\overline{p})\widehat{I}]\overline{p} = \gamma(\overline{p}^t\overline{p})\overline{r} - \gamma(\overline{r}^t\overline{p})\overline{p} = \overline{r} \end{aligned} \tag{1.77}$$

Since $\overline{r}$ and $\overline{p}$ are orthogonal, $\overline{r}^t\overline{p} = 0$. Therefore, Eq. (1.77) gives γ as

$$\gamma = 1/(\overline{p}^t\overline{p}) \tag{1.78}$$

Hence, $\overline{q}^{\circ}$ and $\overline{q}$ are obtained as shown below.

$$\overline{q}^{\circ} = \widetilde{r}\overline{p}/(\overline{p}^t\overline{p}) \tag{1.79}$$

$$\overline{q} = [\widetilde{r}/(\overline{p}^t\overline{p}) + \lambda\widehat{I}]\overline{p} \ ; \ -\infty < \lambda < \infty \tag{1.80}$$

1.8.3 Example 1.3

Consider the following 3 × 3 matrix equation, which is to be solved for $\overline{x}$.

$$\widehat{A}\overline{x} = \overline{y} \tag{1.81}$$

The matrix $\widehat{A}$ can be expressed as follows in terms of its columns.

$$\widehat{A} = \begin{bmatrix}\overline{a}_1 & \overline{a}_2 & \overline{a}_3\end{bmatrix} \tag{1.82}$$

Along with $\widehat{A}$, $\overline{x}$ can be expressed as follows in terms of its elements.

$$\overline{x} = \begin{bmatrix} x_1 \\ x_2 \\ x_3 \end{bmatrix} \tag{1.83}$$

Hence, Eq. (1.81) can be written in a more detailed form as

$$\overline{a}_1 x_1 + \overline{a}_2 x_2 + \overline{a}_3 x_3 = \overline{y} \tag{1.84}$$

Equation (1.84) leads to the following scalar equations with the indicated premultiplications.

$$\overline{a}_2^t\widetilde{a}_3\overline{a}_1 x_1 + \overline{a}_2^t\widetilde{a}_3\overline{a}_2 x_2 + \overline{a}_2^t\widetilde{a}_3\overline{a}_3 x_3 = \overline{a}_2^t\widetilde{a}_3\overline{y} \tag{1.85}$$

$$\overline{a}_3^t\widetilde{a}_1\overline{a}_1 x_1 + \overline{a}_3^t\widetilde{a}_1\overline{a}_2 x_2 + \overline{a}_3^t\widetilde{a}_1\overline{a}_3 x_3 = \overline{a}_3^t\widetilde{a}_1\overline{y} \tag{1.86}$$

$$\overline{a}_1^t\widetilde{a}_2\overline{a}_1 x_1 + \overline{a}_1^t\widetilde{a}_2\overline{a}_2 x_2 + \overline{a}_1^t\widetilde{a}_2\overline{a}_3 x_3 = \overline{a}_1^t\widetilde{a}_2\overline{y} \tag{1.87}$$

Note that, for $i \in \{1, 2, 3\}$ and $j \in \{1, 2, 3\}$,

$$\tilde{a}_i \bar{a}_i = \bar{0} \text{ and } \bar{a}_i^t \tilde{a}_j \bar{a}_i = -\bar{a}_i^t \tilde{a}_i \bar{a}_j = 0 \tag{1.88}$$

Thus, Eqs. (1.85)–(1.87) reduce to the following equations.

$$\bar{a}_2^t \tilde{a}_3 \bar{a}_1 x_1 = \bar{a}_2^t \tilde{a}_3 \bar{y} \tag{1.89}$$

$$\bar{a}_3^t \tilde{a}_1 \bar{a}_2 x_2 = \bar{a}_3^t \tilde{a}_1 \bar{y} \tag{1.90}$$

$$\bar{a}_1^t \tilde{a}_2 \bar{a}_3 x_3 = \bar{a}_1^t \tilde{a}_2 \bar{y} \tag{1.91}$$

Equations (1.89)–(1.91) imply that

$$\det(\widehat{A}) = \bar{a}_1^t \tilde{a}_2 \bar{a}_3 = \bar{a}_2^t \tilde{a}_3 \bar{a}_1 = \bar{a}_3^t \tilde{a}_1 \bar{a}_2 \tag{1.92}$$

Therefore, if $\det(\widehat{A}) \neq 0$, Eqs. (1.89)–(1.91) give the elements of $\bar{x}$ as follows:

$$x_1 = (\bar{a}_2^t \tilde{a}_3 \bar{y})/(\bar{a}_2^t \tilde{a}_3 \bar{a}_1) \tag{1.93}$$

$$x_2 = (\bar{a}_3^t \tilde{a}_1 \bar{y})/(\bar{a}_3^t \tilde{a}_1 \bar{a}_2) \tag{1.94}$$

$$x_3 = (\bar{a}_1^t \tilde{a}_2 \bar{y})/(\bar{a}_1^t \tilde{a}_2 \bar{a}_3) \tag{1.95}$$

Note that the solution obtained above is the same as the solution provided by *Cramer's rule*.

2

Rotation of Vectors and Rotation Matrices

Synopsis

This chapter is devoted to the rotation of vectors and the rotation operators that rotate vectors. The rotation of a vector is expressed both as a vector equation and as a matrix equation written in a selected reference frame. The vector equation is obtained as the Rodrigues formula. The matrix equation is written in terms of the rotation matrix, which is the matrix representation of the rotation operator in the selected reference frame. The expression of the rotation matrix is obtained in terms of the angle of rotation and the unit vector along the axis of rotation. It is shown that the rotation matrix can be expressed very compactly in the exponential form. This chapter also presents the salient mathematical properties of the rotation matrices that can be used conveniently in the symbolic manipulations concerning rotational kinematics. Demonstrative examples are also included.

2.1 Vector Equation of Rotation and the Rodrigues Formula

Figure 2.1 illustrates the rotation of a vector $\vec{p} = \overrightarrow{OA}$ into another vector $\vec{r} = \overrightarrow{OB}$ by an angle θ about an axis described by a unit vector $\vec{n}$.

Incidentally, a vector may be acted upon, simultaneously or successively, by two kinds of *displacement operators*. One of them is a *rotation operator*, which is defined as an operator that changes only the orientation of a vector *irrespective* of any possible change in its location. The other one is a *translation operator*, which is defined as an operator that changes only the location of a vector without changing its orientation.

As mentioned above, a rotation operator is not affected by any translational displacement. Therefore, without any loss of generality, the rotation of $\vec{p}$ is illustrated in Figure 2.1 so that the point O (i.e. the tail point of $\vec{p}$) is assumed to be fixed and the rotation axis is assumed to pass through that point. Moreover, as illustrated on the left-hand side of Figure 2.1, the vector $\vec{p}$ moves on the surface of a cone while it is rotated into the vector $\vec{r}$. The projected appearance of this rotation on the base of the mentioned cone is illustrated on the right-hand side of Figure 2.1.

Kinematics of General Spatial Mechanical Systems, First Edition. M. Kemal Ozgoren.

Companion Website: www.wiley.com/go/ozgoren/spatialmechanicalsystems

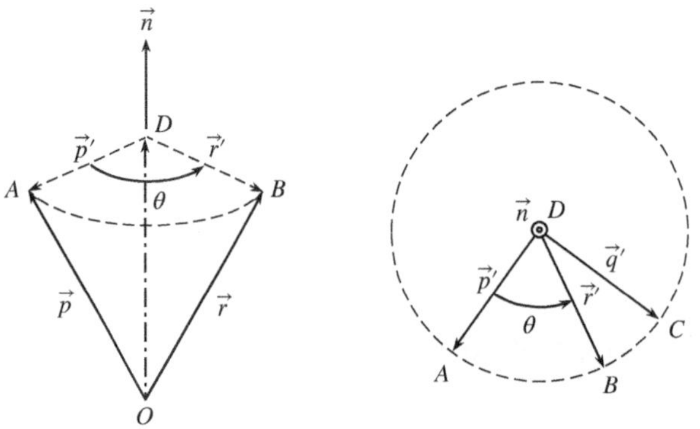

Figure 2.1 Rotation of a vector about an axis.

The rotation illustrated in Figure 2.1 can be described briefly in one of the following ways, in which the rotation operator is denoted by $\text{rot}(\vec{n}, \theta)$.

$$\vec{p} \xrightarrow{\text{rot}(\vec{n},\theta)} \vec{r} \tag{2.1}$$

$$\text{rot}(\vec{n}, \theta) : \vec{p} \to \vec{r} \tag{2.2}$$

The resultant vector $\vec{r}$ can be expressed in terms of the initial vector $\vec{p}$ and the rotation parameters $\vec{n}$ and θ as explained below.

Referring to Figure 2.1, the vectors $\vec{p}' = \overrightarrow{DA}$ and $\vec{r}' = \overrightarrow{DB}$ can be expressed as follows:

$$\vec{p}' = \vec{p} - \vec{s} \tag{2.3}$$

$$\vec{r}' = \vec{r} - \vec{s} \tag{2.4}$$

In Eqs. (2.3) and (2.4), $\vec{s} = \overrightarrow{OD}$ is the common projection of $\vec{p}$ and $\vec{r}$ on the axis of rotation. Note that $\vec{s}$ is not affected by the rotation operator $\text{rot}(\vec{n}, \theta)$ because $\vec{s}$ lies on the rotation axis. The vector $\vec{s}$ is related to $\vec{n}$ and $\vec{p}$ as expressed below.

$$\vec{s} = (\vec{p} \cdot \vec{n})\vec{n} \tag{2.5}$$

Referring again to Figure 2.1, the vector $\vec{q}' = \overrightarrow{DC}$, which is perpendicular to $\vec{p}'$ and $\vec{n}$, is obtained by the following cross product.

$$\vec{q}' = \vec{n} \times \vec{p}' \tag{2.6}$$

Upon substituting Eq. (2.3) and noting that $\vec{n} \times \vec{s} = (\vec{p} \cdot \vec{n})(\vec{n} \times \vec{n}) = \vec{0}$, Eq. (2.6) becomes

$$\vec{q}' = \vec{n} \times \vec{p} \tag{2.7}$$

On the other hand, the projectional view on the right-hand side of Figure 2.1 implies that the vector $\vec{r}'$, which is *coplanar* with the vectors $\vec{p}'$ and $\vec{q}'$, can be expressed as the following linear combination of $\vec{p}'$ and $\vec{q}'$.

$$\vec{r}' = \vec{p}' \cos\theta + \vec{q}' \sin\theta \tag{2.8}$$

After the previously obtained equations concerning $\vec{r}'$, $\vec{p}'$, and $\vec{q}'$ are substituted, Eq. (2.8) becomes

$$\vec{r} - (\vec{p} \cdot \vec{n})\vec{n} = [\vec{p} - (\vec{p} \cdot \vec{n})\vec{n}] \cos\theta + (\vec{n} \times \vec{p}) \sin\theta$$

The preceding equation can be arranged as follows:

$$\vec{r} = \vec{p}\cos\theta + (\vec{n}\times\vec{p})\sin\theta + \vec{n}(\vec{n}\cdot\vec{p})(1-\cos\theta) \tag{2.9}$$

Equation (2.9) is known as the *Rodrigues formula*, which is named after the French mathematician Benjamin Olinde Rodrigues (1795–1851).

2.2 Matrix Equation of Rotation and the Rotation Matrix

The matrix form of the Rodrigues formula, i.e. Eq. (2.9), can be written as follows in a selected reference frame $\mathcal{F}_a$:

$$\overline{r}^{(a)} = \overline{p}^{(a)}\cos\theta + \widetilde{n}^{(a)}\overline{p}^{(a)}\sin\theta + \overline{n}^{(a)}\overline{n}^{(a)t}\overline{p}^{(a)}(1-\cos\theta) \tag{2.10}$$

By noting that $\overline{p}^{(a)} = \widehat{I}\overline{p}^{(a)}$, Eq. (2.10) can be factorized so that

$$\overline{r}^{(a)} = [\widehat{I}\cos\theta + \widetilde{n}^{(a)}\sin\theta + \overline{n}^{(a)}\overline{n}^{(a)t}(1-\cos\theta)]\overline{p}^{(a)} \tag{2.11}$$

Equation (2.11) can be written compactly as

$$\overline{r}^{(a)} = \widehat{R}^{(a)}\overline{p}^{(a)} \tag{2.12}$$

In Eq. (2.12), $\widehat{R}^{(a)}$ is defined as the *rotation matrix* expressed in $\mathcal{F}_a$. It is the matrix representation of the rotation operator $\mathrm{rot}(\vec{n},\theta)$ in $\mathcal{F}_a$. In other words,

$$\widehat{R}^{(a)} = [\mathrm{rot}(\vec{n},\theta)]^{(a)} \tag{2.13}$$

Equations (2.11) and (2.12) show that $\widehat{R}^{(a)}$ is a function of $\overline{n}^{(a)}$ and θ as expressed below.

$$\widehat{R}^{(a)} = \widehat{R}(\overline{n}^{(a)},\theta) = \widehat{I}\cos\theta + \widetilde{n}^{(a)}\sin\theta + \overline{n}^{(a)}\overline{n}^{(a)t}(1-\cos\theta) \tag{2.14}$$

In Eq. (2.14), $\widehat{R}(\overline{n},\theta)$ is defined as the *rotation matrix function* of the arguments $\overline{n}$ and θ. In other words, $\widehat{R}(\overline{n},\theta)$ generates a rotation matrix out of the arguments $\overline{n}$ and θ as shown below, where $\overline{n}$ may be any *column matrix* such that $\overline{n}^t\overline{n} = 1$.

$$\widehat{R}(\overline{n},\theta) = \widehat{I}\cos\theta + \widetilde{n}\sin\theta + \overline{n}\,\overline{n}^t(1-\cos\theta) \tag{2.15}$$

An alternative expression can be derived for the function $\widehat{R}(\overline{n},\theta)$ by recalling the following equation from Section 1.6.

$$\widetilde{n}^2 = \overline{n}\,\overline{n}^t - \widehat{I} \tag{2.16}$$

Upon substituting Eq. (2.16) into Eq. (2.15), the alternative expression is obtained as

$$\widehat{R}(\overline{n},\theta) = \widehat{I} + \widetilde{n}\sin\theta + \widetilde{n}^2(1-\cos\theta) \tag{2.17}$$

Hence, with $\widetilde{n} = \widetilde{n}^{(a)}$, $\widehat{R}^{(a)}$ can also be expressed as

$$\widehat{R}^{(a)} = \widehat{R}(\overline{n}^{(a)},\theta) = \widehat{I} + \widetilde{n}^{(a)}\sin\theta + \widetilde{n}^{(a)2}(1-\cos\theta) \tag{2.18}$$

2.3 Exponentially Expressed Rotation Matrix

The expression of the rotation matrix function $\widehat{R}(\overline{n}, \theta)$, which is given by Eq. (2.17), can be manipulated further as explained below.

Recall that $\sin\theta$ and $\cos\theta$ can be expressed as follows by using their Taylor series expansions:

$$\sin\theta = \theta - \frac{1}{3!}\theta^3 + \frac{1}{5!}\theta^5 - \frac{1}{7!}\theta^7 + \cdots \tag{2.19}$$

$$\cos\theta = 1 - \frac{1}{2!}\theta^2 + \frac{1}{4!}\theta^4 - \frac{1}{6!}\theta^6 + \cdots \tag{2.20}$$

When Eqs. (2.19) and (2.20) are substituted into Eq. (2.17), the Taylor series expansion of $\widehat{R}(\widetilde{n}, \theta)$ is obtained as

$$\widehat{R}(\overline{n}, \theta) = \widehat{I} + \widetilde{n}\theta + \frac{1}{2!}\widetilde{n}^2\theta^2 - \frac{1}{3!}\widetilde{n}\theta^3 - \frac{1}{4!}\widetilde{n}^2\theta^4 + \frac{1}{5!}\widetilde{n}\theta^5 + \frac{1}{6!}\widetilde{n}^2\theta^6 - \cdots \tag{2.21}$$

On the other hand, as shown in Section 1.7, $\widetilde{n}$ has the following exponentiation properties.

$$\widetilde{n}^3 = -\widetilde{n},\ \ \widetilde{n}^4 = -\widetilde{n}^2,\ \ \widetilde{n}^5 = \widetilde{n},\ \ \widetilde{n}^6 = \widetilde{n}^2,\ \ \widetilde{n}^7 = -\widetilde{n},\ \ \text{etc.} \tag{2.22}$$

Hence, Eq. (2.21) can be written again as

$$\widehat{R}(\overline{n}, \theta) = \widehat{I} + (\widetilde{n}\theta) + \frac{1}{2!}(\widetilde{n}\theta)^2 + \frac{1}{3!}(\widetilde{n}\theta)^3 + \frac{1}{4!}(\widetilde{n}\theta)^4 + \frac{1}{5!}(\widetilde{n}\theta)^5 + \cdots \tag{2.23}$$

Note that the series expression on the right-hand side of Eq. (2.23) is analogous to the Taylor series expansion of the *exponential function* of a scalar x, which is written below.

$$e^x = 1 + x + \frac{1}{2!}x^2 + \frac{1}{3!}x^3 + \frac{1}{4!}x^4 + \frac{1}{5!}x^5 + \cdots \tag{2.24}$$

Based on the above analogy, the function $\widehat{R}(\overline{n}, \theta)$ can be expressed exponentially as

$$\widehat{R}(\overline{n}, \theta) = e^{\widetilde{n}\theta} \tag{2.25}$$

Hence, the rotation matrix $\widehat{R}^{(a)}$ can also be expressed exponentially as

$$\widehat{R}^{(a)} = \widehat{R}(\overline{n}^{(a)}, \theta) = e^{\widetilde{n}^{(a)}\theta} \tag{2.26}$$

Note that the exponential expression of $\widehat{R}^{(a)}$ is not only very compact but it also indicates the angle and axis of the rotation explicitly. Therefore, the exponentially expressed rotation matrices can be used quite conveniently in the symbolic manipulations required in the analytical treatments within the scope of rotational kinematics. For the sake of verbal brevity, an exponentially expressed rotation matrix is simply called here an *exponential rotation matrix*.

2.4 Basic Rotation Matrices

A rotation may be carried out about one of the *coordinate axes* of a reference frame $\mathcal{F}_a$. Such a rotation is defined as a *basic rotation* with respect to $\mathcal{F}_a$. More specifically, the kth basic rotation with respect to $\mathcal{F}_a$ takes place about the kth coordinate axis of $\mathcal{F}_a$.

Therefore, the unit vector of the rotation axis of this basic rotation is the kth basis vector of $\mathcal{F}_a$, i.e. $\vec{n} = \vec{u}_k^{(a)}$. The operator of this basic rotation is denoted as

$$\text{rot}[\vec{u}_k^{(a)}, \theta] \tag{2.27}$$

The kth basic rotation operator associated with $\mathcal{F}_a$ is represented in $\mathcal{F}_a$ by the matrix $\hat{R}_k(\theta)$, which is designated as the kth *basic rotation matrix*. It is expressed as follows:

$$\hat{R}_k(\theta) = \{\text{rot}[\vec{u}_k^{(a)}, \theta]\}^{(a)} = \hat{R}(\overline{u}_k^{(a/a)}, \theta) \Rightarrow$$

$$\hat{R}_k(\theta) = \hat{R}(\overline{u}_k, \theta) \tag{2.28}$$

Referring to Section 1.5 for the discussion about the basic column matrix $\overline{u}_k$, it is to be noted that, just like $\overline{u}_k$, the basic rotation matrix $\hat{R}_k(\theta)$ is also an entity that is *not* associated with any reference frame. This is because $\hat{R}_k(\theta)$ represents the rotation operator $\text{rot}[\vec{u}_k^{(x)}, \theta]$ in its own frame $\mathcal{F}_x$, whatever $\mathcal{F}_x$ is. In other words,

$$\hat{R}(\overline{u}_k^{(a/a)}, \theta) = \hat{R}(\overline{u}_k^{(b/b)}, \theta) = \cdots = \hat{R}(\overline{u}_k^{(z/z)}, \theta) = \hat{R}(\overline{u}_k, \theta) = \hat{R}_k(\theta) \tag{2.29}$$

By using Eqs. (2.15), (2.17), and (2.25), $\hat{R}_k(\theta)$ can be expressed in three equivalent ways as shown in the following equations.

$$\hat{R}_k(\theta) = \hat{I}\cos\theta + \tilde{u}_k \sin\theta + \overline{u}_k \overline{u}_k^t (1 - \cos\theta) \tag{2.30}$$

$$\hat{R}_k(\theta) = \hat{I} + \tilde{u}_k \sin\theta + \tilde{u}_k^2 (1 - \cos\theta) \tag{2.31}$$

$$\hat{R}_k(\theta) = e^{\tilde{u}_k \theta} \tag{2.32}$$

Upon inserting the expressions of the basic column matrices into Eq. (2.30), the basic rotation matrices can be expressed element by element as shown below.

$$\overline{u}_1 = \begin{bmatrix} 1 \\ 0 \\ 0 \end{bmatrix} \Rightarrow \hat{R}_1(\theta) = \hat{R}(\overline{u}_1, \theta) = e^{\tilde{u}_1 \theta} = \begin{bmatrix} 1 & 0 & 0 \\ 0 & \cos\theta & -\sin\theta \\ 0 & \sin\theta & \cos\theta \end{bmatrix} \tag{2.33}$$

$$\overline{u}_2 = \begin{bmatrix} 0 \\ 1 \\ 0 \end{bmatrix} \Rightarrow \hat{R}_2(\theta) = \hat{R}(\overline{u}_2, \theta) = e^{\tilde{u}_2 \theta} = \begin{bmatrix} \cos\theta & 0 & \sin\theta \\ 0 & 1 & 0 \\ -\sin\theta & 0 & \cos\theta \end{bmatrix} \tag{2.34}$$

$$\overline{u}_3 = \begin{bmatrix} 0 \\ 0 \\ 1 \end{bmatrix} \Rightarrow \hat{R}_3(\theta) = \hat{R}(\overline{u}_3, \theta) = e^{\tilde{u}_3 \theta} = \begin{bmatrix} \cos\theta & -\sin\theta & 0 \\ \sin\theta & \cos\theta & 0 \\ 0 & 0 & 1 \end{bmatrix} \tag{2.35}$$

2.5 Successive Rotations

Suppose a vector $\vec{p}_0$ is first rotated into a vector $\vec{p}_1$ and then $\vec{p}_1$ is rotated into another vector $\vec{p}_2$. These two successive rotations can be described as indicated below.

$$\vec{p}_0 \xrightarrow{\text{rot}(\vec{n}_{01}, \theta_{01})} \vec{p}_1 \xrightarrow{\text{rot}(\vec{n}_{12}, \theta_{12})} \vec{p}_2 \tag{2.36}$$

On the other hand, according to *Euler's theorem*, the rotation of $\vec{p}_0$ into $\vec{p}_2$ can also be achieved directly in one step. That is,

$$\vec{p}_0 \xrightarrow{\text{rot}(\vec{n}_{02},\theta_{02})} \vec{p}_2 \tag{2.37}$$

The following matrix equations can be written for the rotational steps described above as observed in a reference frame $\mathcal{F}_a$.

$$\overline{p}_1^{(a)} = \widehat{R}_{01}^{(a)}\overline{p}_0^{(a)} \tag{2.38}$$

$$\overline{p}_2^{(a)} = \widehat{R}_{12}^{(a)}\overline{p}_1^{(a)} = \widehat{R}_{12}^{(a)}\widehat{R}_{01}^{(a)}\overline{p}_0^{(a)} \tag{2.39}$$

$$\overline{p}_2^{(a)} = \widehat{R}_{02}^{(a)}\overline{p}_0^{(a)} \tag{2.40}$$

Equations (2.39) and (2.40) show that the *overall* rotation matrix $\widehat{R}_{02}^{(a)}$ is obtained as the following *multiplicative* combination of the *intermediate* rotation matrices $\widehat{R}_{12}^{(a)}$ and $\widehat{R}_{01}^{(a)}$.

$$\widehat{R}_{02}^{(a)} = \widehat{R}_{12}^{(a)}\widehat{R}_{01}^{(a)} \tag{2.41}$$

As a general notational feature, the rotation matrix between $\overline{p}_i^{(a)}$ and $\overline{p}_j^{(a)}$ can be denoted by two alternative but equivalent symbols, which are shown below.

$$\widehat{R}_{ij}^{(a)} = \widehat{R}_{j/i}^{(a)} \tag{2.42}$$

Although $\widehat{R}_{ij}^{(a)}$ and $\widehat{R}_{j/i}^{(a)}$ are mathematically equivalent, their verbal descriptions are not the same. $\widehat{R}_{ij}^{(a)}$ is called a *rotation matrix* that describes the rotation of $\vec{p}_i$ into $\vec{p}_j$, whereas $\widehat{R}_{j/i}^{(a)}$ is called an *orientation matrix* that describes the relative orientation of $\vec{p}_j$ with respect to $\vec{p}_i$.

In a case of m successive rotational steps, the following equations can be written by using the alternative notations described above.

$$\overline{p}_m^{(a)} = \widehat{R}_{0m}^{(a)}\overline{p}_0^{(a)} = \widehat{R}_{m/0}^{(a)}\overline{p}_0^{(a)} \tag{2.43}$$

$$\widehat{R}_{0m}^{(a)} = \widehat{R}_{(m-1)m}^{(a)}\widehat{R}_{(m-2)(m-1)}^{(a)}(\cdots)\widehat{R}_{12}^{(a)}\widehat{R}_{01}^{(a)} \tag{2.44}$$

$$\widehat{R}_{m/0}^{(a)} = \widehat{R}_{m/m-1}^{(a)}\widehat{R}_{m-1/m-2}^{(a)}(\cdots)\widehat{R}_{2/1}^{(a)}\widehat{R}_{1/0}^{(a)} \tag{2.45}$$

2.6 Orthonormality of the Rotation Matrices

Suppose a vector $\vec{p}$ is rotated into another vector $\vec{r}$. This operation is expressed by the following equation as observed in a reference frame $\mathcal{F}_a$.

$$\overline{r}^{(a)} = \widehat{R}^{(a)}\overline{p}^{(a)} \tag{2.46}$$

Since a rotation operator does not change the magnitude of the vector it rotates, the following equations can be written.

$$|\vec{r}| = |\vec{p}| \Rightarrow |\vec{r}|^2 = |\vec{p}|^2 \Rightarrow \vec{r}\cdot\vec{r} = \vec{p}\cdot\vec{p} \Rightarrow$$

$$\overline{r}^{(a)t}\overline{r}^{(a)} = \overline{p}^{(a)t}\overline{p}^{(a)} \Rightarrow [\widehat{R}^{(a)}\overline{p}^{(a)}]^t[\widehat{R}^{(a)}\overline{p}^{(a)}] = \overline{p}^{(a)t}\overline{p}^{(a)} \Rightarrow$$

$$\overline{p}^{(a)t}[\widehat{R}^{(a)t}\widehat{R}^{(a)}]\overline{p}^{(a)} = \overline{p}^{(a)t}\overline{p}^{(a)} \tag{2.47}$$

Equation (2.47) implies that

$$\widehat{R}^{(a)t}\widehat{R}^{(a)} = \widehat{I} \tag{2.48}$$

Hence, Eq. (2.48) implies further that

$$[\widehat{R}^{(a)}]^{-1} = \widehat{R}^{(a)t} \tag{2.49}$$

Here, as a reminder from the matrix algebra, a matrix $\widehat{M}$ is defined to be *orthonormal* if its inverse is equal to its transpose, i.e. if $\widehat{M}^{-1} = \widehat{M}^t$. Therefore, according to Eq. (2.49), $\widehat{R}^{(a)}$ is an orthonormal matrix.

Owing to its orthonormality, the inverse of $\widehat{R}^{(a)}$ can be obtained by using Eq. (2.14), which is repeated below for the sake of convenience.

$$\widehat{R}^{(a)} = \widehat{I}\cos\theta + \widetilde{n}^{(a)}\sin\theta + \overline{n}^{(a)}\overline{n}^{(a)t}(1-\cos\theta) \tag{2.50}$$

Note that $\widehat{I}$ and $\overline{n}^{(a)}\overline{n}^{(a)t}$ are symmetric matrices having their transposes equal to themselves but $\widetilde{n}^{(a)}$ is a skew symmetric matrix, i.e. $\widetilde{n}^{(a)t} = -\widetilde{n}^{(a)}$. Therefore, Eqs. (2.49) and (2.50) lead to the following result.

$$[\widehat{R}^{(a)}]^{-1} = \widehat{R}^{(a)t} = \widehat{I}\cos\theta - \widetilde{n}^{(a)}\sin\theta + \overline{n}^{(a)}\overline{n}^{(a)t}(1-\cos\theta) \tag{2.51}$$

Referring to Eq. (2.26), Eq. (2.51) implies the following exponential expression.

$$[\widehat{R}^{(a)}]^{-1} = \widehat{R}^{(a)t} = \widehat{R}(-\overline{n}^{(a)},\theta) = e^{-\widetilde{n}^{(a)}\theta} \tag{2.52}$$

Equations (2.51) and (2.52) show that the inverse of $\widehat{R}^{(a)}$ always exists. Therefore, its determinant never vanishes. In fact, it happens that

$$\det[\widehat{R}^{(a)}] = 1 \tag{2.53}$$

Equation (2.53) can be verified as explained below.

The matrix $\widehat{R}^{(a)}$ can be expressed as follows in terms of its columns:

$$\widehat{R}^{(a)} = [\overline{\rho}_1^{(a)} \ \ \overline{\rho}_2^{(a)} \ \ \overline{\rho}_3^{(a)}] \tag{2.54}$$

Since the kth basic column matrix $\overline{u}_k$ picks up the kth column of the matrix it multiplies, $\overline{\rho}_k^{(a)}$ can be obtained as follows together with the interpretation that $\overline{u}_k = \overline{u}_k^{(a/a)}$.

$$\overline{\rho}_k^{(a)} = \widehat{R}^{(a)}\overline{u}_k = \widehat{R}^{(a)}\overline{u}_k^{(a/a)} \tag{2.55}$$

According to Eq. (2.55), $\overline{\rho}_k^{(a)}$ represents a vector $\vec{\rho}_k$, which is obtained by rotating the basis vector $\vec{u}_k^{(a)}$ of the reference frame $\mathcal{F}_a$ by means of the rotation operator represented by $\widehat{R}^{(a)}$.

Since $\mathcal{F}_a$ is assumed to be a *right-handed* reference frame, its basis vectors satisfy the following set of cross product equations.

$$\vec{u}_1^{(a)} \times \vec{u}_2^{(a)} = \vec{u}_3^{(a)}, \quad \vec{u}_2^{(a)} \times \vec{u}_3^{(a)} = \vec{u}_1^{(a)}, \quad \vec{u}_3^{(a)} \times \vec{u}_1^{(a)} = \vec{u}_2^{(a)} \tag{2.56}$$

Here, it must be pointed out that a rotation operator does not only retain the magnitudes of the vectors it rotates but it also retains the right-hand rule for their cross products. Therefore, the vectors $\vec{\rho}_1$, $\vec{\rho}_2$, and $\vec{\rho}_3$, which are obtained by rotating $\vec{u}_1^{(a)}$, $\vec{u}_2^{(a)}$, and $\vec{u}_3^{(a)}$, also satisfy a set of equations similar to Eq. Set (2.56). That is,

$$\vec{\rho}_1 \times \vec{\rho}_2 = \vec{\rho}_3, \quad \vec{\rho}_2 \times \vec{\rho}_3 = \vec{\rho}_1, \quad \vec{\rho}_3 \times \vec{\rho}_1 = \vec{\rho}_2 \tag{2.57}$$

In the frame $\mathcal{F}_a$, the matrix equivalent of Eq. Set (2.57) can be written as follows:

$$\tilde{\rho}_1^{(a)}\overline{\rho}_2^{(a)} = \overline{\rho}_3^{(a)}, \quad \tilde{\rho}_2^{(a)}\overline{\rho}_3^{(a)} = \overline{\rho}_1^{(a)}, \quad \tilde{\rho}_3^{(a)}\overline{\rho}_1^{(a)} = \overline{\rho}_2^{(a)} \tag{2.58}$$

On the other hand, as shown in Example 1.3 of Section 1.8, the determinant of a matrix, e.g. $\hat{R}^{(a)}$, can be expressed as follows in terms of its columns, i.e. $\overline{\rho}_1^{(a)}$, $\overline{\rho}_2^{(a)}$, and $\overline{\rho}_3^{(a)}$:

$$\det[\hat{R}^{(a)}] = \overline{\rho}_1^{(a)t}\tilde{\rho}_2^{(a)}\overline{\rho}_3^{(a)} \tag{2.59}$$

According to Eq. Set (2.58), $\tilde{\rho}_2^{(a)}\overline{\rho}_3^{(a)} = \overline{\rho}_1^{(a)}$. Therefore, Eq. (2.59) becomes

$$\det[\hat{R}^{(a)}] = \overline{\rho}_1^{(a)t}\overline{\rho}_1^{(a)} \tag{2.60}$$

At this point, by using Eq. (2.55), Eq. (2.60) can also be written as

$$\det[\hat{R}^{(a)}] = [\hat{R}^{(a)}\overline{u}_1]^t[\hat{R}^{(a)}\overline{u}_1] = \overline{u}_1^t[\hat{R}^{(a)t}\hat{R}^{(a)}]\overline{u}_1 \tag{2.61}$$

Since $\hat{R}^{(a)}$ is an orthonormal matrix, it obeys Eq. (2.48). That is,

$$\hat{R}^{(a)t}\hat{R}^{(a)} = \hat{I} \tag{2.62}$$

Therefore, Eq. (2.61) leads to the verification that

$$\det[\hat{R}^{(a)}] = \overline{u}_1^t\hat{I}\overline{u}_1 = \overline{u}_1^t\overline{u}_1 = 1 \tag{2.63}$$

2.7 Mathematical Properties of the Rotation Matrices

The rotation matrices have several mathematical properties that turn out to be quite useful especially in the symbolic matrix manipulations required in the analytical treatments within the scope of rotational kinematics. These properties are shown and explained below in Sections 2.7.1 and 2.7.2. In both sections, the rotation matrices are expressed in exponential form and the unit vectors of the rotation axes are represented by plain column matrices, such as $\overline{n}$, without explicit frame indication.

2.7.1 Mathematical Properties of General Rotation Matrices

∗ **Determinant of a Rotation Matrix**

As verified in Section 2.6,

$$\det(e^{\tilde{n}\theta}) = 1 \tag{2.64}$$

∗ **Inversion of a Rotation Matrix**

As also verified in Section 2.6,

$$(e^{\tilde{n}\theta})^{-1} = (e^{\tilde{n}\theta})^t = e^{-\tilde{n}\theta} = e^{\tilde{n}(-\theta)} = e^{(-\tilde{n})\theta} \tag{2.65}$$

Equation (2.65) shows that a rotation can be reversed either by reversing the rotation angle or by reversing the unit vector of the rotation axis.

* **Combination of Successive Rotation Matrices**

Two successive rotations about skew axes are neither commutative nor additive. In other words, if $\overline{m} \neq \overline{n}$,

$$e^{\tilde{n}\theta} e^{\tilde{m}\phi} \neq e^{\tilde{m}\phi} e^{\tilde{n}\theta} \neq e^{\tilde{n}\theta+\tilde{m}\phi} \tag{2.66}$$

Two successive rotations about parallel or coincident axes are both commutative and additive. In other words, if $\overline{m} = \overline{n}$,

$$e^{\tilde{n}\theta} e^{\tilde{n}\phi} = e^{\tilde{n}\phi} e^{\tilde{n}\theta} = e^{\tilde{n}\theta+\tilde{n}\phi} = e^{\tilde{n}(\theta+\phi)} \tag{2.67}$$

* **Additional Full, Half, and Quarter Rotations**

The effect of a full additional rotation is nil. That is,

$$e^{\tilde{n}(\theta+2\sigma\pi)} = e^{\tilde{n}\theta} \Rightarrow e^{\tilde{n}(2\sigma\pi)} = e^{\tilde{n}0} = \hat{I} \tag{2.68}$$

The effect of a half additional rotation can be expressed as follows:

$$e^{\tilde{n}(\theta+\sigma\pi)} = -e^{\tilde{n}\theta} + 2\overline{n}\,\overline{n}^t \Rightarrow e^{\tilde{n}(\sigma\pi)} = -\hat{I} + 2\overline{n}\,\overline{n}^t = \hat{I} + 2\tilde{n}^2 \tag{2.69}$$

The effect of a quarter additional rotation can be expressed as follows:

$$e^{\tilde{n}(\theta+\sigma\pi/2)} = \sigma\tilde{n}e^{\tilde{n}\theta} + \overline{n}\,\overline{n}^t \Rightarrow e^{\tilde{n}(\sigma\pi/2)} = \sigma\tilde{n} + \overline{n}\,\overline{n}^t \tag{2.70}$$

In the preceding formulas, σ is an arbitrary sign variable, i.e. $\sigma = \pm 1$.

* **Effectivity of a Rotation Operator**

A rotation operator is ineffective on the unit vector of its own axis. That is,

$$e^{\tilde{n}\theta}\overline{n} = \overline{n} \text{ and } \overline{n}^t e^{\tilde{n}\theta} = \overline{n}^t \tag{2.71}$$

However, one must be careful that

$$e^{\tilde{n}\theta}\tilde{n} = \tilde{n}e^{\tilde{n}\theta} \neq \tilde{n} \tag{2.72}$$

* **Angular Differentiation of a Rotation Matrix**

A rotation matrix $e^{\tilde{n}\theta}$ can be differentiated with respect to θ as follows:

$$\partial(e^{\tilde{n}\theta})/\partial\theta = \tilde{n}e^{\tilde{n}\theta} = e^{\tilde{n}\theta}\tilde{n} \tag{2.73}$$

* **Rotation About Rotated Axis**

Let $\vec{n}$ be rotated into $\vec{m}$ by $\text{rot}(\vec{u}, \beta)$ so that

$$\overline{m} = e^{\tilde{u}\beta}\overline{n} \tag{2.74}$$

Then, it can be shown that Eq. (2.74) leads to the following equations.

$$\tilde{m} = e^{\tilde{u}\beta}\tilde{n}e^{-\tilde{u}\beta} \tag{2.75}$$

$$e^{\tilde{m}\theta} = e^{\tilde{u}\beta} e^{\tilde{n}\theta} e^{-\tilde{u}\beta} \tag{2.76}$$

Equation (2.76) is the expression of the *rotation about rotated axis formula.*

* **Shifting Formulas for the Rotation Matrices**

The following two formulas, which are called *shifting formulas,* can be obtained as two consequences of Eq. (2.76).

$$e^{\tilde{u}\beta} e^{\tilde{n}\theta} = e^{\tilde{m}\theta} e^{\tilde{u}\beta}, \text{ where } \overline{m} = e^{\tilde{u}\beta}\overline{n} \tag{2.77}$$

$$e^{\tilde{n}\theta} e^{\tilde{u}\beta} = e^{\tilde{u}\beta} e^{\tilde{m}'\theta}, \text{ where } \overline{m}' = e^{-\tilde{u}\beta}\overline{n} \tag{2.78}$$

2.7.2 Mathematical Properties of the Basic Rotation Matrices

In the following formulas, σ_{ijk} is defined as in Chapter 1. That is, for distinct indices only,

$$\sigma_{ijk} = \begin{cases} +1 \text{ if } ijk = 123,231,312 \\ -1 \text{ if } ijk = 321,132,213 \end{cases} \tag{2.79}$$

$*$ **Expansion Formulas**

If $j \neq i$,

$$e^{\tilde{u}_i\theta}\overline{u}_j = \overline{u}_j \cos\theta + (\tilde{u}_i\overline{u}_j)\sin\theta = \overline{u}_j \cos\theta + \sigma_{ijk}\overline{u}_k \sin\theta \tag{2.80}$$

$$\overline{u}_j^t e^{\tilde{u}_i\theta} = \overline{u}_j^t \cos\theta + (\tilde{u}_j\overline{u}_i)^t \sin\theta = \overline{u}_j^t \cos\theta + \sigma_{jik}\overline{u}_k^t \sin\theta \tag{2.81}$$

If $j = i$,

$$e^{\tilde{u}_i\theta}\overline{u}_i = \overline{u}_i \text{ and } \overline{u}_i^t e^{\tilde{u}_i\theta} = \overline{u}_i^t \tag{2.82}$$

$*$ **Shifting Formulas with Quarter and Half Rotations**

If $j \neq i$,

$$e^{\tilde{u}_i\pi/2}e^{\tilde{u}_j\theta} = e^{\sigma_{ijk}\tilde{u}_k\theta}e^{\tilde{u}_i\pi/2} \tag{2.83}$$

$$e^{\tilde{u}_i\pi}e^{\tilde{u}_j\theta} = e^{-\tilde{u}_j\theta}e^{\tilde{u}_i\pi} \tag{2.84}$$

If $j = i$,

$$e^{\tilde{u}_i\pi/2}e^{\tilde{u}_i\theta} = e^{\tilde{u}_i\theta}e^{\tilde{u}_i\pi/2} = e^{\tilde{u}_i(\theta+\pi/2)} \tag{2.85}$$

$$e^{\tilde{u}_i\pi}e^{\tilde{u}_i\theta} = e^{\tilde{u}_i\theta}e^{\tilde{u}_i\pi} = e^{\tilde{u}_i(\theta+\pi)} \tag{2.86}$$

$*$ **Three Successive Half Rotations About Mutually Orthogonal Axes**

Provided that $i \neq j \neq k$,

$$e^{\pm\tilde{u}_i\pi}e^{\pm\tilde{u}_j\pi}e^{\pm\tilde{u}_k\pi} = \hat{I} \tag{2.87}$$

2.8 Examples Involving Rotation Matrices

2.8.1 Example 2.1

The first basis vector of a reference frame $\mathcal{F}_a$ is rotated successively in two different sequences, which are indicated below. It is required to express the resultant vectors in $\mathcal{F}_a$.

$$\text{First sequence: } \vec{u}_1^{(a)} \xrightarrow{\text{rot}[\vec{u}_2^{(a)},\phi]} \vec{n}_2 \xrightarrow{\text{rot}[\vec{u}_3^{(a)},\theta]} \vec{n}_3 \tag{2.88}$$

$$\text{Second sequence: } \vec{u}_1^{(a)} \xrightarrow{\text{rot}[\vec{u}_3^{(a)},\theta]} \vec{m}_2 \xrightarrow{\text{rot}[\vec{u}_2^{(a)},\phi]} \vec{m}_3 \tag{2.89}$$

In the first sequence, $\vec{n}_2$ and $\vec{n}_3$ are obtained as described below.

$$\overline{n}_2^{(a)} = e^{\tilde{u}_2^{(a/a)}\phi}\overline{u}_1^{(a/a)} = e^{\tilde{u}_2\phi}\overline{u}_1 = \overline{u}_1\cos\phi - \overline{u}_3\sin\phi \tag{2.90}$$

$$\overline{n}_3^{(a)} = e^{\tilde{u}_3^{(a/a)}\theta}\overline{n}_2^{(a)} = e^{\tilde{u}_3\theta}e^{\tilde{u}_2\phi}\overline{u}_1 = e^{\tilde{u}_3\theta}(\overline{u}_1\cos\phi - \overline{u}_3\sin\phi) \Rightarrow$$

$$\overline{n}_3^{(a)} = (e^{\tilde{u}_3\theta}\overline{u}_1)\cos\phi - (e^{\tilde{u}_3\theta}\overline{u}_3)\sin\phi \Rightarrow$$

$$\overline{n}_3^{(a)} = (\overline{u}_1 \cos\theta + \overline{u}_2 \sin\theta)\cos\phi - (\overline{u}_3)\sin\phi \Rightarrow$$
$$\overline{n}_3^{(a)} = \overline{u}_1 \cos\theta\cos\phi + \overline{u}_2 \sin\theta\cos\phi - \overline{u}_3 \sin\phi \quad (2.91)$$

Noting that $\overline{u}_k = \overline{u}_k^{(a/a)} = [\vec{u}_k^{(a)}]^{(a)}$, the vector equations corresponding to Eqs. (2.90) and (2.91) can be written as follows:

$$\vec{n}_2 = \vec{u}_1^{(a)} \cos\phi - \vec{u}_3^{(a)} \sin\phi \quad (2.92)$$
$$\vec{n}_3 = \vec{u}_1^{(a)} \cos\theta\cos\phi + \vec{u}_2^{(a)} \sin\theta\cos\phi - \vec{u}_3^{(a)} \sin\phi \quad (2.93)$$

In the second sequence, $\vec{m}_2$ and $\vec{m}_3$ are obtained as described below.

$$\overline{m}_2^{(a)} = e^{\tilde{u}_3^{(a/a)}\theta}\overline{u}_1^{(a/a)} = e^{\tilde{u}_3\theta}\overline{u}_1 = \overline{u}_1 \cos\theta + \overline{u}_2 \sin\theta \quad (2.94)$$
$$\overline{m}_3^{(a)} = e^{\tilde{u}_2^{(a/a)}\phi}\overline{m}_2^{(a)} = e^{\tilde{u}_2\phi}e^{\tilde{u}_3\theta}\overline{u}_1 = e^{\tilde{u}_2\phi}(\overline{u}_1 \cos\theta + \overline{u}_2 \sin\theta) \Rightarrow$$
$$\overline{m}_3^{(a)} = (e^{\tilde{u}_2\phi}\overline{u}_1)\cos\theta + (e^{\tilde{u}_2\phi}\overline{u}_2)\sin\theta \Rightarrow$$
$$\overline{m}_3^{(a)} = (\overline{u}_1 \cos\phi - \overline{u}_3 \sin\phi)\cos\theta + (\overline{u}_2)\sin\theta \Rightarrow$$
$$\overline{m}_3^{(a)} = \overline{u}_1 \cos\phi\cos\theta + \overline{u}_2 \sin\theta - \overline{u}_3 \sin\phi\cos\theta \quad (2.95)$$

The corresponding vector equations can be written as follows, like those written for $\vec{n}_2$ and $\vec{n}_3$:

$$\vec{m}_2 = \vec{u}_1^{(a)} \cos\theta + \vec{u}_2^{(a)} \sin\theta \quad (2.96)$$
$$\vec{m}_3 = \vec{u}_1^{(a)} \cos\phi\cos\theta + \vec{u}_2^{(a)} \sin\theta - \vec{u}_3^{(a)} \sin\phi\cos\theta \quad (2.97)$$

2.8.2 Example 2.2

Equation (2.75) is verified here as an example.

Consider three vectors represented by the column matrices $\overline{p}$, $\overline{q}$, and $\overline{r}$ in a reference frame $\mathcal{F}_a$. Let these vectors be such that

$$\overline{r} = \tilde{p}\overline{q} \quad (2.98)$$

Let the same vectors be rotated by the rotation operator represented by $e^{\tilde{u}\beta}$ so that

$$\overline{p}' = e^{\tilde{u}\beta}\overline{p} \quad (2.99)$$
$$\overline{q}' = e^{\tilde{u}\beta}\overline{q} \quad (2.100)$$
$$\overline{r}' = e^{\tilde{u}\beta}\overline{r} \quad (2.101)$$

Since a rotation operator retains the cross product relationship, the new vectors also satisfy an equation similar to Eq. (2.98). That is,

$$\overline{r}' = \tilde{p}'\overline{q}' \quad (2.102)$$

Using Eqs. (2.99)–(2.101), Eq. (2.102) can be manipulated as follows:

$$e^{\tilde{u}\beta}\overline{r} = \tilde{p}' e^{\tilde{u}\beta}\overline{q} \Rightarrow \overline{r} = (e^{-\tilde{u}\beta}\tilde{p}' e^{\tilde{u}\beta})\overline{q} = \tilde{p}\overline{q} \quad (2.103)$$

Hence, it is seen that

$$e^{-\tilde{u}\beta}\tilde{p}' e^{\tilde{u}\beta} = \tilde{p} \Rightarrow \tilde{p}' = e^{\tilde{u}\beta}\tilde{p}e^{-\tilde{u}\beta} \quad (2.104)$$

Equations (2.99) and (2.104) imply the following mutual correspondence, which verifies Eq. (2.75) when $\overline{p}$ and $\overline{p}'$ are replaced with $\overline{n}$ and $\overline{m}$, respectively.

$$\overline{p}' = e^{\tilde{u}\beta}\overline{p} \Leftrightarrow \tilde{p}' = \text{ssm}(\overline{p}') = \text{ssm}(e^{\tilde{u}\beta}\overline{p}) = e^{\tilde{u}\beta}\tilde{p}e^{-\tilde{u}\beta} \quad (2.105)$$

2.8.3 Example 2.3

Equation (2.76) is verified here as a follow-up to Example 2.2.

Starting with Eq. (2.75), the powers of $\widetilde{m}$ can be expressed as follows:

$$\widetilde{m} = e^{\widetilde{u}\beta}\widetilde{n}e^{-\widetilde{u}\beta} \Rightarrow$$

$$\widetilde{m}^2 = (e^{\widetilde{u}\beta}\widetilde{n}e^{-\widetilde{u}\beta})(e^{\widetilde{u}\beta}\widetilde{n}e^{-\widetilde{u}\beta}) = e^{\widetilde{u}\beta}\widetilde{n}^2e^{-\widetilde{u}\beta} \Rightarrow$$

$$\widetilde{m}^3 = \widetilde{m}^2\widetilde{m} = (e^{\widetilde{u}\beta}\widetilde{n}^2e^{-\widetilde{u}\beta})(e^{\widetilde{u}\beta}\widetilde{n}e^{-\widetilde{u}\beta}) = e^{\widetilde{u}\beta}\widetilde{n}^3e^{-\widetilde{u}\beta}$$

Thus, for all $k \geq 0$, it happens that

$$\widetilde{m}^k = e^{\widetilde{u}\beta}\widetilde{n}^ke^{-\widetilde{u}\beta} \tag{2.106}$$

On the other hand, the following Taylor series expansion can be written for $e^{\widetilde{m}\theta}$.

$$e^{\widetilde{m}\theta} = \hat{I} + \widetilde{m}\theta + \frac{1}{2!}(\widetilde{m}\theta)^2 + \frac{1}{3!}(\widetilde{m}\theta)^3 + \frac{1}{4!}(\widetilde{m}\theta)^4 + \cdots \tag{2.107}$$

Using Eq. (2.106) and noting that $e^{\widetilde{u}\beta}e^{-\widetilde{u}\beta} = e^{\widetilde{u}\beta}\hat{I}e^{-\widetilde{u}\beta} = \hat{I}$, Eq. (2.107) can be written in the following factorized form.

$$e^{\widetilde{m}\theta} = e^{\widetilde{u}\beta}\left[\hat{I} + \widetilde{n}\theta + \frac{1}{2!}(\widetilde{n}\theta)^2 + \frac{1}{3!}(\widetilde{n}\theta)^3 + \frac{1}{4!}(\widetilde{n}\theta)^4 + \cdots\right]e^{-\widetilde{u}\beta} \tag{2.108}$$

Hence, as the verification of Eq. (2.76), Eq. (2.108) implies that

$$e^{\widetilde{m}\theta} = e^{\widetilde{u}\beta}e^{\widetilde{n}\theta}e^{-\widetilde{u}\beta} \tag{2.109}$$

2.8.4 Example 2.4

In this example, Eq. (2.87) will be verified for $ijk = 123$ as explained below. However, it can be verified similarly for any other distinct (unrepeated) triplet of indices as well.

For $ijk = 123$, Eq. (2.87) becomes

$$e^{\widetilde{u}_1\pi}e^{\widetilde{u}_2\pi}e^{\widetilde{u}_3\pi} = \hat{I} \tag{2.110}$$

Let $\hat{J}$ be the product of the three matrices on the left-hand side of Eq. (2.110). It can be manipulated as follows:

$$\hat{J} = e^{\widetilde{u}_1\pi}e^{\widetilde{u}_2\pi}e^{\widetilde{u}_3\pi} = e^{\widetilde{u}_1\pi}e^{\widetilde{u}_2\pi/2}e^{\widetilde{u}_2\pi/2}e^{\widetilde{u}_3\pi} \tag{2.111}$$

According to the *shifting formulas* expressed by Eqs. (2.83)–(2.86), Eq. (2.111) can be manipulated further as shown below.

$$\hat{J} = e^{\widetilde{u}_1\pi}e^{\widetilde{u}_2\pi/2}(e^{\widetilde{u}_2\pi/2}e^{\widetilde{u}_3\pi}) = e^{\widetilde{u}_1\pi}e^{\widetilde{u}_2\pi/2}(e^{\widetilde{u}_1\pi}e^{\widetilde{u}_2\pi/2}) \Rightarrow$$

$$\hat{J} = e^{\widetilde{u}_1\pi}(e^{\widetilde{u}_2\pi/2}e^{\widetilde{u}_1\pi})e^{\widetilde{u}_2\pi/2} = e^{\widetilde{u}_1\pi}(e^{\widetilde{u}_1\pi}e^{-\widetilde{u}_2\pi/2})e^{\widetilde{u}_2\pi/2} \Rightarrow$$

$$\hat{J} = (e^{\widetilde{u}_1\pi}e^{\widetilde{u}_1\pi})(e^{-\widetilde{u}_2\pi/2}e^{\widetilde{u}_2\pi/2}) = (e^{2\widetilde{u}_1\pi})(\hat{I}) \tag{2.112}$$

According to Eq. (2.68), $e^{2\widetilde{u}_1\pi} = e^{\widetilde{u}_1 0} = \hat{I}$. Therefore, Eq. (2.112) reduces to

$$\hat{J} = \hat{I} \tag{2.113}$$

Hence, it is seen that Eq. (2.110) is verified owing to Eq. (2.113).

2.9 Determination of the Angle and Axis of a Specified Rotation Matrix

2.9.1 Scalar Equations of Rotation

Consider the following equation, in which $\hat{R} = \hat{R}^{(a)}$ is a specified rotation matrix expressed in an observation frame $\mathcal{F}_a$.

$$\hat{R} = e^{\tilde{n}\theta} = \hat{I}\cos\theta + \tilde{n}\sin\theta + \overline{n}\,\overline{n}^t(1 - \cos\theta) \tag{2.114}$$

Most typically, $\hat{R}$ is specified as an overall rotation matrix in a case of m successive rotations so that $\hat{R} = \hat{R}_{0m} = \hat{R}_{(m-1)m} \cdots \hat{R}_{12}\hat{R}_{01}$. Here, it is required to find the angle and axis, i.e. θ and $\overline{n} = \overline{n}^{(a)}$, of the overall rotation operator. In order to find them, Eq. (2.114) can be written again in the following detailed form.

$$\begin{bmatrix} r_{11} & r_{12} & r_{13} \\ r_{21} & r_{22} & r_{23} \\ r_{31} & r_{32} & r_{33} \end{bmatrix} = \begin{bmatrix} 1 & 0 & 0 \\ 0 & 1 & 0 \\ 0 & 0 & 1 \end{bmatrix}\cos\theta + \begin{bmatrix} 0 & -n_3 & n_2 \\ n_3 & 0 & -n_1 \\ -n_2 & n_1 & 0 \end{bmatrix}\sin\theta + \begin{bmatrix} n_1^2 & n_1n_2 & n_1n_3 \\ n_2n_1 & n_2^2 & n_2n_3 \\ n_3n_1 & n_3n_2 & n_3^2 \end{bmatrix}(1 - \cos\theta) \tag{2.115}$$

Equation (2.115) provides the following scalar equations.

$$r_{11} = \cos\theta + n_1^2(1 - \cos\theta) \tag{2.116}$$

$$r_{22} = \cos\theta + n_2^2(1 - \cos\theta) \tag{2.117}$$

$$r_{33} = \cos\theta + n_3^2(1 - \cos\theta) \tag{2.118}$$

$$r_{12} = n_1n_2(1 - \cos\theta) - n_3\sin\theta \tag{2.119}$$

$$r_{21} = n_2n_1(1 - \cos\theta) + n_3\sin\theta \tag{2.120}$$

$$r_{23} = n_2n_3(1 - \cos\theta) - n_1\sin\theta \tag{2.121}$$

$$r_{32} = n_3n_2(1 - \cos\theta) + n_1\sin\theta \tag{2.122}$$

$$r_{31} = n_3n_1(1 - \cos\theta) - n_2\sin\theta \tag{2.123}$$

$$r_{13} = n_1n_3(1 - \cos\theta) + n_2\sin\theta \tag{2.124}$$

Furthermore, the following three equations can be obtained from Eqs. (2.119)–(2.124).

$$n_1\sin\theta = (r_{32} - r_{23})/2 \tag{2.125}$$

$$n_2\sin\theta = (r_{13} - r_{31})/2 \tag{2.126}$$

$$n_3\sin\theta = (r_{21} - r_{12})/2 \tag{2.127}$$

2.9.2 Determination of the Angle of Rotation

Note that $|\overline{n}| = 1$, i.e.

$$n_1^2 + n_2^2 + n_3^2 = 1 \tag{2.128}$$

Therefore, the side-by-side addition of Eqs. (2.116)–(2.118) leads to the following equation.

$$r_{11} + r_{22} + r_{33} = \text{trace}(\widehat{R}) = 2\cos\theta + 1 \tag{2.129}$$

Hence, $\cos\theta$ and $\sin\theta$ are found as follows:

$$\cos\theta = \xi = (r_{11} + r_{22} + r_{33} - 1)/2 \tag{2.130}$$

$$\sin\theta = \eta = \sigma\sqrt{1 - \xi^2} \tag{2.131}$$

In Eq. (2.131), σ is an arbitrary sign variable, i.e.

$$\sigma = \pm 1 \tag{2.132}$$

With the availability of $\sin\theta$ and $\cos\theta$, θ can be found by means of the *double argument arctangent function*. That is,

$$\theta = \text{atan}_2(\eta, \xi) = \text{atan}_2(\sigma\sqrt{1 - \xi^2}, \xi) = \sigma\text{atan}_2(\sqrt{1 - \xi^2}, \xi) \tag{2.133}$$

The definition and the properties of the atan_2 function can be seen in Section 2.10.

2.9.3 Determination of the Axis of Rotation

(a) General Case with $\sin\theta \neq 0$

In a general case such that $\sin\theta \neq 0$, Eqs. (2.125)–(2.127) give the components of $\overline{n}$ as follows:

$$n_1 = (r_{32} - r_{23})/(2\sin\theta) = \sigma(r_{32} - r_{23})/(2\sqrt{1 - \xi^2}) \tag{2.134}$$

$$n_2 = (r_{13} - r_{31})/(2\sin\theta) = \sigma(r_{13} - r_{31})/(2\sqrt{1 - \xi^2}) \tag{2.135}$$

$$n_3 = (r_{21} - r_{12})/(2\sin\theta) = \sigma(r_{21} - r_{12})/(2\sqrt{1 - \xi^2}) \tag{2.136}$$

(b) Special Cases with $\sin\theta = 0$

However, if $\sin\theta = 0$, then Eqs. (2.125)–(2.127) cannot give the components of $\overline{n}$. There are three distinct cases, in which $\sin\theta = 0$. These cases are discussed below by considering θ to be in the *minimal rotation range*, without any loss of generality. This range is defined so that $0 \leq |\theta| \leq 2\pi$.

$*$ **Special Case with No Rotation**

In this case, $\theta = 0$ and $\overline{n}$ becomes indefinite. Indeed, if there is not any rotation at the moment, then $\overline{n}$ represents any arbitrary unit vector that forms the axis of a prospective rotation, which is yet unknown.

* Special Case with Full Rotation

In this case, $\theta = \pm 2\pi$ and $\overline{n}$ becomes indefinite again. Indeed, after a full rotation, a rotated vector comes back to its initial orientation no matter what the rotation axis is. In other words, according to the mathematical property expressed by Eq. (2.68) in Section 2.7, the following equation is satisfied for any arbitrary $\overline{n}$.

$$e^{\tilde{n}(\pm 2\pi)} = e^{\tilde{n}0} = \hat{I} \tag{2.137}$$

* Special Case with Half Rotation

In this case, $\theta = \pm\pi$. However, unlike the first two cases, $\overline{n}$ does not become indefinite in this case, even though it cannot be found from Eqs. (2.125)–(2.127). Indeed, after a half rotation of a vector, the bisector of the angle between the initial and final orientations of the vector coincides with the rotation axis. This fact is illustrated in Figure 2.2 for two different half rotations of the same vector. Owing to this fact, $\overline{n}$ happens to be definite and therefore it can be found somehow, e.g. as explained below, if not from Eqs. (2.125)–(2.127).

In the case of a half rotation, with $\sin\theta = 0$ and $\cos\theta = -1$, Eqs. (2.116)–(2.124) reduce to the following forms.

$$r_{11} = 2n_1^2 - 1 \tag{2.138}$$

$$r_{22} = 2n_2^2 - 1 \tag{2.139}$$

$$r_{33} = 2n_3^2 - 1 \tag{2.140}$$

$$r_{12} = r_{21} = 2n_1n_2 \tag{2.141}$$

$$r_{23} = r_{32} = 2n_2n_3 \tag{2.142}$$

$$r_{31} = r_{13} = 2n_3n_1 \tag{2.143}$$

Equations (2.138)–(2.140) provide the components of $\overline{n}$ as follows with three arbitrary sign variables, which are $\sigma_1 = \pm 1$, $\sigma_2 = \pm 1$, and $\sigma_3 = \pm 1$:

$$n_1 = \sigma_1\sqrt{(1 + r_{11})/2} \tag{2.144}$$

$$n_2 = \sigma_2\sqrt{(1 + r_{22})/2} \tag{2.145}$$

$$n_3 = \sigma_3\sqrt{(1 + r_{33})/2} \tag{2.146}$$

However, the three sign variables are *not* independent from each other because the components of $\overline{n}$ obtained above must also satisfy Eqs. (2.141)–(2.143). Here, another set of three cases arises depending on whether all the components of $\overline{n}$ are nonzero or not.

Case 1: All the components of $\overline{n}$ are nonzero.

In this case, Eqs. (2.141)–(2.146) imply the following sign variable equations.

$$\sigma_1\sigma_2 = \text{sgn}(r_{12}) = \text{sgn}(r_{21}) \tag{2.147}$$

$$\sigma_2\sigma_3 = \text{sgn}(r_{23}) = \text{sgn}(r_{32}) \tag{2.148}$$

$$\sigma_3\sigma_1 = \text{sgn}(r_{31}) = \text{sgn}(r_{13}) \tag{2.149}$$

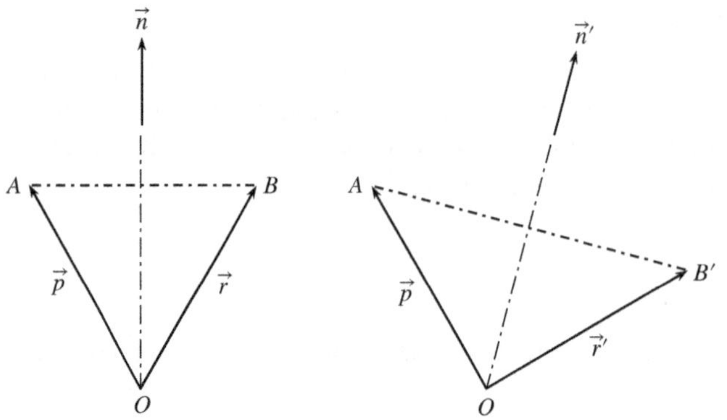

Figure 2.2 Half rotations of the same vector about different axes.

When two of the preceding three equations are multiplied side by side, the following additional equations are obtained by noting that $\sigma_1^2 = \sigma_2^2 = \sigma_3^2 = 1$.

$$\sigma_1\sigma_2 = \text{sgn}(r_{23})\text{sgn}(r_{31}) \tag{2.150}$$

$$\sigma_2\sigma_3 = \text{sgn}(r_{31})\text{sgn}(r_{12}) \tag{2.151}$$

$$\sigma_3\sigma_1 = \text{sgn}(r_{12})\text{sgn}(r_{23}) \tag{2.152}$$

Hence, it is seen that Eqs. (2.150)–(2.152) provide the three sign variables as follows in terms of a single arbitrary sign variable $\sigma' = \pm 1$.

$$\sigma_1 = \sigma'\text{sgn}(r_{23}) \tag{2.153}$$

$$\sigma_2 = \sigma'\text{sgn}(r_{31}) \tag{2.154}$$

$$\sigma_3 = \sigma'\text{sgn}(r_{12}) \tag{2.155}$$

Case 2: One of the components of $\overline{n}$ is zero.

Let $ijk \in \{123, 231, 312\}$ and suppose that $n_k = 0$, while $n_i \neq 0$ and $n_j \neq 0$. In such a case, Eqs. (2.138)–(2.143) imply that $r_{ik} = r_{ki} = r_{kj} = r_{jk} = 0$, $r_{kk} = -1$, and

$$r_{ij} = r_{ji} = 2n_in_j \tag{2.156}$$

Equation (2.156) can be satisfied if σ_i and σ_j are expressed as follows in terms of a single arbitrary sign variable $\sigma' = \pm 1$:

$$\sigma_i = \sigma' \tag{2.157}$$

$$\sigma_j = \sigma'\text{sgn}(r_{ij}) \tag{2.158}$$

Case 3: Two of the components of $\overline{n}$ are zero.

Let $ijk \in \{123, 231, 312\}$ again and this time suppose that $n_i = n_j = 0$, while $n_k \neq 0$. In such a case, Eqs. (2.138)–(2.143) imply that $r_{kk} = +1$, $r_{ij} = r_{ji} = r_{jk} = r_{kj} = r_{ki} = r_{ik} = 0$, and $r_{ii} = r_{jj} = -1$. Then, as the only nonzero component,

$$n_k = \sigma_k = \sigma' = \pm 1 \tag{2.159}$$

2.9.4 Discussion About the Optional Sign Variables

Considering the sign variables σ and σ' that occur in Sections 2.9.2 and 2.9.3, it is possible to select them as $\sigma = +1$ and $\sigma' = +1$ without much loss of generality. A discussion about this statement is presented below.

(a) General Case with $\sin\theta \neq 0$

In such a case, according to Eqs. (2.131)–(2.136), if $\sigma = +1$ leads to θ and $\overline{n}$, then $\sigma = -1$ leads to $\theta' = -\theta$ and $\overline{n}' = -\overline{n}$. However, the pair $\{\overline{n}, \theta\}$ is equivalent to the pair $\{-\overline{n}, -\theta\}$ as confirmed by the following equation.

$$e^{\tilde{n}\theta} = e^{(-\tilde{n})(-\theta)} = \widehat{R} \tag{2.160}$$

Due to Eq. (2.160), σ has no effect on the rotation matrix $\widehat{R}$. Therefore, in a case such that θ and $\overline{n}$ are required to be determined only once in a while or in a somewhat special case such that θ and $\overline{n}$ are required to be determined frequently but the successive values of θ never become zero, the sign ambiguity may be eliminated by selecting the option with $\sigma = +1$ so that $\theta > 0$. However, in a case such that θ and $\overline{n}$ are required to be determined frequently and the successive values of θ turn out to be fluctuating in the vicinity of zero, it may be more appropriate to have θ change its sign (i.e. to have σ switching between $+1$ and -1) rather than having $\vec{n}$ change its orientation from one direction to the opposite one abruptly and frequently. In other words, it may be more preferable to have $\{\overline{n}, \pm|\theta|\}$ rather than $\{\pm\overline{n}, |\theta|\}$.

(b) Special Case with a Half Rotation

In such a case, $\theta = \sigma\pi$ with $\sigma = \pm 1$. On the other hand, as observed in Eqs. (2.153–2.159), σ' (i.e. the sense of $\vec{n}$) is *not* related to $\sigma = \text{sgn}(\theta)$. As a matter of fact, the angle-axis pair $\{\sigma\pi, \sigma'\overline{n}\}$ leads to the same rotation matrix, whatever σ and σ' are. This statement is confirmed as shown below.

$$e^{(\sigma'\tilde{n})(\sigma\pi)} = e^{\tilde{n}(\sigma'\sigma\pi)} = e^{\tilde{n}(\pm\pi)} = e^{\tilde{n}\pi} = \widehat{R} \tag{2.161}$$

Owing to Eq. (2.161), whatever σ is, the sign ambiguity caused by σ' can again be eliminated by selecting the option with $\sigma' = +1$, if θ and $\overline{n}$ are required to be determined only once in a while. However, if θ and $\overline{n}$ are required to be determined frequently in a case such that the values of θ happen to be $\sigma\pi$ at certain successive instants, then it may be more appropriate to prevent the possibility that $\vec{n}$ changes its direction abruptly as soon as θ becomes $\sigma\pi$. In other words, instead of insisting on the choice $\sigma' = +1$, it may be preferable to choose σ' so that $\overline{n}(t)$ with $\theta(t) = \sigma\pi$ and $\overline{n}(t - \Delta t)$ with $\theta(t - \Delta t) = (\sigma\pi - \dot{\theta}\Delta t)$ are almost codirectional, i.e. $\overline{n}^t(t)\overline{n}(t - \Delta t) > 0$.

2.10 Definition and Properties of the Double Argument Arctangent Function

The double argument arctangent function, which is denoted as $\text{atan}_2(\eta, \xi)$, is defined so that it satisfies the following identity for any $\rho > 0$.

$$\theta \equiv \text{atan}_2(\rho\sin\theta, \rho\cos\theta) \tag{2.162}$$

The most characteristic feature of the function $\text{atan}_2(\eta, \xi)$ is that it gives θ without any quadrant ambiguity in the following interval, whenever $\eta \neq 0$ and $\xi \neq 0$.

$$-\pi < \theta < \pi$$

More specifically, the function $\text{atan}_2(\eta, \xi)$ gives the outcome θ as follows depending on the values of the arguments η and ξ:

$$\eta > 0, \quad \xi > 0 \Rightarrow 0 < \theta < \pi/2$$

(θ is in Quadrant I.)

$$\eta > 0, \quad \xi < 0 \Rightarrow \pi/2 < \theta < \pi$$

(θ is in Quadrant II.)

$$\eta < 0, \quad \xi < 0 \Rightarrow -\pi < \theta < -\pi/2$$

(θ is in Quadrant III.)

$$\eta < 0, \quad \xi > 0 \Rightarrow -\pi/2 < \theta < 0$$

(θ is in Quadrant IV.)

$$\eta = 0, \quad \xi > 0 \Rightarrow \theta = 0$$

$$\eta = 0, \quad \xi < 0 \Rightarrow \theta = \pm\pi$$

(θ has a *sign* ambiguity if $\eta = 0$ when $\xi < 0$.)

$$\eta > 0, \quad \xi = 0 \Rightarrow \theta = \pi/2$$

$$\eta < 0, \quad \xi = 0 \Rightarrow \theta = -\pi/2$$

$$\eta = 0, \quad \xi = 0 \Rightarrow \theta = ?$$

(θ becomes indefinite if $\eta = \xi = 0$.)

3

Matrix Representations of Vectors in Different Reference Frames and the Component Transformation Matrices

Synopsis

As mentioned in Chapter 1, the vectors are independent of the reference frames in which they are observed. However, their components are naturally dependent on the selected observation frames. In other words, they have different components and matrix representations in different reference frames. Their matrix representations in different reference frames are related to each other by means of the *transformation matrices*. The transformation matrix between two reference frames can be expressed in various ways. It can be expressed as a *rotation matrix*, or as a *direction cosine matrix*, or as a function of the *Euler angles* of a selected sequence, or as a function of the *basis vectors* of the relevant reference frames. All these expressions are studied in this chapter together with several examples. Moreover, the matrix representations of the *position vectors* of a point in differently oriented and/or located reference frames can be related by means of either *affine* or *homogeneous* transformations. So, the affine and homogeneous transformations together with the 4×4 *homogeneous transformation matrices* are also studied in this chapter.

3.1 Matrix Representations of a Vector in Different Reference Frames

Consider a vector $\vec{r}$ and two different reference frames $\mathcal{F}_a$ and $\mathcal{F}_b$, which are shown in Figure 3.1. Both $\mathcal{F}_a$ and $\mathcal{F}_b$ are assumed to be orthonormal, right-handed, and equally scaled on their axes. They have different orientations described by the following basis vector triads.

$$\mathcal{U}_a = \{\vec{u}_k^{(a)} : k = 1, 2, 3\} \text{ and } \mathcal{U}_b = \{\vec{u}_k^{(b)} : k = 1, 2, 3\} \tag{3.1}$$

In $\mathcal{F}_a$ and $\mathcal{F}_b$, the observed vector $\vec{r}$ is resolved differently as shown below.

$$\vec{r} = \sum_{k=1}^{3} r_k^{(a)} \vec{u}_k^{(a)} = \sum_{k=1}^{3} r_k^{(b)} \vec{u}_k^{(b)} \tag{3.2}$$

In Eq. (3.2), the components of $\vec{r}$ in $\mathcal{F}_a$ and $\mathcal{F}_b$ are obtained as follows for $k \in \{1, 2, 3\}$:

$$r_k^{(a)} = \vec{r} \cdot \vec{u}_k^{(a)} \tag{3.3}$$

$$r_k^{(b)} = \vec{r} \cdot \vec{u}_k^{(b)} \tag{3.4}$$

Kinematics of General Spatial Mechanical Systems, First Edition. M. Kemal Ozgoren.

Companion Website: www.wiley.com/go/ozgoren/spatialmechanicalsystems

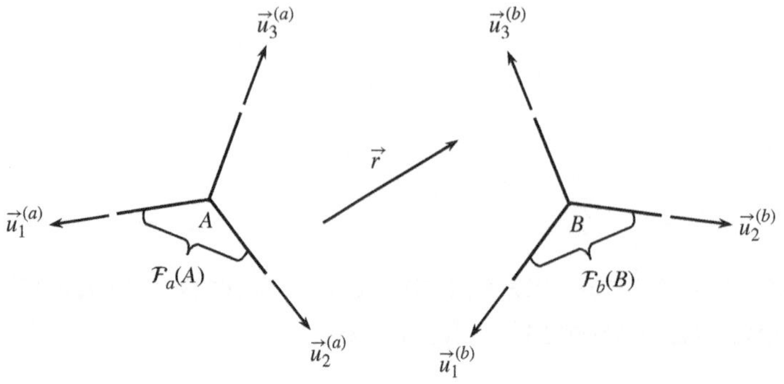

Figure 3.1 A vector observed in two different reference frames.

The components of $\vec{r}$ that are obtained above can be stacked into the following column matrices, which are defined as the matrix representations of $\vec{r}$ in $\mathcal{F}_a$ and $\mathcal{F}_b$.

$$\overline{r}^{(a)} = [\vec{r}]^{(a)} = \begin{bmatrix} r_1^{(a)} \\ r_2^{(a)} \\ r_3^{(a)} \end{bmatrix} \tag{3.5}$$

$$\overline{r}^{(b)} = [\vec{r}]^{(b)} = \begin{bmatrix} r_1^{(b)} \\ r_2^{(b)} \\ r_3^{(b)} \end{bmatrix} \tag{3.6}$$

On the other hand, by recalling the definition of the *basic column matrices* from Chapter 1, the following equations can be written.

$$\overline{u}_k^{(a/a)} = [\vec{u}_k^{(a)}]^{(a)} = \overline{u}_k \tag{3.7}$$

$$\overline{u}_k^{(b/b)} = [\vec{u}_k^{(b)}]^{(b)} = \overline{u}_k \tag{3.8}$$

Hence, referring to Eq. (3.2), $\overline{r}^{(a)}$ and $\overline{r}^{(b)}$ can also be expressed as shown below.

$$\overline{r}^{(a)} = \sum_{k=1}^{3} r_k^{(a)} \overline{u}_k^{(a/a)} = \sum_{k=1}^{3} r_k^{(a)} \overline{u}_k \tag{3.9}$$

$$\overline{r}^{(b)} = \sum_{k=1}^{3} r_k^{(b)} \overline{u}_k^{(b/b)} = \sum_{k=1}^{3} r_k^{(b)} \overline{u}_k \tag{3.10}$$

3.2 Transformation Matrices Between Reference Frames

3.2.1 Definition and Usage of a Transformation Matrix

Consider two reference frames $\mathcal{F}_a$ and $\mathcal{F}_b$. If they are *differently oriented* with respect to each other, a vector $\vec{r}$ appears differently in $\mathcal{F}_a$ and $\mathcal{F}_b$. In other words, $\overline{r}^{(a)} \neq \overline{r}^{(b)}$.

However, since $\overline{r}^{(a)}$ and $\overline{r}^{(b)}$ represent the same vector $\vec{r}$, they are nonetheless related so that

$$\overline{r}^{(a)} = \hat{C}^{(a,b)}\overline{r}^{(b)} \tag{3.11}$$

In Eq. (3.11), $\hat{C}^{(a,b)}$ is defined as the *transformation matrix* between $\mathcal{F}_a$ and $\mathcal{F}_b$. It can be considered as an operator that transforms the components of a vector from $\mathcal{F}_b$ to $\mathcal{F}_a$. So, to be more specific, it may also be called a *component transformation matrix.*

3.2.2 Basic Properties of a Transformation Matrix

(a) Inversion Property

Equation (3.11) can also be written in the following two ways: first by inverting $\hat{C}^{(a,b)}$; and then by interchanging the frame indicators a and b.

$$\overline{r}^{(b)} = [\hat{C}^{(a,b)}]^{-1}\overline{r}^{(a)} \tag{3.12}$$

$$\overline{r}^{(b)} = \hat{C}^{(b,a)}\overline{r}^{(a)} \tag{3.13}$$

Equations (3.11)–(3.13) imply the following equalities.

$$[\hat{C}^{(a,b)}]^{-1} = \hat{C}^{(b,a)} \text{ or } [\hat{C}^{(b,a)}]^{-1} = \hat{C}^{(a,b)} \tag{3.14}$$

$$\hat{C}^{(a,b)}\hat{C}^{(b,a)} = \hat{C}^{(b,a)}\hat{C}^{(a,b)} = \hat{I} \tag{3.15}$$

(b) Orthonormality Property

As also mentioned in Chapter 1, all the reference frames considered in this book are assumed to be orthonormal, right-handed, and equally scaled on their axes. Therefore, the magnitude of a vector $\vec{r}$ appears to be the same in every reference frame. This fact is expressed as follows:

$$|\vec{r}|^2 = \vec{r}\cdot\vec{r} = \overline{r}^{(a)t}\overline{r}^{(a)} = \overline{r}^{(b)t}\overline{r}^{(b)} \tag{3.16}$$

After substituting Eq. (3.11), Eq. (3.16) can be manipulated as shown below.

$$\overline{r}^{(a)t}\overline{r}^{(a)} = [\hat{C}^{(a,b)}\overline{r}^{(b)}]^t[\hat{C}^{(a,b)}\overline{r}^{(b)}] = \overline{r}^{(b)t}\overline{r}^{(b)} \Rightarrow$$

$$\overline{r}^{(b)t}[\hat{C}^{(a,b)t}\hat{C}^{(a,b)}]\overline{r}^{(b)} = \overline{r}^{(b)t}\overline{r}^{(b)} = \overline{r}^{(b)t}\hat{I}\overline{r}^{(b)} \tag{3.17}$$

Equation (3.17) implies the following successive equations.

$$\hat{C}^{(a,b)t}\hat{C}^{(a,b)} = \hat{I} \tag{3.18}$$

$$[\hat{C}^{(a,b)}]^{-1} = \hat{C}^{(a,b)t} \tag{3.19}$$

Furthermore, when Eqs. (3.19) and (3.14) are compared, it is seen that

$$\hat{C}^{(b,a)} = \hat{C}^{(a,b)t} \tag{3.20}$$

As seen above, the inverse of a transformation matrix is equal to its transpose. This property makes a transformation matrix an element of the set of *orthonormal matrices* just like a rotation matrix. The orthonormality of a rotation matrix was shown in Chapter 2.

(c) Combination Property

If a vector $\vec{r}$ is observed in three differently oriented reference frames, such as $\mathcal{F}_a$, $\mathcal{F}_b$, and $\mathcal{F}_c$, the following equations can be written to relate its column matrix representations.

$$\overline{r}^{(a)} = \widehat{C}^{(a,b)}\overline{r}^{(b)} \tag{3.21}$$

$$\overline{r}^{(b)} = \widehat{C}^{(b,c)}\overline{r}^{(c)} \tag{3.22}$$

$$\overline{r}^{(a)} = \widehat{C}^{(a,c)}\overline{r}^{(c)} \tag{3.23}$$

On the other hand, the combination of Eqs. (3.21) and (3.22) results in

$$\overline{r}^{(a)} = \widehat{C}^{(a,b)}\widehat{C}^{(b,c)}\overline{r}^{(c)} \tag{3.24}$$

Equations (3.23) and (3.24) imply that

$$\widehat{C}^{(a,c)} = \widehat{C}^{(a,b)}\widehat{C}^{(b,c)} \tag{3.25}$$

Equation (3.25) can be extended to more than three reference frames as follows:

$$\widehat{C}^{(a,z)} = \widehat{C}^{(a,b)}\widehat{C}^{(b,c)}\widehat{C}^{(c,d)} \dots \widehat{C}^{(x,y)}\widehat{C}^{(y,z)} \tag{3.26}$$

3.3 Expression of a Transformation Matrix in Terms of Basis Vectors

3.3.1 Column-by-Column Expression

Consider two reference frames $\mathcal{F}_a$ and $\mathcal{F}_b$. The kth basis vector of $\mathcal{F}_b$ can be represented by the following column matrix in $\mathcal{F}_a$ for $k \in \{1, 2, 3\}$.

$$\overline{u}_k^{(b/a)} = [\vec{u}_k^{(b)}]^{(a)} \tag{3.27}$$

Using the transformation matrix between $\mathcal{F}_a$ and $\mathcal{F}_b$, $\overline{u}_k^{(b/a)}$ can be expressed as follows:

$$[\vec{u}_k^{(b)}]^{(a)} = \widehat{C}^{(a,b)}[\vec{u}_k^{(b)}]^{(b)} \Rightarrow$$

$$\overline{u}_k^{(b/a)} = \widehat{C}^{(a,b)}\overline{u}_k^{(b/b)} = \widehat{C}^{(a,b)}\overline{u}_k \tag{3.28}$$

Note that $\overline{u}_k$ picks up the kth column of the matrix, by which it is postmultiplied. Thus, in Eq. (3.28), $\overline{u}_k^{(b/a)}$ happens to be the kth column of $\widehat{C}^{(a,b)}$. Therefore, $\widehat{C}^{(a,b)}$ can be expressed *column by column* as shown below.

$$\widehat{C}^{(a,b)} = [\overline{u}_1^{(b/a)} \;\; \overline{u}_2^{(b/a)} \;\; \overline{u}_3^{(b/a)}] \tag{3.29}$$

3.3.2 Row-by-Row Expression

Alternatively, Eq. (3.28) can also be written as follows by interchanging a and b:

$$[\vec{u}_k^{(a)}]^{(b)} = \widehat{C}^{(b,a)}[\vec{u}_k^{(a)}]^{(a)} \Rightarrow$$

$$\overline{u}_k^{(a/b)} = \widehat{C}^{(b,a)}\overline{u}_k^{(a/a)} = \widehat{C}^{(b,a)}\overline{u}_k \tag{3.30}$$

Recalling that $\hat{C}^{(b,a)t} = \hat{C}^{(a,b)}$, Eq. (3.30) leads to the following equation when it is transposed on both sides.

$$\overline{u}_k^{(a/b)t} = \overline{u}_k^t \hat{C}^{(b,a)t} = \overline{u}_k^t \hat{C}^{(a,b)} \tag{3.31}$$

Note that $\overline{u}_k^t$ picks up the kth row of the matrix, by which it is premultiplied. Thus, in Eq. (3.31), $\overline{u}_k^{(a/b)t}$ happens to be the kth row of $\hat{C}^{(a,b)}$. Therefore, as an alternative to Eq. (3.29), $\hat{C}^{(a,b)}$ can also be expressed *row by row* as shown below.

$$\hat{C}^{(a,b)} = \begin{bmatrix} \overline{u}_1^{(a/b)t} \\ \overline{u}_2^{(a/b)t} \\ \overline{u}_3^{(a/b)t} \end{bmatrix} \tag{3.32}$$

Here, it is worth paying attention that the column-by-column expression of $\hat{C}^{(a,b)}$ requires the column matrix expressions of the basis vectors of $\mathcal{F}_b$ in $\mathcal{F}_a$, whereas the row-by-row expression of the same $\hat{C}^{(a,b)}$ requires the row matrix expressions of the basis vectors of $\mathcal{F}_a$ in $\mathcal{F}_b$.

3.3.3 Remark 3.1

Equation (3.29) shows that the *columns* of a transformation matrix, e.g. $\hat{C}^{(a,b)}$, represent the members an *orthonormal* vector triad, i.e. $\mathcal{U}_b = \{\vec{u}_1^{(b)}, \vec{u}_2^{(b)}, \vec{u}_3^{(b)}\}$. This fact poses *six* independent scalar constraint equations on the *nine* elements of $\hat{C}^{(a,b)}$. The independent constraint equations can be expressed as follows for $i \in \{1, 2, 3\}$ and $j \in \{1, 2, 3\}$:

$$\overline{u}_i^{(b/a)t}\overline{u}_j^{(b/a)} = \vec{u}_i^{(b)} \cdot \vec{u}_j^{(b)} = \delta_{ij} = \begin{cases} 1 & \text{if } i = j \\ 0 & \text{if } i \neq j \end{cases} \tag{3.33}$$

Therefore, $\hat{C}^{(a,b)}$ can be expressed completely in terms of only *three* independent parameters.

3.3.4 Remark 3.2

The basis vector triad $\mathcal{U}_b$ associated with $\hat{C}^{(a,b)}$ is not only orthonormal but also *right-handed*. Therefore, one of the columns of $\hat{C}^{(a,b)}$ can be obtained from its other two columns by using the cross product operation. For example, $\vec{u}_3^{(b)} = \vec{u}_1^{(b)} \times \vec{u}_2^{(b)}$ and hence $\overline{u}_3^{(b/a)}$ can be obtained as follows by using the matrix equivalent of the cross product operation:

$$\overline{u}_3^{(b/a)} = \tilde{u}_1^{(b/a)}\overline{u}_2^{(b/a)} \tag{3.34}$$

Note that Eq. (3.34) gives three elements of $\hat{C}^{(a,b)}$ readily in terms of its other six elements. In other words, it already provides three independent constraint equations on the nine elements of $\hat{C}^{(a,b)}$. Therefore, in the presence of Eq. (3.34), only three additional independent constraint equations can be posed on the elements of $\hat{C}^{(a,b)}$. These three additional constraint equations are written as follows involving the first two columns of $\widehat{C}^{(a,b)}$:

$$\overline{u}_1^{(b/a)t}\overline{u}_1^{(b/a)} = 1 \tag{3.35}$$

$$\overline{u}_2^{(b/a)t}\overline{u}_2^{(b/a)} = 1 \tag{3.36}$$

$$\overline{u}_1^{(b/a)t}\overline{u}_2^{(b/a)} = 0 \tag{3.37}$$

3.3.5 Remark 3.3

Remarks 3.1 and 3.2 can also be stated similarly for the rows of $\widehat{C}^{(a,b)}$.

3.3.6 Example 3.1

In this example, the three independent parameters of $\widehat{C}^{(a,b)}$, which is briefly denoted here as $\widehat{C}$, are selected as its three elements, which are c_{11}, c_{21}, and c_{12}. The following values are specified for them.

$$c_{11} = 0.2,\ \ c_{21} = 0.4,\ \ c_{12} = 0.6 \tag{3.38}$$

It is required to determine $\widehat{C}^{(a,b)} = \widehat{C}$ by finding its remaining six elements.

The column-by-column expression of $\widehat{C}$ can be written as

$$\widehat{C} = \begin{bmatrix}\overline{c}_1 & \overline{c}_2 & \overline{c}_3\end{bmatrix} \tag{3.39}$$

By the given information, the first two columns of $\widehat{C}$ are determined partially as shown below.

$$\overline{c}_1 = \begin{bmatrix} c_{11} \\ c_{21} \\ c_{31} \end{bmatrix} = \begin{bmatrix} 0.2 \\ 0.4 \\ c_{31} \end{bmatrix},\ \ \overline{c}_2 = \begin{bmatrix} c_{12} \\ c_{22} \\ c_{32} \end{bmatrix} = \begin{bmatrix} 0.6 \\ c_{22} \\ c_{32} \end{bmatrix} \tag{3.40}$$

Since $\overline{c}_1^t\overline{c}_1 = 1$, c_{31} is found as follows with a sign ambiguity represented by $\sigma_1 = \pm 1$:

$$c_{31} = \sigma_1\sqrt{1 - (0.2)^2 - (0.4)^2} = \sigma_1(0.89443) \tag{3.41}$$

The fact that $\overline{c}_2^t\overline{c}_2 = 1$ leads to the following equation.

$$c_{22}^2 + c_{32}^2 + (0.6)^2 = 1 \Rightarrow$$
$$c_{22}^2 + c_{32}^2 = 0.64 \tag{3.42}$$

The fact that $\overline{c}_1^t\overline{c}_2 = 0$ leads to the following additional equation.

$$0.2 \times 0.6 + 0.4c_{22} + \sigma_1(0.89443)c_{32} = 0 \Rightarrow$$
$$c_{22} = -0.3 - \sigma_1(2.236075)c_{32} \tag{3.43}$$

When c_{22} is eliminated between Eqs. (3.42) and (3.43), the following quadratic equation is obtained for c_{32}.

$$c_{32}^2 + \sigma_1(0.2236075)c_{32} - 0.0916667 = 0 \tag{3.44}$$

Hence, c_{32} is found as follows with an additional sign ambiguity represented by $\sigma_2 = \pm 1$:

$$c_{32} = -\sigma_1(0.111804) + \sigma_2\sqrt{(0.111804)^2 + 0.0916667} \Rightarrow$$
$$c_{32} = \sigma_2(0.322749) - \sigma_1(0.111804) \tag{3.45}$$

Upon inserting c_{32}, Eq. (3.43) gives c_{22} as

$$c_{22} = -0.05 - \sigma_1\sigma_2(0.72169) \tag{3.46}$$

Finally, the third column $\overline{c}_3$ is found as follows according to Eq. (3.34):

$$\overline{c}_3 = \widetilde{c}_1\overline{c}_2 \Rightarrow$$

$$\begin{bmatrix} c_{13} \\ c_{23} \\ c_{33} \end{bmatrix} = \begin{bmatrix} 0 & -c_{31} & c_{21} \\ c_{31} & 0 & -c_{11} \\ -c_{21} & c_{11} & 0 \end{bmatrix} \begin{bmatrix} c_{12} \\ c_{22} \\ c_{32} \end{bmatrix} = \begin{bmatrix} c_{21}c_{32} - c_{31}c_{22} \\ c_{31}c_{12} - c_{11}c_{32} \\ c_{11}c_{22} - c_{21}c_{12} \end{bmatrix} \tag{3.47}$$

Note that the procedure described above provides four different outcomes for $\widehat{C}$ due to the independent sign variables σ_1 and σ_2. To pick up one of these solutions, let $\sigma_1 = \sigma_2 = +1$. This particular choice of σ_1 and σ_2 leads to $\widehat{C}^{(a,b)}$, which is shown below.

$$\widehat{C}^{(a,b)} = \widehat{C} = \begin{bmatrix} 0.2 & 0.6 & 0.7746 \\ 0.4 & -0.77169 & 0.494469 \\ 0.89443 & 0.210945 & -0.394338 \end{bmatrix} \tag{3.48}$$

As a check for the validity of the above solution, it can be shown that $\det(\widehat{C}) = 1$.

3.4 Expression of a Transformation Matrix as a Direction Cosine Matrix

3.4.1 Definitions of Direction Angles and Direction Cosines

The rotational deviation between two reference frames, e.g. $\mathcal{F}_a$ and $\mathcal{F}_b$, can be represented by the *direction angles* as shown in Figure 3.2. In that figure, only six of the nine direction angles are illustrated for the sake of neatness. The direction angles between $\mathcal{F}_a$ and $\mathcal{F}_b$ are denoted and defined as follows for all $i \in \{1, 2, 3\}$ and $j \in \{1, 2, 3\}$:

$$\theta_{ij}^{(a,b)} = \sphericalangle[\vec{u}_i^{(a)} \rightarrow \vec{u}_j^{(b)}] \tag{3.49}$$

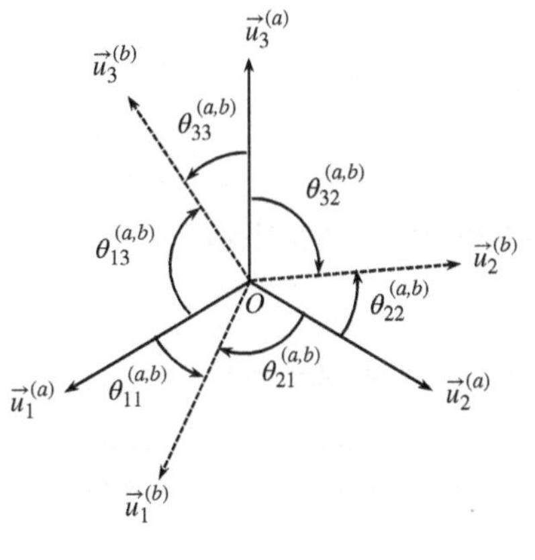

Figure 3.2 Direction angles between two reference frames.

Without any loss of generality, the direction angles can be defined to be positive angles that are confined to the range $[0, \pi]$. That is,

$$0 \le \theta_{ij}^{(a,b)} \le \pi$$

In a direct association with the direction angles, the *direction cosines* between $\mathcal{F}_a$ and $\mathcal{F}_b$ are denoted and defined as follows:

$$c_{ij}^{(a,b)} = \cos\theta_{ij}^{(a,b)} \tag{3.50}$$

3.4.2 Transformation Matrix Formed as a Direction Cosine Matrix

Since the basis vectors of $\mathcal{F}_a$ and $\mathcal{F}_b$ are unit vectors, the direction cosines can also be defined by the following dot product equation written for all $i \in \{1, 2, 3\}$ and $j \in \{1, 2, 3\}$.

$$c_{ij}^{(a,b)} = \vec{u}_i^{(a)} \cdot \vec{u}_j^{(b)} \tag{3.51}$$

Using the transformation matrix $\hat{C}^{(a,b)}$ and the matrix representations of $\vec{u}_i^{(a)}$ and $\vec{u}_j^{(b)}$ in one of the reference frames $\mathcal{F}_a$ and $\mathcal{F}_b$, say $\mathcal{F}_a$, Eq. (3.51) can also be written and manipulated as shown below.

$$c_{ij}^{(a,b)} = \overline{u}_i^{(a/a)t}\overline{u}_j^{(b/a)} = \overline{u}_i^{(a/a)t}[\hat{C}^{(a,b)}\overline{u}_j^{(b/b)}] \Rightarrow$$

$$c_{ij}^{(a,b)} = \overline{u}_i^t\hat{C}^{(a,b)}\overline{u}_j \tag{3.52}$$

As mentioned before, $\overline{u}_i^t$ and $\overline{u}_j$ pick up the ith row and jth column of the matrix they multiply. Therefore, $c_{ij}^{(a,b)}$ happens to be the i-j element of $\hat{C}^{(a,b)}$ according to Eq. (3.52). Owing to this fact, $\hat{C}^{(a,b)}$ can be constructed as a *direction cosine matrix*, i.e. as a matrix constructed as follows by stacking the direction cosines between $\mathcal{F}_a$ and $\mathcal{F}_b$.

$$\hat{C}^{(a,b)} = \begin{bmatrix} c_{11}^{(a,b)} & c_{12}^{(a,b)} & c_{13}^{(a,b)} \\ c_{21}^{(a,b)} & c_{22}^{(a,b)} & c_{23}^{(a,b)} \\ c_{31}^{(a,b)} & c_{32}^{(a,b)} & c_{33}^{(a,b)} \end{bmatrix} = \begin{bmatrix} c\theta_{11}^{(a,b)} & c\theta_{12}^{(a,b)} & c\theta_{13}^{(a,b)} \\ c\theta_{21}^{(a,b)} & c\theta_{22}^{(a,b)} & c\theta_{23}^{(a,b)} \\ c\theta_{31}^{(a,b)} & c\theta_{32}^{(a,b)} & c\theta_{33}^{(a,b)} \end{bmatrix} \tag{3.53}$$

In Eq. (3.53), $c\theta$ is used as an abbreviation for $\cos\theta$.

3.5 Expression of a Transformation Matrix as a Rotation Matrix

3.5.1 Correlation Between the Rotation and Transformation Matrices

Since the reference frames $\mathcal{F}_a$ and $\mathcal{F}_b$ are both orthonormal, right-handed, and equally scaled on their axes, it can be imagined that $\mathcal{F}_b$ is obtained by rotating $\mathcal{F}_a$ as indicated below.

$$\mathcal{F}_a \xrightarrow{\text{rot}(a,b)} \mathcal{F}_b \tag{3.54}$$

The rotation of $\mathcal{F}_a$ into $\mathcal{F}_b$ is actually achieved by rotating the basis vector $\vec{u}_k^{(a)}$ of $\mathcal{F}_a$ into the basis vector $\vec{u}_k^{(b)}$ of $\mathcal{F}_b$ for all $k \in \{1, 2, 3\}$. That is,

$$\vec{u}_k^{(a)} \xrightarrow{\text{rot}(a,b)} \vec{u}_k^{(b)} \tag{3.55}$$

As indicated above, the considered rotation is achieved by means of the rotation operator rot(a, b), which is shown below with its angle-axis detail.

$$\text{rot}(a, b) = \text{rot}(\vec{n}_{ab}, \theta_{ab}) \tag{3.56}$$

Consequently, for $k \in \{1, 2, 3\}$, the basis vectors of $\mathcal{F}_a$ and $\mathcal{F}_b$ are related to each other as follows according to the Rodrigues formula:

$$\vec{u}_k^{(b)} = \vec{u}_k^{(a)} \cos\theta_{ab} + \vec{n}_{ab} \times \vec{u}_k^{(a)} \sin\theta_{ab} + \vec{n}_{ab}[\vec{n}_{ab} \cdot \vec{u}_k^{(a)}](1 - \cos\theta_{ab}) \tag{3.57}$$

Suppose that the rotation of $\mathcal{F}_a$ into $\mathcal{F}_b$ is observed in a third reference frame $\mathcal{F}_c$. Then, the following matrix equation can be written in $\mathcal{F}_c$ in correspondence to Eq. (3.57).

$$\overline{u}_k^{(b/c)} = \widehat{R}_{ab}^{(c)} \overline{u}_k^{(a/c)} \tag{3.58}$$

In Eq. (3.58), $\widehat{R}_{ab}^{(c)}$ is the *rotation matrix* that represents the rotation operator rot(a, b) in $\mathcal{F}_c$. In other words,

$$\widehat{R}_{ab}^{(c)} = [\text{rot}(a, b)]^{(c)} = \widehat{I} \cos\theta_{ab} + \widetilde{n}_{ab}^{(c)} \sin\theta_{ab} + \overline{n}_{ab}^{(c)} \overline{n}_{ab}^{(c)t}(1 - \cos\theta_{ab}) \tag{3.59}$$

By introducing the transformation matrices $\widehat{C}^{(c,a)}$ and $\widehat{C}^{(c,b)}$, Eq. (3.58) can also be written as shown below.

$$\widehat{C}^{(c,b)} \overline{u}_k^{(b/b)} = \widehat{R}_{ab}^{(c)} \widehat{C}^{(c,a)} \overline{u}_k^{(a/a)} \tag{3.60}$$

Recalling that $\overline{u}_k^{(a/a)} = \overline{u}_k^{(b/b)} = \overline{u}_k$ and $[\widehat{C}^{(c,b)}]^{-1} = \widehat{C}^{(b,c)}$, Eq. (3.60) can be manipulated into

$$\overline{u}_k = [\widehat{C}^{(b,c)} \widehat{R}_{ab}^{(c)} \widehat{C}^{(c,a)}] \overline{u}_k \tag{3.61}$$

Equation (3.61) implies that $\widehat{C}^{(b,c)} \widehat{R}_{ab}^{(c)} \widehat{C}^{(c,a)} = \widehat{I}$ or

$$\widehat{R}_{ab}^{(c)} = \widehat{C}^{(c,b)} \widehat{C}^{(a,c)} \tag{3.62}$$

On the other hand,

$$\widehat{C}^{(a,a)} = \widehat{C}^{(b,b)} = \widehat{I} \tag{3.63}$$

Therefore, in either of the two special cases with $\mathcal{F}_c = \mathcal{F}_a$ and $\mathcal{F}_c = \mathcal{F}_b$, Eq. (3.62) leads to the result that

$$\widehat{C}^{(a,b)} = \widehat{R}_{ab}^{(a)} = \widehat{R}_{ab}^{(b)} \tag{3.64}$$

However, unless $\mathcal{F}_c = \mathcal{F}_a$ or $\mathcal{F}_c = \mathcal{F}_b$, Eq. (3.62) implies the inequality that

$$\widehat{C}^{(a,b)} \neq \widehat{R}_{ab}^{(c)} \tag{3.65}$$

3.5.2 Distinction Between the Rotation and Transformation Matrices

Here, it has been shown that the transformation matrix between two reference frames $\mathcal{F}_a$ and $\mathcal{F}_b$ can be expressed as a rotation matrix if that rotation matrix represents the rotation operator rot(a, b) in either $\mathcal{F}_a$ or $\mathcal{F}_b$ but *not* in a third different reference frame $\mathcal{F}_c$.

3.6 Relationship Between the Matrix Representations of a Rotation Operator in Different Reference Frames

Consider the following rotation.

$$\vec{p} \xrightarrow{\text{rot}(\vec{n},\theta)} \vec{r} \tag{3.66}$$

This rotation can be described by the following matrix equations expressed in two different reference frames $\mathcal{F}_a$ and $\mathcal{F}_b$.

$$\overline{r}^{(a)} = \widehat{R}^{(a)}\overline{p}^{(a)} \tag{3.67}$$

$$\overline{r}^{(b)} = \widehat{R}^{(b)}\overline{p}^{(b)} \tag{3.68}$$

In the above equations, $\widehat{R}^{(a)}$ and $\widehat{R}^{(b)}$ are the rotation matrices that represent the same rotation operator rot($\vec{n}$, θ) in $\mathcal{F}_a$ and $\mathcal{F}_b$, respectively. That is,

$$\widehat{R}^{(a)} = [\text{rot}(\vec{n},\theta)]^{(a)} = \widehat{I}\cos\theta + \widetilde{n}^{(a)}\sin\theta + \overline{n}^{(a)}\overline{n}^{(a)t}(1-\cos\theta) \tag{3.69}$$

$$\widehat{R}^{(b)} = [\text{rot}(\vec{n},\theta)]^{(b)} = \widehat{I}\cos\theta + \widetilde{n}^{(b)}\sin\theta + \overline{n}^{(b)}\overline{n}^{(b)t}(1-\cos\theta) \tag{3.70}$$

As for the vectors $\vec{p}$ and $\vec{r}$, their matrix representations are related to each other as follows by the transformation matrix $\widehat{C}^{(a,b)}$:

$$\overline{r}^{(b)} = \widehat{C}^{(b,a)}\overline{r}^{(a)} \tag{3.71}$$

$$\overline{p}^{(b)} = \widehat{C}^{(b,a)}\overline{p}^{(a)} \tag{3.72}$$

When Eqs. (3.71) and (3.72) are substituted, Eq. (3.68) becomes

$$\widehat{C}^{(b,a)}\overline{r}^{(a)} = \widehat{R}^{(b)}\widehat{C}^{(b,a)}\overline{p}^{(a)} \Rightarrow$$

$$\overline{r}^{(a)} = [\widehat{C}^{(a,b)}\widehat{R}^{(b)}\widehat{C}^{(b,a)}]\overline{p}^{(a)} \tag{3.73}$$

When Eqs. (3.67) and (3.73) are compared, it is seen that the matrices $\widehat{R}^{(a)}$ and $\widehat{R}^{(b)}$ are related to each other as follows by using both $\widehat{C}^{(a,b)}$ and $\widehat{C}^{(b,a)}$:

$$\widehat{R}^{(a)} = \widehat{C}^{(a,b)}\widehat{R}^{(b)}\widehat{C}^{(b,a)} \tag{3.74}$$

3.7 Expression of a Transformation Matrix in a Case of Several Successive Rotations

Suppose a reference frame $\mathcal{F}_a$ is rotated into another reference frame $\mathcal{F}_z$ through several successive rotations as described below.

$$\mathcal{F}_a \xrightarrow{\text{rot}(a,b)} \mathcal{F}_b \xrightarrow{\text{rot}(b,c)} \mathcal{F}_c \xrightarrow{\text{rot}(c,d)} \cdots \xrightarrow{\text{rot}(x,y)} \mathcal{F}_y \xrightarrow{\text{rot}(y,z)} \mathcal{F}_z \tag{3.75}$$

In such a rotational sequence, the overall transformation matrix can be obtained as follows:

$$\widehat{C}^{(a,z)} = \widehat{C}^{(a,b)}\widehat{C}^{(b,c)}\widehat{C}^{(cd)}\ldots\widehat{C}^{(x,y)}\widehat{C}^{(y,z)} \tag{3.76}$$

The matrix $\widehat{C}^{(a,z)}$ can be related to the rotation operators that appear in Description (3.75) through two different commonly used formulations depending on the reference frames in which the rotation operators are expressed. These two formulations are explained below.

3.7.1 Rotated Frame Based (RFB) Formulation

In this case, rot(p, q) is expressed as a rotation matrix in one of the two relevant reference frames, i.e. either in the *pre-rotation* frame $\mathcal{F}_p$ or in the *post-rotation* frame $\mathcal{F}_q$. Thus, the transformation matrix $\hat{C}^{(p,q)}$ is related to rot(p, q) as follows according to Eq. (3.64):

$$\hat{C}^{(p,q)} = \hat{R}^{(p)}_{pq} = \hat{R}^{(q)}_{pq} \tag{3.77}$$

Then, Eq. (3.76) gives $\hat{C}^{(a,z)}$ with one of the following equivalent expressions.

$$\hat{C}^{(a,z)} = \hat{R}^{(a)}_{ab}\hat{R}^{(b)}_{bc}\hat{R}^{(c)}_{cd} \cdots \hat{R}^{(x)}_{xy}\hat{R}^{(y)}_{yz} \tag{3.78}$$

$$\hat{C}^{(a,z)} = \hat{R}^{(b)}_{ab}\hat{R}^{(c)}_{bc}\hat{R}^{(d)}_{cd} \cdots \hat{R}^{(y)}_{xy}\hat{R}^{(z)}_{yz} \tag{3.79}$$

On the other hand, $\hat{C}^{(a,z)}$ is also equal to the overall rotation matrix expressed in either $\mathcal{F}_a$ or $\mathcal{F}_z$. That is,

$$\hat{R}^{(a)}_{az} = \hat{R}^{(z)}_{az} = \hat{C}^{(a,z)} \tag{3.80}$$

As noted above, in the rotated frame based (RFB) formulation, the rotation matrices are multiplied in the *same* order as the order of the rotation sequence indicated in Description (3.75).

3.7.2 Initial Frame Based (IFB) Formulation

In this case, all the rotation operators are expressed as the following rotation matrices in the initial reference frame $\mathcal{F}_a$.

$$\hat{R}^{(a)}_{ab} = [\mathrm{rot}(a,b)]^{(a)}, \hat{R}^{(a)}_{bc} = [\mathrm{rot}(b,c)]^{(a)}, \ldots, \hat{R}^{(a)}_{yz} = [\mathrm{rot}(y,z)]^{(a)} \tag{3.81}$$

Of course, in such a formulation, except $\hat{R}^{(a)}_{ab}$, none of the above rotation matrices is a transformation matrix. Therefore, the transformation matrices required in Eq. (3.76) can be obtained as shown below by means of Eq. (3.74).

$$\hat{C}^{(a,b)} = \hat{R}^{(a)}_{ab} \tag{3.82}$$

$$\hat{C}^{(b,c)} = \hat{R}^{(b)}_{bc} = \hat{C}^{(b,a)}\hat{R}^{(a)}_{bc}\hat{C}^{(a,b)} = \hat{R}^{(a)t}_{ab}\hat{R}^{(a)}_{bc}\hat{R}^{(a)}_{ab} \tag{3.83}$$

$$\hat{C}^{(a,c)} = \hat{C}^{(a,b)}\hat{C}^{(b,c)} = [\hat{R}^{(a)}_{ab}][\hat{R}^{(a)t}_{ab}\hat{R}^{(a)}_{bc}\hat{R}^{(a)}_{ab}] \Rightarrow$$

$$\hat{C}^{(a,c)} = \hat{R}^{(a)}_{bc}\hat{R}^{(a)}_{ab} \tag{3.84}$$

$$\hat{C}^{(c,d)} = \hat{R}^{(c)}_{cd} = \hat{C}^{(c,a)}\hat{R}^{(a)}_{cd}\hat{C}^{(a,c)} = \hat{R}^{(a)t}_{ab}\hat{R}^{(a)t}_{bc}\hat{R}^{(a)}_{cd}\hat{R}^{(a)}_{bc}\hat{R}^{(a)}_{ab} \tag{3.85}$$

$$\hat{C}^{(a,d)} = \hat{C}^{(a,c)}\hat{C}^{(c,d)} = [\hat{R}^{(a)}_{bc}\hat{R}^{(a)}_{ab}][\hat{R}^{(a)t}_{ab}\hat{R}^{(a)t}_{bc}\hat{R}^{(a)}_{cd}\hat{R}^{(a)}_{bc}\hat{R}^{(a)}_{ab}] \Rightarrow$$

$$\hat{C}^{(a,d)} = \hat{R}^{(a)}_{cd}\hat{R}^{(a)}_{bc}\hat{R}^{(a)}_{ab} \tag{3.86}$$

The pattern observed in Eqs. (3.84) and (3.86) implies that

$$\hat{C}^{(a,z)} = \hat{R}^{(a)}_{yz}\hat{R}^{(a)}_{xy} \cdots \hat{R}^{(a)}_{cd}\hat{R}^{(a)}_{bc}\hat{R}^{(a)}_{ab} \tag{3.87}$$

As noted above, in the initial frame based (IFB) formulation, the rotation matrices are multiplied in an order *opposite* to the order of the rotation sequence indicated in Description (3.75).

On the other hand, the *rotation matrix* $\hat{R}^{(a)}_{pq}$ (that describes the rotation of $\mathcal{F}_p$ into $\mathcal{F}_q$) is mathematically equivalent to the *orientation matrix* $\hat{R}^{(a)}_{q/p}$ (that describes the orientation of $\mathcal{F}_q$ with respect to $\mathcal{F}_p$). Based on this equivalence, Eq. (3.87) can also be written as follows:

$$\hat{C}^{(a,z)} = \hat{R}^{(a)}_{az} = \hat{R}^{(a)}_{z/a} = \hat{R}^{(a)}_{z/y}\hat{R}^{(a)}_{y/x} \cdots \hat{R}^{(a)}_{d/c}\hat{R}^{(a)}_{c/b}\hat{R}^{(a)}_{b/a} \tag{3.88}$$

3.8 Expression of a Transformation Matrix in Terms of Euler Angles

3.8.1 General Definition of Euler Angles

The Euler angles are named after the Swiss mathematician Leonhard Euler (1707–1783). With a modification of what Euler originally introduced, the definition of the Euler angles was later generalized so that they consist of three rotation angles (ϕ_1, ϕ_2, ϕ_3) about three *specified* rotation axes. The three axes must be specified so that they are neither coplanar nor successively parallel or coincident. Thus, the Euler angles constitute a set of three independent parameters for the transformation matrix $\hat{C}^{(a,b)}$. When a set of Euler angles is used, the reference frame $\mathcal{F}_b$ is obtained by rotating the reference frame $\mathcal{F}_a$ through the following sequence of three rotations.

$$\mathcal{F}_a \xrightarrow{\text{rot}(\vec{n}_1,\phi_1)} \mathcal{F}_m \xrightarrow{\text{rot}(\vec{n}_2,\phi_2)} \mathcal{F}_n \xrightarrow{\text{rot}(\vec{n}_3,\phi_3)} \mathcal{F}_b \tag{3.89}$$

In Description (3.89), $\mathcal{F}_m$ and $\mathcal{F}_n$ are the *intermediate* frames on the way from the *initial* frame $\mathcal{F}_a$ to the *final* or *terminal* frame $\mathcal{F}_b$. As for the unit vectors $\vec{n}_1$, $\vec{n}_2$, and $\vec{n}_3$, they represent the specified rotation axes. They may be specified arbitrarily as desired in the generalized definition of the Euler angles, whereas they are specified as $\vec{n}_1 = \vec{u}_3^{(a)}$, $\vec{n}_2 = \vec{u}_1^{(m)}$, and $\vec{n}_3 = \vec{u}_3^{(n)}$ in the original definition of the Euler angles.

Although $\vec{n}_1$, $\vec{n}_2$, and $\vec{n}_3$ may be specified arbitrarily in general, in almost all the practical cases, each of them is specified as a selected basis vector of a selected reference frame. Thus, different Euler angle sequences arise depending on the selected reference frames and their selected basis vectors. All such Euler angle sequences are grouped into two main categories, which are designated as IFB and RFB sequences. These sequences are described and explained below.

3.8.2 IFB (Initial Frame Based) Euler Angle Sequences

In an IFB sequence, e.g. the IFB *i-j-k* sequence, each of the unit vectors of the rotation axes is specified as one of the basis vectors of the initial reference frame $\mathcal{F}_a$. That is,

$$\vec{n}_1 = \vec{u}_i^{(a)}, \ \vec{n}_2 = \vec{u}_j^{(a)}, \ \vec{n}_3 = \vec{u}_k^{(a)} \tag{3.90}$$

The specified unit vectors must be such that $j \neq i$ and $j \neq k$. Such a rotation sequence can be described as shown below.

$$\mathcal{F}_a \xrightarrow{\text{rot}[\vec{u}_i^{(a)},\ \phi_1]} \mathcal{F}_m \xrightarrow{\text{rot}[\vec{u}_j^{(a)},\ \phi_2]} \mathcal{F}_n \xrightarrow{\text{rot}[\vec{u}_k^{(a)},\ \phi_3]} \mathcal{F}_b \tag{3.91}$$

In such a sequence, the matrix representations of all the rotation operators are expressed naturally in $\mathcal{F}_a$. In other words,

$$[\text{rot}(a,m)]^{(a)} = \hat{R}_{am}^{(a)} = \hat{R}(\tilde{u}_i^{(a/a)}, \phi_1) = \hat{R}_i(\phi_1) = e^{\tilde{u}_i\phi_1} \tag{3.92}$$

$$[\text{rot}(m,n)]^{(a)} = \hat{R}_{mn}^{(a)} = \hat{R}(\tilde{u}_j^{(a/a)}, \phi_2) = \hat{R}_j(\phi_2) = e^{\tilde{u}_j\phi_2} \tag{3.93}$$

$$[\text{rot}(m,n)]^{(a)} = \hat{R}_{nb}^{(a)} = \hat{R}(\tilde{u}_k^{(a/a)}, \phi_3) = \hat{R}_k(\phi_3) = e^{\tilde{u}_k\phi_3} \tag{3.94}$$

Hence, according to the IFB formulation explained in Section 3.7, $\hat{C}^{(a,b)}$ is obtained as follows:

$$\hat{C}^{(a,b)} = \hat{R}_{nb}^{(a)}\hat{R}_{mn}^{(a)}\hat{R}_{am}^{(a)} = \hat{R}_k(\phi_3)\hat{R}_j(\phi_2)\hat{R}_i(\phi_1) \Rightarrow$$

$$\hat{C}^{(a,b)} = e^{\tilde{u}_k\phi_3}e^{\tilde{u}_j\phi_2}e^{\tilde{u}_i\phi_1} \tag{3.95}$$

3.8.3 RFB (Rotated Frame Based) Euler Angle Sequences

In an RFB sequence, e.g. the RFB *i-j-k* sequence, each of the unit vectors of the rotation axes is specified as one of the basis vectors of the reference frames $\mathcal{F}_a$, $\mathcal{F}_m$, and $\mathcal{F}_n$, respectively. That is,

$$\vec{n}_1 = \vec{u}_i^{(a)}, \ \vec{n}_2 = \vec{u}_j^{(m)}, \ \vec{n}_3 = \vec{u}_k^{(n)} \tag{3.96}$$

The specified unit vectors must be such that $j \neq i$ and $j \neq k$. However, since the rotation axes between the pre-rotation and post-rotation frames are common, the following equations can also be written for the unit vectors of the rotation axes.

$$\vec{n}_1 = \vec{u}_i^{(a)} = \vec{u}_i^{(m)}, \ \vec{n}_2 = \vec{u}_j^{(m)} = \vec{u}_j^{(n)}, \ \vec{n}_3 = \vec{u}_k^{(n)} = \vec{u}_k^{(b)} \tag{3.97}$$

Such a rotation sequence can be described as shown below.

$$\mathcal{F}_a \xrightarrow{\text{rot}[\vec{u}_i^{(a)}, \phi_1]} \mathcal{F}_m \xrightarrow{\text{rot}[\vec{u}_j^{(m)}, \phi_2]} \mathcal{F}_n \xrightarrow{\text{rot}[\vec{u}_k^{(n)}, \phi_3]} \mathcal{F}_b \tag{3.98}$$

In a sequence that has the axis unit vectors specified as shown above, i.e. as the basis vectors of the pre-rotation frames, the matrix representations of the rotation operators are also expressed naturally in the pre-rotation frames. In other words,

$$[\text{rot}(a,m)]^{(a)} = \hat{R}_{am}^{(a)} = \hat{R}(\tilde{u}_i^{(a/a)}, \phi_1) = \hat{R}_i(\phi_1) = e^{\tilde{u}_i\phi_1} \tag{3.99}$$

$$[\text{rot}(m,n)]^{(m)} = \hat{R}_{mn}^{(m)} = \hat{R}(\tilde{u}_j^{(m/m)}, \phi_2) = \hat{R}_j(\phi_2) = e^{\tilde{u}_j\phi_2} \tag{3.100}$$

$$[\text{rot}(m,n)]^{(n)} = \hat{R}_{nb}^{(n)} = \hat{R}(\tilde{u}_k^{(n/n)}, \phi_3) = \hat{R}_k(\phi_3) = e^{\tilde{u}_k\phi_3} \tag{3.101}$$

Hence, according to the RFB formulation explained in Section 3.7, $\hat{C}^{(a,b)}$ is obtained as follows:

$$\hat{C}^{(a,b)} = \hat{R}_{am}^{(a)}\hat{R}_{mn}^{(m)}\hat{R}_{nb}^{(n)} = \hat{R}_i(\phi_1)\hat{R}_j(\phi_2)\hat{R}_k(\phi_3) \Rightarrow$$

$$\hat{C}^{(a,b)} = e^{\tilde{u}_i\phi_1}e^{\tilde{u}_j\phi_2}e^{\tilde{u}_k\phi_3} \tag{3.102}$$

3.8.4 Remark 3.4

In an Euler angle sequence, irrespective of whether it is an IFB *i-j-k* or an RFB *i-j-k* sequence, the indices must be such that $j \neq i$ and $j \neq k$ in order to keep the angles ϕ_1, ϕ_2, and ϕ_3 independent. Otherwise, these angles can no longer be independent.

For example, if $j = i$, the three-factor expression in Eq. (3.102) degenerates into the following two-factor expression.

$$\widehat{C}^{(a,b)} = e^{\tilde{u}_i\phi_1} e^{\tilde{u}_i\phi_2} e^{\tilde{u}_k\phi_3} = e^{\tilde{u}_i(\phi_1+\phi_2)} e^{\tilde{u}_k\phi_3} = e^{\tilde{u}_i\phi_4} e^{\tilde{u}_k\phi_3} \tag{3.103}$$

Similarly, if $j = k$, the three-factor expression in Eq. (3.102) degenerates this time into the following two-factor expression.

$$\widehat{C}^{(a,b)} = e^{\tilde{u}_i\phi_1} e^{\tilde{u}_k\phi_2} e^{\tilde{u}_k\phi_3} = e^{\tilde{u}_i\phi_1} e^{\tilde{u}_k(\phi_2+\phi_3)} = e^{\tilde{u}_i\phi_1} e^{\tilde{u}_k\phi_5} \tag{3.104}$$

Equation (3.103) shows that $\widehat{C}^{(a,b)}$ has a missing parameter and it is expressed in terms of only two independent parameters, which are ϕ_4 and ϕ_3. This is because ϕ_1 and ϕ_2 happen to be indistinguishable and indefinite rotation angles about the same axis with a combined effect that can actually be achieved by a single rotation angle ϕ_4. In other words, ϕ_1 and ϕ_2 happen to be dependent on each other because they complement each other to the effective rotation angle ϕ_4, that is, $\phi_1 + \phi_2 = \phi_4$.

Equation (3.104) shows a similar situation with a different effective rotation angle ϕ_5. In this case, ϕ_2 and ϕ_3 happen to be indistinguishable and indefinite rotation angles, which are dependent because they complement each other to the effective rotation angle ϕ_5, that is, $\phi_2 + \phi_3 = \phi_5$.

On the other hand, it is possible to have $k = i \neq j$. Based on this possibility, an Euler angle sequence is called *symmetric* if $k = i$ and *asymmetric* if $k \neq i$. For example, the RFB 1-2-3 sequence is asymmetric, whereas the RFB 3-1-3 sequence is symmetric.

3.8.5 Remark 3.5

The comparison of Eqs. (3.95) and (3.102) shows that any transformation matrix obtained by an IFB sequence can also be obtained by an RFB sequence applied in the reversed order.

For example, the IFB 1-2-3 sequence (with the Euler angles ϕ_1, ϕ_2, and ϕ_3) and the RFB 3-2-1 sequence (with the Euler angles ϕ_1', ϕ_2', and ϕ_3') give the same transformation matrix with the following relationships between the Euler angles.

$$\phi_1' = \phi_3, \quad \phi_2' = \phi_2, \quad \phi_3' = \phi_1$$

The IFB and RFB sequences mentioned above can be described as shown below.

$$\mathcal{F}_a \xrightarrow{\text{rot}[\vec{u}_1^{(a)},\, \phi_1]} \mathcal{F}_p \xrightarrow{\text{rot}[\vec{u}_2^{(a)},\, \phi_2]} \mathcal{F}_q \xrightarrow{\text{rot}[\vec{u}_3^{(a)},\, \phi_3]} \mathcal{F}_b$$

$$\mathcal{F}_a \xrightarrow{\text{rot}[\vec{u}_3^{(a)},\, \phi_3]} \mathcal{F}_m \xrightarrow{\text{rot}[\vec{u}_2^{(m)},\, \phi_2]} \mathcal{F}_n \xrightarrow{\text{rot}[\vec{u}_1^{(n)},\, \phi_1]} \mathcal{F}_b$$

Both of the above sequences lead to the same transformation matrix, which is

$$\widehat{C}^{(a,b)} = e^{\tilde{u}_3\phi_3} e^{\tilde{u}_2\phi_2} e^{\tilde{u}_1\phi_1} \tag{3.105}$$

Note that, although $\mathcal{F}_a$ and $\mathcal{F}_b$ are the same in the two sequences described above, the corresponding intermediate frames are obviously different. That is, $\mathcal{F}_p \neq \mathcal{F}_m$ and $\mathcal{F}_q \neq \mathcal{F}_n$.

3.8.6 Remark 3.6: Preference Between IFB and RFB Sequences

Relying on Remark 3.5, the IFB sequences are almost never used in practice. One reason for this may be the difficulty of visualizing the rotational steps of an IFB sequence while $\mathcal{F}_a$ is rotated into $\mathcal{F}_b$. The visualization of the rotational steps of an RFB sequence happens to be much easier. Another reason why the IFB sequences are not preferred may be the reversal in the order of the rotational steps and the order of multiplying the corresponding rotation matrices.

On the other hand, since the RFB sequences are used almost always in practice, the qualifier RFB is often omitted. In other words, an RFB *i-j-k* sequence is often referred to simply as an *i-j-k* sequence.

3.8.7 Commonly Used Euler Angle Sequences

(a) RFB 1-2-3 Sequence

This sequence is generally known as a *roll-pitch-yaw sequence.* The angles of this sequence are generally named and denoted as *roll angle* ($\phi_1 = \phi$), *pitch angle* ($\phi_2 = \theta$), and *yaw angle* ($\phi_3 = \psi$). As such, the transformation matrix is formed as follows:

$$\hat{C}^{(a,b)} = e^{\tilde{u}_1\phi_1} e^{\tilde{u}_2\phi_2} e^{\tilde{u}_3\phi_3} = e^{\tilde{u}_1\phi} e^{\tilde{u}_2\theta} e^{\tilde{u}_3\psi} \tag{3.106}$$

This sequence is not used very often with the general designations indicated above.

On the other hand, it is used quite often in the area of robotics especially for the purpose of describing the orientation of the end-effector of a manipulator with respect to the base frame. However, when it is used for this purpose, it is designated differently as a *yaw-pitch-roll sequence.* The angles are also named and denoted differently as *yaw or swing angle* ($\phi_1 = \psi$), *pitch or bent angle* ($\phi_2 = \theta$), and *roll or twist angle* ($\phi_3 = \phi$). With these designations, the transformation matrix is formed differently as follows:

$$\hat{C}^{(a,b)} = e^{\tilde{u}_1\phi_1} e^{\tilde{u}_2\phi_2} e^{\tilde{u}_3\phi_3} = e^{\tilde{u}_1\psi} e^{\tilde{u}_2\theta} e^{\tilde{u}_3\phi} \tag{3.107}$$

(b) RFB 3-2-1 Sequence

This sequence is generally known as a *yaw-pitch-roll sequence.* The angles of this sequence are conventionally named and denoted as *yaw angle* ($\phi_1 = \psi$), *pitch angle* ($\phi_2 = \theta$), and *roll angle* ($\phi_3 = \phi$). For this sequence, the transformation matrix is formed as follows:

$$\hat{C}^{(a,b)} = e^{\tilde{u}_3\phi_1} e^{\tilde{u}_2\phi_2} e^{\tilde{u}_1\phi_3} = e^{\tilde{u}_3\psi} e^{\tilde{u}_2\theta} e^{\tilde{u}_1\phi} \tag{3.108}$$

This sequence is used very commonly in the area of vehicle dynamics in order to describe the orientations of all sorts of land, sea, and air vehicles with respect to selected reference frames.

(c) RFB 3-1-3 Sequence

This sequence is generally known as a *precession-nutation-spin sequence.* The angles of this sequence are conventionally named and denoted as *precession angle* ($\phi_1 = \phi$), *nutation angle* ($\phi_2 = \theta$), and *spin angle* ($\phi_3 = \psi$). For this sequence, the transformation matrix is formed as follows:

$$\hat{C}^{(a,b)} = e^{\tilde{u}_3\phi_1} e^{\tilde{u}_1\phi_2} e^{\tilde{u}_3\phi_3} = e^{\tilde{u}_3\phi} e^{\tilde{u}_1\theta} e^{\tilde{u}_3\psi} \tag{3.109}$$

This sequence is used very commonly in the kinematic and dynamic studies that involve spinning bodies such as tops, rotors of gyroscopes, celestial bodies, etc. Actually, this is the sequence that was originally introduced by Leonhard Euler.

(d) RFB 3-2-3 Sequence

This sequence is sometimes used as an alternative to the 3-1-3 sequence in the studies involving spinning bodies. When it is used so, it is also designated as a *precession-nutation-spin sequence.* The angles of this sequence are then similarly named and denoted as *precession angle* ($\phi_1 = \phi$), *nutation angle* ($\phi_2 = \theta$), and *spin angle* ($\phi_3 = \psi$). In such a usage, the transformation matrix is formed as

$$\hat{C}^{(a,b)} = e^{\tilde{u}_3\phi_1} e^{\tilde{u}_2\phi_2} e^{\tilde{u}_3\phi_3} = e^{\tilde{u}_3\phi} e^{\tilde{u}_2\theta} e^{\tilde{u}_3\psi} \tag{3.110}$$

This sequence is also used in the area of robotics as an alternative to the RFB 1-2-3 sequence in order to describe the orientation of an end-effector with respect to the base frame. When it is used so, it is generally designated as a *yaw-declination-roll sequence.* The angles are then named and denoted as *yaw or swing angle* ($\phi_1 = \psi$), *declination angle* ($\phi_2 = \theta$), and *roll or twist angle* ($\phi_3 = \phi$). In such a usage, the transformation matrix is formed as follows:

$$\hat{C}^{(a,b)} = e^{\tilde{u}_3\phi_1} e^{\tilde{u}_2\phi_2} e^{\tilde{u}_3\phi_3} = e^{\tilde{u}_3\psi} e^{\tilde{u}_2\theta} e^{\tilde{u}_3\phi} \tag{3.111}$$

3.8.8 Extraction of Euler Angles from a Given Transformation Matrix

Suppose a transformation matrix is somehow given as

$$\hat{C}^{(a,b)} = \hat{C} \tag{3.112}$$

Then, the Euler angles of a selected sequence can be extracted from $\hat{C}$ by using the procedure explained here. The procedure is explained here for two typical sequences. One of them is the RFB 3-2-3 sequence, which is symmetric, and the other one is the 1-2-3 sequence, which is asymmetric. However, the same procedure can be used similarly for any other sequence, too.

(a) Extraction of the 3-2-3 Euler Angles

If the RFB 3-2-3 sequence is used, $\hat{C}$ is expressed as

$$\hat{C} = e^{\tilde{u}_3\phi_1} e^{\tilde{u}_2\phi_2} e^{\tilde{u}_3\phi_3} \tag{3.113}$$

By using the formulas presented in Chapter 2 about the mathematical properties of the rotation matrices, the following set of five scalar equations can be derived from Eq. (3.113) by picking up the appropriate elements of $\hat{C}$.

$$c_{33} = \overline{u}_3^t \hat{C} \overline{u}_3 = \overline{u}_3^t e^{\tilde{u}_3\phi_1} e^{\tilde{u}_2\phi_2} e^{\tilde{u}_3\phi_3} \overline{u}_3 = \overline{u}_3^t e^{\tilde{u}_2\phi_2} \overline{u}_3 \Rightarrow$$
$$c_{33} = \overline{u}_3^t (e^{\tilde{u}_2\phi_2} \overline{u}_3) = \overline{u}_3^t (\overline{u}_3 \cos\phi_2 + \overline{u}_1 \sin\phi_2) \Rightarrow$$
$$c_{33} = \cos\phi_2 \tag{3.114}$$
$$c_{13} = \overline{u}_1^t \hat{C} \overline{u}_3 = \overline{u}_1^t e^{\tilde{u}_3\phi_1} e^{\tilde{u}_2\phi_2} e^{\tilde{u}_3\phi_3} \overline{u}_3 = (\overline{u}_1^t e^{\tilde{u}_3\phi_1})(e^{\tilde{u}_2\phi_2} \overline{u}_3) \Rightarrow$$
$$c_{13} = (\overline{u}_1^t \cos\phi_1 - \overline{u}_2^t \sin\phi_1)(\overline{u}_3 \cos\phi_2 + \overline{u}_1 \sin\phi_2) \Rightarrow$$
$$c_{13} = \cos\phi_1 \sin\phi_2 \tag{3.115}$$

$$c_{23} = \overline{u}_2^t \widehat{C} \overline{u}_3 = \overline{u}_2^t e^{\widetilde{u}_3\phi_1} e^{\widetilde{u}_2\phi_2} e^{\widetilde{u}_3\phi_3} \overline{u}_3 = (\overline{u}_2^t e^{\widetilde{u}_3\phi_1})(e^{\widetilde{u}_2\phi_2} \overline{u}_3) \Rightarrow$$
$$c_{23} = (\overline{u}_2^t \cos\phi_1 + \overline{u}_1^t \sin\phi_1)(\overline{u}_3 \cos\phi_2 + \overline{u}_1 \sin\phi_2) \Rightarrow$$
$$c_{23} = \sin\phi_1 \sin\phi_2 \tag{3.116}$$
$$c_{31} = \overline{u}_3^t \widehat{C} \overline{u}_1 = \overline{u}_3^t e^{\widetilde{u}_3\phi_1} e^{\widetilde{u}_2\phi_2} e^{\widetilde{u}_3\phi_3} \overline{u}_1 = (\overline{u}_3^t e^{\widetilde{u}_2\phi_2})(e^{\widetilde{u}_3\phi_3} \overline{u}_1) \Rightarrow$$
$$c_{31} = (\overline{u}_3^t \cos\phi_2 - \overline{u}_1^t \sin\phi_2)(\overline{u}_1 \cos\phi_3 + \overline{u}_2 \sin\phi_3) \Rightarrow$$
$$c_{31} = -\sin\phi_2 \cos\phi_3 \tag{3.117}$$
$$c_{32} = \overline{u}_3^t \widehat{C} \overline{u}_2 = \overline{u}_3^t e^{\widetilde{u}_3\phi_1} e^{\widetilde{u}_2\phi_2} e^{\widetilde{u}_3\phi_3} \overline{u}_2 = (\overline{u}_3^t e^{\widetilde{u}_2\phi_2})(e^{\widetilde{u}_3\phi_3} \overline{u}_2) \Rightarrow$$
$$c_{32} = (\overline{u}_3^t \cos\phi_2 - \overline{u}_1^t \sin\phi_2)(\overline{u}_2 \cos\phi_3 - \overline{u}_1 \sin\phi_3) \Rightarrow$$
$$c_{32} = \sin\phi_2 \sin\phi_3 \tag{3.118}$$

From Eq. (3.114), $\sin\phi_2$ and ϕ_2 can be found as follows with an arbitrary sign variable σ:

$$\sin\phi_2 = \sigma\sqrt{1 - c_{33}^2} = \sigma d_{33} \tag{3.119}$$
$$\sigma = \pm 1 \tag{3.120}$$
$$\phi_2 = \operatorname{atan}_2(\sigma d_{33}, c_{33}) = \sigma \operatorname{atan}_2(d_{33}, c_{33}) \tag{3.121}$$

If $\sin\phi_2 \neq 0$, i.e. if $d_{33} > 0$, ϕ_1 and ϕ_3 can be found as follows, respectively, from Eq. Pairs (3.115)–(3.116) and (3.117)–(3.118) consistently with σ, without introducing any additional sign variable.

$$\cos\phi_1 = c_{13}/\sin\phi_2 = \sigma c_{13}/d_{33} \tag{3.122}$$
$$\sin\phi_1 = c_{23}/\sin\phi_2 = \sigma c_{23}/d_{33} \tag{3.123}$$
$$\phi_1 = \operatorname{atan}_2(\sigma c_{23}, \sigma c_{13}) \tag{3.124}$$
$$\cos\phi_3 = -c_{31}/\sin\phi_2 = -\sigma c_{31}/d_{33} \tag{3.125}$$
$$\sin\phi_3 = c_{32}/\sin\phi_2 = \sigma c_{32}/d_{33} \tag{3.126}$$
$$\phi_3 = \operatorname{atan}_2(\sigma c_{32}, -\sigma c_{31}) \tag{3.127}$$

* Selection of the Sign Variable

Based on the solution obtained above for $d_{33} > 0$, the following analysis can be made concerning the sign variable σ.
If $\sigma = +1$ leads to $S = \{\phi_1, \phi_2, \phi_3\}$, then $\sigma = -1$ leads to $S' = \{\phi_1', \phi_2', \phi_3'\}$, where

$$\phi_2' = -\phi_2, \quad \phi_1' = \phi_1 + \sigma_1'\pi, \quad \phi_3' = \phi_3 + \sigma_3'\pi \tag{3.128}$$

Here, σ_1' and σ_3' are two independent sign variables, that is, $\sigma_1' = \pm 1$ and $\sigma_3' = \pm 1$ but they are not necessarily equal. Although S and S' look different, they are actually completely equivalent because they both provide the same transformation matrix as shown below by using the rotation matrix formulas given in Chapter 2.

$$\widehat{C}' = e^{\widetilde{u}_3\phi_1'} e^{\widetilde{u}_2\phi_2'} e^{\widetilde{u}_3\phi_3'} = e^{\widetilde{u}_3(\phi_1+\sigma_1\pi)} e^{-\widetilde{u}_2\phi_2} e^{\widetilde{u}_3(\phi_3+\sigma_3\pi)} \Rightarrow$$
$$\widehat{C}' = e^{\widetilde{u}_3\phi_1} e^{\sigma_1\widetilde{u}_3\pi} e^{-\widetilde{u}_2\phi_2} e^{\sigma_3\widetilde{u}_3\pi} e^{\widetilde{u}_3\phi_3} = e^{\widetilde{u}_3\phi_1} e^{\widetilde{u}_2\phi_2} e^{\sigma_1\widetilde{u}_3\pi} e^{\sigma_3\widetilde{u}_3\pi} e^{\widetilde{u}_3\phi_3} \Rightarrow$$
$$\widehat{C}' = e^{\widetilde{u}_3\phi_1} e^{\widetilde{u}_2\phi_2} e^{\widetilde{u}_3(\sigma_1+\sigma_3)\pi} e^{\widetilde{u}_3\phi_3} = e^{\widetilde{u}_3\phi_1} e^{\widetilde{u}_2\phi_2} \widehat{I} e^{\widetilde{u}_3\phi_3} \Rightarrow$$
$$\widehat{C}' = e^{\widetilde{u}_3\phi_1} e^{\widetilde{u}_2\phi_2} e^{\widetilde{u}_3\phi_3} = \widehat{C} \tag{3.129}$$

Equation (3.129) suggests that σ can be selected as $\sigma = +1$ without a significant loss of generality.

* Singularity Analysis

If $\sin\phi_2 = 0$, i.e. if $d_{33} = 0$, then the 3-2-3 sequence becomes singular and the angles ϕ_1 and ϕ_3 cannot be found from Eq. Pairs (3.115–3.118), which all reduce to $0 = 0$. Such a singularity occurs either if $\phi_2 = 0$ or if $\phi_2 = \sigma_2'\pi$ with $\sigma_2' = \pm 1$. In either case, ϕ_1 and ϕ_3 become indefinite and indistinguishable. So, they cannot be found separately. Nevertheless, their combinations denoted as $\phi_{13} = \phi_1 + \phi_3$ and $\phi_{13}' = \phi_1 - \phi_3$ can still be found depending on whether $\phi_2 = 0$ or $\phi_2 = \sigma_2'\pi$. The way of finding them is explained below.

If $\phi_2 = 0$, Eq. (3.113) can be manipulated as follows:

$$\hat{C} = e^{\tilde{u}_3\phi_1}\hat{I}e^{\tilde{u}_3\phi_3} = e^{\tilde{u}_3\phi_1}e^{\tilde{u}_3\phi_3} = e^{\tilde{u}_3(\phi_1+\phi_3)} = e^{\tilde{u}_3\phi_{13}} \tag{3.130}$$

Equation (3.130) implies that

$$c_{11} = \cos\phi_{13} \text{ and } c_{21} = \sin\phi_{13} \tag{3.131}$$

Hence, ϕ_{13} is found as

$$\phi_{13} = \operatorname{atan}_2(c_{21}, c_{11}) \tag{3.132}$$

If $\phi_2 = \sigma_2'\pi$, Eq. (3.113) can be manipulated as follows:

$$\hat{C} = e^{\tilde{u}_3\phi_1}e^{\sigma_2'\tilde{u}_2\pi}e^{\tilde{u}_3\phi_3} = e^{\tilde{u}_3\phi_1}e^{-\tilde{u}_3\phi_3}e^{\sigma_2'\tilde{u}_2\pi} = e^{\tilde{u}_3(\phi_1-\phi_3)}e^{\sigma_2'\tilde{u}_2\pi} \Rightarrow$$

$$e^{\tilde{u}_3\phi_{13}'} = \hat{C}^* = \hat{C}e^{-\sigma_2'\tilde{u}_2\pi} \tag{3.133}$$

Equation (3.133) implies that

$$\cos\phi_{13}' = c_{22}^* = c_{22} \text{ and } \sin\phi_{13}' = -c_{12}^* = -c_{12} \tag{3.134}$$

Hence, ϕ_{13}' is found as

$$\phi_{13}' = \operatorname{atan}_2(-c_{12}, c_{22}) \tag{3.135}$$

In order to visualize the singularity of the 3-2-3 sequence, the unit vectors of the first and third rotation axes can be expressed as follows in the initial reference frame $\mathcal{F}_a$:

$$\vec{n}_1 = \vec{u}_3^{(a)} \tag{3.136}$$

$$\overline{n}_3^{(a)} = \overline{u}_3^{(b/a)} = \hat{C}^{(a,b)}\overline{u}_3^{(b/b)} = \hat{C}\overline{u}_3 = e^{\tilde{u}_3\phi_1}e^{\tilde{u}_2\phi_2}e^{\tilde{u}_3\phi_3}\overline{u}_3 \Rightarrow$$

$$\overline{n}_3^{(a)} = e^{\tilde{u}_3\phi_1}e^{\tilde{u}_2\phi_2}\overline{u}_3 = e^{\tilde{u}_3\phi_1}(\overline{u}_3\cos\phi_2 + \overline{u}_1\sin\phi_2) \Rightarrow$$

$$\overline{n}_3^{(a)} = e^{\tilde{u}_3\phi_1}\overline{u}_3\cos\phi_2 + e^{\tilde{u}_3\phi_1}\overline{u}_1\sin\phi_2 = \overline{u}_3\cos\phi_2 + e^{\tilde{u}_3\phi_1}\overline{u}_1\sin\phi_2 \Rightarrow$$

$$\overline{n}_3^{(a)} = \overline{u}_3\cos\phi_2 + (\overline{u}_1\cos\phi_1 + \overline{u}_2\sin\phi_1)\sin\phi_2 \Rightarrow$$

$$\vec{n}_3 = \vec{u}_1^{(a)}\cos\phi_1\sin\phi_2 + \vec{u}_2^{(a)}\sin\phi_1\sin\phi_2 + \vec{u}_3^{(a)}\cos\phi_2 \tag{3.137}$$

When the singularity occurs with $\phi_2 = 0$, $\vec{n}_3$ becomes

$$\vec{n}_3 = \vec{u}_3^{(a)} = \vec{n}_1 \tag{3.138}$$

In this singularity, according to Eq. (3.138), the rotations by the angles ϕ_1 and ϕ_3 take place about two axes that have become codirectional. Therefore, only the resultant rotation by the angle $\phi_{13} = \phi_1 + \phi_3$ can be recognized but the angles ϕ_1 and ϕ_3 become obscure and they cannot be distinguished from each other.

When the singularity occurs with $\phi_2 = \sigma_2'\pi$, $\vec{n}_3$ becomes

$$\vec{n}_3 = -\vec{u}_3^{(a)} = -\vec{n}_1 \tag{3.139}$$

In this singularity, according to Eq. (3.139), the rotations by the angles ϕ_1 and ϕ_3 take place about two axes that have become oppositely directed. Therefore, only the resultant rotation by the angle $\phi'_{13} = \phi_1 - \phi_3$ can be recognized but the angles ϕ_1 and ϕ_3 become obscure and they cannot be distinguished from each other.

(b) Extraction of the 1-2-3 Euler Angles

If the RFB 1-2-3 sequence is used, $\hat{C}$ is expressed as

$$\hat{C} = e^{\tilde{u}_1\phi_1} e^{\tilde{u}_2\phi_2} e^{\tilde{u}_3\phi_3} \tag{3.140}$$

Similarly as done above for the 3-2-3 Euler angles, the following five scalar equations can be derived from Eq. (3.140) by picking up the appropriate elements of $\hat{C}$.

$$c_{13} = \overline{u}_1^t \hat{C} \overline{u}_3 = \overline{u}_1^t e^{\tilde{u}_1\phi_1} e^{\tilde{u}_2\phi_2} e^{\tilde{u}_3\phi_3} \overline{u}_3 = \overline{u}_1^t e^{\tilde{u}_2\phi_2} \overline{u}_3 \Rightarrow$$
$$c_{13} = \overline{u}_1^t (e^{\tilde{u}_2\phi_2} \overline{u}_3) = \overline{u}_1^t (\overline{u}_3 \cos\phi_2 + \overline{u}_1 \sin\phi_2) \Rightarrow$$
$$c_{13} = \sin\phi_2 \tag{3.141}$$

$$c_{23} = \overline{u}_2^t \hat{C} \overline{u}_3 = \overline{u}_2^t e^{\tilde{u}_1\phi_1} e^{\tilde{u}_2\phi_2} e^{\tilde{u}_3\phi_3} \overline{u}_3 = (\overline{u}_2^t e^{\tilde{u}_1\phi_1})(e^{\tilde{u}_2\phi_2} \overline{u}_3) \Rightarrow$$
$$c_{23} = (\overline{u}_2^t \cos\phi_1 - \overline{u}_3^t \sin\phi_1)(\overline{u}_3 \cos\phi_2 + \overline{u}_1 \sin\phi_2) \Rightarrow$$
$$c_{23} = -\sin\phi_1 \cos\phi_2 \tag{3.142}$$

$$c_{33} = \overline{u}_3^t \hat{C} \overline{u}_3 = \overline{u}_3^t e^{\tilde{u}_1\phi_1} e^{\tilde{u}_2\phi_2} e^{\tilde{u}_3\phi_3} \overline{u}_3 = (\overline{u}_3^t e^{\tilde{u}_1\phi_1})(e^{\tilde{u}_2\phi_2} \overline{u}_3) \Rightarrow$$
$$c_{33} = (\overline{u}_3^t \cos\phi_1 + \overline{u}_2^t \sin\phi_1)(\overline{u}_3 \cos\phi_2 + \overline{u}_1 \sin\phi_2) \Rightarrow$$
$$c_{33} = \cos\phi_1 \cos\phi_2 \tag{3.143}$$

$$c_{11} = \overline{u}_1^t \hat{C} \overline{u}_1 = \overline{u}_1^t e^{\tilde{u}_1\phi_1} e^{\tilde{u}_2\phi_2} e^{\tilde{u}_3\phi_3} \overline{u}_1 = (\overline{u}_1^t e^{\tilde{u}_2\phi_2})(e^{\tilde{u}_3\phi_3} \overline{u}_1) \Rightarrow$$
$$c_{11} = (\overline{u}_1^t \cos\phi_2 + \overline{u}_3^t \sin\phi_2)(\overline{u}_1 \cos\phi_3 + \overline{u}_2 \sin\phi_3) \Rightarrow$$
$$c_{11} = \cos\phi_2 \cos\phi_3 \tag{3.144}$$

$$c_{12} = \overline{u}_1^t \hat{C} \overline{u}_2 = \overline{u}_1^t e^{\tilde{u}_1\phi_1} e^{\tilde{u}_2\phi_2} e^{\tilde{u}_3\phi_3} \overline{u}_2 = (\overline{u}_1^t e^{\tilde{u}_2\phi_2})(e^{\tilde{u}_3\phi_3} \overline{u}_2) \Rightarrow$$
$$c_{12} = (\overline{u}_1^t \cos\phi_2 + \overline{u}_3^t \sin\phi_2)(\overline{u}_2 \cos\phi_3 - \overline{u}_1 \sin\phi_3) \Rightarrow$$
$$c_{12} = -\cos\phi_2 \sin\phi_3 \tag{3.145}$$

From Eq. (3.141), $\cos\phi_2$ and ϕ_2 can be found as follows with an arbitrary sign variable σ:

$$\cos\phi_2 = \sigma\sqrt{1 - c_{13}^2} = \sigma d_{13} \tag{3.146}$$

$$\sigma = \pm 1 \tag{3.147}$$

$$\phi_2 = \text{atan}_2(c_{13}, \sigma d_{13}) = \sigma_2 \text{atan}_2(|c_{13}|, \sigma d_{13}) \tag{3.148}$$

In Eq. (3.148), σ_2 is different from σ. It is defined as follows if $c_{13} \neq 0$.

$$\sigma_2 = \text{sgn}(\phi_2) = \text{sgn}(c_{13}) \tag{3.149}$$

If $\cos\phi_2 \neq 0$, i.e. if $d_{13} > 0$, ϕ_1 and ϕ_3 can be found as follows, respectively, from Eq. Pairs (3.142–3.145) consistently with σ, without introducing any additional sign variable.

$$\sin\phi_1 = -c_{23}/\cos\phi_2 = -\sigma c_{23}/d_{13} \tag{3.150}$$

$$\cos\phi_1 = c_{33}/\cos\phi_2 = \sigma c_{33}/d_{13} \tag{3.151}$$

$$\phi_1 = \text{atan}_2(-\sigma c_{23}, \sigma c_{33}) \tag{3.152}$$

$$\cos\phi_3 = c_{11}/\cos\phi_2 = \sigma c_{11}/d_{13} \tag{3.153}$$

$$\sin\phi_3 = -c_{12}/\cos\phi_2 = -\sigma c_{12}/d_{13} \tag{3.154}$$

$$\phi_3 = \operatorname{atan}_2(-\sigma c_{12}, \sigma c_{11}) \tag{3.155}$$

∗ **Selection of the Sign Variable**

Based on the solution obtained above for $d_{13} > 0$, the following analysis can be made concerning the sign variable σ.

If $\sigma = +1$ leads to $S = \{\phi_1, \phi_2, \phi_3\}$, then $\sigma = -1$ leads to $S' = \{\phi_1', \phi_2', \phi_3'\}$, where

$$\phi_2' = \sigma_2\pi - \phi_2 = \sigma_2(\pi - |\phi_2|),\ \ \phi_1' = \phi_1 + \sigma_1'\pi,\ \ \phi_3' = \phi_3 + \sigma_3'\pi \tag{3.156}$$

Here, $\sigma_2 = \operatorname{sgn}(\phi_2)$ as introduced before. As for σ_1' and σ_3', they are two independent sign variables, that is, $\sigma_1' = \pm 1$ and $\sigma_3' = \pm 1$ but they are not necessarily equal. Although S and S' look different, they are actually completely equivalent because they both provide the same transformation matrix as shown below similarly as done before for the 3-2-3 sequence.

$$\hat{C}' = e^{\tilde{u}_1\phi_1'}e^{\tilde{u}_2\phi_2'}e^{\tilde{u}_3\phi_3'} = e^{\tilde{u}_1(\phi_1+\sigma_1'\pi)}e^{\tilde{u}_2(\sigma_2\pi-\phi_2)}e^{\tilde{u}_3(\phi_3+\sigma_3'\pi)} \Rightarrow$$

$$\hat{C}' = e^{\tilde{u}_1\phi_1}e^{\sigma_1'\tilde{u}_1\pi}e^{-\tilde{u}_2\phi_2}e^{\sigma_2\tilde{u}_2\pi}e^{\sigma_1'\tilde{u}_3\pi}e^{\tilde{u}_3\phi_3} \Rightarrow$$

$$\hat{C}' = e^{\tilde{u}_1\phi_1}e^{\tilde{u}_2\phi_2}(e^{\sigma_1'\tilde{u}_1\pi}e^{\sigma_2\tilde{u}_2\pi}e^{\sigma_3'\tilde{u}_3\pi})e^{\tilde{u}_3\phi_3} \tag{3.157}$$

According to Eq. (2.87) of Chapter 2 about the *three successive half rotations,*

$$e^{\sigma_1'\tilde{u}_1\pi}e^{\sigma_2\tilde{u}_2\pi}e^{\sigma_3'\tilde{u}_3\pi} = \hat{I}$$

Hence, Eq. (3.157) reduces to

$$\hat{C}' = e^{\tilde{u}_1\phi_1}e^{\tilde{u}_2\phi_2}e^{\tilde{u}_3\phi_3} = \hat{C} \tag{3.158}$$

Equation (3.158) suggests that σ can again be selected as $\sigma = +1$ without a significant loss of generality.

∗ **Singularity Analysis**

If $\cos\phi_2 = 0$, i.e. if $d_{13} = 0$, the 1-2-3 sequence becomes singular and the angles ϕ_1 and ϕ_3 cannot be found from Eq. Pairs (3.142)–(3.143) and (3.144)–(3.145), which all reduce to $0 = 0$. Such a singularity occurs if $\phi_2 = \sigma_2'\pi/2$ with $\sigma_2' = \pm 1$. When it occurs, ϕ_1 and ϕ_3 become indefinite and indistinguishable. So, they cannot be found separately. Nevertheless, their combination denoted as $\phi_{13} = \phi_1 + \sigma_2'\phi_3$ can still be found. The way of finding ϕ_{13} is explained below.

In the singularity with $\phi_2 = \sigma_2'\pi/2$, Eq. (3.140) can be manipulated as follows by using the *shifting formula* given in Chapter 2.

$$\hat{C} = e^{\tilde{u}_1\phi_1}e^{\tilde{u}_2\sigma_2'\pi/2}e^{\tilde{u}_3\phi_3} = e^{\tilde{u}_1\phi_1}e^{\tilde{u}_1\sigma_2'\phi_3}e^{\tilde{u}_2\sigma_s\pi/2} = e^{\tilde{u}_1(\phi_1+\sigma_2'\phi_3)}e^{\tilde{u}_2\sigma_2'\pi/2} \Rightarrow$$

$$e^{\tilde{u}_1(\phi_1+\sigma_2'\phi_3)} = e^{\tilde{u}_1\phi_{13}} = \hat{C}^* = \hat{C}e^{-\tilde{u}_2\sigma_2'\pi/2} \tag{3.159}$$

Equation (3.159) implies that

$$\cos\phi_{13} = c_{22}^* = c_{22} \text{ and } \sin\phi_{13} = c_{32}^* = c_{32} \tag{3.160}$$

Hence, ϕ_{13} is found as

$$\phi_{13} = \operatorname{atan}_2(c_{32}, c_{22}) \tag{3.161}$$

In order to visualize the singularity of the 1-2-3 sequence, the unit vectors of the first and third rotation axes can be expressed as follows in the initial reference frame $\mathcal{F}_a$:

$$\vec{n}_1 = \vec{u}_1^{(a)} \tag{3.162}$$

$$\overline{n}_3^{(a)} = \overline{u}_3^{(b/a)} = \widehat{C}^{(a,b)}\overline{u}_3^{(b/b)} = \widehat{C}\overline{u}_3 = e^{\tilde{u}_1\phi_1}e^{\tilde{u}_2\phi_2}e^{\tilde{u}_3\phi_3}\overline{u}_3 \Rightarrow$$

$$\overline{n}_3^{(a)} = e^{\tilde{u}_1\phi_1}e^{\tilde{u}_2\phi_2}\overline{u}_3 = e^{\tilde{u}_1\phi_1}(\overline{u}_3 \cos\phi_2 + \overline{u}_1 \sin\phi_2) \Rightarrow$$

$$\overline{n}_3^{(a)} = e^{\tilde{u}_1\phi_1}\overline{u}_3 \cos\phi_2 + e^{\tilde{u}_1\phi_1}\overline{u}_1 \sin\phi_2 = e^{\tilde{u}_1\phi_1}\overline{u}_3 \cos\phi_2 + \overline{u}_1 \sin\phi_2 \Rightarrow$$

$$\overline{n}_3^{(a)} = (\overline{u}_3 \cos\phi_1 - \overline{u}_2 \sin\phi_1)\cos\phi_2 + \overline{u}_1 \sin\phi_2 \Rightarrow$$

$$\vec{n}_3 = \vec{u}_1^{(a)} \sin\phi_2 - \vec{u}_2^{(a)} \sin\phi_1 \cos\phi_2 + \vec{u}_3^{(a)} \cos\phi_1 \cos\phi_2 \tag{3.163}$$

When the singularity occurs with $\phi_2 = \sigma_2'\pi/2$, $\vec{n}_3$ becomes

$$\vec{n}_3 = \sigma_2'\vec{u}_1^{(a)} = \sigma_2'\vec{n}_1 \tag{3.164}$$

In this singularity, according to Eq. (3.164), the rotations by the angles ϕ_1 and ϕ_3 take place about two axes that have become parallel, either codirectionally if $\sigma_2' = +1$ or oppositely if $\sigma_2' = -1$. Therefore, only the resultant rotation by the angle $\phi_{13} = \phi_1 + \sigma_2'\phi_3$ can be recognized but the angles ϕ_1 and ϕ_3 become obscure and they cannot be distinguished from each other.

3.9 Position of a Point Expressed in Different Reference Frames and Homogeneous Transformation Matrices

3.9.1 Position of a Point Expressed in Different Reference Frames

Figure 3.3 shows a point P, which is observed in two different reference frames $\mathcal{F}_a(A)$ and $\mathcal{F}_b(B)$. The reference frame $\mathcal{F}_b(B)$ has a general (translating and rotating) displacement with respect to $\mathcal{F}_a(A)$. This general displacement is represented by the translation vector $\vec{r}_{AB}$ and the rotation operator rot(a, b), which functions to rotate $\vec{u}_k^{(a)}$ into $\vec{u}_k^{(b)}$ for $k \in \{1, 2, 3\}$. The position vectors of P appear as $\vec{r}_{AP}$ and $\vec{r}_{BP}$, respectively, in $\mathcal{F}_a(A)$ and $\mathcal{F}_b(B)$. The components of $\vec{r}_{AP}$ in $\mathcal{F}_a(A)$ and $\vec{r}_{BP}$ in $\mathcal{F}_b(B)$ are the *coordinates* of P in $\mathcal{F}_a(A)$ and $\mathcal{F}_b(B)$. In other words, the column matrices $\overline{r}_{AP}^{(a)}$ and $\overline{r}_{BP}^{(b)}$ consist of the coordinates of

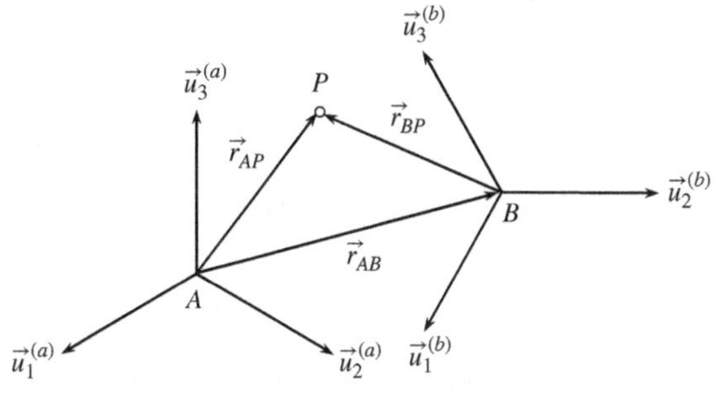

Figure 3.3 A point observed in two different reference frames.

P in $\mathcal{F}_a(A)$ and $\mathcal{F}_b(B)$. On the other hand, as explained in Section 3.5, the transformation matrix between $\mathcal{F}_a(A)$ and $\mathcal{F}_b(B)$ can be expressed in terms of the matrix representations of the rotation operator as $\hat{C}^{(a,b)} = \hat{R}^{(a)}_{ab} = \hat{R}^{(b)}_{ab}$.

As seen in Figure 3.3, the position vectors of P are related to each other as follows:

$$\vec{r}_{AP} = \vec{r}_{AB} + \vec{r}_{BP} \tag{3.165}$$

Equation (3.165), which is a vector equation, can be written as the following matrix equation in one of the involved reference frames, say $\mathcal{F}_a(A)$.

$$\bar{r}^{(a)}_{AP} = \bar{r}^{(a)}_{AB} + \bar{r}^{(a)}_{BP} \tag{3.166}$$

However, it is more convenient to express $\vec{r}_{BP}$ in $\mathcal{F}_b(B)$ rather than in $\mathcal{F}_a(A)$. By doing so, Eq. (3.166) can be written again as follows:

$$\bar{r}^{(a)}_{AP} = \hat{C}^{(a,b)}\bar{r}^{(b)}_{BP} + \bar{r}^{(a)}_{AB} \tag{3.167}$$

3.9.2 Homogeneous, Nonhomogeneous, Linear, Nonlinear, and Affine Relationships

Consider two column matrices $\bar{x} \in \mathcal{R}^n$ and $\bar{y} \in \mathcal{R}^m$. Suppose they are related to each other by means of a function $\bar{f}$ so that

$$\bar{y} = \bar{f}(\bar{x}) \tag{3.168}$$

Depending on the mathematical features of the function $\bar{f}$, the relationship described by Eq. (3.168) is characterized by various designations, which are explained below.

(a) Homogeneous Versus Nonhomogeneous Relationships

The relationship set up by $\bar{f}$ is called *homogeneous* if

$$\bar{f}(\bar{0}) = \bar{0} \tag{3.169}$$

It is called *nonhomogeneous* if

$$\bar{f}(\bar{0}) = \bar{y}_0 \neq \bar{0} \tag{3.170}$$

(b) Linear Versus Nonlinear Relationships

The relationship set up by $\bar{f}$ is called *linear* if, for a scalar k and for all $\bar{x} \in \mathcal{R}^m$,

$$\bar{f}(k\bar{x}) = k\bar{f}(\bar{x}) \tag{3.171}$$

It is called *nonlinear* if

$$\bar{f}(k\bar{x}) \neq k\bar{f}(\bar{x}) \tag{3.172}$$

In the case of a linear relationship, $\bar{f}$ is expressed as follows in terms of an $m \times n$ matrix $\hat{M}$, which does not depend on $\bar{x}$:

$$\bar{y} = \bar{f}(\bar{x}) = \hat{M}\bar{x} \tag{3.173}$$

Note that a linear relationship is also homogeneous, but a nonlinear relationship may or may not be homogeneous.

(c) Affine Relationship

The relationship set up by $\bar{f}$ is called *affine*, if $\bar{f}$ is expressed as follows in terms of an $m \times n$ matrix $\hat{M}$ and an $m \times 1$ matrix $\bar{b}$, which are both independent of $\bar{x}$.

$$\bar{y} = \bar{f}(\bar{x}) = \hat{M}\bar{x} + \bar{b} \tag{3.174}$$

In Eq. (3.174), $\bar{b}$ is defined as the *bias term*. It may or may not be zero, i.e. $|\bar{b}| \geq 0$.

Note that a general affine relationship with $\bar{b} \neq \bar{0}$ is both nonhomogeneous and nonlinear. However, a special affine relationship with $\bar{b} = \bar{0}$ happens to be both homogeneous and linear.

3.9.3 Affine Coordinate Transformation Between Two Reference Frames

As mentioned before in Section 3.9.1, the column matrices $\bar{r}_{AP}^{(a)}$ and $\bar{r}_{BP}^{(b)}$ consist of the coordinates of P in $\mathcal{F}_a(A)$ and $\mathcal{F}_b(B)$. So, the following equation, which is the repetition of Eq. (3.167), constitutes an *affine transformation* between $\bar{r}_{AP}^{(a)}$ and $\bar{r}_{BP}^{(b)}$, i.e. between the coordinates of P in $\mathcal{F}_a(A)$ and $\mathcal{F}_b(B)$.

$$\bar{r}_{AP}^{(a)} = \hat{C}^{(a,b)}\bar{r}_{BP}^{(b)} + \bar{r}_{AB}^{(a)} \tag{3.175}$$

In Eq. (3.175), $\bar{r}_{AB}^{(a)}$ is the *bias term* of the affine transformation. It represents the *translational displacement* of $\mathcal{F}_b(B)$ with respect to $\mathcal{F}_a(A)$. As for $\hat{C}^{(a,b)}$, it represents the *rotational displacement* of $\mathcal{F}_b(B)$ with respect to $\mathcal{F}_a(A)$.

If the point P is observed in several different reference frames such as $\mathcal{F}_a(A)$, $\mathcal{F}_b(B)$, $\mathcal{F}_c(C)$, … , $\mathcal{F}_z(Z)$, then the following affine transformation equations can be written between the successive reference frames.

$$\bar{r}_{AP}^{(a)} = \hat{C}^{(a,b)}\bar{r}_{BP}^{(b)} + \bar{r}_{AB}^{(a)} \tag{3.176}$$

$$\bar{r}_{BP}^{(b)} = \hat{C}^{(b,c)}\bar{r}_{CP}^{(c)} + \bar{r}_{BC}^{(b)} \tag{3.177}$$

$$\bar{r}_{CP}^{(c)} = \hat{C}^{(c,d)}\bar{r}_{DP}^{(d)} + \bar{r}_{CD}^{(c)} \tag{3.178}$$

$$\vdots$$

$$\bar{r}_{YP}^{(y)} = \hat{C}^{(y,z)}\bar{r}_{ZP}^{(z)} + \bar{r}_{YZ}^{(y)} \tag{3.179}$$

As for the overall affine transformation equation, it can be written as

$$\bar{r}_{AP}^{(a)} = \hat{C}^{(a,z)}\bar{r}_{ZP}^{(z)} + \bar{r}_{AZ}^{(a)} \tag{3.180}$$

Upon successive substitutions, the preceding equations lead to the following combined equations.

$$\bar{r}_{AP}^{(a)} = [\hat{C}^{(a,b)}\hat{C}^{(b,c)}]\bar{r}_{CP}^{(c)} + [\bar{r}_{AB}^{(a)} + \hat{C}^{(a,b)}\bar{r}_{BC}^{(b)}] \tag{3.181}$$

$$\bar{r}_{AP}^{(a)} = [\hat{C}^{(a,b)}\hat{C}^{(b,c)}\hat{C}^{(c,d)}]\bar{r}_{DP}^{(d)} + [\bar{r}_{AB}^{(a)} + \hat{C}^{(a,b)}\bar{r}_{BC}^{(b)} + \hat{C}^{(a,b)}\hat{C}^{(b,c)}\bar{r}_{CD}^{(c)}] \tag{3.182}$$

$$\vdots$$

$$\begin{aligned}\bar{r}_{AP}^{(a)} &= [\hat{C}^{(a,b)}\hat{C}^{(b,c)}\hat{C}^{(c,d)} \cdots \hat{C}^{(x,y)}\hat{C}^{(y,z)}]\bar{r}_{ZP}^{(z)} \\ &+ [\bar{r}_{AB}^{(a)} + \hat{C}^{(a,b)}\bar{r}_{BC}^{(b)} + \hat{C}^{(a,b)}\hat{C}^{(b,c)}\bar{r}_{CD}^{(c)} + \cdots + \hat{C}^{(a,b)}\hat{C}^{(b,c)} \cdots \hat{C}^{(x,y)}\bar{r}_{YZ}^{(y)}]\end{aligned} \tag{3.183}$$

Equations (3.180) and (3.183) show the necessity of using the following set of equations in order to obtain the combined rotation and translation matrices.

$$\left.\begin{aligned}\hat{C}^{(a,c)} &= \hat{C}^{(a,b)}\hat{C}^{(b,c)}\\ \hat{C}^{(a,d)} &= \hat{C}^{(a,c)}\hat{C}^{(c,d)}\\ &\vdots\\ \hat{C}^{(a,y)} &= \hat{C}^{(a,x)}\hat{C}^{(x,y)}\\ \hat{C}^{(a,z)} &= \hat{C}^{(a,y)}\hat{C}^{(y,z)}\end{aligned}\right\} \tag{3.184}$$

$$\overline{r}_{AZ}^{(a)} = \overline{r}_{AB}^{(a)} + \hat{C}^{(a,b)}\overline{r}_{BC}^{(b)} + \hat{C}^{(a,c)}\overline{r}_{CD}^{(c)} + \hat{C}^{(a,d)}\overline{r}_{DE}^{(d)} + \cdots + \hat{C}^{(a,y)}\overline{r}_{YZ}^{(y)} \tag{3.185}$$

3.9.4 Homogeneous Coordinate Transformation Between Two Reference Frames

Referring to Eqs. (3.184) and (3.185), it is seen that the result of a combination of several affine transformations necessitates carrying out a considerable number of addition and multiplication operations involving 3×1 and 3×3 matrices. However, a large number of matrix operations is not desirable of course especially from the viewpoint of computational efficiency.

On the other hand, if the transformations are expressed homogeneously, the number of necessary matrix operations reduces considerably to such an extent that only a minimal number of multiplications are required without any additions. However, this reduction in the number of operations necessitates the introduction of 4×1 and 4×4 augmented matrices in return. Even so, the advantage of the reduction in the number of operations emphatically overcomes the disadvantage of the increased dimension of the matrices.

The affine transformation expressed by Eq. (3.175) can be converted into a homogeneous transformation as explained below.

Equation (3.175) can be combined with the trivial equation $1 = 1$ in order to set up the following system of equations.

$$\left.\begin{aligned}\overline{r}_{AP}^{(a)} &= \hat{C}^{(a,b)}\overline{r}_{BP}^{(b)} + \overline{r}_{AB}^{(a)}\\ 1 &= 1\end{aligned}\right\} \tag{3.186}$$

The preceding system of equations can be written as the following single matrix equation.

$$\begin{bmatrix}\overline{r}_{AP}^{(a)}\\ 1\end{bmatrix} = \begin{bmatrix}\hat{C}^{(a,b)} & \overline{r}_{AB}^{(a)}\\ \overline{0}^{t} & 1\end{bmatrix}\begin{bmatrix}\overline{r}_{BP}^{(b)}\\ 1\end{bmatrix} \tag{3.187}$$

Equation (3.187) suggests the following definitions.

$\overline{R}_{AP}^{(a)}$ is defined as the *augmented position matrix* of P in $\mathcal{F}_a(A)$. It is a 4×1 matrix formed as

$$\overline{R}_{AP}^{(a)} = \begin{bmatrix}\overline{r}_{AP}^{(a)}\\ 1\end{bmatrix} \tag{3.188}$$

$\overline{R}_{BP}^{(b)}$ is defined as the *augmented position matrix* of P in $\mathcal{F}_b(B)$. It is a 4×1 matrix formed as

$$\overline{R}_{BP}^{(b)} = \begin{bmatrix} \overline{r}_{BP}^{(b)} \\ 1 \end{bmatrix} \tag{3.189}$$

$\widehat{H}_{AB}^{(a,b)}$ is defined as the *homogeneous transformation matrix* (HTM) between $\mathcal{F}_a(A)$ and $\mathcal{F}_b(B)$. It is a 4×4 matrix formed as

$$\widehat{H}_{AB}^{(a,b)} = \begin{bmatrix} \widehat{C}^{(a,b)} & \overline{r}_{AB}^{(a)} \\ \overline{0}^t & 1 \end{bmatrix} \tag{3.190}$$

Note that the HTM defined above has three major partitions. Its invariant trivial partition is its last row, which is $\begin{bmatrix}\overline{0}^t & 1\end{bmatrix} = \begin{bmatrix}0 & 0 & 0 & 1\end{bmatrix}$. Its *rotational partition* is the 3×3 matrix $\widehat{C}^{(a,b)}$ and its *translational partition* is the 3×1 matrix $\overline{r}_{AB}^{(a)}$.

By using the preceding definitions, Eq. (3.187) can be written in the following compact and linear form, which is known as the *homogeneous transformation equation.*

$$\overline{R}_{AP}^{(a)} = \widehat{H}_{AB}^{(a,b)}\overline{R}_{BP}^{(b)} \tag{3.191}$$

If there are several different reference frames such as $\mathcal{F}_a(A)$, $\mathcal{F}_b(B)$, $\mathcal{F}_c(C)$, … , $\mathcal{F}_z(Z)$, then the following successive homogeneous transformation equations can be written.

$$\left.\begin{aligned} \overline{R}_{AP}^{(a)} &= \widehat{H}_{AB}^{(a,b)}\overline{R}_{BP}^{(b)} \\ \overline{R}_{BP}^{(b)} &= \widehat{H}_{BC}^{(b,c)}\overline{R}_{CP}^{(c)} \\ &\vdots \\ \overline{R}_{YP}^{(y)} &= \widehat{H}_{YZ}^{(y,z)}\overline{R}_{ZP}^{(z)} \end{aligned}\right\} \tag{3.192}$$

As for the overall homogeneous transformation equation, it can be written as

$$\overline{R}_{AP}^{(a)} = \widehat{H}_{AZ}^{(a,z)}\overline{R}_{ZP}^{(z)} \tag{3.193}$$

Upon successive substitutions, the preceding equations lead to the following equation for the combined HTM.

$$\widehat{H}_{AZ}^{(a,z)} = \widehat{H}_{AB}^{(a,b)}\widehat{H}_{BC}^{(b,c)}\widehat{H}_{CD}^{(c,d)}\cdots\widehat{H}_{XY}^{(x,y)}\widehat{H}_{YZ}^{(y,z)} \tag{3.194}$$

As noticed above, the expression of the overall HTM given by Eq. (3.194) involves only matrix multiplications and thus it is much more compact and easier to compute as compared with the accumulation of the consecutive expressions given by Eqs. (3.184) and (3.185) for the rotation matrix and the bias term of the overall affine transformation expressed by Eq. (3.180).

3.9.5 Mathematical Properties of the Homogeneous Transformation Matrices

(a) Determinant of an HTM

Referring to Eq. (3.190), it can be shown that

$$\det[\widehat{H}_{AB}^{(a,b)}] = \det[\widehat{C}^{(a,b)}] = 1 \tag{3.195}$$

(b) Inverse of an HTM

Equation (3.191) can be written in the following two ways: first by interchanging $\mathcal{F}_a(A)$ and $\mathcal{F}_b(B)$; and then by inverting $\widehat{H}_{AB}^{(a,b)}$.

$$\overline{R}_{BP}^{(b)} = \widehat{H}_{BA}^{(b,a)} \overline{R}_{AP}^{(a)} \tag{3.196}$$

$$\overline{R}_{BP}^{(b)} = [\widehat{H}_{AB}^{(a,b)}]^{-1} \overline{R}_{AP}^{(a)} \tag{3.197}$$

Equations (3.196) and (3.197) imply that the inverse of $\widehat{H}_{AB}^{(a,b)}$ can be taken as follows:

$$[\widehat{H}_{AB}^{(a,b)}]^{-1} = \widehat{H}_{BA}^{(b,a)} = \begin{bmatrix} \widehat{C}^{(b,a)} & \overline{r}_{BA}^{(b)} \\ \overline{0}^t & 1 \end{bmatrix} = \begin{bmatrix} \widehat{C}^{(a,b)t} & -\widehat{C}^{(a,b)t} \overline{r}_{AB}^{(a)} \\ \overline{0}^t & 1 \end{bmatrix} \tag{3.198}$$

(c) Decomposition of an HTM

The overall displacement of $\mathcal{F}_b(B)$ with respect to $\mathcal{F}_a(A)$ consists of translational and rotational displacements. So, it can be described in the following two alternative ways.

$$\mathcal{F}_a(A) \xrightarrow{\text{trans}(A \to B)} \mathcal{F}_a(B) \xrightarrow{\text{rot}(a \to b)} \mathcal{F}_b(B) \tag{3.199}$$

$$\mathcal{F}_a(A) \xrightarrow{\text{rot}(a \to b)} \mathcal{F}_b(A) \xrightarrow{\text{trans}(A \to B)} \mathcal{F}_b(B) \tag{3.200}$$

According to the above descriptions, $\widehat{H}_{AB}^{(a,b)}$ can be factorized as shown below.

(i) First translation and then rotation:

$$\widehat{H}_{AB}^{(a,b)} = \widehat{H}_{AB}^{(a,a)} \widehat{H}_{BB}^{(a,b)} = \widehat{H}_{AB}^{(a)} \widehat{H}_{B}^{(a,b)} \tag{3.201}$$

(ii) First rotation and then translation:

$$\widehat{H}_{AB}^{(a,b)} = \widehat{H}_{AA}^{(a,b)} \widehat{H}_{AB}^{(b,b)} = \widehat{H}_{A}^{(a,b)} \widehat{H}_{AB}^{(b)} \tag{3.202}$$

The factorizations described above suggest the following definitions of pure rotational and translational displacements and the associated homogeneous transformation matrices.

(d) HTM of a Pure Rotation

A pure rotational displacement of $\mathcal{F}_b(A)$ with respect to $\mathcal{F}_a(A)$ or a pure rotational displacement of $\mathcal{F}_b(B)$ with respect to $\mathcal{F}_a(B)$ can be achieved by pivoting about either of the origins A and B. In either case, the resultant HTM will be the same. That is,

$$\widehat{H}_{A}^{(a,b)} = \widehat{H}_{B}^{(a,b)} = \begin{bmatrix} \widehat{C}^{(a,b)} & \overline{0} \\ \overline{0}^t & 1 \end{bmatrix} \tag{3.203}$$

In Eq. (3.203), $\widehat{H}_{A}^{(a,b)}$ and $\widehat{H}_{B}^{(a,b)}$ are the abbreviated symbols that stand for $\widehat{H}_{AA}^{(a,b)}$ and $\widehat{H}_{BB}^{(a,b)}$.

Note that Eq. (3.203) is actually valid for any pivot point whatsoever. Therefore, the HTM of a pure rotational displacement does not actually need a subscript and thus it may be denoted even in the following simplest form.

$$\hat{H}^{(a,b)} = \begin{bmatrix} \hat{C}^{(a,b)} & \overline{0} \\ \overline{0}^t & 1 \end{bmatrix} \tag{3.204}$$

Note also that Eqs. (3.203) and (3.204) verify the well-known fact that a rotation operator is indifferent to the location of the pivot point.

(e) HTM of a Pure Translation

A pure translational displacement of $\mathcal{F}_a(B)$ with respect to $\mathcal{F}_a(A)$ or a pure translational displacement of $\mathcal{F}_b(B)$ with respect to $\mathcal{F}_b(A)$ can be expressed by one of the following HTM expressions, depending on the selected one of $\mathcal{F}_a(A)$ and $\mathcal{F}_b(A)$, in which the translation vector $\vec{r}_{AB}$ is observed.

$$\hat{H}_{AB}^{(a)} = \hat{H}_{AB}^{(a,a)} = \begin{bmatrix} \hat{I} & \overline{r}_{AB}^{(a)} \\ \overline{0}^t & 1 \end{bmatrix} \tag{3.205}$$

$$\hat{H}_{AB}^{(b)} = \hat{H}_{AB}^{(b,b)} = \begin{bmatrix} \hat{I} & \overline{r}_{AB}^{(b)} \\ \overline{0}^t & 1 \end{bmatrix} \tag{3.206}$$

(f) Observation in a Third Different Reference Frame

In general, the point P and the reference frames $\mathcal{F}_a(A)$ and $\mathcal{F}_b(B)$ may be observed in a different reference frame $\mathcal{F}_o(O)$. In such a case, considering that the vectors $\vec{r}_{AP}$ and $\vec{r}_{BP}$ are conveniently resolved in $\mathcal{F}_a(A)$ and $\mathcal{F}_b(B)$, Eq. (3.175) can be written as follows:

$$\hat{C}^{(o,a)}\overline{r}_{AP}^{(a)} = \hat{C}^{(o,b)}\overline{r}_{BP}^{(b)} + \overline{r}_{AB}^{(o)} \tag{3.207}$$

The above affine relationship can be expressed in the following homogeneous form.

$$\begin{bmatrix} \hat{C}^{(o,a)} & \overline{0} \\ \overline{0}^t & 1 \end{bmatrix} \begin{bmatrix} \overline{r}_{AP}^{(a)} \\ 1 \end{bmatrix} = \begin{bmatrix} \hat{C}^{(o,b)} & \overline{r}_{AB}^{(o)} \\ \overline{0}^t & 1 \end{bmatrix} \begin{bmatrix} \overline{r}_{BP}^{(b)} \\ 1 \end{bmatrix} \tag{3.208}$$

In Eq. (3.208), the coefficient matrix on the left-hand side is the HTM of a pure rotation from $\mathcal{F}_o(O)$ to $\mathcal{F}_a(A)$ and the coefficient matrix on the right-hand side is the HTM that expresses the overall rotation from $\mathcal{F}_o(O)$ to $\mathcal{F}_b(B)$ together with the translation from A to B as observed in $\mathcal{F}_o(O)$. Thus, Eq. (3.208) can be written compactly as

$$\hat{H}^{(o,a)}\overline{R}_{AP}^{(a)} = \hat{H}_{AB}^{(o,b)}\overline{R}_{BP}^{(b)} \tag{3.209}$$

In case of a *pure rotation* with $B = A$, Eq. (3.209) takes the following form that involves two pure-rotation HTMs.

$$\hat{H}^{(o,a)}\overline{R}_{AP}^{(a)} = \hat{H}^{(o,b)}\overline{R}_{AP}^{(b)} \tag{3.210}$$

In case of a *pure translation* with $b = a$, Eq. (3.209) takes the following form.

$$\hat{H}^{(o,a)}\overline{R}_{AP}^{(a)} = \hat{H}_{AB}^{(o,a)}\overline{R}_{BP}^{(a)} \tag{3.211}$$

As another point of concern, note that Eq. (3.209) can also be written as

$$\overline{R}_{AP}^{(a)} = \widehat{H}^{(a,o)}\widehat{H}_{AB}^{(o,b)}\overline{R}_{BP}^{(b)} \tag{3.212}$$

In Eq. (3.212),

$$\widehat{H}^{(a,o)} = [\widehat{H}^{(o,a)}]^{-1} = \widehat{H}^{(o,a)t} \tag{3.213}$$

When Eqs. (3.212) and (3.193) are compared, it is seen that

$$\widehat{H}_{AB}^{(a,b)} = \widehat{H}^{(a,o)}\widehat{H}_{AB}^{(o,b)} \Rightarrow \widehat{H}_{AB}^{(o,b)} = \widehat{H}^{(o,a)}\widehat{H}_{AB}^{(a,b)} \tag{3.214}$$

Equation (3.214) shows how an HTM can be adapted to the selected observation frame.

3.9.6 Example 3.2

Figure 3.4 shows the initial and final positions of a cube. The length of each edge of the cube is $L = 10$ cm. In the first position of the cube, the edge BC coincides with the first axis of the base frame $\mathcal{F}_b(O)$ so that $OC = 20$ cm. In the second position of the cube, the edge GF coincides with the second axis of $\mathcal{F}_b(O)$ so that $OG = 15$ cm. The reference frame that is fixed to the cube is $\mathcal{F}_c(A)$. It is oriented in such a way that $\overrightarrow{AB} = L\vec{u}_1^{(c)}$, $\overrightarrow{AD} = L\vec{u}_2^{(c)}$, and $\vec{AE} = L\vec{u}_3^{(c)}$.

It is required to express the HTM $\widehat{H}_{A_1A_2}^{(c_1,c_2)}$ between the two positions of the cube.

The translation vector can be expressed in $\mathcal{F}_b(O)$ as follows:

$$\vec{r}_{A_1A_2} = \vec{r}_{A_1B_1} + \vec{r}_{B_1O} + \vec{r}_{OF_2} + \vec{r}_{F_2B_2} + \vec{r}_{B_2A_2} \Rightarrow$$

$$\vec{r}_{A_1A_2} = 10\vec{u}_2^{(b)} - 30\vec{u}_1^{(b)} + 25\vec{u}_2^{(b)} + 10\vec{u}_3^{(b)} - 10\vec{u}_1^{(b)} \Rightarrow$$

$$\vec{r}_{A_1A_2} = -40\vec{u}_1^{(b)} + 35\vec{u}_2^{(b)} + 10\vec{u}_3^{(b)} \tag{3.215}$$

On the other hand, $\mathcal{F}_b(O)$ is oriented with respect to $\mathcal{F}_{c_1}(A_1)$ so that

$$\vec{u}_1^{(b)} = -\vec{u}_2^{(c_1)},\ \vec{u}_2^{(b)} = \vec{u}_1^{(c_1)},\ \vec{u}_3^{(b)} = \vec{u}_3^{(c_1)} \tag{3.216}$$

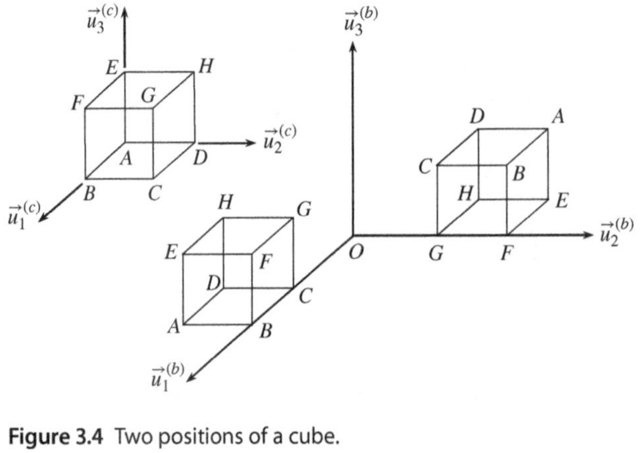

Figure 3.4 Two positions of a cube.

Hence, in $\mathcal{F}_{c_1}(A_1)$, the expression of the translation vector becomes

$$\vec{r}_{A_1A_2} = 35\vec{u}_1^{(c_1)} + 40\vec{u}_2^{(c_1)} + 10\vec{u}_3^{(c_1)} \tag{3.217}$$

Then, the column matrix representation of $\vec{r}_{A_1A_2}$ in $\mathcal{F}_{c_1}(A_1)$ is obtained as

$$\overline{r}_{A_1A_2}^{(c_1)} = 35\overline{u}_1 + 40\overline{u}_2 + 10\overline{u}_3 \tag{3.218}$$

As for the rotation of the cube, Figure 3.4 implies that

$$\mathcal{F}_b \xrightarrow{\text{rot}[\vec{u}_3^{(b)},\ \pi/2]} \mathcal{F}_{c_1} \xrightarrow{\text{rot}[\vec{u}_3^{(b)},\ -\pi/2]} \mathcal{F}_{c_0} \xrightarrow{\text{rot}[\vec{u}_1^{(b)},\ \pi]} \mathcal{F}_{c_2} \tag{3.219}$$

Note that Description (3.219) describes an IFB rotation sequence. Therefore, referring to Section 3.7, the relevant transformation matrices can be obtained as shown below.

$$\widehat{C}^{(b,c_1)} = \widehat{R}_{bc_1}^{(b)} = e^{\widetilde{u}_3\pi/2} \tag{3.220}$$

$$\widehat{C}^{(b,c_0)} = \widehat{R}_{bc_0}^{(b)} = \widehat{R}_{c_1c_0}^{(b)}\widehat{R}_{bc_1}^{(b)} = e^{-\widetilde{u}_3\pi/2}e^{\widetilde{u}_3\pi/2} = \widehat{I} \tag{3.221}$$

$$\widehat{C}^{(b,c_2)} = \widehat{R}_{bc_2}^{(b)} = \widehat{R}_{c_0c_2}^{(b)}\widehat{R}_{c_1c_0}^{(b)}\widehat{R}_{bc_1}^{(b)} = e^{\widetilde{u}_1\pi}e^{-\widetilde{u}_3\pi/2}e^{\widetilde{u}_3\pi/2} = e^{\widetilde{u}_1\pi} \tag{3.222}$$

Hence,

$$\widehat{C}^{(c_1,c_2)} = \widehat{C}^{(c_1,b)}\widehat{C}^{(b,c_2)} = e^{-\widetilde{u}_3\pi/2}e^{\widetilde{u}_1\pi} \tag{3.223}$$

Having found the rotational and translational displacement matrices, i.e. $\widehat{C}^{(c_1,c_2)}$ and $\overline{r}_{A_1A_2}^{(c_1)}$, the HTM $\widehat{H}_{A_1A_2}^{(c_1,c_2)}$ can then be constructed as follows:

$$\widehat{H}_{A_1A_2}^{(c_1,c_2)} = \begin{bmatrix} \widehat{C}^{(c_1,c_2)} & \overline{r}_{A_1A_2}^{(c_1)} \\ \overline{0}^t & 1 \end{bmatrix} \Rightarrow$$

$$\widehat{H}_{A_1A_2}^{(c_1,c_2)} = \begin{bmatrix} e^{-\widetilde{u}_3\pi/2}e^{\widetilde{u}_1\pi} & 35\overline{u}_1 + 40\overline{u}_2 + 10\overline{u}_3 \\ \overline{0}^t & 1 \end{bmatrix} \tag{3.224}$$

In order to have a detailed expression, the rotational partition $\widehat{C}^{(c_1,c_2)}$ can be written as shown below.

$$\widehat{C}^{(c_1,c_2)} = e^{-\widetilde{u}_3\pi/2}e^{\widetilde{u}_1\pi} = \begin{bmatrix} 0 & 1 & 0 \\ -1 & 0 & 0 \\ 0 & 0 & 1 \end{bmatrix}\begin{bmatrix} 1 & 0 & 0 \\ 0 & -1 & 0 \\ 0 & 0 & -1 \end{bmatrix} = \begin{bmatrix} 0 & -1 & 0 \\ -1 & 0 & 0 \\ 0 & 0 & -1 \end{bmatrix} \tag{3.225}$$

Hence,

$$\widehat{H}_{A_1A_2}^{(c_1,c_2)} = \begin{bmatrix} 0 & -1 & 0 & 35 \\ -1 & 0 & 0 & 40 \\ 0 & 0 & -1 & 10 \\ 0 & 0 & 0 & 1 \end{bmatrix} \tag{3.226}$$

As a verification of the expression of $\hat{H}^{(c_1,c_2)}_{A_1A_2}$ obtained above, the coordinates of the points A_2 and G_2 are obtained as follows in $\mathcal{F}_{c_1}(A_1)$:

$$\overline{R}^{(c_1)}_{A_1A_2} = \hat{H}^{(c_1,c_2)}_{A_1A_2}\overline{R}^{(c_2)}_{A_2A_2} = \begin{bmatrix} 0 & -1 & 0 & 35 \\ -1 & 0 & 0 & 40 \\ 0 & 0 & -1 & 10 \\ 0 & 0 & 0 & 1 \end{bmatrix}\begin{bmatrix} 0 \\ 0 \\ 0 \\ 1 \end{bmatrix} = \begin{bmatrix} 35 \\ 40 \\ 10 \\ 1 \end{bmatrix} \Rightarrow$$

$$\overline{r}^{(c_1)}_{A_1A_2} = 35\overline{u}_1 + 40\overline{u}_2 + 10\overline{u}_3 \Rightarrow$$

$$\vec{r}_{A_1A_2} = 35\vec{u}^{(c_1)}_1 + 40\vec{u}^{(c_1)}_2 + 10\vec{u}^{(c_1)}_3 \tag{3.227}$$

$$\overline{R}^{(c_1)}_{A_1G_2} = \hat{H}^{(c_1,c_2)}_{A_1A_2}\overline{R}^{(c_2)}_{A_2G_2} = \begin{bmatrix} 0 & -1 & 0 & 35 \\ -1 & 0 & 0 & 40 \\ 0 & 0 & -1 & 10 \\ 0 & 0 & 0 & 1 \end{bmatrix}\begin{bmatrix} 10 \\ 10 \\ 10 \\ 1 \end{bmatrix} = \begin{bmatrix} 25 \\ 30 \\ 0 \\ 1 \end{bmatrix} \Rightarrow$$

$$\overline{r}^{(c_1)}_{A_1G_2} = 25\overline{u}_1 + 30\overline{u}_2 \Rightarrow$$

$$\vec{r}_{A_1G_2} = 25\vec{u}^{(c_1)}_1 + 30\vec{u}^{(c_1)}_2 \tag{3.228}$$

Note that, by referring to Figure 3.4, both Eqs. (3.227) and (3.228) can be verified by inspection, too. Moreover, Eq. (3.227) matches with Eq. (3.217), as it is supposed to do.

As an additional manipulation with the homogeneous transformation matrices, the coordinates of the points A_2 and G_2 can also be obtained in $\mathcal{F}_b(O)$ as explained below.

The HTM between $\mathcal{F}_b(O)$ and $\mathcal{F}_{c_1}(A_1)$ can be expressed as follows:

$$\hat{H}^{(b,c_1)}_{OA_1} = \begin{bmatrix} \hat{C}^{(b,c_1)} & \overline{r}^{(b)}_{OA_1} \\ \overline{0}^t & 1 \end{bmatrix} \tag{3.229}$$

Recall that $\hat{C}^{(b,c_1)}$ has already been expressed by Eq. (3.220) as

$$\hat{C}^{(b,c_1)} = e^{\tilde{u}_3\pi/2} \tag{3.230}$$

On the other hand, Figure 3.4 implies that

$$\vec{r}_{OA_1} = \vec{r}_{OB_1} + \vec{r}_{B_1A_1} = 30\vec{u}^{(b)}_1 - 10\vec{u}^{(b)}_2 \Rightarrow$$

$$\overline{r}^{(b)}_{OA_1} = 30\overline{u}_1 - 10\overline{u}_2 \tag{3.231}$$

Hence,

$$\hat{H}^{(b,c_1)}_{OA_1} = \begin{bmatrix} e^{\tilde{u}_3\pi/2} & 30\overline{u}_1 - 10\overline{u}_2 \\ \overline{0}^t & 1 \end{bmatrix} = \begin{bmatrix} 0 & -1 & 0 & 30 \\ 1 & 0 & 0 & -10 \\ 0 & 0 & 1 & 0 \\ 0 & 0 & 0 & 1 \end{bmatrix} \tag{3.232}$$

As for the coordinates of A_2 and G_2 in $\mathcal{F}_b(O)$, they are obtained as follows with the help of Eqs. (3.227) and (3.228):

$$\overline{R}_{OA_2}^{(b)} = \widehat{H}_{OA_1}^{(b,c_1)} \overline{R}_{A_1A_2}^{(c_1)} = \begin{bmatrix} 0 & -1 & 0 & 30 \\ 1 & 0 & 0 & -10 \\ 0 & 0 & 1 & 0 \\ 0 & 0 & 0 & 1 \end{bmatrix} \begin{bmatrix} 35 \\ 40 \\ 10 \\ 1 \end{bmatrix} = \begin{bmatrix} -10 \\ 25 \\ 10 \\ 1 \end{bmatrix} \Rightarrow$$

$$\overline{r}_{OA_2}^{(b)} = -10\overline{u}_1 + 25\overline{u}_2 + 10\overline{u}_3 \Rightarrow$$

$$\vec{r}_{OA_2} = -10\vec{u}_1^{(b)} + 25\vec{u}_2^{(b)} + 10\vec{u}_3^{(b)} \tag{3.233}$$

$$\overline{R}_{OG_2}^{(b)} = \widehat{H}_{OA_1}^{(b,c_1)} \overline{R}_{A_1G_2}^{(c_1)} = \begin{bmatrix} 0 & -1 & 0 & 30 \\ 1 & 0 & 0 & -10 \\ 0 & 0 & 1 & 0 \\ 0 & 0 & 0 & 1 \end{bmatrix} \begin{bmatrix} 25 \\ 30 \\ 0 \\ 1 \end{bmatrix} = \begin{bmatrix} 0 \\ 15 \\ 0 \\ 1 \end{bmatrix} \Rightarrow$$

$$\overline{r}_{OG_2}^{(b)} = 15\overline{u}_2 \Rightarrow$$

$$\vec{r}_{OG_2} = 15\vec{u}_2^{(b)} \tag{3.234}$$

Note that, by referring to Figure 3.4, both Eqs. (3.233) and (3.234) can again be verified by inspection, too.

4

Vector Differentiation Accompanied by Velocity and Acceleration Expressions

Synopsis

This chapter is concerned mainly with the differentiation of vectors. As mentioned before, a vector is independent of the reference frame, in which it is observed. However, the derivative of a vector does depend on the reference frame of observation. In other words, vector differentiation is a relative operation that can be carried out only with respect to a selected reference frame. Therefore, the same vector happens to have different derivatives with respect to different reference frames that are rotating differently with respect to each other. The derivatives of the same vector with respect to different reference frames are related to each other through the Coriolis transport theorem. This theorem brings in the definition of relative angular velocity between two reference frames, which is obtained out of the relative orientation matrix. As for the relative angular acceleration, it is obtained by differentiating the relative angular velocity with respect to not any but only one of the involved reference frames. Another concern of this chapter is the relative velocities and accelerations of a point with respect to different reference frames that are translating and rotating with respect to each other. These relative velocities and accelerations are also related to each other through the Coriolis transport theorem.

4.1 Derivatives of a Vector with Respect to Different Reference Frames

4.1.1 Differentiation and Resolution Frames

Vector differentiation is an operation that is relative to a reference frame. In other words, the derivative of a vector $\vec{r}$ with respect to a reference frame $\mathcal{F}_a$ is defined so that $\vec{r}$ is differentiated as if $\mathcal{F}_a$ is stationary. In this definition, $\mathcal{F}_a$ is designated as the *differentiation frame*. The derivative of a vector $\vec{r}$ with respect to a differentiation frame $\mathcal{F}_a$ can be denoted in various ways as shown below.

$$D_a\vec{r} = d_a\vec{r}/dt = [d\vec{r}/dt]_{\mathcal{F}_a} \tag{4.1}$$

The most obvious implication of Eq. (4.1) is the following equation concerning the basis vector triad $\mathcal{U}_a = \{\vec{u}_k^{(a)} : k = 1, 2, 3\}$ of $\mathcal{F}_a$.

$$D_a\vec{u}_k^{(a)} = \vec{0} \tag{4.2}$$

Kinematics of General Spatial Mechanical Systems, First Edition. M. Kemal Ozgoren.

Companion Website: www.wiley.com/go/ozgoren/spatialmechanicalsystems

For a vector $\vec{r}$, if the selected differentiation frame $\mathcal{F}_a$ is also used as the resolution frame, then the following resolution and differentiation equations can be written.

$$\vec{r} = \vec{u}_1^{(a)} r_1^{(a)} + \vec{u}_2^{(a)} r_2^{(a)} + \vec{u}_3^{(a)} r_3^{(a)} \tag{4.3}$$

$$D_a\vec{r} = \vec{u}_1^{(a)} \dot{r}_1^{(a)} + \vec{u}_2^{(a)} \dot{r}_2^{(a)} + \vec{u}_3^{(a)} \dot{r}_3^{(a)} \tag{4.4}$$

In Eq. (4.4),

$$\dot{r}_k^{(a)} = dr_k^{(a)}/dt \tag{4.5}$$

In Eq. (4.5), the component $r_k^{(a)}$ is just a scalar and the derivative of a scalar is not associated with any reference frame. That is why the derivative of $r_k^{(a)}$ can simply be denoted by an overhead *dot* without the necessity of indicating a differentiation frame. In other words, the value of a scalar s is invariant with respect to different reference frames and therefore its derivative can simply be denoted as $\dot{s} = ds/dt$ regardless of any reference frame.

The vectors $\vec{r}$ and $D_a\vec{r}$ can be represented by the following column matrices in the reference frame $\mathcal{F}_a$, when $\mathcal{F}_a$ happens to be both the differentiation and resolution frames.

$$[\vec{r}]^{(a)} = \overline{r}^{(a)} = \begin{bmatrix} r_1^{(a)} \\ r_2^{(a)} \\ r_3^{(a)} \end{bmatrix} \tag{4.6}$$

$$[D_a\vec{r}]^{(a)} = \dot{\overline{r}}^{(a)} = \begin{bmatrix} \dot{r}_1^{(a)} \\ \dot{r}_2^{(a)} \\ \dot{r}_3^{(a)} \end{bmatrix} \tag{4.7}$$

Equation (4.7) has the following points to be noted.

(i) The column matrix $\overline{r}^{(a)}$ is nothing but a collection of three scalars. Actually, any $m \times n$ matrix $\hat{M}$ is a collection of scalar elements such as m_{ij} for $i \in \{1, 2, \ldots, m\}$ and $j \in \{1, 2, \ldots, n\}$. Therefore, just like any of its scalar elements, e.g. m_{ij}, the matrix $\hat{M}$ is also invariant with respect to different reference frames. Based on this fact, the derivative of $\hat{M}$ can simply be denoted as $\dot{\hat{M}} = d\hat{M}/dt$ regardless of any reference frame. The notation $\dot{\hat{M}}$ implies that the elements of $\dot{\hat{M}}$ are $\dot{m}_{ij} = dm_{ij}/dt$ for all $i \in \{1, 2, \ldots, m\}$ and $j \in \{1, 2, \ldots, n\}$. This is the reason why the derivative of $\overline{r}^{(a)}$ is denoted simply as $\dot{\overline{r}}^{(a)}$ in Eq. (4.7).

(ii) As long as the differentiation and resolution frames are the same, the column matrix representation of a differentiated vector is equal to the derivative of the column matrix representation of that vector. That is,

$$[D_a\vec{r}]^{(a)} = [d_a\vec{r}/dt]^{(a)} = d\overline{r}^{(a)}/dt = \dot{\overline{r}}^{(a)} \tag{4.8}$$

4.1.2 Components in Different Differentiation and Resolution Frames

In general, the resolution frame may be different from the differentiation frame. If that is the case, Eq. (4.8) must be modified as follows by using the relevant transformation matrix:

$$[D_a\vec{r}]^{(b)} = \hat{C}^{(b,a)}[D_a\vec{r}]^{(a)} = \hat{C}^{(b,a)}\dot{\overline{r}}^{(a)} \tag{4.9}$$

Concerning Eq. (4.9), it is critical to note that

$$\hat{C}^{(b,a)}\dot{\bar{r}}^{(a)} \neq \dot{\bar{r}}^{(b)} = [D_b\vec{r}]^{(b)} \tag{4.10}$$

In other words,

$$\begin{aligned} &\bar{r}^{(b)} = \hat{C}^{(b,a)}\bar{r}^{(a)} \Rightarrow \dot{\bar{r}}^{(b)} = \hat{C}^{(b,a)}\dot{\bar{r}}^{(a)} + \dot{\hat{C}}^{(b,a)}\bar{r}^{(a)} \Rightarrow \\ &\hat{C}^{(b,a)}\dot{\bar{r}}^{(a)} = \dot{\bar{r}}^{(b)} - \dot{\hat{C}}^{(b,a)}\bar{r}^{(a)} \neq \dot{\bar{r}}^{(b)} \end{aligned} \tag{4.11}$$

Equations (4.9)–(4.11) lead to the following conclusions, as long as $\dot{\hat{C}}^{(b,a)} \neq \hat{0}$.

(i) Unless $\mathcal{F}_b$ is rotationally stationary with respect to $\mathcal{F}_a$, the derivatives of $\vec{r}$ with respect to $\mathcal{F}_a$ and $\mathcal{F}_b$ are not equal. That is,

$$\begin{aligned} &\hat{C}^{(b,a)}\dot{\bar{r}}^{(a)} \neq \dot{\bar{r}}^{(b)} \Rightarrow [D_a\vec{r}]^{(b)} \neq [D_b\vec{r}]^{(b)} \Rightarrow \\ &\quad D_a\vec{r} \neq D_b\vec{r} \end{aligned} \tag{4.12}$$

(ii) Let $\vec{v}_a = D_a\vec{r}$. Then, as seen before,

$$\bar{v}_a^{(a)} = \dot{\bar{r}}^{(a)} \tag{4.13}$$

$$\bar{v}_a^{(b)} \neq \dot{\bar{r}}^{(b)} \tag{4.14}$$

$$\bar{v}_a^{(b)} = \hat{C}^{(b,a)}\dot{\bar{r}}^{(a)} \tag{4.15}$$

Equation (4.13) states that the components of $\vec{v}_a = D_a\vec{r}$ in the frame $\mathcal{F}_a$ are directly equal to the derivatives of the components of the vector $\vec{r}$ in the same frame. On the other hand, Inequality (4.14) states that the components of $\vec{v}_a = D_a\vec{r}$ in a frame $\mathcal{F}_b$ (other than $\mathcal{F}_a$) are *not* equal to the derivatives of the components of a vector in $\mathcal{F}_b$. In other words, as stated by Eq. (4.15), the components of $\vec{v}_a$ in $\mathcal{F}_b$ are not derivatives per se but they are obtained as linear combinations of certain derivatives, which are the components of $\vec{v}_a$ in $\mathcal{F}_a$.

4.1.3 Example

Consider Figure 4.1 showing an aircraft flying with respect to an earth fixed reference frame $\mathcal{F}_e(O)$. As usual, $\mathcal{F}_e(O)$ is taken as an NED (north-east-down) reference frame. Let $\mathcal{F}_b(C)$ be the body fixed reference frame attached to the aircraft, where C is the

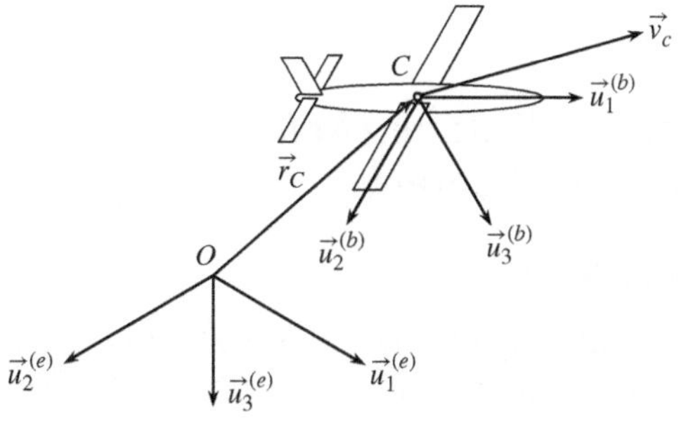

Figure 4.1 An aircraft observed in an earth fixed reference frame.

mass center of the aircraft. The location and orientation of the aircraft, i.e. $\mathcal{F}_b(C)$, with respect to $\mathcal{F}_e(O)$ are described by the following equations.

$$\vec{r}_C = \vec{u}_1^{(e)} x + \vec{u}_2^{(e)} y + \vec{u}_3^{(e)} z \tag{4.16}$$

$$\overline{r}_C^{(e)} = \overline{u}_1 x + \overline{u}_2 y + \overline{u}_3 z \tag{4.17}$$

$$\widehat{C}^{(e,b)} = \widehat{C} = e^{\tilde{u}_3\psi} e^{\tilde{u}_2\theta} e^{\tilde{u}_1\phi} \tag{4.18}$$

The velocity $\vec{v}_C$ of the mass center of the aircraft with respect to $\mathcal{F}_e(O)$ is obtained by differentiating the position vector $\vec{r}_C$ with respect to $\mathcal{F}_e(O)$. That is,

$$\vec{v}_C = D_e\vec{r}_C = \vec{u}_1^{(e)} \dot{x} + \vec{u}_2^{(e)} \dot{y} + \vec{u}_3^{(e)} \dot{z} \tag{4.19}$$

Equation (4.19) gives $\vec{v}_C$ naturally as resolved in $\mathcal{F}_e(O)$. However, especially for the purposes of measurement, navigation, and flight control, it turns out to be more convenient to resolve $\vec{v}_C$ in $\mathcal{F}_b(C)$ as shown below.

$$\vec{v}_C = \vec{u}_1^{(b)} u + \vec{u}_2^{(b)} v + \vec{u}_3^{(b)} w \tag{4.20}$$

The matrix representations of $\vec{v}_C$ in $\mathcal{F}_e(O)$ and $\mathcal{F}_b(C)$ can be expressed as follows:

$$\overline{v}_C^{(e)} = [D_e\vec{r}_C]^{(e)} = \dot{\overline{r}}_C^{(e)} = \overline{u}_1\dot{x} + \overline{u}_2\dot{y} + \overline{u}_3\dot{z} \tag{4.21}$$

$$\overline{v}_C^{(b)} = [D_e\vec{r}_C]^{(b)} = \overline{u}_1 u + \overline{u}_2 v + \overline{u}_3 w \tag{4.22}$$

The preceding matrix representations are related by the following transformation equation.

$$\overline{v}_C^{(b)} = \widehat{C}^{(b,e)}\overline{v}_C^{(e)} = \widehat{C}^{(b,e)}\dot{\overline{r}}_C^{(e)} \tag{4.23}$$

By using Eqs. (4.18), (4.21), and (4.22), Eq. (4.23) can also be written as

$$\overline{u}_1 u + \overline{u}_2 v + \overline{u}_3 w = \widehat{C}^t(\overline{u}_1\dot{x} + \overline{u}_2\dot{y} + \overline{u}_3\dot{z}) \tag{4.24}$$

Note that $\overline{u}_i^t\widehat{C}^t\overline{u}_j = \overline{u}_j^t\widehat{C}\overline{u}_i = c_{ji}$. Thus, the following component relationships are obtained from Eq. (4.24).

$$\left.\begin{aligned} u &= c_{11}\dot{x} + c_{21}\dot{y} + c_{31}\dot{z} \\ v &= c_{12}\dot{x} + c_{22}\dot{y} + c_{32}\dot{z} \\ w &= c_{13}\dot{x} + c_{23}\dot{y} + c_{33}\dot{z} \end{aligned}\right\} \tag{4.25}$$

Equation (4.25) shows that the components of $\vec{v}_C = D_e\vec{r}_C$ in the resolution frame $\mathcal{F}_b(C)$, which is different from the differentiation frame $\mathcal{F}_e(O)$, are not equal to the derivatives of the components of a vector, but they are equal to the indicated linear combinations of the derivatives of the components (x, y, z) of the position vector $\vec{r}_C$ in the differentiation frame $\mathcal{F}_e(O)$.

4.2 Vector Derivatives with Respect to Different Reference Frames and the Coriolis Transport Theorem

4.2.1 First Derivatives and the Relative Angular Velocity

Consider a vector $\vec{r}$, which is observed in two different reference frames $\mathcal{F}_a$ and $\mathcal{F}_b$ that are rotating with respect to each other. The matrix representations of $\vec{r}$ in $\mathcal{F}_a$ and $\mathcal{F}_b$ are related to each other as follows:

$$\overline{r}^{(a)} = \widehat{C}^{(a,b)}\overline{r}^{(b)} \tag{4.26}$$

In Eq. (4.26), $\hat{C}^{(a,b)}$ is the *transformation matrix* between $\mathcal{F}_a$ and $\mathcal{F}_b$, which also happens to be the *relative orientation matrix* of $\mathcal{F}_b$ with respect to $\mathcal{F}_a$. In other words, it is related to the *rotation or orientation matrix* expressed in $\mathcal{F}_a$ or $\mathcal{F}_b$ so that

$$\hat{C}^{(a,b)} = \hat{R}_{ab}^{(a)} = \hat{R}_{ab}^{(b)} \text{ or } \hat{C}^{(a,b)} = \hat{R}_{b/a}^{(a)} = \hat{R}_{b/a}^{(b)} \tag{4.27}$$

The following equation is obtained when Eq. (4.26) is differentiated side by side.

$$\dot{\overline{r}}^{(a)} = \hat{C}^{(a,b)}\dot{\overline{r}}^{(b)} + \dot{\hat{C}}^{(a,b)}\overline{r}^{(b)} \tag{4.28}$$

The expression on the right-hand side of Eq. (4.28) can be factorized as follows:

$$\dot{\overline{r}}^{(a)} = \hat{C}^{(a,b)}[\dot{\overline{r}}^{(b)} + \hat{C}^{(b,a)}\dot{\hat{C}}^{(a,b)}\overline{r}^{(b)}] \tag{4.29}$$

The above factorization is based on the fact that

$$\hat{C}^{(a,b)}\hat{C}^{(b,a)} = \hat{C}^{(b,a)}\hat{C}^{(a,b)} = \hat{I} \tag{4.30}$$

In Eq. (4.29), the matrix product $\hat{C}^{(b,a)}\dot{\hat{C}}^{(a,b)}$ requires special attention. In order to see the kinematic significance of this matrix product through a simpler notation, let

$$\hat{C} = \hat{C}^{(a,b)} \tag{4.31}$$

Then, $\hat{C}^{(b,a)} = \hat{C}^t$ and Eq. (4.30) can be written as

$$\hat{C}\hat{C}^t = \hat{C}^t\hat{C} = \hat{I} \tag{4.32}$$

Equation (4.32) can be differentiated side by side and then manipulated as shown below.

$$\hat{C}^t\dot{\hat{C}} + \dot{\hat{C}}^t\hat{C} = \dot{\hat{I}} = \hat{0} \Rightarrow$$

$$\hat{C}^t\dot{\hat{C}} = -\dot{\hat{C}}^t\hat{C} \Rightarrow \hat{C}^t\dot{\hat{C}} = -(\hat{C}^t\dot{\hat{C}})^t \tag{4.33}$$

Equation (4.33) shows that $\hat{S} = \hat{C}^t\dot{\hat{C}}$ is a *skew symmetric matrix*, because $\hat{S} = -\hat{S}^t$. Therefore, there exists a column matrix $\overline{\omega}$ such that

$$\tilde{\omega} = \hat{C}^t\dot{\hat{C}} \tag{4.34}$$

In Eq. (4.34), $\tilde{\omega}$ is the skew symmetric matrix generated from $\overline{\omega}$, i.e. $\tilde{\omega} = \text{ssm}(\overline{\omega})$.

Recalling that $\hat{C} = \hat{C}^{(a,b)}$, Eq. (4.34) can be written with the currently used elaborate notation as

$$\tilde{\omega}_{b/a}^{(b)} = \hat{C}^{(b,a)}\dot{\hat{C}}^{(a,b)} \tag{4.35}$$

According to Eq. (4.35), there exists a vector $\vec{\omega}_{b/a}$ such that

$$[\vec{\omega}_{b/a}]^{(b)} = \overline{\omega}_{b/a}^{(b)} = \text{colm}[\tilde{\omega}_{b/a}^{(b)}] = \text{colm}[\hat{C}^{(b,a)}\dot{\hat{C}}^{(a,b)}] \tag{4.36}$$

The vector $\vec{\omega}_{b/a}$, which arises from Eqs. (4.35) and (4.36), is defined as the *relative angular velocity* of the reference frame $\mathcal{F}_b$ with respect to the reference frame $\mathcal{F}_a$.

Note that Eq. (4.35) gives $\vec{\omega}_{b/a}$ with its matrix representation in $\mathcal{F}_b$. As for the matrix representation of $\vec{\omega}_{b/a}$ in $\mathcal{F}_a$, it can be obtained as follows:

$$\overline{\omega}_{b/a}^{(a)} = \hat{C}^{(a,b)}\overline{\omega}_{b/a}^{(b)} \Rightarrow$$

$$\tilde{\omega}_{b/a}^{(a)} = \hat{C}^{(a,b)}\tilde{\omega}_{b/a}^{(b)}\hat{C}^{(b,a)} = \hat{C}^{(a,b)}\hat{C}^{(b,a)}\dot{\hat{C}}^{(a,b)}\hat{C}^{(b,a)} = \hat{I}\dot{\hat{C}}^{(a,b)}\hat{C}^{(b,a)} \Rightarrow$$

$$\tilde{\omega}_{b/a}^{(a)} = \dot{\hat{C}}^{(a,b)}\hat{C}^{(b,a)} \tag{4.37}$$

When Eq. (4.35) is substituted, Eq. (4.29) becomes

$$\dot{\bar{r}}^{(a)} = \hat{C}^{(a,b)}[\dot{\bar{r}}^{(b)} + \tilde{\omega}_{b/a}^{(b)} \bar{r}^{(b)}] \tag{4.38}$$

Note that $\dot{\bar{r}}^{(a)} = [D_a\vec{r}]^{(a)}$ and $\dot{\bar{r}}^{(b)} = [D_b\vec{r}]^{(b)}$. Therefore, Eq. (4.38) can also be written as

$$[D_a\vec{r}]^{(a)} = \hat{C}^{(a,b)}[D_b\vec{r} + \vec{\omega}_{b/a} \times \vec{r}]^{(b)} = [D_b\vec{r} + \vec{\omega}_{b/a} \times \vec{r}]^{(a)} \tag{4.39}$$

Equation (4.39) implies the following vector equation.

$$D_a\vec{r} = D_b\vec{r} + \vec{\omega}_{b/a} \times \vec{r} \tag{4.40}$$

Equation (4.40) is the mathematical statement of the *Coriolis transport theorem.*

Here, it is to be pointed out that the relative angular velocity of $\mathcal{F}_b$ with respect to $\mathcal{F}_a$ comes out as a vector but *not* as the derivative of a vector. In other words, there does not exist a relative angular position vector, whose derivative is the relative angular velocity vector. This is because, the relative angular position (i.e. the relative orientation) of $\mathcal{F}_b$ with respect to $\mathcal{F}_a$ is described not by a vector but by a matrix, which is $\hat{C}^{(a,b)}$.

Sometimes, it becomes desirable to arrange the angular velocity equations derived above as the following differential equations, which can be integrated over time in order to determine the relative orientation matrix $\hat{C}^{(a,b)}$, when $\vec{\omega}_{b/a}$ happens to be available with its components in either $\mathcal{F}_a$ or $\mathcal{F}_b$.

$$\dot{\hat{C}}^{(a,b)} = \tilde{\omega}_{b/a}^{(a)} \hat{C}^{(a,b)} \tag{4.41}$$

$$\dot{\hat{C}}^{(a,b)} = \hat{C}^{(a,b)} \tilde{\omega}_{b/a}^{(b)} \tag{4.42}$$

4.2.2 Second Derivatives and the Relative Angular Acceleration

The mathematical statement of the Coriolis transport theorem, i.e. Eq. (4.40), can be differentiated side by side in order to relate the second derivatives of the vector $\vec{r}$ with respect to $\mathcal{F}_a$ and $\mathcal{F}_b$ as shown below.

$$D_a^2\vec{r} = D_a(D_b\vec{r} + \vec{\omega}_{b/a} \times \vec{r}) \Rightarrow$$
$$D_a^2\vec{r} = D_b(D_b\vec{r} + \vec{\omega}_{b/a} \times \vec{r}) + \vec{\omega}_{b/a} \times (D_b\vec{r} + \vec{\omega}_{b/a} \times \vec{r}) \Rightarrow$$
$$D_a^2\vec{r} = D_b^2\vec{r} + 2\vec{\omega}_{b/a} \times (D_b\vec{r}) + \vec{\alpha}_{b/a} \times \vec{r} + \vec{\omega}_{b/a} \times (\vec{\omega}_{b/a} \times \vec{r}) \tag{4.43}$$

In Eq. (4.43), $\vec{\alpha}_{b/a}$ is the *relative angular acceleration* of $\mathcal{F}_b$ with respect to $\mathcal{F}_a$. It is defined by the following equation.

$$\vec{\alpha}_{b/a} = D_b\vec{\omega}_{b/a} \tag{4.44}$$

On the other hand, according to the Coriolis transport theorem,

$$D_a\vec{\omega}_{b/a} = D_b\vec{\omega}_{b/a} + \vec{\omega}_{b/a} \times \vec{\omega}_{b/a} = D_b\vec{\omega}_{b/a} + \vec{0} = D_b\vec{\omega}_{b/a} \tag{4.45}$$

Owing to Eq. (4.45), the relative angular acceleration $\vec{\alpha}_{b/a}$ can be obtained by differentiating the relative angular velocity $\vec{\omega}_{b/a}$ with respect to either $\mathcal{F}_a$ or $\mathcal{F}_b$. In other words,

$$\vec{\alpha}_{b/a} = D_a\vec{\omega}_{b/a} = D_b\vec{\omega}_{b/a} \tag{4.46}$$

However, if $\vec{\omega}_{b/a}$ is differentiated with respect to a third different reference frame $\mathcal{F}_c$, the result will no longer be $\vec{\alpha}_{b/a}$. In other words,

$$\vec{\alpha}_{b/a} \neq D_c\vec{\omega}_{b/a} \tag{4.47}$$

This fact can be shown by the following applications of the Coriolis transport theorem.

$$D_c\vec{\omega}_{b/a} = D_a\vec{\omega}_{b/a} + \vec{\omega}_{a/c} \times \vec{\omega}_{b/a} = \vec{\alpha}_{b/a} + \vec{\omega}_{a/c} \times \vec{\omega}_{b/a} \neq \vec{\alpha}_{b/a} \tag{4.48}$$

$$D_c\vec{\omega}_{b/a} = D_b\vec{\omega}_{b/a} + \vec{\omega}_{b/c} \times \vec{\omega}_{b/a} = \vec{\alpha}_{b/a} + \vec{\omega}_{b/c} \times \vec{\omega}_{b/a} \neq \vec{\alpha}_{b/a} \tag{4.49}$$

As for the relevant matrix equations, Eq. (4.46) implies the following equations for the matrix representations of $\vec{\alpha}_{b/a}$ in $\mathcal{F}_a$ and $\mathcal{F}_b$.

$$\overline{\alpha}^{(a)}_{b/a} = [\vec{\alpha}_{b/a}]^{(a)} = [D_a\vec{\omega}_{b/a}]^{(a)} = \dot{\overline{\omega}}^{(a)}_{b/a} \tag{4.50}$$

$$\overline{\alpha}^{(b)}_{b/a} = [\vec{\alpha}_{b/a}]^{(b)} = [D_b\vec{\omega}_{b/a}]^{(b)} = \dot{\overline{\omega}}^{(b)}_{b/a} \tag{4.51}$$

On the other hand, as seen before,

$$\widetilde{\omega}^{(a)}_{b/a} = \dot{\hat{C}}^{(a,b)}\hat{C}^{(b,a)} \tag{4.52}$$

Upon differentiation, Eq. (4.52) leads to the following formula that relates $\widetilde{\alpha}^{(a)}_{b/a}$ to $\hat{C}^{(a,b)}$ and its derivatives.

$$\widetilde{\alpha}^{(a)}_{b/a} = \dot{\widetilde{\omega}}^{(a)}_{b/a} = \ddot{\hat{C}}^{(a,b)}\hat{C}^{(b,a)} + \dot{\hat{C}}^{(a,b)}\dot{\hat{C}}^{(b,a)} \tag{4.53}$$

Equation (4.53) can be manipulated further in order to obtain the following alternative formula.

$$\widetilde{\alpha}^{(a)}_{b/a} = \ddot{\hat{C}}^{(a,b)}\hat{C}^{(b,a)} + \dot{\hat{C}}^{(a,b)}\hat{C}^{(b,a)}\hat{C}^{(a,b)}\dot{\hat{C}}^{(b,a)} \Rightarrow$$

$$\widetilde{\alpha}^{(a)}_{b/a} = \ddot{\hat{C}}^{(a,b)}\hat{C}^{(b,a)} + \widetilde{\omega}^{(a)}_{b/a}\widetilde{\omega}^{(a)}_{a/b}; \quad \widetilde{\omega}^{(a)}_{a/b} = -\widetilde{\omega}^{(a)}_{b/a} \Rightarrow$$

$$\widetilde{\alpha}^{(a)}_{b/a} = \ddot{\hat{C}}^{(a,b)}\hat{C}^{(b,a)} - \widetilde{\omega}^{(a)2}_{b/a} \tag{4.54}$$

After obtaining the matrix expressions of $\vec{\alpha}_{b/a}$, the matrix equation corresponding to Eq. (4.43) in $\mathcal{F}_a$ can be derived as follows starting from Eq. (4.38), which is repeated below for the sake of convenience:

$$\dot{\overline{r}}^{(a)} = \hat{C}^{(a,b)}[\dot{\overline{r}}^{(b)} + \widetilde{\omega}^{(b)}_{b/a}\overline{r}^{(b)}] \tag{4.55}$$

Before differentiating Eq. (4.55), note that

$$\ddot{\overline{r}}^{(a)} = [D_a^2\vec{r}]^{(a)} \text{ and } \ddot{\overline{r}}^{(b)} = [D_b^2\vec{r}]^{(b)} \tag{4.56}$$

Then, the differentiation of Eq. (4.55) and the subsequent manipulations shown below leads to Eq. (4.57) as the resulting relationship between the second derivatives of $\overline{r}^{(a)}$ and $\overline{r}^{(b)}$.

$$\ddot{\overline{r}}^{(a)} = \hat{C}^{(a,b)}[\ddot{\overline{r}}^{(b)} + \widetilde{\omega}^{(b)}_{b/a}\dot{\overline{r}}^{(b)} + \widetilde{\alpha}^{(b)}_{b/a}\overline{r}^{(b)}] + \dot{\hat{C}}^{(a,b)}[\dot{\overline{r}}^{(b)} + \widetilde{\omega}^{(b)}_{b/a}\overline{r}^{(b)}] \Rightarrow$$

$$\ddot{\overline{r}}^{(a)} = \hat{C}^{(a,b)}[\ddot{\overline{r}}^{(b)} + \widetilde{\omega}^{(b)}_{b/a}\dot{\overline{r}}^{(b)} + \widetilde{\alpha}^{(b)}_{b/a}\overline{r}^{(b)}] + \hat{C}^{(a,b)}\widetilde{\omega}^{(b)}_{b/a}[\dot{\overline{r}}^{(b)} + \widetilde{\omega}^{(b)}_{b/a}\overline{r}^{(b)}] \Rightarrow$$

$$\ddot{\overline{r}}^{(a)} = \hat{C}^{(a,b)}\{\ddot{\overline{r}}^{(b)} + 2\widetilde{\omega}^{(b)}_{b/a}\dot{\overline{r}}^{(b)} + [\widetilde{\alpha}^{(b)}_{b/a} + \widetilde{\omega}^{(b)2}_{b/a}]\overline{r}^{(b)}\} \tag{4.57}$$

In Eq. (4.57), it is worth noting that $\widetilde{\alpha}^{(b)}_{b/a}$ and $\widetilde{\omega}^{(b)2}_{b/a}$ appear as additive terms.

4.3 Combination of Relative Angular Velocities and Accelerations

4.3.1 Combination of Relative Angular Velocities

When a vector $\vec{r}$ is observed in three different reference frames, the following three equations can be written according to the Coriolis transport theorem.

$$D_a\vec{r} = D_b\vec{r} + \vec{\omega}_{b/a} \times \vec{r} \tag{4.58}$$

$$D_b\vec{r} = D_c\vec{r} + \vec{\omega}_{c/b} \times \vec{r} \tag{4.59}$$

$$D_a\vec{r} = D_c\vec{r} + \vec{\omega}_{c/a} \times \vec{r} \tag{4.60}$$

Equations (4.58) and (4.59) can be combined as follows:

$$D_a\vec{r} = (D_c\vec{r} + \vec{\omega}_{c/b} \times \vec{r}) + \vec{\omega}_{b/a} \times \vec{r} = D_c\vec{r} + (\vec{\omega}_{c/b} + \vec{\omega}_{b/a}) \times \vec{r} \tag{4.61}$$

Equations (4.60) and (4.61) imply that

$$\vec{\omega}_{c/a} = \vec{\omega}_{c/b} + \vec{\omega}_{b/a} \tag{4.62}$$

Equation (4.62) states an important fact of the spatial kinematics that the relative angular velocities are represented by vectors and they are combined directly by means of pure vector addition.

Quite differently though, the relative angular positions are represented by matrices and they are combined by means of matrix multiplication. That is,

$$\hat{C}^{(a,c)} = \hat{C}^{(a,b)}\hat{C}^{(b,c)} \tag{4.63}$$

Actually, Eq. (4.62) can also be obtained from Eq. (4.63). For this purpose, Eq. (4.63) is differentiated side by side so that

$$\dot{\hat{C}}^{(a,c)} = \dot{\hat{C}}^{(a,b)}\hat{C}^{(b,c)} + \hat{C}^{(a,b)}\dot{\hat{C}}^{(b,c)} \tag{4.64}$$

Equation (4.64) can be manipulated as follows:

$$\dot{\hat{C}}^{(a,c)}\hat{C}^{(c,a)} = \dot{\hat{C}}^{(a,b)}\hat{C}^{(b,c)}\hat{C}^{(c,a)} + \hat{C}^{(a,b)}\dot{\hat{C}}^{(b,c)}\hat{C}^{(c,a)} \Rightarrow$$

$$\dot{\hat{C}}^{(a,c)}\hat{C}^{(c,a)} = \dot{\hat{C}}^{(a,b)}\hat{C}^{(b,a)} + \hat{C}^{(a,b)}\dot{\hat{C}}^{(b,c)}\hat{C}^{(c,b)}\hat{C}^{(b,a)} \tag{4.65}$$

Referring to Eq. (4.35), note that

$$\dot{\hat{C}}^{(a,c)}\hat{C}^{(c,a)} = \tilde{\omega}^{(a)}_{c/a}, \quad \dot{\hat{C}}^{(a,b)}\hat{C}^{(b,a)} = \tilde{\omega}^{(a)}_{b/a}, \quad \dot{\hat{C}}^{(b,c)}\hat{C}^{(c,b)} = \tilde{\omega}^{(b)}_{c/b}$$

Hence, Eq. (4.65) becomes

$$\tilde{\omega}^{(a)}_{c/a} = \tilde{\omega}^{(a)}_{b/a} + \hat{C}^{(a,b)}\tilde{\omega}^{(b)}_{c/b}\hat{C}^{(b,a)} = \tilde{\omega}^{(a)}_{b/a} + \tilde{\omega}^{(a)}_{c/b} \tag{4.66}$$

Equation (4.66) implies that

$$\overline{\omega}^{(a)}_{c/a} = \overline{\omega}^{(a)}_{c/b} + \overline{\omega}^{(a)}_{b/a} \tag{4.67}$$

Thus, Eq. (4.62) has been verified, because Eq. (4.67) is nothing but the matrix equation corresponding to the vector Eq. (4.62) in $\mathcal{F}_a$.

Equation (4.62) can be generalized to more than three reference frames as shown below.

$$\vec{\omega}_{z/a} = \vec{\omega}_{z/y} + \vec{\omega}_{y/x} + \cdots + \vec{\omega}_{d/c} + \vec{\omega}_{c/b} + \vec{\omega}_{b/a} \tag{4.68}$$

4.3.2 Combination of Relative Angular Accelerations

When Eq. (4.62) is differentiated side by side with respect to one of the three involved reference frames, say $\mathcal{F}_a$, the following equation is obtained.

$$D_a\vec{\omega}_{c/a} = D_a\vec{\omega}_{c/b} + D_a\vec{\omega}_{b/a} \tag{4.69}$$

According to Eq. (4.46) and Inequality (4.47), $D_a\vec{\omega}_{c/a}$ and $D_a\vec{\omega}_{b/a}$ are angular accelerations, but $D_a\vec{\omega}_{c/b}$ is not an angular acceleration. In other words,

$$D_a\vec{\omega}_{c/a} = \vec{\alpha}_{c/a},\ D_a\vec{\omega}_{b/a} = \vec{\alpha}_{b/a};\ D_a\vec{\omega}_{c/b} \neq \vec{\alpha}_{c/b} \tag{4.70}$$

However, $D_a\vec{\omega}_{c/b}$ can be manipulated as follows by using the Coriolis transport theorem.

$$D_a\vec{\omega}_{c/b} = D_b\vec{\omega}_{c/b} + \vec{\omega}_{b/a} \times \vec{\omega}_{c/b} = \vec{\alpha}_{c/b} + \vec{\omega}_{b/a} \times \vec{\omega}_{c/b} \tag{4.71}$$

Hence, Eq. (4.69) becomes

$$\vec{\alpha}_{c/a} = \vec{\alpha}_{c/b} + \vec{\alpha}_{b/a} + \vec{\omega}_{b/a} \times \vec{\omega}_{c/b} \tag{4.72}$$

Equation (4.72) states another important fact of the spatial kinematics that the relative angular accelerations are also represented by vectors but they cannot be combined directly by means of pure vector addition. They are combined indirectly with the addition of an extra term that involves the relative angular velocities.

Equation (4.72) can be generalized to more than three reference frames as shown below.

$$\begin{aligned}\vec{\alpha}_{z/a} = {} & \vec{\alpha}_{z/y} + \vec{\alpha}_{y/x} + \cdots + \vec{\alpha}_{d/c} + \vec{\alpha}_{c/b} + \vec{\alpha}_{b/a} \\ & + \vec{\omega}_{y/a} \times \vec{\omega}_{z/y} + \vec{\omega}_{x/a} \times \vec{\omega}_{y/x} + \cdots + \vec{\omega}_{b/a} \times \vec{\omega}_{c/b}\end{aligned} \tag{4.73}$$

4.4 Angular Velocities and Accelerations Associated with Rotation Sequences

4.4.1 Relative Angular Velocities and Accelerations about Relatively Fixed Axes

Consider two reference frames $\mathcal{F}_a$ and $\mathcal{F}_b$ such that $\mathcal{F}_b$ is obtained by rotating $\mathcal{F}_a$ about an axis that appears fixed in $\mathcal{F}_a$ and $\mathcal{F}_b$. The rotation of $\mathcal{F}_a$ into $\mathcal{F}_b$ is described as shown below.

$$\mathcal{F}_a \xrightarrow{\text{rot}(\vec{n},\theta)} \mathcal{F}_b \tag{4.74}$$

In the rotation described above, $\vec{n}$ is such that

$$\overline{n}^{(a)} = \overline{n}^{(b)} = \overline{n} = \text{constant} \tag{4.75}$$

In such a case, the transformation matrix between $\mathcal{F}_a$ and $\mathcal{F}_b$ happens to be

$$\hat{C}^{(a,b)} = \hat{R}(\overline{n},\theta) = e^{\tilde{n}\theta} \tag{4.76}$$

Hence, the relative angular velocity of $\mathcal{F}_b$ with respect to $\mathcal{F}_a$ is obtained as follows according to Eq. (4.37) and the formula about the differentiation of rotation matrices given in Section 2.7:

$$\tilde{\omega}_{b/a}^{(a)} = \dot{\hat{C}}^{(a,b)}\hat{C}^{(b,a)} = (\dot{\theta}\tilde{n}e^{\tilde{n}\theta})(e^{-\tilde{n}\theta}) = \dot{\theta}\tilde{n} = \dot{\theta}\tilde{n}^{(a)} \tag{4.77}$$

Equation (4.77) implies the following vector equation.

$$\vec{\omega}_{b/a} = \dot{\theta}\vec{n} \tag{4.78}$$

As for the relative angular acceleration, it is found as follows by noting that $D_a\vec{n} = \vec{0}$ because $\vec{n}$ is fixed with respect to $\mathcal{F}_a$.

$$\vec{\alpha}_{b/a} = D_a\vec{\omega}_{b/a} = D_a(\dot{\theta}\vec{n}) = \ddot{\theta}\vec{n} + \dot{\theta}(D_a\vec{n}) \Rightarrow$$
$$\vec{\alpha}_{b/a} = \ddot{\theta}\vec{n} \tag{4.79}$$

Equations (4.78) and (4.79) happen to be very convenient in many practical situations, because the members of a mechanical system are connected mostly with single axis joints (i.e. revolute, prismatic, cylindrical, and screw joints) and the axis of such a joint appears fixed with respect to the members it connects.

4.4.2 Example

Consider the serial manipulator shown in Figure 4.2. It comprises six revolute joints. The unit vector along the axis of the kth joint is $\vec{n}_k$. The reference frame attached to the link $\mathcal{L}_k$ is $\mathcal{F}_k$. The base is the zeroth link (i.e. $\mathcal{L}_0$). The last link $\mathcal{L}_6$ is the end-effector, which is a gripper for this manipulator. The orientations of the successive links are described as shown below.

$$\mathcal{F}_0 \xrightarrow{\text{rot}(\vec{n}_1,\theta_1)} \mathcal{F}_1 \xrightarrow{\text{rot}(\vec{n}_2,\theta_2)} \mathcal{F}_2 \xrightarrow{\text{rot}(\vec{n}_3,\theta_3)} \mathcal{F}_3 \cdots \mathcal{F}_5 \xrightarrow{\text{rot}(\vec{n}_6,\theta_6)} \mathcal{F}_6 \tag{4.80}$$

In Description (4.80), the unit vector $\vec{n}_k$ of the joint $\mathcal{J}_k$ between the links $\mathcal{L}_{k-1}$ and $\mathcal{L}_k$ appears fixed with respect to $\mathcal{F}_{k-1}$ and $\mathcal{F}_k$. In other words,

$$\overline{n}_k^{(k)} = \overline{n}_k^{(k-1)} = \overline{n}_k = \text{constant} \tag{4.81}$$

Hence, the relative orientation matrix $\hat{C}^{(k-1,k)}$ is expressed as

$$\hat{C}^{(k-1,k)} = \hat{R}(\overline{n}_k, \theta_k) = e^{\tilde{n}_k\theta_k} \tag{4.82}$$

By using Eq. (4.82), the orientation of the end-effector ($\mathcal{L}_6$) of the manipulator with respect to the base ($\mathcal{L}_0$) is expressed by the following combined matrix.

$$\hat{C}^{(0,6)} = \hat{C}^{(0,1)}\hat{C}^{(1,2)}\hat{C}^{(2,3)}\hat{C}^{(3,4)}\hat{C}^{(4,5)}\hat{C}^{(5,6)} \Rightarrow$$
$$\hat{C}^{(0,6)} = e^{\tilde{n}_1\theta_1}e^{\tilde{n}_2\theta_2}e^{\tilde{n}_3\theta_3}e^{\tilde{n}_4\theta_4}e^{\tilde{n}_5\theta_5}e^{\tilde{n}_6\theta_6} \tag{4.83}$$

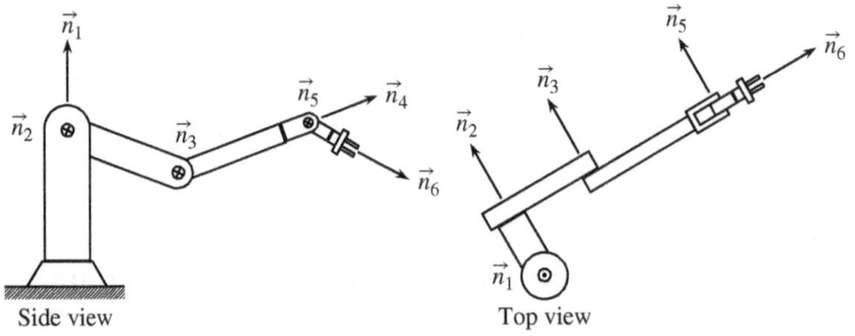

Figure 4.2 A manipulator with six revolute joints.

The relative angular velocities and accelerations between successive links are obtained as follows according to Eqs. (4.78) and (4.79):

$$\vec{\omega}_{k/k-1} = \dot{\theta}_k \vec{n}_k \tag{4.84}$$

$$\vec{\alpha}_{k/k-1} = \ddot{\theta}_k \vec{n}_k \tag{4.85}$$

The angular velocity of the end-effector ($\mathcal{L}_6$) with respect to the base ($\mathcal{L}_0$) is obtained as the following combination of the relative angular velocities according to Eq. (4.68).

$$\vec{\omega}_{6/0} = \vec{\omega}_{6/5} + \vec{\omega}_{5/4} + \vec{\omega}_{4/3} + \vec{\omega}_{3/2} + \vec{\omega}_{2/1} + \vec{\omega}_{1/0} \Rightarrow$$
$$\vec{\omega}_{6/0} = \dot{\theta}_6 \vec{n}_6 + \dot{\theta}_5 \vec{n}_5 + \dot{\theta}_4 \vec{n}_4 + \dot{\theta}_3 \vec{n}_3 + \dot{\theta}_2 \vec{n}_2 + \dot{\theta}_1 \vec{n}_1 \Rightarrow$$
$$\vec{\omega}_{6/0} = \dot{\theta}_1 \vec{n}_1 + \dot{\theta}_2 \vec{n}_2 + \dot{\theta}_3 \vec{n}_3 + \dot{\theta}_4 \vec{n}_4 + \dot{\theta}_5 \vec{n}_5 + \dot{\theta}_6 \vec{n}_6 \tag{4.86}$$

The matrix representation of $\vec{\omega}_{6/0}$ in $\mathcal{F}_0$ is obtained as follows:

$$\overline{\omega}^{(0)}_{6/0} = \dot{\theta}_1 \overline{n}^{(0)}_1 + \dot{\theta}_2 \overline{n}^{(0)}_2 + \dot{\theta}_3 \overline{n}^{(0)}_3 + \dot{\theta}_4 \overline{n}^{(0)}_4 + \dot{\theta}_5 \overline{n}^{(0)}_5 + \dot{\theta}_6 \overline{n}^{(0)}_6$$

Note that $\overline{n}^{(0)}_k = \hat{C}^{(0,k-1)} \overline{n}^{(k-1)}_k = \hat{C}^{(0,k-1)} \overline{n}_k$. Hence, the preceding equation can be manipulated so that

$$\overline{\omega}^{(0)}_{6/0} = \dot{\theta}_1 \overline{n}_1 + \dot{\theta}_2 \hat{C}^{(0,1)} \overline{n}_2 + \dot{\theta}_3 \hat{C}^{(0,2)} \overline{n}_3 + \dot{\theta}_4 \hat{C}^{(0,3)} \overline{n}_4 + \dot{\theta}_5 \hat{C}^{(0,4)} \overline{n}_5 + \dot{\theta}_6 \hat{C}^{(0,5)} \overline{n}_6 \Rightarrow$$
$$\overline{\omega}^{(0)}_{6/0} = \dot{\theta}_1 \overline{n}_1 + \dot{\theta}_2 e^{\tilde{n}_1 \theta_1} \overline{n}_2 + \dot{\theta}_3 e^{\tilde{n}_1 \theta_1} e^{\tilde{n}_2 \theta_2} \overline{n}_3 + \dot{\theta}_4 e^{\tilde{n}_1 \theta_1} e^{\tilde{n}_2 \theta_2} e^{\tilde{n}_3 \theta_3} \overline{n}_4 \Rightarrow$$
$$+ \dot{\theta}_5 e^{\tilde{n}_1 \theta_1} e^{\tilde{n}_2 \theta_2} e^{\tilde{n}_3 \theta_3} e^{\tilde{n}_4 \theta_4} \overline{n}_5 + \dot{\theta}_6 e^{\tilde{n}_1 \theta_1} e^{\tilde{n}_2 \theta_2} e^{\tilde{n}_3 \theta_3} e^{\tilde{n}_4 \theta_4} e^{\tilde{n}_5 \theta_5} \overline{n}_6 \tag{4.87}$$

Equation (4.87) can also be obtained directly from Eq. (4.83) by using Eq. (4.37), which is written as shown below for the present example.

$$\tilde{\omega}^{(0)}_{6/0} = \dot{\hat{C}}^{(0,6)} \hat{C}^{(6,0)} \tag{4.88}$$

When Eq. (4.83) is substituted, Eq. (4.88) becomes

$$\tilde{\omega}^{(0)}_{6/0} = [\dot{\theta}_1 \tilde{n}_1 e^{\tilde{n}_1 \theta_1} e^{\tilde{n}_2 \theta_2} e^{\tilde{n}_3 \theta_3} e^{\tilde{n}_4 \theta_4} e^{\tilde{n}_5 \theta_5} e^{\tilde{n}_6 \theta_6}$$
$$+ \dot{\theta}_2 e^{\tilde{n}_1 \theta_1} \tilde{n}_2 e^{\tilde{n}_2 \theta_2} e^{\tilde{n}_3 \theta_3} e^{\tilde{n}_4 \theta_4} e^{\tilde{n}_5 \theta_5} e^{\tilde{n}_6 \theta_6}$$
$$+ \dot{\theta}_3 e^{\tilde{n}_1 \theta_1} e^{\tilde{n}_2 \theta_2} \tilde{n}_3 e^{\tilde{n}_3 \theta_3} e^{\tilde{n}_4 \theta_4} e^{\tilde{n}_5 \theta_5} e^{\tilde{n}_6 \theta_6} + \cdots$$
$$+ \dot{\theta}_6 e^{\tilde{n}_1 \theta_1} e^{\tilde{n}_2 \theta_2} \cdots e^{\tilde{n}_5 \theta_5} \tilde{n}_6 e^{\tilde{n}_6 \theta_6}][e^{-\tilde{n}_6 \theta_6} e^{-\tilde{n}_5 \theta_5} \cdots e^{-\tilde{n}_2 \theta_2} e^{-\tilde{n}_1 \theta_1}] \Rightarrow$$
$$\tilde{\omega}^{(0)}_{6/0} = \dot{\theta}_1 \tilde{n}_1 + \dot{\theta}_2 e^{\tilde{n}_1 \theta_1} \tilde{n}_2 e^{-\tilde{n}_1 \theta_1} + \dot{\theta}_3 e^{\tilde{n}_1 \theta_1} e^{\tilde{n}_2 \theta_2} \tilde{n}_3 e^{-\tilde{n}_2 \theta_2} e^{-\tilde{n}_1 \theta_1} + \cdots$$
$$+ \dot{\theta}_6 e^{\tilde{n}_1 \theta_1} e^{\tilde{n}_2 \theta_2} \cdots e^{\tilde{n}_5 \theta_5} \tilde{n}_6 e^{-\tilde{n}_5 \theta_5} \cdots e^{-\tilde{n}_2 \theta_2} e^{-\tilde{n}_1 \theta_1} \tag{4.89}$$

Referring to Eq. (1.7.10), it is seen that

$$\overline{y} = \hat{R} \overline{x} \Leftrightarrow \tilde{y} = \hat{R} \tilde{x} \hat{R}^t$$

Therefore, Eq. (4.89) implies Eq. (4.87).

At this point, it is worth paying attention to the particular pattern of Eq. (4.87), which is directly associated with the pattern of Eq. (4.83). Owing to this particular pattern, for any given orientation expression similar to that in Eq. (4.83), it is possible to write down the corresponding angular velocity expression similarly as in Eq. (4.87) directly by inspection.

As for the angular acceleration of the end-effector ($\mathcal{L}_6$) with respect to the base ($\mathcal{L}_0$), it is obtained as the following combination of the relative angular accelerations and velocities according to Eq. (4.73).

$$\begin{aligned}
\vec{\alpha}_{6/0} &= \vec{\alpha}_{6/5} + \vec{\alpha}_{5/4} + \vec{\alpha}_{4/3} + \vec{\alpha}_{3/2} + \vec{\alpha}_{2/1} + \vec{\alpha}_{1/0} \\
&\quad + \vec{\omega}_{5/0} \times \vec{\omega}_{6/5} + \vec{\omega}_{4/0} \times \vec{\omega}_{5/4} + \vec{\omega}_{3/0} \times \vec{\omega}_{4/3} \\
&\quad + \vec{\omega}_{2/0} \times \vec{\omega}_{3/2} + \vec{\omega}_{1/0} \times \vec{\omega}_{2/1} \Rightarrow \\
\vec{\alpha}_{6/0} &= \ddot{\theta}_6 \vec{n}_6 + \ddot{\theta}_5 \vec{n}_5 + \ddot{\theta}_4 \vec{n}_4 + \ddot{\theta}_3 \vec{n}_3 + \ddot{\theta}_2 \vec{n}_2 + \ddot{\theta}_1 \vec{n}_1 \\
&\quad + \dot{\theta}_6\, \vec{\omega}_{5/0} \times \vec{n}_6 + \dot{\theta}_5 \vec{\omega}_{4/0} \times \vec{n}_5 + \dot{\theta}_4 \vec{\omega}_{3/0} \times \vec{n}_4 \\
&\quad + \dot{\theta}_3\, \vec{\omega}_{2/0} \times \vec{n}_3 + \dot{\theta}_2 \vec{\omega}_{1/0} \times \vec{n}_2 \Rightarrow \\
\vec{\alpha}_{6/0} &= \ddot{\theta}_1 \vec{n}_1 + \ddot{\theta}_2 \vec{n}_2 + \ddot{\theta}_3 \vec{n}_3 + \ddot{\theta}_4 \vec{n}_4 + \ddot{\theta}_5 \vec{n}_5 + \ddot{\theta}_6 \vec{n}_6 \\
&\quad + \dot{\theta}_1 \vec{n}_1 \times (\dot{\theta}_2 \vec{n}_2 + \dot{\theta}_3 \vec{n}_3 + \dot{\theta}_4 \vec{n}_4 + \dot{\theta}_5 \vec{n}_5 + \dot{\theta}_6 \vec{n}_6) \\
&\quad + \dot{\theta}_2 \vec{n}_2 \times (\dot{\theta}_3 \vec{n}_3 + \dot{\theta}_4 \vec{n}_4 + \dot{\theta}_5 \vec{n}_5 + \dot{\theta}_6 \vec{n}_6) \\
&\quad + \dot{\theta}_3 \vec{n}_3 \times (\dot{\theta}_4 \vec{n}_4 + \dot{\theta}_5 \vec{n}_5 + \dot{\theta}_6 \vec{n}_6) \\
&\quad + \dot{\theta}_4 \vec{n}_4 \times (\dot{\theta}_5 \vec{n}_5 + \dot{\theta}_6 \vec{n}_6) + \dot{\theta}_5 \dot{\theta}_6 \vec{n}_5 \times \vec{n}_6
\end{aligned} \tag{4.90}$$

4.4.3 Angular Velocities Associated with the Euler Angle Sequences

(a) Rotated Frame Based Euler Angle Sequences

Consider a rotated frame based (RFB) *i-j-k* sequence described as

$$\mathcal{F}_a \xrightarrow{\text{rot}[\vec{u}_i^{(a)},\ \phi_1]} \mathcal{F}_m \xrightarrow{\text{rot}[\vec{u}_j^{(m)},\ \phi_2]} \mathcal{F}_n \xrightarrow{\text{rot}[\vec{u}_k^{(n)},\ \phi_3]} \mathcal{F}_b \tag{4.91}$$

As explained in the previous sections, the orientation matrix and the angular velocity of $\mathcal{F}_b$ with respect to $\mathcal{F}_a$ are expressed as follows:

$$\hat{C}^{(a,b)} = \hat{C}^{(a,m)}\hat{C}^{(m,n)}\hat{C}^{(n,b)} = e^{\tilde{u}_i\phi_1} e^{\tilde{u}_j\phi_2} e^{\tilde{u}_k\phi_3} \tag{4.92}$$

$$\vec{\omega}_{b/a} = \vec{\omega}_{b/n} + \vec{\omega}_{n/m} + \vec{\omega}_{m/a} = \vec{u}_k^{(n)}\dot{\phi}_3 + \vec{u}_j^{(m)}\dot{\phi}_2 + \vec{u}_i^{(a)}\dot{\phi}_1 \tag{4.93}$$

In Eq. (4.93), $\vec{\omega}_{b/a}$ has a hybrid expression, i.e. it is expressed in terms of the basis vectors of different reference frames. However, most of the time, it is required to express $\vec{\omega}_{b/a}$ in a selected reference frame, e.g. $\mathcal{F}_a$. To fulfill this requirement, the matrix representation of $\vec{\omega}_{b/a}$ in $\mathcal{F}_a$ can be obtained as shown below.

$$\begin{aligned}
\overline{\omega}_{b/a}^{(a)} &= \dot{\phi}_1 \overline{u}_i^{(a/a)} + \dot{\phi}_2 \overline{u}_j^{(m/a)} + \dot{\phi}_3 \overline{u}_k^{(n/a)} \Rightarrow \\
\overline{\omega}_{b/a}^{(a)} &= \dot{\phi}_1 \overline{u}_i^{(a/a)} + \dot{\phi}_2 \hat{C}^{(a,m)} \overline{u}_j^{(m/m)} + \dot{\phi}_3 \hat{C}^{(a,n)} \overline{u}_k^{(n/n)} \Rightarrow \\
\overline{\omega}_{b/a}^{(a)} &= \dot{\phi}_1 \overline{u}_i + \dot{\phi}_2 e^{\tilde{u}_i\phi_1} \overline{u}_j + \dot{\phi}_3 e^{\tilde{u}_i\phi_1} e^{\tilde{u}_j\phi_2} \overline{u}_k
\end{aligned} \tag{4.94}$$

As an example, for the RFB 1-2-3 sequence, $\overline{\omega}_{b/a}^{(a)}$ can be written out as follows:

$$\begin{aligned}
\overline{\omega}_{b/a}^{(a)} &= \dot{\phi}_1 \overline{u}_1 + \dot{\phi}_2 e^{\tilde{u}_1\phi_1} \overline{u}_2 + \dot{\phi}_3 e^{\tilde{u}_1\phi_1} e^{\tilde{u}_2\phi_2} \overline{u}_3 \Rightarrow \\
\overline{\omega}_{b/a}^{(a)} &= \overline{u}_1(\dot{\phi}_1 + \dot{\phi}_3 s\phi_2) + \overline{u}_2(\dot{\phi}_2 c\phi_1 - \dot{\phi}_3 s\phi_1 c\phi_2) \\
&\quad + \overline{u}_3(\dot{\phi}_2 s\phi_1 + \dot{\phi}_3 c\phi_1 c\phi_2)
\end{aligned} \tag{4.95}$$

Equation (4.95) implies the following vector equation.

$$\vec{\omega}_{b/a} = \vec{u}_1^{(a)}(\dot{\phi}_1 + \dot{\phi}_3 s\phi_2) + \vec{u}_2^{(a)}(\dot{\phi}_2 c\phi_1 - \dot{\phi}_3 s\phi_1 c\phi_2) + \vec{u}_3^{(a)}(\dot{\phi}_2 s\phi_1 + \dot{\phi}_3 c\phi_1 c\phi_2) \quad (4.96)$$

Coming back to Eq. (4.95), let it be written as follows for the sake of notational brevity:

$$\overline{\omega}_{b/a}^{(a)} = \overline{\omega} = \overline{u}_1\omega_1 + \overline{u}_2\omega_2 + \overline{u}_3\omega_3 \quad (4.97)$$

Meanwhile, let the Euler angles be collected into the following column matrix.

$$\overline{\phi} = \overline{u}_1\phi_1 + \overline{u}_2\phi_2 + \overline{u}_3\phi_3 \quad (4.98)$$

Then, Eq. (4.95) can be written compactly as

$$\overline{\omega} = \widehat{E}\dot{\overline{\phi}} \quad (4.99)$$

In Eq. (4.99), the matrix $\widehat{E}$ is defined as

$$\widehat{E} = \widehat{E}(\overline{\phi}) = \begin{bmatrix} 1 & 0 & s\phi_2 \\ 0 & c\phi_1 & -s\phi_1 c\phi_2 \\ 0 & s\phi_1 & c\phi_1 c\phi_2 \end{bmatrix} \quad (4.100)$$

It can be shown that

$$\det(\widehat{E}) = c\phi_2 \quad (4.101)$$

If $\det(\widehat{E}) \neq 0$, i.e. if $\phi_2 \neq \sigma_2\pi/2$ with $\sigma_2 = \pm 1$, then the following differential equation can be obtained in order to find the Euler angles from the components of the angular velocity.

$$\dot{\overline{\phi}} = \widehat{E}^{-1}\overline{\omega} \quad (4.102)$$

In Eq. (4.102), $\widehat{E}^{-1}$ is expressed as follows with the additional abbreviation that $t\phi_2 = \tan\phi_2$:

$$\widehat{E}^{-1} = \begin{bmatrix} 1 & s\phi_1 t\phi_2 & -c\phi_1 t\phi_2 \\ 0 & c\phi_1 & s\phi_1 \\ 0 & -s\phi_1/c\phi_2 & c\phi_1/c\phi_2 \end{bmatrix} \quad (4.103)$$

If $\det(\widehat{E}) = 0$, i.e. if $\phi_2 = \sigma_2\pi/2$ with $\sigma_2 = \pm 1$, then the RFB 1-2-3 sequence becomes singular. In a case of singularity, Eq. (4.95) reduces to

$$\overline{\omega} = \overline{u}_1(\dot{\phi}_1 + \sigma_2\dot{\phi}_3) + \overline{u}_2\dot{\phi}_2 c\phi_1 + \overline{u}_3\dot{\phi}_2 s\phi_1 \quad (4.104)$$

Equation (4.104) leads to the following set of scalar equations.

$$\left.\begin{aligned} \omega_1 &= \dot{\phi}_1 + \sigma_2\dot{\phi}_3 \\ \omega_2 &= \dot{\phi}_2 c\phi_1 \\ \omega_3 &= \dot{\phi}_2 s\phi_1 \end{aligned}\right\} \quad (4.105)$$

Equation Set (4.105) implies that

$$\dot{\phi}_1 + \sigma_2\dot{\phi}_3 = \omega_1 \quad (4.106)$$

$$\dot{\phi}_2 = \omega_2 c\phi_1 + \omega_3 s\phi_1 \quad (4.107)$$

$$\omega_2 s\phi_1 = \omega_3 c\phi_1 \quad (4.108)$$

Equations (4.106)–(4.108) show the following features of the singularity:

(i) Equation (4.107) shows that $\dot{\phi}_2$ can still be found with a definite value.
(ii) Equation (4.106) shows that $\dot{\phi}_1$ and $\dot{\phi}_3$ become indefinite but the combination $(\dot{\phi}_1 + \sigma_2\dot{\phi}_3)$ can nonetheless be found with a definite value.
(iii) Equation (4.108) shows the constraint that arises concerning the components ω_2 and ω_3.

(b) Initial Frame Based Euler Angle Sequences

Consider an initial frame based (IFB) *i-j-k* sequence described as

$$\mathcal{F}_a \xrightarrow{\text{rot}[\vec{u}_i^{(a)},\ \phi_1]} \mathcal{F}_m \xrightarrow{\text{rot}[\vec{u}_j^{(a)},\ \phi_2]} \mathcal{F}_n \xrightarrow{\text{rot}[\vec{u}_k^{(a)},\ \phi_3]} \mathcal{F}_b \tag{4.109}$$

As explained in Sections 3.7 and 3.8, the IFB *i-j-k* sequence described above is equivalent to the RFB *k-j-i* sequence described below.

$$\mathcal{F}_a \xrightarrow{\text{rot}[\vec{u}_k^{(a)},\ \phi_3]} \mathcal{F}_p \xrightarrow{\text{rot}[\vec{u}_j^{(p)},\ \phi_2]} \mathcal{F}_q \xrightarrow{\text{rot}[\vec{u}_i^{(q)},\ \phi_1]} \mathcal{F}_b \tag{4.110}$$

Based on this equivalence, the orientation matrix and the angular velocity of $\mathcal{F}_b$ with respect to $\mathcal{F}_a$ can be expressed as follows:

$$\hat{C}^{(a,b)} = \hat{C}^{(a,p)}\hat{C}^{(p,q)}\hat{C}^{(q,b)} = e^{\tilde{u}_k\phi_3}e^{\tilde{u}_j\phi_2}e^{\tilde{u}_i\phi_1} \tag{4.111}$$

$$\vec{\omega}_{b/a} = \vec{\omega}_{b/q} + \vec{\omega}_{q/p} + \vec{\omega}_{p/a} = \vec{u}_i^{(q)}\dot{\phi}_1 + \vec{u}_j^{(p)}\dot{\phi}_2 + \vec{u}_k^{(a)}\dot{\phi}_3 \tag{4.112}$$

As for the matrix representation of $\vec{\omega}_{b/a}$ in $\mathcal{F}_a$, it can be obtained as shown below.

$$\overline{\omega}_{b/a}^{(a)} = \dot{\phi}_1\overline{u}_i^{(q/a)} + \dot{\phi}_2\overline{u}_j^{(p/a)} + \dot{\phi}_3\overline{u}_k^{(a/a)} \Rightarrow$$
$$\overline{\omega}_{b/a}^{(a)} = \dot{\phi}_1\hat{C}^{(a,q)}\overline{u}_i^{(q/q)} + \dot{\phi}_2\hat{C}^{(a,p)}\overline{u}_j^{(p/p)} + \dot{\phi}_3\overline{u}_k^{(a/a)} \Rightarrow$$
$$\overline{\omega}_{b/a}^{(a)} = \dot{\phi}_1 e^{\tilde{u}_k\phi_3}e^{\tilde{u}_j\phi_2}\overline{u}_i + \dot{\phi}_2 e^{\tilde{u}_k\phi_3}\overline{u}_j + \dot{\phi}_3\overline{u}_k \tag{4.113}$$

As an example, for the IFB 1-2-3 sequence, $\overline{\omega}_{b/a}^{(a)}$ can be written out as follows:

$$\overline{\omega}_{b/a}^{(a)} = \dot{\phi}_1 e^{\tilde{u}_3\phi_3}e^{\tilde{u}_2\phi_2}\overline{u}_1 + \dot{\phi}_2 e^{\tilde{u}_3\phi_3}\overline{u}_2 + \dot{\phi}_3\overline{u}_3 \Rightarrow$$
$$\overline{\omega}_{b/a}^{(a)} = \overline{u}_1(\dot{\phi}_1 c\phi_3 c\phi_2 - \dot{\phi}_2 s\phi_3) + \overline{u}_2(\dot{\phi}_1 s\phi_3 c\phi_2 + \dot{\phi}_2 c\phi_3)$$
$$+ \overline{u}_3(\dot{\phi}_3 - \dot{\phi}_1 c\phi_2) \tag{4.114}$$

Equation (4.114) implies the following vector equation.

$$\vec{\omega}_{b/a} = \vec{u}_1^{(a)}(\dot{\phi}_1 c\phi_3 c\phi_2 - \dot{\phi}_2 s\phi_3) + \vec{u}_2^{(a)}(\dot{\phi}_1 s\phi_3 c\phi_2 + \dot{\phi}_2 c\phi_3)$$
$$+ \vec{u}_3^{(a)}(\dot{\phi}_3 - \dot{\phi}_1 c\phi_2) \tag{4.115}$$

For the IFB 1-2-3 sequence, the matrix $\hat{E}$ in the equation $\overline{\omega}_{b/a}^{(a)} = \overline{\omega} = \hat{E}\dot{\overline{\phi}}$ takes the following different form as compared with the matrix $\hat{E}$ of the RFB 1-2-3 sequence expressed by Eq. (4.100).

$$\hat{E} = \hat{E}(\overline{\phi}) = \begin{bmatrix} c\phi_3 c\phi_2 & -s\phi_3 & 0 \\ s\phi_3 c\phi_2 & c\phi_3 & 0 \\ -c\phi_2 & 0 & 1 \end{bmatrix} \tag{4.116}$$

4.5 Velocity and Acceleration of a Point with Respect to Different Reference Frames

4.5.1 Velocity of a Point with Respect to Different Reference Frames

Figure 4.3 illustrates a case in which a point P is observed in two different reference frames, $\mathcal{F}_a(A)$ and $\mathcal{F}_b(B)$. The frame $\mathcal{F}_b(B)$ is both translating and rotating with respect to the frame $\mathcal{F}_a(A)$. The relative position vectors of the point P with respect to $\mathcal{F}_a(A)$ and $\mathcal{F}_b(B)$ are related to each other by the following equation.

$$\vec{r}_{P/A} = \vec{r}_{P/B} + \vec{r}_{B/A} \tag{4.117}$$

In order to set up the relationship between the relative velocities of the point P with respect to $\mathcal{F}_a(A)$ and $\mathcal{F}_b(B)$, Eq. (4.117) can be differentiated side by side with respect to one of $\mathcal{F}_a(A)$ and $\mathcal{F}_b(B)$ as the first operation. If $\mathcal{F}_a(A)$ is chosen for this purpose, the following equation is obtained.

$$D_a\vec{r}_{P/A} = D_a\vec{r}_{P/B} + D_a\vec{r}_{B/A} \tag{4.118}$$

The terms in Eq. (4.118) suggest the following definition of *relative velocity*.

$$\vec{v}_{P/B/\mathcal{F}_a} = D_a\vec{r}_{P/B} \tag{4.119}$$

The relative velocity $\vec{v}_{P/B/\mathcal{F}_a}$, which is defined by Eq. (4.119) and denoted with three subscripts, is described verbally as follows: the relative velocity of the point P with respect to the point B and the differentiation frame $\mathcal{F}_a(A)$.

As noticed above, the definition of a relative velocity involves three basic items, which are:

(i) The point of interest, e.g. P.
(ii) The point of reference, e.g. B.
(iii) The frame of differentiation, e.g. $\mathcal{F}_a(A)$.

By using the notation introduced in Eq. (4.119), Eq. (4.118) can be written as the following relative velocity relationship.

$$\vec{v}_{P/A/\mathcal{F}_a} = \vec{v}_{P/B/\mathcal{F}_a} + \vec{v}_{B/A/\mathcal{F}_a} \tag{4.120}$$

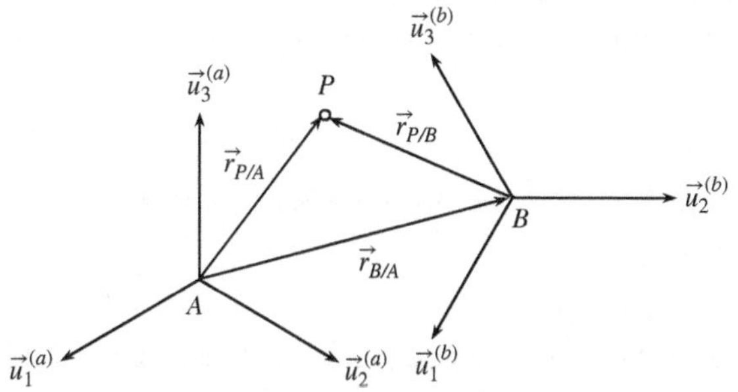

Figure 4.3 A point observed in two different reference frames.

However, it is normally more convenient and preferable to use $\mathcal{F}_b(B)$ as the differentiation frame for the relative position vector $\vec{r}_{P/B}$ because its components are available naturally in $\mathcal{F}_b(B)$. With this in mind, Eq. (4.119) can be written as follows by using the Coriolis transport theorem:

$$\vec{v}_{P/B/\mathcal{F}_a} = D_a\vec{r}_{P/B} = D_b\vec{r}_{P/B} + \vec{\omega}_{b/a} \times \vec{r}_{P/B} = \vec{v}_{P/B/\mathcal{F}_b} + \vec{\omega}_{b/a} \times \vec{r}_{P/B} \tag{4.121}$$

Hence, Eq. (4.120) can be modified to the following equation, which can be used in order to relate the relative velocities of a point P observed in $\mathcal{F}_a(A)$ and $\mathcal{F}_b(B)$.

$$\vec{v}_{P/A/\mathcal{F}_a} = \vec{v}_{P/B/\mathcal{F}_b} + \vec{\omega}_{b/a} \times \vec{r}_{P/B} + \vec{v}_{B/A/\mathcal{F}_a} \tag{4.122}$$

With a slight notational modification, Eq. (4.122) can also be written as

$$\vec{v}_{P/\mathcal{F}_a(A)} = \vec{v}_{P/\mathcal{F}_b(B)} + \vec{\omega}_{b/a} \times \vec{r}_{P/B} + \vec{v}_{B/\mathcal{F}_a(A)} \tag{4.123}$$

In Eq. (4.123), the relative velocity denoted as $\vec{v}_{P/\mathcal{F}_a(A)}$ is described verbally as follows: the relative velocity of the point P with respect to the reference frame $\mathcal{F}_a(A)$. The other relative velocities in the same equation are described similarly.

4.5.2 Acceleration of a Point with Respect to Different Reference Frames

If the vectors in Eq. (4.118) are differentiated once again with respect to $\mathcal{F}_a(A)$, the following equation is obtained.

$$D_a^2\vec{r}_{P/A} = D_a^2\vec{r}_{P/B} + D_a^2\vec{r}_{B/A} \tag{4.124}$$

The terms in Eq. (4.123) suggest the following definition of *relative acceleration*.

$$\vec{a}_{P/B/\mathcal{F}_a} = D_a^2\vec{r}_{P/B} \tag{4.125}$$

Similarly as the relative velocity $\vec{v}_{P/B/\mathcal{F}_a}$, the relative acceleration $\vec{a}_{P/B/\mathcal{F}_a}$, which is defined by Eq. (4.125) and denoted also with three subscripts, is described verbally as follows: the relative acceleration of the point P with respect to the point B and the differentiation frame $\mathcal{F}_a(A)$.

Note that, like the definition of a relative velocity, the definition of a relative acceleration also involves the same three basic items.

By using the notation introduced in Eq. (4.125), Eq. (4.124) can be written as the following relative acceleration relationship.

$$\vec{a}_{P/A/\mathcal{F}_a} = \vec{a}_{P/B/\mathcal{F}_a} + \vec{a}_{B/A/\mathcal{F}_a} \tag{4.126}$$

However, the criterion mentioned before about the relative velocities concerning the preference of the differentiation frame is valid for the relative accelerations, too. So, according to this criterion, Eq. (4.125) can be written as follows by using the Coriolis transport theorem twice:

$$\begin{aligned}
\vec{a}_{P/B/\mathcal{F}_a} &= D_a^2\vec{r}_{P/B} \Rightarrow \\
\vec{a}_{P/B/\mathcal{F}_a} &= D_b^2\vec{r}_{P/B} + 2\vec{\omega}_{b/a} \times D_b\vec{r}_{P/B} + \vec{\alpha}_{b/a} \times \vec{r}_{P/B} + \vec{\omega}_{b/a} \times (\vec{\omega}_{b/a} \times \vec{r}_{P/B}) \Rightarrow \\
\vec{a}_{P/B/\mathcal{F}_a} &= \vec{a}_{P/B/\mathcal{F}_b} + 2\vec{\omega}_{b/a} \times \vec{v}_{P/B/\mathcal{F}_b} \\
&\quad + \vec{\alpha}_{b/a} \times \vec{r}_{P/B} + \vec{\omega}_{b/a} \times (\vec{\omega}_{b/a} \times \vec{r}_{P/B})
\end{aligned} \tag{4.127}$$

Hence, Eq. (4.126) can be modified to the following equation, which can be used in order to relate the relative accelerations of a point P observed in $\mathcal{F}_a(A)$ and $\mathcal{F}_b(B)$.

$$\begin{aligned}
\vec{a}_{P/A/\mathcal{F}_a} &= \vec{a}_{P/B/\mathcal{F}_b} + 2\vec{\omega}_{b/a} \times \vec{v}_{P/B/\mathcal{F}_b} \\
&\quad + \vec{\alpha}_{b/a} \times \vec{r}_{P/B} + \vec{\omega}_{b/a} \times (\vec{\omega}_{b/a} \times \vec{r}_{P/B}) + \vec{a}_{B/A/\mathcal{F}_a}
\end{aligned} \tag{4.128}$$

Note that two of the terms in Eq. (4.128) depend only on the relative velocities ($\vec{v}_{P/B/\mathcal{F}_b}$ and/or $\vec{\omega}_{b/a}$). They are known by the following special names.

$$* \text{ Coriolis acceleration}: \quad 2\vec{\omega}_{b/a} \times \vec{v}_{P/B/\mathcal{F}_b} \tag{4.129}$$

$$* \text{ Centripetal acceleration}: \quad \vec{\omega}_{b/a} \times (\vec{\omega}_{b/a} \times \vec{r}_{P/B}) \tag{4.130}$$

With the same slight notational modification made before for the relative velocities, Eq. (4.128) can also be written as

$$\begin{aligned} \vec{a}_{P/\mathcal{F}_a(A)} &= \vec{a}_{P/\mathcal{F}_b(B)} + 2\vec{\omega}_{b/a} \times \vec{v}_{P/\mathcal{F}_b(B)} \\ &\quad + \vec{\alpha}_{b/a} \times \vec{r}_{P/B} + \vec{\omega}_{b/a} \times (\vec{\omega}_{b/a} \times \vec{r}_{P/B}) + \vec{a}_{B/\mathcal{F}_a(A)} \end{aligned} \tag{4.131}$$

In Eq. (4.131), the relative acceleration denoted as $\vec{a}_{P/\mathcal{F}_a(A)}$ is described verbally as follows: the relative acceleration of the point P with respect to the reference frame $\mathcal{F}_a(A)$. The other relative accelerations in the same equation are described similarly.

4.5.3 Velocity and Acceleration Expressions with Simplified Notations

If only two reference frames such as $\mathcal{F}_a(A)$ and $\mathcal{F}_b(B)$ are used in a particular kinematic study, then a simpler notational scheme may be used instead of the exact and unambiguous but rather cumbersome notational scheme presented above. Such a simpler notational scheme is described below.

$$\vec{r}_{P/A} = \vec{r}, \quad \vec{r}_{P/B} = \vec{r}', \quad \vec{r}_{B/A} = \vec{r}^{\circ} \tag{4.132}$$

$$\vec{v}_{P/A/\mathcal{F}_a} = \vec{v}, \quad \vec{v}_{P/B/\mathcal{F}_b} = \vec{v}', \quad \vec{v}_{B/A/\mathcal{F}_a} = \vec{v}^{\circ} \tag{4.133}$$

$$\vec{a}_{P/A/\mathcal{F}_a} = \vec{a}, \quad \vec{a}_{P/B/\mathcal{F}_b} = \vec{a}', \quad \vec{a}_{B/A/\mathcal{F}_a} = \vec{a}^{\circ} \tag{4.134}$$

$$\vec{\omega}_{b/a} = \vec{\omega}, \quad \vec{\alpha}_{b/a} = \vec{\alpha} \tag{4.135}$$

In the notational scheme described above, the primed symbols denote the relative position, velocity, and acceleration of the observed point with respect to the second reference frame that is moving with respect to the first reference frame and the symbols with the superscript $^{\circ}$ denote the position, velocity, and acceleration of the *origin* of the second reference frame with respect to the first one.

By using the simpler notational scheme, the relationships involving the relative positions, velocities, and accelerations with respect to the first and second reference frames are expressed by the following equations.

$$\vec{r} = \vec{r}' + \vec{r}^{\circ} \tag{4.136}$$

$$\vec{v} = \vec{v}' + \vec{\omega} \times \vec{r}' + \vec{v}^{\circ} \tag{4.137}$$

$$\vec{a} = \vec{a}' + 2\vec{\omega} \times \vec{v}' + \vec{\alpha} \times \vec{r}' + \vec{\omega} \times (\vec{\omega} \times \vec{r}') + \vec{a}^{\circ} \tag{4.138}$$

The preceding vector equations can also be written as matrix equations by introducing the following definitions.

$$\bar{r} = [\vec{r}]^{(a)}, \quad \bar{v} = [\vec{v}]^{(a)}, \quad \bar{a} = [\vec{a}]^{(a)} \tag{4.139}$$

$$\bar{r}^{\circ} = [\vec{r}^{\circ}]^{(a)}, \quad \bar{v}^{\circ} = [\vec{v}^{\circ}]^{(a)}, \quad \bar{a}^{\circ} = [\vec{a}^{\circ}]^{(a)} \tag{4.140}$$

$$\bar{r}' = [\vec{r}']^{(b)}, \quad \bar{v}' = [\vec{v}']^{(b)}, \quad \bar{a}' = [\vec{a}']^{(b)} \tag{4.141}$$

$$\bar{\omega} = [\vec{\omega}]^{(b)}, \quad \bar{\alpha} = [\vec{\alpha}]^{(b)} \tag{4.142}$$

$$\hat{C} = \hat{C}^{(a,b)} \tag{4.143}$$

By means of the above definitions, the vector equations can be converted into the following matrix equations.

$$\bar{r} = \hat{C}\bar{r}' + \bar{r}^{\circ} \tag{4.144}$$

$$\bar{v} = \hat{C}(\bar{v}' + \tilde{\omega}\bar{r}') + \bar{v}^{\circ} \tag{4.145}$$

$$\bar{a} = \hat{C}[\bar{a}' + 2\tilde{\omega}\bar{v}' + (\tilde{\alpha} + \tilde{\omega}^2)\bar{r}'] + \bar{a}^{\circ} \tag{4.146}$$

5

Kinematics of Rigid Body Systems

Synopsis

This chapter presents a systematic formulation for expressing the kinematic relationships concerning a mechanical system such as a manipulator or a mechanism. It is assumed that the members of the system are rigid bodies. In order to facilitate the kinematic description of such a system, the system is divided into several subsystems, which are in the form of kinematic chains. A kinematic chain is such that its members are connected serially. The kinematic chains are also divided into two categories as kinematic branches (i.e. open kinematic chains) and kinematic loops (i.e. closed kinematic chains). The joint or the kinematic pair that connects two members consists of two mating kinematic elements, which are special formations on the connected members. In order to write the equations that express the kinematic relationships, special reference frames are attached not only to the members but also to the kinematic elements on the members. The equations are written in order to express the relative position, velocity, and acceleration of a member first with respect to its immediate neighbor and then with respect to the base (i.e. the reference member) of the kinematic chain.

The main purpose of this chapter is to derive the necessary equations in suitable forms as a preparation for a subsequent detailed kinematic analysis of the system. A detailed kinematic analysis is necessarily based on the analytical or at least semi-analytical solutions of the derived equations for the unspecified joint variables and their rates in accordance with the specifications. The analytical or semi-analytical solutions also facilitate the identification and analysis of the possible multiple poses of the system for the same specifications as well as its poses of position and motion singularities. In order to carry out a detailed kinematic analysis, the necessary equations are derived separately for the orientations and locations of the members. These separate orientation and location equations convey the kinematic details explicitly and they are convenient for further symbolic manipulations in order to be processed into more suitable forms that are ready for the next stages of solution, differentiation, and analysis.

This chapter also includes two comprehensive examples. In the first example, the three-joint arm of a spatial serial manipulator is studied as a typical open kinematic chain. In the second example, a spatial slider-crank mechanism is studied as a typical closed kinematic chain. In both examples, the necessary kinematic equations are derived in order to express the position, velocity, and acceleration relationships. The examples also include the solutions of the kinematic equations for the unspecified joint

Kinematics of General Spatial Mechanical Systems, First Edition. M. Kemal Ozgoren.

Companion Website: www.wiley.com/go/ozgoren/spatialmechanicalsystems

variables together with their first and second rates that correspond to the position and motion specifications. The solutions are accompanied by the discussions about the multiple poses and the poses of position and motion singularities.

5.1 Kinematic Description of a Rigid Body System

5.1.1 Body Frames and Joint Frames

Consider a rigid body system that contains the rigid bodies $\mathcal{B}_o$, $\mathcal{B}_a$, $\mathcal{B}_b$, ..., $\mathcal{B}_z$ as its members. Two of the members ($\mathcal{B}_a$ and $\mathcal{B}_b$) of the system are shown in Figure 5.1. The members $\mathcal{B}_a$ and $\mathcal{B}_b$ are connected to each other with a *joint* denoted as $\mathcal{J}_{ab}$ or $\mathcal{J}_{ba}$, which is also called a *kinematic pair*. The kinematic pair $\mathcal{J}_{ab} = \mathcal{J}_{ba}$ consists of two mating *kinematic elements*, which are $\mathcal{E}_{ab}$ fixed on $\mathcal{B}_a$ and $\mathcal{E}_{ba}$ fixed on $\mathcal{B}_b$.

Usually, a member $\mathcal{B}_a$ of the system is also called a *link* by conceiving it as a bridge that links two or more joints. When it is called so, it is denoted as $\mathcal{L}_a$. However, it is to be noted that, from a logical point of view, a rigid body may be called a link if it contains at least two joints. Using a special Latin-based terminology, a link is called *binary* if it contains *two* joints, *ternary* if it contains *three* joints, *quaternary* if it contains *four* joints, and so on. On the other hand, for the sake of linguistic uniformity and practicality, a rigid body with a single joint may also be sometimes called a link, as in the instance of using the terms *base link* and *last link*, respectively, for the grounded base and the end-effector of a serial manipulator.

The reference frames attached to $\mathcal{B}_a$ and $\mathcal{B}_b$ are $\mathcal{F}_a$ and $\mathcal{F}_b$. These reference frames are called *body frames*. Considering a body frame, such as $\mathcal{F}_a$, the origin and the unit basis vector triad of $\mathcal{F}_a$ are denoted as O_a and $\mathcal{U}_a = \{\vec{u}_1^{(a)}, \vec{u}_2^{(a)}, \vec{u}_3^{(a)}\}$.

In addition to the body frame $\mathcal{F}_a$, the body $\mathcal{B}_a$ carries *joint frames* such as $\mathcal{F}_{ab}$, $\mathcal{F}_{ac}$, $\mathcal{F}_{ad}$, etc. Each of these frames is dedicated to one of the joints, with which $\mathcal{B}_a$ is connected to its neighbors $\mathcal{B}_b$, $\mathcal{B}_c$, $\mathcal{B}_d$, etc. For example, $\mathcal{F}_{ab}$ is dedicated to the joint $\mathcal{J}_{ab}$. The origin and the unit basis vector triad of $\mathcal{F}_{ab}$ are denoted as O_{ab} and $\mathcal{U}_{ab} = \{\vec{u}_1^{(ab)}, \vec{u}_2^{(ab)}, \vec{u}_3^{(ab)}\}$.

The joint frame $\mathcal{F}_{ab}$ is attached to the kinematic element $\mathcal{E}_{ab}$ of the joint $\mathcal{J}_{ab}$ on the body $\mathcal{B}_a$. Similarly, the joint frame $\mathcal{F}_{ba}$ is attached to the kinematic element $\mathcal{E}_{ba}$ of the joint $\mathcal{J}_{ba}$ on the body $\mathcal{B}_b$. The kinematic elements $\mathcal{E}_{ab}$ and $\mathcal{E}_{ba}$ are particular geometric formations that take place at specific locations on $\mathcal{B}_a$ and $\mathcal{B}_b$, respectively. They are shaped

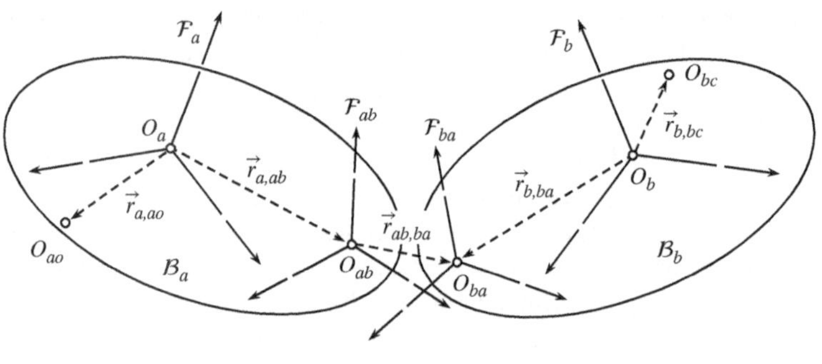

Figure 5.1 Two rigid bodies connected with a joint.

so that they mate each other in order to form the kinematic pair $\mathcal{J}_{ab}$. The companion joint frames $\mathcal{F}_{ab}$ and $\mathcal{F}_{ba}$ are attached to the kinematic elements $\mathcal{E}_{ab}$ and $\mathcal{E}_{ba}$ in such a way that the relative motion of $\mathcal{E}_{ba}$ with respect to $\mathcal{E}_{ab}$ (or vice versa) is expressed in the simplest possible way.

5.1.2 Kinematic Chains, Kinematic Branches, and Kinematic Loops

A general rigid body system may consist of one or several *kinematic chains*. A kinematic chain is defined as a set of *serially* connected rigid bodies. A kinematic chain may be in the form of either a *kinematic branch* or a *kinematic loop*.

A kinematic branch is an *open* kinematic chain. In other words, the bodies of a kinematic branch do not form a loop. The *first* and *last* bodies of a branch are called its *terminal* bodies. The other bodies of the same branch are called its *intermediate* bodies.

On the other hand, a kinematic loop is a *closed* kinematic chain. A kinematic loop can be generated by connecting the terminal bodies of a kinematic branch with some kind of a joint.

5.1.3 Joints or Kinematic Pairs

As mentioned before, the joint $\mathcal{J}_{ab}$ or $\mathcal{J}_{ba}$ between the bodies $\mathcal{B}_a$ and $\mathcal{B}_b$ is composed of two mating kinematic elements, which are $\mathcal{E}_{ab}$ on $\mathcal{B}_a$ and $\mathcal{E}_{ba}$ on $\mathcal{B}_b$. Due to the movable but constrained mating between them, the kinematic elements $\mathcal{E}_{ab}$ and $\mathcal{E}_{ba}$ can get positioned with respect to each other with a *relative mobility* $\mu_{ab} = \mu_{ba}$, which may be such that $1 \leq \mu_{ab} \leq 5$. Accordingly, the relative motion of $\mathcal{E}_{ba}$ with respect to $\mathcal{E}_{ab}$ (or vice versa) can be described by means of μ_{ab} independent variable parameters, which are called *joint variables*.

The most frequently encountered joints are the *revolute* and *prismatic* joints. For both of them, $\mu_{ab} = 1$. They are both characterized by an axis represented by a unit vector $\vec{n}_{ab}$, which is fixed with respect to both $\mathcal{E}_{ab}$ and $\mathcal{E}_{ba}$. The joint frames $\mathcal{F}_{ab}$ and $\mathcal{F}_{ba}$ are attached to $\mathcal{E}_{ab}$ and $\mathcal{E}_{ba}$ in such a way that $\vec{n}_{ab} = \vec{n}_{ba} = \vec{u}_k^{(ab)} = \vec{u}_k^{(ba)}$. Here, k may be 1, 2, or 3 depending on the situation. Then, the other basis vectors of $\mathcal{F}_{ab}$ will be $\vec{u}_i^{(ab)}$ and $\vec{u}_j^{(ab)}$. Similarly, the other basis vectors of $\mathcal{F}_{ba}$ will be $\vec{u}_i^{(ba)}$ and $\vec{u}_j^{(ba)}$. Here and hereafter, the subscript indices i, j, and k are ordered so that $ijk \in \{123, 231, 312\}$.

If $\mathcal{J}_{ab}$ is a *revolute joint*, the relative orientation of $\mathcal{E}_{ba}$ with respect $\mathcal{E}_{ab}$ is described by one joint variable, which is the angle θ_{ab} that rotates $\vec{u}_i^{(ab)}$ and $\vec{u}_j^{(ab)}$, respectively, into $\vec{u}_i^{(ba)}$ and $\vec{u}_j^{(ba)}$ about $\vec{n}_{ab} = \vec{u}_k^{(ab)} = \vec{u}_k^{(ba)}$. The relative location of $\mathcal{E}_{ba}$ with respect to $\mathcal{E}_{ab}$ does not change. In other words, the distance $d_{ab} = O_{ab}O_{ba}$ remains constant. Usually, a revolute joint can be arranged so that $d_{ab} = 0$. A revolute joint is illustrated in Figure 5.2 with its mating kinematic elements.

If $\mathcal{J}_{ab}$ is a *prismatic joint*, the relative location of $\mathcal{E}_{ba}$ with respect to $\mathcal{E}_{ab}$ is described by one joint variable, which is the *sliding distance* $s_{ab} = O_{ab}O_{ba}$ along $\vec{n}_{ab} = \vec{u}_k^{(ab)} = \vec{u}_k^{(ba)}$. The relative orientation of $\mathcal{E}_{ba}$ with respect to $\mathcal{E}_{ab}$ does not change. In other words, the angle δ_{ab} remains constant. This angle describes the constant orientation of $\vec{u}_i^{(ba)}$ or $\vec{u}_j^{(ba)}$ with respect to $\vec{u}_i^{(ab)}$ or $\vec{u}_j^{(ab)}$. Usually, a prismatic joint can be arranged so that $\delta_{ab} = 0$. A prismatic joint is illustrated in Figure 5.2 with its mating kinematic elements.

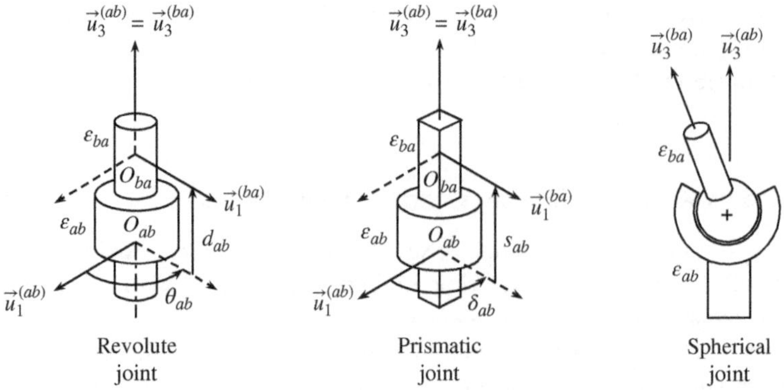

Figure 5.2 Revolute, prismatic, and spherical joints.

Another joint, which is encountered quite frequently, is a *spherical joint*. Its mobility is $\mu_{ab} = 3$. At a spherical joint, the kinematic elements $\mathcal{E}_{ab}$ and $\mathcal{E}_{ba}$ share $O_{ab} = O_{ba}$ as their common center. Therefore, the relative location of $\mathcal{E}_{ba}$ with respect to $\mathcal{E}_{ab}$ does not change. On the other hand, since $\mu_{ab} = 3$, the relative orientation of $\mathcal{E}_{ba}$ with respect to $\mathcal{E}_{ab}$ is described by three independent joint variables, which are usually selected as the Euler angles $(\phi_{ab}, \theta_{ab}, \psi_{ab})$ of an appropriate sequence. The appropriate sequence for a particular spherical joint is case dependent and it is selected so that it helps simplify the expressions of the relevant kinematic relationships. A spherical joint is illustrated in Figure 5.2 with its mating kinematic elements. The appropriate sequences selected for two different spherical joints can be seen as demonstrated for a spatial slider-crank mechanism studied as an example in Section 5.6.

A large collection of joints will be considered later in Chapter 6 with all the details of their kinematic descriptions and the relevant equations.

5.2 Position Equations for a Kinematic Chain of Rigid Bodies

Consider a kinematic chain $\mathcal{K}$ that consists of the rigid bodies $\mathcal{B}_o, \mathcal{B}_a, \mathcal{B}_b, \ldots, \mathcal{B}_z$. In this kinematic chain, $\mathcal{B}_o$ is indicated as the leading member because it is selected as the *base* (i.e. the *reference body*) of the chain. The *relative position* of a member $\mathcal{B}_b$ of the chain with respect to one of its neighbors, e.g. $\mathcal{B}_a$, or with respect to the base $\mathcal{B}_o$ can be expressed by means of the equations written in this section.

Here, concerning a rigid body such as $\mathcal{B}_a$, the word "position" is used in a *general sense* to mean *either* or *both* of the words "location" and "orientation." This usage is based on the fact that saying "the location of $\mathcal{B}_a$" is an abbreviated way of saying "the position of a representative point of $\mathcal{B}_a$" and saying "the orientation of $\mathcal{B}_a$" is equivalent to saying "the angular position of $\mathcal{B}_a$."

In a similar sense, the word "pose" is also used sometimes as an alternative to the word "position." However, they are actually not quite the same, because the word "pose" implies *only the combination* of the words "location" and "orientation." Besides, the word "pose" has also a meaning close to that of the word "posture."

5.2.1 Relative Orientation Equation Between Successive Bodies

Referring to Figure 5.1, the relative orientation of $\mathcal{B}_b$ with respect to $\mathcal{B}_a$ can be expressed by the following equation.

$$\hat{C}^{(a,b)} = \hat{C}^{(a,ab)}\hat{C}^{(ab,ba)}\hat{C}^{(ba,b)} \tag{5.1}$$

In Eq. (5.1), $\hat{C}^{(a,ab)}$ and $\hat{C}^{(ba,b)} = \hat{C}^{(b,ba)t}$ are constant matrices. The matrices $\hat{C}^{(a,ab)}$ and $\hat{C}^{(b,ba)}$ describe the orientations of the kinematic elements $\mathcal{E}_{ab}$ and $\mathcal{E}_{ba}$ with respect to the body frames $\mathcal{F}_a$ and $\mathcal{F}_b$, respectively. As for $\hat{C}^{(ab,ba)}$, it happens to be a variable matrix, if the joint $\mathcal{J}_{ab}$ has rotational mobility. For such a joint, $\hat{C}^{(ab,ba)}$ is a function of the joint variable or variables that describe the relative orientation of $\mathcal{E}_{ba}$ with respect to $\mathcal{E}_{ab}$.

5.2.2 Relative Location Equation Between Successive Bodies

Referring to Figure 5.1 again, the relative location of the origin O_b with respect to the other origin O_a can be expressed by the following equation.

$$\vec{r}_{a,b} = \vec{r}_{a,ab} + \vec{r}_{ab,ba} + \vec{r}_{ba,b} \tag{5.2}$$

In Eq. (5.2), the relative position vectors are defined so that $\vec{r}_{a,b} = \overrightarrow{O_aO_b}$, $\vec{r}_{a,ab} = \overrightarrow{O_aO_{ab}}$, etc. As seen in Figure 5.1, it happens that $\vec{r}_{ba,b} = -\vec{r}_{b,ba}$.

Equation (5.2), which is a vector equation, can also be written as a matrix equation in one of the relevant body frames $\mathcal{F}_a$ and $\mathcal{F}_b$. Here, it is written in $\mathcal{F}_a$ as the following matrix equation.

$$\bar{r}^{(a)}_{a,b} = \bar{r}^{(a)}_{a,ab} + \bar{r}^{(a)}_{ab,ba} + \bar{r}^{(a)}_{ba,b} \Rightarrow$$

$$\bar{r}^{(a)}_{a,b} = \bar{r}^{(a)}_{a,ab} + \hat{C}^{(a,ab)}\bar{r}^{(ab)}_{ab,ba} + \hat{C}^{(a,b)}\bar{r}^{(b)}_{ba,b} \tag{5.3}$$

Equation (5.3) can be written with more detail as follows:

$$\bar{r}^{(a)}_{a,b} = \bar{r}^{(a)}_{a,ab} + \hat{C}^{(a,ab)}[\bar{r}^{(ab)}_{ab,ba} + \hat{C}^{(ab,ba)}\hat{C}^{(ba,b)}\bar{r}^{(b)}_{ba,b}] \tag{5.4}$$

In Eqs. (5.3) and (5.4), $\bar{r}^{(a)}_{a,ab}$ and $\bar{r}^{(b)}_{ba,b} = -\bar{r}^{(b)}_{b,ba}$ are constant column matrices. They describe the locations of the origins O_{ab} and O_{ba} with respect to the body frames $\mathcal{F}_a$ and $\mathcal{F}_b$, respectively. As for $\bar{r}^{(ab)}_{ab,ba}$, it happens to be a variable column matrix, if the joint $\mathcal{J}_{ab}$ has translational mobility. For such a joint, $\bar{r}^{(ab)}_{ab,ba}$ is a function of the joint variable or variables that describe the relative location of O_{ba} with respect to O_{ab} as expressed in the frame $\mathcal{F}_{ab}$.

5.2.3 Orientation of a Body with Respect to the Base of the Kinematic Chain

The orientation of the body $\mathcal{B}_q$ ($q = a, b, c, \ldots, z$) of the kinematic chain $\mathcal{K}$ can be expressed as follows with respect to the base frame $\mathcal{F}_o(O_o)$:

$$\hat{C}^{(o,q)} = \hat{C}^{(o,a)}\hat{C}^{(a,b)}\hat{C}^{(b,c)}\cdots\hat{C}^{(p,q)} \tag{5.5}$$

In Eq. (5.5), all the matrices on the right-hand side are expressed similarly as in Eq. (5.1).

5.2.4 Location of a Body with Respect to the Base of the Kinematic Chain

The location of the origin O_q of the body frame $\mathcal{F}_q(O_q)$ $(q = a, b, c, \ldots, z)$ can be expressed as follows with respect to the base frame $\mathcal{F}_o(O_o)$:

$$\vec{r}_{o,q} = \vec{r}_{o,a} + \vec{r}_{a,b} + \vec{r}_{b,c} + \cdots + \vec{r}_{p,q} \tag{5.6}$$

Equation (5.6) can be written as the following matrix equation in the base frame $\mathcal{F}_o(O_o)$.

$$\overline{r}_{o,q}^{(o)} = \overline{r}_{o,a}^{(o)} + \widehat{C}^{(o,a)}\overline{r}_{a,b}^{(a)} + \widehat{C}^{(o,b)}\overline{r}_{b,c}^{(b)} + \cdots + \widehat{C}^{(o,p)}\overline{r}_{p,q}^{(p)} \tag{5.7}$$

In Eq. (5.7), all the orientation and location matrices on the right-hand side are expressed similarly as in Eqs. (5.1) and (5.3).

5.2.5 Loop Closure Equations for a Kinematic Loop

As mentioned before, a kinematic loop $\mathcal{L}$ is a closed kinematic chain, which is formed by connecting the terminal bodies ($\mathcal{B}_o$ and $\mathcal{B}_z$) of an open kinematic chain $\mathcal{K}$ by a joint $\mathcal{J}_{oz}$. The closure of the resulting kinematic loop can be expressed by the following equations.

(a) Orientation Equation for the Closure at Joint $\mathcal{J}_{oz}$

$$\widehat{C}^{(o,a)}\widehat{C}^{(a,b)}\cdots\widehat{C}^{(x,y)}\widehat{C}^{(y,z)}\widehat{C}^{(z,o)} = \widehat{C}^{(o,o)} = \widehat{I} \Rightarrow$$

$$\widehat{C}^{(o,a)}\widehat{C}^{(a,b)}\cdots\widehat{C}^{(x,y)}\widehat{C}^{(y,z)} = \widehat{C}^{(o,z)} = \widehat{C}^{(o,oz)}\widehat{C}^{(oz,zo)}\widehat{C}^{(zo,z)} \tag{5.8}$$

(b) Location Equation for the Closure at Joint $\mathcal{J}_{oz}$

$$\vec{r}_{o,a} + \vec{r}_{a,b} + \cdots + \vec{r}_{x,y} + \vec{r}_{y,z} + \vec{r}_{z,o} = \vec{r}_{o,o} = \vec{0} \Rightarrow$$

$$\vec{r}_{o,a} + \vec{r}_{a,b} + \cdots + \vec{r}_{x,y} + \vec{r}_{y,z} = \vec{r}_{o,z} = \vec{r}_{o,oz} + \vec{r}_{oz,zo} + \vec{r}_{zo,z} \tag{5.9}$$

Eq. (5.9) can be written as the following matrix equation in the base frame $\mathcal{F}_o(O_o)$.

$$\overline{r}_{o,a}^{(o)} + \widehat{C}^{(o,a)}\overline{r}_{a,b}^{(a)} + \cdots + \widehat{C}^{(o,y)}\overline{r}_{y,z}^{(y)} = \overline{r}_{o,z}^{(o)} \Rightarrow$$

$$\overline{r}_{o,a}^{(o)} + \widehat{C}^{(o,a)}\overline{r}_{a,b}^{(a)} + \cdots + \widehat{C}^{(o,y)}\overline{r}_{y,z}^{(y)} = \overline{r}_{o,oz}^{(o)} + \widehat{C}^{(o,oz)}\overline{r}_{oz,zo}^{(oz)} + \widehat{C}^{(o,z)}\overline{r}_{zo,z}^{(z)} \tag{5.10}$$

Equations (5.8)–(5.10) express that the closure of the kinematic loop is realized by the formation of the joint $\mathcal{J}_{oz}$ between $\mathcal{B}_o$ and $\mathcal{B}_z$. However, it is not necessary to do so. The closure can as well be realized by the formation of an intermediate joint such as $\mathcal{J}_{pq}$ between $\mathcal{B}_p$ and $\mathcal{B}_q$. To do this, it is assumed that $\mathcal{J}_{oz}$ already exists but $\mathcal{J}_{pq}$ is hypothetically disconnected and then reconnected. In such a case, the loop closure equations can be written as shown below.

(c) Orientation Equation for the Closure at Joint $\mathcal{J}_{pq}$

$$\widehat{C}^{(o,a)}\widehat{C}^{(a,b)}\cdots\widehat{C}^{(n,p)}\widehat{C}^{(p,q)} = \widehat{C}^{(o,z)}\widehat{C}^{(z,y)}\cdots\widehat{C}^{(s,r)}\widehat{C}^{(r,q)} \tag{5.11}$$

(d) Location Equation for the Closure at Joint $\mathcal{J}_{pq}$

$$\vec{r}_{o,a} + \vec{r}_{a,b} + \cdots + \vec{r}_{n,p} + \vec{r}_{p,q} = \vec{r}_{o,z} + \vec{r}_{z,y} + \cdots + \vec{r}_{s,r} + \vec{r}_{r,q} \tag{5.12}$$

Equation (5.12) can be written as the following matrix equation in the base frame $\mathcal{F}_o(O_o)$.

$$\overline{r}^{(o)}_{o,a} + \widehat{C}^{(o,a)}\overline{r}^{(a)}_{a,b} + \cdots + \widehat{C}^{(o,n)}\overline{r}^{(n)}_{n,p} + \widehat{C}^{(o,p)}\overline{r}^{(p)}_{p,q}$$

$$= \overline{r}^{(o)}_{o,z} + \widehat{C}^{(o,z)}\overline{r}^{(z)}_{z,y} + \cdots + \widehat{C}^{(o,s)}\overline{r}^{(s)}_{s,r} + \widehat{C}^{(o,r)}\overline{r}^{(r)}_{r,q} \tag{5.13}$$

5.3 Velocity Equations for a Kinematic Chain of Rigid Bodies

This section is concerned with the relative velocities of the members of a kinematic chain $\mathcal{K} = \{\mathcal{B}_o, \mathcal{B}_a, \mathcal{B}_b \ldots, \mathcal{B}_z\}$ with respect to their neighbors and also with respect to the base $\mathcal{B}_o$ of the chain.

5.3.1 Relative Angular Velocity between Successive Bodies

As seen in Chapter 4, the relative angular velocity ($\vec{\omega}_{a,b} = \vec{\omega}_{b/a}$) of a body $\mathcal{B}_b$ with respect to another body $\mathcal{B}_a$ is represented by the column matrix $\overline{\omega}^{(a)}_{a,b}$ in the body frame $\mathcal{F}_a$ of $\mathcal{B}_a$ according to the following equations.

$$\overline{\omega}^{(a)}_{a,b} = \text{colm}[\widetilde{\omega}^{(a)}_{a,b}] \tag{5.14}$$

$$\widetilde{\omega}^{(a)}_{a,b} = \dot{\widehat{C}}^{(a,b)}\widehat{C}^{(b,a)} \tag{5.15}$$

In Eq. (5.15), the matrix $\widehat{C}^{(a,b)}$ describes the relative orientation of $\mathcal{B}_b$ with respect to $\mathcal{B}_a$. It can be expressed by the following equation as explained in Section 5.2.

$$\widehat{C}^{(a,b)} = \widehat{C}^{(a,ab)}\widehat{C}^{(ab,ba)}\widehat{C}^{(ba,b)} \tag{5.16}$$

As mentioned in Section 5.2, $\widehat{C}^{(a,ab)}$ and $\widehat{C}^{(ba,b)} = \widehat{C}^{(b,ba)t}$ are constant matrices. So, Eqs. (5.15) and (5.16) can be combined as shown below.

$$\widetilde{\omega}^{(a)}_{a,b} = \dot{\widehat{C}}^{(a,b)}\widehat{C}^{(b,a)} = [\widehat{C}^{(a,ab)}\dot{\widehat{C}}^{(ab,ba)}\widehat{C}^{(ba,b)}][\widehat{C}^{(b,ba)}\widehat{C}^{(ba,ab)}\widehat{C}^{(ab,a)}] \Rightarrow$$

$$\widetilde{\omega}^{(a)}_{a,b} = \widehat{C}^{(a,ab)}\dot{\widehat{C}}^{(ab,ba)}\widehat{C}^{(ba,a)} \tag{5.17}$$

Here, it must be noted that, although $\widehat{C}^{(ba,b)}$ and $\widehat{C}^{(ab,a)}$ are constant matrices, $\widehat{C}^{(ba,a)}$ is *not* a constant matrix. Indeed,

$$\widehat{C}^{(ba,a)} = \widehat{C}^{(ba,ab)}\widehat{C}^{(ab,a)} \tag{5.18}$$

Keeping Eq. (5.18) in mind, Eq. (5.17) can be manipulated so that

$$\widetilde{\omega}^{(a)}_{a,b} = \widehat{C}^{(a,ab)}\dot{\widehat{C}}^{(ab,ba)}[\widehat{I}]\widehat{C}^{(ba,a)} = \widehat{C}^{(a,ab)}\dot{\widehat{C}}^{(ab,ba)}[\widehat{C}^{(ba,ab)}\widehat{C}^{(ab,ba)}]\widehat{C}^{(ba,a)}$$

$$= \widehat{C}^{(a,ab)}[\dot{\widehat{C}}^{(ab,ba)}\widehat{C}^{(ba,ab)}]\widehat{C}^{(ab,ba)}\widehat{C}^{(ba,a)} \tag{5.19}$$

On the other hand, like the pattern of Eq. (5.15), $\widetilde{\omega}^{(ab)}_{ab,ba}$ is related to $\widehat{C}^{(ab,ba)}$ as follows:

$$\widetilde{\omega}^{(ab)}_{ab,ba} = \dot{\widehat{C}}^{(ab,ba)}\widehat{C}^{(ba,ab)} \tag{5.20}$$

Hence, in addition to Eq. (5.17), $\tilde{\omega}_{a,b}^{(a)}$ can be expressed by the following equation, too.

$$\tilde{\omega}_{a,b}^{(a)} = \hat{C}^{(a,ab)}\tilde{\omega}_{ab,ba}^{(ab)}\hat{C}^{(ab,a)} \tag{5.21}$$

Equation (5.21) implies that

$$\overline{\omega}_{a,b}^{(a)} = \hat{C}^{(a,ab)}\overline{\omega}_{ab,ba}^{(ab)} \tag{5.22}$$

Note that $\hat{C}^{(a,ab)}\overline{\omega}_{ab,ba}^{(ab)} = \overline{\omega}_{ab,ba}^{(a)}$. Therefore, Eq. (5.22) implies in turn the following vector equation.

$$\vec{\omega}_{a,b} = \vec{\omega}_{ab,ba} \tag{5.23}$$

Actually, Eq. (5.23) is an expected equation because $\vec{\omega}_{a,b} = \vec{\omega}_{a,ab} + \vec{\omega}_{ab,ba} + \vec{\omega}_{ba,b}$ and $\vec{\omega}_{a,ab} = \vec{\omega}_{ba,b} = \vec{0}$.

Moreover, if required, $\overline{\omega}_{a,b}^{(b)}$ and $\tilde{\omega}_{a,b}^{(b)}$ can be expressed by the following formulas, which can be obtained from the formulas that give $\overline{\omega}_{a,b}^{(a)}$ and $\tilde{\omega}_{a,b}^{(a)}$.

$$\overline{\omega}_{a,b}^{(b)} = \hat{C}^{(b,a)}\overline{\omega}_{a,b}^{(a)} \tag{5.24}$$

$$\tilde{\omega}_{a,b}^{(b)} = \hat{C}^{(b,a)}\tilde{\omega}_{a,b}^{(a)}\hat{C}^{(a,b)} = \hat{C}^{(b,a)}\dot{\hat{C}}^{(a,b)} \tag{5.25}$$

$$\tilde{\omega}_{a,b}^{(b)} = \hat{C}^{(b,ab)}\dot{\hat{C}}^{(ab,ba)}\hat{C}^{(ba,b)} \tag{5.26}$$

5.3.2 Relative Translational Velocity Between Successive Bodies

As seen in Section 5.2, the origins O_b and O_a of the body frames $\mathcal{F}_b$ and $\mathcal{F}_a$ are located with respect to each other according to the following equation.

$$\vec{r}_{a,b} = \vec{r}_{a,ab} + \vec{r}_{ab,ba} + \vec{r}_{ba,b} \tag{5.27}$$

Equation (5.27) can be differentiated as follows with respect to the body frame $\mathcal{F}_a$.

$$D_a\vec{r}_{a,b} = D_a\vec{r}_{a,ab} + D_a\vec{r}_{ab,ba} + D_a\vec{r}_{ba,b} \tag{5.28}$$

By using the Coriolis transport theorem and recalling that $\vec{\omega}_{b/a} = \vec{\omega}_{a,b}$, Eq. (5.28) can be modified as follows:

$$D_a\vec{r}_{a,b} = D_a\vec{r}_{a,ab} + D_{ab}\vec{r}_{ab,ba} + \vec{\omega}_{a,ab} \times \vec{r}_{ab,ba} + D_b\vec{r}_{ba,b} + \vec{\omega}_{a,b} \times \vec{r}_{ba,b} \tag{5.29}$$

Note that $\mathcal{F}_{ab}$ and $\vec{r}_{a,ab}$ are fixed with respect to $\mathcal{F}_a$ and $\vec{r}_{ba,b}$ is fixed with respect to $\mathcal{F}_b$. Therefore, $D_a\vec{r}_{a,ab} = \vec{0}$, $D_b\vec{r}_{ba,b} = \vec{0}$, and $\vec{\omega}_{a,ab} = \vec{0}$. As for the left-hand side of Eq. (5.27), $D_a\vec{r}_{a,b}$ is the relative velocity of O_b with respect to the body frame $\mathcal{F}_a(O_a)$. In other words,

$$D_a\vec{r}_{a,b} = \vec{v}_{a,b/a} \tag{5.30}$$

According to the preceding explanations, Eq. (5.29) reduces to

$$\vec{v}_{a,b/a} = D_a\vec{r}_{a,b} = D_{ab}\vec{r}_{ab,ba} + \vec{\omega}_{a,b} \times \vec{r}_{ba,b} \tag{5.31}$$

On the other hand, as seen in Chapter 4, the following matrix equation can be written for any vector $\vec{r}$ as long as the differentiation and resolution frames are the same.

$$[D_a\vec{r}]^{(a)} = \dot{\overline{r}}^{(a)} \text{ where } \overline{r}^{(a)} = [\vec{r}]^{(a)} \tag{5.32}$$

Thus, Eq. (5.31) can be written as the following matrix equation in $\mathcal{F}_a$.

$$\overline{v}^{(a)}_{a,b/a} = \dot{\overline{r}}^{(a)}_{a,b} = \widehat{C}^{(a,ab)}\dot{\overline{r}}^{(ab)}_{ab,ba} + \widetilde{\omega}^{(a)}_{a,b}\widehat{C}^{(a,b)}\overline{r}^{(b)}_{ba,b} \tag{5.33}$$

Furthermore, Eqs. (5.33) and (5.17) can be combined into the following equation, which can also be obtained in a much simpler way by differentiating Eq. (5.4) directly.

$$\overline{v}^{(a)}_{a,b/a} = \dot{\overline{r}}^{(a)}_{a,b} = \widehat{C}^{(a,ab)}[\dot{\overline{r}}^{(ab)}_{ab,ba} + \dot{\widehat{C}}^{(ab,ba)}\widehat{C}^{(ba,b)}\overline{r}^{(b)}_{ba,b}] \tag{5.34}$$

5.3.3 Angular Velocity of a Body with Respect to the Base

The angular velocity of the body $\mathcal{B}_q$ ($q = a, b, c, \ldots, z$) can be expressed as follows with respect to the base frame $\mathcal{F}_o(O_o)$:

$$\vec{\omega}_{o,q} = \vec{\omega}_{o,a} + \vec{\omega}_{a,b} + \vec{\omega}_{b,c} + \cdots + \vec{\omega}_{p,q} \tag{5.35}$$

Equation (5.35) can be written as the following matrix equation in the base frame $\mathcal{F}_o(O_o)$.

$$\overline{\omega}^{(o)}_{o,q} = \overline{\omega}^{(o)}_{o,a} + \widehat{C}^{(o,a)}\overline{\omega}^{(a)}_{a,b} + \widehat{C}^{(o,b)}\overline{\omega}^{(b)}_{b,c} + \cdots + \widehat{C}^{(o,p)}\overline{\omega}^{(p)}_{p,q} \tag{5.36}$$

In Eq. (5.36), all the terms on the right-hand side are provided by the previous equations.

5.3.4 Translational Velocity of a Body with Respect to the Base

The velocity of the origin O_q of the body frame $\mathcal{F}_q(O_q)$ ($q = a, b, c, \ldots, z$) with respect to the base frame $\mathcal{F}_o(O_o)$ is defined and denoted with an abbreviated notation ($\vec{v}_{o,q}$) as follows:

$$\vec{v}_{o,q} = \vec{v}_{o,q/o} = D_o\vec{r}_{o,q} \tag{5.37}$$

The velocity $\vec{v}_{o,q}$ can be obtained by differentiating the terms on the right-hand side of Eq. (5.6) with the help of the Coriolis transport theorem. Each term is differentiated as shown below.

$$D_o\vec{r}_{a,b} = D_a\vec{r}_{a,b} + \vec{\omega}_{o,a} \times \vec{r}_{a,b} \tag{5.38}$$

Equation (5.38) can also be written as

$$\vec{v}_{a,b/o} = \vec{v}_{a,b/a} + \vec{\omega}_{o,a} \times \vec{r}_{a,b} \tag{5.39}$$

Hence, $\vec{v}_{o,q}$ is obtained as follows:

$$\begin{aligned}\vec{v}_{o,q} = D_o\vec{r}_{o,q} &= D_o\vec{r}_{o,a} + D_a\vec{r}_{a,b} + D_b\vec{r}_{b,c} + \cdots + D_p\vec{r}_{p,q}\\ &+\vec{\omega}_{o,a} \times \vec{r}_{a,b} + \vec{\omega}_{o,b} \times \vec{r}_{b,c} + \cdots + \vec{\omega}_{o,p} \times \vec{r}_{p,q}\end{aligned} \tag{5.40}$$

Equation (5.40) can be written as the following matrix equation in the base frame $\mathcal{F}_o(O_o)$.

$$\begin{aligned}\overline{v}^{(o)}_{o,q} = \dot{\overline{r}}^{(o)}_{o,q} &= \dot{\overline{r}}^{(o)}_{o,a} + \widehat{C}^{(o,a)}\dot{\overline{r}}^{(a)}_{a,b} + \widehat{C}^{(o,b)}\dot{\overline{r}}^{(b)}_{b,c} + \cdots + \widehat{C}^{(o,p)}\dot{\overline{r}}^{(p)}_{p,q}\\ &+\widetilde{\omega}^{(o)}_{o,a}\widehat{C}^{(o,a)}\overline{r}^{(a)}_{a,b} + \widetilde{\omega}^{(o)}_{o,b}\widehat{C}^{(o,b)}\overline{r}^{(b)}_{b,c} + \cdots + \widetilde{\omega}^{(o)}_{o,p}\widehat{C}^{(o,p)}\overline{r}^{(p)}_{p,q}\end{aligned} \tag{5.41}$$

In Eq. (5.41), all the terms on the right-hand side are provided by the previous equations.

5.3.5 Velocity Equations for a Kinematic Loop

As mentioned in Section 5.2.5, the closure of a kinematic loop $\mathcal{L}$ can be realized by reconnecting a hypothetically disconnected joint $\mathcal{J}_{pq}$. In other words, the kinematic loop $\mathcal{L}$ formed this way consists of two open kinematic chains, which are

$$\mathcal{K}_1 = \{\mathcal{B}_o, \mathcal{B}_a, \mathcal{B}_b, \dots, \mathcal{B}_n, \mathcal{B}_p, \mathcal{B}_q\} \text{ and } \mathcal{K}_2 = \{\mathcal{B}_o, \mathcal{B}_z, \mathcal{B}_y, \dots, \mathcal{B}_s, \mathcal{B}_r, \mathcal{B}_q\}$$

Therefore, the velocity equations associated with $\mathcal{L}$ can be obtained by equating the velocity expressions written for $\mathcal{B}_q$ with respect to the base frame $\mathcal{F}_o(O_o)$ through the open kinematic chains $\mathcal{K}_1$ and $\mathcal{K}_2$. The resulting equations are shown below.

(a) Angular Velocity Equation for the Kinematic Loop

The vector equation can be written as

$$\vec{\omega}_{o,a} + \vec{\omega}_{a,b} + \cdots + \vec{\omega}_{p,q} = \vec{\omega}_{o,z} + \vec{\omega}_{z,y} + \cdots + \vec{\omega}_{r,q} \tag{5.42}$$

Equation (5.42) corresponds to the following matrix equation in the base frame $\mathcal{F}_o(O_o)$.

$$\begin{aligned} &\overline{\omega}_{o,a}^{(o)} + \widehat{C}^{(o,a)}\overline{\omega}_{a,b}^{(a)} + \cdots + \widehat{C}^{(o,p)}\overline{\omega}_{p,q}^{(p)} \\ &= \overline{\omega}_{o,z}^{(o)} + \widehat{C}^{(o,z)}\overline{\omega}_{z,y}^{(z)} + \cdots + \widehat{C}^{(o,r)}\overline{\omega}_{r,q}^{(r)} \end{aligned} \tag{5.43}$$

(b) Translational Velocity Equation for the Kinematic Loop

Referring to Eq. (5.12), the following vector equation can be written.

$$D_o\vec{r}_{o,a} + D_o\vec{r}_{a,b} + \cdots + D_o\vec{r}_{p,q} = D_o\vec{r}_{o,z} + D_o\vec{r}_{z,y} + \cdots + D_o\vec{r}_{r,q} \Rightarrow$$

$$\begin{aligned} &D_o\vec{r}_{o,a} + (D_a\vec{r}_{a,b} + \vec{\omega}_{o,a} \times \vec{r}_{a,b}) + \cdots + (D_p\vec{r}_{p,q} + \vec{\omega}_{o,p} \times \vec{r}_{p,q}) \\ &= D_o\vec{r}_{o,z} + (D_z\vec{r}_{z,y} + \vec{\omega}_{o,z} \times \vec{r}_{z,y}) + \cdots + (D_r\vec{r}_{r,q} + \vec{\omega}_{o,r} \times \vec{r}_{r,q}) \end{aligned} \tag{5.44}$$

Equation (5.44) can be written as the following matrix equation in the base frame $\mathcal{F}_o(O_o)$.

$$\begin{aligned} &\dot{\overline{r}}_{o,a}^{(o)} + \widehat{C}^{(o,a)}\dot{\overline{r}}_{a,b}^{(a)} + \widehat{C}^{(o,b)}\dot{\overline{r}}_{b,c}^{(b)} + \cdots + \widehat{C}^{(o,p)}\dot{\overline{r}}_{p,q}^{(p)} \\ &+ \widetilde{\omega}_{o,a}^{(o)}\widehat{C}^{(o,a)}\overline{r}_{a,b}^{(a)} + \widetilde{\omega}_{o,b}^{(o)}\widehat{C}^{(o,b)}\overline{r}_{b,c}^{(b)} + \cdots + \widetilde{\omega}_{o,p}^{(o)}\widehat{C}^{(o,p)}\overline{r}_{p,q}^{(p)} \\ &= \dot{\overline{r}}_{o,z}^{(o)} + \widehat{C}^{(o,z)}\dot{\overline{r}}_{z,y}^{(z)} + \widehat{C}^{(o,y)}\dot{\overline{r}}_{y,x}^{(y)} + \cdots + \widehat{C}^{(o,r)}\dot{\overline{r}}_{r,q}^{(r)} \\ &+ \widetilde{\omega}_{o,z}^{(o)}\widehat{C}^{(o,z)}\overline{r}_{z,y}^{(z)} + \widetilde{\omega}_{o,y}^{(o)}\widehat{C}^{(o,y)}\overline{r}_{y,x}^{(y)} + \cdots + \widetilde{\omega}_{o,r}^{(o)}\widehat{C}^{(o,r)}\overline{r}_{r,q}^{(r)} \end{aligned} \tag{5.45}$$

5.4 Acceleration Equations for a Kinematic Chain of Rigid Bodies

This section is concerned with the relative accelerations of the members of a kinematic chain $\mathcal{K} = \{\mathcal{B}_o, \mathcal{B}_a, \mathcal{B}_b \dots, \mathcal{B}_z\}$ with respect to their neighbors and also with respect to the base $\mathcal{B}_o$ of the chain.

5.4.1 Relative Angular Acceleration Between Successive Bodies

As seen in Chapter 4, the relative angular acceleration ($\vec{\alpha}_{a,b} = \vec{\alpha}_{b/a}$) of $\mathcal{B}_b$ with respect to $\mathcal{B}_a$ can be obtained by differentiating $\vec{\omega}_{a,b} = \vec{\omega}_{b/a}$ with respect to $\mathcal{F}_a$. That is,

$$\vec{\alpha}_{a,b} = D_a\vec{\omega}_{a,b} = D_a\vec{\omega}_{ab,ba} = D_{ab}\vec{\omega}_{ab,ba} = \vec{\alpha}_{ab,ba} \tag{5.46}$$

In Eq. (5.46), $D_a = D_{ab}$ because the joint frame $\mathcal{F}_{ab}$ is fixed with respect to $\mathcal{F}_a$.

On the other hand, as mentioned before in Section 5.3.2, the following matrix equation can be written for any vector $\vec{w}$ as long as the differentation and resolution frames are the same.

$$[D_a\vec{w}]^{(a)} = \dot{\overline{w}}^{(a)} \text{ where } \overline{w}^{(a)} = [\vec{w}]^{(a)} \tag{5.47}$$

Therefore, Eq. (5.46) corresponds to the following matrix equation in $\mathcal{F}_a$.

$$\overline{\alpha}^{(a)}_{a,b} = \dot{\overline{\omega}}^{(a)}_{a,b} = \hat{C}^{(a,ab)}\dot{\overline{\omega}}^{(ab)}_{ab,ba} \tag{5.48}$$

Equation (5.48) implies that

$$\tilde{\alpha}^{(a)}_{a,b} = \dot{\tilde{\omega}}^{(a)}_{a,b} = \hat{C}^{(a,ab)}\dot{\tilde{\omega}}^{(ab)}_{ab,ba}\hat{C}^{(ab,a)} \tag{5.49}$$

Using the relevant equations of Section 5.3.1, $\tilde{\alpha}^{(a)}_{a,b}$ can also be related to $\ddot{\hat{C}}^{(ab,ba)}$ as shown below.

$$\tilde{\alpha}^{(a)}_{a,b} = \dot{\tilde{\omega}}^{(a)}_{a,b} = \hat{C}^{(a,ab)}\ddot{\hat{C}}^{(ab,ba)}\hat{C}^{(ba,a)} + \hat{C}^{(a,ab)}\dot{\hat{C}}^{(ab,ba)}\dot{\hat{C}}^{(ba,a)} \Rightarrow$$

$$\tilde{\alpha}^{(a)}_{a,b} = \hat{C}^{(a,ab)}\ddot{\hat{C}}^{(ab,ba)}\hat{C}^{(ba,a)} + \hat{C}^{(a,ab)}\dot{\hat{C}}^{(ab,ba)}\dot{\hat{C}}^{(ba,ab)}\hat{C}^{(ab,a)} \Rightarrow$$

$$\tilde{\alpha}^{(a)}_{a,b} = \hat{C}^{(a,ab)}\ddot{\hat{C}}^{(ab,ba)}\hat{C}^{(ba,a)}$$
$$+\hat{C}^{(a,ab)}[\dot{\hat{C}}^{(ab,ba)}\hat{C}^{(ba,ab)}][\hat{C}^{(ab,ba)}\dot{\hat{C}}^{(ba,ab)}]\hat{C}^{(ab,a)} \Rightarrow$$

$$\tilde{\alpha}^{(a)}_{a,b} = \hat{C}^{(a,ab)}\ddot{\hat{C}}^{(ab,ba)}\hat{C}^{(ba,a)} + \hat{C}^{(a,ab)}\tilde{\omega}^{(ab)}_{ab,ba}\tilde{\omega}^{(ab)t}_{ab,ba}\hat{C}^{(ab,a)} \Rightarrow$$

$$\tilde{\alpha}^{(a)}_{a,b} = \hat{C}^{(a,ab)}\ddot{\hat{C}}^{(ab,ba)}\hat{C}^{(ba,a)} - \hat{C}^{(a,ab)}\tilde{\omega}^{(ab)2}_{ab,ba}\hat{C}^{(ab,a)} \Rightarrow$$

$$\tilde{\alpha}^{(a)}_{a,b} = \hat{C}^{(a,ab)}\ddot{\hat{C}}^{(ab,ba)}\hat{C}^{(ba,a)} - \hat{C}^{(a,ab)}\tilde{\omega}^{(ab)}_{ab,ba}\hat{C}^{(ab,a)}\hat{C}^{(a,ab)}\tilde{\omega}^{(ab)}_{ab,ba}\hat{C}^{(ab,a)} \Rightarrow$$

$$\tilde{\alpha}^{(a)}_{a,b} = \hat{C}^{(a,ab)}\ddot{\hat{C}}^{(ab,ba)}\hat{C}^{(ba,a)} - \tilde{\omega}^{(a)2}_{a,b} \tag{5.50}$$

Equation (5.50) can also be written as

$$\hat{C}^{(a,ab)}\ddot{\hat{C}}^{(ab,ba)}\hat{C}^{(ba,a)} = \tilde{\alpha}^{(a)}_{a,b} + \tilde{\omega}^{(a)2}_{a,b} \tag{5.51}$$

5.4.2 Relative Translational Acceleration Between Successive Bodies

The relative acceleration of O_b with respect to the body frame $\mathcal{F}_a(O_a)$ is obtained as

$$\vec{a}_{a,b/a} = D_a\vec{v}_{a,b/a} = D_a^2\vec{r}_{a,b} \tag{5.52}$$

Equation (5.52) corresponds to the following matrix equation in $\mathcal{F}_a(O_a)$.

$$\overline{a}^{(a)}_{a,b/a} = \dot{\overline{v}}^{(a)}_{a,b/a} = \ddot{\overline{r}}^{(a)}_{a,b} \tag{5.53}$$

When Eq. (5.34) is substituted, Eq. (5.53) becomes

$$\overline{a}^{(a)}_{a,b/a} = \ddot{\overline{r}}^{(a)}_{a,b} = \widehat{C}^{(a,ab)}[\ddot{\overline{r}}^{(ab)}_{ab,ba} + \ddot{\widehat{C}}^{(ab,ba)}\widehat{C}^{(ba,b)}\overline{r}^{(b)}_{ba,b}] \tag{5.54}$$

Equation (5.54) can be modified as follows by using Eq. (5.51).

$$\overline{a}^{(a)}_{a,b/a} = \ddot{\overline{r}}^{(a)}_{a,b} = \widehat{C}^{(a,ab)}\ddot{\overline{r}}^{(ab)}_{ab,ba} + [\widetilde{\alpha}^{(a)}_{a,b} + \widetilde{\omega}^{(a)2}_{a,b}]\widehat{C}^{(a,b)}\overline{r}^{(b)}_{ba,b} \tag{5.55}$$

Equation (5.55) corresponds to the following vector equation.

$$\vec{a}_{a,b/a} = D_a^2\vec{r}_{a,b} = D_{ab}^2\vec{r}_{ab,ba} + \vec{\alpha}_{a,b} \times \vec{r}_{ba,b} + \vec{\omega}_{a,b} \times (\vec{\omega}_{a,b} \times \vec{r}_{ba,b}) \tag{5.56}$$

5.4.3 Angular Acceleration of a Body with Respect to the Base

The angular acceleration of the body $\mathcal{B}_q$ ($q = a, b, c, \ldots, z$) with respect to the base frame $\mathcal{F}_o(O_o)$ can be obtained as follows by differentiating Eq. (5.42):

$$\vec{\alpha}_{o,q} = D_o\vec{\omega}_{o,q} = D_o\vec{\omega}_{o,a} + D_o\vec{\omega}_{a,b} + D_o\vec{\omega}_{b,c} + \cdots + D_o\vec{\omega}_{p,q} \tag{5.57}$$

On the other hand, as seen in Chapter 1, if $\mathcal{F}_o \neq \mathcal{F}_a \neq \mathcal{F}_b$,

$$\vec{\alpha}_{a,b} = D_a\vec{\omega}_{a,b} = D_b\vec{\omega}_{a,b} \neq D_o\vec{\omega}_{a,b} \tag{5.58}$$

Nevertheless, the Coriolis transport theorem relates $D_o\vec{\omega}_{a,b}$ to $\vec{\alpha}_{a,b}$ as follows:

$$D_o\vec{\omega}_{a,b} = D_a\vec{\omega}_{a,b} + \vec{\omega}_{o,a} \times \vec{\omega}_{a,b} = \vec{\alpha}_{a,b} + \vec{\omega}_{o,a} \times \vec{\omega}_{a,b} \tag{5.59}$$

Hence, $\vec{\alpha}_{o,q}$ is obtained as follows by combining Eqs. (5.57) and (5.59):

$$\begin{aligned}\vec{\alpha}_{o,q} &= \vec{\alpha}_{o,a} + \vec{\alpha}_{a,b} + \vec{\alpha}_{b,c} + \cdots + \vec{\alpha}_{p,q}\\ &\quad + \vec{\omega}_{o,a} \times \vec{\omega}_{a,b} + \vec{\omega}_{o,b} \times \vec{\omega}_{b,c} + \cdots + \vec{\omega}_{o,p} \times \vec{\omega}_{p,q}\end{aligned} \tag{5.60}$$

Equation (5.60) can be written as the following matrix equation in the base frame $\mathcal{F}_o(O_o)$.

$$\begin{aligned}\overline{\alpha}^{(o)}_{o,q} &= \overline{\alpha}^{(o)}_{o,a} + \widehat{C}^{(o,a)}\overline{\alpha}^{(a)}_{a,b} + \widehat{C}^{(o,b)}\overline{\alpha}^{(b)}_{b,c} + \cdots + \widehat{C}^{(o,p)}\overline{\alpha}^{(p)}_{p,q}\\ &\quad + \widetilde{\omega}^{(o)}_{o,a}\widehat{C}^{(o,a)}\overline{\omega}^{(a)}_{a,b} + \widetilde{\omega}^{(o)}_{o,b}\widehat{C}^{(o,b)}\overline{\omega}^{(b)}_{b,c} + \cdots + \widetilde{\omega}^{(o)}_{o,p}\widehat{C}^{(o,p)}\overline{\omega}^{(p)}_{p,q}\end{aligned} \tag{5.61}$$

In Eq. (5.61), all the terms on the right-hand side are provided by the previous equations.

5.4.4 Translational Acceleration of a Body with Respect to the Base

The acceleration of the origin O_q of the body frame $\mathcal{F}_q(O_q)$ ($q = a, b, c, \ldots, z$) with respect to the base frame $\mathcal{F}_o(O_o)$ is defined and denoted with an abbreviated notation ($\vec{a}_{o,q}$) as follows:

$$\vec{a}_{o,q} = \vec{a}_{o,q/o} = D_o\vec{v}_{o,q/o} = D_o\vec{v}_{o,q} = D_o^2\vec{r}_{o,q} \tag{5.62}$$

The acceleration $\vec{a}_{o,q}$ can be obtained by differentiating the terms on the right-hand side of Eq. (5.40) with the help of the Coriolis transport theorem. The derivatives of the terms associated with $\vec{r}_{a,b}$ lead to the following expression for $D_o^2\vec{r}_{a,b}$.

$$D_o^2\vec{r}_{a,b} = D_o(D_a\vec{r}_{a,b} + \vec{\omega}_{o,a} \times \vec{r}_{a,b}) \Rightarrow$$

$$D_o^2\vec{r}_{a,b} = D_a(D_a\vec{r}_{a,b} + \vec{\omega}_{o,a} \times \vec{r}_{a,b}) + \vec{\omega}_{o,a} \times (D_a\vec{r}_{a,b} + \vec{\omega}_{o,a} \times \vec{r}_{a,b}) \Rightarrow$$

$$D_o^2\vec{r}_{a,b} = D_a^2\vec{r}_{a,b} + 2\vec{\omega}_{o,a} \times (D_a\vec{r}_{a,b}) + \vec{\alpha}_{o,a} \times \vec{r}_{a,b} + \vec{\omega}_{o,a} \times (\vec{\omega}_{o,a} \times \vec{r}_{a,b}) \tag{5.63}$$

Hence, $\vec{a}_{o,q}$ is obtained as follows:

$$\begin{aligned}
\vec{a}_{o,q} = D_o^2\vec{r}_{o,q} &= D_o^2\vec{r}_{o,a} + D_a^2\vec{r}_{a,b} + D_b^2\vec{r}_{b,c} + \cdots + D_p^2\vec{r}_{p,q} \\
&+ 2[\vec{\omega}_{o,a} \times (D_a\vec{r}_{a,b}) + \vec{\omega}_{o,b} \times (D_b\vec{r}_{b,c}) + \cdots + \vec{\omega}_{o,p} \times (D_p\vec{r}_{p,q})] \\
&+ \vec{\alpha}_{o,a} \times \vec{r}_{a,b} + \vec{\alpha}_{o,b} \times \vec{r}_{b,c} + \cdots + \vec{\alpha}_{o,p} \times \vec{r}_{p,q} \\
&+ \vec{\omega}_{o,a} \times (\vec{\omega}_{o,a} \times \vec{r}_{a,b}) + \cdots + \vec{\omega}_{o,p} \times (\vec{\omega}_{o,p} \times \vec{r}_{p,q})
\end{aligned} \tag{5.64}$$

Equation (5.64) can be written as the following matrix equation in the base frame $\mathcal{F}_o(O_o)$.

$$\begin{aligned}
\bar{a}_{o,q}^{(o)} = \ddot{\bar{r}}_{o,q}^{(o)} &= \ddot{\bar{r}}_{o,a}^{(o)} + \hat{C}^{(o,a)}\ddot{\bar{r}}_{a,b}^{(a)} + \hat{C}^{(o,b)}\ddot{\bar{r}}_{b,c}^{(b)} + \cdots + \hat{C}^{(o,p)}\ddot{\bar{r}}_{p,q}^{(p)} \\
&+ 2[\tilde{\omega}_{o,a}^{(o)}\hat{C}^{(o,a)}\dot{\bar{r}}_{a,b}^{(a)} + \tilde{\omega}_{o,b}^{(o)}\hat{C}^{(o,b)}\dot{\bar{r}}_{b,c}^{(b)} + \cdots + \tilde{\omega}_{o,p}^{(o)}\hat{C}^{(o,p)}\dot{\bar{r}}_{p,q}^{(p)}] \\
&+ [\tilde{\alpha}_{o,a}^{(o)} + \tilde{\omega}_{o,a}^{(o)2}]\hat{C}^{(o,a)}\bar{r}_{a,b}^{(a)} + \cdots + [\tilde{\alpha}_{o,p}^{(o)} + \tilde{\omega}_{o,p}^{(o)2}]\hat{C}^{(o,p)}\bar{r}_{p,q}^{(p)}
\end{aligned} \tag{5.65}$$

In Eq. (5.65), all the terms on the right-hand side are provided by the previous equations.

5.4.5 Acceleration Equations for a Kinematic Loop

Considering the kinematic loop $\mathcal{L}$ described in Section 5.3.5, the relevant acceleration equations are obtained by differentiating the relevant velocity equations similarly as done in Sections 5.4.3 and 5.4.4. The resulting equations are shown below.

(a) Angular Acceleration Equation for the Kinematic Loop

The vector equation is obtained as follows:

$$\begin{aligned}
&\vec{\alpha}_{o,a} + \vec{\alpha}_{a,b} + \vec{\alpha}_{b,c} + \cdots + \vec{\alpha}_{p,q} \\
&\quad + \vec{\omega}_{o,a} \times \vec{\omega}_{a,b} + \vec{\omega}_{o,b} \times \vec{\omega}_{b,c} + \cdots + \vec{\omega}_{o,p} \times \vec{\omega}_{p,q} \\
&= \vec{\alpha}_{o,z} + \vec{\alpha}_{z,y} + \vec{\alpha}_{y,x} + \cdots + \vec{\alpha}_{r,q} \\
&\quad + \vec{\omega}_{o,z} \times \vec{\omega}_{z,y} + \vec{\omega}_{o,y} \times \vec{\omega}_{y,x} + \cdots + \vec{\omega}_{o,r} \times \vec{\omega}_{r,q}
\end{aligned} \tag{5.66}$$

Equation (5.66) corresponds to the following matrix equation in $\mathcal{F}_o(O_o)$.

$$\begin{aligned}
&\overline{\alpha}_{o,a}^{(o)} + \widehat{C}^{(o,a)}\overline{\alpha}_{a,b}^{(a)} + \widehat{C}^{(o,b)}\overline{\alpha}_{b,c}^{(b)} + \cdots + \widehat{C}^{(o,p)}\overline{\alpha}_{p,q}^{(p)} \\
&\quad + \widetilde{\omega}_{o,a}^{(o)}\widehat{C}^{(o,a)}\overline{\omega}_{a,b}^{(a)} + \widetilde{\omega}_{o,b}^{(o)}\widehat{C}^{(o,b)}\overline{\omega}_{b,c}^{(b)} + \cdots + \widetilde{\omega}_{o,p}^{(o)}\widehat{C}^{(o,p)}\overline{\omega}_{p,q}^{(p)} \\
&= \overline{\alpha}_{o,z}^{(o)} + \widehat{C}^{(o,z)}\overline{\alpha}_{z,y}^{(z)} + \widehat{C}^{(o,y)}\overline{\alpha}_{y,x}^{(y)} + \cdots + \widehat{C}^{(o,r)}\overline{\alpha}_{r,q}^{(r)} \\
&\quad + \widetilde{\omega}_{o,z}^{(o)}\widehat{C}^{(o,z)}\overline{\omega}_{z,y}^{(z)} + \widetilde{\omega}_{o,y}^{(o)}\widehat{C}^{(o,y)}\overline{\omega}_{y,x}^{(y)} + \cdots + \widetilde{\omega}_{o,r}^{(o)}\widehat{C}^{(o,r)}\overline{\omega}_{r,q}^{(r)}
\end{aligned} \tag{5.67}$$

(b) Translational Acceleration Equation for the Kinematic Loop

The vector equation is obtained as follows:

$$\begin{aligned}
&D_o^2\vec{r}_{o,a} + D_a^2\vec{r}_{a,b} + D_b^2\vec{r}_{b,c} + \cdots + D_p^2\vec{r}_{p,q} \\
&\quad + 2[\vec{\omega}_{o,a} \times (D_a\vec{r}_{a,b}) + \vec{\omega}_{o,b} \times (D_b\vec{r}_{b,c}) + \cdots + \vec{\omega}_{o,p} \times (D_p\vec{r}_{p,q})] \\
&\quad + \vec{\alpha}_{o,a} \times \vec{r}_{a,b} + \vec{\alpha}_{o,b} \times \vec{r}_{b,c} + \cdots + \vec{\alpha}_{o,p} \times \vec{r}_{p,q} \\
&\quad + \vec{\omega}_{o,a} \times (\vec{\omega}_{o,a} \times \vec{r}_{a,b}) + \cdots + \vec{\omega}_{o,p} \times (\vec{\omega}_{o,p} \times \vec{r}_{p,q}) \\
&= D_o^2\vec{r}_{o,z} + D_z^2\vec{r}_{z,y} + D_y^2\vec{r}_{y,x} + \cdots + D_r^2\vec{r}_{r,q} \\
&\quad + 2[\vec{\omega}_{o,z} \times (D_z\vec{r}_{z,y}) + \vec{\omega}_{o,y} \times (D_y\vec{r}_{y,x}) + \cdots + \vec{\omega}_{o,r} \times (D_r\vec{r}_{r,q})] \\
&\quad + \vec{\alpha}_{o,z} \times \vec{r}_{z,y} + \vec{\alpha}_{o,y} \times \vec{r}_{y,x} + \cdots + \vec{\alpha}_{o,r} \times \vec{r}_{r,q} \\
&\quad + \vec{\omega}_{o,z} \times (\vec{\omega}_{o,z} \times \vec{r}_{z,y}) + \cdots + \vec{\omega}_{o,r} \times (\vec{\omega}_{o,r} \times \vec{r}_{r,q})
\end{aligned} \tag{5.68}$$

Equation (5.68) corresponds to the following matrix equation in $\mathcal{F}_o(O_o)$.

$$\begin{aligned}
&\ddot{\overline{r}}_{o,a}^{(o)} + \widehat{C}^{(o,a)}\ddot{\overline{r}}_{a,b}^{(a)} + \widehat{C}^{(o,b)}\ddot{\overline{r}}_{b,c}^{(b)} + \cdots + \widehat{C}^{(o,p)}\ddot{\overline{r}}_{p,q}^{(p)} \\
&\quad + 2[\widetilde{\omega}_{o,a}^{(o)}\widehat{C}^{(o,a)}\dot{\overline{r}}_{a,b}^{(a)} + \widetilde{\omega}_{o,b}^{(o)}\widehat{C}^{(o,b)}\dot{\overline{r}}_{b,c}^{(b)} + \cdots + \widetilde{\omega}_{o,p}^{(o)}\widehat{C}^{(o,p)}\dot{\overline{r}}_{p,q}^{(p)}] \\
&\quad + [\widetilde{\alpha}_{o,a}^{(o)} + \widetilde{\omega}_{o,a}^{(o)2}]\widehat{C}^{(o,a)}\overline{r}_{a,b}^{(a)} + \cdots + [\widetilde{\alpha}_{o,p}^{(o)} + \widetilde{\omega}_{o,p}^{(o)2}]\widehat{C}^{(o,p)}\overline{r}_{p,q}^{(p)} \\
&= \ddot{\overline{r}}_{o,z}^{(o)} + \widehat{C}^{(o,z)}\ddot{\overline{r}}_{z,y}^{(z)} + \widehat{C}^{(o,y)}\ddot{\overline{r}}_{y,x}^{(y)} + \cdots + \widehat{C}^{(o,r)}\ddot{\overline{r}}_{r,q}^{(r)} \\
&\quad + 2[\widetilde{\omega}_{o,z}^{(o)}\widehat{C}^{(o,z)}\dot{\overline{r}}_{z,y}^{(z)} + \widetilde{\omega}_{o,y}^{(o)}\widehat{C}^{(o,y)}\dot{\overline{r}}_{y,x}^{(y)} + \cdots + \widetilde{\omega}_{o,r}^{(o)}\widehat{C}^{(o,r)}\dot{\overline{r}}_{r,q}^{(r)}] \\
&\quad + [\widetilde{\alpha}_{o,z}^{(o)} + \widetilde{\omega}_{o,z}^{(o)2}]\widehat{C}^{(o,z)}\overline{r}_{z,y}^{(z)} + \cdots + [\widetilde{\alpha}_{o,r}^{(o)} + \widetilde{\omega}_{o,r}^{(o)2}]\widehat{C}^{(o,r)}\overline{r}_{r,q}^{(r)}
\end{aligned} \tag{5.69}$$

5.5 Example 5.1 : A Serial Manipulator with an RRP Arm

5.5.1 Kinematic Description of the System

Figure 5.3 shows the side and top views of a serial manipulator, which is a typical example of an open kinematic chain. The arm of the manipulator consists of the first three links that extend up to the wrist point R. The end-effector of the manipulator is a gripper and it is attached to the arm at the wrist point with three successive revolute joints in a spherical arrangement so that their axes are concurrent at the wrist point. This example is concerned with the kinematic analysis of the arm only, excluding the end-effector. Complete kinematic analyses of serial manipulators with all the links will be considered later in Chapters 7–9. The links of the arm are connected with two successive revolute

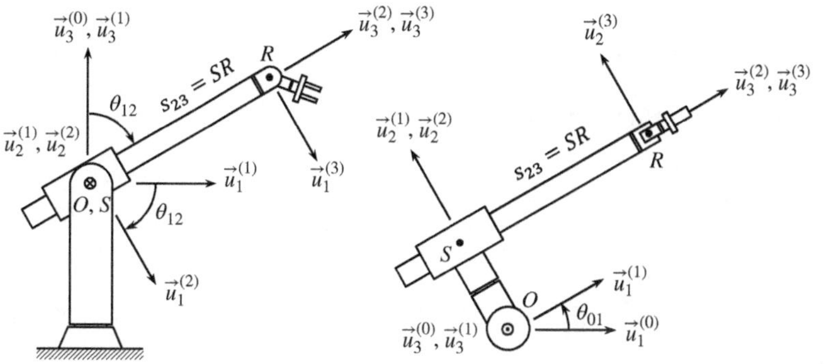

Figure 5.3 A serial manipulator with an RRP arm.

(R) joints and one prismatic (P) joint. The axes of the arm joints are represented by the following unit vectors.

$$\vec{n}_{01} = \vec{n}_{10} = \vec{u}_3^{(0)} = \vec{u}_3^{(1)}, \vec{n}_{12} = \vec{n}_{21} = \vec{u}_2^{(1)} = \vec{u}_2^{(2)}, \vec{n}_{23} = \vec{n}_{32} = \vec{u}_3^{(2)} = \vec{u}_3^{(3)}$$

The following reference frames are attached to the base and the moving links. Their basis vectors are indicated in Figure 5.3.

$$\mathcal{F}_0(O), \mathcal{F}_1(O), \mathcal{F}_2(S), \mathcal{F}_3(R)$$

As noticed, the origins of the link frames are

$$O_0 = O_1 = O, O_2 = S, O_3 = R$$

For this system, the joint frames are assigned as indicated below.

$$\mathcal{F}_{01}(O_{01}) = \mathcal{F}_0(O), \mathcal{F}_{10}(O_{10}) = \mathcal{F}_1(O)$$
$$\mathcal{F}_{12}(O_{12}) = \mathcal{F}_1(S) \parallel \mathcal{F}_1(O), \mathcal{F}_{21}(O_{21}) = \mathcal{F}_2(S)$$
$$\mathcal{F}_{23}(O_{23}) = \mathcal{F}_2(S), \mathcal{F}_{32}(O_{32}) = \mathcal{F}_3(R) \parallel \mathcal{F}_2(S)$$

The base frame origin O is selected on the axis of the first joint ($\mathcal{J}_{01}$) at the level of the *shoulder point* S in order to minimize the number of the nonzero constant parameters. Thus, the only nonzero constant parameter happens to be the *shoulder offset* defined below.

$$d_{12} = O_1O_2 = OS \text{ along } \vec{n}_{12} = \vec{u}_2^{(1)} = \vec{u}_2^{(2)} \tag{5.70}$$

The joint variables are defined as follows:

$$\theta_{01} = \sphericalangle[\vec{u}_1^{(0)} \rightarrow \vec{u}_1^{(1)}] \text{ about } \vec{n}_{01} = \vec{u}_3^{(0)} = \vec{u}_3^{(1)} \tag{5.71}$$
$$\theta_{12} = \sphericalangle[\vec{u}_1^{(1)} \rightarrow \vec{u}_1^{(2)}] \text{ about } \vec{n}_{12} = \vec{u}_2^{(1)} = \vec{u}_2^{(2)} \tag{5.72}$$
$$s_{23} = O_2O_3 = SR \text{ along } \vec{n}_{23} = \vec{u}_3^{(2)} = \vec{u}_3^{(3)} \tag{5.73}$$

5.5.2 Position Analysis

(a) Orientations of the Links

$$\hat{C}^{(0,1)} = \hat{R}_{0,1}^{(0)} = \hat{R}[\overline{u}_3^{(0/0)}, \theta_{01}] = e^{\tilde{u}_3\theta_{01}} \tag{5.74}$$
$$\hat{C}^{(1,2)} = \hat{R}_{1,2}^{(1)} = \hat{R}[\overline{u}_2^{(1/1)}, \theta_{12}] = e^{\tilde{u}_2\theta_{12}} \tag{5.75}$$
$$\hat{C}^{(2,3)} = \hat{R}_{2,3}^{(2)} = \hat{R}[\overline{u}_3^{(2/2)}, 0] = \hat{I} \tag{5.76}$$

$$\hat{C}^{(0,2)} = \hat{C}^{(0,1)}\hat{C}^{(1,2)} = e^{\tilde{u}_3\theta_{01}}e^{\tilde{u}_2\theta_{12}} \tag{5.77}$$

$$\hat{C}^{(0,3)} = \hat{C}^{(0,2)}\hat{C}^{(2,3)} = e^{\tilde{u}_3\theta_{01}}e^{\tilde{u}_2\theta_{12}} \tag{5.78}$$

(b) Locations of the Link Frame Origins

$$\vec{r}_{0,1} = \overrightarrow{O_0O_1} = \overrightarrow{OO} = \vec{0} \Rightarrow$$

$$\overline{r}_{0,1}^{(0)} = \overline{0} \tag{5.79}$$

$$\vec{r}_{1,2} = \overrightarrow{O_1O_2} = \overrightarrow{OS} = d_{12}\vec{u}_2^{(1)} \Rightarrow$$

$$\overline{r}_{1,2}^{(0)} = d_{12}\hat{C}^{(0,1)}\overline{u}_2^{(1/1)} = d_{12}e^{\tilde{u}_3\theta_{01}}\overline{u}_2 \tag{5.80}$$

$$\vec{r}_{2,3} = \overrightarrow{O_2O_3} = \overrightarrow{SR} = s_{23}\vec{u}_3^{(2)} \Rightarrow$$

$$\overline{r}_{2,3}^{(0)} = s_{23}\hat{C}^{(0,2)}\overline{u}_3^{(2/2)} = s_{23}e^{\tilde{u}_3\theta_{01}}e^{\tilde{u}_2\theta_{12}}\overline{u}_3 \tag{5.81}$$

$$\overline{r}_{0,2}^{(0)} = \overline{r}_{0,1}^{(0)} + \overline{r}_{1,2}^{(0)} = d_{12}e^{\tilde{u}_3\theta_{01}}\overline{u}_2 \tag{5.82}$$

$$\overline{r}_{0,3}^{(0)} = \overline{r}_{0,2}^{(0)} + \overline{r}_{2,3}^{(0)} = d_{12}e^{\tilde{u}_3\theta_{01}}\overline{u}_2 + s_{23}e^{\tilde{u}_3\theta_{01}}e^{\tilde{u}_2\theta_{12}}\overline{u}_3 \Rightarrow$$

$$\overline{r}_{0,3}^{(0)} = e^{\tilde{u}_3\theta_{01}}(d_{12}\overline{u}_2 + s_{23}e^{\tilde{u}_2\theta_{12}}\overline{u}_3) \tag{5.83}$$

For the sake of direct association with the characteristic points of the arm, let $\vec{r}_S = \vec{r}_{S/O} = \vec{r}_{0,2}$ and $\vec{r}_R = \vec{r}_{R/O} = \vec{r}_{0,3}$. Then, Eqs. (5.82) and (5.83) can be written again as follows:

$$\overline{r}_S^{(0)} = \overline{r}_{0,2}^{(0)} = d_{12}e^{\tilde{u}_3\theta_{01}}\overline{u}_2 \tag{5.84}$$

$$\overline{r}_R^{(0)} = \overline{r}_{0,3}^{(0)} = e^{\tilde{u}_3\theta_{01}}(d_{12}\overline{u}_2 + s_{23}e^{\tilde{u}_2\theta_{12}}\overline{u}_3) \tag{5.85}$$

By expanding the second term inside the parentheses, Eq. (5.85) can also be written as

$$\overline{r}_R^{(0)} = e^{\tilde{u}_3\theta_{01}}(\overline{u}_1 s_{23}s\theta_{12} + \overline{u}_2 d_{12} + \overline{u}_3 s_{23}c\theta_{12}) \tag{5.86}$$

In Eq. (5.86), it is to be noted that the terms inside the parentheses constitute the matrix representation of the vector $\vec{r}_R$ in the body frame $\mathcal{F}_1(O)$. That is,

$$\overline{r}_R^{(1)} = \overline{u}_1 s_{23}s\theta_{12} + \overline{u}_2 d_{12} + \overline{u}_3 s_{23}c\theta_{12} \tag{5.87}$$

In other words,

$$\overline{r}_R^{(0)} = \hat{C}^{(0,1)}\overline{r}_R^{(1)} = e^{\tilde{u}_3\theta_{01}}\overline{r}_R^{(1)} \tag{5.88}$$

By carrying out a further expansion, Eq. (5.86) can be worked as follows into Eq. (5.89) that shows the components of $\vec{r}_R$ in $\mathcal{F}_0(O)$:

$$\overline{r}_R^{(0)} = (e^{\tilde{u}_3\theta_{01}}\overline{u}_1)s_{23}s\theta_{12} + (e^{\tilde{u}_3\theta_{01}}\overline{u}_2)d_{12} + (e^{\tilde{u}_3\theta_{01}}\overline{u}_3)s_{23}c\theta_{12} \Rightarrow$$

$$\overline{r}_R^{(0)} = (\overline{u}_1 c\theta_{01} + \overline{u}_2 s\theta_{01})s_{23}s\theta_{12} + (\overline{u}_2 c\theta_{01} - \overline{u}_1 s\theta_{01})d_{12} + \overline{u}_3 s_{23}c\theta_{12} \Rightarrow$$

$$\begin{aligned}\overline{r}_R^{(0)} = {} & \overline{u}_1(s_{23}c\theta_{01}s\theta_{12} - d_{12}s\theta_{01}) \\ & + \overline{u}_2(s_{23}s\theta_{01}s\theta_{12} + d_{12}c\theta_{01}) + \overline{u}_3 s_{23}c\theta_{12}\end{aligned} \tag{5.89}$$

(c) Inverse Kinematic Solution

For a serial manipulator, the first stage of the position analysis is the *forward kinematics* (a.k.a. *direct kinematics*), in which the necessary equations are derived in order to express the positions of the moving links in terms of the joint variables. This stage has been completed above as explained in Parts (a) and (b). The next stage of the position analysis is the *inverse kinematics*, in which the equations derived in the forward kinematics stage are solved in order to find the joint variables that correspond to a specified pose of the end-effector. The solution can be obtained either numerically or analytically. The analytical solution has the advantage that it readily indicates the *multiple solutions* and the *position singularities* (if any). It thus can be used to carry out a detailed kinematic analysis subsequently. It also becomes an effective tool for motion planning. The inverse kinematic solution concerning the arm of the present manipulator is obtained analytically as explained below.

The desired pose of the end-effector can be indicated by specifying its orientation matrix $\hat{C} = \hat{C}^{(0,6)}$ and the coordinates of the wrist point R with respect to the base frame $\mathcal{F}_0(O)$. Let the specified coordinates of R be denoted simply as r_1, r_2, and r_3 so that

$$\overline{r}_R^{(0)} = \overline{r} = \overline{u}_1 r_1 + \overline{u}_2 r_2 + \overline{u}_3 r_3 \tag{5.90}$$

Upon the above specification, the inverse kinematic solution for the arm of the manipulator can be started more conveniently from Eq. (5.86) rather than Eq. (5.89). This is because Eq. (5.86) can be written as arranged below so that the unknowns are more or less equally distributed on both sides of the equation, e.g. two to one instead of three to zero.

$$\overline{u}_1 s_{23} s\theta_{12} + \overline{u}_2 d_{12} + \overline{u}_3 s_{23} c\theta_{12} = e^{-\tilde{u}_3 \theta_{01}} \overline{r} \tag{5.91}$$

Equation (5.91) leads to the following three scalar equations, when it is premultiplied on both sides by $\overline{u}_1^t$, $\overline{u}_2^t$, and $\overline{u}_3^t$.

$$s_{23} s\theta_{12} = \overline{u}_1^t e^{-\tilde{u}_3 \theta_{01}} \overline{r} = (\overline{u}_1^t c\theta_{01} + \overline{u}_2^t s\theta_{01}) \overline{r} \Rightarrow$$

$$s_{23} s\theta_{12} = r_1 c\theta_{01} + r_2 s\theta_{01} \tag{5.92}$$

$$d_{12} = \overline{u}_2^t e^{-\tilde{u}_3 \theta_{01}} \overline{r} = (\overline{u}_2^t c\theta_{01} - \overline{u}_1^t s\theta_{01}) \overline{r} \Rightarrow$$

$$r_2 c\theta_{01} - r_1 s\theta_{01} = d_{12} \tag{5.93}$$

$$s_{23} c\theta_{12} = \overline{u}_3^t e^{-\tilde{u}_3 \theta_{01}} \overline{r} = \overline{u}_3^t \overline{r} \Rightarrow$$

$$s_{23} c\theta_{12} = r_3 \tag{5.94}$$

If $r_1^2 + r_2^2 \neq 0$, the angle θ_{01} can be determined after obtaining $s\theta_{01}$ and $c\theta_{01}$ from Eq. (5.93) as follows:

$$(r_1 s\theta_{01} + d_{12})^2 = (r_2 c\theta_{01})^2 = r_2^2[1 - (s\theta_{01})^2] \Rightarrow$$

$$(r_1^2 + r_2^2)(s\theta_{01})^2 + 2d_{12} r_1 (s\theta_{01}) - (r_2^2 - d_{12}^2) = 0 \Rightarrow$$

$$s\theta_{01} = (\sigma r_2 q_{12} - d_{12} r_1)/(r_1^2 + r_2^2) \tag{5.95}$$

$$(r_2 c\theta_{01} - d_{12})^2 = (r_1 s\theta_{01})^2 = r_1^2[1 - (c\theta_{01})^2] \Rightarrow$$

$$(r_1^2 + r_2^2)(c\theta_{01})^2 - 2d_{12} r_2 (c\theta_{01}) - (r_1^2 - d_{12}^2) = 0 \Rightarrow$$

$$c\theta_{01} = (\sigma r_1 q_{12} + d_{12} r_2)/(r_1^2 + r_2^2) \tag{5.96}$$

In Eqs. (5.95) and (5.96),

$$q_{12} = \sqrt{r_1^2 + r_2^2 - d_{12}^2} \tag{5.97}$$

$$\sigma = \pm 1 \tag{5.98}$$

Equations (5.95) and (5.96) together lead to

$$\theta_{01} = \text{atan}_2[(\sigma r_2 q_{12} - d_{12} r_1), (\sigma r_1 q_{12} + d_{12} r_2)] \tag{5.99}$$

Equation (5.97) implies that θ_{01} comes out as a real angle if $r_1^2 + r_2^2 \geq d_{12}^2$. Therefore, the condition that $r_1^2 + r_2^2 \neq 0$, which is required for the validity of Eqs. (5.95) and (5.96), is naturally satisfied as long as $d_{12} \neq 0$.

In Eq. (5.99), $\text{atan}_2(y, x)$ is the *double argument arctangent function*. Although its definition and properties have already been given in Section 2.10 of Chapter 2, they are nonetheless repeated here briefly for the sake of convenience.

The function $\text{atan}_2(y, x)$ is defined according to the following identity for any arbitrary nonzero and positive coefficient r.

$$\theta \equiv \text{atan}_2(rs\theta, rc\theta) \tag{5.100}$$

For the arguments y and x, which are such that $x = rc\theta$ and $y = rs\theta$ with $r = \sqrt{x^2 + y^2}$, $\text{atan}_2(y, x)$ gives θ uniquely in the interval $\{\theta : -\pi < \theta < \pi\}$ as long as $y \neq 0$ and $x \neq 0$. If $x = 0$ with $y \neq 0$, $\text{atan}_2(y, x)$ gives $\theta = (\pi/2)\,\text{sgn}(y)$. If $y = 0$ with $x > 0$, $\text{atan}_2(y, x)$ gives $\theta = 0$; but if $y = 0$ with $x < 0$, $\text{atan}_2(y, x)$ gives θ with a sign ambiguity as $\theta = \pm\pi$. If $y = x = 0$, i.e. if $r = 0$, then $\text{atan}_2(y, x)$ cannot give a definite value for θ.

When Eqs. (5.95) and (5.96) are substituted, Eq. (5.92) becomes

$$s_{23} s\theta_{12} = \sigma q_{12} = \sigma \sqrt{r_1^2 + r_2^2 - d_{12}^2} \tag{5.101}$$

Since $s_{23} > 0$ due to the *joint limitation* that $0 < d_{min} \leq s_{23} \leq d_{max}$, the sum of the squares of the two sides of Eqs. (5.94) and (5.101) leads to

$$s_{23} = \sqrt{r_1^2 + r_2^2 + r_3^2 - d_{12}^2} \tag{5.102}$$

Finally, based on the fact that $s_{23} > 0$, the angle θ_{12} can be determined from Eqs. (5.94) and (5.101) as follows:

$$\theta_{12} = \text{atan}_2(\sigma q_{12}, r_3) \tag{5.103}$$

Meanwhile, Eqs. (5.97) and (5.102) imply the following *workspace limitations* that must be obeyed for an appropriate specification of the wrist point location.

$$r_1^2 + r_2^2 \geq d_{12}^2 \text{ and } d_{12}^2 + d_{min}^2 \leq r_1^2 + r_2^2 + r_3^2 \leq d_{12}^2 + d_{max}^2$$

(d) Multiple Solutions

In an analytical inverse kinematic solution, the *multiple solutions* or the *solution multiplicities* can be represented by sign variables such as σ_1, σ_2, etc. For the present manipulator, the sign variable σ, which appears in the expressions of θ_{01} and θ_{12}, represents two distinct optional inverse kinematic solutions that correspond to the same specified location of the wrist point R. The optional poses of the manipulator

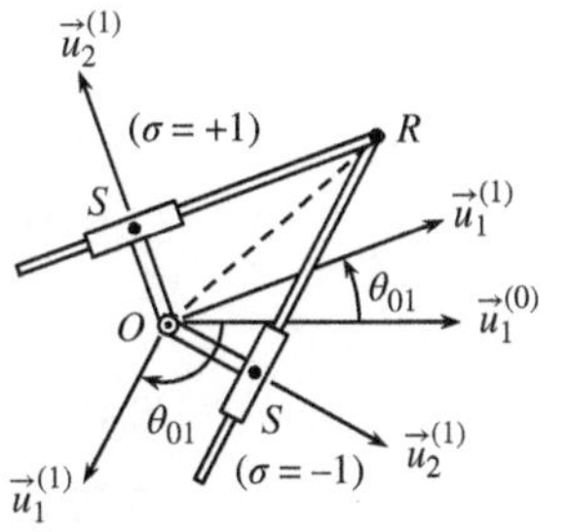

Figure 5.4 Two optional poses of the manipulator for the same location of *R*.

described by $\sigma = +1$ (left-shouldered pose) and $\sigma = -1$ (right-shouldered pose) are sketched in Figure 5.4.

Note that the availability of such optional poses becomes advantageous especially when there are obstacles to be avoided in the vicinity of the manipulator.

(e) Position Singularity

Consider a special version of the present manipulator such that $d_{12} = 0$. It is shown in Figure 5.5 in its side and front views. Regarding Eq. (5.93) for this special manipulator, it is possible to have $r_1^2 + r_2^2 = 0$ or $r_1 = r_2 = 0$. Such a particular pose of the manipulator is called a *position singularity*. In such a pose, Eq. (5.92) implies that the third link assumes a vertical pose so that $s\theta_{12} = 0$, i.e. $\theta_{12} = 0$ or $\theta_{12} = \pi$. However, $\theta_{12} = 0$ is the only physically possible value due to the shapes of the relevant links. On the other hand, Eq. (5.93) degenerates into $0 = 0$ at the singularity and therefore θ_{01} becomes indefinite. In other words, θ_{01} may assume an arbitrary value but, whatever value it assumes, it cannot change the position of the wrist point *R*, which happens to be on the axis along $\vec{u}_3^{(1)}$ at a location determined by s_{23}. The manipulator remains in this pose of position singularity, with the lost effectiveness of θ_{01}, as long as θ_{12} is kept fixed at its singularity value of $\theta_{12} = 0$.

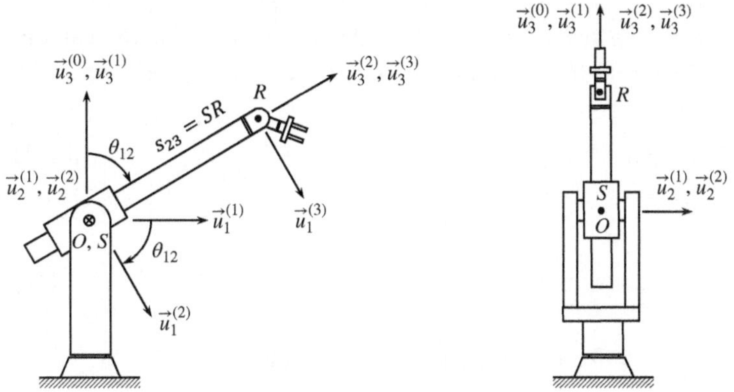

Figure 5.5 Zero-offset version of the manipulator with an RRP arm.

5.5.3 Velocity Analysis

(a) Angular Velocities of the Links

$$\overline{\omega}_{0,1}^{(0)} = \text{colm}[\dot{\widehat{C}}^{(0,1)}\widehat{C}^{(1,0)}] = \dot{\theta}_{01}\overline{u}_3 \tag{5.104}$$

$$\overline{\omega}_{0,2}^{(0)} = \text{colm}[\dot{\widehat{C}}^{(0,2)}\widehat{C}^{(2,0)}] = \dot{\theta}_{01}\overline{u}_3 + \dot{\theta}_{12}e^{\tilde{u}_3\theta_{01}}\overline{u}_2 \Rightarrow$$

$$\overline{\omega}_{0,2}^{(0)} = -\overline{u}_1\dot{\theta}_{12}s\theta_{01} + \overline{u}_2\dot{\theta}_{12}c\theta_{01} + \overline{u}_3\dot{\theta}_{01} \tag{5.105}$$

$$\overline{\omega}_{0,3}^{(0)} = \overline{\omega}_{0,2}^{(0)} \tag{5.106}$$

(b) Velocities of the Link Frame Origins

$$\overline{v}_{0,2}^{(0)} = \overline{v}_{0,2/0}^{(0)} = \dot{\overline{r}}_{0,2}^{(0)} = \dot{\overline{r}}_S^{(0)} = \overline{v}_S^{(0)} \Rightarrow$$

$$\overline{v}_S^{(0)} = -d_{12}\dot{\theta}_{01}(\overline{u}_1c\theta_{01} + \overline{u}_2s\theta_{01}) \tag{5.107}$$

$$\overline{v}_{0,3}^{(0)} = \overline{v}_{0,3/0}^{(0)} = \dot{\overline{r}}_{0,3}^{(0)} = \dot{\overline{r}}_R^{(0)} = \overline{v}_R^{(0)} \Rightarrow$$

$$\overline{v}_R^{(0)} = \dot{\theta}_{01}e^{\tilde{u}_3\theta_{01}}\tilde{u}_3(\overline{u}_1s_{23}s\theta_{12} + \overline{u}_2d_{12} + \overline{u}_3s_{23}c\theta_{12})$$
$$+ e^{\tilde{u}_3\theta_{01}}[(\overline{u}_1s\theta_{12} + \overline{u}_3c\theta_{12})\dot{s}_{23} + (\overline{u}_1c\theta_{12} - \overline{u}_3s\theta_{12})s_{23}\dot{\theta}_{12}] \Rightarrow$$

$$\overline{v}_R^{(0)} = \dot{\theta}_{01}e^{\tilde{u}_3\theta_{01}}(\overline{u}_2s_{23}s\theta_{12} - \overline{u}_1d_{12})$$
$$+ e^{\tilde{u}_3\theta_{01}}[(\overline{u}_1s\theta_{12} + \overline{u}_3c\theta_{12})\dot{s}_{23} + (\overline{u}_1c\theta_{12} - \overline{u}_3s\theta_{12})s_{23}\dot{\theta}_{12}] \Rightarrow$$

$$\overline{v}_R^{(0)} = e^{\tilde{u}_3\theta_{01}}[\overline{u}_1(\dot{s}_{23}s\theta_{12} + s_{23}\dot{\theta}_{12}c\theta_{12} - d_{12}\dot{\theta}_{01}) + \overline{u}_2(s_{23}\dot{\theta}_{01}s\theta_{12})$$
$$+ \overline{u}_3(\dot{s}_{23}c\theta_{12} - s_{23}\dot{\theta}_{12}s\theta_{12})] \tag{5.108}$$

(c) Wrist Point Jacobian Matrix

For the sake of brevity, let

$$\overline{v}_R^{(0)} = \overline{w} \tag{5.109}$$

Then, Eq. (5.108) can also be written as

$$\overline{w} = \overline{W}_1\dot{q}_1 + \overline{W}_2\dot{q}_2 + \overline{W}_3\dot{q}_3 \tag{5.110}$$

In Eq. (5.110), q_k is a general symbol that represents the kth joint variable and $\overline{W}_k$ is the kth *velocity influence coefficient* that represents the influence of $\dot{q}_k$ on $\overline{w}$. The column matrix $\overline{W}_k$ can be obtained from the position matrix $\overline{r}$ or the velocity matrix $\overline{w} = \dot{\overline{r}}$ by taking the following partial derivatives.

$$\overline{W}_k = \partial\overline{r}/\partial q_k = \partial\overline{w}/\partial\dot{q}_k \tag{5.111}$$

For the present manipulator,

$$q_1 = \theta_{01},\ \ q_2 = \theta_{12},\ \ q_3 = s_{23} \tag{5.112}$$

Hence, for $k \in \{1, 2, 3\}$, $\overline{W}_k$ is obtained from Eq. (5.108) as follows:

$$\overline{W}_1 = \partial\overline{w}/\partial\dot{\theta}_{01} = e^{\tilde{u}_3\theta_{01}}(\overline{u}_2s_{23}s\theta_{12} - \overline{u}_1d_{12}) \tag{5.113}$$

$$\overline{W}_2 = \partial\overline{w}/\partial\dot{\theta}_{12} = e^{\tilde{u}_3\theta_{01}}(\overline{u}_1c\theta_{12} - \overline{u}_3s\theta_{12})s_{23} \tag{5.114}$$

$$\overline{W}_3 = \partial\overline{w}/\partial\dot{s}_{23} = e^{\tilde{u}_3\theta_{01}}(\overline{u}_1s\theta_{12} + \overline{u}_3c\theta_{12}) \tag{5.115}$$

Let the joint variables be stacked into the column matrix $\overline{q}$. Then, Eq. (5.110) can be written compactly as

$$\overline{w} = \hat{J}\dot{\overline{q}} \tag{5.116}$$

In Eq. (5.116), $\hat{J}$ is defined as the *Jacobian matrix* between $\overline{w}$ and $\dot{\overline{q}}$. In this example, it is a 3 × 3 matrix, which is made up of the velocity influence coefficients as shown below.

$$\hat{J} = [\overline{W}_1 \ \overline{W}_2 \ \overline{W}_3] \tag{5.117}$$

Note that the Jacobian matrix is a function of position only. That is, $\hat{J} = \hat{J}(\overline{q})$. Note also that the Jacobian matrix functions as a *transformation matrix* that converts the *joint space* velocity information represented by $\dot{\overline{q}}$ to the *task space* velocity information represented by $\overline{w}$.

(d) Inverse Velocity Solution

After obtaining the *forward velocity equations,* as done above in Parts (a), (b), and (c), the next stage of the velocity analysis can be carried out in order to obtain the *inverse velocity solution,* which provides the *joint velocities* (i.e. the first derivatives of the joint variables) that correspond to a specified velocity state of the manipulator in a certain pose it assumes. The solution can be obtained either numerically or analytically. Here, it is to be noted that the velocity relationships happen to be linear. Therefore, the inverse velocity solution provides the joint velocities uniquely unless the manipulator is in a pose of *motion singularity.*

The numerical solution can be obtained from Eq. (5.116) by inverting the Jacobian matrix $\hat{J}$ if the manipulator is not in a pose of motion singularity, in which $\det(\hat{J}) = 0$. In other words, if $\det(\hat{J}) \neq 0$,

$$\dot{\overline{q}} = \hat{J}^{-1}\overline{w} \tag{5.118}$$

The numerical solution has the advantage that it provides the current values of the joint velocities all at once. However, it is not very suitable for a detailed kinematic analysis.

On the other hand, the analytical solution can be obtained through several successive steps but at the end it provides certain formulas for the joint velocities. These formulas eliminate the generally cumbersome task of expressing the determinant of the Jacobian matrix because they readily indicate the motion singularities with their denominator factors. Therefore, the analytical solution has the advantage that it is very suitable for detailed kinematic analyses both in the ordinary and singular poses of the manipulator. Besides, for many manipulators, the formulas obtained through the analytical solution turn out to be rather simple and thus they can be used conveniently for numerical evaluations as well.

The inverse velocity solution for the present manipulator is obtained analytically as explained below.

In a certain pose of the manipulator, let the velocity of the wrist point R be specified with its base frame components (w_1, w_2, w_3) as indicated below.

$$\overline{v}_R^{(0)} = \overline{w} = \overline{u}_1 w_1 + \overline{u}_2 w_2 + \overline{u}_3 w_3 \tag{5.119}$$

Upon this specification, the inverse velocity solution can be started by writing Eq. (5.108) as follows so that the expression on the left-hand side that contains the unknown joint velocities becomes as simple as possible.

$$\overline{u}_1(\dot{s}_{23}s\theta_{12} + s_{23}\dot{\theta}_{12}c\theta_{12} - d_{12}\dot{\theta}_{01}) + \overline{u}_2(s_{23}\dot{\theta}_{01}s\theta_{12}) + \overline{u}_3(\dot{s}_{23}c\theta_{12} - s_{23}\dot{\theta}_{12}s\theta_{12}) = e^{-\tilde{u}_3\theta_{01}}\overline{w} \quad (5.120)$$

Equation (5.120) leads to the following three scalar equations, when it is premultiplied on both sides by $\overline{u}_1^t$, $\overline{u}_2^t$, and $\overline{u}_3^t$.

$$\dot{s}_{23}s\theta_{12} + s_{23}\dot{\theta}_{12}c\theta_{12} - d_{12}\dot{\theta}_{01} = \overline{u}_1^t e^{-\tilde{u}_3\theta_{01}}\overline{w} = w_1c\theta_{01} + w_2s\theta_{01} \quad (5.121)$$

$$s_{23}\dot{\theta}_{01}s\theta_{12} = \overline{u}_2^t e^{-\tilde{u}_3\theta_{01}}\overline{w} = w_2c\theta_{01} - w_1s\theta_{01} \quad (5.122)$$

$$\dot{s}_{23}c\theta_{12} - s_{23}\dot{\theta}_{12}s\theta_{12} = \overline{u}_3^t e^{-\tilde{u}_3\theta_{01}}\overline{w} = w_3 \quad (5.123)$$

If $s\theta_{12} \neq 0$, Eq. (5.122) gives $\dot{\theta}_{01}$ as

$$\dot{\theta}_{01} = (w_2c\theta_{01} - w_1s\theta_{01})/(s_{23}s\theta_{12}) \quad (5.124)$$

With the availability of $\dot{\theta}_{01}$, Eqs. (5.121) and (5.123) can be written together as the following matrix equation.

$$\begin{bmatrix} c\theta_{12} & s\theta_{12} \\ -s\theta_{12} & c\theta_{12} \end{bmatrix} \begin{bmatrix} s_{23}\dot{\theta}_{12} \\ \dot{s}_{23} \end{bmatrix} = \begin{bmatrix} w_{12}^* \\ w_3 \end{bmatrix} \quad (5.125)$$

In Eq. (5.125), w_{12}^* is known as expressed below.

$$w_{12}^* = w_{12} + d_{12}\dot{\theta}_{01} = (w_1c\theta_{01} + w_2s\theta_{01}) + d_{12}\dot{\theta}_{01} \quad (5.126)$$

Note that the way Eq. (5.125) is written makes the coefficient matrix on the left-hand side an orthonormal matrix, whose inverse is equal to its transpose. Therefore, Eq. (5.125) can be solved simply as follows to obtain $\dot{\theta}_{12}$ and $\dot{s}_{23}$ without any singularity problem.

$$\begin{bmatrix} s_{23}\dot{\theta}_{12} \\ \dot{s}_{23} \end{bmatrix} = \begin{bmatrix} c\theta_{12} & -s\theta_{12} \\ s\theta_{12} & c\theta_{12} \end{bmatrix} \begin{bmatrix} w_{12}^* \\ w_3 \end{bmatrix} \Rightarrow$$

$$\dot{\theta}_{12} = (w_{12}^*c\theta_{12} - w_3s\theta_{12})/s_{23} \quad (5.127)$$

$$\dot{s}_{23} = w_{12}^*s\theta_{12} + w_3c\theta_{12} \quad (5.128)$$

(e) Motion Singularities Involving Velocities

According to the condition stated for Eq. (5.124), a motion singularity occurs if $s\theta_{12} = 0$, i.e. if $\theta_{12} = 0$ or $\theta_{12} = \pm\pi$. If such a singularity occurs, Eqs. (5.121)–(5.123) degenerate into the following forms.

$$\sigma' s_{23}\dot{\theta}_{12} - d_{12}\dot{\theta}_{01} = w_1c\theta_{01} + w_2s\theta_{01} = w_{12} \quad (5.129)$$

$$w_2c\theta_{01} - w_1s\theta_{01} = 0 \quad (5.130)$$

$$\sigma'\dot{s}_{23} = w_3 \quad (5.131)$$

In Eqs. (5.129) and (5.131), $\sigma' = \text{sign}(c\theta_{12})$, i.e. $\sigma' = +1$ if $\theta_{12} = 0$ and $\sigma' = -1$ if $\theta_{12} = \pm\pi$.

In a pose of singularity, Eq. (5.131) gives $\dot{s}_{23}$ as

$$\dot{s}_{23} = \sigma' w_3 \quad (5.132)$$

However, $\dot{\theta}_{01}$ and $\dot{\theta}_{12}$ become indefinite because they do not appear in Eq. (5.130) and they cannot be found separately only from Eq. (5.129).

(f) Consequences of the Motion Singularities

A motion singularity, such as the one considered above, is characterized by two typical consequences that are explained below.

∗ Task Space Restriction

Equation (5.130) indicates the task space restriction to be obeyed in planning a task motion. Due to this restriction, the components w_1 and w_2 of the wrist point velocity $\vec{w} = \vec{v}_R$ cannot be specified arbitrarily but they must be specified consistently with Eq. (5.130) so that

$$w_2 c\theta_{01} = w_1 s\theta_{01}$$

∗ Joint Space Indeterminacy

Equation (5.129) indicates the joint space indeterminacy, due to which $\dot{\theta}_{01}$ and $\dot{\theta}_{12}$ cannot be determined separately. This indeterminacy arises in a pose of motion singularity because the same value of w_{12}, which is the component of $\vec{w}$ along $\vec{u}_1^{(1)}$, can be realized redundantly by either or both of $\dot{\theta}_{01}$ and $\dot{\theta}_{12}$. Thus, by assigning an arbitrary value to one of them, the other one can be found from Eq. (5.129).

Here, it is to be pointed out that, as long as the planned motion of the wrist point is consistent with the task space restriction of the motion singularity, i.e. as long as $w_2 c\theta_{01} = w_1 s\theta_{01}$, the joint velocities can still be obtained with finite values, even if one of them is to be selected with an arbitrary value. In other words, it is *not* necessary to avoid a motion singularity. On the contrary, with a proper motion planning that obeys the task space restriction, it is possible to pass through a pose of motion singularity without any occurrence of unbounded joint velocities.

5.5.4 Acceleration Analysis

(a) Angular Accelerations of the Links

The differentiation of Eqs. (5.104)–(5.106) leads to the following equations.

$$\overline{\alpha}_{0,1}^{(0)} = \dot{\overline{\omega}}_{0,1}^{(0)} = \ddot{\theta}_{01}\overline{u}_3 \tag{5.133}$$

$$\begin{aligned}\overline{\alpha}_{0,2}^{(0)} = \dot{\overline{\omega}}_{0,2}^{(0)} &= -\overline{u}_1(\ddot{\theta}_{12} s\theta_{01} + \dot{\theta}_{01}\dot{\theta}_{12} c\theta_{01}) \\ &\quad + \overline{u}_2(\ddot{\theta}_{12} c\theta_{01} - \dot{\theta}_{01}\dot{\theta}_{12} s\theta_{01}) + \overline{u}_3\ddot{\theta}_{01}\end{aligned} \tag{5.134}$$

$$\overline{\alpha}_{0,3}^{(0)} = \overline{\alpha}_{0,2}^{(0)} \tag{5.135}$$

(b) Accelerations of the Link Frame Origins

The acceleration of the shoulder point is obtained as follows by differentiating Eq. (5.107):

$$\overline{a}_{0,2}^{(0)} = \overline{a}_{0,2/0}^{(0)} = \dot{\overline{v}}_{0,2}^{(0)} = \dot{\overline{v}}_S^{(0)} = \overline{a}_S^{(0)} \Rightarrow$$

$$\overline{a}_S^{(0)} = -d_{12}\ddot{\theta}_{01}(\overline{u}_1 c\theta_{01} + \overline{u}_2 s\theta_{01}) + d_{12}\dot{\theta}_{01}^2(\overline{u}_1 s\theta_{01} - \overline{u}_2 c\theta_{01}) \tag{5.136}$$

Similarly, the acceleration of the wrist point is obtained as follows by differentiating Eq. (5.108).

$$\overline{a}_{0,3}^{(0)} = \overline{a}_{0,3/0}^{(0)} = \dot{\overline{v}}_{0,3}^{(0)} = \dot{\overline{v}}_{R}^{(0)} = \overline{a}_{R}^{(0)} \Rightarrow$$

$$\begin{aligned}
\overline{a}_{R}^{(0)} &= e^{\widetilde{u}_3\theta_{01}}[\overline{u}_1(\ddot{s}_{23}s\theta_{12} + s_{23}\ddot{\theta}_{12}c\theta_{12} - d_{12}\ddot{\theta}_{01} + 2\dot{s}_{23}\dot{\theta}_{12}c\theta_{12} - s_{23}\dot{\theta}_{12}^2 s\theta_{12}) \\
&+ \overline{u}_2(s_{23}\ddot{\theta}_{01}s\theta_{12} + \dot{s}_{23}\dot{\theta}_{01}s\theta_{12} + s_{23}\dot{\theta}_{01}\dot{\theta}_{12}c\theta_{12}) \\
&+ \overline{u}_3(\ddot{s}_{23}c\theta_{12} - s_{23}\ddot{\theta}_{12}s\theta_{12} - 2\dot{s}_{23}\dot{\theta}_{12}s\theta_{12} - s_{23}\dot{\theta}_{12}^2 c\theta_{12})] \\
&+ \dot{\theta}_{01}e^{\widetilde{u}_3\theta_{01}}\widetilde{u}_3[\overline{u}_1(\dot{s}_{23}s\theta_{12} + s_{23}\dot{\theta}_{12}c\theta_{12} - d_{12}\dot{\theta}_{01}) + \overline{u}_2(s_{23}\dot{\theta}_{01}s\theta_{12}) \\
&+ \overline{u}_3(\dot{s}_{23}c\theta_{12} - s_{23}\dot{\theta}_{12}s\theta_{12})] \Rightarrow
\end{aligned}$$

$$\begin{aligned}
\overline{a}_{R}^{(0)} &= e^{\widetilde{u}_3\theta_{01}}\{\overline{u}_1[\ddot{s}_{23}s\theta_{12} + s_{23}\ddot{\theta}_{12}c\theta_{12} - d_{12}\ddot{\theta}_{01} \\
&+ 2\dot{s}_{23}\dot{\theta}_{12}c\theta_{12} - s_{23}(\dot{\theta}_{01}^2 + \dot{\theta}_{12}^2)s\theta_{12}] \\
&+ \overline{u}_2(s_{23}\ddot{\theta}_{01}s\theta_{12} + 2\dot{s}_{23}\dot{\theta}_{01}s\theta_{12} + 2s_{23}\dot{\theta}_{01}\dot{\theta}_{12}c\theta_{12} - d_{12}\dot{\theta}_{01}^2) \\
&+ \overline{u}_3(\ddot{s}_{23}c\theta_{12} - s_{23}\ddot{\theta}_{12}s\theta_{12} - 2\dot{s}_{23}\dot{\theta}_{12}s\theta_{12} - s_{23}\dot{\theta}_{12}^2 c\theta_{12})\}
\end{aligned} \tag{5.137}$$

(c) Wrist Point Acceleration by Using the Wrist Point Jacobian Matrix

Let $\overline{a}_R = \overline{a}_R^{(0)}$ and note that

$$\overline{a}_R = \dot{\overline{v}}_R = \dot{\overline{w}} \tag{5.138}$$

Then, Eq. (5.116) leads to the following compact equation for $\overline{a}_R$.

$$\overline{a}_R = \hat{J}\ddot{\overline{q}} + \overset{\circ}{\overline{a}}_R \tag{5.139}$$

In Eq. (5.139), $\overset{\circ}{\overline{a}}_R$ consists of the *velocity-dependent* acceleration terms. That is,

$$\overset{\circ}{\overline{a}}_R = \overset{\circ}{\overline{a}}_R(\overline{q}, \dot{\overline{q}}) = \dot{\hat{J}}\dot{\overline{q}} \tag{5.140}$$

For the system of this example, the detailed expressions of the velocity-dependent acceleration terms are seen in Eqs. (5.134), (5.136), and (5.137).

In general, among the velocity-dependent *translational* acceleration terms, the ones that contain the products of two different joint velocities are called *Coriolis acceleration terms* and the ones that contain the squares of the joint velocities are called *centripetal acceleration terms*. As for the velocity-dependent *angular* acceleration terms, they happen to contain only the products of two different angular joint velocities, and they are called *gyroscopic acceleration terms*.

(d) Inverse Acceleration Solution

After obtaining the *forward acceleration equations*, as done above in Parts (a), (b), and (c), the next stage of the acceleration analysis can be carried out in order to obtain the *inverse acceleration solution*, which provides the *joint accelerations* (i.e. the second derivatives of the joint variables) that correspond to a specified acceleration state of the manipulator in a certain pose and velocity state it assumes. The solution can be obtained either numerically or analytically. Here, it is to be noted that the acceleration relationships happen to be *affine*. Their affinity is due to the presence of the *bias terms* contained in $\overset{\circ}{\overline{a}}_R$. They would be linear if the bias terms were zero. Therefore, like the inverse velocity solution, the inverse acceleration solution provides the joint accelerations uniquely unless the manipulator is in a pose of *motion singularity*.

The numerical solution can be obtained from Eq. (5.139) by inverting the Jacobian matrix $\widehat{\mathcal{J}}$ if the manipulator is not in a pose of motion singularity, in which $\det(\widehat{\mathcal{J}}) = 0$. In other words, if $\det(\widehat{\mathcal{J}}) \neq 0$,

$$\ddot{\overline{q}} = \widehat{\mathcal{J}}^{-1}(\overline{a}_R - \overline{a}_R^{\circ}) \tag{5.141}$$

The numerical solution has the advantage that it provides the current values of the joint accelerations all at once. However, it is not very suitable for a detailed kinematic analysis. As for the analytical inverse acceleration solution, it is obtained for the present manipulator as explained below.

In a certain pose and velocity state of the manipulator, let the acceleration of the wrist point R be specified with its base frame components $(\dot{w}_1, \dot{w}_2, \dot{w}_3)$ as indicated below.

$$\overline{a}_R = \dot{\overline{w}} = \overline{u}_1\dot{w}_1 + \overline{u}_2\dot{w}_2 + \overline{u}_3\dot{w}_3 \tag{5.142}$$

Upon this specification, the inverse acceleration solution can be started by writing Eq. (5.137) as follows so that the expression on the left-hand side that contains the unknown joint accelerations becomes as simple as possible:

$$\overline{u}_1(\ddot{s}_{23}s\theta_{12} + s_{23}\ddot{\theta}_{12}c\theta_{12} - d_{12}\ddot{\theta}_{01}) + \overline{u}_2(s_{23}\ddot{\theta}_{01}s\theta_{12})$$

$$+ \overline{u}_3(\ddot{s}_{23}c\theta_{12} - s_{23}\ddot{\theta}_{12}s\theta_{12}) = e^{-\tilde{u}_3\theta_{01}}\dot{\overline{w}} + \overline{u}_1 a_1^{\circ} + \overline{u}_2 a_2^{\circ} + \overline{u}_3 a_3^{\circ} \tag{5.143}$$

In Eq. (5.143), the velocity-dependent acceleration terms have the following expressions.

$$a_1^{\circ} = s_{23}(\dot{\theta}_{01}^2 + \dot{\theta}_{12}^2)s\theta_{12} - 2\dot{s}_{23}\dot{\theta}_{12}c\theta_{12} \tag{5.144}$$

$$a_2^{\circ} = d_{12}\dot{\theta}_{01}^2 - 2(\dot{s}_{23}\dot{\theta}_{01}s\theta_{12} + s_{23}\dot{\theta}_{01}\dot{\theta}_{12}c\theta_{12}) \tag{5.145}$$

$$a_3^{\circ} = s_{23}\dot{\theta}_{12}^2c\theta_{12} + 2\dot{s}_{23}\dot{\theta}_{12}s\theta_{12} \tag{5.146}$$

Equation (5.143) leads to the following three scalar equations, when it is premultiplied on both sides by $\overline{u}_1^t$, $\overline{u}_2^t$, and $\overline{u}_3^t$.

$$\ddot{s}_{23}s\theta_{12} + s_{23}\ddot{\theta}_{12}c\theta_{12} - d_{12}\ddot{\theta}_{01} = \dot{w}_1c\theta_{01} + \dot{w}_2s\theta_{01} + a_1^{\circ} \tag{5.147}$$

$$s_{23}\ddot{\theta}_{01}s\theta_{12} = \dot{w}_2c\theta_{01} - \dot{w}_1s\theta_{01} + a_2^{\circ} \tag{5.148}$$

$$\ddot{s}_{23}c\theta_{12} - s_{23}\ddot{\theta}_{12}s\theta_{12} = \dot{w}_3 + a_3^{\circ} \tag{5.149}$$

Note that Eqs. (5.147)–(5.149) are similar to Eqs. (5.121)–(5.123) that are written before for the joint velocities. Therefore, they are solved similarly as described below.

If $s\theta_{12} \neq 0$, Eq. (5.148) gives $\ddot{\theta}_{01}$ as

$$\ddot{\theta}_{01} = (\dot{w}_2c\theta_{01} - \dot{w}_1s\theta_{01} + a_2^{\circ})/(s_{23}s\theta_{12}) \tag{5.150}$$

With the availability of $\ddot{\theta}_{01}$, Eqs. (5.147) and (5.149) can be written together as the following matrix equation.

$$\begin{bmatrix} c\theta_{12} & s\theta_{12} \\ -s\theta_{12} & c\theta_{12} \end{bmatrix} \begin{bmatrix} s_{23}\ddot{\theta}_{12} \\ \ddot{s}_{23} \end{bmatrix} = \begin{bmatrix} a_1^* \\ a_3^* \end{bmatrix} \tag{5.151}$$

In Eq. (5.151), a_1^* and a_3^* are known as expressed below.

$$a_1^* = a_1^{\circ} + \dot{w}_1c\theta_{01} + \dot{w}_2s\theta_{01} + d_{12}\ddot{\theta}_{01} \tag{5.152}$$

$$a_3^* = a_3^{\circ} + \dot{w}_3 \tag{5.153}$$

Equation (5.151) gives $\ddot{\theta}_{12}$ and $\ddot{s}_{23}$ as follows:

$$\ddot{\theta}_{12} = (a_1^* c\theta_{12} - a_3^* s\theta_{12})/s_{23} \tag{5.154}$$

$$\ddot{s}_{23} = a_1^* s\theta_{12} + a_3^* c\theta_{12} \tag{5.155}$$

(e) Motion Singularities Involving Accelerations

The pose of a manipulator at the instant of a motion singularity is the same for the relationships that involve velocities, accelerations, and all the other higher order motion derivatives. The conditions stated for Eqs. (5.141) and (5.150) of this section and Eqs. (5.118) and (5.124) of the previous section are evidence of this fact.

If a motion singularity occurs, i.e. if $s\theta_{12} = 0$, Eqs. (5.147)–(5.149) assume the following forms.

$$\sigma' s_{23}\ddot{\theta}_{12} - d_{12}\ddot{\theta}_{01} = \dot{w}_1 c\theta_{01} + \dot{w}_2 s\theta_{01} + a_1^\circ \tag{5.156}$$

$$\dot{w}_2 c\theta_{01} - \dot{w}_1 s\theta_{01} + a_2^\circ = 0 \tag{5.157}$$

$$\sigma' \ddot{s}_{23} = \dot{w}_3 + a_3^\circ \tag{5.158}$$

In Eqs. (5.156) and (5.158), like it is in the velocity analysis, $\sigma' = +1$ if $\theta_{12} = 0$ and $\sigma' = -1$ if $\theta_{12} = \pm\pi$.

In a pose of singularity, Eq. (5.158) gives $\ddot{s}_{23}$ as

$$\ddot{s}_{23} = \sigma'(\dot{w}_3 + a_3^\circ) \tag{5.159}$$

Equation (5.157) indicates the instantaneous *task space restriction* that

$$\dot{w}_1 s\theta_{01} - \dot{w}_2 c\theta_{01} = a_2^\circ.$$

Equation (5.156) indicates the instantaneous *joint space indeterminacy*, due to which $\ddot{\theta}_{01}$ and $\ddot{\theta}_{12}$ cannot be determined separately. However, if an arbitrary value is assigned to one of them, then the other one can be determined from Eq. (5.156). This indeterminacy arises in a pose of motion singularity because the same value of the term on the right-hand side of Eq. (5.156) can be realized redundantly by either or both of $\ddot{\theta}_{01}$ and $\ddot{\theta}_{12}$. Note that the indeterminacy expressed by Eq. (5.156) between the joint accelerations is of course intrinsically related to the indeterminacy expressed by Eq. (5.129) between the joint velocities. Therefore, the elimination of the indeterminacy must be consistent. For example, the elimination of the indeterminacy may be based on the continuity of $\theta_{01}(t)$. Then, the values to be assigned to $\dot{\theta}_{01}$ and $\ddot{\theta}_{01}$ at the instant t_s of the singularity can be taken as their values immediately before the singularity at $t = t_s - \Delta t$ so that $\dot{\theta}_{01}(t_s) = \dot{\theta}_{01}(t_s - \Delta t)$ and $\ddot{\theta}_{01}(t_s) = \ddot{\theta}_{01}(t_s - \Delta t)$.

5.6 Example 5.2 : A Spatial Slider-Crank (RSSP) Mechanism

5.6.1 Kinematic Description of the Mechanism

Figure 5.6 shows a spatial slider-crank mechanism, which is formed as a *kinematic loop* of four links $\mathcal{B}_1$, $\mathcal{B}_2$, $\mathcal{B}_3$, and $\mathcal{B}_0$ (the base). The joint $\mathcal{J}_{01}$ is revolute, the joint $\mathcal{J}_{03}$ is prismatic, and the joints $\mathcal{J}_{12}$ and $\mathcal{J}_{23}$ are spherical. In this example, the *closure joint* is selected to be $\mathcal{J}_{23}$.

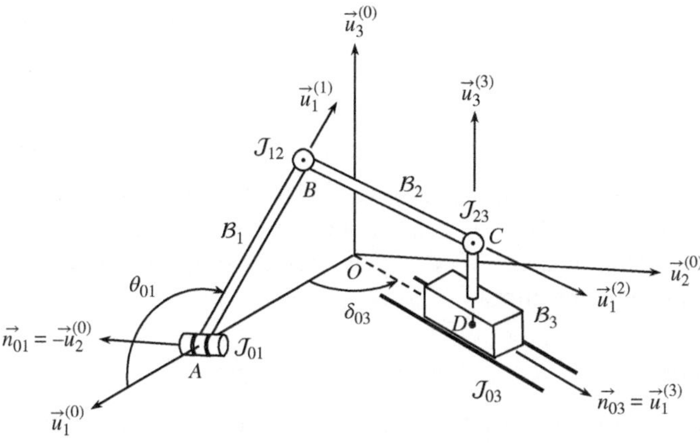

Figure 5.6 A spatial slider-crank mechanism.

The following reference frames are attached to the base and the moving links. Their characterizing basis vectors are indicated in Figure 5.6.

$$\mathcal{F}_0(O_0) : O_0 = O; \mathcal{U}_0 = \{\vec{u}_1^{(0)},\ \vec{u}_2^{(0)},\ \vec{u}_3^{(0)}\}$$
$$\mathcal{F}_1(O_1) : O_1 = A; \mathcal{U}_1 = \{\vec{u}_1^{(1)},\ \vec{u}_2^{(1)} = \vec{u}_2^{(0)},\ \vec{u}_3^{(1)} = \vec{u}_1^{(1)} \times \vec{u}_2^{(1)}\}$$
$$\mathcal{F}_2(O_2) : O_2 = B; \mathcal{U}_2 = \{\vec{u}_1^{(2)},\ \vec{u}_2^{(2)},\ \vec{u}_3^{(2)}\}$$
$$\mathcal{F}_3(O_3) : O_3 = D; \mathcal{U}_3 = \{\vec{u}_1^{(3)},\ \vec{u}_2^{(3)} = \vec{u}_3^{(3)} \times \vec{u}_1^{(3)},\ \vec{u}_3^{(3)} = \vec{u}_3^{(0)}\}$$

The joint frames are attached to the kinematic elements as shown below.

$$\mathcal{F}_{01}(O_{01}) = \mathcal{F}_0(A) \parallel \mathcal{F}_0(O), \mathcal{F}_{10}(O_{10}) = \mathcal{F}_1(A)$$
$$\mathcal{F}_{12}(O_{12}) = \mathcal{F}_1(B) \parallel \mathcal{F}_1(A), \mathcal{F}_{21}(O_{21}) = \mathcal{F}_2(B)$$
$$\mathcal{F}_{23}(O_{23}) = \mathcal{F}_2(C) \parallel \mathcal{F}_2(B), \mathcal{F}_{32}(O_{32}) = \mathcal{F}_3(C) \parallel \mathcal{F}_3(D)$$
$$\mathcal{F}_{03}(O_{03}) = \mathcal{F}_3(O) \parallel \mathcal{F}_3(D), \mathcal{F}_{30}(O_{30}) = \mathcal{F}_3(D)$$

The nonzero constant parameters are listed below.

$$b_{01} = O_0O_{01} = OA \text{ along } \vec{u}_1^{(0)} \tag{5.160}$$
$$b_{12} = O_{10}O_{12} = AB \text{ along } \vec{u}_1^{(1)} \tag{5.161}$$
$$b_{23} = O_{21}O_{23} = BC \text{ along } \vec{u}_1^{(2)} \tag{5.162}$$
$$d_{32} = O_{30}O_{32} = DC \text{ along } \vec{u}_3^{(3)} = \vec{u}_3^{(0)} \tag{5.163}$$
$$\delta_{03} = \sphericalangle[\vec{u}_1^{(0)} \rightarrow \vec{u}_1^{(3)}] \text{ about } \vec{u}_3^{(0)} \tag{5.164}$$

The variables of the revolute and prismatic joints are defined as follows:

$$\theta_{01} = \sphericalangle[\vec{u}_1^{(0)} \rightarrow \vec{u}_1^{(1)}] \text{ about } \vec{n}_{01} = -\vec{u}_2^{(0)} \tag{5.165}$$
$$s_{03} = O_{03}O_{30} = OD \text{ along } \vec{n}_{03} = \vec{u}_1^{(3)} \tag{5.166}$$

Since the joints $\mathcal{J}_{12}$ and $\mathcal{J}_{23}$ are spherical, their joint variables (i.e. three independent angles for each) are embedded into the following rotation matrices that describe

RFB (rotated frame based) rotation sequences, respectively, about the pivot points $B = O_{12} = O_{21}$ and $C = O_{23} = O_{32}$.

$$\widehat{C}^{(1,2)} = \widehat{R}_{12} = e^{\tilde{n}_{121}\phi_{12}} e^{\tilde{n}_{122}\theta_{12}} e^{\tilde{n}_{123}\psi_{12}} \tag{5.167}$$

$$\widehat{C}^{(2,3)} = \widehat{R}_{23} = e^{\tilde{n}_{231}\phi_{23}} e^{\tilde{n}_{232}\theta_{23}} e^{\tilde{n}_{233}\psi_{23}} \tag{5.168}$$

The rotation axis unit vectors for $\widehat{R}_{12}$ and $\widehat{R}_{23}$ are represented by the column matrix triads $\mathcal{N}_{12} = \{\overline{n}_{121}, \overline{n}_{122}, \overline{n}_{123}\}$ and $\mathcal{N}_{23} = \{\overline{n}_{231}, \overline{n}_{232}, \overline{n}_{233}\}$. These triads have not yet been selected. They will be selected later in appropriate ways in order to simplify the expressions of the relevant kinematic relationships.

5.6.2 Loop Closure Equations

The orientational and locational loop closure equations for the considered mechanism can be written as explained below.

(a) Orientational Loop Closure Equation

The rotation matrices between the successive links are written as follows:

$$\widehat{C}^{(0,1)} = \widehat{R}[-\overline{u}_2^{(0/0)}, \theta_{01}] = e^{-\tilde{u}_2\theta_{01}} \tag{5.169}$$

$$\widehat{C}^{(0,3)} = \widehat{R}[\overline{u}_3^{(0/0)}, \delta_{03}] = e^{\tilde{u}_3\delta_{03}} \tag{5.170}$$

$$\widehat{C}^{(1,2)} = \widehat{R}_{12} = e^{\tilde{n}_{121}\phi_{12}} e^{\tilde{n}_{122}\theta_{12}} e^{\tilde{n}_{123}\psi_{12}} \tag{5.171}$$

$$\widehat{C}^{(2,3)} = \widehat{R}_{23} = e^{\tilde{n}_{231}\phi_{23}} e^{\tilde{n}_{232}\theta_{23}} e^{\tilde{n}_{233}\psi_{23}} \tag{5.172}$$

The orientational closure equation is written as shown below in order to express the reconnection of the hypothetically disconnected joint $\mathcal{J}_{23}$.

$$\widehat{C}^{(0,1)}\widehat{C}^{(1,2)}\widehat{C}^{(2,3)} = \widehat{C}^{(0,3)} \tag{5.173}$$

Equation (5.173) can be written in detail as follows:

$$e^{-\tilde{u}_2\theta_{01}}\widehat{R}_{12}\widehat{R}_{23} = e^{\tilde{u}_3\delta_{03}} \Rightarrow \widehat{R}_{12}\widehat{R}_{23} = e^{\tilde{u}_2\theta_{01}} e^{\tilde{u}_3\delta_{03}} \Rightarrow$$

$$(e^{\tilde{n}_{121}\phi_{12}} e^{\tilde{n}_{122}\theta_{12}} e^{\tilde{n}_{123}\psi_{12}})(e^{\tilde{n}_{231}\phi_{23}} e^{\tilde{n}_{232}\theta_{23}} e^{\tilde{n}_{233}\psi_{23}}) = e^{\tilde{u}_2\theta_{01}} e^{\tilde{u}_3\delta_{03}} \tag{5.174}$$

(b) Locational Loop Closure Equation

The locational closure of the mechanism at the joint $\mathcal{J}_{23}$ can be expressed by the following vector equation, which states that the point O_{23} (i.e. the point C_2 of link $\mathcal{B}_2$) and the point O_{32} (i.e. the point C_3 of link $\mathcal{B}_3$) are brought together to be coincident at the joint center denoted simply as C.

$$\begin{aligned} &\overrightarrow{OA} + \overrightarrow{AB} + \overrightarrow{BC} = \overrightarrow{OD} + \overrightarrow{DC} \Rightarrow \\ &\vec{r}_{0,1} + \vec{r}_{1,2} + \vec{r}_{2,23} = \vec{r}_{0,3} + \vec{r}_{3,32} \Rightarrow \\ &b_{01}\vec{u}_1^{(0)} + b_{12}\vec{u}_1^{(1)} + b_{23}\vec{u}_1^{(2)} = s_{03}\vec{u}_1^{(3)} + d_{32}\vec{u}_3^{(0)} \end{aligned} \tag{5.175}$$

Equation (5.175) leads to the following matrix equation in the base frame.

$$\begin{aligned} &b_{01}\overline{u}_1^{(0/0)} + b_{12}\overline{u}_1^{(1/0)} + b_{23}\overline{u}_1^{(2/0)} = s_{03}\overline{u}_1^{(3/0)} + d_{32}\overline{u}_3^{(0/0)} \Rightarrow \\ &b_{01}\overline{u}_1 + b_{12}\widehat{C}^{(0,1)}\overline{u}_1 + b_{23}\widehat{C}^{(0,2)}\overline{u}_1 = s_{03}\widehat{C}^{(0,3)}\overline{u}_1 + d_{32}\overline{u}_3 \Rightarrow \\ &b_{01}\overline{u}_1 + b_{12}e^{-\tilde{u}_2\theta_{01}}\overline{u}_1 + b_{23}e^{-\tilde{u}_2\theta_{01}}\widehat{R}_{12}\overline{u}_1 = s_{03}e^{\tilde{u}_3\delta_{03}}\overline{u}_1 + d_{32}\overline{u}_3 \end{aligned}$$

By inserting the expression of $\widehat{R}_{12}$, the preceding equation can be written as follows:

$$b_{01}\overline{u}_1 + b_{12}e^{-\widetilde{u}_2\theta_{01}}\overline{u}_1 + b_{23}e^{-\widetilde{u}_2\theta_{01}}e^{\widetilde{n}_{121}\phi_{12}}e^{\widetilde{n}_{122}\theta_{12}}e^{\widetilde{n}_{123}\psi_{12}}\overline{u}_1$$

$$= s_{03}e^{\widetilde{u}_3\delta_{03}}\overline{u}_1 + d_{32}\overline{u}_3 \quad (5.176)$$

At this stage, a suitable set of rotation axis unit vectors can be selected for $\widehat{R}_{12}$. Equation (5.176) suggests that a considerable simplification can be achieved if the mentioned unit vectors are selected so that $\overline{n}_{123} = \overline{u}_1$, $\overline{n}_{121} = \overline{u}_2$, and $\overline{n}_{122} = \overline{u}_3$. Thus, $\widehat{R}_{12}$ comes out to be

$$\widehat{R}_{12} = e^{\widetilde{u}_2\phi_{12}}e^{\widetilde{u}_3\theta_{12}}e^{\widetilde{u}_1\psi_{12}} \quad (5.177)$$

The combination of Eqs. (5.176) and (5.177) results in the following simplified equation after the necessary manipulations.

$$b_{01}\overline{u}_1 + b_{12}e^{-\widetilde{u}_2\theta_{01}}\overline{u}_1 + b_{23}e^{-\widetilde{u}_2\theta_{01}}e^{\widetilde{u}_2\phi_{12}}e^{\widetilde{u}_3\theta_{12}}e^{\widetilde{u}_1\psi_{12}}\overline{u}_1$$

$$= s_{03}e^{\widetilde{u}_3\delta_{03}}\overline{u}_1 + d_{32}\overline{u}_3 \Rightarrow$$

$$b_{01}\overline{u}_1 + b_{12}e^{-\widetilde{u}_2\theta_{01}}\overline{u}_1 + b_{23}e^{\widetilde{u}_2(\phi_{12}-\theta_{01})}e^{\widetilde{u}_3\theta_{12}}\overline{u}_1$$

$$= s_{03}e^{\widetilde{u}_3\delta_{03}}\overline{u}_1 + d_{32}\overline{u}_3 \quad (5.178)$$

Note that, owing to the appropriately selected unit vectors for $\widehat{R}_{12}$, the angle ψ_{12} of the joint $\mathcal{J}_{12}$ has been eliminated and the angles ϕ_{12} and θ_{01} of the joints $\mathcal{J}_{12}$ and $\mathcal{J}_{01}$ have become additive in the resulting simplified equation, i.e. Eq. (5.178).

5.6.3 Degree of Freedom or Mobility

The *degree of freedom* (*DoF*) or *mobility* (μ) of an interconnected rigid body system $\mathcal{S}$ with respect to its base $\mathcal{B}_o$ can be determined by using the following *Kutzbach–Gruebler* formula.

$$\mu = \lambda n - \sum_{k=1}^{\lambda-1}(\lambda - k)j_k \quad (5.179)$$

In Eq. (5.179), $\mu = \mu_{\mathcal{S}/\mathcal{B}_o}$ is the mobility of $\mathcal{S}$ with respect to $\mathcal{B}_o$, n is the number of bodies that are *movable* with respect to $\mathcal{B}_o$ (i.e. the number of bodies other than $\mathcal{B}_o$), j_k is the number of the joints with k degrees of relative freedom between their kinematic elements, and λ is the DoF of the *operational space* of the system. More specifically, λ is the number of the independent parameters that are required to describe the position (location and orientation) of an unconstrained rigid body in the operational space of the system. If the operational space of the system is the three-dimensional Euclidean space, then $\lambda = 6$ due to the three locational and three orientational DoFs of a rigid body. Therefore, for a *spatial system*, i.e. for a system that is designed to operate in the three-dimensional Euclidean space, the Kutzbach–Gruebler formula can be written as

$$\mu = 6n - (5j_1 + 4j_2 + 3j_3 + 2j_4 + j_5) \quad (5.180)$$

For the present mechanism, $n = 3$, $j_1 = 2$ (due to $\mathcal{J}_{01}$ and $\mathcal{J}_{03}$), $j_3 = 2$ (due to $\mathcal{J}_{12}$ and $\mathcal{J}_{23}$), and $j_2 = j_4 = j_5 = 0$. Therefore, the mobility of this mechanism (with respect to the base frame $\mathcal{F}_0$ or the link $\mathcal{B}_o$) comes out to be

$$\mu = 2 \tag{5.181}$$

On this occasion, it must be pointed out that, for a general rigid body system $\mathcal{S}$, μ may consist of two terms as expressed below.

$$\mu = \mu^* + \mu' \tag{5.182}$$

In Eq. (5.182), μ^* is the *primary mobility* of $\mathcal{S}$ and μ' is the *secondary mobility* of $\mathcal{S}$. A case with $\mu' > 0$ occurs, if the system $\mathcal{S}$ is divided into two subsystems $\mathcal{S}^*$ and $\mathcal{S}'$ so that

$$\mathcal{S} = \mathcal{S}^* \cup \mathcal{S}' \tag{5.183}$$

The subsystems in Eq. (5.183) are such that $\mu^* = \mu_{\mathcal{S}^*/\mathcal{B}_o}$ is the mobility of $\mathcal{S}^*$ with respect to $\mathcal{B}_o$ and $\mu' = \mu_{\mathcal{S}'/\mathcal{S}^*}$ is the mobility of $\mathcal{S}'$ with respect to $\mathcal{S}^*$. These subsystems are also assumed to be such that the μ' parameters that describe the relative position of $\mathcal{S}'$ with respect to $\mathcal{S}^*$ are independent of the μ^* parameters that describe the relative position of $\mathcal{S}^*$ with respect to $\mathcal{B}_o$.

A typical system with $\mu' > 0$ is such that it has μ^* actuators, which are dedicated to control the position of $\mathcal{S}^*$ with respect to $\mathcal{B}_o$. In other words, the μ^* actuators of $\mathcal{S}$ are used in order to govern the *gross motion* of $\mathcal{S}$, which is defined as the motion of $\mathcal{S}^*$, while the relative motion of $\mathcal{S}'$ with respect to $\mathcal{S}^*$ is ignored and left free because it has no effect on the gross motion of $\mathcal{S}$. Therefore, for such a system, μ^* and μ' are also called *significant mobility* and *insignificant mobility*, respectively.

The most typical example of a system with $\mu' > 0$ is a planar cam mechanism, in which the motion of the cam is transmitted to the follower through a roller. In this mechanism, $\mathcal{S}'$ is the roller and its rotational motion is left free because it is indeed insignificant near the significant gross motion of the mechanism. This gross motion is the same as the motion of the hypothetical mechanism $\mathcal{S}^*$, which consists of two moving links, namely the cam and the follower that is rigidly incorporated with the roller.

The 2-DoF mechanism considered in this section is also a mechanism with $\mu' > 0$. It is such that $\mu^* = 1$ and $\mu' = 1$. The subsystem $\mathcal{S}'$ is the link $\mathcal{B}_2$. The spinning rotation of $\mathcal{B}_2$ about its centerline, i.e. about $\vec{u}_1^{(2)}$, is the insignificant mobility of this mechanism. Indeed, the angle ψ_{12} of this spinning rotation can be left free to change arbitrarily without having any effect on the gross motion of the mechanism, which is the same as the motion of the hypothetical mechanism $\mathcal{S}^*$. As for $\mathcal{S}^*$, it consists of the links $\mathcal{B}_1$, $\mathcal{B}_3$, and an imaginary link $\mathcal{B}_2^\circ$ that is connected to $\mathcal{B}_1$ with a universal joint (i.e. a joint with 2 degrees of relative freedom), instead of a spherical joint, in order to eliminate the arbitrary spinning rotation.

5.6.4 Position Analysis

Since $\mu^* = 1$, if one of the joint variables (other than ψ_{12}) is specified, then the other unspecified joint variables (except the ones that are associated with the spinning motion of $\mathcal{B}_2$) can be found from the loop closure equations.

(a) Solution of the Loop Closure Equations for a Specified Crank Angle

For a specified value of the crank angle θ_{01}, the locational loop closure equation, i.e. Eq. (5.178), can be used to find the joint variables s_{03}, θ_{12}, and ϕ_{12}. For this purpose, Eq. (5.178) can be arranged as

$$b_{23}e^{\tilde{u}_2\phi'_{12}}e^{\tilde{u}_3\theta_{12}}\overline{u}_1 = s_{03}e^{\tilde{u}_3\delta_{03}}\overline{u}_1 - \overline{p} \tag{5.184}$$

In Eq. (5.184), ϕ'_{12} and $\overline{p}$ are defined as follows in terms of θ_{01}.

$$\phi'_{12} = \phi_{12} - \theta_{01} \tag{5.185}$$

$$\overline{p} = b_{12}e^{-\tilde{u}_2\theta_{01}}\overline{u}_1 + b_{01}\overline{u}_1 - d_{32}\overline{u}_3 \Rightarrow$$

$$\overline{p} = \overline{u}_1 p_1 + \overline{u}_3 p_3 = \overline{u}_1(b_{01} + b_{12}c\theta_{01}) + \overline{u}_3(b_{12}s\theta_{01} - d_{32}) \tag{5.186}$$

The unknown angles ϕ'_{12} and θ_{12} can be eliminated by premultiplying both sides of Eq. (5.184) by their transposes. This operation can be worked out as shown below.

$$b_{23}^2(\overline{u}_1^t e^{-\tilde{u}_3\theta_{12}}e^{-\tilde{u}_2\phi'_{12}})(e^{\tilde{u}_2\phi'_{12}}e^{\tilde{u}_3\theta_{12}}\overline{u}_1)$$

$$= (s_{03}\overline{u}_1^t e^{-\tilde{u}_3\delta_{03}} - \overline{p}^t)(s_{03}e^{\tilde{u}_3\delta_{03}}\overline{u}_1 - \overline{p}) \Rightarrow$$

$$b_{23}^2 = s_{03}^2 - 2(\overline{u}_1^t e^{-\tilde{u}_3\delta_{03}}\overline{p})s_{03} + \overline{p}^t\overline{p} \Rightarrow$$

$$s_{03}^2 - 2(p_1 c\delta_{03})s_{03} + (p_1^2 + p_3^2 - b_{23}^2) = 0 \tag{5.187}$$

Equation (5.187) gives s_{03} as follows with an arbitrary closure sign variable, i.e. $\sigma_1 = \pm 1$:

$$s_{03} = p_1 c\delta_{03} + \sigma_1\sqrt{q_{03}} \tag{5.188}$$

In Eq. (5.188),

$$q_{03} = (p_1 c\delta_{03})^2 - (p_1^2 + p_3^2 - b_{23}^2) \Rightarrow$$

$$q_{03} = b_{23}^2 - p_3^2 - (p_1 s\delta_{03})^2 \tag{5.189}$$

With the availability of s_{03}, Eq. (5.184) can be written compactly as

$$e^{\tilde{u}_3\theta_{12}}\overline{u}_1 = e^{-\tilde{u}_2\phi'_{12}}\overline{m} \tag{5.190}$$

In Eq. (5.190), $\overline{m}$ is a known *unit column matrix* defined as follows:

$$\overline{m} = (s_{03}e^{\tilde{u}_3\delta_{03}}\overline{u}_1 - \overline{p})/b_{23} \Rightarrow$$

$$\overline{m} = \overline{u}_1 m_1 + \overline{u}_2 m_2 + \overline{u}_3 m_3$$

$$= \overline{u}_1(s_{03}c\delta_{03} - p_1)/b_{23} + \overline{u}_2(s_{03}s\delta_{03})/b_{23} - \overline{u}_3 p_3/b_{23} \tag{5.191}$$

Equation (5.190) leads to the following three scalar equations, when it is premultiplied on both sides by $\overline{u}_1^t$, $\overline{u}_2^t$, and $\overline{u}_3^t$.

$$\overline{u}_1^t(\overline{u}_1 c\theta_{12} + \overline{u}_2 s\theta_{12}) = (\overline{u}_1^t c\phi'_{12} - \overline{u}_3^t s\phi'_{12})\overline{m} \Rightarrow$$

$$c\theta_{12} = m_1 c\phi'_{12} - m_3 s\phi'_{12} \tag{5.192}$$

$$\overline{u}_2^t(\overline{u}_1 c\theta_{12} + \overline{u}_2 s\theta_{12}) = \overline{u}_2^t\overline{m} \Rightarrow$$

$$s\theta_{12} = m_2 \tag{5.193}$$

$$\overline{u}_3^t\overline{u}_1 = (\overline{u}_3^t c\phi'_{12} + \overline{u}_1^t s\phi'_{12})\overline{m} \Rightarrow$$

$$0 = m_3 c\phi'_{12} + m_1 s\phi'_{12} \tag{5.194}$$

Recall that $\overline{m}$ is defined as a unit column matrix, i.e. $m_1^2 + m_2^2 + m_3^2 = 1$. Thus, Eq. (5.193) gives $c\theta_{12}$ and θ_{12} as follows with another arbitrary closure sign variable, i.e. $\sigma_2 = \pm 1$:

$$c\theta_{12} = \sigma_2\sqrt{1 - m_2^2} = \sigma_2\sqrt{m_1^2 + m_3^2} \tag{5.195}$$

$$\theta_{12} = \text{atan}_2(m_2, \sigma_2\sqrt{m_1^2 + m_3^2}) \tag{5.196}$$

After finding θ_{12} as shown above, Eqs. (5.192) and (5.194) can be combined into the following matrix equation in order to obtain $c\phi'_{12}$ and $s\phi'_{12}$ separately so that ϕ'_{12} can be found without any additional sign variable.

$$\begin{bmatrix} m_1 & -m_3 \\ m_3 & m_1 \end{bmatrix} \begin{bmatrix} c\phi'_{12} \\ s\phi'_{12} \end{bmatrix} = \begin{bmatrix} c\theta_{12} \\ 0 \end{bmatrix} = \begin{bmatrix} \sigma_2\sqrt{m_1^2 + m_3^2} \\ 0 \end{bmatrix} \tag{5.197}$$

If $m_1^2 + m_3^2 \neq 0$, Eq. (5.197) leads to ϕ'_{12} as shown below.

$$\begin{bmatrix} c\phi'_{12} \\ s\phi'_{12} \end{bmatrix} = \frac{1}{m_1^2 + m_3^2} \begin{bmatrix} m_1 & m_3 \\ -m_3 & m_1 \end{bmatrix} \begin{bmatrix} \sigma_2\sqrt{m_1^2 + m_3^2} \\ 0 \end{bmatrix} = \frac{1}{\sqrt{m_1^2 + m_3^2}} \begin{bmatrix} \sigma_2 m_1 \\ -\sigma_2 m_3 \end{bmatrix} \Rightarrow$$

$$\phi'_{12} = \text{atan}_2(-\sigma_2 m_3, \sigma_2 m_1) \tag{5.198}$$

As for the angle ϕ_{12} itself, it is obtained as follows from Eq. (5.185):

$$\phi_{12} = \phi'_{12} + \theta_{01} \tag{5.199}$$

If $m_1^2 + m_3^2 = 0$, i.e. if $m_1 = m_3 = 0$, the mechanism will be in a pose of *position singularity* associated with the specified angle θ_{01}. In such a pose, ϕ'_{12} and hence ϕ_{12} become indefinite. The consequences of this position singularity will be discussed later in Part (d).

In order to find the three joint variables of the spherical joint $\mathcal{J}_{23}$, Eq. (5.174) can be written as follows:

$$\hat{R}_{23} = \hat{R}_{12}^t e^{\tilde{u}_2\theta_{01}} e^{\tilde{u}_3\delta_{03}} = (e^{\tilde{u}_2\phi_{12}} e^{\tilde{u}_3\theta_{12}} e^{\tilde{u}_1\psi_{12}})^t e^{\tilde{u}_2\theta_{01}} e^{\tilde{u}_3\delta_{03}} \Rightarrow$$

$$\hat{R}_{23} = e^{-\tilde{u}_1\psi_{12}} e^{-\tilde{u}_3\theta_{12}} e^{-\tilde{u}_2\phi'_{12}} e^{\tilde{u}_3\delta_{03}} \tag{5.200}$$

The known expression on the right-hand side of Eq. (5.200) can be manipulated as shown below by using the "rotation about rotated axis" formula of the rotation matrices explained previously in Section 2.7 of Chapter 2.

$$\hat{R}_{23} = e^{-\tilde{u}_1\psi_{12}} e^{-\tilde{u}_3\theta_{12}} (e^{\tilde{u}_3\delta_{03}} e^{-\tilde{u}_3\delta_{03}}) e^{-\tilde{u}_2\phi'_{12}} e^{\tilde{u}_3\delta_{03}} \Rightarrow$$

$$\hat{R}_{23} = e^{-\tilde{u}_1\psi_{12}} (e^{-\tilde{u}_3\theta_{12}} e^{\tilde{u}_3\delta_{03}})(e^{-\tilde{u}_3\delta_{03}} e^{-\tilde{u}_2\phi'_{12}} e^{\tilde{u}_3\delta_{03}}) \Rightarrow$$

$$\hat{R}_{23} = e^{-\tilde{u}_1\psi_{12}} e^{-\tilde{u}_3(\theta_{12}-\delta_{03})} e^{-\tilde{n}_2\phi'_{12}} \Rightarrow$$

$$e^{\tilde{n}_{231}\phi_{23}} e^{\tilde{n}_{232}\theta_{23}} e^{\tilde{n}_{233}\psi_{23}} = e^{-\tilde{u}_1\psi_{12}} e^{-\tilde{u}_3(\theta_{12}-\delta_{03})} e^{-\tilde{n}_2\phi'_{12}} \tag{5.201}$$

In Eq. (5.201), $\overline{n}_2$ is defined as

$$\overline{n}_2 = e^{-\tilde{u}_3\delta_{03}} \overline{u}_2 = \overline{u}_2 c\delta_{03} + \overline{u}_1 s\delta_{03} \tag{5.202}$$

Equation (5.201) suggests that the following triad happens to be the most favorable selection for the rotation axis unit vectors of $\hat{R}_{23}$.

$$\{\overline{n}_{231} = \overline{u}_1, \overline{n}_{232} = \overline{u}_3, \overline{n}_{233} = \overline{n}_2\} \tag{5.203}$$

Here, it is to be noted that the axes associated with a rotation matrix need not always be represented by one of the basic column matrices $\overline{u}_1$, $\overline{u}_2$, and $\overline{u}_3$, as seen above. The above selection leads to

$$\widehat{R}_{23} = e^{\widetilde{u}_1 \phi_{23}} e^{\widetilde{u}_3 \theta_{23}} e^{\widetilde{n}_2 \psi_{23}} \tag{5.204}$$

Hence, simply by comparing Eqs. (5.201) and (5.204), it can be directly seen that the three joint variables of $\mathcal{J}_{23}$ assume the following values.

$$\phi_{23} = -\psi_{12} \tag{5.205}$$

$$\theta_{23} = -(\theta_{12} - \delta_{03}) \tag{5.206}$$

$$\psi_{23} = -\phi'_{12} = -(\phi_{12} - \theta_{01}) \tag{5.207}$$

However, as mentioned before, ψ_{12} is indefinite due to the arbitrary spinning motion of $\mathcal{B}_2$. Therefore, ϕ_{23} is also indefinite according to Eq. (5.205).

(b) Existence of Solutions with Continuous Crank Rotation

According to Eq. (5.188), the existence of s_{03} as a real number necessitates that the following condition be satisfied.

$$q_{03} \geq 0 \tag{5.208}$$

When Eq. (5.189) is substituted, Inequality (5.208) becomes

$$\begin{aligned} & b_{23}^2 \geq p_3^2 + p_1^2 (s\delta_{03})^2 \Rightarrow \\ & b_{23}^2 \geq (b_{12} s\theta_{01} - d_{32})^2 + (b_{01} + b_{12} c\theta_{01})^2 (s\delta_{03})^2 \Rightarrow \\ & b_{23}^2 \geq b_{12}^2 [(s\theta_{01})^2 + (s\delta_{03})^2 (c\theta_{01})^2] + 2b_{12}[b_{01}(s\delta_{03})^2 c\theta_{01} - d_{32} s\theta_{01})] \\ & \qquad + d_{32}^2 + b_{01}^2 (s\delta_{03})^2 \end{aligned} \tag{5.209}$$

For a continuous operation of the mechanism, Inequality (5.209) must be satisfied for all values of the crank angle θ_{01} in the interval $[0, 2\pi]$. In other words, b_{23} (the length of the connecting rod BC) must be long enough so that Inequality (5.209) is satisfied even when θ_{01} assumes the critical value θ_{01}^* that maximizes the value of the expression on the right-hand side of Inequality (5.209). This critical value can be found from the following equation, which is obtained by equating the derivative of the right-hand side of Inequality (5.209) to zero.

$$b_{12}(c\delta_{03})^2 s\theta_{01}^* c\theta_{01}^* = b_{01}(s\delta_{03})^2 s\theta_{01}^* + d_{32} c\theta_{01}^* \tag{5.210}$$

(c) Multiple Solutions for the Same Crank Angle

The alternative values of the closure sign variables σ_1 in Eq. (5.188) and σ_2 in Eqs. (5.196) and (5.198) result in multiple solutions, which represent different closure poses of the mechanism corresponding to the same value of the specified joint variable θ_{01}. The multiple solutions are described and discussed below.

* First Kind of Multiplicity

Referring to Eq. (5.188), let $s_{03} = s_{03}^+$ if $\sigma_1 = +1$ and $s_{03} = s_{03}^-$ if $\sigma_1 = -1$. According to Eq. (5.188), $s_{03}^+ > s_{03}^-$ if $q_{03} > 0$. This shows that the alternative values of σ_1 represent two *visually distinct* closure poses of the mechanism whenever $q_{03} > 0$. Interpreting geometrically, the joint centers C^+ (with $\sigma_1 = +1$) and C^- (with $\sigma_1 = -1$) happen to be the intersection points between the sphere with center B and radius BC_2 and the straight line that is parallel to the guideway of the slider and passes through the point C_3. As a special case, if it happens that $q_{03} = 0$ at an instant when $\theta_{01} = \theta_{01}^*$, which is the critical value defined in Part (c), then the two different closure poses mentioned above become the same at that instant. The two distinct closure poses of the mechanism for the same crank angle are illustrated in Figure 5.7 as projected onto the 1–2 plane of the base frame. Here, a due remark is that the desired closures of a mechanism and the corresponding closure signs are selected when the mechanism is assembled. After the assembly, the closures do not change during the operation of the mechanism, unless the geometric parameters of the mechanism allow it to get into a special critical pose that is common for the two closures, e.g. unless $q_{03} = 0$ as discussed in the previous paragraph.

∗ **Second Kind of Multiplicity**

Unlike the visually distinct closure poses of the mechanism represented by the alternative values of σ_1, the closure poses represented by the alternative values of σ_2 are *not* visually distinct. To verify this fact, let the angle pairs $\{\theta_{12}^+, \phi_{12}^+\}$ and $\{\theta_{12}^-, \phi_{12}^-\}$ correspond to $\sigma_2 = +1$ and $\sigma_2 = -1$, respectively. Then, Eq. (5.196) and Eq. Pair (5.198)–(5.199) imply that

$$\theta_{12}^- = \pi - \theta_{12}^+ \tag{5.211}$$

$$\phi_{12}^- = \phi_{12}^+ + \pi \tag{5.212}$$

The angle pairs $\{\theta_{12}^+, \phi_{12}^+\}$ and $\{\theta_{12}^-, \phi_{12}^-\}$ make the unit vector $\vec{u}_1^{(2)}$ along the connecting rod BC assume the orientations that are expressed below for the same value of θ_{01}.

$$\begin{aligned}\overline{u}_1^{(2/0)+} &= \widehat{C}^{(0,2)+}\overline{u}_1^{(2/2)} = \widehat{C}^{(0,1)}\widehat{C}^{(1,2)+}\overline{u}_1 = e^{-\tilde{u}_2\theta_{01}}\widehat{R}_{12}^+\overline{u}_1 \\ &= e^{-\tilde{u}_2\theta_{01}}e^{\tilde{u}_2\phi_{12}^+}e^{\tilde{u}_3\theta_{12}^+}\overline{u}_1 = e^{\tilde{u}_2(\phi_{12}^+-\theta_{01})}e^{\tilde{u}_3\theta_{12}^+}\overline{u}_1\end{aligned} \tag{5.213}$$

$$\begin{aligned}\overline{u}_1^{(2/0)-} &= \widehat{C}^{(0,2)-}\overline{u}_1^{(2/2)} = \widehat{C}^{(0,1)}\widehat{C}^{(1,2)-}\overline{u}_1 = e^{-\tilde{u}_2\theta_{01}}\widehat{R}_{12}^-\overline{u}_1 \\ &= e^{-\tilde{u}_2\theta_{01}}e^{\tilde{u}_2\phi_{12}^-}e^{\tilde{u}_3\theta_{12}^-}\overline{u}_1 = e^{\tilde{u}_2(\phi_{12}^--\theta_{01})}e^{\tilde{u}_3\theta_{12}^-}\overline{u}_1\end{aligned} \tag{5.214}$$

After substituting Eqs. (5.211) and (5.212), Eq. (5.214) can be manipulated as shown below.

$$\begin{aligned}\overline{u}_1^{(2/0)-} &= e^{\tilde{u}_2(\phi_{12}^++\pi-\theta_{01})}e^{\tilde{u}_3(\pi-\theta_{12}^+)}\overline{u}_1 \Rightarrow \\ \overline{u}_1^{(2/0)-} &= e^{\tilde{u}_2(\phi_{12}^+-\theta_{01})}e^{\tilde{u}_2\pi}e^{-\tilde{u}_3\theta_{12}^+}e^{\tilde{u}_3\pi}\overline{u}_1 \Rightarrow \\ \overline{u}_1^{(2/0)-} &= e^{\tilde{u}_2(\phi_{12}^+-\theta_{01})}e^{\tilde{u}_3\theta_{12}^+}e^{\tilde{u}_2\pi}e^{\tilde{u}_3\pi}\overline{u}_1\end{aligned} \tag{5.215}$$

Note that

$$e^{\tilde{u}_3\pi}\overline{u}_1 = -\overline{u}_1 \text{ and } e^{\tilde{u}_2\pi}\overline{u}_1 = -\overline{u}_1$$

Therefore, Eq. (5.215) leads to the result that

$$\overline{u}_1^{(2/0)-} = e^{\tilde{u}_2(\phi_{12}^+-\pi)}e^{\tilde{u}_3\theta_{12}^+}\overline{u}_1 = \overline{u}_1^{(2/0)+} \tag{5.216}$$

Equation (5.216) verifies that the angle pairs $\{\theta_{12}^+, \phi_{12}^+\}$ and $\{\theta_{12}^-, \phi_{12}^-\}$ are geometrically equivalent because both of them lead to the same closure pose of the mechanism. In other words, as verified here, the closure poses represented by the alternative values of σ_2 are not visually distinct. Therefore, the selection $\sigma_2 = +1$ can be used without loss of generality.

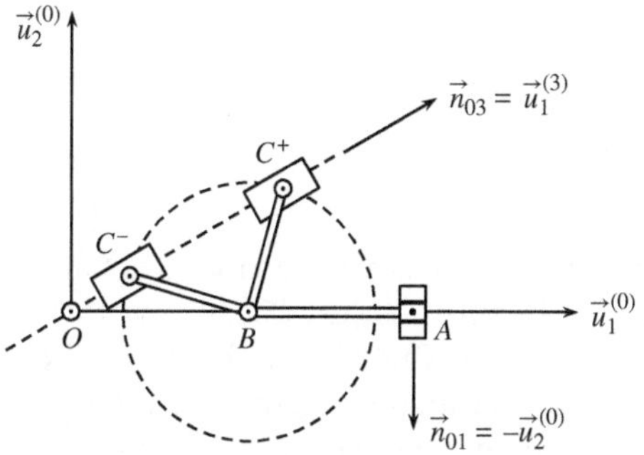

Figure 5.7 Projected views of the two closure poses for the same crank angle.

(d) Position Singularity Associated with the Crank Angle

As mentioned before during the solution for the angle ϕ_{12}, the position singularity of the mechanism occurs if $m_1 = m_3 = 0$. Referring to Eqs. (5.186) and (5.191), the singularity conditions can be expressed in more detail as shown below.

$$m_1 = 0 \Rightarrow p_1 = s_{03}c\delta_{03} \Rightarrow$$

$$s_{03}c\delta_{03} = b_{01} + b_{12}c\theta_{01} \tag{5.217}$$

$$m_3 = 0 \Rightarrow p_3 = 0 \Rightarrow$$

$$b_{12}s\theta_{01} = d_{32} \tag{5.218}$$

Since $|\overline{m}| = 1$, when $m_1 = m_3 = 0$, Eq. (5.191) also implies that

$$m_2 = \sigma_3 = \pm 1 \Rightarrow$$

$$s_{03}s\delta_{03} = \sigma_3 b_{23} \tag{5.219}$$

In the pose of singularity, ϕ_{12} becomes indefinite as implied by Eq. (5.197). In this special pose, Eq. (5.193) and Eqs. (5.217) – (5.219) give the particular value of θ_{01} to be specified and the corresponding values of θ_{12} and s_{03} with the following expressions.

$$s\theta_{12} = \sigma_3 \Rightarrow$$

$$\theta_{12} = \sigma_3\pi/2 \tag{5.220}$$

$$s\theta_{01} = d_{32}/b_{12} \Rightarrow c\theta_{01} = \sigma_0\sqrt{b_{12}^2 - d_{32}^2}/b_{12}; \sigma_0 = \pm 1 \Rightarrow$$

$$\theta_{01} = \text{atan}_2(d_{32}, \sigma_0\sqrt{b_{12}^2 - d_{32}^2}) \tag{5.221}$$

$$s_{03} = \sigma_3 b_{23}s\delta_{03} + (b_{01} + b_{12}c\theta_{01})c\delta_{03} \Rightarrow$$

$$s_{03} = b_{01}c\delta_{03} + \sigma_3 b_{23}s\delta_{03} + \sigma_0 b_{12}c\delta_{03}\sqrt{b_{12}^2 - d_{32}^2} \tag{5.222}$$

With $\theta_{12} = \sigma_3\pi/2$ and $\phi'_{12} = \phi_{12} - \theta_{01}$, the locational loop closure equation, i.e. Eq. (5.178), can be manipulated as follows:

$$b_{01}\overline{u}_1 + b_{12}e^{-\tilde{u}_2\theta_{01}}\overline{u}_1 + b_{23}e^{\tilde{u}_2\phi'_{12}}e^{\tilde{u}_3\sigma_3\pi/2}\overline{u}_1$$
$$= s_{03}e^{\tilde{u}_3\delta_{03}}\overline{u}_1 + d_{32}\overline{u}_3 \Rightarrow$$

$$b_{01}\overline{u}_1 + b_{12}e^{-\tilde{u}_2\theta_{01}}\overline{u}_1 + \sigma_3 b_{23}e^{\tilde{u}_2\phi'_{12}}\overline{u}_2$$
$$= s_{03}e^{\tilde{u}_3\delta_{03}}\overline{u}_1 + d_{32}\overline{u}_3 \Rightarrow$$

$$
\begin{aligned}
&b_{01}\overline{u}_1 + b_{12}e^{-\tilde{u}_2\theta_{01}}\overline{u}_1 + \sigma_3 b_{23}\overline{u}_2 \\
&\quad = s_{03}e^{\tilde{u}_3\delta_{03}}\overline{u}_1 + d_{32}\overline{u}_3 \Rightarrow \\
&\overline{u}_1(b_{01} + b_{12}c\theta_{01}) + \overline{u}_2(\sigma_3 b_{23}) + \overline{u}_3(b_{12}s\theta_{01}) \\
&\quad = \overline{u}_1 s_{03}c\delta_{03} + \overline{u}_2 s_{03}s\delta_{03} + \overline{u}_3 d_{32}
\end{aligned}
\tag{5.223}
$$

Note that, as expected, Eq. (5.223) leads to the same three scalar equations obtained before as Eqs. (5.217)–(5.219). Note also that ϕ'_{12} disappears in Eq. (5.223) and therefore it becomes arbitrary. Thus, ϕ_{12} becomes arbitrary, too. This happens because, when $\theta_{12} = \sigma_3\pi/2$, i.e. when the lines BC and AB become perpendicular, the common axis of the angles θ_{01} and ϕ_{12} gets aligned with the connecting rod, i.e. with the axis of the arbitrary spin angle ψ_{12}. Thus, ϕ_{12} becomes undistinguishable from ψ_{12} and therefore it also becomes arbitrary like ψ_{12}.

(e) Solution of the Loop Closure Equations for a Specified Slider Position

For a specified position of the slider, i.e. for a specified value of s_{03}, the locational loop closure equation, i.e. Eq. (5.178), can be used this time to find the joint variables θ_{01}, θ_{12}, and ϕ_{12}. For this purpose, Eq. (5.178) can be arranged as

$$
b_{23}e^{\tilde{u}_2\phi_{12}}e^{\tilde{u}_3\theta_{12}}\overline{u}_1 = e^{\tilde{u}_2\theta_{01}}\overline{r} - b_{12}\overline{u}_1 \tag{5.224}
$$

In Eq. (5.224), $\overline{r}$ is defined as follows in terms of s_{03}:

$$
\begin{aligned}
&\overline{r} = s_{03}e^{\tilde{u}_3\delta_{03}}\overline{u}_1 + d_{32}\overline{u}_3 - b_{01}\overline{u}_1 \Rightarrow \\
&\overline{r} = \overline{u}_1 r_1 + \overline{u}_2 r_2 + \overline{u}_3 r_3 \\
&\quad = \overline{u}_1(s_{03}c\delta_{03} - b_{01}) + \overline{u}_2 s_{03}s\delta_{03} + \overline{u}_3 d_{32}
\end{aligned}
\tag{5.225}
$$

The unknown angles ϕ_{12} and θ_{12} can be eliminated by premultiplying both sides of Eq. (5.224) by their transposes. This operation can be worked out as shown below.

$$
\begin{aligned}
&b_{23}^2(\overline{u}_1^t e^{-\tilde{u}_3\theta_{12}}e^{-\tilde{u}_2\phi_{12}})(e^{\tilde{u}_2\phi_{12}}e^{\tilde{u}_3\theta_{12}}\overline{u}_1) \\
&\quad = (\overline{r}^t e^{-\tilde{u}_2\theta_{01}} - b_{12}\overline{u}_1^t)(e^{\tilde{u}_2\theta_{01}}\overline{r} - b_{12}\overline{u}_1) \Rightarrow \\
&b_{23}^2 = \overline{r}^t\overline{r} + b_{12}^2 - 2b_{12}(\overline{u}_1^t e^{\tilde{u}_2\theta_{01}}\overline{r}) \Rightarrow \\
&2b_{12}(\overline{u}_1^t c\theta_{01} + \overline{u}_3^t s\theta_{01})\overline{r} = \overline{r}^t\overline{r} + b_{12}^2 - b_{23}^2 \Rightarrow \\
&r_1 c\theta_{01} + r_3 s\theta_{01} = h_{03}
\end{aligned}
\tag{5.226}
$$

In Eq. (5.226),

$$
\begin{aligned}
&h_{03} = (\overline{r}^t\overline{r} + b_{12}^2 - b_{23}^2)/(2b_{12}) \Rightarrow \\
&h_{03} = (s_{03}^2 - 2b_{01}s_{03}c\delta_{03} + b_{01}^2 + d_{32}^2 + b_{12}^2 - b_{23}^2)/(2b_{12}) \Rightarrow \\
&h_{03} = [(s_{03} - b_{01}c\delta_{03})^2 + (b_{01}s\delta_{03})^2 + b_{12}^2 + d_{32}^2 - b_{23}^2]/(2b_{12})
\end{aligned}
\tag{5.227}
$$

If $r_1^2 + r_3^2 \neq 0$, Eq. (5.226) can be solved for θ_{01} by introducing r_{13} and β_{01} so that

$$
r_1 = r_{13}c\beta_{01}, r_3 = r_{13}s\beta_{01} \tag{5.228}
$$

Equation Pair (5.228) implies that

$$
r_{13} = \sqrt{r_1^2 + r_3^2} \tag{5.229}
$$

$$
\beta_{01} = \text{atan}_2(r_3, r_1) \tag{5.230}
$$

By using r_{13} and β_{01}, Eq. (5.226) can be converted into

$$
\cos(\theta_{01} - \beta_{01}) = h_{03}/r_{13} \tag{5.231}
$$

Equation (5.231) implies the following equation with an arbitrary closure sign variable, i.e. $\sigma_1 = \pm 1$.

$$\sin(\theta_{01} - \beta_{01}) = \sigma_1 \sqrt{r_{13}^2 - h_{03}^2}/r_{13} \tag{5.232}$$

Provided that $r_{13}^2 = r_1^2 + r_3^2 \geq h_{03}^2$, Eqs. (5.231) and (5.232) give θ_{01} as follows:

$$\theta_{01} - \beta_{01} = \text{atan}_2(\sigma_1 \sqrt{r_{13}^2 - h_{03}^2}, h_{03}) = \sigma_1 \text{atan}_2(\sqrt{r_{13}^2 - h_{03}^2}, h_{03}) \Rightarrow$$

$$\theta_{01} = \beta_{01} + \sigma_1 \text{atan}_2(\sqrt{r_{13}^2 - h_{03}^2}, h_{03}) \tag{5.233}$$

With the availability of θ_{01}, Eq. (5.224) can be written compactly as

$$e^{\tilde{u}_3\theta_{12}}\overline{u}_1 = e^{-\tilde{u}_2\phi_{12}}\overline{n} \tag{5.234}$$

In Eq. (5.234), $\overline{n}$ is a known *unit column matrix* defined as follows:

$$\begin{aligned}
&\overline{n} = (e^{\tilde{u}_2\theta_{01}}\overline{r} - b_{12}\overline{u}_1)/b_{23} \Rightarrow \\
&\overline{n} = \overline{u}_1 n_1 + \overline{u}_2 n_2 + \overline{u}_3 n_3 \\
&\quad = \overline{u}_1(r_1 c\theta_{01} + r_3 s\theta_{01} - b_{12})/b_{23} + \overline{u}_2(r_2/b_{23}) \\
&\qquad + \overline{u}_3(r_3 c\theta_{01} - r_1 s\theta_{01})/b_{23}
\end{aligned} \tag{5.235}$$

The procedure to obtain ϕ_{12} and θ_{12} from Eq. (5.234) is the same as the procedure described in Part (a) to obtain ϕ'_{12} and θ_{12} from Eq. (5.190). The only difference is that $\overline{m}$ is replaced with $\overline{n}$. As in Part (a), there will be two visually indistinct solutions, which are represented by the two values of another closure sign variable, e.g. $\sigma_2 = \pm 1$, provided that $n_1^2 + n_3^2 \neq 0$.

The treatment of $\hat{R}_{23}$ to find the associated joint variables is also the same as that described in Part (a).

(f) Existence of Solutions

According to Eq. (5.233), the existence of θ_{01} as a real number necessitates that the following condition be satisfied by the specified value of s_{03}.

$$r_{13}^2 = r_1^2 + r_3^2 \geq h_{03}^2 \tag{5.236}$$

When Eqs. (5.225) and (5.227) are used, Inequality (5.236) takes the following detailed form, which contains s_{03} explicitly.

$$\begin{aligned}
&4b_{12}^2[(s_{03}c\delta_{03} - b_{01})^2 + d_{32}^2] \\
&\geq [(s_{03} - b_{01}c\delta_{03})^2 + (b_{01}s\delta_{03})^2 + b_{12}^2 + d_{32}^2 - b_{23}^2]^2
\end{aligned} \tag{5.237}$$

(g) Multiple Solutions for the Same Slider Position

The alternative values of the closure sign variable σ_1 in Eq. (5.233) result in multiple solutions, which represent different closure poses of the mechanism corresponding to the same value of the specified joint variable s_{03} as long as $r_1^2 + r_3^2 > h_{03}^2$. In order to distinguish these closure poses, let $\theta_{01} = \theta_{01}^+$ if $\sigma_1 = +1$ and $\theta_{01} = \theta_{01}^-$ if $\sigma_1 = -1$. According to Eq. (5.233), $\theta_{01}^+ > \theta_{01}^-$. Interpreting geometrically, these visually distinct poses are such

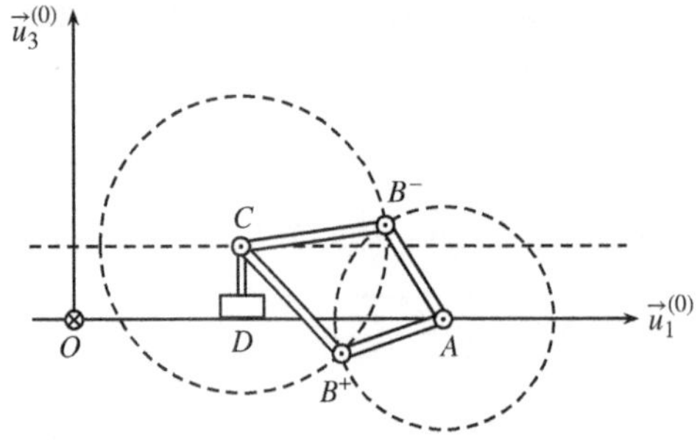

Figure 5.8 Projected views of the two closure poses for the same slider position.

that the joint centers B^+ (with $\sigma_1 = +1$) and B^- (with $\sigma_1 = -1$) happen to be the intersection points between the sphere with center C and radius $CB_2 = b_{23}$ and the circle with center A and radius $AB_1 = b_{12}$. As a special case, if $r_1^2 + r_3^2 = h_{03}^2$, then the two poses become the same, in which $\theta_{01} = \beta_{01} = \text{atan}_2(r_3, r_1)$. In such a case, the above-mentioned sphere and circle intersect each other tangentially at a single point. The two distinct closure poses of the mechanism for the same slider position are illustrated in Figure 5.8 as projected onto the 1–3 plane of the base frame.

On the other hand, the other closure sign variable σ_2 that arises in solving Eq. (5.234) for ϕ_{12} and θ_{12} does not lead to visually distinct closure poses of the mechanism. The fact that these poses are visually indistinct can be verified similarly as in Part (c). Therefore, σ_2 can again be selected as $\sigma_2 = +1$ without loss of generality.

(h) Position Singularities Associated with the Slider Position

For particularly specified values of s_{03}, there may occur two kinds of position singularities.

* First Kind of Position Singularity

This singularity occurs if $n_1^2 + n_3^2 = 0$, i.e. if $n_1 = n_3 = 0$, in Eq. (5.234). If it occurs, it causes ϕ_{12} to become indefinite. Meanwhile, according to Eq. (5.235), θ_{01} assumes the value that satisfies the following equations.

$$n_1 = n_3 = 0 \Rightarrow$$
$$r_1 c\theta_{01} + r_3 s\theta_{01} = b_{12} \tag{5.238}$$
$$r_3 c\theta_{01} - r_1 s\theta_{01} = 0 \tag{5.239}$$

In this pose of singularity, since $|\overline{n}| = 1$, it also happens that

$$n_2 = \sigma_3 = \pm 1 \Rightarrow r_2 = \sigma_3 b_{23} \Rightarrow s_{03} s\delta_{03} = \sigma_3 b_{23} \Rightarrow$$
$$s_{03} = \sigma_3 b_{23} / s\delta_{03} \tag{5.240}$$

Equation (5.240) gives the particular value of s_{03} to be specified in this pose of singularity. On the other hand, Eqs. (5.238) and (5.239) imply that $r_1^2 + r_3^2 = b_{12}^2$. This result in turn implies that $r_1^2 + r_3^2 \neq 0$ because $b_{12} \neq 0$. Therefore, the same equations give $s\theta_{01}$, $c\theta_{01}$, and θ_{01} as follows:

$$s\theta_{01} = b_{12} r_3 / (r_1^2 + r_3^2),\ c\theta_{01} = b_{12} r_1 / (r_1^2 + r_3^2) \Rightarrow$$
$$\theta_{01} = \text{atan}_2(r_3, r_1) \tag{5.241}$$

It is to be recalled from Part (g) that the preceding value of θ_{01} occurs in the special case such that the points B^+ and B^- become coincident. Moreover, by inserting the expressions of r_3 and r_1 given in Eq. (5.225), it can be shown that $s\theta_{01} = d_{32}/b_{12}$ as given by Eq. (5.218). This implies that the present pose of singularity is the same as the pose of singularity described in Part (d). In such a pose, the axis of the angle ϕ_{12} gets aligned with the connecting rod, i.e. with the axis of the arbitrary spin angle ψ_{12}. Thus, ϕ_{12} becomes indistinguishable from ψ_{12} and therefore it also becomes arbitrary like ψ_{12}.

∗ **Second Kind of Position Singularity**

This singularity occurs if $r_1 = r_3 = 0$ in Eq. (5.226). If it occurs, it causes θ_{01} to become indefinite. According to Eq. (5.225), the occurrence of this singularity is possible in a rather special case described by the following equations.

$$d_{32} = 0 \tag{5.242}$$

$$s_{03}c\delta_{03} = b_{01} \tag{5.243}$$

Equation (5.226) implies that $h_{03} = 0$, too. That is,

$$b_{23}^2 = (s_{03} - b_{01}c\delta_{03})^2 + (b_{01}s\delta_{03})^2 + b_{12}^2 + d_{32}^2 \tag{5.244}$$

When Eqs. (5.242) and (5.243) are substituted, Eq. (5.244) can be manipulated so that

$$b_{23}^2 = (s_{03}s\delta_{03})^2 + b_{12}^2 \Rightarrow$$

$$s_{03}s\delta_{03} = \sigma_4\sqrt{b_{23}^2 - b_{12}^2} = \sigma_4 d_{AC}; \sigma_4 = \pm 1 \tag{5.245}$$

Equations (5.243) and (5.245) imply that

$$s_{03} = \sigma_5\sqrt{b_{01}^2 + d_{AC}^2}; \sigma_5 = \pm 1 \tag{5.246}$$

$$\delta_{03} = \text{atan}_2(\sigma_5\sigma_4 d_{AC}, \sigma_5 b_{01}) = \sigma_4\text{atan}_2(\sigma_5 d_{AC}, \sigma_5 b_{01}) \tag{5.247}$$

Note that this singularity can occur only if $d_{32} = 0$, i.e. if the points C and D are coincident. In a pose of this singularity, the line AC becomes aligned with the axis of the first joint according to Eqs. (5.242)–(5.247). Therefore, θ_{01} can change arbitrarily due to the fact that the triangle ABC can rotate freely about the line AC while the slider is kept fixed at the location described by $s_{03} = \sigma_5\sqrt{b_{01}^2 + d_{AC}^2}$, which is given by Eq. (5.246).

5.6.5 Velocity Analysis

The velocity relationships can be obtained most directly by differentiating the loop closure equations that are manipulated into the forms ready for solution. Thus, the following velocity equation can be obtained, when the locational loop closure equation, i.e. Eq. (5.178), is differentiated.

$$-b_{12}\dot{\theta}_{01}e^{-\tilde{u}_2\theta_{01}}\tilde{u}_2\overline{u}_1 + b_{23}\dot{\phi}'_{12}e^{\tilde{u}_2\phi'_{12}}\tilde{u}_2e^{\tilde{u}_3\theta_{12}}\overline{u}_1$$

$$+ b_{23}\dot{\theta}_{12}e^{\tilde{u}_2\phi'_{12}}e^{\tilde{u}_3\theta_{12}}\tilde{u}_3\overline{u}_1 = \dot{s}_{03}e^{\tilde{u}_3\delta_{03}}\overline{u}_1 \tag{5.248}$$

In Eq. (5.248), $\phi'_{12} = \phi_{12} - \theta_{01}$ as defined before. Equation (5.248) can be manipulated as shown below.

$$
\begin{aligned}
&b_{12}\dot{\theta}_{01}e^{-\widetilde{u}_2\theta_{01}}\overline{u}_3 + b_{23}\dot{\phi}'_{12}e^{\widetilde{u}_2\phi'_{12}}\widetilde{u}_2(\overline{u}_1 c\theta_{12} + \overline{u}_2 s\theta_{12}) \\
&\quad + b_{23}\dot{\theta}_{12}e^{\widetilde{u}_2\phi'_{12}}e^{\widetilde{u}_3\theta_{12}}\overline{u}_2 = \dot{s}_{03}e^{\widetilde{u}_3\delta_{03}}\overline{u}_1 \Rightarrow \\
&b_{12}\dot{\theta}_{01}e^{-\widetilde{u}_2\theta_{01}}\overline{u}_3 - b_{23}\dot{\phi}'_{12}e^{\widetilde{u}_2\phi'_{12}}\overline{u}_3 c\theta_{12} \\
&\quad + b_{23}\dot{\theta}_{12}e^{\widetilde{u}_2\phi'_{12}}e^{\widetilde{u}_3\theta_{12}}\overline{u}_2 = \dot{s}_{03}e^{\widetilde{u}_3\delta_{03}}\overline{u}_1 \Rightarrow \\
&b_{12}\dot{\theta}_{01}e^{-\widetilde{u}_2\theta_{01}}\overline{u}_3 + b_{23}e^{\widetilde{u}_2\phi'_{12}}e^{\widetilde{u}_3\theta_{12}}(\overline{u}_2\dot{\theta}_{12} - \overline{u}_3\dot{\phi}'_{12}c\theta_{12}) = \dot{s}_{03}e^{\widetilde{u}_3\delta_{03}}\overline{u}_1
\end{aligned}
$$

The preceding equation can be arranged further as follows:

$$
\begin{aligned}
&b_{12}\dot{\theta}_{01}e^{-\widetilde{u}_3\theta_{12}}e^{-\widetilde{u}_2\phi_{12}}\overline{u}_3 + b_{23}(\overline{u}_2\dot{\theta}_{12} - \overline{u}_3\dot{\phi}'_{12}c\theta_{12}) \\
&\quad = \dot{s}_{03}e^{-\widetilde{u}_3\theta_{12}}e^{-\widetilde{u}_2\phi'_{12}}e^{\widetilde{u}_3\delta_{03}}\overline{u}_1
\end{aligned} \tag{5.249}
$$

When Eq. (5.249) is premultiplied by $\overline{u}_1^t$, $\overline{u}_2^t$, and $\overline{u}_3^t$, the following scalar equations are obtained.

$$(\overline{u}_1^t e^{-\widetilde{u}_3\theta_{12}}e^{-\widetilde{u}_2\phi_{12}}\overline{u}_3)b_{12}\dot{\theta}_{01} = (\overline{u}_1^t e^{-\widetilde{u}_3\theta_{12}}e^{-\widetilde{u}_2\phi'_{12}}e^{\widetilde{u}_3\delta_{03}}\overline{u}_1)\dot{s}_{03} \tag{5.250}$$

$$(\overline{u}_2^t e^{-\widetilde{u}_3\theta_{12}}e^{-\widetilde{u}_2\phi_{12}}\overline{u}_3)b_{12}\dot{\theta}_{01} + b_{23}\dot{\theta}_{12} = (\overline{u}_2^t e^{-\widetilde{u}_3\theta_{12}}e^{-\widetilde{u}_2\phi'_{12}}e^{\widetilde{u}_3\delta_{03}}\overline{u}_1)\dot{s}_{03} \tag{5.251}$$

$$(\overline{u}_3^t e^{-\widetilde{u}_2\phi_{12}}\overline{u}_3)b_{12}\dot{\theta}_{01} - b_{23}\dot{\phi}'_{12}c\theta_{12} = (\overline{u}_3^t e^{-\widetilde{u}_2\phi'_{12}}e^{\widetilde{u}_3\delta_{03}}\overline{u}_1)\dot{s}_{03} \tag{5.252}$$

The preceding equations can be written compactly as follows by denoting the coefficients with complicated expressions simply as f_{11}, f_{13}, etc.

$$f_{13}b_{12}\dot{\theta}_{01} = f_{11}\dot{s}_{03} \tag{5.253}$$

$$f_{23}b_{12}\dot{\theta}_{01} + b_{23}\dot{\theta}_{12} = f_{21}\dot{s}_{03} \tag{5.254}$$

$$f_{33}b_{12}\dot{\theta}_{01} - b_{23}\dot{\phi}'_{12}c\theta_{12} = f_{31}\dot{s}_{03} \tag{5.255}$$

As for the orientational loop closure equation of the present mechanism, it had been simplified into Eqs. (5.205)–(5.207) owing to the special selection of the rotation axis unit vectors for the matrix $\widehat{R}_{23}$. The derivatives of those equations result in the following velocity relationships that give the rates of the angles of $\widehat{R}_{23}$ directly in terms of the rates of the other angles.

$$\dot{\phi}_{23} = -\dot{\psi}_{12} \tag{5.256}$$

$$\dot{\theta}_{23} = -\dot{\theta}_{12} \tag{5.257}$$

$$\dot{\psi}_{23} = -\dot{\phi}'_{12} \tag{5.258}$$

(a) Solution of the Velocity Equations for a Specified Crank Motion

For a specified variation of the crank angle θ_{01}, Eqs. (5.253)–(5.255) can be used to find the joint variable rates $\dot{s}_{03}$, $\dot{\theta}_{12}$, and $\dot{\phi}'_{12}$ in terms of $\dot{\theta}_{01}$ and the current values of the joint variables. The solution is described below.

If $f_{11} \neq 0$, $\dot{s}_{03}$ is found from Eq. (5.253) as

$$\dot{s}_{03} = (f_{13}/f_{11})b_{12}\dot{\theta}_{01} \tag{5.259}$$

With the availability of $\dot{s}_{03}$, $\dot{\theta}_{12}$ is found from Eq. (5.254) and $\dot{\phi}'_{12}$ is found from Eq. (5.255) if $c\theta_{12} \neq 0$. That is,

$$\dot{\theta}_{12} = (f_{21}\dot{s}_{03} - f_{23}b_{12}\dot{\theta}_{01})/b_{23} \tag{5.260}$$

$$\dot{\phi}'_{12} = (f_{33}b_{12}\dot{\theta}_{01} - f_{31}\dot{s}_{03})/(b_{23}c\theta_{12}) \tag{5.261}$$

Afterwards, $\dot{\phi}_{12}$ is obtained as

$$\dot{\phi}_{12} = \dot{\phi}'_{12} + \dot{\theta}_{01} \tag{5.262}$$

Finally, Eqs. (5.256)–(5.258) give the rates of the remaining angles as functions of $\dot{\theta}_{01}$ and the indefinite rate $\dot{\psi}_{12}$.

(b) Motion Singularities Associated with the Crank Motion

There may occur two kinds of motion singularities when the mechanism is attempted to be driven by means of the crank.

*** First Kind of Motion Singularity**

This singularity occurs if

$$\begin{aligned} f_{11} &= \overline{u}_1^t e^{-\tilde{u}_3\theta_{12}} e^{-\tilde{u}_2\phi'_{12}} e^{\tilde{u}_3\delta_{03}} \overline{u}_1 \\ &= c\theta_{12}c\phi'_{12}c\delta_{03} + s\theta_{12}s\delta_{03} = 0 \end{aligned} \tag{5.263}$$

A typical pose of this singularity is described by the following values of ϕ'_{12} and θ_{12}.

$$\phi'_{12} = 0 \tag{5.264}$$

$$\theta_{12} = \delta_{03} \pm \pi/2 \tag{5.265}$$

In such a pose, Eq. (5.253) implies that $\dot{s}_{03}$ becomes indefinite and $\dot{\theta}_{01}$ vanishes. In other words, θ_{01} gets fixed and the mechanism can no longer be driven by means of the crank, but meanwhile the slider gains an uncontrollable mobility, i.e. $\dot{s}_{03}$ becomes indefinite. Due to Eqs. (5.254) and (5.255), $\dot{\theta}_{12}$ and $\dot{\phi}'_{12}$ also become indefinite due to the indefiniteness of $\dot{s}_{03}$.

*** Second Kind of Motion Singularity**

This singularity occurs if

$$c\theta_{12} = 0 \ \text{ or } \ \theta_{12} = \pm\pi/2 \tag{5.266}$$

In such a pose of singularity, Eq. (5.255) implies that $\dot{\phi}'_{12}$ becomes indefinite. However, due to the vanishing term with $\dot{\phi}'_{12}$, Eqs. (5.253) and (5.255) turn into the following consistency equations to be satisfied by $\dot{\theta}_{01}$ and $\dot{s}_{03}$.

$$f_{13}b_{12}\dot{\theta}_{01} = f_{11}\dot{s}_{03} \tag{5.267}$$

$$f_{33}b_{12}\dot{\theta}_{01} = f_{31}\dot{s}_{03} \tag{5.268}$$

Note that, unless $f_{13}/f_{33} = f_{11}/f_{31}$, Eqs. (5.265) and (5.266) can be satisfied only if $\dot{\theta}_{01} = 0$ and $\dot{s}_{03} = 0$. In other words, in this pose of singularity too, the mechanism can no longer be driven by means of the crank and it gets stuck in a pose described by certain fixed values of θ_{01} and s_{03}. Since $\dot{\theta}_{12} = 0$ too due to Eq. (5.254), the only mobility that remains is the uncontrollable spinning rotation of the connecting rod. Note also that Eqs. (5.265) and (5.220) indicate that the mechanism assumes the same pose at the instants of the position singularity and the motion singularity of the second kind that are associated with the crank angle. Therefore, the fixed values of θ_{01} and s_{03} are the same as those given by Eqs. (5.221) and (5.222).

(c) Solution of the Velocity Equations for a Specified Slider Motion

For a specified variation of the slider position described by s_{03}, Eqs. (5.253)–(5.255) can be used to find the joint variable rates $\dot{\theta}_{01}$, $\dot{\theta}_{12}$, and $\dot{\phi}'_{12}$ in terms of $\dot{s}_{03}$ and the current values of the joint variables. The solution is described below.
If $f_{13} \neq 0$, $\dot{\theta}_{01}$ is found from Eq. (5.253) as

$$\dot{\theta}_{01} = (f_{11}\dot{s}_{03})/(f_{13}b_{12}) \tag{5.269}$$

The rest of the solution for the other joint variable rates is very similar to the solution described in Part (a).

(d) Motion Singularities Associated with the Slider Motion

There may occur two kinds of motion singularities when the mechanism is attempted to be driven by means of the slider.

*** First Kind of Motion Singularity**

This singularity occurs if

$$f_{13} = \overline{u}_1^t e^{-\tilde{u}_3\theta_{12}} e^{-\tilde{u}_2\phi_{12}} \overline{u}_3 = -c\theta_{12} s\phi_{12} = 0 \tag{5.270}$$

Equation (5.270) is satisfied if either or both of θ_{12} and ϕ_{12} assume the following values.

$$\theta_{12} = \pm\pi/2 \tag{5.271}$$

$$\phi_{12} = 0 \text{ or } \phi_{12} = \pm\pi \tag{5.272}$$

In a pose of this singularity, Eq. (5.253) implies that $\dot{\theta}_{01}$ becomes indefinite and $\dot{s}_{03}$ vanishes. In other words, s_{03} gets fixed and the mechanism can no longer be driven by means of the slider, but meanwhile the crank gains an uncontrollable mobility. Due to Eqs. (5.254) and (5.255), $\dot{\theta}_{12}$ and $\dot{\phi}'_{12}$ also become indefinite under the effect of $\dot{\theta}_{01}$.

*** Second Kind of Motion Singularity**

This singularity associated with the slider motion is very similar to that associated with the crank motion, which is discussed in Part (b).

5.6.6 Acceleration Analysis

The acceleration relationships can be obtained directly by differentiating the velocity equations that are manipulated into the forms ready for solution. Thus, the following acceleration equations can be obtained, when the velocity equations, i.e. Eqs. (5.253)–(5.258), are differentiated.

$$f_{13}b_{12}\ddot{\theta}_{01} = f_{11}\ddot{s}_{03} + \mathring{a}_1 \tag{5.273}$$

$$f_{23}b_{12}\ddot{\theta}_{01} + b_{23}\ddot{\theta}_{12} = f_{21}\ddot{s}_{03} + \mathring{a}_2 \tag{5.274}$$

$$f_{33}b_{12}\ddot{\theta}_{01} - b_{23}\ddot{\phi}'_{12}c\theta_{12} = f_{31}\ddot{s}_{03} + \mathring{a}_3 \tag{5.275}$$

$$\ddot{\phi}_{23} = -\ddot{\psi}_{12} \tag{5.276}$$

$$\ddot{\theta}_{23} = -\ddot{\theta}_{12} \tag{5.277}$$

$$\ddot{\psi}_{23} = -\ddot{\phi}'_{12} \tag{5.278}$$

In Eqs. (5.273)–(5.275), the velocity-dependent acceleration terms are defined as follows:

$$a_1^\circ = \dot{f}_{11}\dot{s}_{03} - \dot{f}_{13}b_{12}\dot{\theta}_{01} \tag{5.279}$$

$$a_2^\circ = \dot{f}_{21}\dot{s}_{03} - \dot{f}_{23}b_{12}\dot{\theta}_{01} \tag{5.280}$$

$$a_3^\circ = \dot{f}_{31}\dot{s}_{03} + b_{23}\dot{\phi}'_{12}\dot{\theta}_{12}s\theta_{12} - \dot{f}_{33}b_{12}\dot{\theta}_{01} \tag{5.281}$$

Note that Eqs. (5.273)–(5.278) are very similar to the velocity equations. Therefore, the acceleration analysis can be carried out similarly as the velocity analysis.

6

Joints and Their Kinematic Characteristics

Synopsis

This chapter is devoted to the kinematic features of a wide set of joints that are used in various mechanical systems formed by rigid body members. The joints are conceived as kinematic pairs that consist of mating kinematic elements. The kinematic elements are represented by the joint frames attached to them. The joint frames are attached in such a way that the relative position (i.e. location and orientation) of one kinematic element with respect to its mate is expressed in terms of the joint variable or variables by means of the simplest possible equations. These equations are called here characteristic equations. In this treatment, the characteristic equations are preferably written separately for the relative location and orientation between the kinematic elements. By doing so, the effect of the joint variable or variables on the relative location and orientation can be expressed explicitly by displaying all the kinematic details. Thus, the subsequent symbolic manipulations can be performed conveniently in order to obtain analytical or semi-analytical solutions for the unspecified variables and to carry out detailed kinematic analyses involving the multiple and singular poses of the considered system. The last section of this chapter is supplemented with several examples of mechanical systems that demonstrate how some of the typical joints are used with their joint frames and characteristic equations.

6.1 Kinematic Details of the Joints

6.1.1 Description of a Joint as a Kinematic Pair

A joint $\mathcal{J}_{ab}$ (or $\mathcal{J}_{ba}$) between two members $\mathcal{B}_a$ and $\mathcal{B}_b$ of a mechanical system consists of a pair of mating *kinematic elements* $\mathcal{E}_{ab}$ and $\mathcal{E}_{ba}$ that belong to $\mathcal{B}_a$ and $\mathcal{B}_b$, respectively. That is why a joint is also called a *kinematic pair*.

The kinematic elements $\mathcal{E}_{ab}$ and $\mathcal{E}_{ba}$ are represented by the reference frames $\mathcal{F}_{ab}$ and $\mathcal{F}_{ba}$ that are attached to them. The origins of $\mathcal{F}_{ab}$ and $\mathcal{F}_{ba}$ are denoted as O_{ab} and O_{ba}. Their unit basis vector triads are denoted as shown below.

$$\mathcal{U}_{ab} = \{\vec{u}_i^{(ab)},\ \vec{u}_j^{(ab)},\ \vec{u}_k^{(ab)}\} \text{ and } \mathcal{U}_{ba} = \{\vec{u}_i^{(ba)},\ \vec{u}_j^{(ba)},\ \vec{u}_k^{(ba)}\} \tag{6.1}$$

Kinematics of General Spatial Mechanical Systems, First Edition. M. Kemal Ozgoren.

Companion Website: www.wiley.com/go/ozgoren/spatialmechanicalsystems

Here, for the sake of generality, the unit basis vectors are denoted by the indices i, j, and k, which are ordered as follows in order to have the property of right-handedness:

$$ijk \in \{123, 231, 312\} \tag{6.2}$$

However, most of the time, $\mathcal{F}_{ab}$ and $\mathcal{F}_{ba}$ are oriented so that $ijk = 123$. Yet, in some special cases, a different ordering (such as $ijk = 231$ or $ijk = 312$) may simplify the kinematic relationships between the joint frames (e.g. $\mathcal{F}_{ab}$) and the accomodating body frames (e.g. $\mathcal{F}_a$). Such a special case occurs in the first example (which is about the arm of a serial manipulator) presented previously in Chapter 5.

The reference frames $\mathcal{F}_{ab}$ and $\mathcal{F}_{ba}$ are attached to $\mathcal{E}_{ab}$ and $\mathcal{E}_{ba}$ in such a way that the relative position (i.e. location and orientation) of $\mathcal{F}_{ba}$ with respect to $\mathcal{F}_{ab}$ (or vice versa) is expressed in the simplest possible form in terms of the relevant joint variable or variables. The location and orientation equations written for such an arrangement of $\mathcal{F}_{ab}$ and $\mathcal{F}_{ba}$ are called *characteristic equations* of the joint $\mathcal{J}_{ab}$ (or $\mathcal{J}_{ba}$).

6.1.2 Degree of Freedom or Mobility of a Joint

The *degree of freedom (DoF)* or *mobility* of a joint $\mathcal{J}_{ab}$ is denoted here as μ_{ab} and defined as the number of *independent* joint variables that are required to describe the relative location and orientation of $\mathcal{E}_{ba}$ with respect to $\mathcal{E}_{ab}$ completely. In a *spatial mechanical system*, i.e. in a mechanical system that works in the three-dimensional Euclidean space, a movable joint can have a mobility such that $\mu_{ab} = 1$ at least and $\mu_{ab} = 5$ at most.

6.1.3 Number of Distinct Joints Between Two Rigid Bodies

In most of the cases, two members of a mechanical system such as $\mathcal{B}_a$ and $\mathcal{B}_b$ are connected by only one joint denoted as $\mathcal{J}_{ab}$ (or $\mathcal{J}_{ba}$). However, there are also some cases, in which $\mathcal{B}_a$ and $\mathcal{B}_b$ are connected by several (e.g. p) joints, which form a *joint set* as $\mathcal{G}_{ab} = \{\mathcal{J}_{ab}^{(k)} : k = 1, 2, \ldots, p\}$. Typical examples of one-joint and two-joint connections are illustrated in Figure 6.1.

Here, it is to be mentioned that any two elements, e.g. $\mathcal{J}_{ab}^{(i)}$ and $\mathcal{J}_{ab}^{(j)}$, of a joint set $\mathcal{G}_{ab}$ can be considered to be distinct if their kinematic characteristics are independent of each other. Otherwise, $\mathcal{J}_{ab}^{(i)}$ and $\mathcal{J}_{ab}^{(j)}$ can be merged into a single kinematically equivalent joint. For example, two revolute joints that share a common axis can be replaced with a single revolute joint that also has the same axis. Similarly, two prismatic joints that have axes parallel to a unit vector $\vec{n}$ can be replaced with a single prismatic joint whose axis is also parallel to $\vec{n}$.

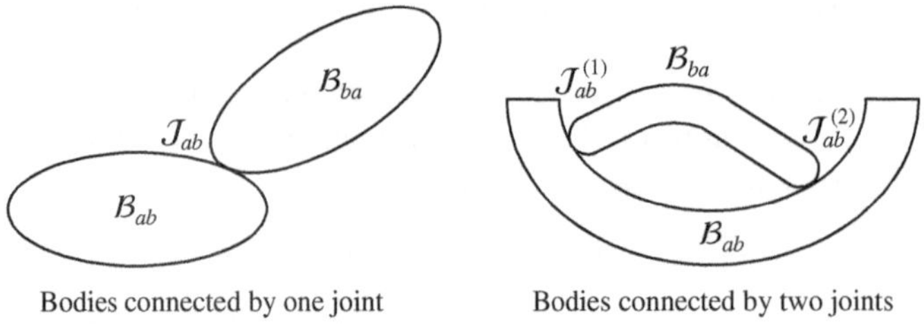

Figure 6.1 Bodies connected by one and two joints.

Another point to be mentioned about a connection formed by several joints is that the resultant relative mobility (μ_{ab}) between $\mathcal{B}_a$ and $\mathcal{B}_b$ must be such that $\mu_{ab} \geq 1$. Referring to Chapter 5 for the Kutzbach–Gruebler formula, i.e. Eq. (5.179), μ_{ab} can be expressed as follows assuming that the mechanical system is spatial with $\lambda = 6$ and considering that there is only one moving body (i.e. $\mathcal{B}_b$) with respect to the reference body (i.e. $\mathcal{B}_a$).

$$\mu_{ab} = \lambda \times 1 - \sum_{k=1}^{p} [\lambda - \mu_{ab}^{(k)}] = 6 - \sum_{k=1}^{p} [6 - \mu_{ab}^{(k)}] \Rightarrow$$

$$\mu_{ab} = \sum_{k=1}^{p} \mu_{ab}^{(k)} - 6(p-1) \tag{6.3}$$

In Eq. (6.3), $\mu_{ab}^{(k)}$ is the mobility of $\mathcal{J}_{ab}^{(k)}$. According to Eq. (6.3), the requirement that $\mu_{ab} \geq 1$ leads to the following condition on the mobilities of the joints in the set $\mathcal{G}_{ab}$.

$$\sum_{k=1}^{p} \mu_{ab}^{(k)} \geq 6p - 5 \tag{6.4}$$

For example, considering the two-joint connection illustrated in Figure 6.1, both $\mathcal{J}_{ab}^{(1)}$ and $\mathcal{J}_{ab}^{(2)}$ are spatial cam joints with $\mu_{ab}^{(1)} = \mu_{ab}^{(2)} = 5$. Therefore, with $p = 2$, Inequality (6.4) is satisfied as $10 > 7$ and Eq. (6.3) gives the resultant relative mobility as $\mu_{ab} = 4$. Note that this result is the same as the expected result that can be obtained by inspection.

6.1.4 Classification of the Joints

The joints can be classified into three main categories as follows according to the nature of the contact between the kinematic elements:

- Surface contact joints
- Line contact joints
- Point contact joints

The joints can also be classified as follows according to the expected level of the contact stresses that arise during a statically or dynamically loaded operation of the system:

- Lower order joints
- Higher order joints

The two classifications described above are interrelated so that the surface contact joints happen to be lower order joints, whereas the line contact and point contact joints happen to be higher order joints.

At a surface contact joint, the kinematic elements contact each other on a surface, which may be planar, cylindrical, or spherical. Typical examples of surface contact joints are prismatic, revolute, cylindrical, spherical, and planar contact joints.

At a line contact joint, the kinematic elements contact each other along a straight line. Typical examples of line contact joints are a cam pair formed by two cylindrical cams, a cam pair formed by a cylindrical cam and a knife-edge follower, a spur gear pair, a bevel gear pair, and the joint between a scraper and a plate.

At a point contact joint, the kinematic elements contact each other at a point. Typical examples of point contact joints are a cam pair formed by an arbitrary cam and a pin-point follower, a cam pair formed by two ellipsoidal cams, and the joint between a pen and a tablet. The line and point contact joints can be classified further into two categories as follows:

- Simple contact joints
- Rolling contact joints

A simple contact joint is such that the contact point or the contact line remains fixed on one of the kinematic elements. Typical examples are a cam pair formed by an arbitrary cam and a pin-point follower, a cam pair formed by a cylindrical cam and a knife-edge follower, the joint between a pen and a tablet, and the joint between a scraper and a plate.

A rolling contact joint is such that the contact point or the contact line does not remain fixed on any of the two kinematic elements. Typical examples are a cam pair formed by two ellipsoidal cams, a cam pair formed by two cylindrical cams, a spur gear pair, a rack-and-pinion pair, a bevel gear pair, the joint between a wheel and the ground, and the joint between a cone and a plate.

At a rolling contact joint, the relative rolling motion between the kinematic elements may be with or without slipping. The slipping can be prevented either by means of friction as in the case of a wheel on a dry road or by means of teeth as in the case of gear pairs.

When the slipping is prevented by means of friction or meshing teeth, the motion restriction induced by such means happens to be a *nonholonomic constraint*. A nonholonomic constraint does not affect the relative positioning of the kinematic elements when they are brought together during the assembly stage, but it becomes effective only when the kinematic elements start to move with respect to each other.

A variety of typical joints from all the categories mentioned above are described in the following sections together with their characteristic equations. The DOFs of the considered joints range from $\mu_{ab} = 1$ to $\mu_{ab} = 5$.

6.2 Typical Lower Order Joints

6.2.1 Single-Axis Joints

Figure 6.2 shows a schematic illustration of a general single-axis joint. At a single-axis joint, the kinematic elements $\mathcal{E}_{ab}$ and $\mathcal{E}_{ba}$ can rotate and/or translate with respect to each other about and along the same axis, which is represented by a unit vector $\vec{n}_{ab}$. The reference frames $\mathcal{F}_{ab}$ and $\mathcal{F}_{ba}$ are attached to the kinematic elements $\mathcal{E}_{ab}$ and $\mathcal{E}_{ba}$ so that both $\vec{u}_k^{(ab)}$ and $\vec{u}_k^{(ba)}$ are aligned with $\vec{n}_{ab}$. Thus, the orientation and location of $\mathcal{F}_{ba}$ with respect to $\mathcal{F}_{ab}$ are expressed as shown below in terms of the joint variables θ_{ab} and s_{ab}.

$$\text{rot}[\vec{u}_k^{(ab)}, \theta_{ab}] : \vec{u}_i^{(ab)} \to \vec{u}_i^{(ba)} \text{ and } \vec{u}_j^{(ab)} \to \vec{u}_j^{(ba)} \tag{6.5}$$

$$\vec{r}_{ab,ba} = \overrightarrow{O_{ab}O_{ba}} = \vec{u}_k^{(ab)} s_{ab} \tag{6.6}$$

Noting that $\overline{u}_k^{(ab/ab)} = \overline{u}_k$, the relative orientation and location of $\mathcal{F}_{ba}$ with respect to $\mathcal{F}_{ab}$ are expressed by the following matrix equations in $\mathcal{F}_{ab}$.

$$\hat{C}^{(ab,ba)} = e^{\tilde{u}_k \theta_{ab}} \tag{6.7}$$

$$\overline{r}_{ab,ba}^{(ab)} = \overline{u}_k s_{ab} \tag{6.8}$$

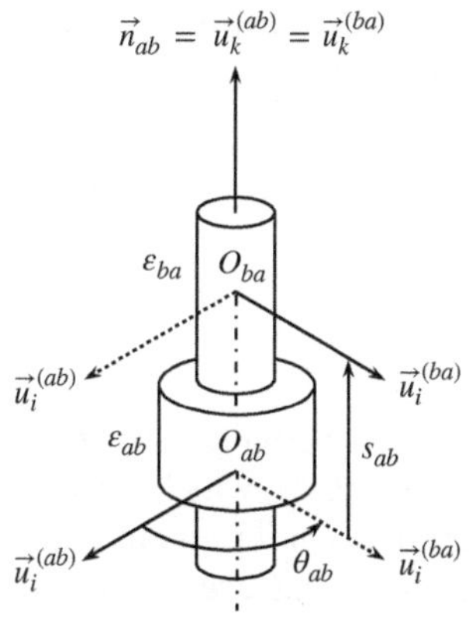

Figure 6.2 A general single-axis joint.

There are four types of single-axis joints. They are described below.

(a) Cylindrical Joint

A cylindrical joint is such that both θ_{ab} and s_{ab} are independent joint variables. Therefore, its mobility is $\mu_{ab} = 2$. As such, it is characterized by Eqs. (6.7) and (6.8) without having any constraint on them.

(b) Revolute Joint

A revolute joint is also known as a *rotary joint* or a *hinge joint*. Its mobility is $\mu_{ab} = 1$ due to having $s_{ab} = d_{ab} =$ constant as caused by appropriate physical constraints. However, this joint can usually be arranged so that $d_{ab} = 0$. Then, with $d_{ab} = 0$ and the only joint variable θ_{ab}, this joint is characterized by the following equations.

$$\widehat{C}^{(ab,ba)} = e^{\widetilde{u}_k \theta_{ab}} \tag{6.9}$$

$$\overline{r}^{(ab)}_{ab,ba} = \overline{0} \tag{6.10}$$

(c) Prismatic Joint

A prismatic joint is also known as a *sliding joint*. Its mobility is also $\mu_{ab} = 1$ due to having $\theta_{ab} = \delta_{ab} =$ constant as caused by appropriate physical constraints. However, this joint can usually be arranged so that $\delta_{ab} = 0$. Then, with $\delta_{ab} = 0$ and the only joint variable s_{ab}, this joint is characterized by the following equations.

$$\widehat{C}^{(ab,ba)} = \widehat{I} \tag{6.11}$$

$$\overline{r}^{(ab)}_{ab,ba} = \overline{u}_k s_{ab} \tag{6.12}$$

(d) Screw Joint

A screw joint is also known as a *helical joint*. It is another joint with mobility $\mu_{ab} = 1$ due to the following relationship between the joint variables s_{ab} and θ_{ab}.

$$s_{ab} = \lambda_{ab}\theta_{ab} + d_{ab} \tag{6.13}$$

In Eq. (6.13), λ_{ab} is a constant coefficient, which is known as the *lead* or *pitch* of the screw joint. As for d_{ab}, it is the constant *offset* between the kinematic elements. Usually, the joint can be arranged so that $d_{ab} = 0$. Then, by using Eq. (6.13) with $d_{ab} = 0$, the screw joint can be characterized by the following equations, in which θ_{ab} is taken as the principal joint variable.

$$\hat{C}^{(ab,ba)} = e^{\tilde{u}_k \theta_{ab}} \tag{6.14}$$

$$\bar{r}^{(ab)}_{ab,ba} = \bar{u}_k \lambda_{ab} \theta_{ab} \tag{6.15}$$

6.2.2 Universal Joint

A universal joint is also known as a *Hooke joint* or a *Cardan joint*. It actually consists of two independent revolute joints that connect three kinematic elements, which are $\mathcal{E}_{ab}$, $\mathcal{E}_{bab}$, and $\mathcal{E}_{ba}$. Here, $\mathcal{E}_{bab}$ is an intermediate kinematic element between the main kinematic elements $\mathcal{E}_{ab}$ and $\mathcal{E}_{ba}$, which belong to the members $\mathcal{B}_a$ and $\mathcal{B}_b$ that are connected by the universal joint. Because of the two independent revolute joints, the mobility of a universal joint is $\mu_{ab} = 2$.

There are basically two versions of a universal joint, which are distinguished by the arrangement of the axes of the first and second revolute joints. The two versions are illustrated in Figure 6.3. For both versions, the characteristic location equation is

$$\bar{r}^{(ab)}_{ab,ba} = \bar{0} \tag{6.16}$$

For the version shown on the left-hand side of Figure 6.3, the rotation sequence is

$$\mathcal{F}_{ab} \xrightarrow{\text{rot}[\vec{u}_k^{(ab)} = \vec{u}_k^{(bab)}, \phi_{ab}]} \mathcal{F}_{bab} \xrightarrow{\text{rot}[\vec{u}_i^{(bab)} = \vec{u}_i^{(ba)}, \theta_{ab}]} \mathcal{F}_{ba} \tag{6.17}$$

The corresponding orientation equation is

$$\hat{C}^{(ab,ba)} = e^{\tilde{u}_k \phi_{ab}} e^{\tilde{u}_i \theta_{ab}} \tag{6.18}$$

For the version shown on the right-hand side of Figure 6.3, the rotation sequence is

$$\mathcal{F}_{ab} \xrightarrow{\text{rot}[\vec{u}_i^{(ab)} = \vec{u}_i^{(bab)}, \theta_{ab}]} \mathcal{F}_{bab} \xrightarrow{\text{rot}[\vec{u}_k^{(bab)} = \vec{u}_k^{(ba)}, \phi_{ab}]} \mathcal{F}_{ba} \tag{6.19}$$

The corresponding orientation equation is

$$\hat{C}^{(ab,ba)} = e^{\tilde{u}_i \theta_{ab}} e^{\tilde{u}_k \phi_{ab}} \tag{6.20}$$

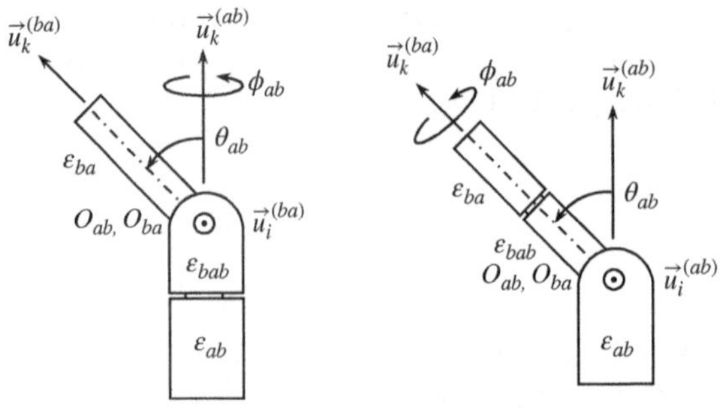

Figure 6.3 Two versions of a universal joint.

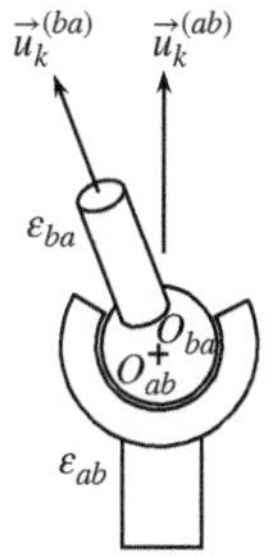

Figure 6.4 A spherical joint.

6.2.3 Spherical Joint

A spherical joint is also known as a *globular joint* or a *ball-and-socket joint*. Indeed, it consists of two kinematic elements that have the shapes of a ball and a socket as illustrated in Figure 6.4. Within the encasement of the socket, the ball can rotate freely about any arbitrary direction. Therefore, the mobility of a spherical joint is $\mu_{ab} = 3$, which is only due to the rotational freedom. Due to the encasement of the socket, the characteristic location equation is

$$\bar{r}_{ab,ba}^{(ab)} = \bar{0} \tag{6.21}$$

As for the characteristic orientation equation, it can be written in the following general form in terms of three independent angular joint variables $(\phi_{ab}, \theta_{ab}, \psi_{ab})$.

$$\hat{C}^{(ab,ba)} = e^{\tilde{n}_{ab1}\phi_{ab}} e^{\tilde{n}_{ab2}\theta_{ab}} e^{\tilde{n}_{ab3}\psi_{ab}} \tag{6.22}$$

Equation (6.22) is based on the following rotation sequence with two hypothetical intermediate kinematic elements $\mathcal{E}_{bc}$ and $\mathcal{E}_{cb}$.

$$\mathcal{F}_{ab} \xrightarrow{\text{rot}[\vec{n}_{ab1},\phi_{ab}]} \mathcal{F}_{bc} \xrightarrow{\text{rot}[\vec{n}_{ab2},\theta_{ab}]} \mathcal{F}_{cb} \xrightarrow{\text{rot}[\vec{n}_{ab3},\psi_{ab}]} \mathcal{F}_{ba} \tag{6.23}$$

In Description (6.23), $\vec{n}_{ab1}$, $\vec{n}_{ab2}$, and $\vec{n}_{ab3}$ are the unit vectors along the relevant rotation axes. In writing Eq. (6.22), these unit vectors are specified (without any loss of generality) according to the RFB (rotating frame based) convention so that

$$\left.\begin{aligned} \bar{n}_{ab1} &= \bar{n}_{ab1}^{(ab)} = \bar{n}_{ab1}^{(bc)} = \text{constant} \\ \bar{n}_{ab2} &= \bar{n}_{ab2}^{(bc)} = \bar{n}_{ab2}^{(cb)} = \text{constant} \\ \bar{n}_{ab3} &= \bar{n}_{ab3}^{(cb)} = \bar{n}_{ab3}^{(ba)} = \text{constant} \end{aligned}\right\} \tag{6.24}$$

Here, it is worth noting that it is not a favorable practice to prespecify $\bar{n}_{ab1}$, $\bar{n}_{ab2}$, and $\bar{n}_{ab3}$ when the spherical joint stands alone by itself. Actually, it happens to be much more favorable to specify them judiciously in order to simplify the expressions of the orientation relationships that involve the considered spherical joint together with the members it connects. The advantage of doing so is demonstrated for the spherical joint of the RRRSP mechanism studied in Example 6.1 (Section 6.7.1) of this chapter. It is also demonstrated for the two spherical joints of the spatial slider-crank mechanism studied in Example 5.2 (Section 5.6) of Chapter 5.

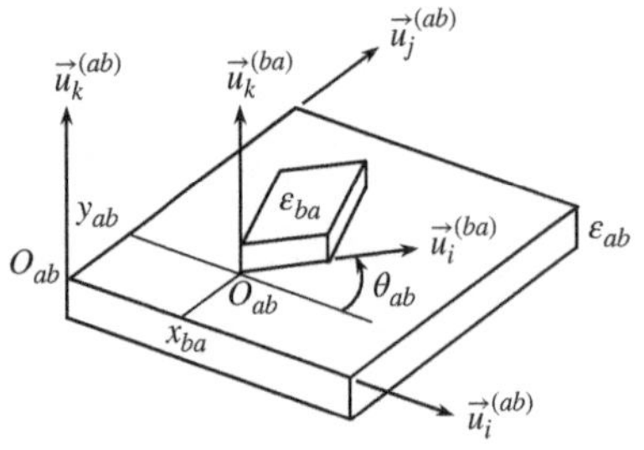

Figure 6.5 A plane-on-plane joint.

6.2.4 Plane-on-Plane Joint

A plane-on-plane joint is illustrated in Figure 6.5. It is also known as a *planar contact joint*. It is called so because the kinematic elements $\mathcal{E}_{ab}$ and $\mathcal{E}_{ba}$ are in contact with each other on their $i-j$ planes. As such, its mobility is $\mu_{ab} = 3$ and it is characterized by the following orientation and location equations, which are written in terms of the joint variables x_{ab}, y_{ab}, and θ_{ab}.

$$\hat{C}^{(ab,ba)} = e^{\tilde{u}_k \theta_{ab}} \tag{6.25}$$

$$\overline{r}_{ab,ba}^{(ab)} = \overline{u}_i x_{ab} + \overline{u}_j y_{ab} \tag{6.26}$$

6.3 Higher Order Joints with Simple Contacts

6.3.1 Line-on-Plane Joint

A line-on-plane joint is illustrated in Figure 6.6. It is so called because the kinematic elements $\mathcal{E}_{ab}$ and $\mathcal{E}_{ba}$ are in contact with each other on a straight line along $\vec{u}_i^{(ba)}$. It may also be called a *line-plane contact joint*. With respect to $\mathcal{F}_{ab}$, O_{ba} moves in the plane formed by $\vec{u}_i^{(ab)}$ and $\vec{u}_j^{(ab)}$. Meanwhile, $\mathcal{F}_{ba}$ rotates sequentially first about $\vec{u}_k^{(ab)}$ and then about $\vec{u}_i^{(ba)}$ as if there is a universal joint centered at O_{ba}. Therefore, the mobility of a line-on-plane joint is $\mu_{ab} = 4$ and it is characterized by the following orientation and location equations, which are written in terms of a set of four joint variables such as $\{\phi_{ab}, \theta_{ab}, x_{ab}, y_{ab}\}$.

$$\hat{C}^{(ab,ba)} = e^{\tilde{u}_k \phi_{ab}} e^{\tilde{u}_i \theta_{ab}} \tag{6.27}$$

$$\overline{r}_{ab,ba}^{(ab)} = \overline{u}_i x_{ab} + \overline{u}_j y_{ab} \tag{6.28}$$

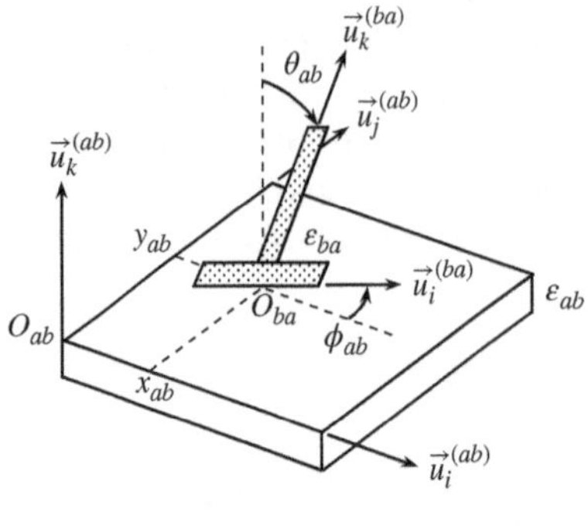

Figure 6.6 A line-on-plane joint.

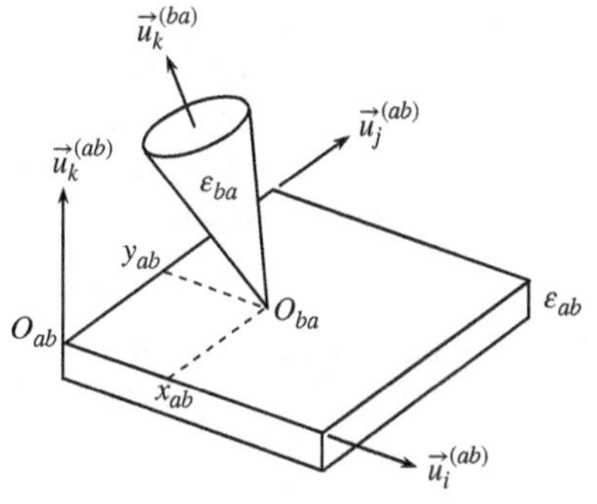

Figure 6.7 A point-on-plane joint.

6.3.2 Point-on-Plane Joint

A point-on-plane joint is illustrated in Figure 6.7. It is so called because the kinematic elements $\mathcal{E}_{ab}$ and $\mathcal{E}_{ba}$ are in contact with each other at a single point O_{ba}. It may also be called a *point-plane contact joint*. With respect to $\mathcal{F}_{ab}$, O_{ba} moves in the plane formed by $\vec{u}_i^{(ab)}$ and $\vec{u}_j^{(ab)}$. Meanwhile, $\mathcal{F}_{ba}$ rotates as if there is a spherical joint centered at O_{ba}. Therefore, the mobility of a point-on-plane joint is $\mu_{ab} = 5$ and it is characterized by the following orientation and location equations, which involve a set of five joint variables such as $\{\phi_{ab}, \theta_{ab}, \psi_{ab}, x_{ab}, y_{ab}\}$.

$$\widehat{C}^{(ab,ba)} = e^{\widetilde{n}_{ab1}\phi_{ab}} e^{\widetilde{n}_{ab2}\theta_{ab}} e^{\widetilde{n}_{ab3}\psi_{ab}} \tag{6.29}$$

$$\overline{r}_{ab,ba}^{(ab)} = \overline{u}_i x_{ab} + \overline{u}_j y_{ab} \tag{6.30}$$

The unit column matrices $\overline{n}_{ab1}$, $\overline{n}_{ab2}$, and $\overline{n}_{ab3}$ that take place in Eq. (6.29) are selected according to similar considerations explained in Section 6.2.3 for a spherical joint.

6.3.3 Point-on-Surface Joint

A point-on-surface joint is illustrated in Figure 6.8. It is so called because $\mathcal{E}_{ab}$ and $\mathcal{E}_{ba}$ are in contact with each other at a single point O_{ba} and $\mathcal{E}_{ab}$ has the shape of a specified surface S_{ab}. The reference frame $\mathcal{F}_{ab}$ is to be selected in such a way that the function $f_{ab}(x, y, z)$ that describes the surface S_{ab} has the simplest expression. For example, if the

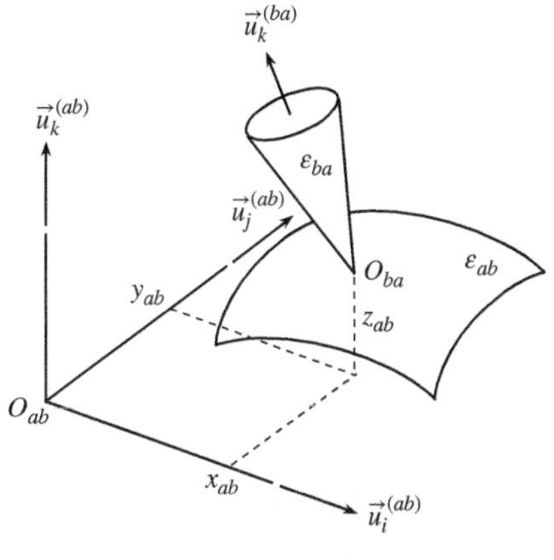

Figure 6.8 A point-on-surface joint.

surface is ellipsoidal, then O_{ab} must be placed at the center of the ellipsoid and the basis vectors of $\mathcal{F}_{ab}$ must be aligned with the principal axes of the ellipsoid. With respect to $\mathcal{F}_{ab}$, the contact point O_{ba} moves so as to remain on the surface S_{ab}. Meanwhile, $\mathcal{F}_{ba}$ rotates as if there is a spherical joint centered at O_{ba}. Therefore, the mobility of this joint is $\mu_{ab} = 5$ even though it is represented by the following set of six joint variables.

$$\{x_{ab}, y_{ab}, z_{ab}, \phi_{ab}, \theta_{ab}, \psi_{ab}\} \tag{6.31}$$

This joint is characterized by the following orientation and location equations together with the surface equation, which constitutes a constraint among the three locational joint variables.

$$\widehat{C}^{(ab,ba)} = e^{\widetilde{n}_{ab1}\phi_{ab}} e^{\widetilde{n}_{ab2}\theta_{ab}} e^{\widetilde{n}_{ab3}\psi_{ab}} \tag{6.32}$$

$$\overline{r}_{ab,ba}^{(ab)} = \overline{u}_i x_{ab} + \overline{u}_j y_{ab} + \overline{u}_k z_{ab} \tag{6.33}$$

$$f_{ab}(x_{ab}, y_{ab}, z_{ab}) = 0 \tag{6.34}$$

The unit column matrices $\overline{n}_{ab1}$, $\overline{n}_{ab2}$, and $\overline{n}_{ab3}$ that take place in Eq. (6.32) are selected according to the similar considerations explained in Section 6.2.3 for a spherical joint.

As a typical surface example, an ellipsoidal surface is described by the following equation.

$$f_{ab}(x_{ab}, y_{ab}, z_{ab}) = (x_{ab}/a_{ab})^2 + (y_{ab}/b_{ab})^2 + (z_{ab}/c_{ab})^2 - 1 = 0 \tag{6.35}$$

In Eq. (6.35), a_{ab}, b_{ab}, and c_{ab} are the semi-axis lengths of the ellipsoidal surface.

6.4 Typical Multi-Joint Connections

6.4.1 Fork-on-Surface Joint

A fork-on-surface joint is illustrated in Figure 6.9. It is so called because $\mathcal{E}_{ab}$ has the shape of a specified surface S_{ab} and $\mathcal{E}_{ba}$ has the shape of a fork with two prongs. Thus, $\mathcal{E}_{ab}$ and $\mathcal{E}_{ba}$ are connected to each other with two point-on-surface joints. The contact points

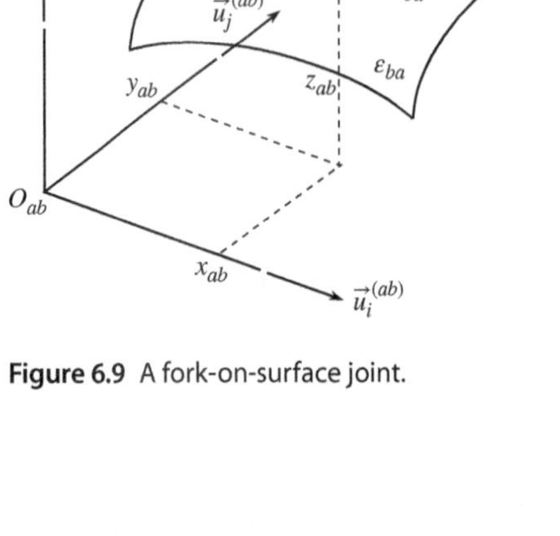

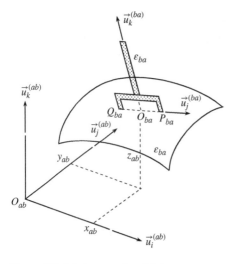

Figure 6.9 A fork-on-surface joint.

on $\mathcal{E}_{ab}$ and $\mathcal{E}_{ba}$ are denoted as $\{P_{ab}, Q_{ab}\}$ and $\{P_{ba}, Q_{ba}\}$. The reference frame $\mathcal{F}_{ab}$ is to be selected in such a way that the function $f_{ab}(x, y, z)$ that describes the surface S_{ab} has the simplest expression. As for the reference frame $\mathcal{F}_{ba}$, it is selected here so that $\vec{u}_i^{(ba)}$ is directed from Q_{ba} to P_{ba} and $\vec{u}_k^{(ba)}$ is aligned with the handle of the fork, i.e. with the line that is normal to the line segment between Q_{ba} to P_{ba} at its midpoint O_{ba}. The contact points on $\mathcal{E}_{ba}$ are located by the position vectors $\vec{r}_{bap}$ and $\vec{r}_{baq}$ with respect to O_{ba}. Let the distance between the prongs of the fork be $2d_{ba}$. Then, $\vec{r}_{bap}$ and $\vec{r}_{baq}$ are represented by the following column matrices in $\mathcal{F}_{ba}$.

$$\left.\begin{aligned} \bar{r}_{bap}^{(ba)} &= +\, d_{ba}\bar{u}_i \\ \bar{r}_{baq}^{(ba)} &= -\, d_{ba}\bar{u}_i \end{aligned}\right\} \tag{6.36}$$

The relative position of $\mathcal{F}_{ba}$ with respect to $\mathcal{F}_{ab}$ is represented by six joint variables, which are the coordinates of O_{ba} (i.e. x_{ab}, y_{ab}, z_{ab}) and the Euler angles (i.e. $\psi_{ab}, \theta_{ab}, \phi_{ab}$) of an appropriately selected sequence. However, these six variables are subject to two constraint equations, which state that the coordinates of the contact points must satisfy the surface equation specified for $\mathcal{E}_{ab}$. Therefore, the mobility of this joint is $\mu_{ab} = 4$. Its characteristic equations are written as follows by expressing the relative orientation matrix with the k–j–i sequence, which will be seen to be an appropriate selection later in Eqs. (6.41) and (6.42).

$$\hat{C}^{(ab,ba)} = e^{\tilde{u}_k\psi_{ab}} e^{\tilde{u}_j\theta_{ab}} e^{\tilde{u}_i\phi_{ab}} \tag{6.37}$$

$$\bar{r}_{ab,ba}^{(ab)} = \bar{u}_i x_{ab} + \bar{u}_j y_{ab} + \bar{u}_k z_{ab} \tag{6.38}$$

$$\left.\begin{aligned} f_{ab}(x_{abp}, y_{abp}, z_{abp}) &= 0 \\ f_{ab}(x_{abq}, y_{abq}, z_{abq}) &= 0 \end{aligned}\right\} \tag{6.39}$$

In Eq. Set (6.39), the arguments of the surface function $f_{ab}(x, y, z)$ are the i, j, k components of the position vectors $\vec{r}_{abp}$ and $\vec{r}_{abq}$, which locate the companion contact points P_{ab} and Q_{ab} of $\mathcal{E}_{ab}$ with respect to O_{ab}. These vectors are represented by the following column matrices in $\mathcal{F}_{ab}$.

$$\left.\begin{aligned} \bar{r}_{abp}^{(ab)} &= \bar{r}_{ab,ba}^{(ab)} + \hat{C}^{(ab,ba)}\bar{r}_{bap}^{(ba)} \\ \bar{r}_{abq}^{(ab)} &= \bar{r}_{ab,ba}^{(ab)} + \hat{C}^{(ab,ba)}\bar{r}_{baq}^{(ba)} \end{aligned}\right\} \tag{6.40}$$

When Eqs. (6.36)–(6.38) are substituted, Eq. Set (6.40) provides the following detailed equations.

$$\bar{r}_{abp}^{(ab)} = \bar{u}_i x_{ab} + \bar{u}_j y_{ab} + \bar{u}_k z_{ab} + d_{ba} e^{\tilde{u}_k\psi_{ab}} e^{\tilde{u}_j\theta_{ab}} \bar{u}_i \tag{6.41}$$

$$\bar{r}_{abq}^{(ab)} = \bar{u}_i x_{ab} + \bar{u}_j y_{ab} + \bar{u}_k z_{ab} - d_{ba} e^{\tilde{u}_k\psi_{ab}} e^{\tilde{u}_j\theta_{ab}} \bar{u}_i \tag{6.42}$$

Equations (6.41) and (6.42) show that the selected $k-j-i$ sequence is appropriate, because it has provided a simplification owing to the fact that $e^{\tilde{u}_i\phi_{ab}}\bar{u}_i = \bar{u}_i$. Incidentally, considering this simplification criterion, the $j-k-i$ sequence could also be selected as another appropriate sequence, if desired.

Note that $\tilde{u}_i\bar{u}_j = \bar{u}_k$, $\tilde{u}_j\bar{u}_k = \bar{u}_i$, and $\tilde{u}_k\bar{u}_i = \bar{u}_j$ according to the right-handedness property. Therefore, referring to Section 2.7 of Chapter 2, it can be shown that

$$\left.\begin{aligned} e^{\tilde{u}_i\theta}\bar{u}_j &= \bar{u}_j c\theta + \bar{u}_k s\theta \\ \bar{u}_j^t e^{\tilde{u}_i\theta} &= \bar{u}_j^t c\theta - \bar{u}_k^t s\theta \end{aligned}\right\} \tag{6.43}$$

Note also that

$$\left.\begin{array}{l} x_{abp} = \overline{u}_i^t \overline{r}_{abp}^{(ab)},\ y_{abp} = \overline{u}_j^t \overline{r}_{abp}^{(ab)},\ z_{abp} = \overline{u}_k^t \overline{r}_{abp}^{(ab)} \\ x_{abq} = \overline{u}_i^t \overline{r}_{abq}^{(ab)},\ y_{abq} = \overline{u}_j^t \overline{r}_{abq}^{(ab)},\ z_{abq} = \overline{u}_k^t \overline{r}_{abq}^{(ab)} \end{array}\right\} \tag{6.44}$$

The preceding equations can be used to obtain the following contact point coordinates in $\mathcal{F}_{ab}$, which are required in Eq. Set (6.39).

$$\left.\begin{array}{l} x_{abp} = x_{ab} + d_{ba} c\psi_{ab} c\theta_{ab} \\ y_{abp} = y_{ab} + d_{ba} s\psi_{ab} c\theta_{ab} \\ z_{abp} = z_{ab} - d_{ba} s\theta_{ab} \end{array}\right\} \tag{6.45}$$

$$\left.\begin{array}{l} x_{abq} = x_{ab} - d_{ba} c\psi_{ab} c\theta_{ab} \\ y_{abq} = y_{ab} - d_{ba} s\psi_{ab} c\theta_{ab} \\ z_{abq} = z_{ab} + d_{ba} s\theta_{ab} \end{array}\right\} \tag{6.46}$$

As a special case, if the surface is the $i-j$ plane of $\mathcal{F}_{ab}$, i.e. if $f_{ab}(x, y, z) = z = 0$, then Eq. Sets (6.45), (6.46), and (6.39) imply that $z_{abp} = z_{abq} = z_{ab} = 0$ and $\theta_{ab} = 0$. Consequently, Eqs. (6.37) and (6.38) reduce to

$$\widehat{C}^{(ab,ba)} = e^{\tilde{u}_k \psi_{ab}} e^{\tilde{u}_i \phi_{ab}} \tag{6.47}$$

$$\overline{r}_{ab,ba}^{(ab)} = \overline{u}_i x_{ab} + \overline{u}_j y_{ab} \tag{6.48}$$

Equations (6.47) and (6.48) imply in turn that a fork-on-surface joint reduces as expected to a line-on-plane joint (with the joint variables x_{ab}, y_{ab}, ψ_{ab} and ϕ_{ab}) if the surface S_{ab} flattens down to a plane.

6.4.2 Triangle-on-Surface Joint

A triangle-on-surface joint is illustrated in Figure 6.10. It is so called because $\mathcal{E}_{ab}$ has the shape of a specified surface S_{ab} and $\mathcal{E}_{ba}$ has the shape of the base triangle of a trivet or a pedestal with three supports. Thus, $\mathcal{E}_{ab}$ and $\mathcal{E}_{ba}$ are connected to each other with three point-on-surface joints. The contact points on $\mathcal{E}_{ab}$ and $\mathcal{E}_{ba}$ are denoted as $\{P_{ab}, Q_{ab}, R_{ab}\}$ and $\{P_{ba}, Q_{ba}, R_{ba}\}$, respectively. The reference frame $\mathcal{F}_{ab}$ is to be selected in such a way

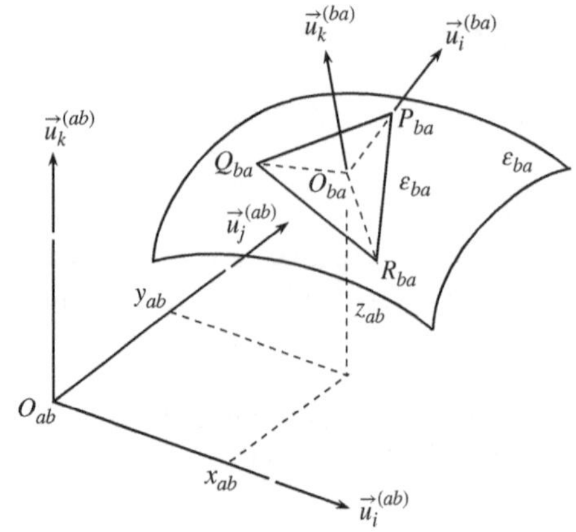

Figure 6.10 A triangle-on-surface joint.

that the function $f_{ab}(x, y, z)$ that describes the surface $\mathcal{S}_{ab}$ has the simplest expression. As for the reference frame $\mathcal{F}_{ba}$, it is selected here so that $\vec{u}_k^{(ba)}$ is normal to the plane of the triangle and $\vec{u}_i^{(ba)}$ is directed from O_{ba} to P_{ba} as shown in Figure 6.10. The contact points on $\mathcal{E}_{ba}$ are located by the position vectors $\vec{r}_{bap}$, $\vec{r}_{baq}$, and $\vec{r}_{bar}$ with respect to O_{ba}.

Here, $\mathcal{E}_{ba}$ is assumed to be shaped like an isosceles triangle. However, if it were shaped like an arbitrary triangle, a similar but naturally more complicated formulation could also be obtained. The geometric features of an isosceles triangle are described by a length d_{ba} and an angle γ_{ba}. Referring to Figure 6.10, they are defined as follows:

$$d_{ba} = O_{ba}P_{ba} = O_{ba}Q_{ba} = O_{ba}R_{ba} \tag{6.49}$$

$$\gamma_{ba} = [\sphericalangle(Q_{ba}O_{ba}R_{ba})]/2 \tag{6.50}$$

By using the geometric parameters defined above, the position vectors of the contact points are represented by the following column matrices in $\mathcal{F}_{ba}$.

$$\left.\begin{aligned} \overline{r}_{bap}^{(ba)} &= d_{ba}\overline{u}_i \\ \overline{r}_{baq}^{(ba)} &= -d_{ba}(\overline{u}_i c\gamma_{ba} - \overline{u}_j s\gamma_{ba}) \\ \overline{r}_{bar}^{(ba)} &= -d_{ba}(\overline{u}_i c\gamma_{ba} + \overline{u}_j s\gamma_{ba}) \end{aligned}\right\} \tag{6.51}$$

The relative position of $\mathcal{F}_{ba}$ with respect to $\mathcal{F}_{ab}$ is represented by six joint variables, which are the coordinates of O_{ba} (i.e. x_{ab}, y_{ab}, z_{ab}) and the Euler angles (i.e. ψ_{ab}, θ_{ab}, ϕ_{ab}) of an appropriate sequence. However, these six variables are subject to three constraint equations, which state that the coordinates of the contact points must satisfy the surface equation specified for $\mathcal{E}_{ab}$. Therefore, the mobility of this joint is $\mu_{ab} = 3$. Its characteristic equations are written as shown below by expressing the relative orientation matrix with the $k-j-i$ sequence, which turns out to be appropriate according to the simplification criterion mentioned previously in Section 6.4.1 concerning the fork-on-surface joint.

$$\widehat{C}^{(ab,ba)} = e^{\tilde{u}_k\psi_{ab}}e^{\tilde{u}_j\theta_{ab}}e^{\tilde{u}_i\phi_{ab}} \tag{6.52}$$

$$\overline{r}_{ab,ba}^{(ab)} = \overline{u}_i x_{ab} + \overline{u}_j y_{ab} + \overline{u}_k z_{ab} \tag{6.53}$$

$$\left.\begin{aligned} f_{ab}(x_{abp}, y_{abp}, z_{abp}) &= 0 \\ f_{ab}(x_{abq}, y_{abq}, z_{abq}) &= 0 \\ f_{ab}(x_{abr}, y_{abr}, z_{abr}) &= 0 \end{aligned}\right\} \tag{6.54}$$

In Eq. Set (6.54), the arguments of the surface function $f_{ab}(x, y, z)$ are the i, j, k components of the position vectors $\vec{r}_{abp}$, $\vec{r}_{abq}$, and $\vec{r}_{abr}$, which locate the companion contact points P_{ab}, Q_{ab}, and R_{ab} of $\mathcal{E}_{ab}$ with respect to O_{ab}. These vectors are represented by the following column matrices in $\mathcal{F}_{ab}$.

$$\left.\begin{aligned} \overline{r}_{abp}^{(ab)} &= \overline{r}_{ab,ba}^{(ab)} + \widehat{C}^{(ab,ba)}\overline{r}_{bap}^{(ba)} \\ \overline{r}_{abq}^{(ab)} &= \overline{r}_{ab,ba}^{(ab)} + \widehat{C}^{(ab,ba)}\overline{r}_{baq}^{(ba)} \\ \overline{r}_{abr}^{(ab)} &= \overline{r}_{ab,ba}^{(ab)} + \widehat{C}^{(ab,ba)}\overline{r}_{bar}^{(ba)} \end{aligned}\right\} \tag{6.55}$$

When Eqs. (6.52), (6.53), and Eq. Set (6.51) are substituted, Eq. Set (6.55) provides the following detailed equations.

$$\overline{r}_{abp}^{(ab)} = \overline{u}_i x_{ab} + \overline{u}_j y_{ab} + \overline{u}_k z_{ab} + d_{ba} e^{\tilde{u}_k \psi_{ab}} e^{\tilde{u}_j \theta_{ab}} \overline{u}_i \tag{6.56}$$

$$\begin{aligned}\overline{r}_{abq}^{(ab)} &= \overline{u}_i x_{ab} + \overline{u}_j y_{ab} + \overline{u}_k z_{ab} \\ &\quad - d_{ba} e^{\tilde{u}_k \psi_{ab}} e^{\tilde{u}_j \theta_{ab}} [\overline{u}_i c\gamma_{ba} - (\overline{u}_j c\phi_{ab} + \overline{u}_k s\phi_{ab}) s\gamma_{ba}]\end{aligned} \tag{6.57}$$

$$\begin{aligned}\overline{r}_{abr}^{(ab)} &= \overline{u}_i x_{ab} + \overline{u}_j y_{ab} + \overline{u}_k z_{ab} \\ &\quad - d_{ba} e^{\tilde{u}_k \psi_{ab}} e^{\tilde{u}_j \theta_{ab}} [\overline{u}_i c\gamma_{ba} + (\overline{u}_j c\phi_{ab} + \overline{u}_k s\phi_{ab}) s\gamma_{ba}]\end{aligned} \tag{6.58}$$

Note that $x_{abp} = \overline{u}_i^t \overline{r}_{abp}^{(ab)}$, $y_{abp} = \overline{u}_j^t \overline{r}_{abp}^{(ab)}$, $z_{abp} = \overline{u}_k^t \overline{r}_{abp}^{(ab)}$, and so on. Thus, similarly as done before for the fork-on-surface joint, the preceding equations can be manipulated to obtain the following contact point coordinates in $\mathcal{F}_{ab}$.

$$\left.\begin{aligned} x_{abp} &= x_{ab} + d_{ba} c\psi_{ab} c\theta_{ab} \\ y_{abp} &= y_{ab} + d_{ba} s\psi_{ab} c\theta_{ab} \\ z_{abp} &= z_{ab} - d_{ba} s\theta_{ab} \end{aligned}\right\} \tag{6.59}$$

$$\left.\begin{aligned} x_{abq} &= x_{ab} - d_{ba}[c\psi_{ab} c\theta_{ab} c\gamma_{ba} + (s\psi_{ab} c\phi_{ab} - c\psi_{ab} s\theta_{ab} s\phi_{ab}) s\gamma_{ba}] \\ y_{abq} &= y_{ab} - d_{ba}[s\psi_{ab} c\theta_{ab} c\gamma_{ba} - (c\psi_{ab} c\phi_{ab} + s\psi_{ab} s\theta_{ab} s\phi_{ab}) s\gamma_{ba}] \\ z_{abq} &= z_{ab} + d_{ba}(s\theta_{ab} c\gamma_{ba} + c\theta_{ab} s\phi_{ab} s\gamma_{ba}) \end{aligned}\right\} \tag{6.60}$$

$$\left.\begin{aligned} x_{abr} &= x_{ab} - d_{ba}[c\psi_{ab} c\theta_{ab} c\gamma_{ba} - (s\psi_{ab} c\phi_{ab} - c\psi_{ab} s\theta_{ab} s\phi_{ab}) s\gamma_{ba}] \\ y_{abr} &= y_{ab} - d_{ba}[s\psi_{ab} c\theta_{ab} c\gamma_{ba} + (c\psi_{ab} c\phi_{ab} + s\psi_{ab} s\theta_{ab} s\phi_{ab}) s\gamma_{ba}] \\ z_{abr} &= z_{ab} + d_{ba}(s\theta_{ab} c\gamma_{ba} - c\theta_{ab} s\phi_{ab} s\gamma_{ba}) \end{aligned}\right\} \tag{6.61}$$

As a special case, if the surface is the $i-j$ plane of $\mathcal{F}_{ab}$, i.e. if $f_{ab}(x,y,z) = z = 0$, then Eq. Sets (6.59)–(6.61) and (6.54) imply that $z_{abp} = z_{abq} = z_{abr} = z_{ab} = 0$ and also $\theta_{ab} = \phi_{ab} = 0$. Consequently, Eqs. (6.52) and (6.53) reduce to:

$$\hat{C}^{(ab,ba)} = e^{\tilde{u}_k \psi_{ab}} \tag{6.62}$$

$$\overline{r}_{ab,ba}^{(ab)} = \overline{u}_i x_{ab} + \overline{u}_j y_{ab} \tag{6.63}$$

Equations (6.62) and (6.63) imply in turn that a triangle-on-surface joint reduces as expected to a plane-on-plane joint (with the joint variables x_{ab}, y_{ab}, and ψ_{ab}) if the surface S_{ab} flattens down to a plane.

An example that involves a triangle-on-surface joint is presented as Example 6.2 in Section 6.7.2. The surface in that example consists of three mutually orthogonal planes and each point of the triangle touches a different one of the planes.

6.5 Rolling Contact Joints with Point Contacts

6.5.1 Surface-on-Surface Joint

Figure 6.11 illustrates a joint formed by two kinematic elements $\mathcal{E}_{ab}$ and $\mathcal{E}_{ba}$, which are in rolling contact at their coincident points Q_{ab} and Q_{ba}. The kinematic elements have the shapes of specified surfaces S_{ab} and S_{ba}, which are described by functions $f_{ab}(x,y,z)$ and $f_{ba}(x,y,z)$. Both S_{ab} and S_{ba} are assumed to be smooth. In other words, the functions that describe them are assumed to be continuous and differentiable with respect to

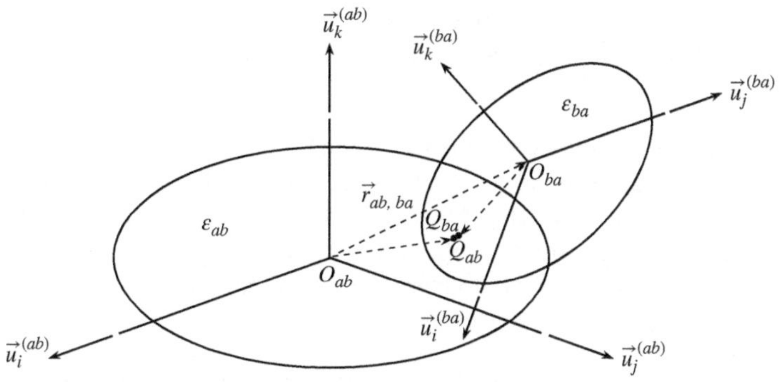

Figure 6.11 A surface-on-surface joint.

their arguments at all the possible contact points. The surfaces can have a point contact if one of them is a *double-curvature* surface such as an ellipsoidal surface. The other surface may also be a double-curvature surface. However, it may also be a *single-curvature* surface, e.g. a conical or cylindrical surface, or a *zero-curvature* surface, i.e. a plane. A point contact can also occur even if both of $\mathcal{S}_{ab}$ and $\mathcal{S}_{ba}$ are single-curvature surfaces. However, in that case, $\mathcal{S}_{ab}$ and $\mathcal{S}_{ba}$ must not be sharing a common contact line.

(a) Relative Position Equations

The following equations can be written in order to express the relative position relationships between $\mathcal{E}_{ab}$ and $\mathcal{E}_{ba}$.

The fact that Q_{ab} and Q_{ba} are coincident points can be expressed as follows:

$$\overrightarrow{O_{ab}Q_{ab}} = \overrightarrow{O_{ab}O_{ba}} + \overrightarrow{O_{ba}Q_{ba}} \Rightarrow$$

$$\vec{r}_{abq} = \vec{r}_{ab,ba} + \vec{r}_{baq} \tag{6.64}$$

Equation (6.64) can be replaced with the following matrix equation.

$$\bar{r}_{abq}^{(ab)} = \bar{r}_{ab,ba}^{(ab)} + \hat{C}^{(ab,ba)}\bar{r}_{baq}^{(ba)} \tag{6.65}$$

The column matrices in Eq. (6.65) can be expressed as follows in $\mathcal{F}_{ab}$ and $\mathcal{F}_{ba}$:

$$\bar{r}_{ab,ba}^{(ab)} = \bar{u}_i x_{ab} + \bar{u}_j y_{ab} + \bar{u}_k z_{ab} \tag{6.66}$$

$$\bar{r}_{abq}^{(ab)} = \bar{u}_i x_{abq} + \bar{u}_j y_{abq} + \bar{u}_k z_{abq} \tag{6.67}$$

$$\bar{r}_{baq}^{(ba)} = \bar{u}_i x_{baq} + \bar{u}_j y_{baq} + \bar{u}_k z_{baq} \tag{6.68}$$

The orientation matrix that occurs in Eq. (6.65) can be expressed as follows with suitably selected constant unit column matrices $\bar{n}_{ab1}$, $\bar{n}_{ab2}$, and $\bar{n}_{ab3}$:

$$\hat{C}^{(ab,ba)} = e^{\tilde{n}_{ab1}\phi_{ab}} e^{\tilde{n}_{ab2}\theta_{ab}} e^{\tilde{n}_{ab3}\psi_{ab}} \tag{6.69}$$

In addition to Eq. (6.65), the coordinates of the contact points must also satisfy the following surface equations.

$$f_{ab}(x_{abq}, y_{abq}, z_{abq}) = 0 \tag{6.70}$$

$$f_{ba}(x_{baq}, y_{baq}, z_{baq}) = 0 \tag{6.71}$$

Moreover, the surfaces S_{ab} and S_{ba} must be tangent to each other at the contact points. This fact can be expressed by the following *gradient alignment equation*.

$$\vec{g}_{ba} = -\lambda_{ab}\vec{g}_{ab} \tag{6.72}$$

In Eq. (6.72), $\vec{g}_{ab}$ and $\vec{g}_{ba}$ are the gradient vectors of the surfaces S_{ab} and S_{ba}. Here, it must be pointed out that the surface functions f_{ab} and f_{ba} are conventionally expressed in terms of their arguments in such a way that $\vec{g}_{ab}$ and $\vec{g}_{ba}$ are generally oriented away from the convex sides of the respective surfaces S_{ab} and S_{ba}. For example, the equation that describes a *spherical surface* of radius r is conventionally written as

$$f(x, y, z) = x^2 + y^2 + z^2 - r^2 = 0$$

On the other hand, it would be unconventional to write the equation of the same surface as

$$f'(x, y, z) = -f(x, y, z) = r^2 - x^2 - y^2 - z^2 = 0$$

As for the scalar coefficient λ_{ab}, it is defined as the *gradient ratio* between the two surfaces. In most of the cases, the contacting kinematic elements have convex shapes and therefore the gradients $\vec{g}_{ab}$ and $\vec{g}_{ba}$ are oriented oppositely at the coincident contact points. That is why Eq. (6.72) is written with a minus sign so that λ_{ab} happens to be positive in the most common case of convex-convex contacts. According to this sign convention, λ_{ab} will be negative in the less common case of concave-convex or convex-concave contacts. If one of the surfaces, say S_{ab}, happens to be planar within the contact region, then $\vec{g}_{ab}$ becomes $\vec{n}_{ab}$, which is the unit vector normal to S_{ab}. In that case, the sign of λ_{ab} depends on how $\vec{n}_{ab}$ is oriented optionally (i.e. arbitrarily as desired) with respect to S_{ab}, which is neither convex nor concave.

The gradient vectors $\vec{g}_{ab}$ and $\vec{g}_{ba}$ are represented by the following column matrices, respectively, in $\mathcal{F}_{ab}$ and $\mathcal{F}_{ba}$.

$$\overline{g}_{ab}^{(ab)} = \overline{u}_i f_{abx} + \overline{u}_j f_{aby} + \overline{u}_k f_{abz} \tag{6.73}$$

$$\overline{g}_{ba}^{(ba)} = \overline{u}_i f_{bax} + \overline{u}_j f_{bay} + \overline{u}_k f_{baz} \tag{6.74}$$

In Eqs. (6.73) and (6.74),

$$\left.\begin{aligned} f_{abx} &= \partial f_{ab}/\partial x \\ f_{aby} &= \partial f_{ab}/\partial y \\ f_{abz} &= \partial f_{ab}/\partial z \end{aligned}\right\} \tag{6.75}$$

$$\left.\begin{aligned} f_{bax} &= \partial f_{ba}/\partial x \\ f_{bay} &= \partial f_{ba}/\partial y \\ f_{baz} &= \partial f_{ba}/\partial z \end{aligned}\right\} \tag{6.76}$$

By using the matrix representations of $\vec{g}_{ab}$ and $\vec{g}_{ba}$ in $\mathcal{F}_{ab}$ and $\mathcal{F}_{ba}$, the gradient alignment equation, i.e. Eq. (6.72), can be written as

$$\hat{C}^{(ab,ba)}\overline{g}_{ba}^{(ba)} = -\lambda_{ab}\overline{g}_{ab}^{(ab)} \tag{6.77}$$

As seen above, a surface-on-surface joint involves 13 variables, which are listed below in two groups.

Joint variables:

$$x_{ab}, y_{ab}, z_{ab}; \phi_{ab}, \theta_{ab}, \psi_{ab}; \lambda_{ab}$$

Contact point coordinates:

$$x_{abq}, y_{abq}, z_{abq}; x_{baq}, y_{baq}, z_{baq}$$

The 13 variables listed above are interrelated by eight independent scalar equations. Six of them are contained in Eqs. (6.65) and (6.77). The remaining two are Eqs. (6.70) and (6.71). Therefore, the mobility of a surface-on-surface joint is $\mu_{ab} = 13 - 8 = 5$.

Thus, if five of the variables (e.g. $\phi_{ab}, \theta_{ab}, \psi_{ab}, x_{ab}, y_{ab}$) are specified, then the remaining eight variables (i.e. z_{ab}, λ_{ab}, and the six contact point coordinates) can be determined by using the mentioned equations.

(b) Relative Velocity Equations

The relative velocity equations of $\mathcal{E}_{ba}$ with respect to $\mathcal{E}_{ab}$ can be obtained by differentiating the relative position equations. As a part of this process, Eqs. (6.70), (6.71), and (6.65) lead to the following equations that relate the velocities of the contact points Q_{ab} and Q_{ba} as their locations change on their respective surfaces S_{ab} and S_{ba}.

$$f_{abx}\dot{x}_{abq} + f_{aby}\dot{y}_{abq} + f_{abz}\dot{z}_{abq} = 0 \Rightarrow$$

$$\overline{g}_{ab}^{(ab)t}\dot{\overline{r}}_{abq}^{(ab)} = 0 \tag{6.78}$$

$$f_{bax}\dot{x}_{baq} + f_{bay}\dot{y}_{baq} + f_{baz}\dot{z}_{baq} = 0 \Rightarrow$$

$$\overline{g}_{ba}^{(ba)t}\dot{\overline{r}}_{baq}^{(ba)} = 0 \tag{6.79}$$

$$\dot{\overline{r}}_{abq}^{(ab)} = \dot{\overline{r}}_{ab,ba}^{(ab)} + \hat{C}^{(ab,ba)}\dot{\overline{r}}_{baq}^{(ba)} + \dot{\hat{C}}^{(ab,ba)}\overline{r}_{baq}^{(ba)} \Rightarrow$$

$$\dot{\overline{r}}_{abq}^{(ab)} = \overline{v}_{ab,ba}^{(ab)} + \hat{C}^{(ab,ba)}\dot{\overline{r}}_{baq}^{(ba)} + \tilde{\omega}_{ab,ba}^{(ab)}\hat{C}^{(ab,ba)}\overline{r}_{baq}^{(ba)} \tag{6.80}$$

In Eq. (6.80), the relative translational and rotational velocities of $\mathcal{E}_{ba}$ with respect to $\mathcal{E}_{ab}$ are defined so that

$$\overline{v}_{ab,ba}^{(ab)} = \dot{\overline{r}}_{ab,ba}^{(ab)} \tag{6.81}$$

$$\tilde{\omega}_{ab,ba}^{(ab)} = \dot{\hat{C}}^{(ab,ba)}\hat{C}^{(ba,ab)} \Rightarrow \overline{\omega}_{ab,ba}^{(ab)} = \mathrm{colm}[\tilde{\omega}_{ab,ba}^{(ab)}] \tag{6.82}$$

By using Eqs. (6.66) and (6.69), the relative velocities $\overline{v}_{ab,ba}^{(ab)}$ and $\overline{\omega}_{ab,ba}^{(ab)}$ can be expressed as follows in terms of the joint variables and their derivatives:

$$\overline{v}_{ab,ba}^{(ab)} = \overline{u}_i\dot{x}_{ab} + \overline{u}_j\dot{y}_{ab} + \overline{u}_k\dot{z}_{ab} \tag{6.83}$$

$$\overline{\omega}_{ab,ba}^{(ab)} = \dot{\phi}_{ab}\overline{n}_{ab1} + \dot{\theta}_{ab}e^{\tilde{n}_{ab1}\phi_{ab}}\overline{n}_{ab2} + \dot{\psi}_{ab}e^{\tilde{n}_{ab1}\phi_{ab}}e^{\tilde{n}_{ab2}\theta_{ab}}\overline{n}_{ab3} \tag{6.84}$$

On the other hand, the differentiation of the gradient alignment equation, i.e. Eq. (6.77), leads to the following equation.

$$\hat{C}^{(ab,ba)}\dot{\overline{g}}_{ba}^{(ba)} + \dot{\hat{C}}^{(ab,ba)}\overline{g}_{ba}^{(ba)} = -\lambda_{ab}\dot{\overline{g}}_{ab}^{(ab)} - \dot{\lambda}_{ab}\overline{g}_{ab}^{(ab)} \Rightarrow$$

$$\hat{C}^{(ab,ba)}\dot{\overline{g}}_{ba}^{(ba)} + \tilde{\omega}_{ab,ba}^{(ab)}\hat{C}^{(ab,ba)}\overline{g}_{ba}^{(ba)} = -\lambda_{ab}\dot{\overline{g}}_{ab}^{(ab)} - \dot{\lambda}_{ab}\overline{g}_{ab}^{(ab)} \tag{6.85}$$

Referring to Eqs. (6.73) and (6.74), the terms $\dot{\overline{g}}_{ab}^{(ab)}$ and $\dot{\overline{g}}_{ba}^{(ba)}$ can be written out as shown below by using the same differentiation notation used in Eqs. (6.75) and (6.76).

$$\begin{aligned}\dot{\overline{g}}_{ab}^{(ab)} = {} & \overline{u}_i(f_{abxx}\dot{x}_{abq} + f_{abxy}\dot{y}_{abq} + f_{abxz}\dot{z}_{abq}) \\ & + \overline{u}_j(f_{abyx}\dot{x}_{abq} + f_{abyy}\dot{y}_{abq} + f_{abyz}\dot{z}_{abq}) \\ & + \overline{u}_k(f_{abzx}\dot{x}_{abq} + f_{abzy}\dot{y}_{abq} + f_{abzz}\dot{z}_{abq})\end{aligned} \tag{6.86}$$

$$\dot{\overline{g}}_{ba}^{(ba)} = \overline{u}_i(f_{baxx}\dot{x}_{baq} + f_{baxy}\dot{y}_{baq} + f_{baxz}\dot{z}_{baq})$$
$$+ \overline{u}_j(f_{bayx}\dot{x}_{baq} + f_{bayy}\dot{y}_{baq} + f_{bayz}\dot{z}_{baq})$$
$$+ \overline{u}_k(f_{bazx}\dot{x}_{baq} + f_{bazy}\dot{y}_{baq} + f_{bazz}\dot{z}_{baq}) \quad (6.87)$$

Equations (6.86) and (6.87) can be written compactly as follows:

$$\dot{\overline{g}}_{ab}^{(ab)} = \widehat{H}_{ab}^{(ab)}\dot{\overline{r}}_{abq}^{(ab)} \quad (6.88)$$

$$\dot{\overline{g}}_{ba}^{(ba)} = \widehat{H}_{ba}^{(ba)}\dot{\overline{r}}_{baq}^{(ba)} \quad (6.89)$$

In Eqs. (6.88) and (6.89), $\widehat{H}_{ab}^{(ab)}$ and $\widehat{H}_{ba}^{(ba)}$ are known as the *Hessian matrices* associated with the surface functions $f_{ab}(x, y, z)$ and $f_{ba}(x, y, z)$. Equations (6.86) and (6.87) imply that the Hessian matrices have the following detailed expressions in terms of their elements.

$$\widehat{H}_{ab}^{(ab)} = \begin{bmatrix} f_{abxx} & f_{abxy} & f_{abxz} \\ f_{abyx} & f_{abyy} & f_{abyz} \\ f_{abzx} & f_{abzy} & f_{abzz} \end{bmatrix} \quad (6.90)$$

$$\widehat{H}_{ba}^{(ba)} = \begin{bmatrix} f_{baxx} & f_{baxy} & f_{baxz} \\ f_{bayx} & f_{bayy} & f_{bayz} \\ f_{bazx} & f_{bazy} & f_{bazz} \end{bmatrix} \quad (6.91)$$

By using the Hessian matrices and Eq. (6.77), Eq. (6.85) can be written again as follows:

$$\widehat{C}^{(ab,ba)}\widehat{H}_{ba}^{(ba)}\dot{\overline{r}}_{baq}^{(ba)} + \lambda_{ab}\widehat{H}_{ab}^{(ab)}\dot{\overline{r}}_{abq}^{(ab)} = [\lambda_{ab}\widetilde{\omega}_{ab,ba}^{(ab)} - \dot{\lambda}_{ab}\widehat{I}]\overline{g}_{ab}^{(ab)} \quad (6.92)$$

Similarly as in the position equations, Eqs. (6.78), (6.79), (6.80), and (6.92) provide eight independent scalar equations that relate the following 13 derivatives.

$$\dot{x}_{ab}, \dot{y}_{ab}, \dot{z}_{ab}; \dot{\phi}_{ab}, \dot{\theta}_{ab}, \dot{\psi}_{ab}; \dot{\lambda}_{ab}; \dot{x}_{abq}, \dot{y}_{abq}, \dot{z}_{abq}; \dot{x}_{abq}, \dot{y}_{abq}, \dot{z}_{abq}$$

Hence, it is seen again that the mobility of a surface-on-surface joint is $\mu_{ab} = 5$. Thus, if five of the derivatives (e.g. $\dot{\phi}_{ab}$, $\dot{\theta}_{ab}$, $\dot{\psi}_{ab}$, $\dot{x}_{ab}$, $\dot{y}_{ab}$) are specified along with the position specifications, then the remaining eight derivatives (i.e. $\dot{z}_{ab}$, $\dot{\lambda}_{ab}$, and the derivatives of the six contact point coordinates) can be determined by using the mentioned velocity equations.

(c) Relative Velocity Equations in the Case of Rolling Without Slipping

If $\mathcal{E}_{ba}$ rolls over $\mathcal{E}_{ab}$ without slipping due to the presence of sticking friction, then the instantaneous relative velocity of the contact point Q_{ba} with respect to its coincident companion Q_{ab} vanishes. In other words, the derivatives $\dot{\overline{r}}_{abq}^{(ab)}$ and $\dot{\overline{r}}_{baq}^{(ba)}$ become related so that

$$\dot{\overline{r}}_{abq}^{(ab)} = \widehat{C}^{(ab,ba)}\dot{\overline{r}}_{baq}^{(ba)} \quad (6.93)$$

Note that the relationship expressed by Eq. (6.93) arises only if $\mathcal{E}_{ba}$ is in motion with respect to $\mathcal{E}_{ab}$, i.e. only if $\dot{\overline{r}}_{baq}^{(ba)} \neq \overline{0}$ and hence $\dot{\overline{r}}_{abq}^{(ab)} \neq \overline{0}$ either. In other words, Eq. (6.93) constitutes an instantaneous constraint on the velocity relationships. However, it does not affect the position relationships immediately at the same instant. In other words, it cannot be integrated instantaneously into a relationship directly between $\overline{r}_{abq}^{(ab)}$ and $\overline{r}_{baq}^{(ba)}$. As such, it describes a *nonholonomic* (i.e. unintegrable) constraint.

Another point to be noted about Eq. (6.93) is that it contains only two nontrivial independent scalar constraint equations on the components of $\dot{\bar{r}}_{abq}^{(ab)}$ and $\dot{\bar{r}}_{baq}^{(ba)}$. This is because the velocity vectors of the contact points lie naturally on the *common tangent plane* of the surfaces S_{ab} and S_{ba}, which is normal to the aligned gradient vectors. This fact is already implied by Eqs. (6.78) and (6.79), which state that $\bar{g}_{ab}^{(ab)t}\dot{\bar{r}}_{abq}^{(ab)} = \bar{g}_{ba}^{(ba)t}\dot{\bar{r}}_{baq}^{(ba)} = 0$. Due to the two additional independent velocity constraints contained in Eq. (6.93), the mobility of a surface-on-surface joint reduces to $\mu_{ab} = 5 - 2 = 3$ during the rolling-without-slipping motion.

Actually, during the rolling-without-slipping motion, Eq. (6.93) gives $\dot{\bar{r}}_{abq}^{(ab)}$ directly in terms of $\dot{\bar{r}}_{baq}^{(ba)}$ and this fact makes Eq. (6.79) redundant because it becomes obtainable from Eq. (6.78) as shown below with the assistance of Eqs. (6.77) and (6.93).

$$\bar{g}_{ab}^{(ab)t}\dot{\bar{r}}_{abq}^{(ab)} = 0 \Rightarrow -\lambda_{ab}\bar{g}_{ab}^{(ab)t}\dot{\bar{r}}_{abq}^{(ab)} = 0 \Rightarrow$$

$$[\widehat{C}^{(ab,ba)}\bar{g}_{ba}^{(ba)}]^t[\widehat{C}^{(ab,ba)}\dot{\bar{r}}_{baq}^{(ba)}] = \bar{g}_{ba}^{(ba)t}[\widehat{C}^{(ba,ab)}\widehat{C}^{(ab,ba)}]\dot{\bar{r}}_{baq}^{(ba)} = 0 \Rightarrow$$

$$\bar{g}_{ba}^{(ba)t}\dot{\bar{r}}_{baq}^{(ba)} = 0$$

In this mode of motion, Eq. (6.93) changes Eqs. (6.80) and (6.92) into the following forms.

$$\bar{v}_{ab,ba}^{(ab)} = -\widetilde{\omega}_{ab,ba}^{(ab)}\widehat{C}^{(ab,ba)}\bar{r}_{baq}^{(ba)} \tag{6.94}$$

$$\widehat{K}_{ab}\dot{\bar{r}}_{abq}^{(ab)} = [\lambda_{ab}\widetilde{\omega}_{ab,ba}^{(ab)} - \dot{\lambda}_{ab}\widehat{I}]\bar{g}_{ab}^{(ab)} \tag{6.95}$$

In Eq. (6.95), the matrix $\widehat{K}_{ab}$ is defined as

$$\widehat{K}_{ab} = \widehat{C}^{(ab,ba)}\widehat{H}_{ba}^{(ba)}\widehat{C}^{(ba,ab)} + \lambda_{ab}\widehat{H}_{ab}^{(ab)} \tag{6.96}$$

Furthermore, Eq. (6.94) can also be written as follows by using Eq. (6.65):

$$\bar{v}_{ab,ba}^{(ab)} = \widetilde{\omega}_{ab,ba}^{(ab)}[\bar{r}_{ab,ba}^{(ab)} - \bar{r}_{abq}^{(ab)}] \tag{6.97}$$

Incidentally, Eq. (6.97) shows the most typical feature of the rolling-without-slipping motion by implying that the instantaneous relative motion of $\mathcal{E}_{ba}$ with respect to $\mathcal{E}_{ab}$ appears as a rotation about the contact point Q_{ab} as if Q_{ab} is fixed instantaneously.

During this motion, Eq. (6.93) already relates $\dot{\bar{r}}_{baq}^{(ab)}$ to $\dot{\bar{r}}_{abq}^{(ba)}$. Therefore, there remains 10 derivatives to be related by the other equations. These derivatives are $\dot{\phi}_{ab}$, $\dot{\theta}_{ab}$, $\dot{\psi}_{ab}$, $\dot{x}_{ab}$, $\dot{y}_{ab}$, $\dot{z}_{ab}$, $\dot{x}_{abq}$, $\dot{y}_{abq}$, $\dot{z}_{abq}$, and $\dot{\lambda}_{ab}$. Thus, if three of these derivatives (e.g. $\dot{\phi}_{ab}$, $\dot{\theta}_{ab}$, $\dot{\psi}_{ab}$) are specified along with the position specifications, then the other seven (i.e. $\dot{x}_{ab}$, $\dot{y}_{ab}$, $\dot{z}_{ab}$, $\dot{x}_{abq}$, $\dot{y}_{abq}$, $\dot{z}_{abq}$, $\dot{\lambda}_{ab}$) can be determined by using the seven independent scalar equations contained in Eqs. (6.95), (6.97), and (6.78). The solution procedure is explained below.

If the derivatives $\dot{\phi}_{ab}$, $\dot{\theta}_{ab}$, and $\dot{\psi}_{ab}$ are specified as mentioned above, then $\bar{\omega}_{ab,ba}^{(ab)}$ becomes available according to Eq. (6.84). Then, $\bar{v}_{ab,ba}^{(ab)}$ is also provided readily by Eq. (6.97). Hence, noting that $\bar{v}_{ab,ba}^{(ab)} = \bar{u}_i\dot{x}_{ab} + \bar{u}_j\dot{y}_{ab} + \bar{u}_k\dot{z}_{ab}$, three of the derivatives become determined. On the other hand, Eqs. (6.95) and (6.78) can be used to determine $\dot{\lambda}_{ab}$ and $\dot{\bar{r}}_{abq}^{(ab)}$. For this purpose, Eq. (6.96) is used first in order to express $\dot{\bar{r}}_{abq}^{(ab)}$ as follows in terms of $\dot{\lambda}_{ab}$ as long as the matrix $\widehat{K}_{ab}$ is not singular:

$$\dot{\bar{r}}_{abq}^{(ab)} = \widehat{K}_{ab}^{-1}[\lambda_{ab}\widetilde{\omega}_{ab,ba}^{(ab)} - \dot{\lambda}_{ab}\widehat{I}]\bar{g}_{ab}^{(ab)} \tag{6.98}$$

Then, recalling that Eq. (6.78) is $\overline{g}_{ab}^{(ab)t}\dot{\overline{r}}_{abq}^{(ab)} = 0$, $\dot{\lambda}_{ab}$ is obtained as shown below if $\overline{g}_{ab}^{(ab)t}\widehat{K}_{ab}^{-1}\overline{g}_{ab}^{(ab)} \neq 0$.

$$\dot{\lambda}_{ab} = [\overline{g}_{ab}^{(ab)t}\widehat{K}_{ab}^{-1}\widetilde{\omega}_{ab,ba}^{(ab)}\overline{g}_{ab}^{(ab)}\lambda_{ab}]/[\overline{g}_{ab}^{(ab)t}\widehat{K}_{ab}^{-1}\overline{g}_{ab}^{(ab)}] \tag{6.99}$$

Finally, $\dot{\overline{r}}_{abq}^{(ab)}$ is also obtained by substituting Eq. (6.99) into Eq. (6.98). Hence, the last three derivatives $(\dot{x}_{abq}, \dot{y}_{abq}, \dot{z}_{abq})$ also become determined.

6.5.2 Curve-on-Surface Joint

Figure 6.12 illustrates a joint formed by two kinematic elements $\mathcal{E}_{ab}$ and $\mathcal{E}_{ba}$, which are in rolling contact at their coincident points Q_{ab} and Q_{ba}. $\mathcal{E}_{ab}$ has the shape of a specified surface S_{ab}, which is described by a function $f_{ab}(x, y, z)$. As for $\mathcal{E}_{ba}$, it has a sharp edge as the contact interface and its edge contour has the shape of a specified curve C_{ba}, which is described parametrically as shown below.

$$C_{ba} = \{x = f_{bai}(p), y = f_{baj}(p), z = f_{bak}(p) : p_{min} \leq p \leq p_{max}\} \tag{6.100}$$

(a) Relative Position Equations

The following equations can be written in order to express the relative position relationships between $\mathcal{E}_{ab}$ and $\mathcal{E}_{ba}$.

The fact that Q_{ab} and Q_{ba} are coincident points can be expressed as follows:

$$\vec{r}_{abq} = \vec{r}_{ab,ba} + \vec{r}_{baq} \tag{6.101}$$

The vectors in Eq. (6.101) can be represented by the following column matrices in $\mathcal{F}_{ab}$ and $\mathcal{F}_{ba}$.

$$\overline{r}_{ab,ba}^{(ab)} = \overline{u}_i x_{ab} + \overline{u}_j y_{ab} + \overline{u}_k z_{ab} \tag{6.102}$$

$$\overline{r}_{abq}^{(ab)} = \overline{u}_i x_{abq} + \overline{u}_j y_{abq} + \overline{u}_k z_{abq} \tag{6.103}$$

$$\overline{r}_{baq}^{(ba)} = \overline{u}_i f_{bai}(p) + \overline{u}_j f_{baj}(p) + \overline{u}_k f_{bak}(p) \tag{6.104}$$

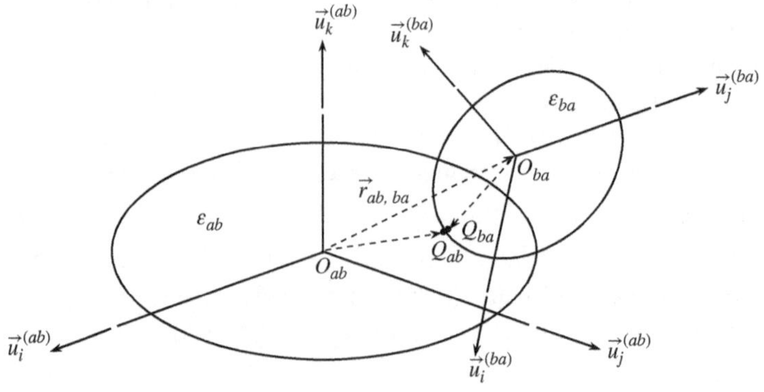

Figure 6.12 A curve-on-surface joint.

Hence, Eq. (6.101) can be replaced with the following matrix equation.

$$\overline{u}_i(x_{abq} - x_{ab}) + \overline{u}_j(y_{abq} - y_{ab}) + \overline{u}_k(z_{abq} - z_{ab}) = \widehat{C}^{(ab,ba)}[\overline{u}_i f_{bai}(p) + \overline{u}_j f_{baj}(p) + \overline{u}_k f_{bak}(p)] \tag{6.105}$$

In addition to Eq. (6.105), the coordinates of the contact point Q_{ab} must also satisfy the following surface equation.

$$f_{ab}(x_{abq}, y_{abq}, z_{abq}) = 0 \tag{6.106}$$

In Eq. (6.105), the orientation matrix can be expressed as follows with suitably selected constant unit column matrices $\overline{n}_{ab1}$, $\overline{n}_{ab2}$, and $\overline{n}_{ab3}$:

$$\widehat{C}^{(ab,ba)} = e^{\widetilde{n}_{ab1}\phi_{ab}} e^{\widetilde{n}_{ab2}\theta_{ab}} e^{\widetilde{n}_{ab3}\psi_{ab}} \tag{6.107}$$

Moreover, the curve C_{ba} must be tangent to the surface S_{ba} at the contact point. This fact can be expressed by the following equation.

$$\vec{g}_{ab} \cdot \vec{u}_{bat} = 0 \tag{6.108}$$

In Eq. (6.108), $\vec{g}_{ab}$ is the gradient vector of the surface S_{ab} and $\vec{u}_{bat}$ is the tangential unit vector of the curve C_{ba}. These vectors are represented by the following column matrices, respectively, in $\mathcal{F}_{ab}$ and $\mathcal{F}_{ba}$.

$$\overline{g}_{ab}^{(ab)} = \overline{u}_i f_{abx} + \overline{u}_j f_{aby} + \overline{u}_k f_{abz} \tag{6.109}$$

$$\overline{u}_{bat}^{(ba)} = (\overline{u}_i f'_{bai} + \overline{u}_j f'_{baj} + \overline{u}_k f'_{bak}) / [(f'_{bai})^2 + (f'_{baj})^2 + (f'_{bak})^2]^{1/2} \tag{6.110}$$

In Eqs. (6.109) and (6.110),

$$\left.\begin{aligned} f_{abx} &= \partial f_{ab}/\partial x \\ f_{aby} &= \partial f_{ab}/\partial y \\ f_{abz} &= \partial f_{ab}/\partial z \end{aligned}\right\} \tag{6.111}$$

$$\left.\begin{aligned} f'_{bai} &= df_{bai}/dp \\ f'_{baj} &= df_{baj}/dp \\ f'_{bak} &= df_{bak}/dp \end{aligned}\right\} \tag{6.112}$$

By using the matrix representations of $\vec{g}_{ab}$ and $\vec{u}_{bat}$ in $\mathcal{F}_{ab}$ and $\mathcal{F}_{ba}$, the tangency equation, i.e. Eq. (6.108), can be written as follows:

$$\overline{g}_{ab}^{(ab)t}\widehat{C}^{(ab,ba)}\overline{u}_{bat}^{(ba)} = 0 \Rightarrow$$
$$(\overline{u}_i f_{abx} + \overline{u}_j f_{aby} + \overline{u}_k f_{abz})^t \widehat{C}^{(ab,ba)}(\overline{u}_i f'_{bai} + \overline{u}_j f'_{baj} + \overline{u}_k f'_{bak}) = 0 \tag{6.113}$$

As seen above, the curve-on-surface joint involves 10 variables, which are

$$x_{ab}, y_{ab}, z_{ab}; \phi_{ab}, \theta_{ab}, \psi_{ab}; x_{abq}, y_{abq}, z_{abq}; p$$

The 10 variables listed above are interrelated by five independent scalar equations. Three of them are contained in Eq. (6.105) and two of them are Eqs. (6.106) and (6.113). Therefore, the mobility of a curve-on-surface joint is $\mu_{ab} = 10 - 5 = 5$. Thus, if five of the variables (e.g. ϕ_{ab}, θ_{ab}, ψ_{ab}, x_{ab}, y_{ab}) are specified, then the other five variables (i.e. z_{ab}, x_{abq}, y_{abq}, z_{abq}, and p) can be determined by using the mentioned equations.

(b) Relative Velocity Equations

The differentiation of Eqs. (6.105)–(6.107) and (6.113) leads to the following relative velocity equations.

$$\begin{aligned}&\overline{u}_i(\dot{x}_{abq}-\dot{x}_{ab})+\overline{u}_j(\dot{y}_{abq}-\dot{y}_{ab})+\overline{u}_k(\dot{z}_{abq}-\dot{z}_{ab})\\&=\widetilde{\omega}^{(ab)}_{ab,ba}[\overline{u}_i(x_{abq}-x_{ab})+\overline{u}_j(y_{abq}-y_{ab})+\overline{u}_k(z_{abq}-z_{ab})]\\&+\dot{p}\widehat{C}^{(ab,ba)}(\overline{u}_if'_{bai}+\overline{u}_jf'_{baj}+\overline{u}_kf'_{bak})\end{aligned}\tag{6.114}$$

$$f_{abx}\dot{x}_{abq}+f_{aby}\dot{y}_{abq}+f_{abz}\dot{z}_{abq}=0\Rightarrow$$

$$\overline{g}^{(ab)t}_{ab}\dot{\overline{r}}^{(ab)}_{abq}=0\tag{6.115}$$

$$\begin{aligned}&\dot{\overline{r}}^{(ab)t}_{abq}\widehat{H}^{(ab)}_{ab}\widehat{C}^{(ab,ba)}(\overline{u}_if'_{bai}+\overline{u}_jf'_{baj}+\overline{u}_kf'_{bak})\\&+\overline{g}^{(ba)t}_{ba}\widetilde{\omega}^{(ab)}_{ab,ba}\widehat{C}^{(ab,ba)}(\overline{u}_if'_{bai}+\overline{u}_jf'_{baj}+\overline{u}_kf'_{bak})\\&+\dot{p}\overline{g}^{(ba)t}_{ba}\widehat{C}^{(ab,ba)}(\overline{u}_if''_{bai}+\overline{u}_jf''_{baj}+\overline{u}_kf''_{bak})=0\end{aligned}\tag{6.116}$$

In the above equations, $\overline{\omega}^{(ab)}_{ab,ba}$ is expressed as follows:

$$\overline{\omega}^{(ab)}_{ab,ba}=\dot{\phi}_{ab}\overline{n}_{ab1}+\dot{\theta}_{ab}e^{\widetilde{n}_{ab1}\phi_{ab}}\overline{n}_{ab2}+\dot{\psi}_{ab}e^{\widetilde{n}_{ab1}\phi_{ab}}e^{\widetilde{n}_{ab2}\theta_{ab}}\overline{n}_{ab3}\tag{6.117}$$

In Eq. (6.116), the Hessian matrix $\widehat{H}^{(ab)}_{ab}$ is the same as expressed previously in Section 6.5.1. As for the double primes, they indicate the second derivatives of the relevant functions with respect to the curve parameter p.

Similarly as in Part (a), if five of the derivatives (e.g. $\dot{\phi}_{ab}$, $\dot{\theta}_{ab}$, $\dot{\psi}_{ab}$, $\dot{x}_{ab}$, $\dot{y}_{ab}$) are specified, then the other five derivatives (i.e. $\dot{z}_{ab}$, $\dot{x}_{abq}$, $\dot{y}_{abq}$, $\dot{z}_{abq}$, and $\dot{p}$) can be determined by using Eqs. (6.114)–(6.116).

(c) Relative Velocity Equations in the Case of Rolling Without Slipping

If $\mathcal{E}_{ba}$ rolls over $\mathcal{E}_{ab}$ without slipping due to the presence of sticking friction, then, as discussed in Section 6.5.1, the derivatives $\dot{\overline{r}}^{(ab)}_{abq}$ and $\dot{\overline{r}}^{(ba)}_{baq}$ become related by the following nonholonomic constraint.

$$\dot{\overline{r}}^{(ab)}_{abq}=\widehat{C}^{(ab,ba)}\dot{\overline{r}}^{(ba)}_{baq}=\dot{p}\widehat{C}^{(ab,ba)}(\overline{u}_if'_{bai}+\overline{u}_jf'_{baj}+\overline{u}_kf'_{bak})\tag{6.118}$$

Note that Eq. (6.118) has already given $\dot{\overline{r}}^{(ab)}_{abq}$ in terms of $\dot{p}$. Due to the same equation, Eqs. (6.114)–(6.117) reduce to the following equations that contain the remaining seven derivatives, which are $\dot{x}_{ab}$, $\dot{y}_{ab}$, $\dot{z}_{ab}$, $\dot{\phi}_{ab}$, $\dot{\theta}_{ab}$, $\dot{\psi}_{ab}$, and $\dot{p}$.

$$\begin{aligned}&\overline{u}_i\dot{x}_{ab}+\overline{u}_j\dot{y}_{ab}+\overline{u}_k\dot{z}_{ab}\\&=\widetilde{\omega}^{(ab)}_{ab,ba}[\overline{u}_i(x_{ab}-x_{abq})+\overline{u}_j(y_{ab}-y_{abq})+\overline{u}_k(z_{ab}-z_{abq})\end{aligned}\tag{6.119}$$

$$\overline{g}^{(ab)t}_{ab}\dot{\overline{r}}^{(ab)}_{abq}=0\Rightarrow$$

$$\overline{g}^{(ab)t}_{ab}\widehat{C}^{(ab,ba)}\dot{\overline{r}}^{(ba)}_{baq}=\dot{p}\overline{g}^{(ab)t}_{ab}\widehat{C}^{(ab,ba)}(\overline{u}_if'_{bai}+\overline{u}_jf'_{baj}+\overline{u}_kf'_{bak})=0\Rightarrow$$

$$(\overline{u}_if_{abx}+\overline{u}_jf_{aby}+\overline{u}_kf_{abz})^t\widehat{C}^{(ab,ba)}(\overline{u}_if'_{bai}+\overline{u}_jf'_{baj}+\overline{u}_kf'_{bak})=0\tag{6.120}$$

$$\dot{p}(\overline{u}_i f'_{bai} + \overline{u}_j f'_{baj} + \overline{u}_k f'_{bak})^t \hat{C}^{(ba,ab)} \hat{H}^{(ab)}_{ab} \hat{C}^{(ab,ba)} (\overline{u}_i f'_{bai} + \overline{u}_j f'_{baj} + \overline{u}_k f'_{bak})$$
$$+ (\overline{u}_i f_{abx} + \overline{u}_j f_{aby} + \overline{u}_k f_{abz})^t \tilde{\omega}^{(ab)}_{ab,ba} \hat{C}^{(ab,ba)} (\overline{u}_i f'_{bai} + \overline{u}_j f'_{baj} + \overline{u}_k f'_{bak})$$
$$+ \dot{p}(\overline{u}_i f_{abx} + \overline{u}_j f_{aby} + \overline{u}_k f_{abz})^t \hat{C}^{(ab,ba)} (\overline{u}_i f''_{bai} + \overline{u}_j f''_{baj} + \overline{u}_k f''_{bak}) = 0 \tag{6.121}$$

As seen above, Eq. (6.120) has come out to be the same as the tangency equation, i.e. Eq. (6.113), written previously among the relative position equations. So, concerning the velocities, there remain only Eqs. (6.119) and (6.121). Equation (6.121) is already a scalar equation and Eq. (6.119) provides three scalar equations. These four scalar equations contain the seven derivatives ($\dot{x}_{ab}$, $\dot{y}_{ab}$, $\dot{z}_{ab}$, $\dot{\phi}_{ab}$, $\dot{\theta}_{ab}$, $\dot{\psi}_{ab}$, and $\dot{p}$) mentioned before. Therefore, the mobility of a curve-on-surface joint reduces to $\mu_{ab} = 7 - 4 = 3$ during the rolling-without-slipping motion. Thus, if three of the derivatives (e.g. $\dot{\phi}_{ab}$, $\dot{\theta}_{ab}$, $\dot{\psi}_{ab}$) are specified along with the position specifications, then the other four derivatives (i.e. $\dot{x}_{ab}$, $\dot{y}_{ab}$, $\dot{z}_{ab}$, and $\dot{p}$) can be determined by using Eqs. (6.119) and (6.121).

6.5.3 Curve-on-Curve Joint

Figure 6.13 illustrates a joint formed by two kinematic elements $\mathcal{E}_{ab}$ and $\mathcal{E}_{ba}$, which are in rolling contact at their coincident points Q_{ab} and Q_{ba}. The shapes of $\mathcal{E}_{ab}$ and $\mathcal{E}_{ba}$ are such that they both have sharp edges as their contact interfaces. Their edge contours have the shapes of two specified curves C_{ab} and C_{ba}, which are described parametrically as shown below.

$$C_{ab} = \{x = f_{abi}(p_a), y = f_{abj}(p_a), z = f_{abk}(p_a) : p_{amin} \le p_a \le p_{amax}\} \tag{6.122}$$

$$C_{ba} = \{x = f_{bai}(p_b), y = f_{baj}(p_b), z = f_{bak}(p_b) : p_{bmin} \le p_b \le p_{bmax}\} \tag{6.123}$$

(a) Relative Position Equations

The fact that Q_{ab} and Q_{ba} are coincident points can be expressed as follows:

$$\vec{r}_{abq} = \vec{r}_{ab,ba} + \vec{r}_{baq} \tag{6.124}$$

The vectors in Eq. (6.124) can be represented by the following column matrices in $\mathcal{F}_{ab}$ and $\mathcal{F}_{ba}$.

$$\overline{r}^{(ab)}_{ab,ba} = \overline{u}_i x_{ab} + \overline{u}_j y_{ab} + \overline{u}_k z_{ab} \tag{6.125}$$

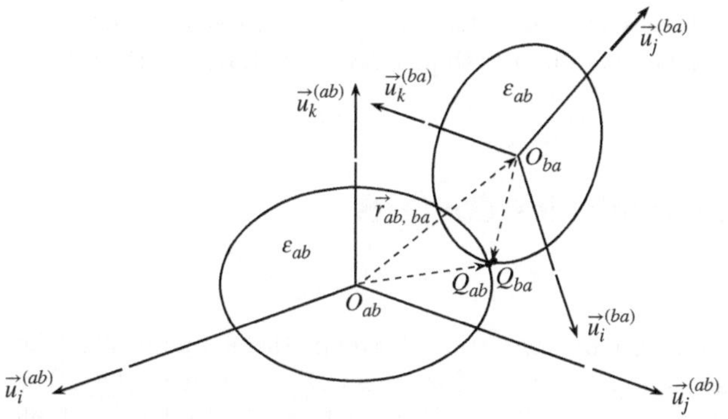

Figure 6.13 A curve-on-curve joint.

$$\overline{r}_{abq}^{(ab)} = \overline{u}_i f_{abi}(p_a) + \overline{u}_j f_{abj}(p_a) + \overline{u}_k f_{abk}(p_a) \tag{6.126}$$

$$\overline{r}_{baq}^{(ba)} = \overline{u}_i f_{bai}(p_b) + \overline{u}_j f_{baj}(p_b) + \overline{u}_k f_{bak}(p_b) \tag{6.127}$$

Hence, Eq. (6.124) can be replaced with the following matrix equation.

$$\begin{aligned}&\overline{u}_i[f_{abi}(p_a) - x_{ab}] + \overline{u}_j[f_{abj}(p_a) - y_{ab}] + \overline{u}_k[f_{abk}(p_a) - z_{ab}]\\&= \widehat{C}^{(ab,ba)}[\overline{u}_i f_{bai}(p_b) + \overline{u}_j f_{baj}(p_b) + \overline{u}_k f_{bak}(p_b)]\end{aligned} \tag{6.128}$$

In Eq. (6.128), the orientation matrix can be expressed as follows with suitably selected constant unit column matrices $\overline{n}_{ab1}$, $\overline{n}_{ab2}$, and $\overline{n}_{ab3}$:

$$\widehat{C}^{(ab,ba)} = e^{\widetilde{n}_{ab1}\phi_{ab}} e^{\widetilde{n}_{ab2}\theta_{ab}} e^{\widetilde{n}_{ab3}\psi_{ab}} \tag{6.129}$$

As seen above, the curve-on-curve joint involves eight variables, which are

$$x_{ab}, y_{ab}, z_{ab}; \phi_{ab}, \theta_{ab}, \psi_{ab}; p_a, p_b$$

The eight variables listed above are interrelated by three independent scalar equations contained in Eq. (6.128). Therefore, the mobility of this joint is $\mu_{ab} = 8 - 3 = 5$. Thus, if five of the variables (e.g. $\phi_{ab}, \theta_{ab}, \psi_{ab}, x_{ab}, y_{ab}$) are specified, then the other three variables (i.e. p_a, p_b, and z_{ab}) can be determined by using Eq. (6.128).

(b) Relative Velocity Equations

The differentiation of Eq. (6.128) leads to the following relative velocity equation.

$$\begin{aligned}&\overline{u}_i(f'_{abi}\dot{p}_a - \dot{x}_{ab}) + \overline{u}_j(f'_{abj}\dot{p}_a - \dot{y}_{ab}) + \overline{u}_k(f'_{abk}\dot{p}_a - \dot{z}_{ab})\\&= \widetilde{\omega}_{ab,ba}^{(ab)}[\overline{u}_i(f_{abi} - x_{ab}) + \overline{u}_j(f_{abj} - y_{ab}) + \overline{u}_k(f_{abk} - z_{ab})]\\&\quad + \dot{p}_b\widehat{C}^{(ab,ba)}(\overline{u}_i f'_{bai} + \overline{u}_j f'_{baj} + \overline{u}_k f'_{bak})\end{aligned} \tag{6.130}$$

In Eq. (6.130), $\overline{\omega}_{ab,ba}^{(ab)}$ is obtained from Eq. (6.129). That is,

$$\overline{\omega}_{ab,ba}^{(ab)} = \dot{\phi}_{ab}\overline{n}_{ab1} + \dot{\theta}_{ab} e^{\widetilde{n}_{ab1}\phi_{ab}}\overline{n}_{ab2} + \dot{\psi}_{ab} e^{\widetilde{n}_{ab1}\phi_{ab}} e^{\widetilde{n}_{ab2}\theta_{ab}}\overline{n}_{ab3} \tag{6.131}$$

Equation (6.130) can be used so that, if five of the derivatives (e.g. $\dot{\phi}_{ab}, \dot{\theta}_{ab}, \dot{\psi}_{ab}, \dot{x}_{ab}, \dot{y}_{ab}$) are specified, then the other three derivatives (i.e. $\dot{p}_a$, $\dot{p}_b$, and $\dot{z}_{ab}$) can be determined.

However, it is to be noted that, although the curve-on-curve joint is a rolling contact joint, it is not possible to have a relative rolling-without-slipping motion between its sharp-edge kinematic elements. If there is sticking friction between these kinematic elements, then the relative motion becomes similar to that between the kinematic elements of a spherical joint.

6.6 Rolling Contact Joints with Line Contacts

6.6.1 Cone-on-Cone Joint

Figure 6.14 illustrates a joint formed by two conical kinematic elements $\mathcal{E}_{ab}$ and $\mathcal{E}_{ba}$, which are in rolling contact along the coincident lines $O_{ab}Q_{ab}$ and $O_{ba}Q_{ba}$. The kinematic elements have the shapes of specified conical surfaces S_{ab} and S_{ba}, which can be described by the *base functions* $f_{ab}(\xi, \eta)$ and $f_{ba}(\xi, \eta)$ in the associated reference

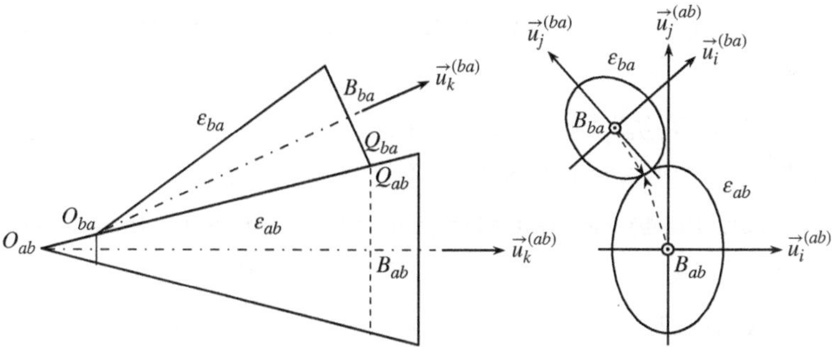

Figure 6.14 A cone-on-cone joint.

frames $\mathcal{F}_{ab}$ and $\mathcal{F}_{ba}$. These functions are also called *directrix functions*. They are assumed to be continuous and differentiable with respect to their arguments. As noticed, the origins O_{ab} and O_{ba} of the frames $\mathcal{F}_{ab}$ and $\mathcal{F}_{ba}$ are placed at the vertices of the conical surfaces.

(a) Relative Position Equations

The conical surfaces are described by the following equations.

$$f_{ab}(\xi_{abq}, \eta_{abq}) = 0 \tag{6.132}$$

$$f_{ba}(\xi_{baq}, \eta_{baq}) = 0 \tag{6.133}$$

In Eqs. (6.132) and (6.133), the arguments of the base functions are the *coordinate ratios* that are defined as follows:

$$\left.\begin{aligned} \xi_{abq} &= x_{abq}/z_{abq} \\ \eta_{abq} &= y_{abq}/z_{abq} \end{aligned}\right\} \tag{6.134}$$

$$\left.\begin{aligned} \xi_{baq} &= x_{baq}/z_{baq} \\ \eta_{baq} &= y_{baq}/z_{baq} \end{aligned}\right\} \tag{6.135}$$

For the function $f_{ab}(\xi_{abq}, \eta_{abq})$, the derivatives required for the gradient vector $\vec{g}_{ab}$ are taken as shown below.

$$f_{abx} = \partial f_{ab}/\partial x_{abq} = (\partial f_{ab}/\partial \xi_{abq})(\partial \xi_{abq}/\partial x_{abq}) = f_{ab\xi}/z_{abq} \tag{6.136}$$

$$f_{aby} = \partial f_{ab}/\partial y_{abq} = (\partial f_{ab}/\partial \eta_{abq})(\partial \eta_{abq}/\partial y_{abq}) = f_{ab\eta}/z_{abq} \tag{6.137}$$

$$f_{abz} = \partial f_{ab}/\partial z_{abq} \Rightarrow$$

$$f_{abz} = (\partial f_{ab}/\partial \xi_{abq})(\partial \xi_{abq}/\partial z_{abq}) + (\partial f_{ab}/\partial \eta_{abq})(\partial \eta_{abq}/\partial z_{abq}) \Rightarrow$$

$$f_{abz} = -(f_{ab\xi}\xi_{abq} + f_{ab\eta}\eta_{abq})/z_{abq} \tag{6.138}$$

In order to obtain the companion gradient vector $\vec{g}_{ba}$, similar derivates are taken for the function $f_{ba}(\xi_{baq}, \eta_{baq})$, too.

Thus, the gradient vectors can be represented by the following column matrices in their respective frames $\mathcal{F}_{ab}$ and $\mathcal{F}_{ba}$.

$$\left.\begin{aligned} \overline{g}_{ab}^{(ab)} &= \overline{\gamma}_{ab}^{(ab)}/z_{abq} \\ \overline{g}_{ba}^{(ba)} &= \overline{\gamma}_{ba}^{(ba)}/z_{baq} \end{aligned}\right\} \tag{6.139}$$

In Eq. Set (6.139), $\overline{\gamma}_{ab}^{(ab)}$ and $\overline{\gamma}_{ba}^{(ba)}$ are defined as the *modified gradient functions* with the following expressions.

$$\left.\begin{aligned}\overline{\gamma}_{ab}^{(ab)} &= \overline{u}_i f_{ab\xi} + \overline{u}_j f_{ab\eta} - \overline{u}_k (f_{ab\xi}\xi_{abq} + f_{ab\eta}\eta_{abq}) \\ \overline{\gamma}_{ba}^{(ba)} &= \overline{u}_i f_{ba\xi} + \overline{u}_j f_{ba\eta} - \overline{u}_k (f_{ba\xi}\xi_{baq} + f_{ba\eta}\eta_{baq})\end{aligned}\right\} \tag{6.140}$$

Equation Sets (6.139) and (6.140) can be used to express the alignment of the gradient vectors as shown below.

$$\widehat{C}^{(ab,ba)}\overline{g}_{ba}^{(ba)} = -\lambda_{ab}\overline{g}_{ab}^{(ab)} \Rightarrow$$

$$\widehat{C}^{(ab,ba)}\overline{\gamma}_{ba}^{(ba)} = -\lambda'_{ab}\overline{\gamma}_{ab}^{(ab)} \tag{6.141}$$

In Eq. (6.141), λ'_{ab} is the *modified gradient ratio*, which is related to the *regular gradient ratio* λ_{ab} as follows:

$$\lambda'_{ab} = (z_{baq}/z_{abq})\lambda_{ab} \tag{6.142}$$

The position vectors of the contact points along the contact lines (i.e. the coincident *generatrix lines*) can be expressed in the frames $\mathcal{F}_{ab}$ and $\mathcal{F}_{ba}$ as written below.

$$\overline{r}_{abq}^{(ab)} = (\overline{u}_i\xi_{abq} + \overline{u}_j\eta_{abq} + \overline{u}_k)z_{abq} = \overline{w}_{abl}^{(ab)} z_{abq} \tag{6.143}$$

$$\overline{r}_{baq}^{(ba)} = (\overline{u}_i\xi_{baq} + \overline{u}_j\eta_{baq} + \overline{u}_k)z_{baq} = \overline{w}_{bal}^{(ba)} z_{baq} \tag{6.144}$$

In Eqs. (6.143) and (6.144), $\overline{w}_{abl}^{(ab)}$ and $\overline{w}_{bal}^{(ba)}$ are the matrix representations of the *generatrix line vectors* $\vec{w}_{abl}$ and $\vec{w}_{bal}$ of the conical surfaces. They are defined as follows:

$$\left.\begin{aligned}\overline{w}_{abl}^{(ab)} &= \overline{u}_i\xi_{abq} + \overline{u}_j\eta_{abq} + \overline{u}_k = \overline{\rho}_{abq}^{(ab)} + \overline{u}_k \\ \overline{w}_{bal}^{(ba)} &= \overline{u}_i\xi_{baq} + \overline{u}_j\eta_{baq} + \overline{u}_k = \overline{\rho}_{baq}^{(ba)} + \overline{u}_k\end{aligned}\right\} \tag{6.145}$$

As for $\overline{\rho}_{abq}^{(ab)}$ and $\overline{\rho}_{baq}^{(ba)}$ in Eq. Set (6.145), they are the matrix representations of the base plane position vectors $\vec{\rho}_{abq}$ and $\vec{\rho}_{baq}$ of the conical surfaces. They are the vectors illustrated by dotted lines on the right-hand side of Figure 6.14. Considering the base plane origins B_{ab} and B_{ba}, they are defined as indicated below.

$$\vec{\rho}_{abq} = \overrightarrow{B_{ab}Q_{ab}}/z_{abq} \text{ and } \vec{\rho}_{baq} = \overrightarrow{B_{ba}Q_{ba}}/z_{baq}$$

By using the preceding definitions, the coincidence of the contact lines $O_{ab}Q_{ab}$ and $O_{ba}Q_{ba}$ can be expressed as follows:

$$\overline{r}_{abq}^{(ab)} = \overline{r}_{ab,ba}^{(ab)} + \widehat{C}^{(ab,ba)}\overline{r}_{baq}^{(ba)} \Rightarrow$$

$$\overline{w}_{abl}^{(ab)} z_{abq} = \overline{r}_{ab,ba}^{(ab)} + \widehat{C}^{(ab,ba)}\overline{w}_{bal}^{(ba)} z_{baq} \tag{6.146}$$

Due to the coincidence of the contact lines, O_{ba} is also located on the line $O_{ab}Q_{ab}$. That is,

$$\overline{r}_{ab,ba}^{(ab)} = \overline{u}_i x_{ab} + \overline{u}_j y_{ab} + \overline{u}_k z_{ab} = \overline{w}_{abl}^{(ab)} z_{ab} = [\overline{\rho}_{abq}^{(ab)} + \overline{u}_k] z_{ab} \Rightarrow$$

$$\overline{r}_{ab,ba}^{(ab)} = (\overline{u}_i\xi_{abq} + \overline{u}_j\eta_{abq} + \overline{u}_k)z_{ab} \tag{6.147}$$

Equation (6.147) implies the following consistency relationships.

$$\left.\begin{aligned} x_{ab} &= \xi_{abq} z_{ab} \\ y_{ab} &= \eta_{abq} z_{ab} \end{aligned}\right\} \tag{6.148}$$

Equations (6.146) and (6.147) can be combined into the following *contact line equation.*

$$\overline{w}_{abl}^{(ab)}(\zeta_{abq} - \zeta_{ab}) = \widehat{C}^{(ab,ba)} \overline{w}_{bal}^{(ba)} \tag{6.149}$$

In Eq. (6.149),

$$\left.\begin{aligned} \zeta_{abq} &= z_{abq}/z_{baq} \\ \zeta_{ab} &= z_{ab}/z_{baq} \end{aligned}\right\} \tag{6.150}$$

At a general cone-on-cone joint, one of z_{abq} and z_{baq} can be taken as a constant. However, in the case shown in Figure 6.14, z_{abq} depends on the coordinate z_{ab} of O_{ba}, which is not necessarily constant. Therefore, in this case, z_{baq} is taken as a constant. That is,

$$z_{baq} = h_{ba} = \text{constant} \tag{6.151}$$

As for the relative orientation matrix, it is expressed here as follows according to the $k-i-k$ sequence, which is suitable for the conical geometry. However, if desired, the $k-j-k$ sequence may also be used as an alternative.

$$\widehat{C}^{(ab,ba)} = e^{\tilde{u}_k \phi_{ab}} e^{\tilde{u}_i \theta_{ab}} e^{\tilde{u}_k \psi_{ab}} \tag{6.152}$$

Considering the orientation angles and all the coordinate ratios defined above together with the gradient ratio, a cone-on-cone joint involves 10 principal variables that are listed below.

$$\phi_{ab}, \theta_{ab}, \psi_{ab}; \lambda'_{ab}; \zeta_{ab}, \zeta_{abq}; \xi_{abq}, \eta_{abq}; \xi_{baq}, \eta_{baq}$$

The variables listed above must satisfy two scalar equations, i.e. Eqs. (6.132) and (6.133), and two column matrix equations, i.e. Eqs. (6.141) and (6.149). However, the gradient vectors are naturally perpendicular to the contact line vectors. This fact can be seen in the following identically satisfied equations, which are derived by using Eq. Set (6.140) together with Eqs. (6.143) and (6.144).

$$\begin{aligned} \overline{r}_{abq}^{(ab)t} \overline{g}_{ab}^{(ab)} &= \overline{w}_{abl}^{(ab)t} \overline{\gamma}_{ab}^{(ab)} \\ &= (\overline{u}_i \xi_{abq} + \overline{u}_j \eta_{abq} + \overline{u}_k)^t [\overline{u}_i f_{ab\xi} + \overline{u}_j f_{ab\eta} - \overline{u}_k (f_{ab\xi} \xi_{abq} + f_{ab\eta} \eta_{abq})] \equiv 0 \end{aligned} \tag{6.153}$$

$$\begin{aligned} \overline{r}_{baq}^{(ba)t} \overline{g}_{ba}^{(ba)} &= \overline{w}_{bal}^{(ba)t} \overline{\gamma}_{ba}^{(ba)} \\ &= (\overline{u}_i \xi_{baq} + \overline{u}_j \eta_{baq} + \overline{u}_k)^t [\overline{u}_i f_{ba\xi} + \overline{u}_j f_{ba\eta} - \overline{u}_k (f_{ba\xi} \xi_{baq} + f_{ba\eta} \eta_{baq})] \equiv 0 \end{aligned} \tag{6.154}$$

Because of the above-mentioned feature of the gradient vectors, Eq. (6.141) contains only two independent scalar equations. Consequently, the total number of independent scalar equations becomes seven and the mobility of a cone-on-cone joint turns out to be $\mu_{ab} = 10 - 7 = 3$. Thus, if three of the variables (e.g. $\zeta_{ab}, \xi_{abq}, \xi_{baq}$) are specified, then the remaining seven variables (i.e. $\phi_{ab}, \theta_{ab}, \psi_{ab}, \lambda'_{ab}, \zeta_{abq}, \eta_{abq}, \eta_{baq}$) can be determined by using the mentioned equations. Afterwards, the values of the actual coordinates (x_{ab}, y_{ab}, z_{ab}, and so on) can also be determined by using Eq. (6.151) with the specified value of h_{ba}.

The two independent scalar equations contained in Eq. (6.141) can be picked up by premultiplying both sides by the transposes of $\overline{w}_{abt}^{(ab)}$ and $\overline{w}_{abr}^{(ab)}$, which are the orthogonal companions of $\overline{w}_{abl}^{(ab)}$. This operation results in the following equations.

$$\overline{w}_{abt}^{(ab)t}\widehat{C}^{(ab,ba)}\overline{\gamma}_{ba}^{(ba)} = -\lambda'_{ab}\overline{w}_{abt}^{(ab)t}\overline{\gamma}_{ab}^{(ab)} \tag{6.155}$$

$$\overline{w}_{abr}^{(ab)t}\widehat{C}^{(ab,ba)}\overline{\gamma}_{ba}^{(ba)} = -\lambda'_{ab}\overline{w}_{abr}^{(ab)t}\overline{\gamma}_{ab}^{(ab)} \tag{6.156}$$

In Eqs. (6.155) and (6.156), $\overline{w}_{abt}^{(ab)}$ and $\overline{w}_{abr}^{(ab)}$ are defined as the *transversal* and *radial* orthogonal companions of $\overline{w}_{abl}^{(ab)}$. They are defined so that $\overline{w}_{abl}^{(ab)t}\overline{w}_{abt}^{(ab)} = \overline{w}_{abl}^{(ab)t}\overline{w}_{abr}^{(ab)} = 0$ and $\overline{w}_{abr}^{(ab)} = \widetilde{w}_{abl}^{(ab)}\overline{w}_{abt}^{(ab)}$. Similar definitions can be made for the orthogonal companions of $\overline{w}_{bal}^{(ba)}$, too. Referring to Eq. (6.145), the triads $\{\overline{w}_{abl}^{(ab)}, \overline{w}_{abt}^{(ab)}, \overline{w}_{abr}^{(ab)}\}$ and $\{\overline{w}_{bal}^{(ba)}, \overline{w}_{bat}^{(ba)}, \overline{w}_{bar}^{(ba)}\}$ can be expressed together as shown below.

$$\left.\begin{aligned}
\overline{w}_{abl}^{(ab)} &= \overline{u}_i\xi_{abq} + \overline{u}_j\eta_{abq} + \overline{u}_k \\
\overline{w}_{abt}^{(ab)} &= \overline{u}_i\eta_{abq} - \overline{u}_j\xi_{abq} \\
\overline{w}_{abr}^{(ab)} &= \overline{u}_i\xi_{abq} + \overline{u}_j\eta_{abq} - \overline{u}_k(\xi_{abq}^2 + \eta_{abq}^2)
\end{aligned}\right\} \tag{6.157}$$

$$\left.\begin{aligned}
\overline{w}_{bal}^{(ba)} &= \overline{u}_i\xi_{baq} + \overline{u}_j\eta_{baq} + \overline{u}_k \\
\overline{w}_{bat}^{(ba)} &= \overline{u}_i\eta_{baq} - \overline{u}_j\xi_{baq} \\
\overline{w}_{bar}^{(ba)} &= \overline{u}_i\xi_{baq} + \overline{u}_j\eta_{baq} - \overline{u}_k(\xi_{baq}^2 + \eta_{baq}^2)
\end{aligned}\right\} \tag{6.158}$$

(b) Relative Velocity Equations

The differentiation of Eqs. (6.132), (6.133), (6.155), (6.156), and (6.149) leads to the following relative velocity equations.

$$f_{ab\xi}\dot{\xi}_{abq} + f_{ab\eta}\dot{\eta}_{abq} = 0 \tag{6.159}$$

$$f_{ba\xi}\dot{\xi}_{baq} + f_{ba\eta}\dot{\eta}_{baq} = 0 \tag{6.160}$$

$$\begin{aligned}
&\dot{\overline{w}}_{abt}^{(ab)t}\widehat{C}^{(ab,ba)}\overline{\gamma}_{ba}^{(ba)} + \overline{w}_{abt}^{(ab)t}\left[\widehat{C}^{(ab,ba)}\dot{\overline{\gamma}}_{ba}^{(ba)} + \widetilde{\omega}_{ab,ba}^{(ab)}\widehat{C}^{(ab,ba)}\overline{\gamma}_{ba}^{(ba)}\right] \\
&= -\dot{\lambda}'_{ab}\overline{w}_{abt}^{(ab)t}\overline{\gamma}_{ab}^{(ab)} - \lambda'_{ab}\left[\dot{\overline{w}}_{abt}^{(ab)t}\overline{\gamma}_{ab}^{(ab)} + \overline{w}_{abt}^{(ab)t}\dot{\overline{\gamma}}_{ab}^{(ab)}\right]
\end{aligned} \tag{6.161}$$

$$\begin{aligned}
&\dot{\overline{w}}_{abr}^{(ab)t}\widehat{C}^{(ab,ba)}\overline{\gamma}_{ba}^{(ba)} + \overline{w}_{abr}^{(ab)t}\left[\widehat{C}^{(ab,ba)}\dot{\overline{\gamma}}_{ba}^{(ba)} + \widetilde{\omega}_{ab,ba}^{(ab)}\widehat{C}^{(ab,ba)}\overline{\gamma}_{ba}^{(ba)}\right] \\
&= -\dot{\lambda}'_{ab}\overline{w}_{abr}^{(ab)t}\overline{\gamma}_{ab}^{(ab)} - \lambda'_{ab}\left[\dot{\overline{w}}_{abr}^{(ab)t}\overline{\gamma}_{ab}^{(ab)} + \overline{w}_{abr}^{(ab)t}\dot{\overline{\gamma}}_{ab}^{(ab)}\right]
\end{aligned} \tag{6.162}$$

$$\begin{aligned}
&\dot{\overline{w}}_{abl}^{(ab)}(\zeta_{abq} - \zeta_{ab}) + \overline{w}_{abl}^{(ab)}(\dot{\zeta}_{abq} - \dot{\zeta}_{ab}) \\
&= \widehat{C}^{(ab,ba)}\dot{\overline{w}}_{bal}^{(ba)} + \widetilde{\omega}_{ab,ba}^{(ab)}\widehat{C}^{(ab,ba)}\overline{w}_{bal}^{(ba)}
\end{aligned} \tag{6.163}$$

In the preceding equations, $\overline{\omega}_{ab,ba}^{(ab)}$ is derived from Eq. (6.152) as follows:

$$\overline{\omega}_{ab,ba}^{(ab)} = \dot{\phi}_{ab}\overline{u}_k + \dot{\theta}_{ab}e^{\widetilde{u}_k\phi_{ab}}\overline{u}_i + \dot{\psi}_{ab}e^{\widetilde{u}_k\phi_{ab}}e^{\widetilde{u}_i\theta_{ab}}\overline{u}_k \tag{6.164}$$

The derivatives $\dot{\overline{w}}_{abl}^{(ab)}$, $\dot{\overline{w}}_{abt}^{(ab)}$, $\dot{\overline{w}}_{abr}^{(ab)}$, and $\dot{\overline{\gamma}}_{ab}^{(ab)}$ can be expressed in terms of $\dot{\xi}_{abq}$ and $\dot{\eta}_{abq}$. Similarly, the derivatives $\dot{\overline{w}}_{bal}^{(ba)}$ and $\dot{\overline{\gamma}}_{ba}^{(ba)}$ can be expressed in terms of $\dot{\xi}_{baq}$ and $\dot{\eta}_{baq}$.

As such, the relative velocity equations written above contain 10 derivatives, which are the derivatives of the principal variables listed in Part (a). So, if three of them (e.g. $\dot{\zeta}_{ab}$, $\dot{\xi}_{abq}$, $\dot{\xi}_{baq}$) are specified along with the position specifications, the others (i.e. $\dot{\phi}_{ab}$, $\dot{\theta}_{ab}$, $\dot{\psi}_{ab}$, $\dot{\lambda}'_{ab}$, $\dot{\zeta}_{abq}$, $\dot{\eta}_{abq}$, $\dot{\eta}_{baq}$) can be determined from Eqs. (6.159)–(6.163).

(c) Relative Velocity Equations in the Case of Rolling Without Slipping

If $\mathcal{E}_{ba}$ rolls over $\mathcal{E}_{ab}$ without slipping due to the presence of sticking friction, then, as discussed in Section 6.5, the derivatives $\dot{\overline{r}}^{(ab)}_{abq}$ and $\dot{\overline{r}}^{(ba)}_{baq}$ become related by the following nonholonomic constraint.

$$\dot{\overline{r}}^{(ab)}_{abq} = \widehat{C}^{(ab,ba)}\dot{\overline{r}}^{(ba)}_{baq} \tag{6.165}$$

With $z_{baq} = h_{ba} = \text{constant}$, Eqs. (6.143), (6.144), and (6.165) lead to the following equation.

$$\dot{\overline{w}}^{(ab)}_{abl}\zeta_{abq} + \overline{w}^{(ab)}_{abl}\dot{\zeta}_{abq} = \widehat{C}^{(ab,ba)}\dot{\overline{w}}^{(ba)}_{bal} \tag{6.166}$$

By using Eq. Set (6.145), Eq. (6.166) can be written in more detail as follows:

$$\begin{aligned}&(\overline{u}_i\dot{\xi}_{abq} + \overline{u}_j\dot{\eta}_{abq})\zeta_{abq} + (\overline{u}_i\xi_{abq} + \overline{u}_j\eta_{abq} + \overline{u}_k)\dot{\zeta}_{abq}\\ &= \widehat{C}^{(ab,ba)}(\overline{u}_i\dot{\xi}_{baq} + \overline{u}_j\dot{\eta}_{baq})\end{aligned} \tag{6.167}$$

When Eqs. (6.166) and (6.149) are substituted, Eq. (6.163) reduces to

$$\dot{\overline{w}}^{(ab)}_{abl}\zeta_{ab} + \overline{w}^{(ab)}_{abl}\dot{\zeta}_{ab} + \widetilde{\omega}^{(ab)}_{ab,ba}\overline{w}^{(ab)}_{abl}(\zeta_{abq} - \zeta_{ab}) = \overline{0} \tag{6.168}$$

On the other hand, Eq. (6.147) leads to the following equation for the velocity of O_{ba} with respect to $\mathcal{F}_{ab}(O_{ab})$.

$$\overline{v}^{(ab)}_{ab,ba} = \dot{\overline{r}}^{(ab)}_{ab,ba} = \dot{\overline{w}}^{(ab)}_{abl}z_{ab} + \overline{w}^{(ab)}_{abl}\dot{z}_{ab} = [\dot{\overline{w}}^{(ab)}_{abl}\zeta_{ab} + \overline{w}^{(ab)}_{abl}\dot{\zeta}_{ab}]h_{ba} \tag{6.169}$$

Using Eq. (6.169), Eq. (6.168) can be written again as

$$\overline{v}^{(ab)}_{ab,ba} = \widetilde{\omega}^{(ab)}_{ab,ba}\overline{w}^{(ab)}_{abl}(z_{ab} - z_{abq}) \tag{6.170}$$

Note that O_{ba} is also a contact point on the contact line $O_{ba}Q_{ba}$. In other words, O_{ba} is such a special contact point that $z_{abq} = z_{ab}$. Therefore, Eq. (6.170) reduces to

$$\overline{v}^{(ab)}_{ab,ba} = \overline{0} \tag{6.171}$$

Note also that $\dot{\zeta}_{ab} = 0$ due to the lack of slipping. Therefore, Eqs. (6.171) and (6.169) imply that $z_{ab} = \zeta_{ab}h_{ba} = 0$, which implies further that

$$O_{ba} = O_{ab} \tag{6.172}$$

According to this result, the two cones can roll over each other without slipping only if their vertices remain permanently coincident.

Moreover, with $z_{ab} = 0$, $\overline{v}^{(ab)}_{ab,ba} = \overline{0}$, $\overline{w}^{(ab)}_{abl}z_{abq} = \widehat{C}^{(ab,ba)}\overline{w}^{(ba)}_{bal}h_{ba}$, and $\overline{w}^{(ba)}_{bal} = \overline{\rho}^{(ba)}_{baq} + \overline{u}_k$, Eq. (6.170) can be written as follows by using Eqs. (6.145) and (6.149):

$$\begin{aligned}&\widetilde{\omega}^{(ab)}_{ab,ba}\widehat{C}^{(ab,ba)}\overline{w}^{(ba)}_{bal}h_{ba} = \overline{0} \Rightarrow\\ &\widetilde{\omega}^{(ab)}_{ab,ba}\widehat{C}^{(ab,ba)}\overline{\rho}^{(ba)}_{baq}h_{ba} + \widetilde{\omega}^{(ab)}_{ab,ba}\widehat{C}^{(ab,ba)}\overline{u}_k h_{ba} = \overline{0}\end{aligned} \tag{6.173}$$

Equation (6.173) implies that

$$\vec{\omega}_{ab,ba} \times \overrightarrow{B_{ba}Q_{ba}} + \vec{\omega}_{ab,ba} \times \overrightarrow{O_{ba}B_{ba}} = \vec{0} \Rightarrow$$

$$\vec{\omega}_{ab,ba} \times \overrightarrow{Q_{ba}B_{ba}} = \vec{\omega}_{ab,ba} \times \overrightarrow{O_{ba}B_{ba}} = \vec{v}_{B_{ba}/\mathcal{F}_{ab}(O_{ab})} \tag{6.174}$$

Equation (6.174) implies in turn that, during a rolling-without-slipping motion of $\mathcal{E}_{ba}$ over $\mathcal{E}_{ab}$, the velocity of the base plane origin B_{ba} can be determined in one of the two ways. It can be determined either by using the rotation of the longitudinal line segment $O_{ba}B_{ba}$ about the permanently fixed vertex point O_{ba} or by using the rotation of the radial line segment $Q_{ba}B_{ba}$ about the instantaneously fixed contact point Q_{ba}.

The preceding analysis shows that the rolling-without-slipping motion of $\mathcal{E}_{ba}$ over $\mathcal{E}_{ab}$ is characterized by the following two column matrix equations in addition to the four scalar equations written previously in Part (b), i.e. Eqs. (6.159)–(6.162).

$$\tilde{\omega}^{(ab)}_{ab,ba}\overline{w}^{(ab)}_{abl} = \overline{0} \tag{6.175}$$

$$\dot{\overline{w}}^{(ab)}_{abl}\zeta_{abq} + \overline{w}^{(ab)}_{abl}\dot{\zeta}_{abq} = \widehat{C}^{(ab,ba)}\dot{\overline{w}}^{(ba)}_{bal} \tag{6.176}$$

Equation (6.175) implies that $\vec{\omega}_{ab,ba}$ is aligned with the generatrix line vector $\vec{w}_{abl}$. In other words, as expressed below, $\vec{\omega}_{ab,ba}$ can actually be represented by a single parameter such as ω_{ab}.

$$\overline{\omega}^{(ab)}_{ab,ba} = \omega_{ab}\overline{u}^{(ab)}_{abl} \tag{6.177}$$

In Eq. (6.177), $\vec{u}_{abl}$ is the unit vector along $\vec{w}_{abl}$. That is,

$$\overline{u}^{(ab)}_{abl} = \overline{w}^{(ab)}_{abl} / \mid \overline{w}^{(ab)}_{abl} \mid = (\overline{u}_i\xi_{abq} + \overline{u}_j\eta_{abq} + \overline{u}_k)/(\xi^2_{abq} + \eta^2_{abq} + 1)^{1/2} \tag{6.178}$$

As for Eq. (6.176), it contains only two independent scalar equations, because it is trivially satisfied in the direction of the aligned gradient vectors. This fact can be verified by means of the following manipulations.

When Eq. (6.176) is premultiplied on both sides by $\overline{\gamma}^{(ab)t}_{ab}$, it becomes

$$\overline{\gamma}^{(ab)t}_{ab}\dot{\overline{w}}^{(ab)}_{abl}\zeta_{abq} + \overline{\gamma}^{(ab)t}_{ab}\overline{w}^{(ab)}_{abl}\dot{\zeta}_{abq} = \overline{\gamma}^{(ab)t}_{ab}\widehat{C}^{(ab,ba)}\dot{\overline{w}}^{(ba)}_{bal} \tag{6.179}$$

Note that $\overline{\gamma}^{(ab)t}_{ab}\overline{w}^{(ab)}_{abl} = 0$ and $\widehat{C}^{(ab,ba)}\overline{\gamma}^{(ba)}_{ba} = -\lambda'_{ab}\overline{\gamma}^{(ab)}_{ab}$ due to Eqs. (6.153) and (6.141). Therefore, Eq. (6.179) simplifies to

$$\overline{\gamma}^{(ab)t}_{ab}\dot{\overline{w}}^{(ab)}_{abl}\zeta_{abq} = -\overline{\gamma}^{(ba)t}_{ba}\dot{\overline{w}}^{(ba)}_{bal} / \lambda'_{ab} \tag{6.180}$$

By inserting the expressions given in Eq. Sets (6.140) and (6.145), Eq. (6.180) can be written out and manipulated as shown below.

$$[\overline{u}_i f_{ab\xi} + \overline{u}_j f_{ab\eta} - \overline{u}_k(f_{ab\xi}\xi_{abq} + f_{ab\eta}\eta_{abq})]^t(\overline{u}_i\dot{\xi}_{abq} + \overline{u}_j\dot{\eta}_{abq})\zeta_{abq}$$

$$= -[\overline{u}_i f_{ba\xi} + \overline{u}_j f_{ba\eta} - \overline{u}_k(f_{ba\xi}\xi_{baq} + f_{ba\eta}\eta_{baq})]^t(\overline{u}_i\dot{\xi}_{baq} + \overline{u}_j\dot{\eta}_{baq})/\lambda'_{ab} \Rightarrow$$

$$(f_{ab\xi}\dot{\xi}_{abq} + f_{ab\eta}\dot{\eta}_{abq})\zeta_{abq} = -(f_{ba\xi}\dot{\xi}_{baq} + f_{ba\eta}\dot{\eta}_{baq})/\lambda'_{ab} \tag{6.181}$$

On the other hand, according to Eqs. (6.159) and (6.160),

$$f_{ab\xi}\dot{\xi}_{abq} + f_{ab\eta}\dot{\eta}_{abq} = f_{ba\xi}\dot{\xi}_{baq} + f_{ba\eta}\dot{\eta}_{baq} = 0$$

Therefore, Eq. (6.181) reduces to the trivial equation 0 = 0. Thus, it has been verified that Eq. (6.176) contains only two independent scalar equations. These scalar equations can be obtained as follows by using the orthogonal companions of $\overline{w}_{abl}^{(ab)}$, which are expressed by Eq. Set (6.157):

$$\overline{w}_{abt}^{(ab)t}\dot{\overline{w}}_{abl}^{(ab)}\zeta_{abq} = \overline{w}_{abt}^{(ab)t}\widehat{C}^{(ab,ba)}\dot{\overline{w}}_{bal}^{(ba)} \tag{6.182}$$

$$\overline{w}_{abn}^{(ab)t}\dot{\overline{w}}_{abl}^{(ab)}\zeta_{abq} = \overline{w}_{abn}^{(ab)t}\widehat{C}^{(ab,ba)}\dot{\overline{w}}_{bal}^{(ba)} \tag{6.183}$$

As shown above, there are six independent scalar equations that characterize the rolling-without-slipping motion, which are Eqs. (6.159)–(6.162) and Eqs. (6.182) and (6.183). These equations are to be satisfied by seven velocity parameters, which are ω_{ab}, $\dot{\lambda}'_{ab}$, ζ_{abq}, $\dot{\xi}_{abq}$ $\dot{\eta}_{abq}$, $\dot{\xi}_{baq}$, and $\dot{\eta}_{baq}$. Therefore, the mobility of a cone-on-cone joint becomes $\mu_{ab} = 1$ during a rolling-without-slipping motion. Thus, if one of the velocity parameters (e.g. ω_{ab}) is specified along with the position specifications, the other six (i.e. $\dot{\xi}_{abq}$, $\dot{\eta}_{abq}$, ζ_{abq}; $\dot{\xi}_{baq}$, $\dot{\eta}_{baq}$, and $\dot{\lambda}'_{ab}$) can be determined from the indicated equations.

As for the derivatives $\dot{\phi}_{ab}$, $\dot{\theta}_{ab}$, and $\dot{\psi}_{ab}$, they can be determined from the following equation, which is obtained by combining Eqs. (6.164) and (6.177).

$$\dot{\phi}_{ab}\overline{u}_k + \dot{\theta}_{ab}e^{\tilde{u}_k\phi_{ab}}\overline{u}_i + \dot{\psi}_{ab}e^{\tilde{u}_k\phi_{ab}}e^{\tilde{u}_i\theta_{ab}}\overline{u}_k = \omega_{ab}\overline{u}_{abl}^{(ab)} \tag{6.184}$$

Equation (6.184) can be worked out to the following scalar equations.

$$\dot{\theta}_{ab}c\phi_{ab} + \dot{\psi}_{ab}s\phi_{ab}s\theta_{ab} = \omega_{abi} = \omega_{ab}\overline{u}_i^t\overline{u}_{abl}^{(ab)} \tag{6.185}$$

$$\dot{\theta}_{ab}s\phi_{ab} - \dot{\psi}_{ab}c\phi_{ab}s\theta_{ab} = \omega_{abj} = \omega_{ab}\overline{u}_j^t\overline{u}_{abl}^{(ab)} \tag{6.186}$$

$$\dot{\phi}_{ab} + \dot{\psi}_{ab}c\theta_{ab} = \omega_{abk} = \omega_{ab}\overline{u}_k^t\overline{u}_{abl}^{(ab)} \tag{6.187}$$

Equations (6.185)–(6.187) provide $\dot{\phi}_{ab}$, $\dot{\theta}_{ab}$, and $\dot{\psi}_{ab}$ with the following expressions in a typical situation such that $s\theta_{ab} \neq 0$.

$$\dot{\theta}_{ab} = \omega_{abi}c\phi_{ab} + \omega_{abj}s\phi_{ab} \tag{6.188}$$

$$\dot{\psi}_{ab} = (\omega_{abi}s\phi_{ab} - \omega_{abj}c\phi_{ab})/s\theta_{ab} \tag{6.189}$$

$$\dot{\phi}_{ab} = \omega_{abk} - (\omega_{abi}s\phi_{ab} - \omega_{abj}c\phi_{ab})c\theta_{ab}/s\theta_{ab} \tag{6.190}$$

6.6.2 Cone-on-Cylinder Joint

Figure 6.15 illustrates a joint formed as a pair of conical and cylindrical kinematic elements $\mathcal{E}_{ab}$ and $\mathcal{E}_{ba}$, which are in rolling contact along the coincident lines $O_{ba}Q_{ab}$ and $O_{ba}Q_{ba}$. The kinematic elements have the shapes of specified cylindrical and conical surfaces S_{ab} and S_{ba}. These surfaces are described by the base functions $f_{ab}(x, y)$ and $f_{ba}(\xi, \eta)$, which are assumed to be continuous and differentiable with respect to their arguments.

(a) Relative Position Equations

The conical and cylindrical surfaces are described by the following equations.

$$f_{ab}(x_{abq}, y_{abq}) = 0 \tag{6.191}$$

$$f_{ba}(\xi_{baq}, \eta_{baq}) = 0 \tag{6.192}$$

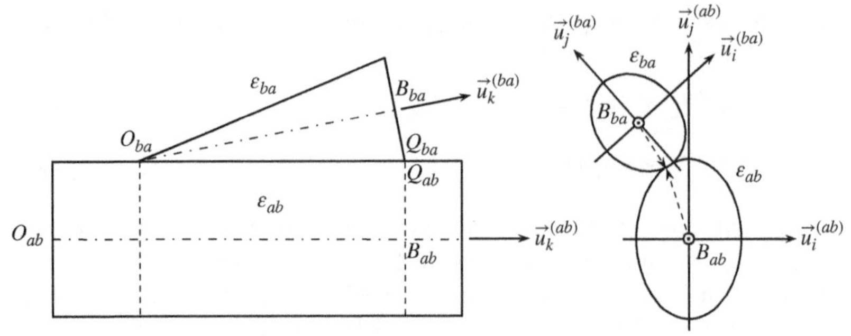

Figure 6.15 A cone-on-cylinder joint.

In Eq. (6.192), the arguments of the conical surface function are the coordinate ratios that are defined as follows:

$$\left.\begin{aligned} \xi_{baq} &= x_{baq}/z_{baq} \\ \eta_{baq} &= y_{baq}/z_{baq} \end{aligned}\right\} \tag{6.193}$$

For f_{ab}, the gradient vector $\vec{g}_{ab}$ has the following expression in $\mathcal{F}_{ab}$.

$$\overline{g}_{ab}^{(ab)} = \overline{u}_i f_{abx} + \overline{u}_j f_{aby} \tag{6.194}$$

In Eq. (6.194),

$$\left.\begin{aligned} f_{abx} &= \partial f_{ab}/\partial x_{abq} \\ f_{aby} &= \partial f_{ab}/\partial y_{abq} \end{aligned}\right\} \tag{6.195}$$

For f_{ba}, the gradient vector $\vec{g}_{ba}$ can be expressed as follows in $\mathcal{F}_{ba}$:

$$\overline{g}_{ba}^{(ba)} = \overline{\gamma}_{ba}^{(ba)}/z_{baq} \tag{6.196}$$

In Eq. (6.196),

$$\overline{\gamma}_{ba}^{(ba)} = \overline{u}_i f_{ba\xi} + \overline{u}_j f_{ba\eta} - \overline{u}_k (f_{ba\xi}\xi_{baq} + f_{ba\eta}\eta_{baq}) \tag{6.197}$$

$$\left.\begin{aligned} f_{ba\xi} &= \partial f_{ba}/\partial \xi_{baq} \\ f_{ba\eta} &= \partial f_{ba}/\partial \eta_{baq} \end{aligned}\right\} \tag{6.198}$$

Equations (6.194) and (6.196) can be combined into the following gradient alignment equation.

$$\widehat{C}^{(ab,ba)}\overline{\gamma}_{ba}^{(ba)} = -\lambda_{ab} z_{baq} (\overline{u}_i f_{abx} + \overline{u}_j f_{aby}) \tag{6.199}$$

The position vectors of the contact points along the contact lines (i.e. the coincident generatrix lines) are expressed in the frames $\mathcal{F}_{ab}$ and $\mathcal{F}_{ba}$ as shown below.

$$\left.\begin{aligned} \overline{r}_{abq}^{(ab)} &= \overline{u}_i x_{abq} + \overline{u}_j y_{abq} + \overline{u}_k z_{abq} \\ \overline{r}_{baq}^{(ba)} &= (\overline{u}_i \xi_{baq} + \overline{u}_j \eta_{baq} + \overline{u}_k) z_{baq} \end{aligned}\right\} \tag{6.200}$$

Hence, the coincidence of the contact lines can be expressed as follows:

$$\begin{aligned} &\overline{u}_i x_{abq} + \overline{u}_j y_{abq} + \overline{u}_k z_{abq} \\ &\quad = \overline{r}_{ab,ba}^{(ab)} + \widehat{C}^{(ab,ba)} (\overline{u}_i \xi_{baq} + \overline{u}_j \eta_{baq} + \overline{u}_k) z_{baq} \end{aligned} \tag{6.201}$$

The coincidence of the contact lines necessitates that

$$\overline{r}^{(ab)}_{ab,ba} = \overline{u}_i x_{abq} + \overline{u}_j y_{abq} + \overline{u}_k z_{ab} \quad (6.202)$$

Equations (6.201) and (6.202) can be combined into the following equation.

$$\overline{u}_k(z_{abq} - z_{ab}) = \widehat{C}^{(ab,ba)}(\overline{u}_i \xi_{baq} + \overline{u}_j \eta_{baq} + \overline{u}_k) z_{baq} \quad (6.203)$$

Since the gradient vectors are naturally perpendicular to the contact lines, which are coincident along $\vec{u}_k^{(ab)}$, Eq. (6.199) contains two independent scalar equations, which can be picked up as shown below.

$$\overline{u}_i^t \widehat{C}^{(ab,ba)} \overline{\gamma}^{(ba)}_{ba} = -\lambda_{ab} z_{baq} f_{abx} \quad (6.204)$$

$$\overline{u}_j^t \widehat{C}^{(ba,ab)} \overline{\gamma}^{(ba)}_{ba} = -\lambda_{ab} z_{baq} f_{aby} \quad (6.205)$$

As for the relative orientation matrix, it can be expressed similarly as in Section 6.6.1. That is,

$$\widehat{C}^{(ab,ba)} = e^{\widetilde{u}_k \phi_{ab}} e^{\widetilde{u}_i \theta_{ab}} e^{\widetilde{u}_k \psi_{ab}} \quad (6.206)$$

For the cone-on-cylinder joint, z_{baq} can be taken as a constant. That is,

$$z_{baq} = h_{ba} = \text{constant} \quad (6.207)$$

Therefore, the cone-on-cylinder joint involves 10 principal variables, which are

$$\phi_{ab}, \theta_{ab}, \psi_{ab}; \lambda_{ab}, z_{ab}; x_{abq}, y_{abq}, z_{abq}; \xi_{baq}, \eta_{baq}$$

The variables indicated above must satisfy four scalar equations, i.e. Eqs. (6.191), (6.192), (6.204), and (6.205), and one column matrix equation, i.e. Eq. (6.203). Therefore, the mobility of the cone-on-cylinder joint turns out to be $\mu_{ab} = 10 - 7 = 3$. Thus, if three of the variables (e.g. $z_{ab}, x_{abq}, \xi_{baq}$) are specified, then the remaining seven variables (i.e. $\phi_{ab}, \theta_{ab}, \psi_{ab}, \lambda_{ab}, z_{abq}, y_{abq}, \eta_{baq}$) can be determined by using the mentioned equations. Afterwards, the values of the actual coordinates ($x_{baq} = \xi_{baq} h_{ba}$ and $y_{baq} = \eta_{baq} h_{ba}$) can also be determined by using the specified value of h_{ba}.

(b) Relative Velocity Equations

They can be obtained upon the differentiation of Eqs. (6.191), (6.192), (6.204), (6.205), and (6.203) by noting that $z_{baq} = h_{ba} =$ constant. However, for a cone-on-cylinder joint, it is *not* possible to have a relative rolling-without-slipping motion. The reason can be explained as follows: The cylindrical kinematic element $\mathcal{E}_{ab}$ can be considered as a degenerate conical element having its vertex at infinity. As for the kinematic element $\mathcal{E}_{ba}$, it is a proper cone having its vertex at a finite distance from its base. Otherwise, it would also degenerate into a cylinder. Therefore, the vertices of $\mathcal{E}_{ab}$ and $\mathcal{E}_{ba}$ can never be coincident. This fact prevents rolling without slipping, because, as shown in Section 6.6.1, a rolling-without-slipping motion can occur only if the vertices of the cones are permanently coincident.

6.6.3 Cone-on-Plane Joint

Figure 6.16 illustrates a joint formed as a pair of conical and planar kinematic elements $\mathcal{E}_{ab}$ and $\mathcal{E}_{ba}$, which are in rolling contact along the coincident lines $O_{ab}Q_{ab}$ and $O_{ba}Q_{ba}$. The kinematic element $\mathcal{E}_{ba}$ has the shape of a specified conical surface S_{ba}. This surface

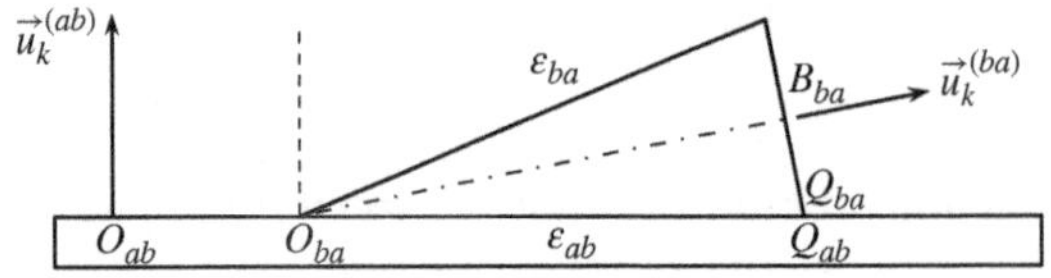

Figure 6.16 A cone-on-plane joint.

is described by the base function $f_{ba}(\xi, \eta)$, which is assumed to be continuous and differentiable with respect to its arguments.

(a) Relative Position Equations

The planar and conical surfaces are described by the following equations.

$$f_{ab}(x_{abq}, y_{abq}, z_{abq}) = z_{abq} = 0 \tag{6.208}$$

$$f_{ba}(\xi_{baq}, \eta_{baq}) = 0 \tag{6.209}$$

In Eq. (6.209), the arguments of the conical surface function are the coordinate ratios that are defined as follows:

$$\left.\begin{aligned} \xi_{baq} &= x_{baq}/z_{baq} \\ \eta_{baq} &= y_{baq}/z_{baq} \end{aligned}\right\} \tag{6.210}$$

For the function f_{ab}, the gradient vector $\vec{g}_{ab}$ has the following expression in $\mathcal{F}_{ab}$.

$$\overline{g}_{ab}^{(ab)} = \overline{u}_k \tag{6.211}$$

For the function f_{ba}, the gradient vector $\vec{g}_{ba}$ has the following expression in $\mathcal{F}_{ba}$.

$$\begin{aligned} \overline{g}_{ba}^{(ba)} &= \overline{\gamma}_{ba}^{(ba)}/z_{baq} \\ &= [\overline{u}_i f_{ba\xi} + \overline{u}_j f_{ba\eta} - \overline{u}_k (f_{ba\xi}\xi_{baq} + f_{ba\eta}\eta_{baq})]/z_{baq} \end{aligned} \tag{6.212}$$

In Eq. (6.212),

$$\left.\begin{aligned} f_{ba\xi} &= \partial f_{ba}/\partial \xi_{baq} \\ f_{ba\eta} &= \partial f_{ba}/\partial \eta_{baq} \end{aligned}\right\} \tag{6.213}$$

Equations (6.211) and (6.212) can be combined into the following gradient alignment equation.

$$\widehat{C}^{(ab,ba)}\overline{\gamma}_{ba}^{(ba)} = -\lambda_{ab} z_{baq} \overline{u}_k \tag{6.214}$$

The coincidence of the contact lines can be expressed as follows:

$$\overline{u}_i x'_{abq} + \overline{u}_j y'_{abq} = \widehat{C}^{(ab,ba)}(\overline{u}_i \xi_{baq} + \overline{u}_j \eta_{baq} + \overline{u}_k) z_{baq} \tag{6.215}$$

In Eq. (6.215),

$$\left.\begin{aligned} x'_{abq} &= x_{abq} - x_{ab} \\ y'_{abq} &= y_{abq} - y_{ab} \end{aligned}\right\} \tag{6.216}$$

Equation (6.215) suggests to introduce the contact line unit vector $\vec{u}_{abl}$, which is expressed as follows in $\mathcal{F}_{ab}$:

$$\overline{u}_{abl}^{(ab)} = (\overline{u}_i x'_{abq} + \overline{u}_j y'_{abq})/d_{ab} \tag{6.217}$$

The orthogonal coplanar companion $\vec{u}_{abn}$ of $\vec{u}_{abl}$ can then also be introduced so that

$$\overline{u}_{abn}^{(ab)} = (\overline{u}_j x'_{abq} - \overline{u}_i y'_{abq})/d_{ab} \tag{6.218}$$

In Eqs. (6.217) and (6.218),

$$d_{ab} = [(x'_{abq})^2 + (y'_{abq})^2]^{1/2} \tag{6.219}$$

On the other hand, the gradient vector $\vec{g}_{ba}$ is naturally perpendicular to the contact line, i.e. $\vec{u}_{abl}$. Therefore, Eq. (6.214) contains only two independent scalar equations. They can be extracted from Eq. (6.214) by premultiplying both sides by $\overline{u}_k^t$ and $\overline{u}_{abn}^{(ab)t}$. This operation results in the following equations.

$$\overline{u}_k^t \widehat{C}^{(ab,ba)} \overline{\gamma}_{ba}^{(ba)} = -\lambda_{ab} z_{baq} \tag{6.220}$$

$$\overline{u}_j^t \widehat{C}^{(ab,ba)} \overline{\gamma}_{ba}^{(ba)} x'_{abq} = \overline{u}_i^t \widehat{C}^{(ab,ba)} \overline{\gamma}_{ba}^{(ba)} y'_{abq} \tag{6.221}$$

For the cone-on-plane joint, the relative orientation matrix can be expressed as follows:

$$\widehat{C}^{(ab,ba)} = e^{\tilde{u}_k \phi_{ab}} e^{\tilde{u}_j \theta_{ab}} e^{\tilde{u}_k \psi_{ab}} \tag{6.222}$$

The above expression happens to be suitable if it is preferable to have $\vec{u}_k^{(ba)}$ become parallel to $\vec{u}_1^{(ab)}$ when $\phi_{ab} = 0$ and $\theta_{ab} = \pi/2$.

As for z_{baq}, it can be taken as a constant as in the previously considered conical kinematic elements. That is,

$$z_{baq} = h_{ba} = \text{constant} \tag{6.223}$$

Therefore, the cone-on-plane joint involves 10 principal variables, which are

$$\phi_{ab}, \theta_{ab}, \psi_{ab}; \lambda_{ab}, x_{ab}, y_{ab}; x_{abq}, y_{abq}; \xi_{baq}, \eta_{baq}$$

The variables listed above must satisfy three scalar equations, i.e. Eqs. (6.209), (6.220), and (6.221), and one column matrix equation, i.e. Eq. (6.215). Therefore, the mobility of the cone-on-plane joint turns out to be $\mu_{ab} = 10 - 6 = 4$. Thus, if four of the variables (e.g. x_{ab}, y_{ab}, ϕ_{ab}, ψ_{ab}) are specified, then the remaining six variables (i.e. θ_{ab}, λ_{ab}, x_{abq}, y_{abq}, ξ_{baq}, η_{baq}) can be determined by using the mentioned equations. Afterwards, the values of the actual coordinates ($x_{baq} = \xi_{baq} h_{ba}$ and $y_{abq} = \eta_{baq} h_{ba}$) can also be determined by using the specified value of h_{ba}.

(b) Relative Velocity Equations

The relative velocity equations can be obtained upon the differentiation of Eqs. (6.209), (6.220), (6.221), and (6.215) by noting that $z_{baq} = h_{ba} = \text{constant}$. This operation results in the following equations.

$$f_{ba\xi} \dot{\xi}_{baq} + f_{ba\eta} \dot{\eta}_{baq} = 0 \tag{6.224}$$

$$\overline{u}_k^t \widehat{C}^{(ab,ba)} \dot{\overline{\gamma}}_{ba}^{(ba)} + \overline{u}_k^t \tilde{\overline{\omega}}_{ab,ba}^{(ab)} \widehat{C}^{(ab,ba)} \overline{\gamma}_{ba}^{(ba)} = -h_{ba} \dot{\lambda}_{ab} \tag{6.225}$$

$$\begin{aligned} &\overline{u}_j^t \widehat{C}^{(ab,ba)} [\dot{\overline{\gamma}}_{ba}^{(ba)} x'_{abq} + \overline{\gamma}_{ba}^{(ba)} \dot{x}'_{abq}] + \overline{u}_j^t \tilde{\overline{\omega}}_{ab,ba}^{(ab)} \widehat{C}^{(ab,ba)} \overline{\gamma}_{ba}^{(ba)} x'_{abq} \\ &= \overline{u}_i^t \widehat{C}^{(ab,ba)} [\dot{\overline{\gamma}}_{ba}^{(ba)} y'_{abq} + \overline{\gamma}_{ba}^{(ba)} \dot{y}'_{abq}] + \overline{u}_i^t \tilde{\overline{\omega}}_{ab,ba}^{(ab)} \widehat{C}^{(ab,ba)} \overline{\gamma}_{ba}^{(ba)} y'_{abq} \end{aligned} \tag{6.226}$$

$$\overline{u}_i(\dot{x}_{abq} - \dot{x}_{ab}) + \overline{u}_j(\dot{y}_{abq} - \dot{y}_{ab}) = h_{ba}\widetilde{\omega}_{ab,ba}^{(ab)}\widehat{C}^{(ab,ba)}(\overline{u}_i\xi_{baq} + \overline{u}_j\eta_{baq} + \overline{u}_k) + h_{ba}\widehat{C}^{(ab,ba)}(\overline{u}_i\dot{\xi}_{baq} + \overline{u}_j\dot{\eta}_{baq}) \tag{6.227}$$

As for the relative angular velocity, the following equation can be derived from Eq. (6.222).

$$\overline{\omega}_{ab,ba}^{(ab)} = \dot{\phi}_{ab}\overline{u}_k + \dot{\theta}_{ab}e^{\widetilde{u}_k\phi_{ab}}\overline{u}_j + \dot{\psi}_{ab}e^{\widetilde{u}_k\phi_{ab}}e^{\widetilde{u}_j\theta_{ab}}\overline{u}_k \tag{6.228}$$

Hence, if four of the derivatives (e.g. $\dot{x}_{ab}$, $\dot{y}_{ab}$, $\dot{\phi}_{ab}$, $\dot{\psi}_{ab}$) are specified along with the position specifications, then the remaining six derivatives (i.e. $\dot{\theta}_{ab}$, $\dot{\lambda}_{ab}$, $\dot{x}_{abq}$, $\dot{y}_{abq}$, $\dot{\xi}_{baq}$, $\dot{\eta}_{baq}$) can be determined by using Eqs. (6.224)–(6.227).

(c) Relative Velocity Equations in the Case of Rolling Without Slipping

If $\mathcal{E}_{ba}$ rolls over $\mathcal{E}_{ab}$ without slipping due to the presence of sticking friction, then, as discussed in Section 6.5, the derivatives $\dot{\overline{r}}_{abq}^{(ab)}$ and $\dot{\overline{r}}_{baq}^{(ba)}$ become related by the following nonholonomic constraint.

$$\dot{\overline{r}}_{abq}^{(ab)} = \widehat{C}^{(ab,ba)}\dot{\overline{r}}_{baq}^{(ba)} \tag{6.229}$$

With $z_{baq} = h_{ba} =$ constant, Eq. (6.229) becomes

$$\overline{u}_i\dot{x}_{abq} + \overline{u}_j\dot{y}_{abq} = h_{ba}\widehat{C}^{(ab,ba)}(\overline{u}_i\dot{\xi}_{baq} + \overline{u}_j\dot{\eta}_{baq}) \tag{6.230}$$

When Eq. (6.230) is substituted, Eq. (6.227) reduces to

$$h_{ba}\widetilde{\omega}_{ab,ba}^{(ab)}\widehat{C}^{(ab,ba)}(\overline{u}_i\xi_{baq} + \overline{u}_j\eta_{baq} + \overline{u}_k) = -(\overline{u}_i\dot{x}_{ab} + \overline{u}_j\dot{y}_{ab}) \tag{6.231}$$

By using Eq.(6.215) with $z_{baq} = h_{ba}$, Eq.(6.231) can also be written as

$$\widetilde{\omega}_{ab,ba}^{(ab)}(\overline{u}_i x'_{abq} + \overline{u}_j y'_{abq}) = -(\overline{u}_i\dot{x}_{ab} + \overline{u}_j\dot{y}_{ab}) \tag{6.232}$$

Since O_{ba} is also a contact point, Eq. (6.232) must be satisfied for $x'_{abq} = y'_{abq} = 0$, too. This can be possible if $\dot{x}_{ab} = \dot{y}_{ab} = 0$. This result implies that a rolling-without-slipping motion can occur only if the origin O_{ba} is stationary on the plane, i.e. if x_{ab} and y_{ab} have constant values such as x°_{ab} and y°_{ab}.

Moreover, with $\dot{x}_{ab} = \dot{y}_{ab} = 0$, Eqs. (6.232) and (6.217) imply that

$$\widetilde{\omega}_{ab,ba}^{(ab)}(\overline{u}_i x'_{abq} + \overline{u}_j y'_{abq}) = \overline{0} \Rightarrow$$

$$\widetilde{\omega}_{ab,ba}^{(ab)}\overline{u}_{abl}^{(ab)} = \overline{0} \tag{6.233}$$

Equation (6.233) implies further that $\mathcal{E}_{ba}$ rolls over $\mathcal{E}_{ab}$ about the contact line. That is,

$$\overline{\omega}_{ab,ba}^{(ab)} = \omega_{ab}\overline{u}_{abl}^{(ab)} = \omega_{ab}[\overline{u}_i(x_{abq} - x^{\circ}_{ab}) + \overline{u}_j(y_{abq} - y^{\circ}_{ab})]/d_{ab} \tag{6.234}$$

In Eq. (6.234), d_{ab} is defined similarly as in Eq. (6.219).

According to the preceding results, the number of velocity parameters reduces to six (i.e. ω_{ab}, $\dot{\lambda}_{ab}$, $\dot{x}_{abq}$, $\dot{y}_{abq}$, $\dot{\xi}_{baq}$, $\dot{\eta}_{baq}$). They have to satisfy five independent scalar equations, which are Eqs. (6.224), (6.225), (6.226), and the following two equations generated from Eq. (6.230).

$$\dot{x}_{abq} = h_{ba}\overline{u}_i^t\widehat{C}^{(ab,ba)}(\overline{u}_i\dot{\xi}_{baq} + \overline{u}_j\dot{\eta}_{baq}) \tag{6.235}$$

$$\dot{y}_{abq} = h_{ba}\overline{u}_j^t\widehat{C}^{(ab,ba)}(\overline{u}_i\dot{\xi}_{baq} + \overline{u}_j\dot{\eta}_{baq}) \tag{6.236}$$

Consequently, the mobility of the cone-on-plane joint reduces to $\mu_{ab} = 6 - 5 = 1$ while the conical kinematic element undergoes a rolling-without-slipping motion on the planar kinematic element. Thus, if one of the above velocity parameters (e.g. ω_{ab}) is specified along with the position specifications, the others (i.e. $\dot{\lambda}_{ab}, \dot{x}_{abq}, \dot{y}_{abq}, \dot{\xi}_{baq}, \dot{\eta}_{baq}$) can be determined from the mentioned equations.

As for the derivatives $\dot{\phi}_{ab}$, $\dot{\theta}_{ab}$, and $\dot{\psi}_{ab}$, they can be determined from the following equation, which is obtained by combining Eqs. (6.228) and (6.234).

$$\dot{\phi}_{ab}\overline{u}_k + \dot{\theta}_{ab}e^{\widetilde{u}_k\phi_{ab}}\overline{u}_j + \dot{\psi}_{ab}e^{\widetilde{u}_k\phi_{ab}}e^{\widetilde{u}_j\theta_{ab}}\overline{u}_k = \omega_{ab}\overline{u}_{abl}^{(ab)} \tag{6.237}$$

By using Eq. (6.217), Eq. (6.237) can be worked out to the following scalar equations.

$$\dot{\psi}_{ab}c\phi_{ab}s\theta_{ab} - \dot{\theta}_{ab}s\phi_{ab} = \omega_{abi} = \omega_{ab}(x_{abq} - x_{ab}^{\circ})/d_{ab} \tag{6.238}$$

$$\dot{\psi}_{ab}s\phi_{ab}s\theta_{ab} + \dot{\theta}_{ab}c\phi_{ab} = \omega_{abj} = \omega_{ab}(y_{abq} - y_{ab}^{\circ})/d_{ab} \tag{6.239}$$

$$\dot{\phi}_{ab} + \dot{\psi}_{ab}c\theta_{ab} = 0 \tag{6.240}$$

Equations (6.238)–(6.240) provide $\dot{\phi}_{ab}$, $\dot{\theta}_{ab}$, and $\dot{\psi}_{ab}$ with the following expressions in a typical situation such that $s\theta_{ab} \neq 0$.

$$\dot{\theta}_{ab} = \omega_{abj}c\phi_{ab} - \omega_{abi}s\phi_{ab} \tag{6.241}$$

$$\dot{\psi}_{ab} = (\omega_{abi}c\phi_{ab} + \omega_{abj}s\phi_{ab})/s\theta_{ab} \tag{6.242}$$

$$\dot{\phi}_{ab} = -(\omega_{abi}c\phi_{ab} + \omega_{abj}s\phi_{ab})c\theta_{ab}/s\theta_{ab} \tag{6.243}$$

6.6.4 Cylinder-on-Cylinder Joint

Figure 6.17 illustrates a joint formed by two cylindrical kinematic elements $\mathcal{E}_{ab}$ and $\mathcal{E}_{ba}$, which are in rolling contact along the coincident lines $P_{ab}Q_{ab}$ and $P_{ba}Q_{ba}$. The kinematic elements have the shapes of specified cylindrical surfaces S_{ab} and S_{ba}. These surfaces are described by the base functions $f_{ab}(x, y)$ and $f_{ba}(x, y)$, which are assumed to be continuous and differentiable with respect to their arguments.

(a) Relative Position Equations

The cylindrical surfaces are described by the following equations.

$$f_{ab}(x_{abq}, y_{abq}) = 0 \tag{6.244}$$

$$f_{ba}(x_{baq}, y_{baq}) = 0 \tag{6.245}$$

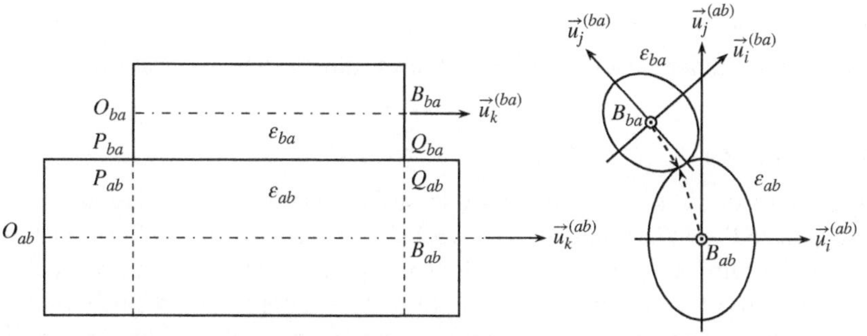

Figure 6.17 A cylinder-on-cylinder joint.

The gradient vectors are expressed by the following column matrices in their respective frames.

$$\overline{g}_{ab}^{(ab)} = \overline{u}_i f_{abx} + \overline{u}_j f_{aby} \tag{6.246}$$

$$\overline{g}_{ba}^{(ba)} = \overline{u}_i f_{bax} + \overline{u}_j f_{bay} \tag{6.247}$$

Since $\vec{u}_k^{(ba)}$ and $\vec{u}_k^{(ab)}$ are parallel, the relative orientation matrix is expressed simply as

$$\hat{C}^{(ab,ba)} = e^{\tilde{u}_k \psi_{ab}} \tag{6.248}$$

The locations of Q_{ab}, Q_{ba}, and O_{ba} are described by the following column matrices in the relevant frames.

$$\overline{r}_{ab,ba}^{(ab)} = \overline{u}_i x_{ab} + \overline{u}_j y_{ab} + \overline{u}_k z_{ab} \tag{6.249}$$

$$\overline{r}_{abq}^{(ab)} = \overline{u}_i x_{abq} + \overline{u}_j y_{abq} + \overline{u}_k (z_{ab} + h_{ba}) \tag{6.250}$$

$$\overline{r}_{baq}^{(ba)} = \overline{u}_i x_{baq} + \overline{u}_j y_{baq} + \overline{u}_k h_{ba} \tag{6.251}$$

In Eqs. (6.250) and (6.251), h_{ba} is the specified constant length of $\mathcal{E}_{ba}$.

The alignment of the contact lines can be expressed by two nontrivial scalar equations, which are derived as shown below.

$$\overline{r}_{abq}^{(ab)} = \overline{r}_{ab,ba}^{(ab)} + \hat{C}^{(ab,ba)} \overline{r}_{baq}^{(ba)} \Rightarrow$$

$$\overline{u}_i x_{abq} + \overline{u}_j y_{abq} + \overline{u}_k (z_{ab} + h_{ba})$$
$$= \overline{u}_i x_{ab} + \overline{u}_j y_{ab} + \overline{u}_k z_{ab} + e^{\tilde{u}_k \psi_{ab}} (\overline{u}_i x_{baq} + \overline{u}_j y_{baq} + \overline{u}_k h_{ba}) \Rightarrow$$

$$\overline{u}_i x_{abq} + \overline{u}_j y_{abq} = \overline{u}_i x_{ab} + \overline{u}_j y_{ab} + e^{\tilde{u}_k \psi_{ab}} (\overline{u}_i x_{baq} + \overline{u}_j y_{baq}) \Rightarrow$$

$$x_{abq} = x_{ab} + x_{baq} c\psi_{ab} - y_{baq} s\psi_{ab} \tag{6.252}$$

$$y_{abq} = y_{ab} + x_{baq} s\psi_{ab} + y_{baq} c\psi_{ab} \tag{6.253}$$

The alignment of the gradient vectors can also be expressed by two nontrivial scalar equations, which are derived as shown below.

$$\hat{C}^{(ab,ba)} \overline{g}_{ba}^{(ba)} = -\lambda_{ab} \overline{g}_{ab}^{(ab)} \Rightarrow$$

$$e^{\tilde{u}_k \psi_{ab}} (\overline{u}_i f_{bax} + \overline{u}_j f_{bay}) = -\lambda_{ab} (\overline{u}_i f_{abx} + \overline{u}_j f_{aby}) \Rightarrow$$

$$f_{bax} c\psi_{ab} - f_{bay} s\psi_{ab} = -\lambda_{ab} f_{abx} \tag{6.254}$$

$$f_{bax} s\psi_{ab} + f_{bay} c\psi_{ab} = -\lambda_{ab} f_{aby} \tag{6.255}$$

As seen above, the cylinder-on-cylinder joint involves nine principal variables that are listed below.

$$z_{ab};\, x_{ab},\, y_{ab},\, \psi_{ab},\, \lambda_{ab};\, x_{abq},\, y_{abq};\, x_{baq},\, y_{baq}$$

The nine variables listed above have to satisfy six scalar equations, which are Eq. Pairs (6.244)–(6.245), (6.252)–(6.253), and (6.254)–(6.255). Therefore, the mobility of the cylinder-on-cylinder joint is $\mu_{ab} = 9 - 6 = 3$. However, z_{ab} and the other eight variables are not interrelated. In other words, one of the mobilities is represented solely by z_{ab} and therefore it must be specified independently of the other variables. Thus, if two of the other variables (e.g. x_{ab} and ψ_{ab}) are specified in addition to z_{ab}, then the remaining six variables (i.e. y_{ab}, λ_{ab}, x_{abq}, y_{abq}, x_{baq}, y_{baq}) can be determined by using the mentioned scalar equations.

(b) Relative Velocity Equations

The relative velocity equations can be obtained upon the differentiation of the scalar equation pairs mentioned in Part (a). This operation results in the following equations.

$$f_{abx}\dot{x}_{abq} + f_{aby}\dot{y}_{abq} = 0 \tag{6.256}$$

$$f_{bax}\dot{x}_{baq} + f_{bay}\dot{y}_{baq} = 0 \tag{6.257}$$

$$\dot{x}_{abq} = \dot{x}_{ab} + \dot{x}_{baq}c\psi_{ab} - \dot{y}_{baq}s\psi_{ab} - (x_{baq}s\psi_{ab} + y_{baq}c\psi_{ab})\dot{\psi}_{ab} \tag{6.258}$$

$$\dot{y}_{abq} = \dot{y}_{ab} + \dot{x}_{baq}s\psi_{ab} + \dot{y}_{baq}c\psi_{ab} + (x_{baq}c\psi_{ab} - y_{baq}s\psi_{ab})\dot{\psi}_{ab} \tag{6.259}$$

$$\begin{aligned}(f_{baxx}c\psi_{ab} - f_{bayx}s\psi_{ab})\dot{x}_{baq} + (f_{baxy}c\psi_{ab} - f_{bayy}s\psi_{ab})\dot{y}_{baq} \\ - (f_{bax}s\psi_{ab} + f_{bay}c\psi_{ab})\dot{\psi}_{ab} \\ = -\dot{\lambda}_{ab}f_{abx} - \lambda_{ab}(f_{abxx}\dot{x}_{abq} + f_{abxy}\dot{y}_{abq})\end{aligned} \tag{6.260}$$

$$\begin{aligned}(f_{baxx}s\psi_{ab} + f_{bayx}c\psi_{ab})\dot{x}_{baq} + (f_{baxy}s\psi_{ab} + f_{bayy}c\psi_{ab})\dot{y}_{baq} \\ + (f_{bax}c\psi_{ab} - f_{bay}s\psi_{ab})\dot{\psi}_{ab} \\ = -\dot{\lambda}_{ab}f_{aby} - \lambda_{ab}(f_{abyx}\dot{x}_{abq} + f_{abyy}\dot{y}_{abq})\end{aligned} \tag{6.261}$$

By using the preceding equations, if two of the derivatives (e.g. $\dot{x}_{ab}$ and $\dot{\psi}_{ab}$) are specified along with $\dot{z}_{ab}$ and the position specifications, then the other six derivatives (i.e. $\dot{y}_{ab}$, $\dot{\lambda}_{ab}$, $\dot{x}_{abq}$, $\dot{y}_{abq}$, $\dot{x}_{baq}$, $\dot{y}_{baq}$) can be determined.

(c) Relative Velocity Equations in the Case of Rolling without Slipping

If $\mathcal{E}_{ba}$ rolls over $\mathcal{E}_{ab}$ without slipping due to the presence of sticking friction, then, as discussed in Section 6.5, the derivatives $\dot{\vec{r}}_{abq}^{(ab)}$ and $\dot{\vec{r}}_{baq}^{(ba)}$ become related by the following nonholonomic constraint.

$$\dot{\overline{r}}_{abq}^{(ab)} = \widehat{C}^{(ab,ba)}\dot{\overline{r}}_{baq}^{(ba)} \tag{6.262}$$

For the cylinder-on-cylinder joint, Eq. (6.262) leads to three scalar equations as shown below.

$$\overline{u}_i\dot{x}_{abq} + \overline{u}_j\dot{y}_{abq} + \overline{u}_k\dot{z}_{ab} = e^{\tilde{u}_k\psi_{ab}}(\overline{u}_i\dot{x}_{baq} + \overline{u}_j\dot{y}_{baq}) \Rightarrow$$

$$\dot{x}_{abq} = \dot{x}_{baq}c\psi_{ab} - \dot{y}_{baq}s\psi_{ab} \tag{6.263}$$

$$\dot{y}_{abq} = \dot{x}_{baq}s\psi_{ab} + \dot{y}_{baq}c\psi_{ab} \tag{6.264}$$

$$\dot{z}_{ab} = 0 \tag{6.265}$$

When Eqs. (6.263) and (6.264) are substituted together with Eqs. (6.252) and (6.253), Eqs. (6.258) and (6.259) become

$$\dot{x}_{ab} = -(y_{ab} - y_{abq})\dot{\psi}_{ab} \tag{6.266}$$

$$\dot{y}_{ab} = +(x_{ab} - x_{abq})\dot{\psi}_{ab} \tag{6.267}$$

Equations (6.266) and (6.267) imply that $\mathcal{E}_{ba}$ rolls over $\mathcal{E}_{ab}$ without slipping by rotating about the common contact line. Besides, Eqs. (6.266) and (6.267) give $\dot{x}_{ab}$ and $\dot{y}_{ab}$ directly in terms of $\dot{\psi}_{ab}$. Furthermore, Eqs. (6.263) and (6.264) give $\dot{x}_{abq}$ and $\dot{y}_{abq}$ directly in terms of $\dot{x}_{baq}$ and $\dot{y}_{baq}$. Therefore, there remains four derivatives ($\dot{x}_{baq}$, $\dot{y}_{baq}$, $\dot{\psi}_{ab}$, $\dot{\lambda}_{ab}$) to be related. These derivatives must satisfy three independent scalar equations, which are Eqs. (6.260), (6.261), and (6.257). In this case, Eq. (6.256) happens to be redundant because it becomes the same as Eq. (6.257) upon substituting Eqs. (6.263) and (6.264) together with Eqs. (6.254) and (6.255). Thus, the mobility of the cylinder-on-cylinder joint reduces to $\mu_{ab} = 4 - 3 = 1$ during a rolling-without-slipping motion. Consequently, if one of the derivatives (e.g. $\dot{\psi}_{ab}$) is specified along with the position specifications, then the other three derivatives (i.e. $\dot{\lambda}_{ab}$, $\dot{x}_{baq}$, $\dot{y}_{baq}$) can be determined by using the mentioned three equations.

6.6.5 Cylinder-on-Plane Joint

Figure 6.18 illustrates a joint formed as a pair of cylindrical and planar kinematic elements $\mathcal{E}_{ab}$ and $\mathcal{E}_{ba}$, which are in rolling contact along the coincident lines $P_{ab}Q_{ab}$ and $P_{ba}Q_{ba}$. The kinematic element $\mathcal{E}_{ba}$ has the shape of a specified cylindrical surface $\mathcal{S}_{ba}$. This surface is described by the base function $f_{ba}(x, y)$, which is assumed to be continuous and differentiable with respect to its arguments.

(a) Relative Position Equations

The planar and cylindrical surfaces are described by the following equations.

$$f_{ab}(x_{abq}, y_{abq}, z_{abq}) = z_{abq} = 0 \tag{6.268}$$

$$f_{ba}(x_{baq}, y_{baq}) = 0 \tag{6.269}$$

The gradient vectors $\vec{g}_{ab}$ and $\vec{g}_{ba}$ have the following matrix expressions, respectively, in the frames $\mathcal{F}_{ab}$ and $\mathcal{F}_{ba}$.

$$\overline{g}_{ab}^{(ab)} = \overline{u}_k \tag{6.270}$$

$$\overline{g}_{ba}^{(ba)} = \overline{u}_i f_{bax} + \overline{u}_j f_{bay} \tag{6.271}$$

The relative orientation matrix can be expressed in the following equivalent forms.

$$\hat{C}^{(ab,ba)} = e^{\tilde{u}_k \phi_{ab}} e^{\tilde{u}_j \pi/2} e^{\tilde{u}_k \psi_{ab}} \Rightarrow \hat{C}^{(ab,ba)} = e^{\tilde{u}_k \phi_{ab}} e^{\tilde{u}_i \psi_{ab}} e^{\tilde{u}_j \pi/2} \tag{6.272}$$

Hence, the alignment of the gradient vectors can be expressed as shown below.

$$\hat{C}^{(ab,ba)} \overline{g}_{ba}^{(ba)} = -\lambda_{ab} \overline{g}_{ab}^{(ab)} \Rightarrow$$

$$e^{\tilde{u}_k \phi_{ab}} e^{\tilde{u}_i \psi_{ab}} e^{\tilde{u}_j \pi/2} (\overline{u}_i f_{bax} + \overline{u}_j f_{bay}) = -\lambda_{ab} \overline{u}_k \Rightarrow$$

$$e^{\tilde{u}_j \pi/2} (\overline{u}_i f_{bax} + \overline{u}_j f_{bay}) = -\lambda_{ab} e^{-\tilde{u}_i \psi_{ab}} e^{-\tilde{u}_k \phi_{ab}} \overline{u}_k = -\lambda_{ab} e^{-\tilde{u}_i \psi_{ab}} \overline{u}_k \Rightarrow$$

$$-\overline{u}_k f_{bax} + \overline{u}_j f_{bay} = -\lambda_{ab} (\overline{u}_k c\psi_{ab} + \overline{u}_j s\psi_{ab}) \Rightarrow$$

$$f_{bax} = \lambda_{ab} c\psi_{ab} \tag{6.273}$$

$$f_{bay} = -\lambda_{ab} s\psi_{ab} \tag{6.274}$$

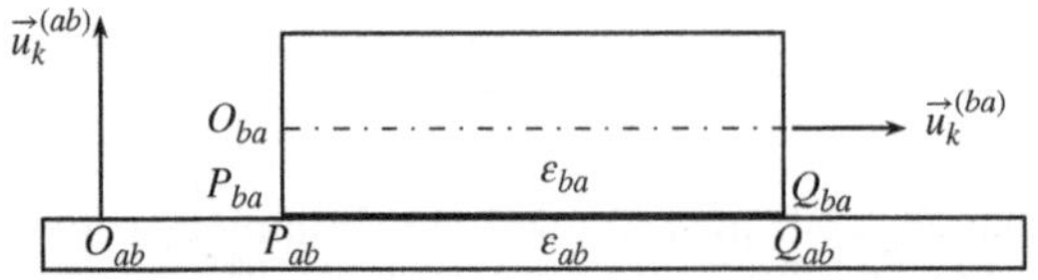

Figure 6.18 A cylinder-on-plane joint.

As for the coincidence of the contact lines, it can be expressed as follows by using h_{ba} as the constant length of the cylinder:

$$\begin{aligned}
&\overline{r}_{abq}^{(ab)} = \overline{r}_{ab,ba}^{(ab)} + \widehat{C}^{(ab,ba)}\overline{r}_{baq}^{(ba)} \Rightarrow \\
&\overline{u}_i x_{abq} + \overline{u}_j y_{abq} = \overline{u}_i x_{ab} + \overline{u}_j y_{ab} + \overline{u}_k z_{ab} \\
&\qquad + e^{\widetilde{u}_k\phi_{ab}} e^{\widetilde{u}_i\psi_{ab}} e^{\widetilde{u}_j\pi/2}(\overline{u}_i x_{baq} + \overline{u}_j y_{baq} + \overline{u}_k h_{ba}) \Rightarrow \\
&\overline{u}_i(x_{abq} - x_{ab}) + \overline{u}_j(y_{abq} - y_{ab}) - \overline{u}_k z_{ab} \\
&\quad = e^{\widetilde{u}_k\phi_{ab}} e^{\widetilde{u}_i\psi_{ab}}(\overline{u}_i h_{ba} + \overline{u}_j y_{baq} - \overline{u}_k x_{baq}) \Rightarrow \\
&\overline{u}_i(x_{abq} - x_{ab}) + \overline{u}_j(y_{abq} - y_{ab}) - \overline{u}_k z_{ab} \\
&\quad = e^{\widetilde{u}_k\phi_{ab}}[\overline{u}_i h_{ba} + \overline{u}_j(y_{baq}c\psi_{ab} + x_{baq}s\psi_{ab}) + \overline{u}_k(y_{baq}s\psi_{ab} - x_{baq}c\psi_{ab})] \Rightarrow \\
&\overline{u}_i(x_{abq} - x_{ab}) + \overline{u}_j(y_{abq} - y_{ab}) - \overline{u}_k z_{ab} \\
&\quad = (\overline{u}_i c\phi_{ab} + \overline{u}_j s\phi_{ab})h_{ba} + (\overline{u}_j c\phi_{ab} - \overline{u}_i s\phi_{ab})(y_{baq}c\psi_{ab} + x_{baq}s\psi_{ab}) \\
&\qquad + \overline{u}_k(y_{baq}s\psi_{ab} - x_{baq}c\psi_{ab}) \Rightarrow
\end{aligned}$$

$$x_{abq} = x_{ab} + h_{ba}c\phi_{ab} - (y_{baq}c\psi_{ab} + x_{baq}s\psi_{ab})s\phi_{ab} \tag{6.275}$$

$$y_{abq} = y_{ab} + h_{ba}s\phi_{ab} + (y_{baq}c\psi_{ab} + x_{baq}s\psi_{ab})c\phi_{ab} \tag{6.276}$$

$$z_{ab} = x_{baq}c\psi_{ab} - y_{baq}s\psi_{ab} \tag{6.277}$$

As seen above, the cylinder-on-plane joint involves 10 principal variables, which are

$$\phi_{ab}, \psi_{ab}; x_{ab}, y_{ab}, z_{ab}, \lambda_{ab}; x_{abq}, y_{abq}; x_{baq}, y_{baq}$$

The variables listed above must satisfy six scalar equations, which are Eqs. (6.269) and (6.273)–(6.277). Therefore, the mobility of the cylinder-on-plane joint is $\mu_{ab} = 10 - 6 = 4$. Thus, if four of the variables (e.g. x_{ab}, y_{ab}, ϕ_{ab}, ψ_{ab}) are specified, then the remaining six variables (i.e. z_{ab}, λ_{ab}, x_{abq}, y_{abq}, x_{baq}, y_{baq}) can be determined by using the mentioned equations.

(b) Relative Velocity Equations

The relative velocity equations can be obtained by differentiating the six relative position equations mentioned in Part (a). This operation results in the following equations.

$$f_{bax}\dot{x}_{baq} + f_{bay}\dot{y}_{baq} = 0 \tag{6.278}$$

$$f_{baxx}\dot{x}_{baq} + f_{baxy}\dot{y}_{baq} = \dot{\lambda}_{ab}c\psi_{ab} - \lambda_{ab}\dot{\psi}_{ab}s\psi_{ab} \tag{6.279}$$

$$f_{bayx}\dot{x}_{baq} + f_{bayy}\dot{y}_{baq} = -\dot{\lambda}_{ab}s\psi_{ab} - \lambda_{ab}\dot{\psi}_{ab}c\psi_{ab} \tag{6.280}$$

$$\begin{aligned}
\dot{x}_{abq} = \dot{x}_{ab} - h_{ba}\dot{\phi}_{ab}s\phi_{ab} - (y_{baq}c\psi_{ab} + x_{baq}s\psi_{ab})\dot{\phi}_{ab}c\phi_{ab} \\
- (\dot{y}_{baq}c\psi_{ab} + \dot{x}_{baq}s\psi_{ab})s\phi_{ab} - (x_{baq}c\psi_{ab} - y_{baq}s\psi_{ab})\dot{\psi}_{ab}s\phi_{ab}
\end{aligned} \tag{6.281}$$

$$\begin{aligned}
\dot{y}_{abq} = \dot{y}_{ab} + h_{ba}\dot{\phi}_{ab}c\phi_{ab} - (y_{baq}c\psi_{ab} + x_{baq}s\psi_{ab})\dot{\phi}_{ab}s\phi_{ab} \\
+ (\dot{y}_{baq}c\psi_{ab} + \dot{x}_{baq}s\psi_{ab})c\phi_{ab} + (x_{baq}c\psi_{ab} - y_{baq}s\psi_{ab})\dot{\psi}_{ab}c\phi_{ab}
\end{aligned} \tag{6.282}$$

$$\dot{z}_{ab} = \dot{x}_{baq}c\psi_{ab} - \dot{y}_{baq}s\psi_{ab} - (x_{baq}s\psi_{ab} + y_{baq}c\psi_{ab})\dot{\psi}_{ab} \tag{6.283}$$

Thus, if four of the derivatives (e.g. $\dot{x}_{ab}$, $\dot{y}_{ab}$, $\dot{\phi}_{ab}$, $\dot{\psi}_{ab}$) are specified along with the position specifications, then the remaining six derivatives (i.e. $\dot{z}_{ab}$, $\dot{\lambda}_{ab}$, $\dot{x}_{abq}$, $\dot{y}_{abq}$, $\dot{x}_{baq}$, $\dot{y}_{baq}$) can be determined by using Eqs. (6.277) – (6.282).

(c) Relative Velocity Equations in the Case of Rolling Without Slipping

If $\mathcal{E}_{ba}$ rolls over $\mathcal{E}_{ab}$ without slipping due to the presence of sticking friction, then, as discussed in Section 6.5, the derivatives $\dot{\overline{r}}_{abq}^{(ab)}$ and $\dot{\overline{r}}_{baq}^{(ba)}$ become related by the following nonholonomic constraint.

$$\dot{\overline{r}}_{abq}^{(ab)} = \widehat{C}^{(ab,ba)}\dot{\overline{r}}_{baq}^{(ba)} \tag{6.284}$$

Equation (6.284) leads to the following equations for the cylinder-on-plane joint.

$$\overline{u}_i\dot{x}_{abq} + \overline{u}_j\dot{y}_{abq} = e^{\tilde{u}_k\phi_{ab}}e^{\tilde{u}_i\psi_{ab}}e^{\tilde{u}_j\pi/2}(\overline{u}_i\dot{x}_{baq} + \overline{u}_j\dot{y}_{baq}) \Rightarrow$$

$$\overline{u}_i\dot{x}_{abq} + \overline{u}_j\dot{y}_{abq} = e^{\tilde{u}_k\phi_{ab}}e^{\tilde{u}_i\psi_{ab}}(\overline{u}_j\dot{y}_{baq} - \overline{u}_k\dot{x}_{baq}) \Rightarrow$$

$$\begin{aligned}\overline{u}_i\dot{x}_{abq} + \overline{u}_j\dot{y}_{abq} &= e^{\tilde{u}_k\phi_{ab}}[(\overline{u}_j c\psi_{ab} + \overline{u}_k s\psi_{ab})\dot{y}_{baq} - (\overline{u}_k c\psi_{ab} - \overline{u}_j s\psi_{ab})\dot{x}_{baq}] \\ &= (e^{\tilde{u}_k\phi_{ab}}\overline{u}_j)(\dot{y}_{baq}c\psi_{ab} + \dot{x}_{baq}s\psi_{ab}) + \overline{u}_k(\dot{y}_{baq}s\psi_{ab} - \dot{x}_{baq}c\psi_{ab}) \\ &= (\overline{u}_j c\phi_{ab} - \overline{u}_i s\phi_{ab})(\dot{y}_{baq}c\psi_{ab} + \dot{x}_{baq}s\psi_{ab}) + \overline{u}_k(\dot{y}_{baq}s\psi_{ab} - \dot{x}_{baq}c\psi_{ab}) \Rightarrow\end{aligned}$$

$$\dot{x}_{abq} = -(\dot{y}_{baq}c\psi_{ab} + \dot{x}_{baq}s\psi_{ab})s\phi_{ab} \tag{6.285}$$

$$\dot{y}_{abq} = (\dot{y}_{baq}c\psi_{ab} + \dot{x}_{baq}s\psi_{ab})c\phi_{ab} \tag{6.286}$$

$$\dot{y}_{baq}s\psi_{ab} = \dot{x}_{baq}c\psi_{ab} \tag{6.287}$$

When Eqs. (6.285) and (6.286) are substituted, Eqs. (6.281) and (6.282) reduce to the following equations.

$$\begin{aligned}\dot{x}_{ab} = {} & (y_{baq}c\psi_{ab} + x_{baq}s\psi_{ab})\dot{\phi}_{ab}c\phi_{ab} + h_{ba}\dot{\phi}_{ab}s\phi_{ab} \\ & + (x_{baq}c\psi_{ab} - y_{baq}s\psi_{ab})\dot{\psi}_{ab}s\phi_{ab}\end{aligned} \tag{6.288}$$

$$\begin{aligned}\dot{y}_{ab} = {} & (y_{baq}c\psi_{ab} + x_{baq}s\psi_{ab})\dot{\phi}_{ab}s\phi_{ab} - h_{ba}\dot{\phi}_{ab}c\phi_{ab} \\ & - (x_{baq}c\psi_{ab} - y_{baq}s\psi_{ab})\dot{\psi}_{ab}c\phi_{ab}\end{aligned} \tag{6.289}$$

Note that $\dot{x}_{ab}$ and $\dot{y}_{ab}$ are the $\mathcal{F}_{ab}$ components of $\vec{v}_{ab}$, which is the relative velocity of O_{ba} with respect to O_{ab}. That is,

$$\vec{v}_{ab} = \vec{u}_i^{(ab)}\dot{x}_{ab} + \vec{u}_j^{(ab)}\dot{y}_{ab} \tag{6.290}$$

On the other hand, while the cylinder is rolling on the plane without slipping, $\vec{v}_{ab}$ can only be perpendicular to the centerline of the cylinder, i.e. to the unit vector $\vec{u}_{bal} = \vec{u}_k^{(ba)}$, which is expressed as follows in $\mathcal{F}_{ab}$:

$$\vec{u}_{bal} = \vec{u}_i^{(ab)}c\phi_{ab} + \vec{u}_j^{(ab)}s\phi_{ab} \tag{6.291}$$

In other words, $\vec{v}_{ab}$ must be such that

$$\vec{v}_{ab}\cdot\vec{u}_{bal} = \dot{x}_{ab}c\phi_{ab} + \dot{y}_{ab}s\phi_{ab} = 0 \tag{6.292}$$

When Eqs. (6.288) and (6.289) are substituted, Eq. (6.292) becomes

$$(y_{baq}c\psi_{ab} + x_{baq}s\psi_{ab})\dot{\phi}_{ab} = 0 \tag{6.293}$$

Equation (6.293) can be satisfied for any arbitrary value of ψ_{ab}, only if

$$\dot{\phi}_{ab} = 0 \tag{6.294}$$

Equation (6.294) implies that the centerline of the cylinder, i.e. the unit vector $\vec{u}_{bal}$, does not change its orientation during a rolling-without-slipping motion. Due to this fact, Eqs. (6.288) and (6.289) simplify to the following equations.

$$\dot{x}_{ab} = (x_{baq} c\psi_{ab} - y_{baq} s\psi_{ab})\dot{\psi}_{ab} s\phi_{ab} \tag{6.295}$$

$$\dot{y}_{ab} = -(x_{baq} c\psi_{ab} - y_{baq} s\psi_{ab})\dot{\psi}_{ab} c\phi_{ab} \tag{6.296}$$

The velocity $\vec{v}_{ab}$ can also be expressed as

$$\vec{v}_{ab} = v_{ab} \vec{u}_{ban} \tag{6.297}$$

In Eq. (6.297), $\vec{u}_{ban}$ is the normal coplanar companion of $\vec{u}_{bal}$. That is,

$$\vec{u}_{ban} = \vec{u}_j^{(ab)} c\phi_{ab} - \vec{u}_i^{(ab)} s\phi_{ab} \tag{6.298}$$

By using Eqs. (6.297), (6.298), and (6.277), v_{ab} can be obtained as follows:

$$v_{ab} = \vec{v}_{ab} \cdot \vec{u}_{ban} = \dot{y}_{ab} c\phi_{ab} - \dot{x}_{ab} s\phi_{ab} = -(x_{baq} c\psi_{ab} - y_{baq} s\psi_{ab})\dot{\psi}_{ab} \Rightarrow$$

$$v_{ab} = -z_{ab}\dot{\psi}_{ab} \tag{6.299}$$

Equation (6.299) implies that $\mathcal{E}_{ba}$ rolls over $\mathcal{E}_{ab}$ without slipping by rotating about the contact line. During this motion, the principal derivatives to be related happen to be $\dot{\psi}_{ab}$, $\dot{x}_{baq}$, $\dot{y}_{baq}$, and $\dot{\lambda}_{ab}$. These derivatives have to satisfy three scalar equations, which are Eqs. (6.279), (6.280), and *one* of Eqs. (6.278) and (6.287) because they are dependent on each other through Eqs. (6.273) and (6.274). Therefore, the mobility of the cylinder-on-plane joint reduces to $\mu_{ab} = 1$ during the rolling-without-slipping motion. Thus, if $\dot{\psi}_{ab}$ is specified along with the position specifications, then $\dot{x}_{baq}$, $\dot{y}_{baq}$, and $\dot{\lambda}_{ab}$ can be determined by using the mentioned equations.

6.7 Examples

6.7.1 Example 6.1: An RRRSP Mechanism

(a) Kinematic Description of the System

Figure 6.19 shows an RRRSP mechanism with its side and top views on the left-hand and right-hand sides. As its designation implies, it consists of three revolute (R) joints, one spherical (S) joint, and one prismatic (P) joint. It contains four moving links. As such,

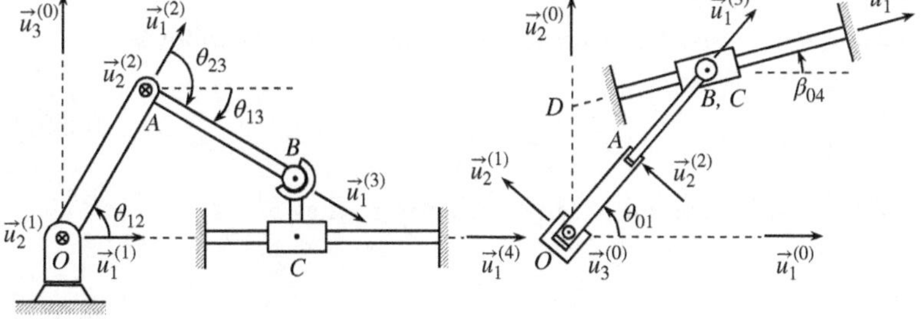

Figure 6.19 An RRRSP mechanism.

i.e. with $\lambda = 6$, $n = 4$, $j_1 = 4$, $j_3 = 1$, and $j_2 = j_4 = j_5 = 0$, it has the following mobility according to the Kutzbach–Gruebler formula.

$$\mu = \lambda n - \sum_{k=1}^{5} (\lambda - k) j_k = 6 \times 4 - (5 \times 4 + 3 \times 1) = 1 \tag{6.300}$$

The joint frames are attached to the relevant kinematic elements so that they are related to the link frames as shown below.

$$\mathcal{F}_{01}(O_{01}) = \mathcal{F}_0(O), \mathcal{F}_{10}(O_{10}) = \mathcal{F}_1(O)$$
$$\mathcal{F}_{12}(O_{12}) = \mathcal{F}_1(O), \mathcal{F}_{21}(O_{21}) = \mathcal{F}_2(O)$$
$$\mathcal{F}_{23}(O_{23}) = \mathcal{F}_2(A), \mathcal{F}_{32}(O_{32}) = \mathcal{F}_3(A)$$
$$\mathcal{F}_{34}(O_{34}) = \mathcal{F}_3(B), \mathcal{F}_{43}(O_{43}) = \mathcal{F}_4(B)$$
$$\mathcal{F}_{40}(O_{40}) = \mathcal{F}_4(C), \mathcal{F}_{04}(O_{04}) = \mathcal{F}_4(D)$$

Hence, the relative orientation matrices between the successive links can be expressed as follows:

$$\hat{C}^{(0,1)} = e^{\tilde{u}_3 \theta_{01}} \tag{6.301}$$

$$\hat{C}^{(1,2)} = e^{\tilde{u}_2(-\theta_{12})} = e^{-\tilde{u}_2 \theta_{12}} \tag{6.302}$$

$$\hat{C}^{(2,3)} = e^{\tilde{u}_2 \theta_{23}} \tag{6.303}$$

$$\hat{C}^{(3,4)} = e^{\tilde{n}_{341} \phi_{34}} e^{\tilde{n}_{342} \theta_{34}} e^{\tilde{n}_{343} \psi_{34}} \tag{6.304}$$

$$\hat{C}^{(0,4)} = e^{\tilde{u}_3 \beta_{04}} \tag{6.305}$$

In Eq. (6.302), θ_{12} is inserted into the rotation matrix formula with a minus sign because it represents a negative rotation about $\vec{u}_2^{(1)}$ according to the right-hand rule. In Eq. (6.304), the three unit vectors $(\vec{n}_{341}, \vec{n}_{342}, \vec{n}_{343})$ associated with the spherical joint $\mathcal{J}_{34}$ have not yet been specified, because they will be selected later in order to simplify the relevant kinematic relationships.

(b) Loop Closure Equations

Equations (6.301)–(6.305) lead to the following orientational closure equation.

$$\hat{C}^{(0,1)}\hat{C}^{(1,2)}\hat{C}^{(2,3)}\hat{C}^{(3,4)}\hat{C}^{(4,0)} = \hat{I} \Rightarrow \hat{C}^{(0,1)}\hat{C}^{(1,2)}\hat{C}^{(2,3)}\hat{C}^{(3,4)} = \hat{C}^{(0,4)} \Rightarrow$$
$$e^{\tilde{u}_3 \theta_{01}} e^{-\tilde{u}_2 \theta_{12}} e^{\tilde{u}_2 \theta_{23}} e^{\tilde{n}_{341} \phi_{34}} e^{\tilde{n}_{342} \theta_{34}} e^{\tilde{n}_{343} \psi_{34}} = e^{\tilde{u}_3 \beta_{04}} \Rightarrow$$
$$e^{\tilde{n}_{341} \phi_{34}} e^{\tilde{n}_{342} \theta_{34}} e^{\tilde{n}_{343} \psi_{34}} = e^{-\tilde{u}_2(\theta_{23}-\theta_{12})} e^{-\tilde{u}_3(\theta_{01}-\beta_{04})} \Rightarrow$$
$$e^{\tilde{n}_{341} \phi_{34}} e^{\tilde{n}_{342} \theta_{34}} e^{\tilde{n}_{343} \psi_{34}} = e^{-\tilde{u}_2 \theta_{13}} e^{-\tilde{u}_3 \theta'_{01}} \tag{6.306}$$

In Eq. (6.306), θ_{13} and θ'_{01} are *combined angles,* which are defined as follows:

$$\left.\begin{aligned} \theta_{13} &= \theta_{23} - \theta_{12} \\ \theta'_{01} &= \theta_{01} - \beta_{04} \end{aligned}\right\} \tag{6.307}$$

Based on Eq. (6.306), $\overline{n}_{341}$, $\overline{n}_{342}$, and $\overline{n}_{343}$ can be selected so that

$$\left.\begin{aligned} \overline{n}_{341} &= \overline{u}_1 \\ \overline{n}_{342} &= \overline{u}_2 \\ \overline{n}_{343} &= \overline{u}_3 \end{aligned}\right\} \tag{6.308}$$

Upon the above judicious selection, the angles of the spherical joint come out simply as follows without any algebraic manipulation:

$$\left.\begin{aligned} \phi_{34} &= 0 \\ \theta_{34} &= -\theta_{13} \\ \psi_{34} &= -\theta'_{01} \end{aligned}\right\} \tag{6.309}$$

The locational closure equation of the mechanism can be written as follows by expressing the closure at the center B of the spherical joint $\mathcal{J}_{34}$:

$$\vec{r}_{OA} + \vec{r}_{AB} = \vec{r}_{OD} + \vec{r}_{DC} + \vec{r}_{CB} \Rightarrow$$

$$b_2\vec{u}_1^{(2)} + b_3\vec{u}_1^{(3)} = d_4\vec{u}_2^{(0)} + s_{14}\vec{u}_1^{(4)} + h_4\vec{u}_3^{(0)} \tag{6.310}$$

In Eq. (6.310), s_{14} is the joint variable of the prismatic joint and the lengths b_2, b_3, d_4, and h_4 are the constant geometric parameters of the mechanism. Equation (6.310) can be written as the following column matrix equation in the base frame $\mathcal{F}_0(O)$.

$$b_2\overline{u}_1^{(2/0)} + b_3\overline{u}_1^{(3/0)} = d_4\overline{u}_2^{(0/0)} + s_{14}\overline{u}_1^{(4/0)} + h_4\overline{u}_3^{(0/0)} \Rightarrow$$

$$b_2\widehat{C}^{(0,2)}\overline{u}_1 + b_3\widehat{C}^{(0,3)}\overline{u}_1 = \mathrm{d}_4\overline{u}_2 + h_4\overline{u}_3 + s_{14}\widehat{C}^{(0,4)}\overline{u}_1 \tag{6.311}$$

The relevant orientation matrices in Eq. (6.311) can be expressed as follows:

$$\widehat{C}^{(0,2)} = \widehat{C}^{(0,1)}\widehat{C}^{(1,2)} = e^{\tilde{u}_3\theta_{01}}e^{-\tilde{u}_2\theta_{12}} \tag{6.312}$$

$$\widehat{C}^{(0,3)} = \widehat{C}^{(0,2)}\widehat{C}^{(2,3)} = e^{\tilde{u}_3\theta_{01}}e^{-\tilde{u}_2\theta_{12}}e^{\tilde{u}_2\theta_{23}} = e^{\tilde{u}_3\theta_{01}}e^{\tilde{u}_2\theta_{13}} \tag{6.313}$$

$$\widehat{C}^{(0,4)} = e^{\tilde{u}_3\beta_{04}} \tag{6.314}$$

Hence, Eq. (6.311) can be manipulated as shown below.

$$b_2 e^{\tilde{u}_3\theta_{01}}e^{-\tilde{u}_2\theta_{12}}\overline{u}_1 + b_3 e^{\tilde{u}_3\theta_{01}}e^{\tilde{u}_2\theta_{13}}\overline{u}_1 = d_4\overline{u}_2 + h_4\overline{u}_3 + s_{14}e^{\tilde{u}_3\beta_{04}}\overline{u}_1 \Rightarrow$$

$$b_2 e^{-\tilde{u}_2\theta_{12}}\overline{u}_1 + b_3 e^{\tilde{u}_2\theta_{13}}\overline{u}_1 = d_4 e^{-\tilde{u}_3\theta_{01}}\overline{u}_2 + h_4\overline{u}_3 + s_{14}e^{-\tilde{u}_3\theta'_{01}}\overline{u}_1 \Rightarrow$$

$$\begin{aligned} &b_2(\overline{u}_1 c\theta_{12} + \overline{u}_3 s\theta_{12}) + b_3(\overline{u}_1 c\theta_{13} - \overline{u}_3 s\theta_{13}) \\ &\quad = d_4(\overline{u}_2 c\theta_{01} + \overline{u}_1 s\theta_{01}) + h_4\overline{u}_3 + \overline{u}_1 s_{14} c\theta'_{01} - \overline{u}_2 s_{14} s\theta'_{01} \end{aligned} \tag{6.315}$$

Equation (6.315) leads to the following scalar equations.

$$b_2 c\theta_{12} + b_3 c\theta_{13} = d_4 s\theta_{01} + s_{14}c\theta'_{01} = d_4 s\theta_{01} + s_{14}c(\theta_{01} - \beta_{04}) \tag{6.316}$$

$$d_4 c\theta_{01} = s_{14}s\theta'_{01} = s_{14}s(\theta_{01} - \beta_{04}) \tag{6.317}$$

$$b_2 s\theta_{12} - b_3 s\theta_{13} = h_4 \tag{6.318}$$

Since $\mu = 1$ for this mechanism, one of the joint variables can be specified and the other three can be determined by solving Eqs. (6.316)–(6.318) for them. The solutions can be obtained analytically as explained below by taking first θ_{01} and then s_{14} as the specified variable.

(c) Determination of the Unspecified Variables for Specified θ_{01}

If θ_{01} is specified, then Eq. (6.317) gives s_{14} as follows, if $\theta'_{01} = \theta_{01} - \beta_{04} \neq 0$:

$$s_{14} = d_4 c\theta_{01}/s\theta'_{01} \tag{6.319}$$

Having found s_{14}, Eqs. (6.316) and (6.318) can be written as shown below.

$$b_2 c\theta_{12} = f_4 - b_3 c\theta_{13} \tag{6.320}$$

$$b_2 s\theta_{12} = h_4 + b_3 s\theta_{13} \tag{6.321}$$

In Eq. (6.320), f_4 is known as

$$f_4 = f_4(\theta_{01}) = d_4 s\theta_{01} + s_{14} c\theta'_{01} = d_4 c\beta_{04}/s\theta'_{01} \tag{6.322}$$

Equations (6.320) and (6.321) can be combined into the following equation.

$$b_2^2 = (f_4 - b_3 c\theta_{13})^2 + (h_4 + b_3 s\theta_{13})^2 \Rightarrow$$

$$b_2^2 = f_4^2 + h_4^2 + b_3^2 + 2b_3(h_4 s\theta_{13} - f_4 c\theta_{13}) \Rightarrow$$

$$f_4 c\theta_{13} - h_4 s\theta_{13} = g_4 \tag{6.323}$$

In Eq. (6.323), g_4 is known as

$$g_4 = g_4(\theta_{01}) = (f_4^2 + h_4^2 + b_3^2 - b_2^2)/(2b_3) \tag{6.324}$$

Equation (6.323) can be solved for θ_{13} as shown below.

$$\left.\begin{aligned} c\theta_{13} &= \xi_{13} = [\sigma h_4 \sqrt{f_4^2 + h_4^2 - g_4^2} + f_4 g_4]/m_4 \\ s\theta_{13} &= \eta_{13} = [\sigma f_4 \sqrt{f_4^2 + h_4^2 - g_4^2} - h_4 g_4]/m_4 \end{aligned}\right\} \tag{6.325}$$

$$\theta_{13} = \mathrm{atan}_2(\eta_{13}, \xi_{13}) \tag{6.326}$$

In Eq. Set (6.325), $\sigma = \pm 1$ is an arbitrarily selectable *closure sign variable* and

$$m_4 = f_4^2 + h_4^2 \tag{6.327}$$

After finding θ_{13} as shown above, θ_{12} can be found as follows without any other sign variable by using both Eqs. (6.320) and (6.321) with the help of Eq. Set (6.325).

$$\theta_{12} = \mathrm{atan}_2[(h_4 + b_3 s\theta_{13}), (f_4 - b_3 c\theta_{13})] \tag{6.328}$$

Finally, referring to Eq. (6.307), θ_{23} is found as

$$\theta_{23} = \theta_{12} + \theta_{13} \tag{6.329}$$

The value of the closure sign variable σ corresponds to one of the two distinct closures of the mechanism. These closures are implied by Eq. Set (6.325). If $\sigma = +1$ is selected, then $s\theta_{13}$ and hence θ_{13} will be positive and point A will be above the line OB as shown in Figure 6.19. On the other hand, if $\sigma = -1$ is selected, then $s\theta_{13}$ and hence θ_{13} will be negative and point A will be below the line OB.

(d) Determination of the Unspecified Variables for Specified s_{14}

If s_{14} is specified, then Eq. (6.317) can be written again as

$$(s_{14} c\beta_{04}) s\theta_{01} = (d_4 + s_{14} s\beta_{04}) c\theta_{01} \tag{6.330}$$

Equation (6.330) implies that

$$\left.\begin{aligned} c\theta_{01} &= \sigma'(s_{14} c\beta_{04})/m_1 \\ s\theta_{01} &= \sigma'(d_4 + s_{14} s\beta_{04})/m_1 \end{aligned}\right\} \tag{6.331}$$

In Eq. Set (6.331), $\sigma' = \pm 1$ is an arbitrarily selectable *closure sign variable* and

$$m_1 = \sqrt{(s_{14}c\beta_{04})^2 + (d_4 + s_{14}s\beta_{04})^2} = \sqrt{s_{14}^2 + 2d_4s_{14}s\beta_{04} + d_4^2} \tag{6.332}$$

Hence, if $m_1 \neq 0$, i.e. if $s_{14} \neq d_4$ and $\beta_{04} \neq -\pi/2$, i.e. if point B is not on the axis along $\vec{u}_3^{(0)}$, θ_{01} can be found as

$$\theta_{01} = \text{atan}_2[\sigma'(d_4 + s_{14}s\beta_{04}), \sigma'(s_{14}c\beta_{04})] \tag{6.333}$$

Afterwards, θ_{12} and θ_{23} can be found similarly as described above in Part (c) for the case of specified θ_{01}. However, it is to be noted that, while Part (c) involves only one closure sign variable σ, which is associated directly with θ_{13}, Part (d) involves two closure sign variables σ' and σ, which are associated with θ_{01} and θ_{13}, respectively. The role of σ has already been discussed in Part (c). The role of σ' is discussed below along with σ.

If $\sigma' = +1$ is selected, then Eq. (6.333) implies that θ_{01} will be an acute angle as shown in Figure 6.19. In that case, as explained in Part (c), $\sigma = +1$ and $\sigma = -1$ will, respectively, lead to the closures designated as A-above-OB and A-below-OB.

If $\sigma' = -1$ is selected, let the resultant angles be denoted as θ_{01}^-, θ_{12}^-, θ_{23}^-, and $\theta_{13}^- = \theta_{23}^- - \theta_{12}^-$. Then, Eq. (6.333) implies that θ_{01}^- will be an obtuse angle, which is related to θ_{01} (caused by $\sigma' = +1$) as follows:

$$\theta_{01}^- = \theta_{01} + \pi \tag{6.334}$$

Consequently, according to Eqs. (6.322) and (6.326), f_4 will be negative and then $s\theta_{13}^-$ and hence θ_{13}^- will also be negative if $\sigma = +1$. This conclusion implies that the closure role of σ will be reversed. In other words, the A-above-OB and A-below-OB closures will now be caused, respectively, by $\sigma = -1$ and $\sigma = +1$.

6.7.2 Example 6.2: A Two-Link Mechanism with Three Point-on-Plane Joints

(a) Kinematic Description of the System

The mechanism shown in Figure 6.20 consists of two links $\mathcal{L}_0$ and $\mathcal{L}_1$, which are also the kinematic elements $\mathcal{E}_{01}$ and $\mathcal{E}_{10}$. $\mathcal{L}_0$ is fixed and $\mathcal{L}_1$ is connected to $\mathcal{L}_0$ with a triangle-on-surface joint. In this example, the surface is the union of the three coordinate planes (i.e. the planes formed by the pairs of coordinate axes) of the frame

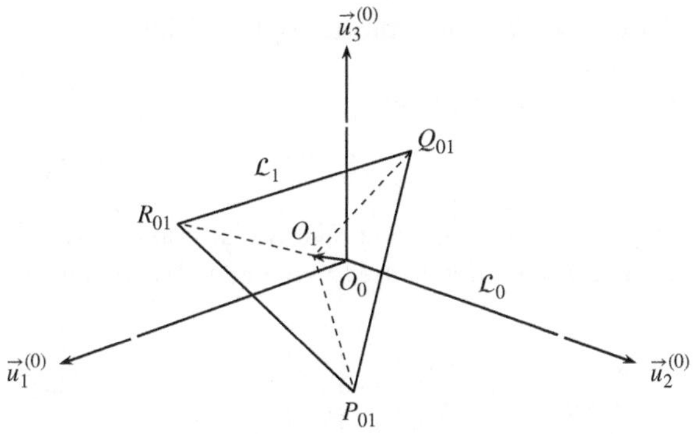

Figure 6.20 A two-link mechanism with three point-on-plane joints.

$\mathcal{F}_0$ attached to $\mathcal{L}_0$. The relative position of $\mathcal{L}_1$ with respect to $\mathcal{L}_0$ can be described by the following location and orientation equations.

$$\overline{r}_{0,1}^{(0)} = \overline{u}_1 x_{01} + \overline{u}_2 y_{01} + \overline{u}_3 z_{01} \tag{6.335}$$

$$\widehat{C}^{(0,1)} = e^{\tilde{u}_3 \psi_{01}} e^{\tilde{u}_2 \theta_{01}} e^{\tilde{u}_1 \phi_{01}} \tag{6.336}$$

The joint $\mathcal{J}_{01}$ is formed in such a way that P_{01} is the contact point with the 1 – 2 plane of $\mathcal{F}_0$, Q_{01} is the contact point with the 2 – 3 plane of $\mathcal{F}_0$, and R_{01} is the contact point with the 3 – 1 plane of $\mathcal{F}_0$. In other words, $z_{P_{01}} = x_{Q_{01}} = y_{R_{01}} = 0$. Therefore, referring to Section 6.4, the third, first, and second respective members of Eq. Sets (6.59)–(6.61) imply the following equations that relate the locational joint variables x_{01}, y_{01}, and z_{01} (which are the coordinates of O_1) to the angular joint variables ψ_{01}, θ_{01}, and ϕ_{01} (which are the 3-2-1 Euler angles of $\mathcal{L}_1$).

$$x_{01} = d_1[c\psi_{01} c\theta_{01} c\gamma_1 + (s\psi_{01} c\phi_{01} - c\psi_{01} s\theta_{01} s\phi_{01}) s\gamma_1] \tag{6.337}$$

$$y_{01} = d_1[s\psi_{01} c\theta_{01} c\gamma_1 + (c\psi_{01} c\phi_{01} + s\psi_{01} s\theta_{01} s\phi_{01}) s\gamma_1] \tag{6.338}$$

$$z_{01} = d_1 s\theta_{01} \tag{6.339}$$

In the above equations, $d_1 = O_1 P_{01} = O_1 Q_{01} = O_1 R_{01}$ and $\gamma_1 = [\sphericalangle(Q_{01} O_1 R_{01})]/2$. Upon adapting to this example and substituting the above equations, the other members of Eq. Sets (6.59)–(6.61) give the nonzero coordinates of the contact points P_{01}, Q_{01}, and R_{01} in terms of the angular joint variables.

(b) Determination of the Unspecified Variables

If the angular joint variables are specified, then Eqs. (6.337)–(6.339) give the locational joint variables directly. However, if the locational joint variables are specified, then the angular joint variables cannot be extracted from Eqs. (6.337)–(6.339) in a completely analytical way. In other words, it is not possible to obtain closed-form expressions for all three of the angular joint variables in terms of the specified locational joint variables. The closed-form expressions and the analytical solutions are explained in Part (c). In this particular example, even if an analytical solution is not attainable, the angular joint variables can be found at least in a *first-order semi-analytical* way. In such a solution, one of the unknown variables can only be found by means of a suitable numerical method, but the others can still be found analytically. The procedure of this semi-analytical solution is explained below.

From Eq. (6.339), θ_{01} can be found as follows with an arbitrary sign variable:

$$s\theta_{01} = \zeta_{01} = z_{01}/d_1 \Rightarrow c\theta_{01} = \sigma\sqrt{1 - \zeta_{01}^2}; \sigma = \pm 1 \Rightarrow$$

$$\theta_{01} = \text{atan}_2(\zeta_{01}, \sigma\sqrt{1 - \zeta_{01}^2}) \tag{6.340}$$

Note that θ_{01} exists if $\zeta_{01}^2 \leq 1$, i.e. if $|z_{01}| \leq d_1$. When θ_{01} becomes available as described above, Eqs. (6.337) and (6.338) can be written together as the following matrix equation in order to find ϕ_{01} and ψ_{01}.

$$\begin{bmatrix} c\theta_{01} c\gamma_1 - s\theta_{01} s\phi_{01} s\gamma_1 & c\phi_{01} s\gamma_1 \\ c\phi_{01} s\gamma_1 & c\theta_{01} c\gamma_1 + s\theta_{01} s\phi_{01} s\gamma_1 \end{bmatrix} \begin{bmatrix} c\psi_{01} \\ s\psi_{01} \end{bmatrix} = \begin{bmatrix} \xi_{01} \\ \eta_{01} \end{bmatrix} \tag{6.341}$$

In Eq. (6.341),

$$\left.\begin{array}{l}\xi_{01} = x_{01}/d_1\\ \eta_{01} = y_{01}/d_1\end{array}\right\} \tag{6.342}$$

The coefficient matrix in Eq. (6.341) has the following determinant.

$$D_{01} = (c\theta_{01}c\gamma_1)^2 - [(s\theta_{01}s\phi_{01})^2 + (c\phi_{01})^2](s\gamma_1)^2 \tag{6.343}$$

If $D_{01} \neq 0$, Eq. (6.341) can be solved as follows for $c\psi_{01}$ and $s\psi_{01}$.

$$c\psi_{01} = [(c\theta_{01}c\gamma_1 + s\theta_{01}s\phi_{01}s\gamma_1)\xi_{01} - (c\phi_{01}s\gamma_1)\eta_{01}]/D_{01} = A_{01}/D_{01} \tag{6.344}$$

$$s\psi_{01} = [(c\theta_{01}c\gamma_1 - s\theta_{01}s\phi_{01}s\gamma_1)\eta_{01} - (c\phi_{01}s\gamma_1)\xi_{01}]/D_{01} = B_{01}/D_{01} \tag{6.345}$$

By adding the squares of Eqs. (6.344) and (6.345), ψ_{01} is eliminated and the following rather complicated equation is obtained for ϕ_{01}.

$$\begin{aligned}&(c\psi_{01})^2 + (s\psi_{01})^2 = 1 \Rightarrow D_{01}^2 = A_{01}^2 + B_{01}^2 \Rightarrow\\ &\{(c\theta_{01}c\gamma_1)^2 - [(s\theta_{01}s\phi_{01})^2 + (c\phi_{01})^2](s\gamma_1)^2\}^2\\ &\quad = \xi_{01}^2(c\theta_{01}c\gamma_1 + s\theta_{01}s\phi_{01}s\gamma_1)^2 + \eta_{01}^2(c\phi_{01}s\gamma_1)^2\\ &\quad + \eta_{01}^2(c\theta_{01}c\gamma_1 - s\theta_{01}s\phi_{01}s\gamma_1)^2 + \xi_{01}^2(c\phi_{01}s\gamma_1)^2\\ &\quad - 4\xi_{01}\eta_{01}s\gamma_1c\gamma_1c\theta_{01}c\phi_{01}\end{aligned} \tag{6.346}$$

Clearly, Eq. (6.346) cannot be solved analytically for ϕ_{01}. However, it can be solved numerically by means of a suitable iterative method. Afterwards, ψ_{01} can be found analytically as follows from Eqs. (6.344) and (6.345):

$$\psi_{01} = \mathrm{atan}_2(B_{01}, A_{01}) \tag{6.347}$$

If $D_{01} = 0$, then a singularity occurs. In that case, ψ_{01} becomes indefinite. Nevertheless, it will still be finite, if $x_{01} = d_1\xi_{01}$ and $y_{01} = d_1\eta_{01}$ are specified according to the consistency relationship imposed by Eqs. (6.344) and (6.345), which may be expressed by either of the following equations.

$$(c\theta_{01}c\gamma_1 + s\theta_{01}s\phi_{01}s\gamma_1)x_{01} = (c\phi_{01}s\gamma_1)y_{01} \tag{6.348}$$

$$(c\theta_{01}c\gamma_1 - s\theta_{01}s\phi_{01}s\gamma_1)y_{01} = (c\phi_{01}s\gamma_1)x_{01} \tag{6.349}$$

(c) Closed-Form Expressions and Analytical Solutions

A *closed-form expression* is formed as a combination of a *finite* number of the *elementary* functions. The combination may involve all the algebraic operations such as addition, subtraction, multiplication, division, exponentiation, etc. The set of elementary functions of an argument x comprise functions such as $f(x) = x$, $f(x) = \exp(x)$, $f(x) = \ln(x)$, $f(x) = \sin(x)$, $f(x) = \cos(x)$, $f(x) = \sin^{-1}(x)$, $f(x) = \cos^{-1}(x)$, $f(x) = \tan^{-1}(x)$, etc.

Consider an equation such as $\phi(y, x) = 0$, which is to be solved so as to obtain y as a function of x. Suppose y can be obtained by means of certain *symbolic manipulations* so that $y = g(x)$, where $g(x)$ is a *closed-form expression*. Then, $y = g(x)$ is called an *analytical solution* of the equation $\phi(y, x) = 0$.

6.7.3 Example 6.3: A Spatial Cam Mechanism

(a) Kinematic Description of the System

Figure 6.21 shows a spatial cam mechanism with its two projected views. The mechanism consists of two moving links $\mathcal{L}_1$ and $\mathcal{L}_2$, which are connected to the fixed link $\mathcal{L}_0$ with revolute joints $\mathcal{J}_{01}$ and $\mathcal{J}_{02}$. The kinematic elements that form the cam joint $\mathcal{J}_{12}$ are elliptical and spherical cams attached, respectively, to $\mathcal{L}_1$ and $\mathcal{L}_2$. The link frames attached to $\mathcal{L}_1$, $\mathcal{L}_2$, and $\mathcal{L}_0$ are defined as follows with their origins and basis vectors:

$$\mathcal{F}_0 = \mathcal{F}_0\{O_0; \vec{u}_1^{(0)}, \ \vec{u}_2^{(0)}, \ \vec{u}_3^{(0)}\}$$
$$\mathcal{F}_1 = \mathcal{F}_1\{O_1; \vec{u}_1^{(1)}, \ \vec{u}_2^{(1)}, \ \vec{u}_3^{(1)}\}; \vec{u}_2^{(1)} = \vec{u}_2^{(0)}$$
$$\mathcal{F}_2 = \mathcal{F}_2\{O_2; \vec{u}_1^{(2)}, \ \vec{u}_2^{(2)}, \ \vec{u}_3^{(2)}\}; \vec{u}_1^{(2)} = \vec{u}_1^{(0)}$$

The joint frames attached to the kinematic elements are related to the link frames as shown below.

$$\mathcal{F}_{01}(O_{01}) = \mathcal{F}_0(O_{01}), \mathcal{F}_{10}(O_{10}) = \mathcal{F}_1(O_{01}); O_{10} = O_{01} = O_0$$
$$\mathcal{F}_{02}(O_{02}) = \mathcal{F}_0(O_{02}), \mathcal{F}_{20}(O_{20}) = \mathcal{F}_2(O_{02}); O_{20} = O_{02}$$
$$\mathcal{F}_{12}(O_{12}) = \mathcal{F}_1(O_1), \mathcal{F}_{21}(O_{21}) = \mathcal{F}_2(O_2); O_{12} = O_1, O_{21} = O_2$$

The fixed geometric parameters are indicated below.

$$d_3 = O_{01}B_{03}, d_2 = B_{03}B_{02}, d_1 = B_{02}O_{02}$$
$$c_1 = O_{01}O_1 = O_0O_1, c_2 = O_{02}O_2$$

b_1, b_2, b_3: semi-axis lengths of the ellipsoidal cam
r_2: radius of the spherical cam

(b) Loop Closure Equations

The orientational closure of the mechanism can be expressed by the following matrix equation.

$$\hat{C}^{(0,1)}\hat{C}^{(1,2)} = \hat{C}^{(0,2)} \tag{6.350}$$

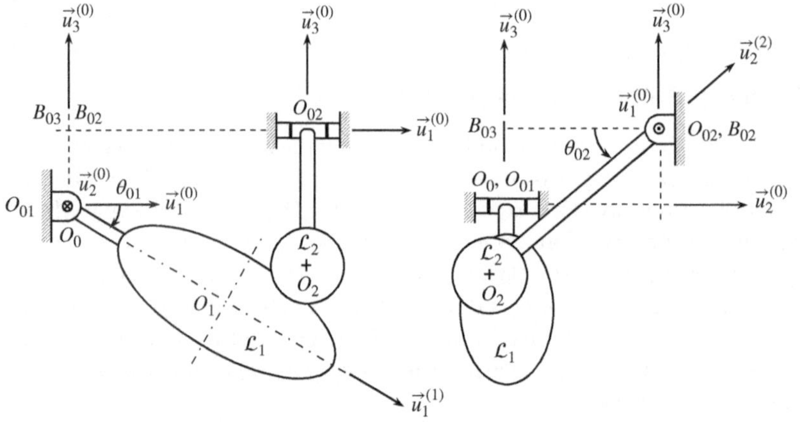

Figure 6.21 A spatial cam mechanism with elliptic and spherical cams.

In Eq. (6.350),

$$\widehat{C}^{(0,1)} = e^{\widetilde{u}_2\theta_{01}} \tag{6.351}$$

$$\widehat{C}^{(0,2)} = e^{\widetilde{u}_1\theta_{02}} \tag{6.352}$$

$$\widehat{C}^{(1,2)} = e^{\widetilde{n}_{121}\phi_{12}} e^{\widetilde{n}_{122}\theta_{12}} e^{\widetilde{n}_{123}\psi_{12}} \tag{6.353}$$

In Eq. (6.353), the unit column matrices $\overline{n}_{121}$, $\overline{n}_{122}$, and $\overline{n}_{123}$ represent the axes of the three successive rotations of $\mathcal{L}_2$ with respect to $\mathcal{L}_1$. On the other hand, Eqs. (6.350) and (6.353) suggest to select the 2-3-1 sequence for the sake of simplified kinematics. Upon this selection, it turns out that $\overline{n}_{121} = \overline{u}_2$, $\overline{n}_{122} = \overline{u}_3$, $\overline{n}_{123} = \overline{u}_1$, and

$$\widehat{C}^{(1,2)} = e^{\widetilde{u}_2\phi_{12}} e^{\widetilde{u}_3\theta_{12}} e^{\widetilde{u}_1\psi_{12}} \tag{6.354}$$

Hence, Eq. (6.350) becomes

$$e^{\widetilde{u}_2\theta_{01}} e^{\widetilde{u}_2\phi_{12}} e^{\widetilde{u}_3\theta_{12}} e^{\widetilde{u}_1\psi_{12}} = e^{\widetilde{u}_1\theta_{02}} \Rightarrow$$

$$e^{\widetilde{u}_2(\phi_{12}+\theta_{01})} e^{\widetilde{u}_3\theta_{12}} e^{\widetilde{u}_1(\psi_{12}-\theta_{02})} = \widehat{I} \tag{6.355}$$

According to Eq. (6.355), the angular variables of the joint $\mathcal{J}_{12}$ are readily determined as

$$\phi_{12} = -\theta_{01},\ \ \theta_{12} = 0,\ \ \psi_{12} = \theta_{02} \tag{6.356}$$

As for the locational closure of the mechanism, the following vector equation can be written.

$$\overrightarrow{O_{01}O_1} + \overrightarrow{O_1Q_{12}} = \overrightarrow{O_{01}B_{03}} + \overrightarrow{B_{03}B_{02}} + \overrightarrow{B_{02}O_{02}} + \overrightarrow{O_{02}O_2} + \overrightarrow{O_2Q_{21}} \tag{6.357}$$

By using the variables (θ_{01} and θ_{02}) of the revolute joints, the coordinates (x_1, x_2, x_3; y_1, y_2, y_3) of the contact points (Q_{12} and Q_{21}) in the link frames of the cams, and the constant geometric parameters indicated in Part (a), Eq. (6.357) can be written again as shown below.

$$\begin{aligned} c_1\vec{u}_1^{(1)} + [x_1\vec{u}_1^{(1)} + x_2\vec{u}_2^{(1)} + x_3\vec{u}_3^{(1)}] &= d_3\vec{u}_3^{(0)} + d_2\vec{u}_2^{(0)} + d_1\vec{u}_1^{(0)} - c_2\vec{u}_2^{(2)} \\ &\quad + [y_1\vec{u}_1^{(2)} + y_2\vec{u}_2^{(2)} + y_3\vec{u}_3^{(2)}] \end{aligned} \tag{6.358}$$

By using the transformation matrices expressed by Eqs. (6.351) and (6.352), the matrix equation corresponding to Eq. (6.358) can be written in $\mathcal{F}_0$ as follows:

$$\begin{aligned} e^{\widetilde{u}_2\theta_{01}}[(x_1 + c_1)\overline{u}_1 + x_2\overline{u}_2 + x_3\overline{u}_3] &= d_3\overline{u}_3 + d_2\overline{u}_2 + d_1\overline{u}_1 \\ &+ e^{\widetilde{u}_1\theta_{02}}[y_1\overline{u}_1 + (y_2 - c_2)\overline{u}_2 + y_3\overline{u}_3] \end{aligned} \tag{6.359}$$

Equation (6.359) can be written more compactly as

$$\overline{x} = e^{-\widetilde{u}_2\theta_{01}}(e^{\widetilde{u}_1\theta_{02}}\overline{y} + \overline{z}) \tag{6.360}$$

In Eq. (6.360),

$$\overline{x} = x_1\overline{u}_1 + x_2\overline{u}_2 + x_3\overline{u}_3 \tag{6.361}$$

$$\overline{y} = y_1\overline{u}_1 + y_2\overline{u}_2 + y_3\overline{u}_3 \tag{6.362}$$

$$\overline{z} = \overline{z}(\theta_{01}, \theta_{02}) = \overline{d} - c_1 e^{\widetilde{u}_2\theta_{01}}\overline{u}_1 - c_2 e^{\widetilde{u}_1\theta_{02}}\overline{u}_2 \Rightarrow$$

$$\overline{z} = \overline{u}_1(d_1 - c_1c\theta_{01}) + \overline{u}_2(d_2 - c_2c\theta_{02}) + \overline{u}_3(d_3 + c_1s\theta_{01} - c_2s\theta_{02}) \tag{6.363}$$

On the other hand, the coordinates of the contact points obey the following surface equations of the elliptical and spherical cams.

$$f_1(x_1, x_2, x_3) = (x_1/b_1)^2 + (x_2/b_2)^2 + (x_3/b_3)^2 - 1 = 0 \tag{6.364}$$

$$f_2(y_1, y_2, y_3) = (y_1/r_2)^2 + (y_2/r_2)^2 + (y_3/r_2)^2 - 1 = 0 \tag{6.365}$$

Equations (6.364) and (6.365) can also be written as follows:

$$f_1(x_1, x_2, x_3) = (x_1/\beta_1)^2 + (x_2/\beta_2)^2 + (x_3/\beta_3)^2 - r_2^2 = 0 \tag{6.366}$$

$$f_2(y_1, y_2, y_3) = y_1^2 + y_2^2 + y_3^2 - r_2^2 = 0 \tag{6.367}$$

Equation (6.366) is written by introducing the following dimensionless parameters.

$$\beta_1 = b_1/r_2, \quad \beta_2 = b_2/r_2, \quad \beta_3 = b_3/r_2 \tag{6.368}$$

Moreover, the tangency of the cams necessitate the following gradient alignment equation.

$$\hat{C}^{(1,2)}\overline{g}_2^{(2)} = -\lambda_{12}\overline{g}_1^{(1)} \tag{6.369}$$

According to Eqs. (6.354) and (6.356),

$$\hat{C}^{(1,2)} = e^{\tilde{u}_2\phi_{12}}e^{\tilde{u}_3\theta_{12}}e^{\tilde{u}_1\psi_{12}} = e^{-\tilde{u}_2\theta_{01}}e^{\tilde{u}_1\theta_{02}} \tag{6.370}$$

Hence, Eq. (6.369) can be written as shown below with the specific features of the mechanism.

$$e^{-\tilde{u}_2\theta_{01}}e^{\tilde{u}_1\theta_{02}}[\overline{u}_1(2y_1) + \overline{u}_2(2y_2) + \overline{u}_3(2y_3)]$$

$$= -\lambda_{12}[\overline{u}_1(2x_1/\beta_1^2) + \overline{u}_2(2x_2/\beta_2^2) + \overline{u}_3(2x_3/\beta_3^2)] \Rightarrow$$

$$e^{\tilde{u}_1\theta_{02}}(\overline{u}_1 y_1 + \overline{u}_2 y_2 + \overline{u}_3 y_3) = -\lambda_{12}e^{\tilde{u}_2\theta_{01}}(\overline{u}_1 x_1/\beta_1^2 + \overline{u}_2 x_2/\beta_2^2 + \overline{u}_3 x_3/\beta_3^2) \Rightarrow$$

$$\overline{y} = -\lambda_{12}e^{-\tilde{u}_1\theta_{02}}e^{\tilde{u}_2\theta_{01}}\hat{B}\overline{x} \tag{6.371}$$

In Eq. (6.371),

$$\hat{B} = \begin{bmatrix} 1/\beta_1^2 & 0 & 0 \\ 0 & 1/\beta_2^2 & 0 \\ 0 & 0 & 1/\beta_3^2 \end{bmatrix} \tag{6.372}$$

Meanwhile, Eqs. (6.366) and (6.367) can also be written compactly as follows:

$$f_1(\overline{x}) = \overline{x}^t\hat{B}\overline{x} - r_2^2 = 0 \tag{6.373}$$

$$f_2(\overline{y}) = \overline{y}^t\overline{y} - r_2^2 = 0 \tag{6.374}$$

(c) Determination of the Unspecified Variables for Specified θ_{01}

Note that the mobility of the mechanism is $\mu = 1$. Therefore, if θ_{01} is specified, the other eight variables ($\theta_{02}, \lambda_{12}; x_1, x_2, x_3; y_1, y_2, y_3$) can be determined as explained below from the equations written in Part (b), i.e. Eqs. (6.360), (6.371), (6.373), and (6.374).

Equations (6.360) and (6.371) can be solved together for $\overline{x}$ and $\overline{y}$ as follows:

$$\overline{x} = (\hat{I} + \lambda_{12}\hat{B})^{-1}e^{-\tilde{u}_2\theta_{01}}\overline{z} \tag{6.375}$$

$$\overline{y} = -\lambda_{12}e^{-\tilde{u}_1\theta_{02}}e^{\tilde{u}_2\theta_{01}}\hat{B}(\hat{I} + \lambda_{12}\hat{B})^{-1}e^{-\tilde{u}_2\theta_{01}}\overline{z} \tag{6.376}$$

When Eqs. (6.375) and (6.376) are substituted into Eqs. (6.373) and (6.374), the following scalar equations are obtained.

$$\overline{z}^t e^{\tilde{u}_2\theta_{01}}(\hat{I} + \lambda_{12}\hat{B})^{-1}\hat{B}(\hat{I} + \lambda_{12}\hat{B})^{-1}e^{-\tilde{u}_2\theta_{01}}\overline{z} = r_2^2 \tag{6.377}$$

$$\lambda_{12}^2\overline{z}^t e^{\tilde{u}_2\theta_{01}}(\hat{I} + \lambda_{12}\hat{B})^{-1}\hat{B}^2(\hat{I} + \lambda_{12}\hat{B})^{-1}e^{-\tilde{u}_2\theta_{01}}\overline{z} = r_2^2 \tag{6.378}$$

Referring to Eq. (6.363), note that $\overline{z} = \overline{z}(\theta_{01}, \theta_{02})$. Therefore, Eqs. (6.377) and (6.378) contain two unknowns, which are λ_{12} and θ_{02}. Note also that Eqs. (6.377) and (6.378) are highly nonlinear and therefore they can be solved only by using a suitable iterative numerical method. However, after λ_{12} and θ_{02} are thus determined, the coordinates of the contact points can be found readily by using Eqs. (6.375) and (6.376).

6.7.4 Example 6.4: A Spatial Cam Mechanism That Allows Rolling Without Slipping

(a) Kinematic Description of the System

Figure 6.22 shows a modified version of the spatial cam mechanism considered in the previous example. The modification is such that the previous revolute joint $\mathcal{J}_{02}$ is converted into a cylindrical joint and the previous link $\mathcal{L}_2$ is divided into two links $\mathcal{L}_2$ and $\mathcal{L}_3$, which are connected by a revolute joint $\mathcal{J}_{23}$. Upon this division, the spherical cam is attached to $\mathcal{L}_3$ and it forms the cam joint $\mathcal{J}_{13}$ together with the elliptical cam, which is still attached to $\mathcal{L}_1$. Thus, the mobility of the modified mechanism becomes $\mu = 3$ as long as there is no sticking friction between $\mathcal{L}_1$ and $\mathcal{L}_3$. However, as seen in Section 6.5.1, if there is sticking friction between $\mathcal{L}_1$ and $\mathcal{L}_3$, then the mobility decreases by two, i.e. it becomes $\mu' = 1$, during a motion of the mechanism such that the cams roll without slipping with respect to each other.

For the mechanism of this example, the link frames attached to $\mathcal{L}_1$, $\mathcal{L}_2$, $\mathcal{L}_3$, and $\mathcal{L}_0$ are defined as follows with their origins and basis vectors:

$$\mathcal{F}_0 = \mathcal{F}_0\{O_0; \vec{u}_1^{(0)}, \vec{u}_2^{(0)}, \vec{u}_3^{(0)}\}$$
$$\mathcal{F}_1 = \mathcal{F}_1\{O_1; \vec{u}_1^{(1)}, \vec{u}_2^{(1)}, \vec{u}_3^{(1)}\}; \vec{u}_2^{(1)} = \vec{u}_2^{(0)}$$
$$\mathcal{F}_2 = \mathcal{F}_2\{O_2; \vec{u}_1^{(2)}, \vec{u}_2^{(2)}, \vec{u}_3^{(2)}\}; \vec{u}_1^{(2)} = \vec{u}_1^{(0)}; O_2 = O_{20}$$
$$\mathcal{F}_3 = \mathcal{F}_3\{O_3; \vec{u}_1^{(3)}, \vec{u}_2^{(3)}, \vec{u}_3^{(3)}\}; \vec{u}_2^{(3)} = \vec{u}_2^{(2)}$$

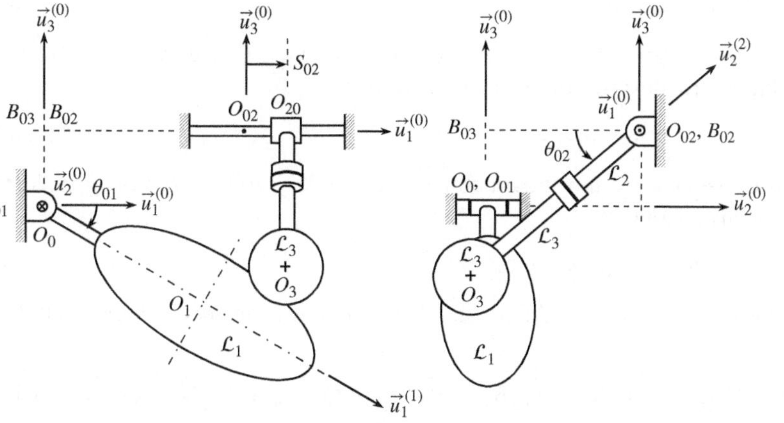

Figure 6.22 A spatial cam mechanism that allows rolling without slipping.

The joint frames attached to the kinematic elements are related to the link frames as shown below.

$$\mathcal{F}_{01}(O_{01}) = \mathcal{F}_0(O_{01}), \mathcal{F}_{10}(O_{10}) = \mathcal{F}_1(O_{01}); O_{10} = O_{01} = O_0$$

$$\mathcal{F}_{02}(O_{02}) = \mathcal{F}_0(O_{02}), \mathcal{F}_{20}(O_{20}) = \mathcal{F}_2(O_{02}); O_{20} \neq O_{02}$$

$$\mathcal{F}_{23}(O_{23}) = \mathcal{F}_2(O_{20}), \mathcal{F}_{32}(O_{32}) = \mathcal{F}_3(O_{20}); O_{23} = O_{32} = O_{20}$$

$$\mathcal{F}_{13}(O_{13}) = \mathcal{F}_1(O_1), \mathcal{F}_{31}(O_{31}) = \mathcal{F}_3(O_3); O_{12} = O_1, O_{31} = O_3$$

The fixed geometric parameters are indicated below.

$$d_3 = O_{01}B_{03}, d_2 = B_{03}B_{02}, d_1 = B_{02}O_{02}$$

$$c_1 = O_{01}O_1, c_3 = O_{20}O_3$$

b_1, b_2, b_3: semi-axis lengths of the ellipsoidal cam
r_2: radius of the spherical cam

(b) Loop Closure Equations

The orientational closure of the mechanism can be expressed by the following matrix equation.

$$\hat{C}^{(0,1)}\hat{C}^{(1,3)} = \hat{C}^{(0,2)}\hat{C}^{(2,3)} \tag{6.379}$$

In Eq. (6.379),

$$\hat{C}^{(0,1)} = e^{\tilde{u}_2\theta_{01}} \tag{6.380}$$

$$\hat{C}^{(0,2)} = e^{\tilde{u}_1\theta_{02}} \tag{6.381}$$

$$\hat{C}^{(2,3)} = e^{\tilde{u}_2\theta_{23}} \tag{6.382}$$

$$\hat{C}^{(1,3)} = e^{\tilde{n}_{131}\phi_{13}}e^{\tilde{n}_{132}\theta_{13}}e^{\tilde{n}_{133}\psi_{13}} \tag{6.383}$$

In Eq. (6.383), the unit column matrices $\overline{n}_{131}$, $\overline{n}_{132}$, and $\overline{n}_{133}$ represent the axes of the three successive rotations of $\mathcal{L}_3$ with respect to $\mathcal{L}_1$. On the other hand, Eqs. (6.380)–(6.383) and Eq. (6.379) can be combined so that

$$\hat{C}^{(1,3)} = \hat{C}^{(1,0)}\hat{C}^{(0,2)}\hat{C}^{(2,3)} \Rightarrow$$

$$e^{\tilde{n}_{131}\phi_{13}}e^{\tilde{n}_{132}\theta_{13}}e^{\tilde{n}_{133}\psi_{13}} = e^{-\tilde{u}_2\theta_{01}}e^{\tilde{u}_1\theta_{02}}e^{\tilde{u}_2\theta_{23}} \tag{6.384}$$

Equation (6.384) suggests the following selection for $\overline{n}_{131}$, $\overline{n}_{132}$, and $\overline{n}_{133}$ in order to simplify the kinematic description of the mechanism.

$$\overline{n}_{131} = \overline{u}_2, \quad \overline{n}_{132} = \overline{u}_1, \quad \overline{n}_{133} = \overline{u}_2 \tag{6.385}$$

Owing to the special selection shown above, the angular variables of the joint $\mathcal{J}_{13}$ are determined readily as follows:

$$\phi_{13} = -\theta_{01}, \quad \theta_{13} = \theta_{02}, \quad \psi_{13} = \theta_{23} \tag{6.386}$$

As for the locational closure of the mechanism, the following vector equation can be written.

$$\overrightarrow{O_{01}O_1} + \overrightarrow{O_1Q_{13}} = \overrightarrow{O_{01}B_{03}} + \overrightarrow{B_{03}B_{02}} + \overrightarrow{B_{02}O_{02}} + \overrightarrow{O_{02}O_{20}} + \overrightarrow{O_{20}O_3} + \overrightarrow{O_3Q_{31}} \tag{6.387}$$

By using the variables (θ_{01}, θ_{02}, s_{02}, and θ_{23}) of the revolute and cylindrical joints, the coordinates (x_1, x_2, x_3; y_1, y_2, y_3) of the contact points (Q_{13} and Q_{31}), and the constant geometric parameters indicated in Part (a), Eq. (6.387) can be written again as shown below.

$$\begin{aligned} c_1\vec{u}_1^{(1)} + [x_1\vec{u}_1^{(1)} + x_2\vec{u}_2^{(1)} + x_3\vec{u}_3^{(1)}] &= d_3\vec{u}_3^{(0)} + d_2\vec{u}_2^{(0)} + d_1\vec{u}_1^{(0)} + s_{02}\vec{u}_1^{(0)} \\ &- c_3\vec{u}_2^{(3)} + [y_1\vec{u}_1^{(3)} + y_2\vec{u}_2^{(3)} + y_3\vec{u}_3^{(3)}] \end{aligned} \quad (6.388)$$

The matrix equation corresponding to Eq. (6.388) can be written in $\mathcal{F}_0$ as follows:

$$\begin{aligned} e^{\tilde{u}_2\theta_{01}}[(x_1 + c_1)\overline{u}_1 + x_2\overline{u}_2 + x_3\overline{u}_3] &= d_3\overline{u}_3 + d_2\overline{u}_2 + (d_1 + s_{02})\overline{u}_1 \\ &+ e^{\tilde{u}_1\theta_{02}}e^{\tilde{u}_2\theta_{23}}[y_1\overline{u}_1 + (y_2 - c_3)\overline{u}_2 + y_3\overline{u}_3] \end{aligned} \quad (6.389)$$

Equation (6.389) can be written more compactly as

$$\overline{x} = e^{-\tilde{u}_2\theta_{01}}(e^{\tilde{u}_1\theta_{02}}e^{\tilde{u}_2\theta_{23}}\overline{y} + \overline{z}) \quad (6.390)$$

In Eq. (6.390),

$$\overline{x} = x_1\overline{u}_1 + x_2\overline{u}_2 + x_3\overline{u}_3 \quad (6.391)$$

$$\overline{y} = y_1\overline{u}_1 + y_2\overline{u}_2 + y_3\overline{u}_3 \quad (6.392)$$

$$\begin{aligned} \overline{z} &= \overline{z}(\theta_{01}, \theta_{02}, \theta_{23}, s_{02}) = \overline{d} + s_{02}\overline{u}_1 - c_1e^{\tilde{u}_2\theta_{01}}\overline{u}_1 - c_3e^{\tilde{u}_1\theta_{02}}\overline{u}_2 \Rightarrow \\ \overline{z} &= \overline{u}_1(d_1 + s_{02} - c_1c\theta_{01}) + \overline{u}_2(d_2 - c_3c\theta_{02}) \\ &\quad + \overline{u}_3(d_3 + c_1s\theta_{01} - c_3s\theta_{02}) \end{aligned} \quad (6.393)$$

On the other hand, the elliptical and spherical cams obey the same surface equations as in the previous example. That is, recalling that $\beta_1 = b_1/r_2$, $\beta_2 = b_2/r_2$, $\beta_3 = b_3/r_2$,

$$\begin{aligned} f_1(x_1, x_2, x_3) &= (x_1/\beta_1)^2 + (x_2/\beta_2)^2 + (x_3/\beta_3)^2 - r_2^2 = 0 \Rightarrow \\ f_1(\overline{x}) &= \overline{x}^t\widehat{B}\overline{x} - r_2^2 = 0 \end{aligned} \quad (6.394)$$

$$\begin{aligned} f_3(y_1, y_2, y_3) &= y_1^2 + y_2^2 + y_3^2 - r_2^2 = 0 \Rightarrow \\ f_3(\overline{y}) &= \overline{y}^t\overline{y} - r_2^2 = 0 \end{aligned} \quad (6.395)$$

Moreover, the tangency of the cams necessitate the following gradient alignment equation.

$$\widehat{C}^{(1,3)}\overline{g}_3^{(3)} = -\lambda_{13}\overline{g}_1^{(1)} \quad (6.396)$$

According to Eqs. (6.385) and (6.386),

$$\widehat{C}^{(1,3)} = e^{-\tilde{u}_2\theta_{01}}e^{\tilde{u}_1\theta_{02}}e^{\tilde{u}_2\theta_{23}} \quad (6.397)$$

Hence, by using the previous definitions, Eq. (6.396) can be written in the following form that shows the specific features of the mechanism.

$$\begin{aligned} e^{-\tilde{u}_2\theta_{01}}e^{\tilde{u}_1\theta_{02}}e^{\tilde{u}_2\theta_{23}}&[\overline{u}_1(2y_1) + \overline{u}_2(2y_2) + \overline{u}_3(2y_3)] \\ &= -\lambda_{13}[\overline{u}_1(2x_1/\beta_1^2) + \overline{u}_2(2x_2/\beta_2^2) + \overline{u}_3(2x_3/\beta_3^2)] \Rightarrow \\ \overline{y} &= -\lambda_{13}e^{-\tilde{u}_2\theta_{23}}e^{-\tilde{u}_1\theta_{02}}e^{\tilde{u}_2\theta_{01}}\widehat{B}\overline{x} \end{aligned} \quad (6.398)$$

(c) Determination of the Unspecified Variables for Specified θ_{01}, θ_{23}, and s_{02}

As mentioned before, the mobility of the modified mechanism is $\mu = 3$, when it is not in a rolling-without-slipping motion. Therefore, if θ_{01}, θ_{23}, and s_{02} are specified, the other eight variables (θ_{02}, λ_{12}; x_1, x_2, x_3; y_1, y_2, y_3) can be determined as explained below from the equations written in Part (b), i.e. Eqs. (6.390), (6.394), (6.395), and (6.398).

Equations (6.390) and (6.398) can be solved together for $\overline{x}$ and $\overline{y}$ as follows:

$$\overline{x} = (\hat{I} + \lambda_{13}\hat{B})^{-1}e^{-\tilde{u}_2\theta_{01}}\overline{z} \tag{6.399}$$

$$\overline{y} = -\lambda_{13}e^{-\tilde{u}_2\theta_{23}}e^{-\tilde{u}_1\theta_{02}}e^{\tilde{u}_2\theta_{01}}\hat{B}(\hat{I} + \lambda_{13}\hat{B})^{-1}e^{-\tilde{u}_2\theta_{01}}\overline{z} \tag{6.400}$$

When Eqs. (6.399) and (6.400) are substituted into Eqs. (6.394) and (6.395), the following scalar equations are obtained.

$$\overline{z}^t e^{\tilde{u}_2\theta_{01}}(\hat{I} + \lambda_{13}\hat{B})^{-1}\hat{B}(\hat{I} + \lambda_{13}\hat{B})^{-1}e^{-\tilde{u}_2\theta_{01}}\overline{z} = r_2^2 \tag{6.401}$$

$$\lambda_{13}^2\overline{z}^t e^{\tilde{u}_2\theta_{01}}(\hat{I} + \lambda_{13}\hat{B})^{-1}\hat{B}^2(\hat{I} + \lambda_{13}\hat{B})^{-1}e^{-\tilde{u}_2\theta_{01}}\overline{z} = r_2^2 \tag{6.402}$$

Referring to Eq. (6.393), note that $\overline{z} = \overline{z}(\theta_{01}, \theta_{02}, \theta_{23}, s_{02})$. Therefore, Eqs. (6.401) and (6.402) contain two unknowns, which are λ_{13} and θ_{02}. Note also that Eqs. (6.401) and (6.402) are highly nonlinear and therefore they can be solved only by using a suitable iterative numerical method. However, after λ_{13} and θ_{02} are thus determined, the coordinates of the contact points can be found readily by using Eqs. (6.399) and (6.400).

(d) Velocity Equations and Their Solution in the Case of Rolling with Slipping

The differentiation of Eqs. (6.394), (6.395), (6.390), and (6.398) results in the following velocity equations, in which the derivatives of the exponential factors are written in a suitable way by using the commutativity property that $e^{\tilde{n}\theta}\tilde{n} = \tilde{n}e^{\tilde{n}\theta}$.

$$\overline{x}^t\hat{B}\dot{\overline{x}} = 0 \tag{6.403}$$

$$\overline{y}^t\dot{\overline{y}} = 0 \tag{6.404}$$

$$\begin{aligned}
&e^{\tilde{u}_2\theta_{01}}\dot{\overline{x}} + \dot{\theta}_{01}e^{\tilde{u}_2\theta_{01}}\tilde{u}_2\overline{x} = \dot{\overline{z}} + e^{\tilde{u}_1\theta_{02}}e^{\tilde{u}_2\theta_{23}}\dot{\overline{y}}\\
&\qquad + \dot{\theta}_{02}e^{\tilde{u}_1\theta_{02}}\tilde{u}_1e^{\tilde{u}_2\theta_{23}}\overline{y} + \dot{\theta}_{23}e^{\tilde{u}_1\theta_{02}}\tilde{u}_2e^{\tilde{u}_2\theta_{23}}\overline{y} \Rightarrow\\
&e^{\tilde{u}_1\theta_{02}}e^{\tilde{u}_2\theta_{23}}\dot{\overline{y}} + e^{\tilde{u}_1\theta_{02}}(\dot{\theta}_{02}\tilde{u}_1 + \dot{\theta}_{23}\tilde{u}_2)e^{\tilde{u}_2\theta_{23}}\overline{y}\\
&\quad = e^{\tilde{u}_2\theta_{01}}\dot{\overline{x}} + \dot{\theta}_{01}e^{\tilde{u}_2\theta_{01}}\tilde{u}_2\overline{x} - \dot{\overline{z}}
\end{aligned} \tag{6.405}$$

$$\begin{aligned}
&e^{\tilde{u}_1\theta_{02}}e^{\tilde{u}_2\theta_{23}}\dot{\overline{y}} + \dot{\theta}_{02}e^{\tilde{u}_1\theta_{02}}\tilde{u}_1e^{\tilde{u}_2\theta_{23}}\overline{y} + \dot{\theta}_{23}e^{\tilde{u}_1\theta_{02}}\tilde{u}_2e^{\tilde{u}_2\theta_{23}}\overline{y}\\
&\quad = -\lambda_{13}e^{\tilde{u}_2\theta_{01}}\hat{B}\dot{\overline{x}} - \lambda_{13}\dot{\theta}_{01}\tilde{u}_2e^{\tilde{u}_2\theta_{01}}\hat{B}\overline{x} - \dot{\lambda}_{13}e^{\tilde{u}_2\theta_{01}}\hat{B}\overline{x} \Rightarrow\\
&e^{\tilde{u}_1\theta_{02}}e^{\tilde{u}_2\theta_{23}}\dot{\overline{y}} + e^{\tilde{u}_1\theta_{02}}(\dot{\theta}_{02}\tilde{u}_1 + \dot{\theta}_{23}\tilde{u}_2)e^{\tilde{u}_2\theta_{23}}\overline{y}\\
&\quad = -\lambda_{13}e^{\tilde{u}_2\theta_{01}}\hat{B}\dot{\overline{x}} - \lambda_{13}\dot{\theta}_{01}\tilde{u}_2e^{\tilde{u}_2\theta_{01}}\hat{B}\overline{x} - \dot{\lambda}_{13}e^{\tilde{u}_2\theta_{01}}\hat{B}\overline{x}
\end{aligned} \tag{6.406}$$

In Eq. (6.405),

$$\dot{\overline{z}} = \overline{u}_1(\dot{s}_{02} + c_1\dot{\theta}_{01}s\theta_{01}) + \overline{u}_2(c_3\dot{\theta}_{02}s\theta_{02}) + \overline{u}_3(c_1\dot{\theta}_{01}c\theta_{01} - c_3\dot{\theta}_{02}c\theta_{02}) \tag{6.407}$$

Whenever there is not enough friction to prevent slipping, the mobility of the mechanism is $\mu = 3$. Therefore, three of the derivatives must be specified in order to determine its motion.

Let the specified derivatives be $\dot{\theta}_{01}$, $\dot{\theta}_{23}$, and $\dot{s}_{02}$. Then, the other eight derivatives ($\dot{\theta}_{02}$, $\dot{\lambda}_{13}$; $\dot{x}_1, \dot{x}_2, \dot{x}_3$; $\dot{y}_1, \dot{y}_2, \dot{y}_3$) can be determined as explained below from the velocity equations written above, i.e. Eqs. (6.403)–(6.407).

As the first step, the right-hand sides of Eqs. (6.405) and (6.406) lead to the following expression for $\dot{\overline{x}}$.

$$\dot{\overline{x}} = (\hat{I} + \lambda_{13}\hat{B})^{-1}[e^{-\tilde{u}_2\theta_{01}}\dot{\overline{z}} - \dot{\theta}_{01}\tilde{u}_2(\hat{I} + \lambda_{13}\hat{B})\overline{x} - \dot{\lambda}_{13}\hat{B}\overline{x}] \tag{6.408}$$

Afterwards, by inserting the above expression of $\dot{\overline{x}}$, $\dot{\overline{y}}$ can be obtained from Eq. (6.405) as shown below.

$$\dot{\overline{y}} = e^{-\tilde{u}_2\theta_{23}}[\dot{\overline{x}} - e^{-\tilde{u}_2\theta_{01}}\dot{\overline{z}} + \dot{\theta}_{01}\tilde{u}_2\overline{x} - (\dot{\theta}_{02}\tilde{u}_1 + \dot{\theta}_{23}\tilde{u}_2)e^{\tilde{u}_2\theta_{23}}\overline{y}] \tag{6.409}$$

On the other hand, when $\dot{\overline{z}}$ is inserted as given by Eq. (6.407), $\dot{\overline{x}}$ and $\dot{\overline{y}}$ can be re-expressed as follows:

$$\dot{\overline{x}} = \overline{\Phi}_{01}\dot{\theta}_{01} + \overline{\Phi}_{02}\dot{\theta}_{02} + \overline{\Phi}_{23}\dot{\theta}_{23} + \overline{V}_{02}\dot{s}_{02} + \overline{M}_{13}\dot{\lambda}_{13} \tag{6.410}$$

$$\dot{\overline{y}} = \overline{\Psi}_{01}\dot{\theta}_{01} + \overline{\Psi}_{02}\dot{\theta}_{02} + \overline{\Psi}_{23}\dot{\theta}_{23} + \overline{W}_{02}\dot{s}_{02} + \overline{N}_{13}\dot{\lambda}_{13} \tag{6.411}$$

In Eqs. (6.410) and (6.411), all the coefficients are functions of the current pose of the mechanism. When Eqs. (6.410) and (6.411) are substituted into Eqs. (6.403) and (6.404), the resulting equations can be combined into the following matrix equation.

$$\begin{bmatrix} \overline{x}^t\hat{B}\overline{\Phi}_{02} & \overline{x}^t\hat{B}\overline{\Phi}_{02} \\ \overline{y}^t\overline{\Psi}_{02} & \overline{y}^t\overline{N}_{13} \end{bmatrix} \begin{bmatrix} \dot{\theta}_{02} \\ \dot{\lambda}_{13} \end{bmatrix} = -\begin{bmatrix} v \\ w \end{bmatrix} \tag{6.412}$$

In Eq. (6.412), v and w are known. That is,

$$\begin{bmatrix} v \\ w \end{bmatrix} = \begin{bmatrix} \overline{x}^t\hat{B}(\overline{\Phi}_{01}\dot{\theta}_{01} + \overline{\Phi}_{23}\dot{\theta}_{23} + \overline{V}_{02}\dot{s}_{02}) \\ \overline{y}^t(\overline{\Psi}_{01}\dot{\theta}_{01} + \overline{\Psi}_{23}\dot{\theta}_{23} + \overline{W}_{02}\dot{s}_{02}) \end{bmatrix} \tag{6.413}$$

From Eq. (6.412), $\dot{\theta}_{02}$ and $\dot{\lambda}_{13}$ can be obtained as follows, if $D \neq 0$.

$$\dot{\theta}_{02} = [(\overline{x}^t\hat{B}\overline{M}_{13})w - (\overline{y}^t\overline{N}_{13})v]/D \tag{6.414}$$

$$\dot{\lambda}_{13} = [(\overline{y}^t\overline{\Psi}_{02})v - (\overline{x}^t\hat{B}\overline{\Phi}_{02})w]/D \tag{6.415}$$

In the above equations, D is the determinant of the coefficient matrix in Eq. (6.412). That is,

$$D = (\overline{y}^t\overline{N}_{13})(\overline{x}^t\hat{B}\overline{\Phi}_{02}) - (\overline{y}^t\overline{\Psi}_{02})(\overline{x}^t\hat{B}\overline{M}_{13}) \tag{6.416}$$

After $\dot{\theta}_{02}$ and $\dot{\lambda}_{13}$ are thus found, $\dot{\overline{x}}$ and $\dot{\overline{y}}$ can also be found readily by using Eqs. (6.410) and (6.411).

(e) Velocity Equations and Their Solution in the Case of Rolling without Slipping

Referring to Section 6.5.1, the rolling without slipping constraint can be written as follows for the mechanism of this example:

$$\dot{\overline{r}}_{13q}^{(1)} = \hat{C}^{(1,3)}\dot{\overline{r}}_{31q}^{(3)} \Rightarrow$$

$$\dot{\overline{x}} = e^{-\tilde{u}_2\theta_{01}}e^{\tilde{u}_1\theta_{02}}e^{\tilde{u}_2\theta_{23}}\dot{\overline{y}} \tag{6.417}$$

With this constraint relationship, Eqs. (6.403) and (6.404) take the following forms.

$$\overline{x}^t \widehat{B} e^{-\widetilde{u}_2\theta_{01}} e^{\widetilde{u}_1\theta_{02}} e^{\widetilde{u}_2\theta_{23}} \dot{\overline{y}} = 0 \tag{6.418}$$

$$\overline{y}^t \dot{\overline{y}} = 0 \tag{6.419}$$

Meanwhile, Eq. (6.398) implies that

$$\overline{x}^t \widehat{B} = -\overline{y}^t e^{-\widetilde{u}_2\theta_{23}} e^{-\widetilde{u}_1\theta_{02}} e^{\widetilde{u}_2\theta_{01}} / \lambda_{13} \tag{6.420}$$

Note that Eq. (6.420) does not have a zero-division problem, because λ_{13} never vanishes. This fact is implied by the following relationships, which are based on Eqs. (6.394)–(6.396).

$$\vec{g}_3 = \vec{u}_1^{(3)}(2y_1) + \vec{u}_2^{(3)}(2y_2) + \vec{u}_3^{(1)}(2y_3) \Rightarrow$$

$$|\vec{g}_3| = 2\sqrt{y_1^2 + y_2^2 + y_3^2} = 2\sqrt{\overline{y}^t\overline{y}} = 2r_2 \neq 0 \tag{6.421}$$

$$\vec{g}_1 = \vec{u}_1^{(1)}(2x_1/\beta_1^2) + \vec{u}_2^{(1)}(2x_2/\beta_2^2) + \vec{u}_3^{(1)}(2x_3/\beta_3^2) \Rightarrow$$

$$|\vec{g}_1| = 2\sqrt{x_1^2/\beta_1^4 + x_2^2/\beta_2^4 + x_3^2/\beta_3^4} = 2\sqrt{\overline{x}^t \widehat{B}^2 \overline{x}};\ \sqrt{\overline{x}^t \widehat{B} \overline{x}} = r_2 \neq 0 \Rightarrow$$

$$|\vec{g}_1| \neq 0 \tag{6.422}$$

$$\vec{g}_3 = -\lambda_{13}\vec{g}_1 \Rightarrow |\vec{g}_3| = |\lambda_{13}| \cdot |\vec{g}_1| \Rightarrow |\lambda_{13}| \neq 0 \tag{6.423}$$

When Eq. (6.420) is substituted, Eq. (6.418) becomes the same as Eq. (6.419). That is,

$$-\overline{y}^t \dot{\overline{y}} / \lambda_{12} = 0 \Rightarrow \overline{y}^t \dot{\overline{y}} = 0$$

Therefore, only Eq. (6.419) can be taken into the set of independent velocity equations.

On the other hand, when Eq. (6.417) is substituted, Eqs. (6.405) and (6.406) take the following forms.

$$e^{\widetilde{u}_1\theta_{02}}(\dot{\theta}_{02}\widetilde{u}_1 + \dot{\theta}_{23}\widetilde{u}_2) e^{\widetilde{u}_2\theta_{23}} \overline{y} = \dot{\theta}_{01} e^{\widetilde{u}_2\theta_{01}} \widetilde{u}_2 \overline{x} - \dot{\overline{z}} \Rightarrow$$

$$e^{\widetilde{u}_1\theta_{02}}(\dot{\theta}_{02}\widetilde{u}_1 + \dot{\theta}_{23}\widetilde{u}_2) e^{\widetilde{u}_2\theta_{23}} \overline{y} = \dot{\theta}_{01} e^{\widetilde{u}_2\theta_{01}} \widetilde{u}_2 \overline{x}$$

$$-\overline{u}_1(\dot{s}_{02} + c_1\dot{\theta}_{01}s\theta_{01}) - \overline{u}_2(c_3\dot{\theta}_{02}s\theta_{02}) - \overline{u}_3(c_1\dot{\theta}_{01}c\theta_{01} - c_3\dot{\theta}_{02}c\theta_{02}) \tag{6.424}$$

$$(\widehat{I} + \lambda_{13}\widehat{B}) e^{-\widetilde{u}_2\theta_{01}} e^{\widetilde{u}_1\theta_{02}} e^{\widetilde{u}_2\theta_{23}} \dot{\overline{y}}$$

$$= -(\dot{\lambda}_{13}\widehat{I} + \lambda_{13}\dot{\theta}_{01}\widetilde{u}_2)\widehat{B}\overline{x} - e^{-\widetilde{u}_2\theta_{01}} e^{\widetilde{u}_1\theta_{02}} (\dot{\theta}_{02}\widetilde{u}_1 + \dot{\theta}_{23}\widetilde{u}_2) e^{\widetilde{u}_2\theta_{23}} \overline{y} \tag{6.425}$$

As noted above, Eqs. (6.417), (6.419), (6.424) and (6.425) constitute a set of 10 independent scalar equations. These equations contain 11 scalar derivatives, which are $\dot{\lambda}_{13}$, $\dot{\theta}_{01}$, $\dot{\theta}_{02}$, $\dot{\theta}_{23}$, $\dot{s}_{02}$, and the six components contained in $\dot{\overline{x}}$ and $\dot{\overline{y}}$. Thus, if one these derivatives (e.g. $\dot{\theta}_{01}$) is specified, then the other 10 can be determined. If $\dot{\theta}_{01}$ is specified, the solution can be expressed as shown below by means of the relevant *velocity influence coefficients*.

$$\dot{\theta}_{02} = \Omega_{02}(\overline{x}, \overline{y}; \lambda_{13}, \theta_{01}, \theta_{02}, \theta_{23}, s_{02})\dot{\theta}_{01} \tag{6.426}$$

$$\dot{\theta}_{23} = \Omega_{23}(\overline{x}, \overline{y}; \lambda_{13}, \theta_{01}, \theta_{02}, \theta_{23}, s_{02})\dot{\theta}_{01} \tag{6.427}$$

$$\dot{s}_{02} = W_{02}(\overline{x}, \overline{y}; \lambda_{13}, \theta_{01}, \theta_{02}, \theta_{23}, s_{02})\dot{\theta}_{01} \tag{6.428}$$

$$\dot{\lambda}_{13} = \Gamma_{13}(\overline{x}, \overline{y}; \lambda_{13}, \theta_{01}, \theta_{02}, \theta_{23}, s_{02})\dot{\theta}_{01} \tag{6.429}$$

$$\dot{\overline{x}} = \overline{W}_x(\overline{x}, \overline{y}; \lambda_{13}, \theta_{01}, \theta_{02}, \theta_{23}, s_{02})\dot{\theta}_{01} \tag{6.430}$$

$$\dot{\overline{y}} = \overline{W}_y(\overline{x}, \overline{y}; \lambda_{13}, \theta_{01}, \theta_{02}, \theta_{23}, s_{02})\dot{\theta}_{01} \tag{6.431}$$

(f) Finding the Pose of the Mechanism During a Rolling-Without-Slipping Motion

For a specified time variation of θ_{01}, the procedure explained in Part (e) gives the derivatives of the other variables as differential equations by means of the velocity influence coefficients that depend on the current pose of the mechanism. All these differential equations can be written compactly as the following matrix equation.

$$\dot{\bar{\xi}} = \overline{\Psi}(\bar{\xi}, \theta_{01})\dot{\theta}_{01} \tag{6.432}$$

In Eq. (6.432),

$$\bar{\xi} = [x_1, x_2, x_3; y_1, y_2, y_3; \lambda_{13}, \theta_{02}, \theta_{23}, s_{02}]^t \tag{6.433}$$

With known $\theta_{01} = \theta_{01}(t)$, Eq. (6.432) can be integrated by using a suitable numerical method starting from a known initial pose of the mechanism at the instant $t = t_0$. Thus, the pose of the mechanism can be predicted at discrete instants of time such as t_1, t_2, t_3, and so on. The accuracy of $\bar{\xi}(t_k)$ predicted this way can be improved by means of sufficient number of corrective iterations on $\bar{\xi}(t_k)$ so that the loop closure equations derived in Part (b) are also satisfied.

7

Kinematic Features of Serial Manipulators

Synopsis

This chapter presents a detailed kinematic description of serial manipulators. A serial manipulator is described as an open kinematic chain that consists of a base and m moving links connected by m joints, which can be either revolute or prismatic. Its terminal link incorporates an end-effector, which is a special device required to execute a planned task. As the usual practice in the relevant literature, the kinematic relationships concerning the links and the joints are formulated according to the Denavit–Hartenberg (D–H) convention. However, in this book, a special version of the D–H convention is used. It is based on the notational scheme that is used throughout the book. The D–H convention is first explained in general terms and then its usage is discussed in the commonly encountered special cases, which involve the first joint, the last joint, and the successive joints with intersecting, parallel, and coincident axes.

7.1 Kinematic Description of a General Serial Manipulator

A serial manipulator is a mechanical system that consists of m bodies that are movable with respect to a base. In this book, the bodies are assumed to be rigid. The bodies are arranged to form an *open kinematic chain* and they are connected to their neighbors by either *revolute* or *prismatic* joints. Therefore, the number of joints is also m. In a serial manipulator, all the joints are actuated. The intermediate bodies between the joints are naturally called links. However, for the sake of terminological uniformity, even the base (i.e. the zeroth body) and the last (i.e. the mth) body are also called links. If the base is fixed on the earth, it is also called *ground*. In order to set up an analogy to the human arm, the part of the manipulator beyond the penultimate revolute joint is called *hand*, the center of the penultimate revolute joint is called *wrist point*, and the part of the manipulator between the first joint and the penultimate revolute joint is called *arm*. It is conceived that the *end-effector* (i.e. the special manipulation tool that is attached to the last link) is a part of the hand. In a general purpose manipulator, the end-effector is a *two-finger gripper*. In a special purpose manipulator, the end-effector may be one of the devices such as a pen, a screwdriver, a drilling tool, a welding tool, a painting gun, a multi-finger gripper, etc. Considering a link or a joint with its position in the kinematic chain, it is called *proximal* if it is closer to the base and *distal* if it is further away from the base.

Kinematics of General Spatial Mechanical Systems, First Edition. M. Kemal Ozgoren.

Companion Website: www.wiley.com/go/ozgoren/spatialmechanicalsystems

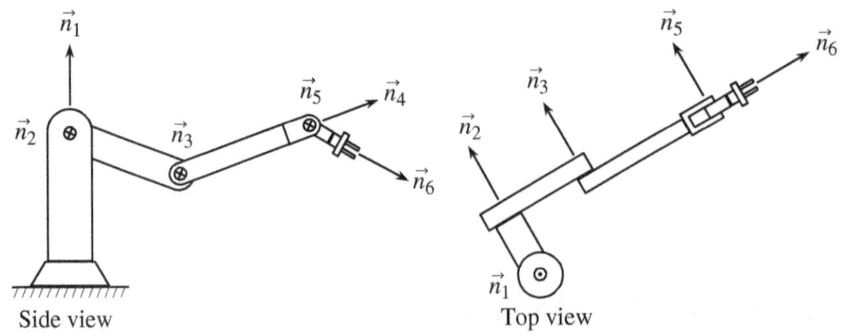

Figure 7.1 A typical serial manipulator with six revolute joints.

A typical serial manipulator with six revolute joints is illustrated in Figure 7.1 showing its side and top views. The axes of the joints are represented by the unit vectors $\vec{n}_1, \vec{n}_2, \ldots, \vec{n}_6$.

7.2 Denavit–Hartenberg Convention

The kinematic features of a serial manipulator can be described conveniently according to the Denavit–Hartenberg (D–H) convention. The D–H convention is based on the *joint axes* and the *common normals* between them. However, there have so far been different formulations proposed for the D–H convention. Aside from the notational differences, these formulations differ from each other basically by the indexing rules. These rules assign an index number to each of the relevant entities such as the joint variables, the constant geometric parameters, the characteristic points, and the unit vectors along the joint axes and the common normals.

The D–H convention was originally introduced in 1955 by Denavit and Hartenberg in order to describe the kinematic features of the spatial mechanisms that comprise *single-axis joints* such as cylindrical, revolute, prismatic, and screw joints. Based on this convention, Denavit and Hartenberg constructed 4×4 homogeneous transformation matrices in order to express the kinematic relationships between the adjacent links of a spatial mechanism and then they used them for the purposes of analysis and synthesis. The definition and details of the homogeneous transformation matrices can be seen in Section 3.9 of Chapter 3.

Although the D–H convention was introduced for the spatial mechanisms with single-axis joints, it can also be extended to other spatial mechanical systems that may contain multi-axis joints, too, but not any kind of rolling contact joints such as cam and gear joints. Of course, such an extension necessitates a virtual decomposition of all the multi-axis joints of the system into single-axis joints. For example, a spherical joint can be virtually decomposed into three revolute joints with noncoincident axes that intersect at a single point.

The original formulation (D–H-0) of the D–H convention is somewhat different from its subsequent formulations (D–H-1 and D–H-2), which have become more popular in the current literature of robotics. The D–H-1 formulation appeared in about 1970 in the publications of Pieper and Paul. Later in about 1980, the D–H-2 formulation was proposed by Craig as an alternative to the precedent D–H-0 and D–H-1 formulations. When these formulations are compared, it is observed as the major difference that the D–H-0 and D–H-1 formulations necessitate taking the third axis of the base frame to be coincident with the axis of the first joint of the manipulator, whereas the D–H-2 formulation eliminates this necessity. In other words, the D–H-2 formulation allows the orientation of the base frame to be selected independently of the manipulator and this feature makes it more convenient to use.

In this book, another formulation (D–H-3) is used for the D–H convention. However, apart from the notational differences and the difference in the index of an angle, the D–H-3 formulation is very similar to the D–H-2 formulation because it is based on the same principle mentioned above about keeping the orientation of the base frame independent of the manipulator.

The D–H convention is explained in the following section with its general features by using the D–H-3 formulation.

7.3 D–H Convention for Successive Intermediate Links and Joints

7.3.1 Assignment and Description of the Link Frames

Figure 7.2 shows a sketch that involves four successive intermediate links ($\mathcal{L}_{k-2}$, $\mathcal{L}_{k-1}$, $\mathcal{L}_k$, $\mathcal{L}_{k+1}$) of a serial manipulator. These links are positioned in the kinematic chain so that $\mathcal{L}_{k-2}$ is the most proximal one. The joint that connects the link $\mathcal{L}_k$ to the preceding link $\mathcal{L}_{k-1}$ is designated as $\mathcal{J}_k$.

The link $\mathcal{L}_k$ is represented kinematically by the reference frame $\mathcal{F}_k$. The origin of $\mathcal{F}_k$ is O_k and its basis vector triad is

$$\mathcal{U}_k = \{\vec{u}_1^{(k)}, \vec{u}_2^{(k)}, \vec{u}_3^{(k)}\} \tag{7.1}$$

The origin and the basis vectors of $\mathcal{F}_k$ are defined as explained below.

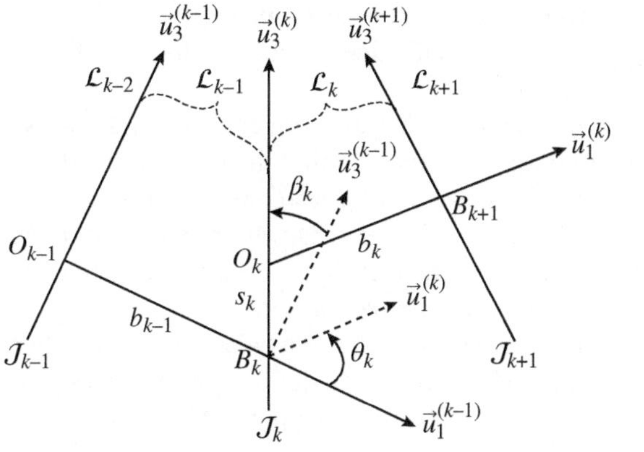

Figure 7.2 D–H convention for successive intermediate links and joints.

The unit vector $\vec{u}_3^{(k)}$ is assigned to the axis of the joint $\mathcal{J}_k$. The sense of $\vec{u}_3^{(k)}$ along the axis of $\mathcal{J}_k$ may be selected arbitrarily.

The unit vector $\vec{u}_1^{(k)}$ is assigned to the *common normal* $\mathcal{N}_k$ between the axes of the joints $\mathcal{J}_k$ and $\mathcal{J}_{k+1}$. The sense of $\vec{u}_1^{(k)}$ is selected so that it is directed from the proximal joint to the distal joint, i.e. from $\mathcal{J}_k$ to $\mathcal{J}_{k+1}$, if the axes of these joints are not intersecting. If their axes are intersecting, i.e. if they have equal proximality, then the sense of $\vec{u}_1^{(k)}$ may be selected arbitrarily.

After the assignment of the basis vectors $\vec{u}_3^{(k)}$ and $\vec{u}_1^{(k)}$ as explained above, the remaining basis vector $\vec{u}_2^{(k)}$ is determined so that the reference frame $\mathcal{F}_k$ is right-handed. That is,

$$\vec{u}_2^{(k)} = \vec{u}_3^{(k)} \times \vec{u}_1^{(k)} \tag{7.2}$$

The origin O_k is defined as the intersection point of the common normal $\mathcal{N}_k$ with the axis of the joint $\mathcal{J}_k$. Another point associated with $\mathcal{N}_k$ is the auxiliary point B_{k+1}, which is defined as the intersection point of $\mathcal{N}_k$ with the axis of the next joint $\mathcal{J}_{k+1}$.

7.3.2 D–H Parameters

The four parameters associated with the D–H convention are defined as follows:

* The *twist angle* (β_k) between the axes of the joints $\mathcal{J}_{k-1}$ and $\mathcal{J}_k$:

$$\beta_k = \sphericalangle[\vec{u}_3^{(k-1)} \rightarrow \vec{u}_3^{(k)}] \text{ about } \vec{u}_1^{(k-1)} \tag{7.3}$$

* The *rotation angle* (θ_k) between the common normals $\mathcal{N}_{k-1}$ and $\mathcal{N}_k$:

$$\theta_k = \sphericalangle[\vec{u}_1^{(k-1)} \rightarrow \vec{u}_1^{(k)}] \text{ about } \vec{u}_3^{(k)} \tag{7.4}$$

* The *offset* (s_k) between the common normals $\mathcal{N}_{k-1}$ and $\mathcal{N}_k$:

$$s_k = B_k O_k \text{ along } \vec{u}_3^{(k)} \tag{7.5}$$

* The *effective length* (b_k) of the link $\mathcal{L}_k$ between the joints $\mathcal{J}_k$ and $\mathcal{J}_{k+1}$:

$$b_k = O_k B_{k+1} \text{ along } \vec{u}_1^{(k)} \tag{7.6}$$

The parameters β_k and b_k are always constant. The parameter b_k may also be called the *joint-to-joint distance* from $\mathcal{J}_k$ to $\mathcal{J}_{k+1}$. If it is especially desired to be called so, then it may also be denoted alternatively by b'_{k+1} so that $b'_{k+1} = b_k$.

If $\mathcal{J}_k$ is a *revolute joint*, then θ_k is the *joint variable* and s_k is a constant. In that case, s_k is replaced with the symbol d_k in order to emphasize the fact that the offset is constant.

If $\mathcal{J}_k$ is a *prismatic joint*, then s_k is the *joint variable* and θ_k is a constant. In that case, the variable offset s_k may also be called the *sliding distance*. On the other hand, θ_k is replaced with the symbol δ_k in order to emphasize the fact that it is constant. The constant parameter δ_k may be called the *deflection angle* or *sway angle*.

A practicable serial manipulator is designed in such a way that the twist angles are either $\beta_k = 0$ or $\beta_k = \pm\pi/2$ for all $k = 1, 2, \ldots, m$. Moreover, if the joint $\mathcal{J}_k$ of a practicable manipulator is prismatic, then the associated deflection angle happens to be either $\delta_k = 0$ or $\delta_k = \pm\pi/2$.

7.3.3 Relative Position Formulas Between Successive Links

By using the four D–H parameters defined above in reference to Figure 7.2, the relative position (orientation and location) of $\mathcal{F}_k(O_k)$ with respect to $\mathcal{F}_{k-1}(O_{k-1})$ can be expressed by the following formulas.

(a) Relative orientation of $\mathcal{F}_k(O_k)$ with respect to $\mathcal{F}_{k-1}(O_{k-1})$:

$$\widehat{C}^{(k-1,k)} = \widehat{R}_1(\beta_k)\widehat{R}_3(\theta_k) = e^{\widetilde{u}_1\beta_k}e^{\widetilde{u}_3\theta_k} \tag{7.7}$$

(b) Relative location of $\mathcal{F}_k(O_k)$ with respect to $\mathcal{F}_{k-1}(O_{k-1})$:

$$\vec{r}_{k-1,k} = \vec{r}_{O_{k-1}O_k} = \vec{r}_{O_{k-1}B_k} + \vec{r}_{B_kO_k} = b_{k-1}\vec{u}_1^{(k-1)} + s_k\vec{u}_3^{(k)}$$

The preceding vector equation can be written as the following column matrix equation in $\mathcal{F}_{k-1}(O_{k-1})$.

$$\overline{r}_{k-1,k}^{(k-1)} = b_{k-1}\overline{u}_1^{(k-1/k-1)} + s_k\overline{u}_3^{(k/k-1)} = b_{k-1}\overline{u}_1 + s_k\widehat{C}^{(k-1,k)}\overline{u}_3^{(k/k)} \Rightarrow$$

$$\overline{r}_{k-1,k}^{(k-1)} = b_{k-1}\overline{u}_1 + s_k\widehat{C}^{(k-1,k)}\overline{u}_3 \tag{7.8}$$

Equation (7.8) can be written in more detail as follows:

$$\overline{r}_{k-1,k}^{(k-1)} = b_{k-1}\overline{u}_1 + s_k e^{\widetilde{u}_1\beta_k}e^{\widetilde{u}_3\theta_k}\overline{u}_3 = b_{k-1}\overline{u}_1 + s_k e^{\widetilde{u}_1\beta_k}\overline{u}_3 \Rightarrow$$

$$\overline{r}_{k-1,k}^{(k-1)} = \overline{u}_1 b_{k-1} - \overline{u}_2 s_k s\beta_k + \overline{u}_3 s_k c\beta_k \tag{7.9}$$

Equations (7.7) and (7.9) can be combined as shown below in order to obtain the *homogeneous transformation matrix* between $\mathcal{F}_{k-1}(O_{k-1})$ and $\mathcal{F}_k(O_k)$.

$$\widehat{H}^{(k-1,k)} = \widehat{H}_{O_{k-1}O_k}^{(k-1,k)} = \begin{bmatrix} \widehat{C}^{(k-1,k)} & \overline{r}_{k-1,k}^{(k-1)} \\ \overline{0}^t & 1 \end{bmatrix} \Rightarrow$$

$$\widehat{H}^{(k-1,k)} = \begin{bmatrix} e^{\widetilde{u}_1\beta_k}e^{\widetilde{u}_3\theta_k} & \overline{u}_1 b_{k-1} - \overline{u}_2 s_k s\beta_k + \overline{u}_3 s_k c\beta_k \\ \overline{0}^t & 1 \end{bmatrix} \Rightarrow$$

$$\widehat{H}^{(k-1,k)} = \begin{bmatrix} c\theta_k & -s\theta_k & 0 & b_{k-1} \\ c\beta_k s\theta_k & c\beta_k c\theta_k & -s\beta_k & -s_k s\beta_k \\ s\beta_k s\theta_k & s\beta_k c\theta_k & c\beta_k & s_k c\beta_k \\ 0 & 0 & 0 & 1 \end{bmatrix} \tag{7.10}$$

7.3.4 Alternative Multi-Index Notation for the D–H Convention

The single-index notation described above for the D–H-3 formulation of the D–H convention can be made more expressive by means of the multi-index notation described in this section. This notation is naturally more cumbersome but it provides concordance between the D–H convention and the systematics of Chapter 6 for the kinematic characterization of all kinds of joints. The multi-index notation is explained below together with its relationship to the single-index notation by considering a succession of three links denoted by the indices $k, j = k-1$, and $i = k-2$. This notation is based on the fact that the kinematic pair $\mathcal{J}_{jk} = \mathcal{J}_{kj}$ is made up of the mating kinematic elements $\mathcal{E}_{jk}$ and $\mathcal{E}_{kj}$ that belong, respectively, to the links $\mathcal{L}_j$ and $\mathcal{L}_k$.

* The joint *between* the links $\mathcal{L}_j$ and $\mathcal{L}_k$:

$$\mathcal{J}_k = \mathcal{J}_{jk} = \mathcal{J}_{kj} \tag{7.11}$$

* The common *normal* between the joints $\mathcal{J}_j = \mathcal{J}_{ij}$ and $\mathcal{J}_k = \mathcal{J}_{jk}$:

$$\mathcal{N}_j = \mathcal{N}_{ijk} \text{ or } \mathcal{N}_j = \mathcal{N}_{ik} \tag{7.12}$$

* The unit vector *along* the axis of the joint $\mathcal{J}_k = \mathcal{J}_{jk} = \mathcal{J}_{kj}$:

$$\vec{u}_3^{(k)} = \vec{u}_3^{(jk)} = \vec{u}_3^{(kj)} \tag{7.13}$$

* The unit vector along the common normal $\mathcal{N}_j = \mathcal{N}_{ijk}$:

$$\vec{u}_1^{(j)} = \vec{u}_1^{(ji)} = \vec{u}_1^{(jk)} \tag{7.14}$$

* The twist *angle* between the unit vectors $\vec{u}_3^{(j)} = \vec{u}_3^{(ji)}$ and $\vec{u}_3^{(k)} = \vec{u}_3^{(jk)}$:

$$\beta_k = \beta_{jk} = \sphericalangle[\vec{u}_3^{(j)} \to \vec{u}_3^{(k)}] \text{ about } \vec{u}_1^{(j)} \tag{7.15}$$

* The *rotation* angle between the unit vectors $\vec{u}_1^{(j)} = \vec{u}_1^{(jk)}$ and $\vec{u}_1^{(k)} = \vec{u}_1^{(kj)}$:

$$\theta_k = \theta_{jk} = \sphericalangle[\vec{u}_1^{(j)} \to \vec{u}_1^{(k)}] \text{ about } \vec{u}_3^{(k)} \tag{7.16}$$

* The *offset* between the unit vectors $\vec{u}_1^{(j)} = \vec{u}_1^{(jk)}$ and $\vec{u}_1^{(k)} = \vec{u}_1^{(kj)}$:

$$s_k = s_{jk} = O_{jk}O_{kj} = B_kO_k \text{ along } \vec{u}_3^{(k)} \tag{7.17}$$

* The *effective link length* of the link $\mathcal{L}_j$ between the joints $\mathcal{J}_{ji}$ and $\mathcal{J}_{jk}$:

$$b_j = b_{ik} = b_{ijk} = O_{ji}O_{jk} = O_jB_k \text{ along } \vec{u}_1^{(j)} \tag{7.18}$$

* The joint frames $\mathcal{F}_{ji}(O_{ji})$ and $\mathcal{F}_{jk}(O_{jk})$ of the kinematic elements $\mathcal{E}_{ji}$ and $\mathcal{E}_{jk}$ on the link $\mathcal{L}_j$:

The joint frame $\mathcal{F}_{ji}(O_{ji})$ and the link frame $\mathcal{F}_j(O_j)$ are taken to be coincident. That is,

$$\mathcal{F}_{ji}(O_{ji}) = \mathcal{F}_j(O_j); \vec{u}_k^{(ji)} = \vec{u}_k^{(j)} \text{ for } k = 1, 2, 3 \tag{7.19}$$

As for the joint frame $\mathcal{F}_{jk}(O_{jk})$, it is characterized as follows:

$$\mathcal{F}_{jk}(O_{jk}) = \mathcal{F}_{jk}(B_k); \vec{u}_1^{(jk)} = \vec{u}_1^{(ji)} = \vec{u}_1^{(j)}, \vec{u}_3^{(jk)} = \vec{u}_3^{(kj)} = \vec{u}_3^{(k)} \tag{7.20}$$

7.4 D–H Convention for the First Joint

Figure 7.3 shows a sketch that involves the base link ($\mathcal{L}_0$) and the first link ($\mathcal{L}_1$) of a serial manipulator. In general, the base frame $\mathcal{F}_0(O_0)$ is attached to $\mathcal{L}_0$ independently of how the manipulator is installed. However, the joint frame $\mathcal{F}_{01}(O_{01})$ is attached to $\mathcal{L}_0$ in accordance with the first axis of the manipulator. So, $\mathcal{F}_{01}(O_{01})$ is also called a *manipulator-specific base frame*. The third and first basis vectors of $\mathcal{F}_{01}(O_{01})$ are such that $\vec{u}_3^{(01)} = \vec{u}_3^{(1)}$ and $\vec{u}_1^{(01)}$ is along the common normal $\mathcal{N}_{01}$ between the axes along $\vec{u}_3^{(0)}$ and $\vec{u}_3^{(1)}$. The origin O_{01} coincides with the auxiliary point B_1, at which $\mathcal{N}_{01}$ intersects the first joint axis along $\vec{u}_3^{(1)}$. Meanwhile, $\mathcal{N}_{01}$ intersects the axis along $\vec{u}_3^{(0)}$ at the point H_1.

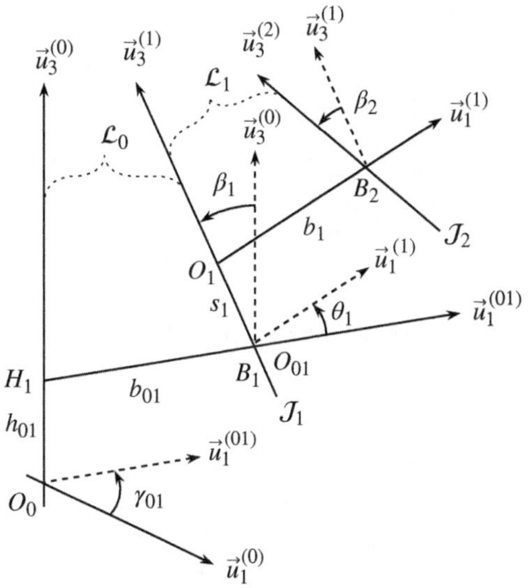

Figure 7.3 D–H convention for the first joint.

The location and orientation of $\mathcal{F}_{01}(O_{01})$ with respect to $\mathcal{F}_0(O_0)$ are described by the following constant parameters.

$$\gamma_{01} = \sphericalangle[\vec{u}_1^{(0)} \to \vec{u}_1^{(01)}] \text{ about } \vec{u}_3^{(0)} \tag{7.21}$$

$$\beta_{01} = \beta_1 = \sphericalangle[\vec{u}_3^{(0)} \to \vec{u}_3^{(1)}] \text{ about } \vec{u}_1^{(01)} \tag{7.22}$$

$$h_{01} = O_0H_1 \text{ along } \vec{u}_3^{(0)} \tag{7.23}$$

$$b_{01} = H_1O_{01} = H_1B_1 \text{ along } \vec{u}_1^{(01)} \tag{7.24}$$

However, in many cases, the demanded task can be carried out by using only one manipulator without considering its coordination with the other manipulators (if there are any) that share the same base. In such a case, the kinematic description of the considered manipulator can be simplified by repositioning the base frame $\mathcal{F}_0(O_0)$ so that

$$O_0 = O_{01} = B_1 \text{ and } \vec{u}_1^{(01)} = \vec{u}_1^{(0)} \tag{7.25}$$

In Eq. Set (7.25), $\vec{u}_1^{(0)}$ can be oriented arbitrarily as desired. Besides, Eq. Set (7.25) implies that

$$\gamma_{01} = 0 \text{ and } h_{01} = b_{01} = 0 \tag{7.26}$$

On the other hand, if the base is the ground as usual, it is customary to orient the base frame $\mathcal{F}_0(O_0)$ so that the acceleration of gravity appears as $\vec{g} = -g\vec{u}_3^{(0)}$. Moreover, a manipulator is usually installed on its base so that the axis of its first joint is parallel to the third axis of $\mathcal{F}_0(O_0)$. With such orientations of $\vec{u}_3^{(0)}$ and $\vec{u}_3^{(1)}$, the following additional simplification is achieved.

$$\vec{u}_3^{(0)} = \vec{u}_3^{(1)} \text{ or } \beta_1 = 0 \tag{7.27}$$

The simplifications mentioned above are illustrated in Figure 7.4.

Moreover, if the first joint of the considered manipulator is a revolute joint, which is so in general, then a further simplification can be achieved by selecting $\mathcal{F}_0(O_0)$ so that

$$\mathcal{F}_0(O_0) = \mathcal{F}_0(O_1) \text{ or } s_1 = d_1 = 0 \tag{7.28}$$

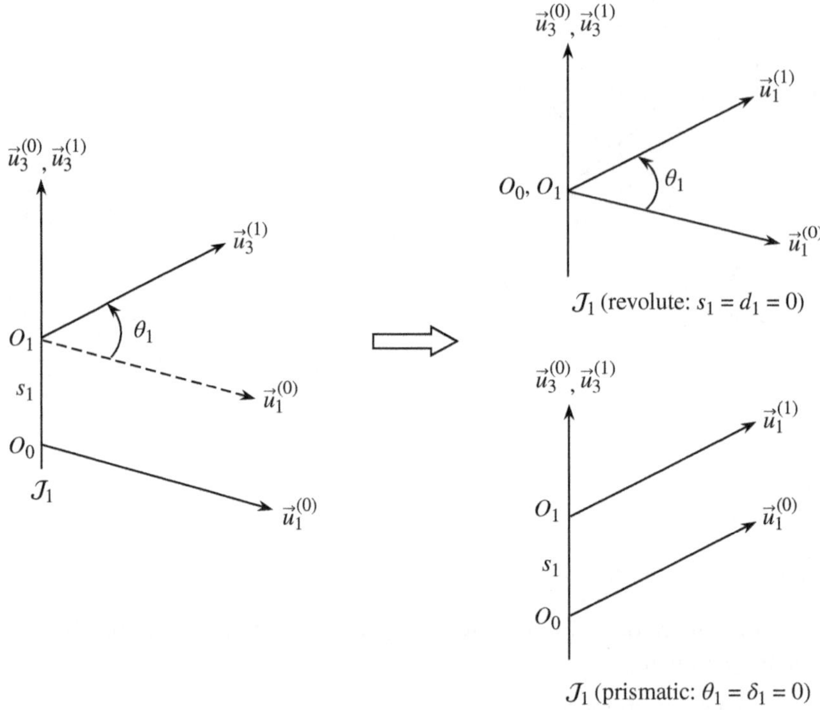

Figure 7.4 D–H convention for the special cases of the first joint.

Note that, as long as $\theta_1 \neq 0$, $\mathcal{F}_1(O_1) \neq \mathcal{F}_0(O_1)$ even though they share the same origin.

On the other hand, if the first joint of the considered manipulator is a prismatic joint, then it is possible to have a further simplification by selecting $\mathcal{F}_0(O_0)$ so that

$$\mathcal{F}_0(O_0) = \mathcal{F}_1(B_1) = \mathcal{F}_1(O_0) \quad \text{or} \quad \theta_1 = \delta_1 = 0 \tag{7.29}$$

Note that, as long as $s_1 \neq 0$, $\mathcal{F}_1(O_1) \neq \mathcal{F}_1(O_0)$ even though they have the same orientation.

The further simplifications achieved by positioning $\mathcal{F}_0(O_0)$ appropriately, which are explained above depending on whether $\mathcal{J}_1$ is revolute or prismatic, are also illustrated in Figure 7.4.

In the most general case, the position of $\mathcal{F}_1(O_1)$ with respect to $\mathcal{F}_0(O_0)$ can be expressed by the following equations.

$$\begin{aligned}
\hat{C}^{(0,1)} &= \hat{C}^{(0,01)}\hat{C}^{(01,1)} = (e^{\tilde{u}_3\gamma_{01}}e^{\tilde{u}_1\beta_1})(e^{\tilde{u}_3\theta_1}) \Rightarrow \\
\hat{C}^{(0,1)} &= e^{\tilde{u}_3\gamma_{01}}e^{\tilde{u}_1\beta_1}e^{\tilde{u}_3\theta_1}
\end{aligned} \tag{7.30}$$

$$\begin{aligned}
\vec{r}_{0,1} &= \vec{r}_{O_0O_1} = \vec{r}_{O_0H_1} + \vec{r}_{H_1B_1} + \vec{r}_{B_1O_1} \Rightarrow \\
\vec{r}_{0,1} &= h_{01}\vec{u}_3^{(0)} + b_{01}\vec{u}_1^{(01)} + s_1\vec{u}_3^{(01)} \Rightarrow \\
\overline{r}_{0,1}^{(0)} &= h_{01}\overline{u}_3^{(0/0)} + b_{01}\overline{u}_1^{(01/0)} + s_1\overline{u}_3^{(01/0)} \Rightarrow \\
\overline{r}_{0,1}^{(0)} &= h_{01}\overline{u}_3 + b_{01}\hat{C}^{(0,01)}\overline{u}_1 + s_1\hat{C}^{(0,01)}\overline{u}_3 \Rightarrow \\
\overline{r}_{0,1}^{(0)} &= h_{01}\overline{u}_3 + b_{01}e^{\tilde{u}_3\gamma_{01}}e^{\tilde{u}_1\beta_1}\overline{u}_1 + s_1e^{\tilde{u}_3\gamma_{01}}e^{\tilde{u}_1\beta_1}\overline{u}_3 \Rightarrow \\
\overline{r}_{0,1}^{(0)} &= \overline{u}_3(h_{01} + s_1c\beta_1) + \overline{u}_1(b_{01}c\gamma_{01} + s_1s\beta_1s\gamma_{01}) + \overline{u}_2(b_{01}s\gamma_{01} - s_1s\beta_1c\gamma_{01})
\end{aligned} \tag{7.31}$$

In the special case expressed by Eqs. (7.26) and (7.27), which allows the joint $\mathcal{J}_1$ to be either revolute or prismatic, Eqs. (7.30) and (7.31) reduce to the following equation pair.

$$\hat{C}^{(0,1)} = e^{\tilde{u}_3\theta_1}, \quad \overline{r}_{0,1}^{(0)} = \overline{u}_3 s_1 \tag{7.32}$$

In the further special cases depending on whether $\mathcal{J}_1$ is revolute or prismatic, Eq. Pair (7.32) takes the following simpler forms.

If $\mathcal{J}_1$ is revolute,

$$\hat{C}^{(0,1)} = e^{\tilde{u}_3\theta_1}, \quad \overline{r}_{0,1}^{(0)} = \overline{0} \tag{7.33}$$

If $\mathcal{J}_1$ is prismatic,

$$\hat{C}^{(0,1)} = \hat{I}, \quad \overline{r}_{0,1}^{(0)} = \overline{u}_3 s_1 \tag{7.34}$$

7.5 D–H Convention for the Last Joint

Figure 7.5 shows a sketch that involves the last link $\mathcal{L}_m$ of a serial manipulator together with an *end-effector* (i.e. the body $\mathcal{B}_e$) that is rigidly fixed to $\mathcal{L}_m$.

Considering $\mathcal{L}_m$, the frame $\mathcal{F}_m(O_m)$ is attached to $\mathcal{L}_m$ according to the same pattern of the D–H convention. However, there is not a next joint such as $\mathcal{J}_{m+1}$. Therefore, the location of O_m on the axis of $\mathcal{J}_m$ and the orientation of $\vec{u}_1^{(m)}$ (which is to be normal to the axis of $\mathcal{J}_m$) can in general be selected arbitrarily. However, this arbitrariness can be eliminated by relating their selection to the geometric features of the affixed end-effector.

Considering the end-effector $\mathcal{B}_e$, the frame $\mathcal{F}_e(O_e)$ attached to it may be quite different from the frame $\mathcal{F}_m(O_m)$ attached to $\mathcal{L}_m$. This is because the assembly position of the end-effector on $\mathcal{L}_m$ is determined by paying attention primarily to its operational characteristics, rather than the position of $\mathcal{J}_m$, and $\mathcal{F}_e(O_e)$ is attached to the end-effector in order to represent its geometric features in the most natural way. For example, the origin O_e is placed at the *tip point P* of the end-effector and the unit vector $\vec{u}_3^{(e)}$ is oriented along the major functioning part of the end-effector, e.g. the muzzle of a painting gun.

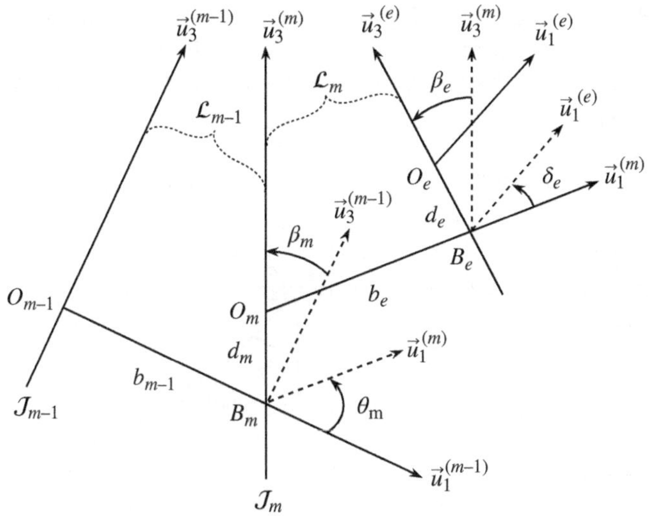

Figure 7.5 D–H convention for the last joint.

The position of $\mathcal{F}_m(O_m)$ with respect to the penultimate frame $\mathcal{F}_{m-1}(O_{m-1})$ and the position of the end-effector frame $\mathcal{F}_e(O_e)$ with respect to $\mathcal{F}_m(O_m)$ are described by the following parameters, which are defined so that the previously mentioned arbitrariness in the location of O_m and the orientation of $\vec{u}_1^{(m)}$ is eliminated by using the common normal between the axes along $\vec{u}_3^{(m)}$ and $\vec{u}_3^{(e)}$ as long as these axes are not coincident.

$$b_{m-1} = O_{m-1}B_m \text{ along } \vec{u}_1^{(m-1)} \tag{7.35}$$

$$d_m = B_m O_m \text{ along } \vec{u}_3^{(m)} \tag{7.36}$$

$$b_e = O_m B_e \text{ along } \vec{u}_1^{(m)} \tag{7.37}$$

$$d_e = B_e O_e \text{ along } \vec{u}_3^{(e)} \tag{7.38}$$

$$\beta_m = \sphericalangle[\vec{u}_3^{(m-1)} \to \vec{u}_3^{(m)}] \text{ about } \vec{u}_1^{(m-1)} \tag{7.39}$$

$$\theta_m = \sphericalangle[\vec{u}_1^{(m-1)} \to \vec{u}_1^{(m)}] \text{ about } \vec{u}_3^{(m)} \tag{7.40}$$

$$\beta_e = \sphericalangle[\vec{u}_3^{(m)} \to \vec{u}_3^{(e)}] \text{ about } \vec{u}_1^{(m)} \tag{7.41}$$

$$\delta_e = \sphericalangle[\vec{u}_1^{(m)} \to \vec{u}_1^{(e)}] \text{ about } \vec{u}_3^{(e)} \tag{7.42}$$

However, in a practicable manipulator, for the sake of simplicity, the end-effector is preferably shaped and affixed to the last link so that it becomes possible to take $\mathcal{F}_e(O_e)$ coincident with $\mathcal{F}_m(O_m)$. Such an end-effector can be designated as a *regular end-effector*. The most typical example of a regular end-effector is a gripper with two fingers, which is illustrated in Figure 7.6.

The relevant definitions concerning the gripper shown in Figure 7.6 are given below.

$$P = O_m\text{: Tip point (origin of the last link frame } \mathcal{F}_\mathrm{m}) \tag{7.43}$$

$$R = O_{m-1}\text{: Wrist point (origin of the penultimate link frame } \mathcal{F}_{\mathrm{m}-1}) \tag{7.44}$$

$$\vec{u}_3^{(m)} = \vec{u}_a\text{: Approach vector} \tag{7.45}$$

$$\vec{u}_1^{(m)} = \vec{u}_n\text{: Normal vector} \tag{7.46}$$

$$\vec{u}_2^{(m)} = \vec{u}_s\text{: Side vector} \tag{7.47}$$

In the preceding definitions, $\vec{u}_n$ is called a *normal vector* because it is perpendicular to the gripper plane formed by the fingers. Similarly, $\vec{u}_s$ is called a *side vector* because it indicates the lateral direction in the gripper plane. The reason why $\vec{u}_a$ is called an *approach vector* is due to imagining that the gripper approaches the object to be gripped essentially in the direction of $\vec{u}_a$.

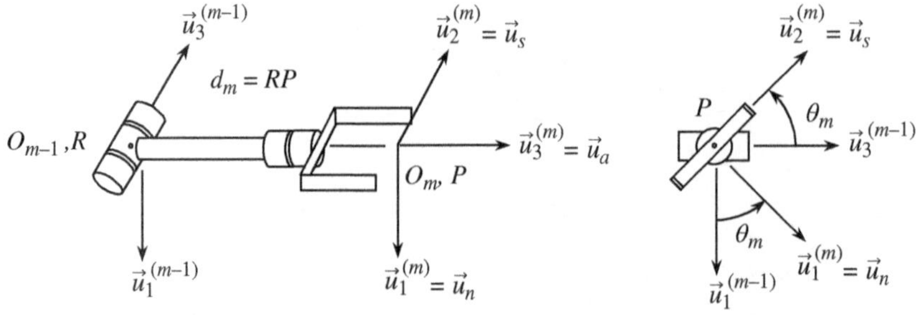

Figure 7.6 D–H convention for a gripper.

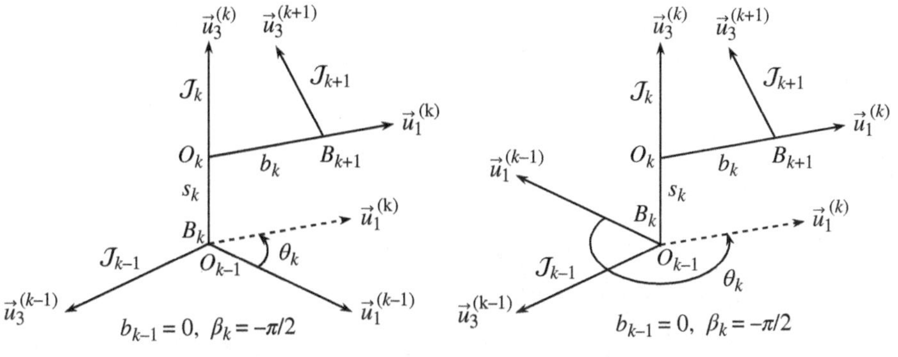

Figure 7.7 D–H convention for perpendicularly intersecting joint axes.

7.6 D–H Convention for Successive Joints with Perpendicularly Intersecting Axes

Figure 7.7 shows a case, in which the axes of the joints $\mathcal{J}_{k-1}$ and $\mathcal{J}_k$ intersect each other perpendicularly. In such a case, the proximality of $\mathcal{J}_{k-1}$ and $\mathcal{J}_k$ become equal and therefore the sense of the unit vector $\vec{u}_1^{(k-1)}$ along the common normal $\mathcal{N}_{k-1}$ becomes arbitrary. In other words, the sense of $\vec{u}_1^{(k-1)}$ can be selected either as seen on the left-hand side of Figure 7.7 (Option 1) or as seen on the right-hand side of Figure 7.7 (Option 2). As noticed, in both options, s_k is the same and $b_{k-1} = 0$. However, if $\theta_k = \theta'_k$ in Option 1 with $\beta_k = -\pi/2$, then in Option 2 with $\beta_k = \pi/2$, it happens that $\theta_k = \theta''_k = \theta'_k + \pi$.

Between the two options, the general tendency is to choose the one in which θ_k appears as an acute angle in the major part of the operational range of the manipulator.

7.7 D–H Convention for Successive Joints with Parallel Axes

Figures 7.8 and 7.9 illustrate the two options of using the D–H convention for two successive joints with parallel axes. In the figure, the indices are such that $l = k + 1, j = k - 1$, and $i = k - 2$. As noticed, the joints with parallel axes are $\mathcal{J}_j$ and $\mathcal{J}_k$. In such a case, the common normal $\mathcal{N}_j$ is not unique. In fact, there are infinitely many common normals, which are all parallel to each other. However, only two of these common normals lead to two options with simplified kinematic expressions, in which the number of nonzero constant parameters is minimized.

Since the axes of $\mathcal{J}_j$ and $\mathcal{J}_k$ are parallel in this joint arrangement, $\beta_k = 0$ in both options. The other vanishing parameter can be either s_j or s_k depending on the selected option. These two options are described and discussed below.

(a) Option 1: In this option, the common normal $\mathcal{N}_j$ is selected so that $O_j = B_j$ and $s_j = d_j = 0$. Therefore, this option can be valid only if $\mathcal{J}_j$ is a revolute joint. The next joint $\mathcal{J}_k$ can be either revolute or prismatic.

(b) Option 2: In this option, the common normal $\mathcal{N}_j$ is selected so that $O_k = B_k$ and $s_k = d_k = 0$. Therefore, this option can be valid only if $\mathcal{J}_k$ is a revolute joint. The preceding joint $\mathcal{J}_j$ can be either revolute or prismatic.

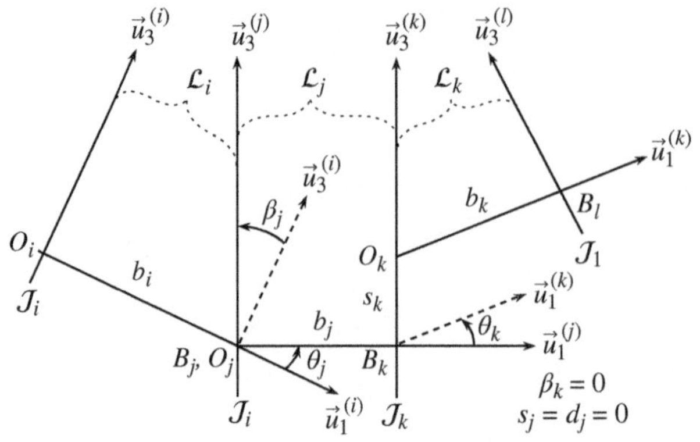

Figure 7.8 D–H convention for parallel joint axes (Option 1).

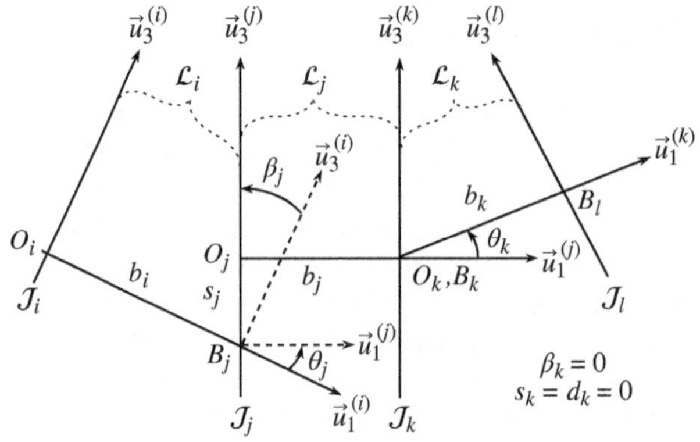

Figure 7.9 D–H convention for parallel joint axes (Option 2).

The selection of one of the two options described above depends on the types of the joints $\mathcal{J}_j$ and $\mathcal{J}_k$ as discussed below.

(i) If $\mathcal{J}_j$ and $\mathcal{J}_k$ are both revolute, then any of the two options can be selected with equal degree of preferability.

(ii) If $\mathcal{J}_j$ is revolute but $\mathcal{J}_k$ is prismatic, then only Option 1 (with $d_j = 0$) can be selected, because s_k is a joint variable and it is not zero in general.

(iii) If $\mathcal{J}_j$ is prismatic but $\mathcal{J}_k$ is revolute, then only Option 2 (with $d_k = 0$) can be selected, because s_j is a joint variable and it is not zero in general.

(iv) If $\mathcal{J}_j$ and $\mathcal{J}_k$ are both prismatic, then there remains no point in comparing the two options, because a manipulator with such a joint arrangement happens to be *kinematically defective*. This defect is due to the fact that both $\mathcal{J}_j$ and $\mathcal{J}_k$ provide translational displacements (s_j and s_k) in the same direction and therefore one of s_j and s_k becomes redundant. In other words, the same translational displacement can be imparted to the end-effector by actuating only one of $\mathcal{J}_j$ and $\mathcal{J}_k$ while the other one is kept fixed. Thus, one of $\mathcal{J}_j$ and $\mathcal{J}_k$ gets effectively lost and therefore the mobility of the end-effector undesirably reduces by one to $m' = m - 1$. The translational displacement redundancy mentioned above can be seen in the following equations that are written to express the location of O_k with respect to O_i.

$$\vec{r}_{i,k} = \vec{r}_{i,j} + \vec{r}_{j,k} = [b_i\vec{u}_1^{(i)} + s_j\vec{u}_3^{(j)}] + [b_j\vec{u}_1^{(j)} + s_k\vec{u}_3^{(k)}] \tag{7.48}$$

Since the axes of $\mathcal{J}_j$ and $\mathcal{J}_k$ are parallel, i.e. since $\vec{u}_3^{(j)} = \vec{u}_3^{(k)}$, Eq. (7.48) can be written as follows:

$$\vec{r}_{i,k} = b_i\vec{u}_1^{(i)} + b_j\vec{u}_1^{(j)} + s_{jk}\vec{u}_3^{(j)} = b_i\vec{u}_1^{(i)} + b_j\vec{u}_1^{(j)} + s_{jk}\vec{u}_3^{(k)} \tag{7.49}$$

In Eq. (7.49),

$$s_{jk} = s_j + s_k \tag{7.50}$$

Equations (7.49) and (7.50) show that the overall translational displacement s_{jk} can effectively be achieved by actuating only one of $\mathcal{J}_j$ and $\mathcal{J}_k$ while the other one is fixed and kept out of the operation.

7.8 D–H Convention for Successive Joints with Coincident Axes

Figures 7.10 and 7.11 illustrate the two cases of coincident joint axes that may be allowed to occur in a serial manipulator. As in the preceding case of parallel joint axes, the indices are again such that $l = k+1$, $j = k-1$, and $i = k-2$. In an allowable case of two successive joints with coincident axes, if one of the joints is revolute, the other one can only be prismatic. Such a pair of revolute and prismatic joints constitute a *cylindrical joint arrangement*. It is called so because the combined effect of the two joints forming this arrangement is equivalent to the effect of a single cylindrical joint. On the other hand, the arrangements that comprise two successive revolute (or prismatic) joints

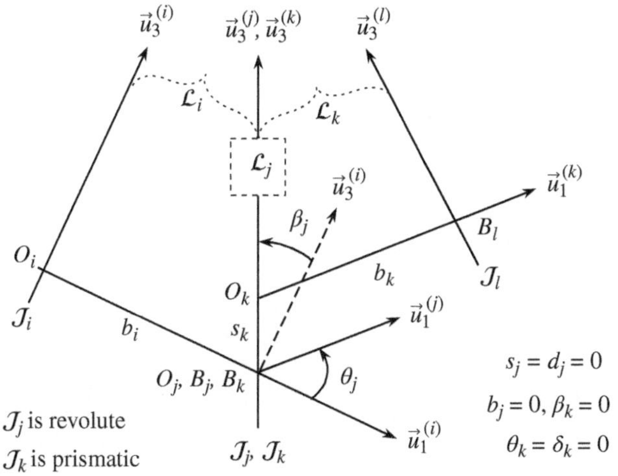

Figure 7.10 D–H convention for coincident joint axes (case 1).

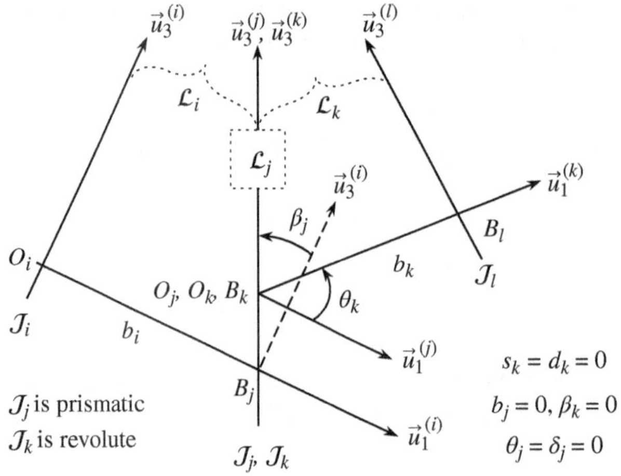

Figure 7.11 D–H convention for coincident joint axes (case 2).

with coincident axes cannot be allowed in a serial manipulator. This is because two revolute (or prismatic) joints that share the same axis can effectively be replaced with a kinematically equivalent single revolute (or prismatic) joint on the same axis. As a consequence of this replacement, one of the joints gets effectively lost and the mobility of the end-effector undesirably reduces by one to $m' = m - 1$.

In the allowable cases illustrated in Figures 7.10 and 7.11, the joints with coincident axes are $\mathcal{J}_j$ and $\mathcal{J}_k$. The joint variables and the constant parameters associated with each case are indicated in the mentioned figures.

8

Position and Motion Analyses of Generic Serial Manipulators

Synopsis

The first main part of this chapter encompasses position analysis. It is concerned with the relationships between the joint variables and the positions (orientations and locations) of the links and the end-effector of a generic serial manipulator with respect to the base frame. These relationships are established in two successive stages.

In the first stage, which is known as *forward kinematics*, the positions of the links (including the end-effector) are expressed in terms of the joint variables. The expressions can be obtained by using two alternative formulations. In the *compact formulation*, which is suitable especially for computational purposes, the orientations and locations of the links are expressed together in a combined way within the *homogeneous transformation matrices*. These expressions turn out to be quite brief but they do not show the joint variables explicitly and they do not reflect the characteristic features of the manipulator. The compact formulation is explained in Section 8.2. In the *detailed formulation*, which is suitable especially for analytical purposes, the orientations and locations of the links are expressed separately. These expressions turn out to be lengthier but they show the joint variables explicitly. They are naturally specific (i.e. tailored) to the particular manipulator of concern but they reflect all the characteristic features clearly. Therefore, they are convenient to be manipulated symbolically in various analytical treatments. The detailed formulation is explained in Sections 8.3 and 8.4.

In the second stage, which is known as *inverse kinematics*, the main concern is the process of finding the joint variables corresponding to a specified position of the end-effector. The result of this process is called *inverse kinematic solution*. The inverse kinematic solution involves multiplicities, because the forward kinematic equations that are to be solved for the joint variables contain the angular joint variables nonlinearly with their trigonometric (sine and cosine) functions. The multiplicities are associated with the *posture modes* of the manipulator. For example, a typical manipulator with six revolute joints may assume a pose either in the elbow-up mode or in the elbow-down mode in order to attain the same position of the end-effector. Besides, the inverse kinematic solution also involves singularities, which are called *position singularities*. The identification and analysis of the multiplicities and singularities are the other concerns in the stage of inverse kinematics. The nonlinear equations of forward kinematics can always be solved, of course, by using an iterative numerical method. However, if a numerical method is used, it becomes quite difficult to identify the multiplicities and singularities. On the other hand, they can be readily identified

Kinematics of General Spatial Mechanical Systems, First Edition. M. Kemal Ozgoren.

Companion Website: www.wiley.com/go/ozgoren/spatialmechanicalsystems

as by-products if an analytical solution is used, whenever it is possible. If a completely analytical solution is not possible, then at least a semi-analytical solution can be used in order to reduce the number of equations to be solved numerically and at the same time identify some (if not all) of the multiplicities and singularities.

Since the analytical and semi-analytical solutions can be obtained only for the manipulators with specified kinematic features, they will be studied in detail in Chapter 9 for certain specific manipulators.

Another aspect of the inverse kinematics is the requirement to obtain inverse kinematic solutions not only for the *regular* manipulators but also for the *redundant* and *deficient* manipulators. A regular manipulator has a necessary and sufficient number of actuated joints in order to execute the required task. A redundant manipulator has a larger number of actuated joints than is necessary to execute the required task. On the other hand, a deficient manipulator has a smaller number of actuated joints than is necessary to execute an arbitrary task. So, it can be used only for some appropriately compromised tasks. All these matters that are studied within the scope of the inverse kinematics are explained and discussed in Sections 8.5–8.8.

The second main part of this chapter encompasses motion analysis. It is concerned with the relationships between the joint motions and the translational and rotational motions of the links and the end-effector with respect to the base frame. The motion relationships are limited here so as to cover velocities and accelerations. However, if desired, they can be extended to cover higher-order motion derivatives (e.g. jerks) as well. Like the position relationships, the motion relationships are also established in two successive stages.

In the first stage, which is known as *forward kinematics of motion*, the motions of the end-effector and the other links are expressed in terms of the joint motions. The motion relationships happen to be linear among the velocities and affine among the accelerations. They are expressed by using two alternative formulations. In the *compact formulation*, which is suitable especially for computational purposes, the translational and rotational motions are expressed together by using augmented column matrices that indicate the velocity and acceleration states of the links. These column matrices are related to the column matrices of the joint velocities and accelerations compactly and briefly by means of *Jacobian matrices*. The generation and usage of Jacobian matrices are explained in Section 8.10. In the *detailed formulation*, which is suitable especially for analytical purposes, the rotational and translational motions of the end-effector and the other links are expressed separately. These expressions are lengthier but they reflect all the characteristic features of the manipulator. Therefore, they are suitable for the symbolic manipulations required in various analytical treatments. The detailed formulation is explained in Sections 8.9.1 and 8.9.2.

In the second stage, which is known as *inverse kinematics of motion* or *inverse motion analysis*, the main concern is the process of finding the joint motions corresponding to a specified motion of the end-effector. The result of this process is called an *inverse motion solution*. However, as mentioned before, an inverse motion solution is limited here to be one of the inverse velocity and inverse acceleration solutions. An inverse motion solution does not involve multiplicities, because the motion relationships are either *linear* (among velocities) or *affine* (among accelerations). However, an inverse motion solution does involve singularities, which are called *motion singularities*. It is to be underlined that the motion singularities are in general different from the position singularities that occur in the inverse kinematic solution concerning the position relationships. The identification and analysis of the motion singularities are the other concerns of the inverse motion analysis. If the compact formulation is preferred, then the motion singularities happen to be the singularities of the relevant Jacobian matrix. In that case, unless the determinant of the Jacobian matrix has a simple expression, it will be rather tedious to find the singularities and analyze their consequences. On the other hand, if the detailed formulation is preferred, then the motion singularities can be readily identified by looking at the denominators of the symbolically obtained expressions for the derivatives of the joint variables. Moreover, the symbolically obtained expressions also facilitate the analysis of the singularities concerning their consequences in the joint and task spaces of the manipulator.

In Section 8.12, the way of obtaining the inverse velocity and acceleration solutions is explained for a generic regular manipulator by using its Jacobian matrix. In Section 8.15, the way of obtaining the inverse velocity and acceleration solutions is explained for another generic manipulator by using the detailed formulation. These solutions imply that the detailed formulation may become especially effective if it is used for a specific manipulator with known characteristic features. This statement is verified later in Chapter 9, which contains several specific manipulators.

Another aspect of the inverse motion analysis is the requirement to obtain inverse motion solutions not only for the regular manipulators but also for the redundant and deficient manipulators. The inverse motion analyses of redundant and deficient manipulators are explained and discussed in Sections 8.13 and 8.14.

The position, velocity, and acceleration expressions of forward kinematics for all the links of a generic serial manipulator including the end-effector can also be obtained through a systematic *recursive* procedure. This procedure is explained in Section 8.11.

8.1 Forward Kinematics

Concerning a serial manipulator, forward kinematics is the process of expressing the position (location and orientation) of the end-effector with respect to the base frame in terms of the joint variables. In a more general sense, the intermediate links are also taken into the scope of forward kinematics. Occasionally, forward kinematics is also referred to as *direct kinematics.*

Forward kinematics can be conceived as a mapping from the *joint space* (S_J) to the *task space* (S_T). For a spatial serial manipulator with m joints, S_J is an m-dimensional space that covers all the possible values of the m joint variables ($q_1, q_2, \ldots, q_m$) and S_T is a six-dimensional space that covers all the possible values of the *three coordinates* of the tip point (p_1, p_2, p_3) and the *three independent orientation parameters* of the end-effector with respect to the base frame. The orientation of the end-effector with respect to the base frame is described by the orientation matrix $\hat{C} = \hat{C}^{(0,m)}$ and the three independent parameters of $\hat{C}$ are usually taken as the three Euler angles (ϕ_1, ϕ_2, ϕ_3) of a suitable sequence such as 1-2-3 or 3-2-3. Sometimes, however, they may be taken alternatively as the primary direction angles ($\theta_{11}, \theta_{22}, \theta_{33}$) or the angle-axis parameters (θ; α, β) of the orienting rotation. Here, θ is the rotation angle about the rotation axis, whose inclination is described by the azimuth angle α and the elevation angle β. The above-mentioned coordinates and orientation parameters describe the position of the end-effector with respect to the base frame within the *working volume* or the *reachable range* of the manipulator.

Depending on the purpose, the expressions of forward kinematics can be generated in one of the two major versions. The first version is a *compact formulation*, in which the orientation and location of each link are treated together by stacking them into the relevant *homogeneous transformation matrix*. The compact formulation contains the joint variables implicitly and it does not reflect the characteristic features of the manipulator. The second version is a *detailed formulation*, in which the orientation and location of each link are treated *separately*. The detailed formulation shows the joint variables explicitly and it reflects all the characteristic features of the manipulator. The compact formulation is convenient especially for computational purposes. On the other hand, the detailed formulation is convenient especially for analytical purposes. More specifically, the detailed formulation becomes indispensable in the stage of inverse kinematics, if the joint variables are desired to be obtained analytically or semi-analytically from the equations of forward kinematics. The detailed formulation is also convenient for differentiation in order to derive the velocity and acceleration relationships.

The main points of the compact and detailed formulations are explained in the following sections. For the sake of avoiding the complications that do not normally occur in the practicable cases, these formulations are obtained here by assuming that the following simplifying features exist for the first joint and the end-effector, which are explained in Chapter 7.

∗ The base frame is positioned so that $\gamma_{01} = 0$, $h_{01} = 0$, and $b_{01} = 0$. Moreover, in most of the cases, it may be oriented so that $\beta_{01} = \beta_1 = 0$, too, as illustrated in Figure 7.4.

∗ The end-effector is regular as illustrated in Figure 7.6. That is, $\mathcal{F}_e(O_e) = \mathcal{F}_m(O_m)$.

8.2 Compact Formulation of Forward Kinematics

In Section 7.3, the link-to-link rotation matrix and the origin-to-origin position vector between two successive links $\mathcal{L}_{k-1}$ and $\mathcal{L}_k$ are expressed as follows:

$$\hat{C}^{(k-1,k)} = e^{\tilde{u}_1\beta_k} e^{\tilde{u}_3\theta_k} \tag{8.1}$$

$$\vec{r}_{k-1,k} = \vec{r}_{O_{k-1}O_k} = b_{k-1}\vec{u}_1^{(k-1)} + s_k\vec{u}_3^{(k)} \tag{8.2}$$

$$\overline{r}_{k-1,k}^{(k-1)} = \overline{u}_1 b_{k-1} - \overline{u}_2 s_k s\beta_k + \overline{u}_3 s_k c\beta_k \tag{8.3}$$

In the first stage of the compact formulation, Eqs. (8.1) and (8.3) are combined as shown below to obtain the *homogeneous transformation matrix* between the links $\mathcal{L}_{k-1}$ and $\mathcal{L}_k$.

$$\hat{H}^{(k-1,k)} = \begin{bmatrix} \hat{C}^{(k-1,k)} & \overline{r}_{k-1,k}^{(k-1)} \\ \overline{0}^t & 1 \end{bmatrix} \tag{8.4}$$

As for the element-by-element expression,

$$\hat{H}^{(k-1,k)} = \begin{bmatrix} c\theta_k & -s\theta_k & 0 & b_{k-1} \\ c\beta_k s\theta_k & c\beta_k c\theta_k & -s\beta_k & -s_k s\beta_k \\ s\beta_k s\theta_k & s\beta_k c\theta_k & c\beta_k & s_k c\beta_k \\ 0 & 0 & 0 & 1 \end{bmatrix} \tag{8.5}$$

After obtaining the link-to-link homogeneous transformation matrices as explained above, the *homogeneous position matrix* of $\mathcal{L}_k$ with respect to $\mathcal{F}_0(O_0)$, i.e. the homogeneous transformation matrix between $\mathcal{F}_k(O_k)$ and $\mathcal{F}_0(O_0)$, can be obtained simply as follows for any $k = 1, 2, \ldots, m$:

$$\hat{H}^{(0,k)} = \hat{H}^{(0,1)}\hat{H}^{(1,2)}\hat{H}^{(2,3)}\cdots\hat{H}^{(k-1,k)} \tag{8.6}$$

For the sake of brevity, $\hat{H}^{(0,k)}$ can also be denoted and expressed as shown below.

$$\hat{H}^{(0,k)} = \begin{bmatrix} \hat{C}^{(0,k)} & \bar{r}^{(0)}_{0,k} \\ \bar{0}^t & 1 \end{bmatrix} \Rightarrow \hat{H}_k = \begin{bmatrix} \hat{C}_k & \bar{r}_k \\ \bar{0}^t & 1 \end{bmatrix} \tag{8.7}$$

Thus, after the computation of $\hat{H}_k = \hat{H}^{(0,k)}$ by using the specified values of the joint variables, the numerical information about the orientation and location of $\mathcal{L}_k$ with respect to $\mathcal{F}_0(O_0)$ can be acquired from $\hat{H}_k$ by looking at its partitions $\hat{C}_k = \hat{C}^{(0,k)}$ and $\bar{r}_k = \bar{r}^{(0)}_{0,k}$.

Considering the end-effector in particular, the notation for its homogeneous position matrix can be abbreviated as shown below for the sake further convenience in its frequent usage.

$$\hat{H}^{(0,m)} = \begin{bmatrix} \hat{C}^{(0,m)} & \bar{r}^{(0)}_{0,m} \\ \bar{0}^t & 1 \end{bmatrix} \Rightarrow \hat{H}_m = \begin{bmatrix} \hat{C}_m & \bar{r}_m \\ \bar{0}^t & 1 \end{bmatrix} \Rightarrow \hat{H} = \begin{bmatrix} \hat{C} & \bar{p} \\ \bar{0}^t & 1 \end{bmatrix} \tag{8.8}$$

The above abbreviation implies the following related abbreviations.

$$\hat{C} = \hat{C}_m = \hat{C}^{(0,m)} \tag{8.9}$$

$$\bar{p} = \bar{r}_m = \bar{r}^{(0)}_{0,m} = \bar{r}^{(0)}_{O_0O_m} = \bar{r}^{(0)}_{OP} \tag{8.10}$$

Once $\hat{C}$ becomes available as explained above, its three independent parameters, e.g. the three Euler angles (ϕ_1, ϕ_2, ϕ_3) of a selected sequence, can also be extracted from $\hat{C}$.

8.3 Detailed Formulation of Forward Kinematics

By using Eq. (8.1), the orientation matrix of a link $\mathcal{L}_k$ with respect to the base link $\mathcal{L}_0$ can be obtained as follows for $k = 1, 2, 3, \ldots, m$:

$$\hat{C}_k = \hat{C}^{(0,k)} = \hat{C}^{(0,1)}\hat{C}^{(1,2)}\hat{C}^{(2,3)}\cdots\hat{C}^{(k-1,k)} \Rightarrow$$

$$\hat{C}_k = e^{\tilde{u}_1\beta_1}e^{\tilde{u}_3\theta_1}e^{\tilde{u}_1\beta_2}e^{\tilde{u}_3\theta_2}\cdots e^{\tilde{u}_1\beta_k}e^{\tilde{u}_3\theta_k} \tag{8.11}$$

By using the *shifting property* of the rotation matrices, which is provided in Section 2.7, $\hat{C}_k$ in Eq. (8.11) can be re-expressed as shown below.

$$\hat{C}_k = \hat{\Phi}_k e^{\tilde{u}_1\gamma_k} \tag{8.12}$$

In Eq. (8.12),

$$\hat{\Phi}_k = e^{\tilde{n}_1\theta_1}e^{\tilde{n}_2\theta_2}\cdots e^{\tilde{n}_{k-1}\theta_{k-1}}e^{\tilde{n}_k\theta_k} \tag{8.13}$$

$$\bar{n}_k = e^{\tilde{u}_1\gamma_k}\bar{u}_3 \tag{8.14}$$

$$\gamma_k = \beta_1 + \beta_2 + \cdots + \beta_k \tag{8.15}$$

In the above equations, γ_k is the *cumulative twist angle* of the joint $\mathcal{J}_k$ and the column matrix $\overline{n}_k$ represents the joint axis unit vector $\vec{u}_3^{(k)}$ of $\mathcal{J}_k$ in the base frame $\mathcal{F}_0$ when the relevant angular joint variables $(\theta_1, \theta_2, ..., \theta_{k-1})$ are all zero. This property of $\overline{n}_k$ can be verified by noting that

$$\overline{u}_3^{(k/0)} = \hat{C}^{(0,k)}\overline{u}_3^{(k/k)} = (\hat{\Phi}_k e^{\tilde{u}_1\gamma_k})\overline{u}_3 = \hat{\Phi}_k(e^{\tilde{u}_1\gamma_k}\overline{u}_3) = \hat{\Phi}_k\overline{n}_k = \hat{\Phi}_{k-1}\overline{n}_k \tag{8.16}$$

As seen above, $\overline{u}_3^{(k/0)} \to \overline{n}_k$ as $\hat{\Phi}_{k-1} \to \hat{I}$, i.e. as $\theta_i \to 0$ for all $i = 1, 2, \ldots, k-1$.

With the availability of the orientation matrices of the links, Eqs. (8.2), (8.11), and (8.12) lead to the following column matrix expressions that represent the location of the origin O_k with respect to $\mathcal{F}_0(O_0)$ for $k = 1, 2, 3, \ldots, m$. The following equations are written by noting that $b_{01} = 0$ as in the common special case mentioned in Section 8.1.

$$\begin{aligned}
&\vec{r}_k = \vec{r}_{0,k} = \vec{r}_{0,1} + \vec{r}_{1,2} + \vec{r}_{2,3} + \cdots + \vec{r}_{k-1,k} \Rightarrow \\
&\overline{r}_k = \overline{r}_k^{(0)} = \overline{r}_{0,k}^{(0)} = \overline{r}_{0,1}^{(0)} + \overline{r}_{1,2}^{(0)} + \overline{r}_{2,3}^{(0)} + \cdots + \overline{r}_{k-1,k}^{(0)} \Rightarrow \\
&\overline{r}_k = [s_1\overline{u}_3^{(1/0)}] + [b_1\overline{u}_1^{(1/0)} + s_2\overline{u}_3^{(2/0)}] + [\, b_2\overline{u}_1^{(2/0)} + s_3\overline{u}_3^{(3/0)}] + \cdots \\
&\qquad + [b_{k-2}\overline{u}_1^{(k-2/0)} + s_{k-1}\overline{u}_3^{(k-1/0)}] + [b_{k-1}\overline{u}_1^{(k-1/0)} + s_k\overline{u}_3^{(k/0)}] \Rightarrow \\
&\overline{r}_k = \hat{C}_1(s_1\overline{u}_3 + b_1\overline{u}_1) + \hat{C}_2(s_2\overline{u}_3 + b_2\overline{u}_1) + \hat{C}_3(s_3\overline{u}_3 + b_3\overline{u}_1) + \cdots \\
&\qquad + \hat{C}_{k-1}(s_{k-1}\overline{u}_3 + b_{k-1}\overline{u}_1) + s_k\hat{C}_k\overline{u}_3
\end{aligned} \tag{8.17}$$

Alternatively, $\overline{r}_k$ can also be expressed as follows by using Eqs. (8.12)–(8.17):

$$\begin{aligned}
&\overline{r}_k = s_1\overline{n}_1 + \hat{\Phi}_1(b_1\overline{u}_1 + s_2\overline{n}_2) + \hat{\Phi}_2(b_2\overline{u}_1 + s_3\overline{n}_3) + \hat{\Phi}_3(b_3\overline{u}_1 + s_4\overline{n}_4) + \cdots \\
&\qquad + \hat{\Phi}_{k-1}(b_{k-1}\overline{u}_1 + s_k\overline{n}_k)
\end{aligned} \tag{8.18}$$

Equation (8.18) is obtained from Eq. (8.17) by using the following equalities.

$$\hat{C}_k\overline{u}_1 = \hat{\Phi}_k e^{\tilde{u}_1\gamma_k}\overline{u}_1 = \hat{\Phi}_k\overline{u}_1 \tag{8.19}$$

$$\hat{C}_k\overline{u}_3 = \hat{\Phi}_k e^{\tilde{u}_1\gamma_k}\overline{u}_3 = \hat{\Phi}_k\overline{n}_k = \hat{\Phi}_{k-1}\overline{n}_k \tag{8.20}$$

As for the end-effector, it is convenient to use the following definitions.

$$\hat{C} = \hat{C}_m = \hat{C}^{(0,m)} : \ \text{Orientation matrix of the end-effector, i.e.}\ \mathcal{L}_m \tag{8.21}$$

$$\overline{p} = \overline{r}_P = \overline{r}_m = \overline{r}_{0,m}^{(0)} : \ \text{Location matrix of the tip point}\ P = O_m \tag{8.22}$$

$$\overline{r} = \overline{r}_R = \overline{r}_{m-1} = \overline{r}_{0,m-1}^{(0)} : \ \text{Location matrix of the wrist point}\ R = O_{m-1} \tag{8.23}$$

According to the assumption that the end-effector is regular such as a gripper shown in Figure 7.6, the column matrices $\overline{r}$ and $\overline{p}$ are related to each other as follows:

$$\begin{aligned}
&\overline{r}_{OP}^{(0)} = \overline{r}_{OR}^{(0)} + \overline{r}_{RP}^{(0)} = \overline{r}_{0,m}^{(0)} + d_m\hat{C}^{(0,m)}\overline{u}_a^{(m)} = \overline{r}_{0,m}^{(0)} + d_m\hat{C}^{(0,m)}\overline{u}_3^{(m/m)} \Rightarrow \\
&\overline{p} = \overline{r} + d_m\hat{C}\overline{u}_3
\end{aligned} \tag{8.24}$$

In Eq. (8.24), d_m is the length of the end-effector between the wrist and tip points. Since it is a length measured along $\vec{u}_3^{(m)}$, it is conceived as an offset and called a *tip point offset*.

By using the previous orientation and location equations derived for a link $\mathcal{L}_k$, the following equations can be written for the position matrices $\hat{C}$ and $\bar{r}$ of the end-effector.

$$\hat{C} = \hat{\Phi}_m e^{\tilde{u}_1 \gamma_m} = (e^{\tilde{n}_1 \theta_1} e^{\tilde{n}_2 \theta_2} \cdots e^{\tilde{n}_m \theta_m}) e^{\tilde{u}_1 \gamma_m} \tag{8.25}$$

$$\begin{aligned} \bar{r} = {} & \hat{C}_1(s_1\bar{u}_3 + b_1\bar{u}_1) + \hat{C}_2(s_2\bar{u}_3 + b_2\bar{u}_1) + \hat{C}_3(s_3\bar{u}_3 + b_3\bar{u}_1) + \cdots \\ & + \hat{C}_{m-2}(s_{m-2}\bar{u}_3 + b_{m-2}\bar{u}_1) + s_{m-1}\hat{C}_{m-1}\bar{u}_3 \end{aligned} \tag{8.26}$$

$$\begin{aligned} \bar{r} = {} & s_1\bar{n}_1 + \hat{\Phi}_1(b_1\bar{u}_1 + s_2\bar{n}_2) + \hat{\Phi}_2(b_2\bar{u}_1 + s_3\bar{n}_3) + \hat{\Phi}_3(b_3\bar{u}_1 + s_4\bar{n}_4) + \cdots \\ & + \hat{\Phi}_{m-2}(b_{m-2}\bar{u}_1 + s_{m-1}\bar{n}_{m-1}) \end{aligned} \tag{8.27}$$

Note that, once $\hat{C}$ and $\bar{r}$ are obtained as expressed by Eqs. (8.25)–(8.27), $\bar{p}$ becomes readily available as expressed by the following modified version of Eq. (8.24).

$$\bar{p} = \bar{r} + d_m \hat{\Phi}_{m-1} \bar{n}_m \tag{8.28}$$

At first glance, the separate orientation and location equations derived above may appear to be lengthy and complicated. However, for almost all the practicable manipulators, the kinematic parameters happen to be such that these equations can be simplified considerably by means of the necessary symbolic manipulations carried out by using the algebraic properties of the rotation matrices presented in Chapter 2. Then, the simplified equations can be used conveniently for the subsequent analytical purposes, such as obtaining the inverse kinematic solution and carrying out the accompanying multiplicity and singularity analyses. They are also convenient to differentiate in order to obtain the velocity and acceleration equations. On the other hand, the simplified equations are naturally tailored to the particular manipulator of concern. In other words, manipulators with different configurations (i.e. with different joint types and joint arrangements) will have different simplified equations. Nevertheless, even though the simplified equations happen to be specific to the manipulator of concern, they turn out to be so handy that they can be used conveniently even for the computational purposes. Such simplified forward kinematic equations will be obtained for several serial manipulators in Chapter 9.

8.4 Manipulators with or without Spherical Wrists

A manipulator with a spherical wrist is such that its hand is connected to its arm with three successive revolute joints, whose axes intersect each other at a single point R, which is defined as the wrist point. Here, as usual, the hand is defined as the part of the manipulator beyond the wrist point and the arm is defined as the part of the manipulator up to the wrist point. The end-effector is integrated with the last link.

A spherical wrist and a nonspherical wrist are illustrated in Figure 8.1. The wrist on the left-hand side is spherical because the axes of the joints $\mathcal{J}_m$, $\mathcal{J}_{m-1}$, and $\mathcal{J}_{m-2}$ intersect each other at the same point R, which is the wrist point as defined before. On the other hand, the wrist on the right-hand side is nonspherical because the axes of the joints $\mathcal{J}_{m-1}$ and $\mathcal{J}_{m-2}$ do not even intersect each other. Yet, for a nonspherical wrist, the point R, which happens to be the center of the penultimate revolute joint $\mathcal{J}_{m-1}$ that precedes the tip point P, is still defined as the wrist point.

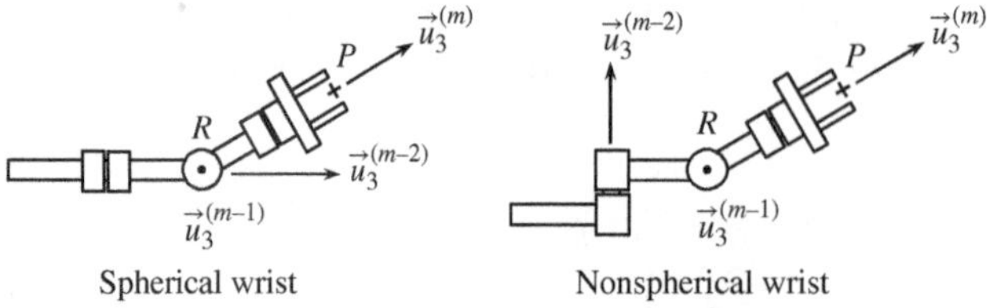

Figure 8.1 A spherical wrist and a nonspherical wrist.

In a manipulator with a spherical wrist, the wrist comprises the last three joints, which are called *wrist joints*. In such a manipulator, the arm comprises the first $m-3$ joints, which are called *arm joints*. The most prominent feature of such a manipulator is that the location and orientation functions of the wrist and arm joints are separated as explained below.

The wrist joints do not have any effect on the location of the wrist point with respect to the base frame. Their sole function is to orient the end-effector with respect to the base frame.

The main function of the arm joints is to locate the wrist point with respect to the base frame. However, as the arm joints function to locate the wrist point, the revolute joints among them also impart a certain orientation to the last link of the arm (i.e. $\mathcal{L}_{m-3}$) with respect to the base frame. Then, the wrist joints actually function to rotate the end-effector (i.e. $\mathcal{L}_m$) with respect to $\mathcal{L}_{m-3}$ about the wrist point in order to provide an overall orientation for $\mathcal{L}_m$ with respect to the base frame. In other words, the overall orientation of the end-effector with respect to the base fame is naturally determined by all the revolute joints of the manipulator.

On the other hand, in a manipulator without a spherical wrist, the functions of the joints cannot be separated distinctly as explained above. In other words, the location of the wrist point with respect to the base frame is determined not only by the first $m-3$ joints but also by the antepenultimate joint $\mathcal{J}_{m-2}$.

Besides, in some special configurations such that the approach vector $\vec{u}_a$ of the end-effector is not aligned with the axis vector $\vec{u}_3^{(m)}$ of the last joint $\mathcal{J}_m$, even the penultimate joint $\mathcal{J}_{m-1}$ becomes effective on the location of the wrist point R. Such a configuration is illustrated in Figure 8.2. Note that, in such a configuration, the wrist point R happens to be located at O_m instead of the usual O_{m-1}.

The preceding statements about the spherical and nonspherical wrists can be expressed by the following equations that show the functional relationships between the joint variables and the orientation of the end-effector and the location of the wrist point with respect to the base frame.

For any manipulator,

$$\hat{C} = \hat{F}(q_1, q_2, \ldots, q_{m-3}, q_{m-2}, q_{m-1}, q_m) \tag{8.29}$$

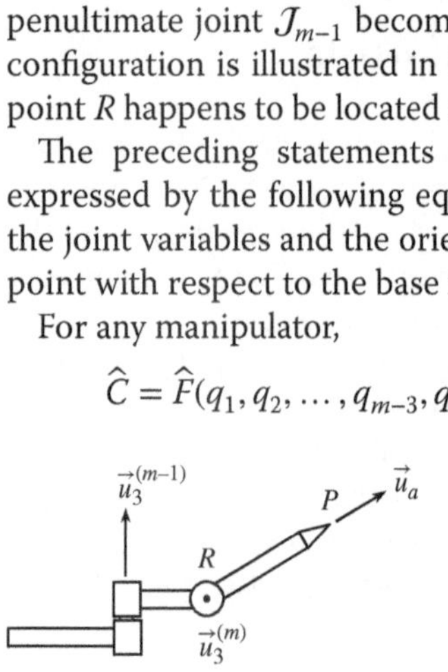

Figure 8.2 A pencil-like end-effector.

For a manipulator with a general nonspherical wrist,

$$\overline{r} = \overline{f}(q_1, q_2, \dots, q_{m-3}, q_{m-2}, q_{m-1}) \tag{8.30}$$

For a manipulator with a spherical wrist,

$$\overline{r} = \overline{f}(q_1, q_2, \dots, q_{m-3}) \tag{8.31}$$

In the preceding equations, q_k is used as a general symbol for the joint variable that belongs to the joint $\mathcal{J}_k$. It is defined as follows:

$$q_k = \begin{cases} \theta_k & \text{if } \mathcal{J}_k \text{ is revolute} \\ s_k & \text{if } \mathcal{J}_k \text{ is prismatic} \end{cases} \tag{8.32}$$

As a note concerning Eq. (8.29), q_k will naturally be absent in the expression of $\widehat{C}$ if $\mathcal{J}_k$ is a prismatic joint. As another note concerning Eq. (8.30), q_{m-1} will be absent in the expression of $\overline{r}$ for a usually encountered end-effector such that $\vec{u}_a = \vec{u}_3^{(m)}$.

8.5 Inverse Kinematics

Inverse kinematics is a mapping from the task space $\mathcal{S}_T$ to the joint space $\mathcal{S}_J$ and as such it is the inverse of the mapping defined previously as forward kinematics. Inverse kinematics is primarily concerned with the process of finding the joint variables corresponding to a specified position of the end-effector. The result of this primary process is called an *inverse kinematic solution*. Inverse kinematics is also concerned with the identification and analysis of the *multiplicities* and *singularities* that occur in the inverse kinematic solution. The specification set required for the inverse kinematics consists of the specified orientation of the end-effector (i.e. the matrix $\widehat{C}$) and the specified location of the tip point (i.e. the column matrix $\overline{p}$) with respect to the base frame. This specification set is the outcome of a *motion planning process* carried out in order to perform a desired task in the task space $\mathcal{S}_T$.

The column matrix $\overline{p}$ is specified in general by specifying the three rectangular coordinates (p_1, p_2, p_3) of the tip point P. In other words, the specification for $\overline{p}$ is presented as

$$\overline{p} = \overline{u}_1 p_1 + \overline{u}_2 p_2 + \overline{u}_3 p_3 \tag{8.33}$$

As for the matrix $\widehat{C}$, it is usually specified by specifying the three Euler angles (ϕ_1, ϕ_2, ϕ_3) according to a selected sequence. In practice, the most commonly selected sequences are 1-2-3 and 3-2-3. In other words, the specification for $\widehat{C}$ is usually presented in one of the following forms.

(i) Specification according to the 1-2-3 sequence:

$$\widehat{C} = \widehat{R}_1(\phi_1)\widehat{R}_2(\phi_2)\widehat{R}_3(\phi_3) = e^{\tilde{u}_1\phi_1}e^{\tilde{u}_2\phi_2}e^{\tilde{u}_3\phi_3} \tag{8.34}$$

(ii) Specification according to the 3-2-3 sequence:

$$\widehat{C} = \widehat{R}_3(\phi_1)\widehat{R}_2(\phi_2)\widehat{R}_3(\phi_3) = e^{\tilde{u}_3\phi_1}e^{\tilde{u}_2\phi_2}e^{\tilde{u}_3\phi_3} \tag{8.35}$$

However, in some particular tasks, instead of specifying three independent orientation parameters, it may be more convenient to specify $\widehat{C}$ by specifying the approach vector, i.e. $\overline{u}_a = \overline{u}_3^{(m/0)}$, and one of the normal and side vectors, i.e. either $\overline{u}_n = \overline{u}_1^{(m/0)}$ or $\overline{u}_s = \overline{u}_2^{(m/0)}$. Since the end-effector frame is right-handed, the unspecified one of the normal or side vectors becomes readily available so that either $\overline{u}_s = \widetilde{u}_a\overline{u}_n$ or $\overline{u}_n = \widetilde{u}_s\overline{u}_a$. In such a task, the specification for $\widehat{C}$ is presented column by column as shown below.

$$\widehat{C} = \left[\overline{u}_n \ \overline{u}_s \ \overline{u}_a\right] \tag{8.36}$$

If the matrices $\widehat{C}$ and $\overline{p}$ are specified completely with all of their elements, then the wrist-point location matrix, i.e. the column matrix $\overline{r}$, also becomes available as readily provided by Eq. (8.24). That is,

$$\overline{r} = \overline{p} - d_m\widehat{C}\overline{u}_3 \tag{8.37}$$

Owing to Eq. (8.37), the pair $\widehat{C}$ and $\overline{r}$ can be treated as the specified data instead of the pair $\widehat{C}$ and $\overline{p}$. This way, the inverse kinematic solution can be facilitated, because the dependence of $\overline{r}$ on the joint variables is in general simpler than that of $\overline{p}$. Moreover, for the special but common class of manipulators with spherical wrists, the matrix $\overline{r}$ depends only on the first $m-3$ of the joint variables and this feature facilitates the inverse kinematic solution to a considerable extent.

8.6 Inverse Kinematic Solution for a Regular Manipulator

A serial manipulator is defined to be *regular* if

$$m = \mu \tag{8.38}$$

In Eq. (8.38), m is the number of the joints and μ is the mobility required for the end-effector in order to execute the planned task.

For example, if a serial manipulator with m joints is to be used to change the location and orientation of an arbitrary object arbitrarily in the three-dimensional space, then the mobility required for the end-effector will be $\mu = 6$. For such a task, the manipulator will be *regular* if $m = \mu = 6$. It will be *redundant* if $m > \mu = 6$ and it will be *deficient* if $m < \mu = 6$.

On the other hand, if the above-mentioned manipulator is to be used to change the location of a uniform spherical object arbitrarily in the three-dimensional space, then the mobility required for the end-effector will be $\mu = 3$. For this task, the manipulator with $m = 6$ will be redundant. It would become regular if the variables of its three selected joints were specified so that its degree of freedom would reduce to $m' = m - 3 = \mu = 3$.

In this section, the inverse kinematic solution will be outlined for a generic six-joint spatial manipulator, which is used for a general task such that $\mu = 6$. The outline will be explained for two possible configurations. In the first configuration, the wrist is spherical, whereas it is nonspherical in the second configuration.

8.6.1 Regular Manipulator with a Spherical Wrist

For a regular manipulator with a spherical wrist, Eqs. (8.29) and (8.31) are written as follows:

$$\widehat{C} = \widehat{F}(q_1, q_2, q_3; q_4, q_5, q_6) \tag{8.39}$$

$$\overline{r} = \overline{f}(q_1, q_2, q_3) \tag{8.40}$$

Since Eq. (8.40) comprises three scalar equations, it can be solved for the arm variables, i.e. the first three joint variables q_1, q_2, and q_3. The solution will not be unique in general and it can be indicated as follows, if the manipulator is not in a pose of *position singularity* associated with the wrist point:

$$q_k = g_k^{(i)}(r_1, r_2, r_3); \quad k = 1, 2, 3; \quad i = 1, 2, \ldots \tag{8.41}$$

In Eq. (8.41), i is the *multiplicity index.* It indicates one of the multiple poses that the manipulator assumes corresponding to the same location of the wrist point. These poses belong to one of the *posture modes* characterized by each i. The posture modes are usually designated with humanoid terms such as *right shouldered, left shouldered, elbow-up, elbow-down,* etc. The posture modes of some selected manipulators can be seen in Chapter 9.

Depending on the configuration of the manipulator, i.e. the types and arrangements of the joints, the solution can be obtained analytically, semi-analytically, or numerically. However, for most of the practicable manipulators, it is fortunate that the solution can be obtained analytically so that the functions in Eq. (8.41) consist of closed-form expressions for all k and i.

As a reminding note, a closed-form expression consists of a finite number of the elementary functions of the involved variables. Concerning a manipulator, the elementary functions encountered in an analytical solution consist of expressions that contain a finite number of terms with sine and cosine functions, double-argument arctangent functions, square roots of such expressions, and reciprocals of such expressions. Therefore, the multiplicities are caused by the sign ambiguities associated with the square roots and the singularities are caused by the reciprocals whenever they turn into vanishing denominators.

If Eq. (8.40) can be solved analytically, then the posture modes can also be indicated by *independent* sign variables such as $\sigma_j = \pm 1$ for $j = 1, 2$. They are called *posture mode sign variables.* Here, it is to be pointed out that Eq. (8.40) leads to at most two posture mode sign variables for a spatial manipulator. This fact is evidenced for several manipulators considered in Chapter 9. By using the posture mode sign variables, Eq. (8.41) can be rendered as

$$q_k = g_k(r_1, r_2, r_3; \sigma_1, \sigma_2); \quad k = 1, 2, 3 \tag{8.42}$$

The posture modes are initially selected according to the desired task and the neighboring obstacles (if there are any). Once the manipulator is started with a selected posture mode, it keeps operating in the same mode, unless it reaches a critical pose, which is called a *posture mode changing pose.* In such a pose, the posture of the manipulator can be changed easily from one mode to its alternative without effecting the location of the wrist point. For example, it can be changed from one of the elbow-up and elbow-down modes to the other, when the two links connected by the elbow joint (i.e. the revolute joint at the elbow point) happen to be aligned. This phenomenon is demonstrated for the applicable ones of the manipulators considered in Chapter 9.

After finding the arm variables (q_1, q_2, q_3) as explained above, the matrix function $\hat{F}$ in Eq. (8.39) can be factorized as shown below.

$$\hat{C} = \hat{F}(q_1, q_2, q_3; q_4, q_5, q_6) = \hat{C}'\hat{C}^* e^{\tilde{u}_1 \gamma_6} \tag{8.43}$$

This factorization is based on Eq. (8.25), which is adapted to Eq. (8.43) as follows:

$$\hat{C} = \hat{\Phi}_6 e^{\tilde{u}_1 \gamma_6} = (e^{\tilde{n}_1 \epsilon_1 \theta_1} e^{\tilde{n}_2 \epsilon_2 \theta_2} e^{\tilde{n}_3 \epsilon_3 \theta_3})(e^{\tilde{n}_4 \theta_4} e^{\tilde{n}_5 \theta_5} e^{\tilde{n}_6 \theta_6}) e^{\tilde{u}_1 \gamma_6} \tag{8.44}$$

In Eq. (8.44), $\varepsilon_k = 1$ if $\mathcal{J}_k$ is a revolute joint and $\varepsilon_k = 0$ if $\mathcal{J}_k$ is a prismatic joint for $k = 1, 2, 3$. By using the factors indicated in Eq. (8.44), Eq. (8.43) can be rearranged as shown below in order to prepare it for finding the wrist joint variables, i.e. $q_4 = \theta_4$, $q_5 = \theta_5$, and $q_6 = \theta_6$.

$$e^{\tilde{n}_4\theta_4} e^{\tilde{n}_5\theta_5} e^{\tilde{n}_6\theta_6} = \hat{C}^* \tag{8.45}$$

In Eq. (8.45), $\hat{C}^*$ is known as a combination of the available matrices $\hat{C}$ and $\hat{C}'$. That is,

$$\hat{C}^* = (\hat{C}')^{-1}\hat{C}e^{-\tilde{u}_1\gamma_6} = e^{-\tilde{n}_3\varepsilon_3\theta_3} e^{-\tilde{n}_2\varepsilon_2\theta_2} e^{-\tilde{n}_1\varepsilon_1\theta_1} \hat{C}e^{-\tilde{u}_1\gamma_6} \tag{8.46}$$

Equation (8.45) implies that the angular variables θ_4, θ_5, and θ_6 can be extracted from $\hat{C}^*$ similarly as the Euler angles are extracted. If the sequence appearing in Eq. (8.45) is not singular, the extracted angles can be expressed as follows with another sign variable σ_3:

$$\theta_k = g_k(c^*_{11}, \ldots, c^*_{33}; \sigma_3); \quad \sigma_3 = \pm 1; \quad k = 4, 5, 6 \tag{8.47}$$

The sign variable σ_3 in Eq. (8.47) leads to two alternative orientation modes of the end-effector. However, as evidenced in Chapter 9, these modes are not visually distinguishable from each other if the wrist is spherical. So, for a spherical wrist, σ_3 can be taken as $\sigma_3 = +1$ without loss of generality.

As for the *position singularities*, there are two major kinds, which are associated with the wrist point location and the end-effector orientation.

A *wrist-point location singularity* occurs, if $\overline{r} = \overline{r}^\circ$, where some or all components of $\overline{r}^\circ$ have certain critical values. In this singularity, one or two of the arm joint variables (q_1, q_2, q_3) become *indefinite* and *ineffective*. In other words, they may be changed arbitrarily without effecting the critical components of $\overline{r}^\circ$. Nevertheless, the remaning ones of the arm joint variables can still be found from Eq. (8.40).

An *end-effector orientation singularity* occurs, if $\hat{C} = \hat{C}^\circ$, where the elements of $\hat{C}^\circ$ have certain critical values. In this singularity, regarding Eq. (8.45), the intermediate angle θ_5 also assumes an associated critical value θ_5° such that the axes of the angles θ_4 and θ_6 become aligned and therefore θ_4 and θ_6 become indistinguishable from each other. This can be seen in the following equation, which is derived from Eq. (8.45) with $\theta_5 = \theta_5^\circ$ by using the *shifting property* of the rotation matrices.

$$e^{\tilde{n}_4\theta_4} e^{\tilde{n}_5\theta_5^\circ} e^{\tilde{n}_6\theta_6} = e^{\tilde{n}_4\theta_4} e^{\sigma'\tilde{n}_4\theta_6} e^{\tilde{n}_5\theta_5^\circ} = e^{\tilde{n}_4(\theta_4+\sigma'\theta_6)} e^{\tilde{n}_5\theta_5^\circ} = \hat{C}^* \tag{8.48}$$

As noted above, θ_5° has such a critical value that $e^{\tilde{n}_5\theta_5^\circ}\overline{n}_6 = \sigma'\overline{n}_4$, where $\sigma' = \pm 1$. As also noted, θ_4 and θ_6 become *indefinite*, as a typical feature of the singularity, because they cannot be found separately from Eq. (8.48). Therefore, they may take arbitrary values, but their values happen to be ineffective in causing any change in $\hat{C}$. On the other hand, the arbitrary values they may take are not independent of each other because their combination $(\theta_4 + \sigma'\theta_6)$ can still be found with a definite value from the following version of Eq. (8.48).

$$e^{\tilde{n}_4(\theta_4+\sigma'\theta_6)} = \hat{C}^\# = \hat{C}^* e^{-\tilde{n}_5\theta_5^\circ} \tag{8.49}$$

The position singularities of several manipulators are discussed in detail in Chapter 9.

8.6.2 Regular Manipulator with a Nonspherical Wrist

For such a manipulator, Eq. (8.29) and the two versions of Eq. (8.30) that correspond to two possible nonspherical wrist configurations are written as follows:

$$\hat{C} = \hat{F}(q_1, q_2, q_3, q_4, q_5, q_6) \tag{8.50}$$

For a nonspherical wrist of the first kind such that $\vec{u}_a = \vec{u}_3^{(m)}$,

$$\bar{r} = \bar{f}(q_1, q_2, q_3, q_4) \tag{8.51}$$

For a nonspherical wrist of the second kind such that $\vec{u}_a \neq \vec{u}_3^{(m)}$,

$$\bar{r} = \bar{f}(q_1, q_2, q_3, q_4, q_5) \tag{8.52}$$

For some special manipulators with particularly arranged parallel and/or coincident joint axes, it may be possible to obtain analytical solutions to Eq. Pairs (8.50)–(8.51) and (8.50)–(8.52). Examples of such manipulators are provided in Chapter 9. However, for a general manipulator with a nonspherical wrist, Eq. Pairs (8.50)–(8.51) and (8.50)–(8.52) can have only semi-analytical solutions.

The semi-analytical inverse kinematic solutions can be classified into three categories designated as *first order*, *second order*, and *third order*. In the first stage of a kth-order semi-analytical solution, closed-form expressions can be obtained for the $k' = 6 - k$ of the variables in terms of the selected k of the variables. Then, in the second stage, the selected k variables are found iteratively by using a suitable numerical method in order to satisfy the k rather complicated scalar equations associated with them. These equations arise usually as some consistency equations.

A first-order semi-analytical solution to Eq. Pair (8.50)–(8.51) and a second-order semi-analytical solution to Eq. Pair (8.50)–(8.52) are outlined below as two typical examples.

(a) First-order semi-analytical solution to Eq. Pair (8.50)–(8.51):

Suppose that Eq. (8.51) can be solved analytically so that

$$q_k = g_k(q_4); \quad k = 1, 2, 3 \tag{8.53}$$

Then, Eq. (8.50) can be written as follows similarly to Eq. (8.45):

$$e^{\tilde{n}_4 \theta_4} e^{\tilde{n}_5 \theta_5} e^{\tilde{n}_6 \theta_6} = \hat{C}^*(q_4) \tag{8.54}$$

Since Eq. (8.54) is a rotation matrix equation, it can also be solved analytically so that

$$q_k = g_k(q_4) \text{ for } k = 4, 5, 6 \tag{8.55}$$

In the preceding solutions, it is assumed that a position singularity does not occur, and the posture and orientation modes have already been selected. Moreover, the known arguments (i.e. the elements of $\bar{r}$ and $\hat{C}$ and the selected sign variables) are not shown for the sake of brevity.

Note that Eq. (8.55) gives q_4 in terms of itself. Thus, the following consistency equation arises.

$$q_4 = g_4(q_4) \tag{8.56}$$

Since Eq. (8.56) is quite complicated, the value of q_4 can be found from Eq. (8.56) only by using a suitable iterative numerical method. Afterwards, the values of the other variables will be readily provided by Eqs. (8.53) and (8.55).

(b) Second-order semi-analytical solution to Eq. Pair (8.50)–(8.52):

Suppose that Eq. (8.52) can also be solved analytically so that

$$q_k = g_k(q_4, q_5) \;;\; k = 1, 2, 3 \tag{8.57}$$

Then, Eq. (8.50) can be written as follows similarly to Eq. (8.54):

$$e^{\tilde{n}_4\theta_4}e^{\tilde{n}_5\theta_5}e^{\tilde{n}_6\theta_6} = \hat{C}^*(q_4, q_5) \tag{8.58}$$

Since Eq. (8.58) is a rotation matrix equation, it can again be solved analytically so that

$$q_k = g_k(q_4, q_5) \;;\; k = 4, 5, 6 \tag{8.59}$$

In the preceding solutions, it is again assumed that a position singularity does not occur, and the posture and orientation modes have already been selected. Moreover, the known arguments are not again shown for the sake of brevity.

Note that Eq. (8.59) gives q_4 and q_5 in terms of themselves. Thus, the following consistency equations arise.

$$q_4 = g_4(q_4, q_5) \tag{8.60}$$

$$q_5 = g_5(q_4, q_5) \tag{8.61}$$

Since Eqs. (8.60) and (8.61) are coupled and quite complicated, the values of q_4 and q_5 can be found from them only by using a suitable iterative numerical method. Afterwards, the values of the other variables will be readily provided by Eqs. (8.57) and (8.59).

8.7 Inverse Kinematic Solution for a Redundant Manipulator

A serial manipulator is defined to be *redundant* if

$$m > \mu \tag{8.62}$$

For a redundant manipulator, the inverse kinematic solution can be obtained by using one of the two methods that are explained below.

8.7.1 Solution by Specifying the Variables of Certain Joints

In this method, the redundant manipulator is turned into a regular manipulator by specifying the variables of appropriately selected $r = m - \mu$ of its m joints. The selected joint or joints and/or the values assigned to their variables may be changed from time to time during the execution of the required task in order to increase the operational dexterity of the manipulator. Once the r joints and their specified variables are selected, the inverse kinematic solution can then be obtained similarly as described in Section 8.6 for a regular manipulator with six joints.

In a special task of manipulating uniform spherical objects with a six-joint manipulator, it happens that $m = 6 > \mu = 3$. In such a case, the orientation of the end-effector is left free, i.e. the matrix $\hat{C}$ is not specified. What is specified is only the location matrix $\overline{p}$ of the tip point, which is assumed to be coincident with the center of the manipulated spherical object. On the other hand, according to Eq. (8.28),

$$\overline{p} = \overline{\phi}(q_1, q_2, q_3, q_4, q_5) \tag{8.63}$$

Let the last three joints be selected as the joints with specified variables and let their variables be kept fixed at some values such as q_4°, q_5°, and q_6°. With these values, Eq. (8.63) becomes

$$\overline{p} = \overline{\phi}(q_1, q_2, q_3, q_4^\circ, q_5^\circ) \tag{8.64}$$

Assuming that no position singularity is encountered, the joint variables of the operational joints can then be found from Eq. (8.64) with appropriately selected posture modes. That is,

$$q_k = g_k(p_1, p_2, p_3, q_4^\circ, q_5^\circ); \;\; k = 1, 2, 3 \tag{8.65}$$

As a by-product of the above solution, the unspecified orientation matrix is also obtained as follows depending on the specified location of the tip point:

$$\hat{C} = \hat{F}(q_1, q_2, q_3, q_4^\circ, q_5^\circ, q_6^\circ) = \hat{G}(p_1, p_2, p_3, q_4^\circ, q_5^\circ, q_6^\circ) \tag{8.66}$$

8.7.2 Solution by Optimization

In this method, a performance function U_p of the joint variables is required to be minimized in addition to the required manipulation task. For example, U_p can be taken as the potential energy of the manipulator. Such a function is shown below generically with its arguments.

$$U_p = U_p(q_1, q_2, \dots, q_{m-3}, q_{m-2}, q_{m-1}, q_m) \tag{8.67}$$

The minimization of the function U_p is naturally subject to the following constraints, which are the forward kinematic equations that must be satisfied to attain the specified matrices $\hat{C}$ and $\overline{r}$.

$$\hat{F}(q_1, q_2, \dots, q_{m-3}, q_{m-2}, q_{m-1}, q_m) - \hat{C} = \hat{0} \tag{8.68}$$

$$\overline{f}(q_1, q_2, \dots, q_{m-3}, q_{m-2}, q_{m-1}) - \overline{r} = \overline{0} \tag{8.69}$$

Equations (8.68) and (8.69) provide the following set of six independent scalar constraint equations for $i = 1, 2, 3$.

$$f_i(q_1, q_2, \dots, q_{m-3}, q_{m-2}, q_{m-1}) - r_i = 0 \tag{8.70}$$

$$F_{ii}(q_1, q_2, \dots, q_{m-3}, q_{m-2}, q_{m-1}, q_m) - C_{ii} = 0 \tag{8.71}$$

As noticed above, Eq. (8.71) involves only the diagonal elements of $\hat{F}$ and $\hat{C}$. The reason for taking only the diagonal elements is that $\hat{F}$ and $\hat{C}$ are orthonormal matrices and therefore each of them contains only three independent elements. Their independent elements are taken here as their diagonal elements for the sake of regularity. However, taking the diagonal elements is not a necessity. This is because, for a specific manipulator of concern, some of the off-diagonal elements may be preferred owing to the convenience that they depend on the joint variables with simpler expressions.

In order to obtain the optimal solution for the joint variables, the function U_p and the scalar constraint equations can be combined into an augmented function U_p^* as follows:

$$U_p^* = U_p + \sum_{i=1}^{3}[\lambda_i(f_i - r_i) + \lambda_i'(F_{ii} - C_{ii})] \tag{8.72}$$

In Eq. (8.72), λ_i and λ_i' are the relevant *Lagrange multipliers.*

The optimality condition leads to the following set of equations written for $j = 1, 2, 3, \ldots, m$.

$$\partial U_p^*/\partial q_j = \partial U_p/\partial q_j + \sum_{i=1}^{3}[\lambda_i(\partial f_i/\partial q_j) + \lambda_i'(\partial F_{ii}/\partial q_j)] = 0 \tag{8.73}$$

Equation Sets (8.70), (8.71), and (8.73) constitute a complete set of $m^* = m + 6$ scalar equations that contain $m^* = m + 6$ unknowns, which are the m joint variables (q_1, q_2, …, q_m) and the six Lagrange multipliers ($\lambda_1, \lambda_2, \lambda_3; \lambda_1', \lambda_2', \lambda_3'$). Thus, these m^* scalar equations can be solved to find the optimal values of the joint variables together with the corresponding values of the Lagrange multipliers.

8.8 Inverse Kinematic Solution for a Deficient Manipulator

A serial manipulator is defined to be *deficient* if

$$m < \mu \leq 6 \tag{8.74}$$

For a deficient manipulator with a regular end-effector such that $\vec{u}_3^{(m)} = \vec{u}_a$ as shown in Figure 8.1, the following forward kinematic equations can be written for the tip point location and the end-effector orientation.

$$\overline{p} = \overline{\phi}(q_1, q_2, q_3, \ldots, q_{m-3}, q_{m-2}, q_{m-1}) \tag{8.75}$$

$$\hat{C} = \hat{F}(q_1, q_2, q_3, \ldots, q_{m-3}, q_{m-2}, q_{m-1}, q_m) \tag{8.76}$$

With a deficient manipulator, even if the intention is to have μ for the task mobility, it is not possible to realize this intention. Therefore, it is necessary to have a compromise by reducing the required task mobility from μ to $\mu' = m$. This compromise makes the manipulator regular and thus the joint variables can be found so as to satisfy the μ' specifications (instead of μ) for the position (location and orientation) of the end-effector.

For a deficient manipulator, the above-mentioned compromise can be made in one of the two ways that are explained below.

8.8.1 Compromise in Orientation in Favor of a Completely Specified Location

According to this compromise, three of the joint variables (e.g. q_1, q_2, and q_3) are obtained from Eq. (8.75) as functions of the specified $\overline{p}$ and the remaining joint variable or variables, i.e. q_4 if $m = 4$ or q_4 and q_5 if $m = 5$. This solution leaves $\mu_e = m - 3 = \mu' - 3$ degrees of freedom for the orientation of the end-effector. Note that $\mu_e = 1$ or $\mu_e = 2$. This result implies that the matrix $\hat{C}$ cannot be specified completely, i.e. its three independent parameters cannot all be specified. Nevertheless, its μ_e parameter or parameters can still be specified. For example, if $m = 4$, then $\mu_e = 1$ and the only joint variable left for $\widehat{C}$ is $q_4 = \theta_4$. Due to this fact, the orientation of the end-effector can only be partially controlled by means of θ_4, which can be found in correspondence to the consistently specified single parameter of $\widehat{C}$. The other two unspecified independent parameters of $\widehat{C}$ can then be found from Eq. (8.76) as part of the solution.

8.8.2 Compromise in Location in Favor of a Completely Specified Orientation

According to this compromise, three of the joint variables (e.g. q_a, q_b, and q_c) are obtained from Eq. (8.76) as functions of the specified $\widehat{C}$ and the remaining joint variable or variables, i.e. q_d if $m = 4$ or q_d and q_e if $m = 5$. This solution leaves $\mu_e = m - 3 = \mu' - 3$ degrees of freedom for the location of the tip point. Note that $\mu_e = 1$ or $\mu_e = 2$. This result implies that the column matrix $\overline{p}$ cannot be specified completely, i.e. the three coordinates of the tip point cannot all be specified. Nevertheless, its μ_e coordinate or coordinates can still be specified. For example, if $m = 4$, then $\mu_e = 1$ and the only joint variable left for $\overline{p}$ is q_d. Due to this fact, the location of the tip point can only be partially controlled by means of q_d, which can be found in correspondence to the consistently specified single coordinate of the tip point. The other two unspecified coordinates of $\overline{p}$ can then be found from Eq. (8.75) as part of the solution.

8.9 Forward Kinematics of Motion

8.9.1 Forward Kinematics of Velocity Relationships

The forward kinematic relationships for the angular velocity of the link $\mathcal{L}_k$ and the translational velocity of the link frame origin O_k can be obtained by using the derivatives of the matrices $\widehat{C}_k = \widehat{C}^{(0,k)}$ and $\overline{r}_k = \overline{r}_{0,k}^{(0)}$, whose expressions are given in Eqs. (8.12) and (8.18).

The *angular velocity* of the link $\mathcal{L}_k$ can be obtained from $\widehat{C}_k$ or $\widehat{\Phi}_k$ as follows:

$$\widetilde{\omega}_{0,k}^{(0)} = \dot{\widehat{C}}^{(0,k)}\widehat{C}^{(k,0)} \Rightarrow \widetilde{\omega}_k = \dot{\widehat{C}}_k\widehat{C}_k^t = (\dot{\widehat{\Phi}}_k e^{\widetilde{u}_1\gamma_k})(\widehat{\Phi}_k e^{\widetilde{u}_1\gamma_k})^t = \dot{\widehat{\Phi}}_k\widehat{\Phi}_k^t \Rightarrow$$

$$\overline{\omega}_k = \mathrm{colm}[\dot{\widehat{C}}_k\widehat{C}_k^t] = \mathrm{colm}[\dot{\widehat{\Phi}}_k\widehat{\Phi}_k^t] \tag{8.77}$$

After Eq. (8.13) of $\widehat{\Phi}_k$ is substituted, Eq. (8.77) can be manipulated into the following equation for the angular velocity of $\mathcal{L}_k$.

$$\overline{\omega}_k = \dot{\theta}_1\overline{n}_1 + \dot{\theta}_2 e^{\widetilde{n}_1\theta_1}\overline{n}_2 + \dot{\theta}_3 e^{\widetilde{n}_1\theta_1}e^{\widetilde{n}_2\theta_2}\overline{n}_3 + \cdots$$
$$+ \dot{\theta}_k e^{\widetilde{n}_1\theta_1}e^{\widetilde{n}_2\theta_2}e^{\widetilde{n}_3\theta_3}\ldots e^{\widetilde{n}_{k-1}\theta_{k-1}}\overline{n}_k \tag{8.78}$$

In particular, let $\overline{\omega}$ denote briefly the angular velocity of the end-effector. That is,

$$\overline{\omega} = \overline{\omega}_m = \overline{\omega}_{0,m}^{(0)} = \mathrm{colm}[\dot{\widehat{\Phi}}_m\widehat{\Phi}_m^t] = \mathrm{colm}[\dot{\widehat{C}}\widehat{C}^t] \tag{8.79}$$

Then, according to Eqs. (8.78) and (8.78),

$$\overline{\omega} = \dot{\theta}_1\overline{n}_1 + \dot{\theta}_2 e^{\widetilde{n}_1\theta_1}\overline{n}_2 + \dot{\theta}_3 e^{\widetilde{n}_1\theta_1}e^{\widetilde{n}_2\theta_2}\overline{n}_3 + \cdots$$
$$+ \dot{\theta}_m e^{\widetilde{n}_1\theta_1}e^{\widetilde{n}_2\theta_2}e^{\widetilde{n}_3\theta_3}\ldots e^{\widetilde{n}_{m-1}\theta_{m-1}}\overline{n}_m \tag{8.80}$$

Equations (8.78) and (8.80) can also be written as follows:

$$\overline{\omega}_k = \dot{\theta}_1\overline{n}_1 + \dot{\theta}_2\widehat{\Phi}_1\overline{n}_2 + \dot{\theta}_3\widehat{\Phi}_2\overline{n}_3 + \cdots + \dot{\theta}_k\widehat{\Phi}_{k-1}\overline{n}_k \tag{8.81}$$

$$\overline{\omega} = \overline{\omega}_m = \dot{\theta}_1\overline{n}_1 + \dot{\theta}_2\widehat{\Phi}_1\overline{n}_2 + \dot{\theta}_3\widehat{\Phi}_2\overline{n}_3 + \cdots + \dot{\theta}_m\widehat{\Phi}_{m-1}\overline{n}_m \tag{8.82}$$

The *translational velocity* of the link frame origin O_k can be obtained from $\overline{r}_k$. That is,

$$\overline{v}^{(0)}_{0,k} = \dot{\overline{r}}^{(0)}_{0,k} \Rightarrow \overline{v}_k = \dot{\overline{r}}_k \tag{8.83}$$

Equation (8.18) of $\overline{r}_k$ can be differentiated so that

$$\begin{aligned}\overline{v}_k &= \dot{\widehat{\Phi}}_1(b_1\overline{u}_1 + s_2\overline{n}_2) + \dot{\widehat{\Phi}}_2(b_2\overline{u}_1 + s_3\overline{n}_3) + \cdots + \dot{\widehat{\Phi}}_{k-1}(b_{k-1}\overline{u}_1 + s_k\overline{n}_k)\\ &\quad + \dot{s}_1\overline{n}_1 + \dot{s}_2\widehat{\Phi}_1\overline{n}_2 + \dot{s}_3\widehat{\Phi}_2\overline{n}_3 + \cdots + \dot{s}_k\widehat{\Phi}_{k-1}\overline{n}_k\end{aligned} \tag{8.84}$$

Note that $\widehat{\Phi}^t_k\widehat{\Phi}_k = \widehat{I}$ and $\dot{\widehat{\Phi}}_k\widehat{\Phi}^t_k = \widetilde{\omega}_k$. Therefore,

$$\dot{\widehat{\Phi}}_k = \dot{\widehat{\Phi}}_k\widehat{\Phi}^t_k\widehat{\Phi}_k = \widetilde{\omega}_k\widehat{\Phi}_k \tag{8.85}$$

Hence, Eq. (8.84) can also be written as follows:

$$\begin{aligned}\overline{v}_k &= \widetilde{\omega}_1\widehat{\Phi}_1(b_1\overline{u}_1 + s_2\overline{n}_2) + \widetilde{\omega}_2\widehat{\Phi}_2(b_2\overline{u}_1 + s_3\overline{n}_3) + \cdots\\ &\quad + \widetilde{\omega}_{k-1}\widehat{\Phi}_{k-1}(b_{k-1}\overline{u}_1 + s_k\overline{n}_k)\\ &\quad + \dot{s}_1\overline{n}_1 + \dot{s}_2\widehat{\Phi}_1\overline{n}_2 + \dot{s}_3\widehat{\Phi}_2\overline{n}_3 + \cdots + \dot{s}_k\widehat{\Phi}_{k-1}\overline{n}_k\end{aligned} \tag{8.86}$$

The velocities of the *wrist point* (R) and the *tip point* (P) can be denoted by the following notations.

$$\overline{w} = \overline{v}_R = \dot{\overline{r}} = \overline{v}_{m-1} = \overline{v}^{(0)}_{0,m-1} = \overline{v}^{(0)}_{R/O} \tag{8.87}$$

$$\overline{v} = \overline{v}_P = \dot{\overline{p}} = \overline{v}_m = \overline{v}^{(0)}_{0,m} = \overline{v}^{(0)}_{P/O} \tag{8.88}$$

By means of Eq. (8.86), $\overline{w}$ can be expressed as shown below.

$$\begin{aligned}\overline{w} &= \widetilde{\omega}_1\widehat{\Phi}_1(b_1\overline{u}_1 + s_2\overline{n}_2) + \widetilde{\omega}_2\widehat{\Phi}_2(b_2\overline{u}_1 + s_3\overline{n}_3) + \cdots\\ &\quad + \widetilde{\omega}_{m-2}\widehat{\Phi}_{m-2}(b_{m-2}\overline{u}_1 + s_{m-1}\overline{n}_{m-1})\\ &\quad + \dot{s}_1\overline{n}_1 + \dot{s}_2\widehat{\Phi}_1\overline{n}_2 + \dot{s}_3\widehat{\Phi}_2\overline{n}_3 + \cdots + \dot{s}_{m-1}\widehat{\Phi}_{m-2}\overline{n}_{m-1}\end{aligned} \tag{8.89}$$

On the other hand, Eqs. (8.24) and (8.28) lead to the following equations for $\overline{v} = \overline{v}_P$.

$$\overline{v} = \overline{w} + d_m\dot{\widehat{C}}\overline{u}_3 = \overline{w} + d_m\widetilde{\omega}\widehat{C}\overline{u}_3 \tag{8.90}$$

$$\overline{v} = \overline{w} + d_m\dot{\widehat{\Phi}}_{m-1}\overline{n}_m = \overline{w} + d_m\widetilde{\omega}_{m-1}\widehat{\Phi}_{m-1}\overline{n}_m \tag{8.91}$$

8.9.2 Forward Kinematics of Acceleration Relationships

The forward kinematic relationships for the angular acceleration of the link $\mathcal{L}_k$ and the translational acceleration of the link frame origin O_k can be obtained by taking the derivatives of the preceding velocity equations. This process leads to the following acceleration equations.

Let $\overline{\alpha}_k = \dot{\overline{\omega}}_k$. Then, upon differentiation, Eq. (8.78) leads to $\overline{\alpha}_k$ as shown below.

$$\begin{aligned}\overline{\alpha}_k &= \ddot{\theta}_1\overline{n}_1 + \ddot{\theta}_2 e^{\widetilde{n}_1\theta_1}\overline{n}_2 + \ddot{\theta}_3 e^{\widetilde{n}_1\theta_1}e^{\widetilde{n}_2\theta_2}\overline{n}_3 + \cdots\\ &\quad + \ddot{\theta}_k e^{\widetilde{n}_1\theta_1}e^{\widetilde{n}_2\theta_2}e^{\widetilde{n}_3\theta_3}\ldots e^{\widetilde{n}_{k-1}\theta_{k-1}}\overline{n}_k + \overline{\alpha}^{\circ}_k\end{aligned} \tag{8.92}$$

Equation (8.92) can also be written in the following more compact form.

$$\overline{\alpha}_k = \ddot{\theta}_1\overline{n}_1 + \ddot{\theta}_2\widehat{\Phi}_1\overline{n}_2 + \ddot{\theta}_3\widehat{\Phi}_2\overline{n}_3 + \cdots + \ddot{\theta}_k\widehat{\Phi}_{k-1}\overline{n}_k + \overline{\alpha}_k^{\circ} \tag{8.93}$$

In Eqs. (8.92) and (8.93), $\overline{\alpha}_k^{\circ}$ is the *velocity-dependent angular acceleration*. It is also called the *angular acceleration bias term*. It consists of the *gyroscopic acceleration* terms that are formed as the binary products of different angular joint variable derivatives. It is defined as follows:

$$\begin{aligned}\overline{\alpha}_k^{\circ} = {} & \dot{\theta}_1\dot{\theta}_2 e^{\tilde{n}_1\theta_1}\tilde{n}_1\overline{n}_2 + \dot{\theta}_1\dot{\theta}_3 e^{\tilde{n}_1\theta_1}\tilde{n}_1 e^{\tilde{n}_2\theta_2}\overline{n}_3 + \dot{\theta}_2\dot{\theta}_3 e^{\tilde{n}_1\theta_1}e^{\tilde{n}_2\theta_2}\tilde{n}_2\overline{n}_3 + \cdots \\ & + \dot{\theta}_1\dot{\theta}_k e^{\tilde{n}_1\theta_1}\tilde{n}_1 e^{\tilde{n}_2\theta_2}e^{\tilde{n}_3\theta_3}\cdots e^{\tilde{n}_{k-1}\theta_{k-1}}\overline{n}_k \\ & + \dot{\theta}_2\dot{\theta}_k e^{\tilde{n}_1\theta_1}e^{\tilde{n}_2\theta_2}\tilde{n}_2 e^{\tilde{n}_3\theta_3}\cdots e^{\tilde{n}_{k-1}\theta_{k-1}}\overline{n}_k + \cdots \\ & + \dot{\theta}_{k-1}\dot{\theta}_k e^{\tilde{n}_1\theta_1}e^{\tilde{n}_2\theta_2}e^{\tilde{n}_3\theta_3}\cdots e^{\tilde{n}_{k-1}\theta_{k-1}}\tilde{n}_{k-1}\overline{n}_k\end{aligned} \tag{8.94}$$

In particular, let $\overline{\alpha}$ denote briefly the angular acceleration of the end-effector. That is,

$$\overline{\alpha} = \overline{\alpha}_m = \overline{\alpha}_{0,m}^{(0)} = \dot{\overline{\omega}} \tag{8.95}$$

Then, $\overline{\alpha}$ can be obtained from Eq. (8.93) as follows with $k = m$ and $\overline{\alpha}^{\circ} = \overline{\alpha}_m^{\circ}$:

$$\overline{\alpha} = \ddot{\theta}_1\overline{n}_1 + \ddot{\theta}_2\widehat{\Phi}_1\overline{n}_2 + \ddot{\theta}_3\widehat{\Phi}_2\overline{n}_3 + \cdots + \ddot{\theta}_m\widehat{\Phi}_{m-1}\overline{n}_m + \overline{\alpha}^{\circ} \tag{8.96}$$

The translational acceleration of the link frame origin O_k can be obtained from Eqs. (8.83) and (8.86). That is,

$$\overline{a}_{0,k}^{(0)} = \dot{\overline{v}}_{0,k}^{(0)} \Rightarrow \overline{a}_k = \dot{\overline{v}}_k \tag{8.97}$$

Equation (8.86) of $\overline{v}_k$ can be differentiated so that

$$\begin{aligned}\overline{a}_k = {} & b_1(\tilde{\alpha}_1 + \tilde{\omega}_1^2)\widehat{\Phi}_1\overline{u}_1 + b_2(\tilde{\alpha}_2 + \tilde{\omega}_2^2)\widehat{\Phi}_2\overline{u}_1 + \cdots \\ & + b_{k-1}(\tilde{\alpha}_{k-1} + \tilde{\omega}_{k-1}^2)\widehat{\Phi}_{k-1}\overline{u}_1 \\ & + \ddot{s}_1\overline{n}_1 + \ddot{s}_2\widehat{\Phi}_1\overline{n}_2 + \ddot{s}_3\widehat{\Phi}_2\overline{n}_3 + \cdots + \ddot{s}_k\widehat{\Phi}_{k-1}\overline{n}_k \\ & + 2\dot{s}_2\tilde{\omega}_1\widehat{\Phi}_1\overline{n}_2 + 2\dot{s}_3\tilde{\omega}_2\widehat{\Phi}_2\overline{n}_3 + \cdots + 2\dot{s}_k\tilde{\omega}_{k-1}\widehat{\Phi}_{k-1}\overline{n}_k\end{aligned} \tag{8.98}$$

Equation (8.98) can also be written more compactly as follows:

$$\begin{aligned}\overline{a}_k = {} & b_1\tilde{\alpha}_1\widehat{\Phi}_1\overline{u}_1 + b_2\tilde{\alpha}_2\widehat{\Phi}_2\overline{u}_1 + \cdots + b_{k-1}\tilde{\alpha}_{k-1}\widehat{\Phi}_{k-1}\overline{u}_1 \\ & + \ddot{s}_1\overline{n}_1 + \ddot{s}_2\widehat{\Phi}_1\overline{n}_2 + \ddot{s}_3\widehat{\Phi}_2\overline{n}_3 + \cdots + \ddot{s}_k\widehat{\Phi}_{k-1}\overline{n}_k + \overline{a}_k^{\circ}\end{aligned} \tag{8.99}$$

In Eq. (8.99), $\overline{a}_k^{\circ}$ is the *velocity-dependent translational acceleration*. It is also called the *translational acceleration bias term*. It is separated into two special terms so that

$$\overline{a}_k^{\circ} = \overline{a}_k^{cf} + \overline{a}_k^{cor} \tag{8.100}$$

In Eq. (8.100), $\overline{a}_k^{cf}$ consists of the *centripetal acceleration* terms and $\overline{a}_k^{cor}$ consists of the *Coriolis acceleration* terms. They are defined as follows:

$$\overline{a}_k^{cf} = b_1\tilde{\omega}_1^2\widehat{\Phi}_1\overline{u}_1 + b_2\tilde{\omega}_2^2\widehat{\Phi}_2\overline{u}_1 + \cdots + b_{k-1}\tilde{\omega}_{k-1}^2\widehat{\Phi}_{k-1}\overline{u}_1 \tag{8.101}$$

$$\overline{a}_k^{cor} = 2\dot{s}_2\tilde{\omega}_1\widehat{\Phi}_1\overline{n}_2 + 2\dot{s}_3\tilde{\omega}_2\widehat{\Phi}_2\overline{n}_3 + \cdots + 2\dot{s}_k\tilde{\omega}_{k-1}\widehat{\Phi}_{k-1}\overline{n}_k \tag{8.102}$$

The accelerations of the wrist and tip points can be denoted by the following notations.

$$\overline{a}_R = \dot{\overline{v}}_R = \dot{\overline{w}} = \overline{a}_{m-1} = \overline{a}_{0,m-1}^{(0)} = \overline{a}_{R/O}^{(0)} \tag{8.103}$$

$$\overline{a}_P = \dot{\overline{v}}_P = \dot{\overline{v}} = \overline{a}_m = \overline{a}_{0,m}^{(0)} = \overline{a}_{P/O}^{(0)} \tag{8.104}$$

Thus, $\overline{a}_R = \overline{a}_{m-1}$ and $\overline{a}_P = \overline{a}_m$ can also be expressed by using Eq. (8.99).

On the other hand, Eq. (8.90) leads to the following relationship between $\overline{a}_P$ and $\overline{a}_R$.

$$\overline{a}_P = \overline{a}_R + d_m(\widetilde{\alpha} + \widetilde{\omega}^2)\widehat{C}\overline{u}_3 \tag{8.105}$$

8.10 Jacobian Matrices Associated with the Wrist and Tip Points

Considering a spatial manipulator with m joints, the *velocity state* of the end-effector can be represented in the task space S_T by one of the following augmented column matrices.

$$\overline{\eta}_P = \begin{bmatrix} \overline{v}_P \\ \overline{\omega}_m \end{bmatrix} = \begin{bmatrix} \overline{v} \\ \overline{\omega} \end{bmatrix} \tag{8.106}$$

$$\overline{\eta}_R = \begin{bmatrix} \overline{v}_R \\ \overline{\omega}_m \end{bmatrix} = \begin{bmatrix} \overline{w} \\ \overline{\omega} \end{bmatrix} \tag{8.107}$$

In the above equations, $\overline{\eta}_P$ and $\overline{\eta}_R$ represent the 6×1 velocity state matrices associated with the tip and wrist points. They are briefly called the *tip point velocity state matrix* and the *wrist point velocity state matrix*.

On this occasion, Eq. (8.90) can be modified as shown below.

$$\overline{v} = \overline{w} + d_m\widetilde{\omega}\widehat{C}\overline{u}_3 = \overline{w} - d_m[\text{cpm}(\widehat{C}\overline{u}_3)]\overline{\omega} \Rightarrow$$

$$\overline{v} = \overline{w} - d_m\widehat{C}\widetilde{u}_3\widehat{C}^t\overline{\omega} \tag{8.108}$$

According to Eqs. (8.106) – (8.108), $\overline{\eta}_P$ can be related to $\overline{\eta}_R$ as follows:

$$\begin{bmatrix} \overline{v} \\ \overline{\omega} \end{bmatrix} = \begin{bmatrix} \overline{w} - d_m\widehat{C}\widetilde{u}_3\widehat{C}^t\overline{\omega} \\ \overline{\omega} \end{bmatrix} = \begin{bmatrix} \widehat{I} & -d_m\widehat{C}\widetilde{u}_3\widehat{C}^t \\ \widehat{0} & \widehat{I} \end{bmatrix} \begin{bmatrix} \overline{w} \\ \overline{\omega} \end{bmatrix} \Rightarrow$$

$$\overline{\eta}_P = \widehat{D}_{PR}\overline{\eta}_R \tag{8.109}$$

In Eq. (8.109), $\widehat{D}_{PR}$ is the *velocity state conversion matrix*. As seen above, it is defined as the following function of $\widehat{C}$.

$$\widehat{D}_{PR} = \begin{bmatrix} \widehat{I} & -d_m\widehat{C}\widetilde{u}_3\widehat{C}^t \\ \widehat{0} & \widehat{I} \end{bmatrix} \tag{8.110}$$

Equations (8.108) – (8.110) imply the following inversion.

$$\overline{\eta}_R = \widehat{D}_{RP}\overline{\eta}_P \tag{8.111}$$

$$\widehat{D}_{RP} = (\widehat{D}_{PR})^{-1} = \begin{bmatrix} \widehat{I} & d_m\widehat{C}\widetilde{u}_3\widehat{C}^t \\ \widehat{0} & \widehat{I} \end{bmatrix} \tag{8.112}$$

On the other hand, the position of the end-effector of the same manipulator can be represented in the joint space S_J by the following m-element column matrix that consists of the joint variables.

$$\overline{q} = \begin{bmatrix} q_1 \\ q_2 \\ \vdots \\ q_m \end{bmatrix} \tag{8.113}$$

In Eq. (8.113), q_k is used as a general symbol for the joint variable that belongs to the joint $\mathcal{J}_k$. As defined previously by Eq. (8.32), $q_k = \theta_k$ if $\mathcal{J}_k$ is revolute and $q_k = s_k$ if $\mathcal{J}_k$ is prismatic.

As for the velocity state of the end-effector in the joint space $\mathcal{S}_J$, it is represented by $\dot{\overline{q}}$. The elements of $\dot{\overline{q}}$ (i.e. $\dot{q}_1, \dot{q}_2, \ldots, \dot{q}_m$) are briefly called *joint velocities.*

The velocity state representations of the end-effector in $\mathcal{S}_T$ and $\mathcal{S}_J$ can be related to each other by one of the following equations.

$$\overline{\eta}_R = \hat{\mathcal{J}}_R \dot{\overline{q}} \tag{8.114}$$

$$\overline{\eta}_P = \hat{\mathcal{J}}_P \dot{\overline{q}} \tag{8.115}$$

In Eqs. (8.114) and (8.115), the $6 \times m$ matrices $\hat{\mathcal{J}}_R$ and $\hat{\mathcal{J}}_P$ are defined as the *Jacobian matrices* associated with the wrist and tip points. So, they are briefly called the *wrist point Jacobian matrix* and the *tip point Jacobian matrix.*

Note that $\hat{\mathcal{J}}_R$ and $\hat{\mathcal{J}}_P$ are related to each other as follows according to Eqs. (8.109) and (8.111):

$$\hat{\mathcal{J}}_P = \hat{D}_{PR} \hat{\mathcal{J}}_R \tag{8.116}$$

$$\hat{\mathcal{J}}_R = \hat{D}_{RP} \hat{\mathcal{J}}_P \tag{8.117}$$

As a preparation for obtaining the Jacobian matrices, the column matrices that represent the wrist and tip point velocities and the angular velocity of the end-effector can be expressed in terms of the joint velocities as shown below.

$$\overline{w} = \sum_{k=1}^{m} \overline{W}_k \dot{q}_k \tag{8.118}$$

$$\overline{v} = \sum_{k=1}^{m} \overline{V}_k \dot{q}_k \tag{8.119}$$

$$\overline{\omega} = \sum_{k=1}^{m} \overline{\Omega}_k \dot{q}_k \tag{8.120}$$

In the above equations, $\overline{W}_k$, $\overline{V}_k$, and $\overline{\Omega}_k$ are called *velocity influence coefficients.* They are so called because they show the influence of $\dot{q}_k$ on $\overline{w}$, $\overline{v}$, and $\overline{\omega}$. They are defined by the following partial derivatives.

$$\overline{W}_k = \partial\overline{w}/\partial\dot{q}_k = \partial\overline{r}/\partial q_k \tag{8.121}$$

$$\overline{V}_k = \partial\overline{v}/\partial\dot{q}_k = \partial\overline{p}/\partial q_k \tag{8.122}$$

$$\overline{\Omega}_k = \partial\overline{\omega}/\partial\dot{q}_k = \mathrm{colm}[(\partial\hat{C}/\partial q_k)\hat{C}^t] \tag{8.123}$$

Note that Eq. (8.123) is deduced directly from Eq. (8.79) because

$$\dot{\hat{C}} = \sum_{k=1}^{m} (\partial\hat{C}/\partial q_k)\dot{q}_k \tag{8.124}$$

The velocity influence coefficients defined above lead to the following detailed expressions of the Jacobian matrices $\hat{\mathcal{J}}_R$ and $\hat{\mathcal{J}}_P$.

$$\hat{\mathcal{J}}_R = \begin{bmatrix} \overline{W}_1 & \overline{W}_2 & \cdots & \overline{W}_m \\ \overline{\Omega}_1 & \overline{\Omega}_2 & \cdots & \overline{\Omega}_m \end{bmatrix} \tag{8.125}$$

$$\hat{\mathcal{J}}_P = \begin{bmatrix} \overline{V}_1 & \overline{V}_2 & \cdots & \overline{V}_m \\ \overline{\Omega}_1 & \overline{\Omega}_2 & \cdots & \overline{\Omega}_m \end{bmatrix} \tag{8.126}$$

As for the detailed expressions of the velocity influence coefficients, they are obtained from Eqs. (8.82), (8.89), and (8.90) as shown below.

$$\overline{\Omega}_k = \begin{cases} \hat{\Phi}_{k-1}\overline{n}_k & \text{if } q_k = \theta_k \\ \overline{0} & \text{if } q_k = s_k \end{cases} \tag{8.127}$$

$$\overline{W}_k = \begin{cases} \overline{W}_k^* & \text{if } q_k = \theta_k \\ \hat{\Phi}_{k-1}\overline{n}_k & \text{if } q_k = s_k \end{cases} \tag{8.128}$$

$$\overline{V}_k = \overline{W}_k + d_m \widetilde{\Omega}_k \hat{C} \overline{u}_3 \tag{8.129}$$

In Eq. (8.128), $\overline{W}_k^*$ is obtained from Eq. (8.89) as follows after substituting Eq. (8.81):

$$\overline{W}_k^* = (\hat{\Phi}_{k-1}\widetilde{n}_k\hat{\Phi}_{k-1}^t)(\overline{z}_k + \overline{z}_{k+1} + \cdots + \overline{z}_{m-2}) \tag{8.130}$$

In Eq. (8.130), $\hat{\Phi}_0 = \hat{I}$ for $k = 1$ and for $i = k, k+1, \ldots, m-2$,

$$\overline{z}_i = \hat{\Phi}_i(b_i\overline{u}_1 + s_{i+1}\overline{n}_{i+1}) \tag{8.131}$$

When the preceding equations are examined, it is seen that all the velocity influence coefficients are functions of position only. In other words, they depend only on the current values of the joint variables. Therefore, the Jacobian matrices are also functions of position only as indicated below.

$$\hat{J}_R = \hat{J}_R(q_1, q_2, \ldots, q_m) \text{ and } \hat{J}_P = \hat{J}_P(q_1, q_2, \ldots, q_m) \tag{8.132}$$

8.11 Recursive Position, Velocity, and Acceleration Formulations

The positions, velocities, and accelerations of the links of a manipulator can also be expressed in terms of the joint variables and their derivatives in a recursive manner. The expressions obtained this way can be used conveniently especially for computational purposes. The derivation of the recursive formulas are explained below for all the links, i.e. for $\{\mathcal{L}_k : k = 1, 2, \ldots, m\}$.

8.11.1 Orientations of the Links

Referring to Eqs. (8.1) and (8.11), the orientations of two successive links are related as shown below.

$$\hat{C}^{(0,k)} = \hat{C}^{(0,k-1)}\hat{C}^{(k-1,k)} \Rightarrow$$

$$\hat{C}_k = \hat{C}_{k-1}e^{\widetilde{u}_1\beta_k}e^{\widetilde{u}_3\theta_k} \tag{8.133}$$

$$\hat{C}_0 = \hat{I} \tag{8.134}$$

On the other hand, recall that

$$\hat{C}_k = \hat{\Phi}_k e^{\widetilde{u}_1\gamma_k} \tag{8.135}$$

Then, by using Eqs. (8.13)–(8.15), Eq. (8.133) can be modified into

$$\hat{\Phi}_k = \hat{\Phi}_{k-1}e^{\widetilde{n}_k\theta_k} \tag{8.136}$$

In Eq. (8.136),

$$\overline{n}_k = e^{\tilde{u}_1 \gamma_k} \overline{u}_3 \tag{8.137}$$

$$\gamma_k = \gamma_{k-1} + \beta_k \tag{8.138}$$

$$\gamma_0 = 0; \quad \hat{\Phi}_0 = \hat{I} \tag{8.139}$$

8.11.2 Locations of the Link Frame Origins

Referring to Section 8.3, recall that $\overline{r}_k = \overline{r}^{(0)}_{0,k}$. Hence, Eq. (8.18) can be written in the following recursive form.

$$\overline{r}_k = \overline{r}_{k-1} + \hat{\Phi}_{k-1}(b_{k-1}\overline{u}_1 + s_k \overline{n}_k) \tag{8.140}$$

$$\overline{r}_0 = \overline{0} \tag{8.141}$$

8.11.3 Locations of the Mass Centers of the Links

Let C_k be the mass center of the link $\mathcal{L}_k$. Let also

$$\overline{r}^*_k = \overline{r}^{(0)}_{O_0 C_k} \tag{8.142}$$

Furthermore, noting that C_k appears as a fixed point with respect to the link frame $\mathcal{F}_k(O_k)$, let the constant column matrix $\overline{c}_k$ be defined so that

$$\overline{c}_k = \overline{r}^{(k)}_{O_k C_k} \tag{8.143}$$

Hence, the following equation can be written for the location of C_k with respect to the base frame.

$$\overline{r}^{(0)}_{O_0 C_k} = \overline{r}^{(0)}_{O_0 O_k} + \overline{r}^{(0)}_{O_k C_k} = \overline{r}^{(0)}_{O_0 O_k} + \hat{C}^{(0,k)} \overline{r}^{(k)}_{O_k C_k} \Rightarrow$$

$$\overline{r}^*_k = \overline{r}_k + \hat{C}_k \overline{c}_k = \overline{r}_k + \hat{\Phi}_k e^{\tilde{u}_1 \gamma_k} \overline{c}_k \tag{8.144}$$

8.11.4 Angular Velocities of the Links

The angular velocities of two successive links are related as follows:

$$\vec{\omega}_{0,k} = \vec{\omega}_{0,k-1} + \vec{\omega}_{k-1,k} \Rightarrow \vec{\omega}_k = \vec{\omega}_{k-1} + \vec{\omega}_{k-1,k} \tag{8.145}$$

Note that $\mathcal{L}_k$ rotates with respect to $\mathcal{L}_{k-1}$ about the joint axis along $\vec{u}^{(k)}_3$ and $\vec{u}^{(k)}_3$ appears fixed with respect to $\mathcal{L}_k$ and $\mathcal{L}_{k-1}$. Therefore,

$$\vec{\omega}_{k-1,k} = \dot{\theta}_k \vec{u}^{(k)}_3 \tag{8.146}$$

Hence, the matrix equivalent of Eq. (8.144) can be written as follows in the base frame:

$$\overline{\omega}^{(0)}_k = \overline{\omega}^{(0)}_{k-1} + \dot{\theta}_k \overline{u}^{(k/0)}_3 = \overline{\omega}^{(0)}_{k-1} + \dot{\theta}_k \hat{C}^{(0,k)} \overline{u}^{(k/k)}_3 \Rightarrow$$

$$\overline{\omega}_k = \overline{\omega}_{k-1} + \dot{\theta}_k \hat{C}_k \overline{u}_3 \tag{8.147}$$

Using Eqs. (8.135) – (8.138), Eq. (8.147) can be modified so that

$$\overline{\omega}_k = \overline{\omega}_{k-1} + \dot{\theta}_k \hat{\Phi}_{k-1} \overline{n}_k \tag{8.148}$$

$$\overline{\omega}_0 = \overline{0} \tag{8.149}$$

8.11.5 Velocities of the Link Frame Origins

Noting that $\overline{v}_k = \dot{\overline{r}}_k$, Eq. (8.140) leads to the following velocity equation upon differentiation.

$$\overline{v}_k = \overline{v}_{k-1} + \dot{\widehat{\Phi}}_{k-1}(b_{k-1}\overline{u}_1 + s_k\overline{n}_k) + \dot{s}_k\widehat{\Phi}_{k-1}\overline{n}_k \quad (8.150)$$

Referring to Eq. (8.77), recall that

$$\dot{\widehat{\Phi}}_{k-1} = \dot{\widehat{\Phi}}_{k-1}\widehat{\Phi}^t_{k-1}\widehat{\Phi}_{k-1} = \widetilde{\omega}_{k-1}\widehat{\Phi}_{k-1} \quad (8.151)$$

Hence, the following formula is obtained.

$$\overline{v}_k = \overline{v}_{k-1} + \dot{s}_k\widehat{\Phi}_{k-1}\overline{n}_k + \widetilde{\omega}_{k-1}\widehat{\Phi}_{k-1}(b_{k-1}\overline{u}_1 + s_k\overline{n}_k) \quad (8.152)$$

$$\overline{v}_0 = \overline{0} \quad (8.153)$$

8.11.6 Velocities of the Mass Centers of the Links

Referring to Eq. (8.142), the velocity of the mass center C_k can be obtained as

$$\overline{v}^*_k = \dot{\overline{r}}^*_k \quad (8.154)$$

Hence, Eq. (8.144) leads to the following velocity equation.

$$\overline{v}^*_k = \overline{v}_k + \widetilde{\omega}_k\widehat{\Phi}_k e^{\widetilde{u}_1\gamma_k}\overline{c}_k \quad (8.155)$$

8.11.7 Angular Accelerations of the Links

Upon differentiation of Eq. (8.148), the angular accelerations of two successive links can be related as follows:

$$\overline{\omega}_k = \overline{\omega}_{k-1} + d(\dot{\theta}_k\widehat{\Phi}_{k-1}\overline{n}_k)/dt \Rightarrow$$

$$\overline{\alpha}_k = \overline{\alpha}_{k-1} + \ddot{\theta}_k\widehat{\Phi}_{k-1}\overline{n}_k + \dot{\theta}_k\dot{\widehat{\Phi}}_{k-1}\overline{n}_k \Rightarrow$$

$$\overline{\alpha}_k = \overline{\alpha}_{k-1} + \ddot{\theta}_k\widehat{\Phi}_{k-1}\overline{n}_k + \dot{\theta}_k\widetilde{\omega}_{k-1}\widehat{\Phi}_{k-1}\overline{n}_k \Rightarrow$$

$$\overline{\alpha}_k = \overline{\alpha}_{k-1} + (\ddot{\theta}_k\widehat{I} + \dot{\theta}_k\widetilde{\omega}_{k-1})\widehat{\Phi}_{k-1}\overline{n}_k \quad (8.156)$$

$$\overline{\alpha}_0 = \overline{0} \quad (8.157)$$

8.11.8 Accelerations of the Link Frame Origins

The differentiation of Eq. (8.152) leads to the following formula that relates the accelerations of two successive link frame origins.

$$\dot{\overline{v}}_k = \dot{\overline{v}}_{k-1} + \ddot{s}_k\widehat{\Phi}_{k-1}\overline{n}_k + \dot{s}_k\dot{\widehat{\Phi}}_{k-1}\overline{n}_k + \widetilde{\omega}_{k-1}\widehat{\Phi}_{k-1}(\dot{s}_k\overline{n}_k)$$
$$+ \widetilde{\alpha}_{k-1}\widehat{\Phi}_{k-1}(b_{k-1}\overline{u}_1 + s_k\overline{n}_k) + \widetilde{\omega}_{k-1}\dot{\widehat{\Phi}}_{k-1}(b_{k-1}\overline{u}_1 + s_k\overline{n}_k) \Rightarrow$$

$$\overline{a}_k = \overline{a}_{k-1} + \ddot{s}_k\widehat{\Phi}_{k-1}\overline{n}_k + 2\dot{s}_k\widetilde{\omega}_{k-1}\widehat{\Phi}_{k-1}\overline{n}_k$$
$$+ (\widetilde{\alpha}_{k-1} + \widetilde{\omega}^2_{k-1})\widehat{\Phi}_{k-1}(b_{k-1}\overline{u}_1 + s_k\overline{n}_k) \quad (8.158)$$

$$\overline{a}_0 = \overline{0} \quad (8.159)$$

8.11.9 Accelerations of the Mass Centers of the Links

Equation (8.155) can be differentiated so that $\overline{a}_k^* = \dot{\overline{v}}_k^*$ is obtained as follows:

$$\overline{a}_k^* = \overline{a}_k + (\widetilde{\alpha}_k + \widetilde{\omega}_k^2)\widehat{\Phi}_k e^{\widetilde{u}_1 \gamma_k} \overline{c}_k \tag{8.160}$$

8.12 Inverse Motion Analysis of a Manipulator Based on the Jacobian Matrix

As seen before in Section 8.10, $\overline{\eta}_R$ and $\widehat{\mathcal{J}}_R$ are related to $\overline{\eta}_P$ and $\widehat{\mathcal{J}}_P$ directly by means of the *velocity state conversion matrix* $\widehat{D}_{RP}$, which depends only on the orientation of the end-effector, i.e. the matrix $\widehat{C}$. Therefore, if $\widehat{C}$ is specified completely, then the inverse motion analysis can be carried out more conveniently by using $\widehat{\mathcal{J}}_R$ rather than $\widehat{\mathcal{J}}_P$, because the relationship between $\overline{\eta}_R$ and $\dot{\overline{q}}$ is simpler than that between $\overline{\eta}_P$ and $\dot{\overline{q}}$. Moreover, this simplicity is enhanced especially if the wrist of the manipulator is spherical. However, $\widehat{C}$ may not be specified completely if a deficient manipulator is used. Then, it becomes necessary to use $\widehat{\mathcal{J}}_P$. In any case, in this section, the subscript R or P will be omitted and the velocity equation will be written simply as follows for the sake of brevity:

$$\overline{\eta} = \widehat{\mathcal{J}}\dot{\overline{q}} \tag{8.161}$$

In Eq. (8.161), depending on the situation,

$$\overline{\eta} = \overline{\eta}_R \text{ or } \overline{\eta} = \overline{\eta}_P \tag{8.162}$$

Similarly and respectively,

$$\widehat{\mathcal{J}} = \widehat{\mathcal{J}}_R \text{ or } \widehat{\mathcal{J}} = \widehat{\mathcal{J}}_P \tag{8.163}$$

Note that $\widehat{\mathcal{J}} = \widehat{\mathcal{J}}(\overline{q})$. Therefore, at this stage, it is assumed that $\overline{q}$ has already been obtained within a selected posture mode as the main outcome of the preceding stage of inverse kinematics. In other words, if the manipulator is not in a pose of *position singularity*, $\overline{q}$ is known definitely. Otherwise, it is assumed that some suitable values have been assigned to those components of $\overline{q}$, which acquire arbitrariness in a pose of position singularity.

Here, it is to be pointed out that the position singularities are associated with the position relationships and therefore they occur irrespective of the Jacobian matrix $\widehat{\mathcal{J}}(\overline{q})$. On the other hand, the motion singularities are associated primarily with the velocity relationships. They are of course associated with the relationships that involve higher order motion derivatives as well. Therefore, the motion singularities are the same as the singularities of $\widehat{\mathcal{J}}(\overline{q})$. This can be seen in Section 8.12.1. In general, the motion singularities are different from the position singularities because they arise from different sources, i.e. from the velocity relationships, which are of course different from the position relationships, even though the two relationships are interrelated via differentiation. The differences between the position and motion singularities will be evident for the manipulators studied in Chapter 9.

8.12.1 Inverse Velocity Analysis of a Regular Manipulator

For a spatial regular manipulator, the Jacobian matrix $\hat{\mathcal{J}}$ happens to be square (i.e. 6×6). Therefore, if the manipulator is not in a pose of *motion singularity*, i.e. if $\det(\hat{\mathcal{J}}) \neq 0$, $\dot{\overline{q}}$ can be obtained from Eq. (8.161) directly as

$$\dot{\overline{q}} = \hat{\mathcal{J}}^{-1}\overline{\eta} \tag{8.164}$$

If the manipulator gets into a pose of motion singularity, i.e. if $\det(\hat{\mathcal{J}}) = 0$, then $\dot{\overline{q}}$ cannot be obtained definitely as in Eq. (8.164). In such a pose, the rank of $\hat{\mathcal{J}}$ reduces from 6 to $r < 6$. Consequently, $\dot{\overline{q}}$ becomes indefinite. Nevertheless, although indefinite, it can still be kept finite in magnitude, if $\overline{\eta}$ is specified in compliance with the *restriction* imposed by the singularity in the *task space*. According to this restriction, only r of the 6 components of $\overline{\eta}$ can be specified arbitrarily as desired. The other $r' = 6 - r$ components of $\overline{\eta}$ must be specified compatibly with the imposed restriction. On the other hand, $\dot{\overline{q}}$ becomes indefinite due to the *arbitrariness* induced by the singularity in the *joint space*. Consequently, r' of the 6 joint velocities (i.e. the joint variable derivatives) can be chosen arbitrarily. Then, the other r of the joint velocities can be obtained depending on the r' joint velocities that are chosen arbitrarily and the r components of $\overline{\eta}$ that are specified arbitrarily.

The above-mentioned consequences of a motion singularity in the task and joint spaces can be verified mathematically as explained below.

Since the rank of $\hat{\mathcal{J}}$ reduces to r in a motion singularity, Eq. (8.161) can be written in the following partitioned form.

$$\begin{bmatrix} \overline{\eta}_a \\ \overline{\eta}_b \end{bmatrix} = \begin{bmatrix} \hat{\mathcal{J}}_{aa} & \hat{\mathcal{J}}_{ab} \\ \hat{\mathcal{J}}_{ba} & \hat{\mathcal{J}}_{bb} \end{bmatrix} \begin{bmatrix} \dot{\overline{q}}_a \\ \dot{\overline{q}}_b \end{bmatrix} \tag{8.165}$$

The partitioning in Eq. (8.165) is not necessarily unique, that is, it can be changed by shuffling the rows and columns of $\hat{\mathcal{J}}$. However, the rank of $\hat{\mathcal{J}}_{aa}$ must be full, i.e. it must be r. Equation (8.165) can be separated into the following pair of equations.

$$\overline{\eta}_a = \hat{\mathcal{J}}_{aa}\dot{\overline{q}}_a + \hat{\mathcal{J}}_{ab}\dot{\overline{q}}_b \tag{8.166}$$

$$\overline{\eta}_b = \hat{\mathcal{J}}_{ba}\dot{\overline{q}}_a + \hat{\mathcal{J}}_{bb}\dot{\overline{q}}_b \tag{8.167}$$

Since $\text{rank}(\hat{\mathcal{J}}_{aa}) = r$, $\det(\hat{\mathcal{J}}_{aa}) \neq 0$ and Eq. (8.166) can be solved for $\dot{\overline{q}}_a$. That is,

$$\dot{\overline{q}}_a = \hat{\mathcal{J}}_{aa}^{-1}(\overline{\eta}_a - \hat{\mathcal{J}}_{ab}\dot{\overline{q}}_b) \tag{8.168}$$

When Eq. (8.168) is substituted, Eq. (8.167) becomes

$$\overline{\eta}_b = (\hat{\mathcal{J}}_{ba}\hat{\mathcal{J}}_{aa}^{-1})\overline{\eta}_a + (\hat{\mathcal{J}}_{bb} - \hat{\mathcal{J}}_{ba}\hat{\mathcal{J}}_{aa}^{-1}\hat{\mathcal{J}}_{ab})\dot{\overline{q}}_b \tag{8.169}$$

The singularity of $\hat{\mathcal{J}}$ implies that

$$\hat{\mathcal{J}}_{bb} = \hat{\mathcal{J}}_{ba}\hat{\mathcal{J}}_{aa}^{-1}\hat{\mathcal{J}}_{ab} \tag{8.170}$$

Thus, Eq. (8.169) reduces to the following *task space compatibility condition*.

$$\overline{\eta}_b = (\hat{\mathcal{J}}_{ba}\hat{\mathcal{J}}_{aa}^{-1})\overline{\eta}_a \tag{8.171}$$

Equation (8.171) expresses the restriction imposed in the task space. Due to this restriction, $\overline{\eta}$ cannot be specified totally as desired. Only the partition $\overline{\eta}_a$ can be specified as desired, but the other partition $\overline{\eta}_b$ must necessarily be specified so as to satisfy the compatibility condition expressed by Eq. (8.171). Provided that $\overline{\eta}_b$ is specified as such, the manipulator can be driven through the motion singularity without causing any unbounded growth in the joint velocities.

On the other hand, Eq. (8.168) expresses the arbitrariness induced in the joint space. Due to this arbitrariness, the joints of the manipulator must be controlled so that $\dot{\overline{q}}_b$ is generated arbitrarily as desired, but the complementary partition $\dot{\overline{q}}_a$ is generated as dictated by $\overline{\eta}_a$ and $\dot{\overline{q}}_b$.

8.12.2 Inverse Acceleration Analysis of a Regular Manipulator

The derivative of Eq. (8.161) can be taken so that

$$\hat{J}\ddot{\overline{q}} = \overline{\gamma} \tag{8.172}$$

Suppose the position and velocity analyses have already been completed. Thus, noting that $\hat{J} = \hat{J}(\overline{q})$, $\overline{\gamma}$ in Eq. (8.172) is available as a known function of $\overline{q}$, $\dot{\overline{q}}$, and $\dot{\overline{\eta}}$. That is,

$$\overline{\gamma} = \overline{\gamma}(\dot{\overline{\eta}}, \overline{q}, \dot{\overline{q}}) = \dot{\overline{\eta}} - \dot{\hat{J}}\dot{\overline{q}} = \dot{\overline{\eta}} - \left[\sum_{k=1}^{6}(\partial\hat{J}/\partial q_k)\dot{q}_k\right]\dot{\overline{q}} \tag{8.173}$$

If the manipulator is not in a pose of motion singularity, $\ddot{\overline{q}}$ can be obtained from Eq. (8.172) directly as

$$\ddot{\overline{q}} = \hat{J}^{-1}\overline{\gamma} \tag{8.174}$$

If the manipulator gets into a pose of motion singularity, then Eq. (8.172) is treated in the same way with the same partitioning as discussed before for Eq. (8.161).

8.13 Inverse Motion Analysis of a Redundant Manipulator

8.13.1 Inverse Velocity Analysis

For a spatial redundant manipulator with $m > 6$, the Jacobian matrix $\hat{J}$ is no longer square. That is, it cannot be inverted. Therefore, the inverse velocity solution can be obtained by using one of the methods explained below.

(a) Solution by Specifying the Motions of Certain Joints

In this method, Eq. (8.161) is written in a partitioned form such as shown below.

$$\overline{\eta} = \hat{J}_c\dot{\overline{q}}_c + \hat{J}_d\dot{\overline{q}}_d \tag{8.175}$$

In Eq. (8.175), $\hat{J}_c$ is a square (6×6) matrix. The indicated partitioning is not necessarily unique, but it must be consistent with the partitioning used in obtaining the inverse kinematic solution as discussed in Section 8.7. The partition $\overline{q}_d$ of $\overline{q}$ is specified either as a constant or as a certain function of time and therefore $\dot{\overline{q}}_d$ is known. If $\hat{J}_c$ is not singular, then Eq. (8.175) can be solved for $\dot{\overline{q}}_c$ so that

$$\dot{\overline{q}}_c = \hat{J}_c^{-1}(\overline{\eta} - \hat{J}_d\dot{\overline{q}}_d) \tag{8.176}$$

If $\hat{\mathcal{J}}_c$ becomes singular, then a different partitioning may be tried in order to obtain a definite solution. However, in some cases, the motion singularity turns out to be so fundamental that $\hat{\mathcal{J}}_c$ becomes singular no matter how the partitioning is made. In that case, $\hat{\mathcal{J}}_c$ can further be partitioned with a main full-rank partition $\hat{\mathcal{J}}_{caa}$ and the solution can be obtained as explained in Section 8.12.

(b) Optimized Solution Based on a Position Criterion

This method is based on the optimization method explained in Section 8.7. When applied at an instant t_k, the method in Section 8.7 gives $\overline{q}(t_k)$ optimally depending on $\overline{p}(t_k)$ and $\hat{C}(t_k)$, which denote the specified location and orientation of the end-effector at the instant t_k. When applied again at a slightly later instant $t_{k+1} = t_k + \Delta t$, the same method gives $\overline{q}(t_{k+1})$ optimally depending on $\overline{p}(t_{k+1})$ and $\hat{C}(t_{k+1})$. Hence, $\dot{\overline{q}}(t_{k+1})$ can be obtained numerically by using the following *discrete backward differentiation*.

$$\dot{\overline{q}}(t_{k+1}) = [\overline{q}(t_{k+1}) - \overline{q}(t_k)]/\Delta t; k = 0, 1, 2, \ldots \tag{8.177}$$

Note that this method cannot give $\dot{\overline{q}}(t_0)$. However, this is not a problem, because a task motion is normally planned so that the manipulator starts with zero velocity, i.e. it happens that $\dot{\overline{q}}(t_0) = \overline{0}$.

(c) Optimized Solution Based on a Motion Criterion

In this method, a positive definite *performance function* U_v of the joint velocities is required to be minimized in order to perform the task with minimal velocities. For example, U_v can be taken as the kinetic energy of the manipulator. The function U_v can be expressed generically as follows:

$$U_v = \frac{1}{2}\dot{\overline{q}}^t\hat{B}\dot{\overline{q}} \tag{8.178}$$

In Eq. (8.178), $\hat{B}$ is the *weighting matrix*. Depending on the task requirements, it is selected as a suitable *positive definite symmetric* matrix. In many cases, it may be selected simply as a diagonal matrix, e.g. $\hat{B} = \mathrm{diag}(B_1, B_2, \ldots, B_m)$. When $\hat{B}$ is diagonal, the diagonal element B_k is called the *weighting coefficient* for the joint velocity $\dot{q}_k$. A larger value of B_k leads to a smaller value of $|\dot{q}_k|$ as the result of the minimization. The minimization of the function U_v is naturally subject to the following constraint, which must be satisfied by $\dot{\overline{q}}$ in order to attain the specified velocity state of the end-effector, i.e. $\overline{\eta}$.

$$\overline{\eta} = \hat{\mathcal{J}}\dot{\overline{q}} \tag{8.179}$$

This optimization problem can be solved by augmenting the performance function so that

$$U_v^* = \frac{1}{2}\dot{\overline{q}}^t\hat{B}\dot{\overline{q}} + \overline{\lambda}^t(\overline{\eta} - \hat{\mathcal{J}}\dot{\overline{q}}) \tag{8.180}$$

In Eq. (8.180), $\overline{\lambda}$ is the column matrix that comprises the *Lagrange multipliers*. That is,

$$\overline{\lambda} = \begin{bmatrix} \lambda_1 & \lambda_2 & \cdots & \lambda_6 \end{bmatrix}^t \tag{8.181}$$

The optimization necessitates that

$$\partial U_v^*/\partial\dot{\overline{q}} = \hat{B}\dot{\overline{q}} - \hat{\mathcal{J}}^t\overline{\lambda} = \overline{0} \tag{8.182}$$

Equation (8.182) gives $\dot{\overline{q}}$ in terms of $\overline{\lambda}$ as follows:

$$\dot{\overline{q}} = \hat{B}^{-1}\hat{J}^t\overline{\lambda} \tag{8.183}$$

The above expression of $\dot{\overline{q}}$ must satisfy Eq. (8.179). That is,

$$\overline{\eta} = \hat{G}\overline{\lambda} \tag{8.184}$$

In Eq. (8.184),

$$\hat{G} = \hat{J}\hat{B}^{-1}\hat{J}^t \tag{8.185}$$

Hence, if $\det(\hat{G}) \neq 0$, i.e. if the manipulator is not in a pose of motion singularity, then $\overline{\lambda}$ and $\dot{\overline{q}}$ are obtained in terms of $\overline{\eta}$ as follows from Eqs. (8.184) and (8.183):

$$\overline{\lambda} = \hat{G}^{-1}\overline{\eta} \tag{8.186}$$

$$\dot{\overline{q}} = (\hat{B}^{-1}\hat{J}^t\hat{G}^{-1})\overline{\eta} = \hat{K}\overline{\eta} \tag{8.187}$$

Here, it must be pointed out that the same optimization criterion must be used for both the inverse position and motion solutions for the sake of consistency. Therefore, the inverse velocity solution expressed by Eq. (8.187) must be used to obtain the inverse position solution as well by means of integration. In other words, noting that $\hat{K} = \hat{K}(\overline{q})$, $\overline{q}(t)$ must be obtained by solving the following differential equation online while the task is executed.

$$\dot{\overline{q}}(t) = \hat{K}(\overline{q}(t))\overline{\eta}(t); \quad t \geq t_0, \quad \overline{q}(t_0) = \overline{q}_0 \tag{8.188}$$

Of course, the initial value $\overline{q}_0$ must be determined in such a way that $\overline{p}_0 = \overline{p}(t_0)$ and $\hat{C}_0 = \hat{C}(t_0)$ must have their specified values. For this purpose, a special optimization criterion can be used only for $t = t_0$. For example, the manipulator can be driven with minimized kinetic energy but it can be started from an initial pose with minimized potential energy.

If $\det(\hat{G}) = 0$, then the manipulator assumes a pose of motion singularity and the rank of $\hat{G}$ reduces to $r < 6$. In that case, a solution with finite but indefinite values of the joint velocities can be obtained through the following partitioning of Eq. (8.184).

$$\overline{\eta}_a = \hat{G}_{aa}\overline{\lambda}_a + \hat{G}_{ab}\overline{\lambda}_b \tag{8.189}$$

$$\overline{\eta}_b = \hat{G}_{ba}\overline{\lambda}_a + \hat{G}_{bb}\overline{\lambda}_b \tag{8.190}$$

In the above equations, $\mathrm{rank}(\hat{G}_{aa}) = r$ and $\overline{\lambda}_b$ is the indefinite partition of $\overline{\lambda}$, which can be selected arbitrarily. Thus, Eq. (8.190) gives $\overline{\lambda}_a$ as follows in terms of $\overline{\eta}_a$ and $\overline{\lambda}_b$.

$$\overline{\lambda}_a = \hat{G}_{aa}^{-1}(\overline{\eta}_a - \hat{G}_{ab}\overline{\lambda}_b) \tag{8.191}$$

Consequently, noting that $\hat{G}_{bb} = \hat{G}_{ba}\hat{G}_{aa}^{-1}\hat{G}_{ab}$ due to the singularity of $\hat{G}$, Eq. (8.190) turns into the following task space restriction relationship between $\overline{\eta}_a$ and $\overline{\eta}_b$.

$$\overline{\eta}_b = (\hat{G}_{ba}\hat{G}_{aa}^{-1})\overline{\eta}_a \tag{8.192}$$

According to Eq. (8.192), only the partition $\overline{\eta}_a$ of $\overline{\eta}$ can be specified arbitrarily as desired in the considered motion singularity. The other partition $\overline{\eta}_b$ must be specified in compliance with Eq. (8.192) in order to pass through the singularity with finite but indefinite joint velocities. The indefiniteness in $\dot{\overline{q}}$ arises due to its dependence on the indefinite partition $\overline{\lambda}_b$ of $\overline{\lambda}$ as expressed by the following version of Eq. (8.183).

$$\dot{\overline{q}} = \widehat{B}^{-1}(\widehat{J}_a^t\overline{\lambda}_a + \widehat{J}_b^t\overline{\lambda}_b) \Rightarrow$$
$$\dot{\overline{q}} = \widehat{B}^{-1}[(\widehat{J}_a^t\widehat{G}_{aa}^{-1})\overline{\eta}_a + (\widehat{J}_b^t - \widehat{J}_a^t\widehat{G}_{aa}^{-1}\widehat{G}_{ab})\overline{\lambda}_b] \tag{8.193}$$

Concerning the solution of the differential equation Eq. (8.188) smoothly so that $\dot{\overline{q}}(t)$ is generated as a continuous function of time even in a singularity, $\overline{\eta}(t)$ must be specified in such a way that it obeys the compatibility condition Eq. (8.192). If $\overline{\eta}(t)$ is specified as such, then, knowing that $\dot{\overline{q}}(t)$ and hence $\overline{q}(t)$ will be continuous at the singularity instant t_s, the discretized solution of the differential equation can be continued simply by skipping t_s, i.e. by choosing the nearest successive integration instants t_k and t_{k+1} so that $t_k < t_s < t_{k+1}$.

8.13.2 Inverse Acceleration Analysis

The inverse acceleration solution can also be obtained by using the continuation of the selected one of the methods that are explained for the inverse velocity solution. The continuation of those methods is explained below.

(a) Solution by Specifying the Motions of Certain Joints

In this method, Eq. (8.175) is differentiated so that

$$\dot{\overline{\eta}} = \widehat{J}_a\ddot{\overline{q}}_a + \widehat{J}_b\ddot{\overline{q}}_b + \dot{\widehat{J}}_a\dot{\overline{q}}_a + \dot{\widehat{J}}_b\dot{\overline{q}}_b \tag{8.194}$$

Since $\overline{q}_b$ is specified, $\dot{\overline{q}}_b$ and $\ddot{\overline{q}}_b$ are known. Furthermore, $\overline{q}_a$ and $\dot{\overline{q}}_a$ are also known from the previously obtained inverse position and velocity solutions. Therefore, if $\widehat{J}_a$ is not singular, Eq. (8.194) can be solved for $\ddot{\overline{q}}_a$ as follows:

$$\ddot{\overline{q}}_a = \widehat{J}_a^{-1}[\dot{\overline{\eta}} - (\widehat{J}_b\ddot{\overline{q}}_b + \dot{\widehat{J}}_a\dot{\overline{q}}_a + \dot{\widehat{J}}_b\dot{\overline{q}}_b)] \tag{8.195}$$

If $\widehat{J}_a$ happens to be singular, then the solution can be obtained similarly as suggested in Part 8.13.1(a) of this section.

(b) Optimized Solution Based on a Position Criterion

Here, at an instant $t_{k+2} = t_{k+1} + \Delta t = t_k + 2\Delta t$, $\ddot{\overline{q}}(t_{k+2})$ can be obtained from Eq. (8.177) as follows by using the discrete backward differentiation once again:

$$\ddot{\overline{q}}(t_{k+2}) = \frac{[\dot{\overline{q}}(t_{k+2}) - \dot{\overline{q}}(t_{k+1})]}{\Delta t}; \quad k = 0, 1, 2, \ldots \tag{8.196}$$

By using Eq. (8.177) successively, Eq. (8.196) can also be written as

$$\ddot{\overline{q}}(t_{k+2}) = [\overline{q}(t_{k+2}) - 2\overline{q}(t_{k+1}) + \overline{q}(t_k)]/(\Delta t)^2; \quad k = 0, 1, 2, \ldots \tag{8.197}$$

Note that the preceding formulas cannot give $\ddot{\overline{q}}(t_1)$ and $\ddot{\overline{q}}(t_0)$. However, as mentioned before, a task motion is normally planned so that the manipulator starts with zero velocity and acceleration, i.e. it happens that $\dot{\overline{q}}(t_0) = \overline{0}$ and $\ddot{\overline{q}}(t_0) = \overline{0}$. As for $\ddot{\overline{q}}(t_1)$, it is obtained as follows:

$$\ddot{\overline{q}}(t_1) = [\dot{\overline{q}}(t_1) - \dot{\overline{q}}(t_0)]/\Delta t = \dot{\overline{q}}(t_1)/\Delta t = [\overline{q}(t_1) - \overline{q}(t_0)]/(\Delta t)^2 \tag{8.198}$$

(c) Optimized Solution Based on a Motion Criterion

If the manipulator is not in a pose of motion singularity, i.e. if $\det(\hat{G}) \neq 0$, then $\ddot{\overline{q}}$ can be obtained by differentiating Eq. (8.187). That is,

$$\ddot{\overline{q}} = \hat{K}\dot{\overline{\eta}} + \dot{\hat{K}}\overline{\eta} = \hat{K}\dot{\overline{\eta}} + \left[\sum_{k=1}^{m}(\partial\hat{K}/\partial q_k)\dot{q}_k\right]\overline{\eta} \tag{8.199}$$

If the manipulator gets into a pose of motion singularity at an instant t_s, $\dot{\overline{q}}$ and $\ddot{\overline{q}}$ will be continuous provided that the partitions of $\overline{\eta}$ be specified so as to satisfy the compatibility condition Eq. (8.192). Then, the computation of $\ddot{\overline{q}}(t_s)$ can be skipped momentarily at $t = t_s$ and a little later, i.e. at $t = t_s + \Delta t$, it can be computed simply as the following average value, instead of using the complicated derivative of Eq. (8.193) at $t = t_s$.

$$\ddot{\overline{q}}(t_s) \cong [\ddot{\overline{q}}(t_s + \Delta t) + \ddot{\overline{q}}(t_s - \Delta t)]/2 \tag{8.200}$$

8.14 Inverse Motion Analysis of a Deficient Manipulator

For a deficient manipulator with $m < 6$, Eqs. (8.161) and (8.172) can be written in the following partitioned forms, which must be consistent with the task compromises discussed in Section 8.8.

$$\begin{bmatrix}\hat{\mathcal{J}}_c \\ \hat{\mathcal{J}}_d\end{bmatrix}\dot{\overline{q}} = \begin{bmatrix}\overline{\eta}_c \\ \overline{\eta}_d\end{bmatrix} \tag{8.201}$$

$$\begin{bmatrix}\hat{\mathcal{J}}_c \\ \hat{\mathcal{J}}_d\end{bmatrix}\ddot{\overline{q}} = \begin{bmatrix}\overline{\gamma}_c \\ \overline{\gamma}_d\end{bmatrix} \tag{8.202}$$

In the above equations, $\hat{\mathcal{J}}_c$ is the $m \times m$ partition of $\hat{\mathcal{J}}$. In Eq. (8.201), $\overline{\eta}_c \in \mathcal{R}^m$ is the specified partition of $\overline{\eta} \in \mathcal{R}^6$ and $\overline{\eta}_d \in \mathcal{R}^{6-m}$ is the partition of $\overline{\eta}$ that is left free, i.e. it will be accepted as a by-product of the solution. In Eq. (8.202), the partitions of $\overline{\gamma}$ are related to the partitions of $\overline{\eta}$ as shown below according to Eq. (8.173).

$$\begin{bmatrix}\overline{\gamma}_c \\ \overline{\gamma}_d\end{bmatrix} = \begin{bmatrix}\dot{\overline{\eta}}_c \\ \dot{\overline{\eta}}_d\end{bmatrix} - \begin{bmatrix}\dot{\hat{\mathcal{J}}}_c \\ \dot{\hat{\mathcal{J}}}_d\end{bmatrix}\dot{\overline{q}} \tag{8.203}$$

If $\det(\hat{\mathcal{J}}_c) \neq 0$, i.e. if the manipulator is not in a pose of motion singularity associated with the indicated partitioning, Eqs. (8.201) and (8.202) yield the following solutions for the pairs $\{\dot{\overline{q}}, \overline{\eta}_d\}$ and $\{\ddot{\overline{q}}, \overline{\gamma}_d\}$.

$$\dot{\overline{q}} = \hat{\mathcal{J}}_c^{-1}\overline{\eta}_c \tag{8.204}$$

$$\overline{\eta}_d = (\hat{\mathcal{J}}_d\hat{\mathcal{J}}_c^{-1})\overline{\eta}_c \tag{8.205}$$

$$\ddot{\overline{q}} = \hat{\mathcal{J}}_c^{-1}\overline{\gamma}_c \tag{8.206}$$

$$\overline{\gamma}_d = (\hat{\mathcal{J}}_d\hat{\mathcal{J}}_c^{-1})\overline{\gamma}_c \tag{8.207}$$

If $\det(\hat{\mathcal{J}}_c) = 0$, then $\hat{\mathcal{J}}_c$ can further be partitioned with a main full-rank partition $\hat{\mathcal{J}}_{caa}$ and the solutions for $\dot{\overline{q}}$ and $\ddot{\overline{q}}$ can be obtained as explained in Section 8.12. Afterwards, $\overline{\eta}_d$ and $\overline{\gamma}_d$ can also be obtained by using the equations $\overline{\eta}_d = \hat{\mathcal{J}}_d\dot{\overline{q}}$ and $\overline{\gamma}_d = \hat{\mathcal{J}}_d\ddot{\overline{q}}$.

8.15 Inverse Motion Analysis of a Regular Manipulator Using the Detailed Formulation

In this section, the inverse motion analysis of a regular generic manipulator is carried out by using the detailed formulation. For the sake of simplicity, the procedure is explained here for a regular manipulator with a spherical wrist. Owing to the spherical wrist, the motion of the wrist point depends only on the motions of the first three joints. Thus, by using the relevant velocity influence coefficients, the forward kinematic equations can be written separately for the translational velocity of the wrist point and the angular velocity of the end-effector. Such separate equations are already available as Eqs. (8.118) and (8.120) in Section 8.10. They are adapted to the present manipulator as follows:

$$\overline{w} = \overline{W}_1\dot{q}_1 + \overline{W}_2\dot{q}_2 + \overline{W}_3\dot{q}_3 \tag{8.208}$$

$$\overline{\omega} = \overline{\Omega}_1\dot{q}_1 + \overline{\Omega}_2\dot{q}_2 + \overline{\Omega}_3\dot{q}_3 + \overline{\Omega}_4\dot{q}_4 + \overline{\Omega}_5\dot{q}_5 + \overline{\Omega}_6\dot{q}_6 \tag{8.209}$$

Here, the velocity influence coefficients are known because they depend only on the joint variables, which are already determined at the stage of inverse kinematics.

8.15.1 Inverse Velocity Solution

The inverse velocity solution can be started from Eq. (8.207) by generating the following three scalar equations, which are obtained by noting that $\widetilde{W}_i\overline{W}_i = \overline{0}$ and $\overline{W}_i^t\widetilde{W}_j\overline{W}_i = 0$.

$$\overline{W}_3^t\widetilde{W}_1\overline{w} = \overline{W}_3^t\widetilde{W}_1\overline{W}_2\dot{q}_2 \tag{8.210}$$

$$\overline{W}_1^t\widetilde{W}_2\overline{w} = \overline{W}_1^t\widetilde{W}_2\overline{W}_3\dot{q}_3 \tag{8.211}$$

$$\overline{W}_2^t\widetilde{W}_3\overline{w} = \overline{W}_2^t\widetilde{W}_3\overline{W}_1\dot{q}_1 \tag{8.212}$$

On the other hand, as seen in Chapter 1,

$$\overline{W}_1^t\widetilde{W}_2\overline{W}_3 = \overline{W}_2^t\widetilde{W}_3\overline{W}_1 = \overline{W}_3^t\widetilde{W}_1\overline{W}_2 = \det\left[\overline{W}_1 \;\; \overline{W}_2 \;\; \overline{W}_3\right] = D_1 \tag{8.213}$$

Hence, the first three joint velocities can be obtained as follows, if a *motion singularity of the first kind* is not encountered, i.e. if $D_1 \neq 0$:

$$\dot{q}_1 = \overline{W}_2^t\widetilde{W}_3\overline{w}/D_1 = N_1/D_1 \tag{8.214}$$

$$\dot{q}_2 = \overline{W}_3^t\widetilde{W}_1\overline{w}/D_1 = N_2/D_1 \tag{8.215}$$

$$\dot{q}_3 = \overline{W}_1^t\widetilde{W}_2\overline{w}/D_1 = N_3/D_1 \tag{8.216}$$

The inverse velocity solution can be continued with Eq. (8.209). Noting that the last three joint variables are angular, Eq. (8.209) can also be written in the following more detailed form by using Eq. (8.127) for the expressions of the angular velocity influence coefficients.

$$\begin{aligned}\overline{\omega} = {} & \varepsilon_1\dot{\theta}_1\overline{n}_1 + \varepsilon_2\dot{\theta}_2 e^{\tilde{n}_1\theta_1}\overline{n}_2 + \varepsilon_3\dot{\theta}_3 e^{\tilde{n}_1\theta_1}e^{\tilde{n}_2\theta_2}\overline{n}_3 + \dot{\theta}_4 e^{\tilde{n}_1\theta_1}e^{\tilde{n}_2\theta_2}e^{\tilde{n}_3\theta_3}\overline{n}_4 \\ & + \dot{\theta}_5 e^{\tilde{n}_1\theta_1}e^{\tilde{n}_2\theta_2}e^{\tilde{n}_3\theta_3}e^{\tilde{n}_4\theta_4}\overline{n}_5 + \dot{\theta}_6 e^{\tilde{n}_1\theta_1}e^{\tilde{n}_2\theta_2}e^{\tilde{n}_3\theta_3}e^{\tilde{n}_4\theta_4}e^{\tilde{n}_5\theta_5}\overline{n}_6.\end{aligned} \tag{8.217}$$

In Eq. (8.217), $\varepsilon_k = 1$ if $q_k = \theta_k$ and $\varepsilon_k = 0$ if $q_k = s_k$ for $k = 1, 2, 3$. Since the first three joint velocities have already been found, Eq. (8.217) can be rearranged as shown below by noting that $e^{\tilde{n}_4\theta_4}\overline{n}_4 = \overline{n}_4$.

$$e^{\tilde{n}_1\theta_1}e^{\tilde{n}_2\theta_2}e^{\tilde{n}_3\theta_3}e^{\tilde{n}_4\theta_4}(\dot{\theta}_4\overline{n}_4 + \dot{\theta}_5\overline{n}_5 + \dot{\theta}_6 e^{\tilde{n}_5\theta_5}\overline{n}_6)$$
$$= \overline{\omega} - (\varepsilon_1\dot{\theta}_1\overline{n}_1 + \varepsilon_2\dot{\theta}_2 e^{\tilde{n}_1\theta_1}\overline{n}_2 + \varepsilon_3\dot{\theta}_3 e^{\tilde{n}_1\theta_1}e^{\tilde{n}_2\theta_2}\overline{n}_3) \tag{8.218}$$

Equation (8.218) can be written very briefly as follows:

$$\dot{\theta}_4\overline{n}_4 + \dot{\theta}_5\overline{n}_5 + \dot{\theta}_6\overline{m}_6 = \overline{\omega}^* \tag{8.219}$$

In Eq. (8.219), $\overline{m}_6$ and $\overline{\omega}^*$ are known because they depend on the previously determined joint positions and velocities and the specified components of $\overline{\omega}$ according to the following equations.

$$\overline{m}_6 = e^{\tilde{n}_5\theta_5}\overline{n}_6 \tag{8.220}$$

$$\overline{\omega}^* = \hat{\Phi}_4^{-1}[\overline{\omega} - (\varepsilon_1\dot{\theta}_1\overline{n}_1 + \varepsilon_2\dot{\theta}_2 e^{\tilde{n}_1\theta_1}\overline{n}_2 + \varepsilon_3\dot{\theta}_3 e^{\tilde{n}_1\theta_1}e^{\tilde{n}_2\theta_2}\overline{n}_3)] \tag{8.221}$$

$$\hat{\Phi}_4 = e^{\tilde{n}_1\theta_1}e^{\tilde{n}_2\theta_2}e^{\tilde{n}_3\theta_3}e^{\tilde{n}_4\theta_4} \tag{8.222}$$

Note that Eq. (8.219) is in the same form as Eq. (8.208). Therefore, it can be solved similarly as shown below, if a *motion singularity of the second kind* is not encountered, i.e. if $D_2 \neq 0$.

$$\dot{\theta}_4 = \overline{n}_5^t\tilde{m}_6\overline{\omega}^*/D_2 = N_4/D_2 \tag{8.223}$$

$$\dot{\theta}_5 = \overline{m}_6^t\tilde{n}_4\overline{\omega}^*/D_2 = N_5/D_2 \tag{8.224}$$

$$\dot{\theta}_6 = \overline{n}_4^t\tilde{n}_5\overline{\omega}^*/D_2 = N_6/D_2 \tag{8.225}$$

In the above equations,

$$D_2 = \det\left[\overline{n}_4\ \overline{n}_5\ \overline{m}_6\right] = \overline{n}_4^t\tilde{n}_5\overline{m}_6 = \overline{n}_5^t\tilde{m}_6\overline{n}_4 = \overline{m}_6^t\tilde{n}_4\overline{n}_5 \tag{8.226}$$

As noticed above, the motion singularities occur if $D_1 = 0$ and/or $D_2 = 0$. These singularities are identified readily during the solution by looking at the common denominators of the relevant expressions in Eqs. (8.213)–(8.215) and Eqs. (8.222)–(8.224). However, since the manipulator considered here is generic, D_1 and D_2 cannot be expressed easily so as to show their dependence on the joint variables explicitly. On the other hand, if a specific manipulator is of concern, then D_1 and D_2 can be expressed rather easily in terms of the joint variables and these expressions happen to be quite simple in general. Thus, it becomes possible to carry out a comprehensive singularity analysis with all the consequences in the task and joint spaces. Such analyses are carried out in Chapter 9 for a set of specific manipulators.

8.15.2 Inverse Acceleration Solution

As seen in Section 8.15.1, when the detailed formulation is used, the inverse velocity solution provides the joint velocities with the following generic formulas for $k = 1, 2, \ldots, m$.

$$\dot{q}_k = N_k/D_k \tag{8.227}$$

In Eq. (8.227), it happens that D_k is a function of the joint variables only, whereas N_k is a function of the joint variables and the specified components of the angular and translational velocities of the end-effector.

Owing to the availability of Eq. (8.227), the inverse acceleration solution can be obtained simply and directly by its differentiation for $k = 1, 2, \ldots, m$. That is,

$$\ddot{q}_k = (\dot{N}_k D_k - N_k \dot{D}_k)/D_k^2 = [\dot{N}_k - (N_k/D_k)\dot{D}_k]/D_k \Rightarrow$$
$$\ddot{q}_k = (\dot{N}_k - \dot{q}_k \dot{D}_k)/D_k \tag{8.228}$$

Note that both $\dot{q}_k$ and $\ddot{q}_k$ have the common denominator D_k in their expressions. Thus, the manipulator gets into a pose of motion singularity when $D_k = 0$, which affects both the velocity and acceleration relationships. Actually, it affects the relationships that involve the higher order motion derivatives, too. For example, by taking another derivative, the inverse jerk solution for the kth joint can be obtained as follows with the same denominator.

$$\dddot{q}_k = (\ddot{N}_k - \ddot{D}_k \dot{q}_k - 2\dot{D}_k \ddot{q}_k)/D_k \tag{8.229}$$

9

Kinematic Analyses of Typical Serial Manipulators

Synopsis

This chapter presents thorough kinematic analyses of some typical serial manipulators. The analyses are carried out by formulating the locations and orientations of the links separately so that the characteristic features of the manipulators are shown explicitly, and the symbolic manipulations required for the analytical treatments are facilitated. Most of the symbolic manipulations are carried out by using the algebraic properties (especially the shifting and expansion properties) of the exponentially expressed rotation matrices presented in Chapter 2. The kinematic descriptions of the manipulators are based on the D–H (Denavit–Hartenberg) convention explained in Chapter 7. Most of the manipulators considered here have certain configurations that are typically encountered in practice. They are known with certain generic names such as Puma manipulator, Stanford manipulator, etc. Each of these generic names represents a class of manipulators with similar configurations, i.e. with similar types and arrangements of the joints. This chapter also includes some conceptual manipulators that do not have generic names. They are not necessarily encountered in practice. However, they have certain kinematic features that make them interesting and instructive to study. Moreover, this chapter contains examples of deficient and redundant manipulators as well in order to demonstrate the particular complications that occur in their inverse kinematic solutions.

9.1 Puma Manipulator

A Puma manipulator is shown in Figure 9.1. Its generic name comes from the acronym PUMA that stands for "Programmable Universal Manipulation Arm." It was developed originally by *Victor Scheinman* at the *Unimation Robot Company*. Typical versions of Puma manipulators are the Puma-560 and Puma-600. A Puma manipulator comprises six revolute joints. So, it is designated symbolically as 6R or R^6. Its wrist is spherical. Its joint axes are also shown in Figure 9.1 together with the relevant unit vectors. Its kinematic details (the joint variables and the constant geometric parameters) are shown in the line diagrams in Figure 9.2. The line diagrams comprise the *side view*, the *top view*, and two *auxiliary views* that show the joint variables that are not seen in the side and top views. The significant points of the manipulator are as follows:

Kinematics of General Spatial Mechanical Systems, First Edition. M. Kemal Ozgoren.

Companion Website: www.wiley.com/go/ozgoren/spatialmechanicalsystems

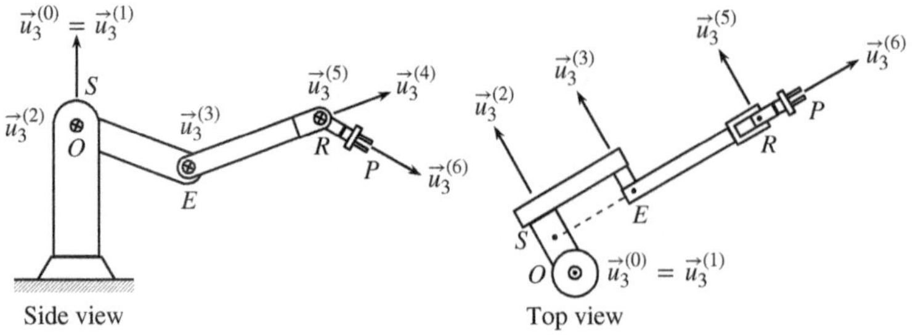

Figure 9.1 A Puma manipulator in its side and top views.

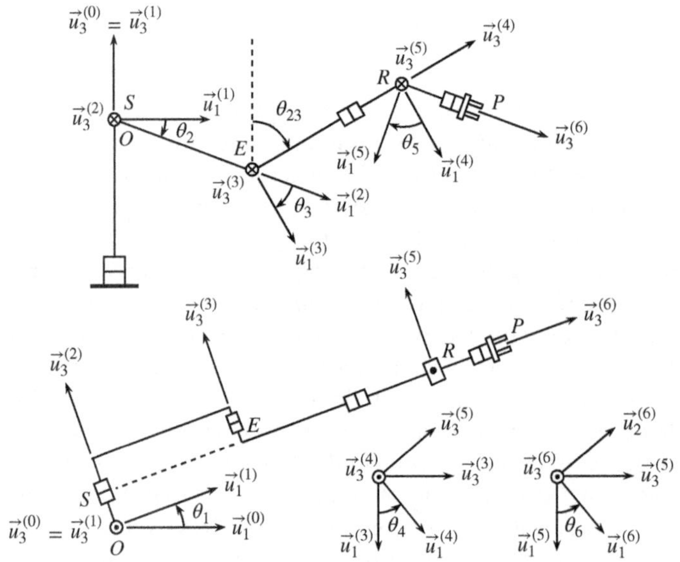

Figure 9.2 Line diagrams that show the kinematic details.

O: Center point or neck point (origin of the base frame)
S: Shoulder point, E: Elbow point, R: Wrist point, P: Tip point

9.1.1 Kinematic Description According to the D–H Convention

(a) Joint Variables

$$\theta_1, \theta_2, \theta_3, \theta_4, \theta_5, \theta_6$$

They are the rotation angles about the joint axes. They are shown in Figure 9.2.

(b) Twist Angles

$$\beta_1 = 0, \beta_2 = -\pi/2, \beta_3 = 0, \beta_4 = \pi/2, \beta_5 = -\pi/2, \beta_6 = \pi/2$$

(c) Offsets

$$d_1 = 0, d_2 = OS, d_3 = 0, d_4 = ER, d_5 = 0, d_6 = RP$$

The nonzero offsets are given the following special names.

d_2: Shoulder offset, d_6: Tip point offset
d_4: Front arm length (i.e. the overall length of Links 3 and 4)

(d) Effective Link Lengths

$$b_0 = 0, b_1 = 0, b_2 = SE, b_3 = 0, b_4 = 0, b_5 = 0$$

The only nonzero effective link length is b_2. It is called the "upper arm length" (i.e. the length of Link 2).

(e) Link Frame Origins

$$O_0 = O, O_1 = O, O_2 = S, O_3 = E, O_4 = R, O_5 = R, O_6 = P$$

9.1.2 Forward Kinematics in the Position Domain

(a) Link-to-Link Orientation Matrices

$$\hat{C}^{(0,1)} = e^{\tilde{u}_1\beta_1}e^{\tilde{u}_3\theta_1} = e^{\tilde{u}_3\theta_1} \quad (9.1)$$

$$\hat{C}^{(1,2)} = e^{\tilde{u}_1\beta_2}e^{\tilde{u}_3\theta_2} = e^{-\tilde{u}_1\pi/2}e^{\tilde{u}_3\theta_2} \quad (9.2)$$

$$\hat{C}^{(2,3)} = e^{\tilde{u}_1\beta_3}e^{\tilde{u}_3\theta_3} = e^{\tilde{u}_3\theta_3} \quad (9.3)$$

$$\hat{C}^{(3,4)} = e^{\tilde{u}_1\beta_4}e^{\tilde{u}_3\theta_4} = e^{\tilde{u}_1\pi/2}e^{\tilde{u}_3\theta_4} \quad (9.4)$$

$$\hat{C}^{(4,5)} = e^{\tilde{u}_1\beta_5}e^{\tilde{u}_3\theta_5} = e^{-\tilde{u}_1\pi/2}e^{\tilde{u}_3\theta_5} \quad (9.5)$$

$$\hat{C}^{(5,6)} = e^{\tilde{u}_1\beta_6}e^{\tilde{u}_3\theta_6} = e^{\tilde{u}_1\pi/2}e^{\tilde{u}_3\theta_6} \quad (9.6)$$

(b) Orientation Matrices of the Links with Respect to the Base Frame

$$\hat{C}_1 = \hat{C}^{(0,1)} = e^{\tilde{u}_3\theta_1} \quad (9.7)$$

$$\hat{C}_2 = \hat{C}^{(0,2)} = \hat{C}^{(0,1)}\hat{C}^{(1,2)} = e^{\tilde{u}_3\theta_1}e^{-\tilde{u}_1\pi/2}e^{\tilde{u}_3\theta_2} \Rightarrow$$

$$\hat{C}_2 = e^{\tilde{u}_3\theta_1}e^{\tilde{u}_2\theta_2}e^{-\tilde{u}_1\pi/2} \quad (9.8)$$

$$\hat{C}_3 = \hat{C}^{(0,3)} = \hat{C}^{(0,2)}\hat{C}^{(2,3)} = e^{\tilde{u}_3\theta_1}e^{\tilde{u}_2\theta_2}e^{-\tilde{u}_1\pi/2}e^{\tilde{u}_3\theta_3} \Rightarrow$$

$$\hat{C}_3 = e^{\tilde{u}_3\theta_1}e^{\tilde{u}_2\theta_2}e^{\tilde{u}_2\theta_3}e^{-\tilde{u}_1\pi/2} = e^{\tilde{u}_3\theta_1}e^{\tilde{u}_2(\theta_2+\theta_3)}e^{-\tilde{u}_1\pi/2} \Rightarrow$$

$$\hat{C}_3 = e^{\tilde{u}_3\theta_1}e^{\tilde{u}_2\theta_{23}}e^{-\tilde{u}_1\pi/2} \quad (9.9)$$

$$\hat{C}_4 = \hat{C}^{(0,4)} = \hat{C}^{(0,3)}\hat{C}^{(3,4)} = e^{\tilde{u}_3\theta_1}e^{\tilde{u}_2\theta_{23}}e^{-\tilde{u}_1\pi/2}e^{\tilde{u}_1\pi/2}e^{\tilde{u}_3\theta_4} \Rightarrow$$

$$\hat{C}_4 = e^{\tilde{u}_3\theta_1}e^{\tilde{u}_2\theta_{23}}e^{\tilde{u}_3\theta_4} \quad (9.10)$$

$$\hat{C}_5 = \hat{C}^{(0,5)} = \hat{C}^{(0,4)}\hat{C}^{(4,5)} = e^{\tilde{u}_3\theta_1}e^{\tilde{u}_2\theta_{23}}e^{\tilde{u}_3\theta_4}e^{-\tilde{u}_1\pi/2}e^{\tilde{u}_3\theta_5} \Rightarrow$$

$$\hat{C}_5 = e^{\tilde{u}_3\theta_1}e^{\tilde{u}_2\theta_{23}}e^{\tilde{u}_3\theta_4}e^{\tilde{u}_2\theta_5}e^{-\tilde{u}_1\pi/2} \quad (9.11)$$

$$\hat{C}_6 = \hat{C}^{(0,6)} = \hat{C}^{(0,5)}\hat{C}^{(5,6)} = e^{\tilde{u}_3\theta_1}e^{\tilde{u}_2\theta_{23}}e^{\tilde{u}_3\theta_4}e^{\tilde{u}_2\theta_5}e^{-\tilde{u}_1\pi/2}e^{\tilde{u}_1\pi/2}e^{\tilde{u}_3\theta_6} \Rightarrow$$

$$\hat{C}_6 = \hat{C} = e^{\tilde{u}_3\theta_1}e^{\tilde{u}_2\theta_{23}}e^{\tilde{u}_3\theta_4}e^{\tilde{u}_2\theta_5}e^{\tilde{u}_3\theta_6} \quad (9.12)$$

In Eq. (9.9), θ_{23} is introduced as a *combined joint variable.* It is defined as follows:

$$\theta_{23} = \theta_2 + \theta_3 \quad (9.13)$$

The angle θ_{23} is also shown in Figure 9.2. It has a particular significance because it indicates the declination of the front arm with respect to the third axis of the base frame.

(c) Location of the Wrist Point with Respect to the Base Frame

$$\vec{r} = \vec{r}_{OR} = \vec{r}_{OS} + \vec{r}_{SE} + \vec{r}_{ER} = d_2\vec{u}_3^{(2)} + b_2\vec{u}_1^{(2)} + d_4\vec{u}_3^{(4)} \tag{9.14}$$

The vector Eq. (9.14) implies the following matrix equation in the base frame.

$$\bar{r} = \bar{r}^{(0)} = d_2\bar{u}_3^{(2/0)} + b_2\bar{u}_1^{(2/0)} + d_4\bar{u}_3^{(4/0)} \tag{9.15}$$

Equation (9.15) can be manipulated as shown below.

$$\bar{r} = d_2\hat{C}^{(0,2)}\bar{u}_3^{(2/2)} + b_2\hat{C}^{(0,2)}\bar{u}_1^{(2/2)} + d_4\hat{C}^{(0,4)}\bar{u}_3^{(4/4)} \Rightarrow$$

$$\bar{r} = d_2 e^{\tilde{u}_3\theta_1} e^{\tilde{u}_2\theta_2} e^{-\tilde{u}_1\pi/2}\bar{u}_3 + b_2 e^{\tilde{u}_3\theta_1} e^{\tilde{u}_2\theta_2} e^{-\tilde{u}_1\pi/2}\bar{u}_1 + d_4 e^{\tilde{u}_3\theta_1} e^{\tilde{u}_2\theta_{23}} e^{\tilde{u}_3\theta_4}\bar{u}_3 \Rightarrow$$

$$\bar{r} = d_2 e^{\tilde{u}_3\theta_1}\bar{u}_2 + b_2 e^{\tilde{u}_3\theta_1} e^{\tilde{u}_2\theta_2}\bar{u}_1 + d_4 e^{\tilde{u}_3\theta_1} e^{\tilde{u}_2\theta_{23}}\bar{u}_3 \Rightarrow$$

$$\bar{r} = e^{\tilde{u}_3\theta_1}(d_2\bar{u}_2 + b_2 e^{\tilde{u}_2\theta_2}\bar{u}_1 + d_4 e^{\tilde{u}_2\theta_{23}}\bar{u}_3) \Rightarrow$$

$$\bar{r} = e^{\tilde{u}_3\theta_1}[d_2\bar{u}_2 + b_2(\bar{u}_1 c\theta_2 - \bar{u}_3 s\theta_2) + d_4(\bar{u}_3 c\theta_{23} + \bar{u}_1 s\theta_{23})] \Rightarrow$$

$$\bar{r} = e^{\tilde{u}_3\theta_1}[\bar{u}_1(d_4 s\theta_{23} + b_2 c\theta_2) + \bar{u}_2 d_2 + \bar{u}_3(d_4 c\theta_{23} - b_2 s\theta_2)] \tag{9.16}$$

Equation (9.16) can also be written as

$$\bar{r} = e^{\tilde{u}_3\theta_1}\bar{r}' \tag{9.17}$$

In Eq. (9.17),

$$\bar{r}' = \bar{u}_1(d_4 s\theta_{23} + b_2 c\theta_2) + \bar{u}_2 d_2 + \bar{u}_3(d_4 c\theta_{23} - b_2 s\theta_2) \tag{9.18}$$

Note that $\bar{r}'$ is the matrix representation of the vector $\vec{r}$ in the link frame $\mathcal{F}_1(O)$, i.e. $\bar{r}' = \bar{r}^{(1)}$. This fact can be verified as follows:

$$\bar{r} = \bar{r}^{(0)} = \hat{C}^{(0,1)}\bar{r}^{(1)} = e^{\tilde{u}_3\theta_1}\bar{r}^{(1)} \Rightarrow \bar{r}^{(1)} = \bar{r}' \tag{9.19}$$

Note further that, referring to Figure 9.2, $\bar{r}' = \bar{u}_1 r_1' + \bar{u}_2 r_2' + \bar{u}_3 r_3'$ can also be written down directly by inspection. In other words, Figure 9.2 readily implies the following coordinates of the wrist point R in the frame $\mathcal{F}_1(O)$.

$$\left.\begin{aligned} r_1' &= d_4 s\theta_{23} + b_2 c\theta_2 \\ r_2' &= d_2 \\ r_3' &= d_4 c\theta_{23} - b_2 s\theta_2 \end{aligned}\right\} \tag{9.20}$$

On the other hand, the expansion of the expression in Eq. (9.16) leads to the coordinates of the wrist point R in the base frame $\mathcal{F}_0(O)$ as follows:

$$\bar{r}^{(0)} = \bar{r} = \bar{u}_1 r_1 + \bar{u}_2 r_2 + \bar{u}_3 r_3 \tag{9.21}$$

$$\bar{r} = (e^{\tilde{u}_3\theta_1}\bar{u}_1)(d_4 s\theta_{23} + b_2 c\theta_2) + (e^{\tilde{u}_3\theta_1}\bar{u}_2)d_2 + (e^{\tilde{u}_3\theta_1}\bar{u}_3)(d_4 c\theta_{23} - b_2 s\theta_2) \Rightarrow$$

$$\bar{r} = (\bar{u}_1 c\theta_1 + \bar{u}_2 s\theta_1)(d_4 s\theta_{23} + b_2 c\theta_2) + (\bar{u}_2 c\theta_1 - \bar{u}_1 s\theta_1)d_2 + (\bar{u}_3)(d_4 c\theta_{23} - b_2 s\theta_2) \Rightarrow$$

$$\begin{aligned}\bar{r} = {} & \bar{u}_1[(d_4 s\theta_{23} + b_2 c\theta_2)c\theta_1 - d_2 s\theta_1] + \bar{u}_2[(d_4 s\theta_{23} + b_2 c\theta_2)s\theta_1 + d_2 c\theta_1] \\ & + \bar{u}_3(d_4 c\theta_{23} - b_2 s\theta_2)\end{aligned} \tag{9.22}$$

Equation (9.22) indicates that

$$\left.\begin{aligned} r_1 &= (d_4 s\theta_{23} + b_2 c\theta_2)c\theta_1 - d_2 s\theta_1 \\ r_2 &= (d_4 s\theta_{23} + b_2 c\theta_2)s\theta_1 + d_2 c\theta_1 \\ r_3 &= d_4 c\theta_{23} - b_2 s\theta_2 \end{aligned}\right\} \tag{9.23}$$

(d) Location of the Tip Point with Respect to the Base Frame

$$\vec{p} = \vec{r}_{OP} = \vec{r}_{OR} + \vec{r}_{RP} = \vec{r} + d_6 \vec{u}_3^{(6)} \tag{9.24}$$

Equation (9.24) implies the following matrix equation in the base frame.

$$\overline{p} = \overline{p}^{(0)} = \overline{r}^{(0)} + d_6 \overline{u}_3^{(6/0)} = \overline{r}^{(0)} + d_6 \widehat{C}^{(0,6)} \overline{u}_3^{(6/6)} \Rightarrow$$

$$\overline{p} = \overline{r} + d_6 \widehat{C} \overline{u}_3 \tag{9.25}$$

When Eq. (9.12) is substituted, Eq. (9.25) becomes

$$\overline{p} = \overline{r} + d_6 e^{\tilde{u}_3 \theta_1} e^{\tilde{u}_2 \theta_{23}} e^{\tilde{u}_3 \theta_4} e^{\tilde{u}_2 \theta_5} \overline{u}_3 \tag{9.26}$$

9.1.3 Inverse Kinematics in the Position Domain

Here, $\overline{p}$ and $\widehat{C}$ are specified. Hence, $\overline{r}$ also becomes specified according to Eq. (9.25). That is,

$$\overline{r} = \overline{p} - d_6 \widehat{C} \overline{u}_3 \tag{9.27}$$

(a) Solution for the Arm Joint Variables

The wrist of a Puma manipulator is spherical. Therefore, the inverse kinematic solution can be started from the wrist location equation, i.e. Eq. (9.16), which can be rearranged as follows in order to distribute the unknowns $(\theta_1, \theta_2, \theta_{23})$ more or less evenly to both sides of the equation:

$$\overline{u}_1(d_4 s\theta_{23} + b_2 c\theta_2) + \overline{u}_2 d_2 + \overline{u}_3(d_4 c\theta_{23} - b_2 s\theta_2) = e^{-\tilde{u}_3 \theta_1} \overline{r} \tag{9.28}$$

Upon premultiplications by $\overline{u}_1^t$, $\overline{u}_2^t$, and $\overline{u}_3^t$, Eq. (9.28) leads to the following scalar equations.

$$d_4 s\theta_{23} + b_2 c\theta_2 = \overline{u}_1^t e^{-\tilde{u}_3 \theta_1} \overline{r} = (\overline{u}_1^t c\theta_1 + \overline{u}_2^t s\theta_1)\overline{r} \Rightarrow$$

$$d_4 s\theta_{23} + b_2 c\theta_2 = r_1 c\theta_1 + r_2 s\theta_1 \tag{9.29}$$

$$d_2 = \overline{u}_2^t e^{-\tilde{u}_3 \theta_1} \overline{r} = (\overline{u}_2^t c\theta_1 - \overline{u}_1^t s\theta_1)\overline{r} \Rightarrow$$

$$r_2 c\theta_1 - r_1 s\theta_1 = d_2 \tag{9.30}$$

$$d_4 c\theta_{23} - b_2 s\theta_2 = \overline{u}_3^t e^{-\tilde{u}_3 \theta_1} \overline{r} = \overline{u}_3^t \overline{r} \Rightarrow$$

$$d_4 c\theta_{23} - b_2 s\theta_2 = r_3 \tag{9.31}$$

Note that Eq. (9.30) contains θ_1 as the only unknown. It is a *sine–cosine equation* and it can be solved for θ_1 by using various methods. One of the most frequently used methods is the *phase angle method*, which is based on the following transformation of r_1 and r_2 into r_{12} and ϕ_1.

$$\left.\begin{aligned} r_1 &= r_{12}\cos\phi_1 \\ r_2 &= r_{12}\sin\phi_1 \end{aligned}\right\} \tag{9.32}$$

By defining the transformation with the positive square root, Eq. (9.32) implies that

$$r_{12} = \sqrt{r_1^2 + r_2^2} \tag{9.33}$$

If $r_{12} \neq 0$, i.e. if r_1 and r_2 are not both zero, Eq. (9.32) implies further that

$$\phi_1 = \text{atan}_2(r_2, r_1) \tag{9.34}$$

After Eq. Set (9.32) is substituted, Eq. (9.30) can be manipulated into the following form.

$$r_{12}\sin(\phi_1 - \theta_1) = d_2 \tag{9.35}$$

Equation (9.35) can further be written as

$$\sin\psi_1 = \eta_1 \tag{9.36}$$

In Eq. (9.36),

$$\psi_1 = \phi_1 - \theta_1 \tag{9.37}$$

$$\eta_1 = d_2 / r_{12} \tag{9.38}$$

Equation (9.36) implies that

$$\cos\psi_1 = \sigma_1\sqrt{1 - \eta_1^2}; \quad \sigma_1 = \pm 1 \tag{9.39}$$

Hence, θ_1 is obtained as follows jointly from Eqs. (9.36) and (9.39).

$$\psi_1 = \text{atan}_2(\eta_1, \sigma_1\sqrt{1 - \eta_1^2}) \Rightarrow$$

$$\theta_1 = \phi_1 - \psi_1 = \text{atan}_2(r_2, r_1) - \text{atan}_2(\eta_1, \sigma_1\sqrt{1 - \eta_1^2}) \tag{9.40}$$

Equation (9.40) implies that θ_1 will be real if $\eta_1^2 = (d_2/r_{12})^2 \leq 1$, i.e. if

$$r_1^2 + r_2^2 \geq d_2^2 \tag{9.41}$$

The above inequality constitutes a *workspace limitation* for the manipulator.
With the availability of θ_1, Eqs. (9.29) and (9.31) can be manipulated as shown below in order to find θ_3 and θ_2.

$$\left.\begin{aligned} &d_4(s\theta_2 c\theta_3 + c\theta_2 s\theta_3) + b_2 c\theta_2 = r_{12}c\phi_1 c\theta_1 + r_{12}s\phi_1 s\theta_1 \\ &d_4(c\theta_2 c\theta_3 - s\theta_2 s\theta_3) - b_2 s\theta_2 = r_3 \end{aligned}\right\} \Rightarrow$$

$$\left.\begin{aligned} &(d_4 c\theta_3)s\theta_2 + (b_2 + d_4 s\theta_3)c\theta_2 = r_{12}\cos(\phi_1 - \theta_1) \\ &(d_4 c\theta_3)c\theta_2 - (b_2 + d_4 s\theta_3)s\theta_2 = r_3 \end{aligned}\right\} \Rightarrow$$

$$(d_4 c\theta_3)s\theta_2 + (b_2 + d_4 s\theta_3)c\theta_2 = \sigma_1 r_{12}\sqrt{1 - \eta_1^2} = \sigma_1\sqrt{r_1^2 + r_2^2 - d_2^2} \tag{9.42}$$

$$(d_4 c\theta_3)c\theta_2 - (b_2 + d_4 s\theta_3)s\theta_2 = r_3 \tag{9.43}$$

The following equation is obtained by adding the squares of Eqs. (9.42) and (9.43).

$$(d_4 c\theta_3)^2 + (b_2 + d_4 s\theta_3)^2 = r_1^2 + r_2^2 + r_3^2 - d_2^2 \Rightarrow$$
$$d_4^2 + b_2^2 + 2b_2 d_4 s\theta_3 = r_1^2 + r_2^2 + r_3^2 - d_2^2 \Rightarrow$$
$$\sin\theta_3 = \eta_3 = [(r_1^2 + r_2^2 + r_3^2) - (d_2^2 + b_2^2 + d_4^2)]/(2b_2 d_4) \tag{9.44}$$

Equation (9.44) implies that

$$\cos\theta_3 = \sigma_3\sqrt{1-\eta_3^2};\quad \sigma_3 = \pm 1 \tag{9.45}$$

Equations (9.44) and (9.45) give θ_3 as

$$\theta_3 = \operatorname{atan}_2(\eta_3, \sigma_3\sqrt{1-\eta_3^2}) \tag{9.46}$$

Equation (9.46) implies that θ_3 will be real if $-1 \le \eta_3 \le 1$, i.e. if

$$-2b_2 d_4 \le (r_1^2 + r_2^2 + r_3^2) - (d_2^2 + b_2^2 + d_4^2) \le 2b_2 d_4 \Rightarrow$$
$$d_2^2 + (b_2 - d_4)^2 \le r_1^2 + r_2^2 + r_3^2 \le d_2^2 + (b_2 + d_4)^2 \tag{9.47}$$

The above inequality constitutes another *workspace limitation* for the manipulator. After finding θ_3, Eqs. (9.42) and (9.43) can be written together as the following matrix equation in order to obtain $s\theta_2$ and $c\theta_2$ separately so that an additional superfluous sign variable is not introduced in finding θ_2.

$$\begin{bmatrix} d_4 c\theta_3 & (b_2 + d_4 s\theta_3) \\ -(b_2 + d_4 s\theta_3) & d_4 c\theta_3 \end{bmatrix} \begin{bmatrix} s\theta_2 \\ c\theta_2 \end{bmatrix} = \begin{bmatrix} \sigma_1\sqrt{r_1^2 + r_2^2 - d_2^2} \\ r_3 \end{bmatrix} \tag{9.48}$$

In Eq. (9.48), the determinant of the coefficient matrix is

$$D_2 = (d_4 c\theta_3)^2 + (b_2 + d_4 s\theta_3)^2 = b_2^2 + d_4^2 + 2b_2 d_4 s\theta_3 \tag{9.49}$$

If $D_2 \ne 0$, i.e. if $D_2 > 0$, Eq. (9.48) gives $s\theta_2$ and $c\theta_2$ as follows:

$$\begin{bmatrix} s\theta_2 \\ c\theta_2 \end{bmatrix} = \frac{1}{D_2}\begin{bmatrix} d_4 c\theta_3 & -(b_2 + d_4 s\theta_3) \\ (b_2 + d_4 s\theta_3) & d_4 c\theta_3 \end{bmatrix} \begin{bmatrix} \sigma_1\sqrt{r_1^2 + r_2^2 - d_2^2} \\ r_3 \end{bmatrix} \Rightarrow$$

$$\left.\begin{aligned} s\theta_2 &= [\sigma_1 d_4\sqrt{r_1^2 + r_2^2 - d_2^2}c\theta_3 - (b_2 + d_4 s\theta_3)r_3]/D_2 = B_2/D_2 \\ c\theta_2 &= [\sigma_1(b_2 + d_4 s\theta_3)\sqrt{r_1^2 + r_2^2 - d_2^2} + d_4 r_3 c\theta_3]/D_2 = A_2/D_2 \end{aligned}\right\} \tag{9.50}$$

Hence, θ_2 is found as

$$\theta_2 = \operatorname{atan}_2(B_2, A_2) \tag{9.51}$$

(b) Solution for the Wrist Joint Variables

In order to find the wrist joint variables $(\theta_4, \theta_5, \theta_6)$ after finding the arm joint variables, the end-effector orientation equation, i.e. Eq. (9.12), can be written as follows:

$$e^{\tilde{u}_3\theta_4} e^{\tilde{u}_2\theta_5} e^{\tilde{u}_3\theta_6} = \hat{C}^* \tag{9.52}$$

In Eq. (9.52), $\hat{C}^*$ is known as

$$\hat{C}^* = e^{-\tilde{u}_2\theta_{23}} e^{-\tilde{u}_3\theta_1}\hat{C} \tag{9.53}$$

Note that the unknown angles in Eq. (9.52) are arranged like the Euler angles of the 3-2-3 sequence. Therefore, they can be extracted by using the similar procedure explained in Chapter 3. This procedure leads to the following scalar equation pairs generated from Eq. (9.52).

$$c_{33}^* = \overline{u}_3^t \hat{C}^* \overline{u}_3 = c\theta_5 \Rightarrow s\theta_5 = \sigma_5\sqrt{1-(c_{33}^*)^2}; \quad \sigma_5 = \pm 1 \tag{9.54}$$

$$c_{23}^* = \overline{u}_2^t \hat{C}^* \overline{u}_3 = s\theta_4 s\theta_5; \quad c_{13}^* = \overline{u}_1^t \hat{C}^* \overline{u}_3 = c\theta_4 s\theta_5 \tag{9.55}$$

$$c_{32}^* = \overline{u}_3^t \hat{C}^* \overline{u}_2 = s\theta_5 s\theta_6; \quad c_{31}^* = \overline{u}_3^t \hat{C}^* \overline{u}_1 = -s\theta_5 c\theta_6 \tag{9.56}$$

Equation Pair (9.54) gives θ_5 as

$$\theta_5 = \text{atan}_2(\sigma_5\sqrt{1-(c_{33}^*)^2}, c_{33}^*) = \sigma_5 \text{atan}_2(\sqrt{1-(c_{33}^*)^2}, c_{33}^*) \tag{9.57}$$

If $s\theta_5 \neq 0$, the other equation pairs give θ_4 and θ_6 as follows without additional superfluous sign variables:

$$\theta_4 = \text{atan}_2(\sigma_5 c_{23}^*, \sigma_5 c_{13}^*) \tag{9.58}$$

$$\theta_6 = \text{atan}_2(\sigma_5 c_{32}^*, -\sigma_5 c_{31}^*) \tag{9.59}$$

9.1.4 Multiplicity Analysis

(a) First Kind of Multiplicity

The first kind of multiplicity is associated with the sign variable σ_1 that arises in the process of finding θ_1 by using Eqs. (9.39) and (9.40). The manipulator attains the same location of the wrist point by assuming one of the two alternative poses that correspond to $\sigma_1 = +1$ and $\sigma_1 = -1$. These poses are illustrated in Figure 9.3. The poses corresponding to $\sigma_1 = +1$ and $\sigma_1 = -1$ are designated, respectively, as *left shouldered pose* and *right shouldered pose*. In both poses, as implied by Eq. (9.34), the angle ϕ_1 denotes the projected direction of the line *OR* on the 1–2 plane of the base frame. Meanwhile, Eq. (9.39) implies that the angle ψ_1 becomes *acute* (i.e. $\psi_1 < \pi/2$ with $c\psi_1 > 0$) if $\sigma_1 = +1$ and it becomes *obtuse* (i.e. $\psi_1 > \pi/2$ with $c\psi_1 < 0$) if $\sigma_1 = -1$. On the other hand, $\theta_1 = \phi_1 - \psi_1$ according to Eq. (9.40). That is why θ_1 assumes the positive and negative values shown

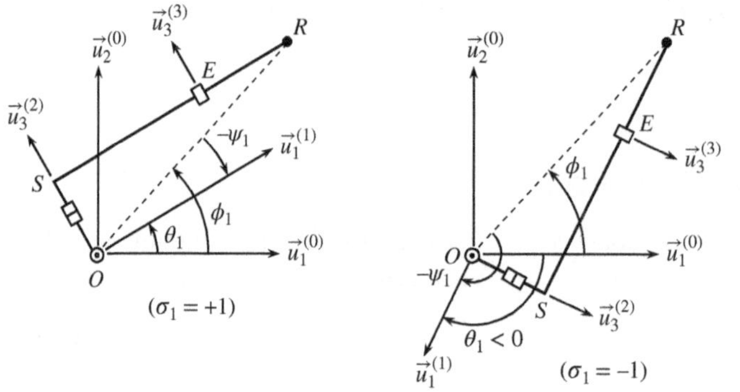

Figure 9.3 Left and right shouldered poses of a Puma manipulator.

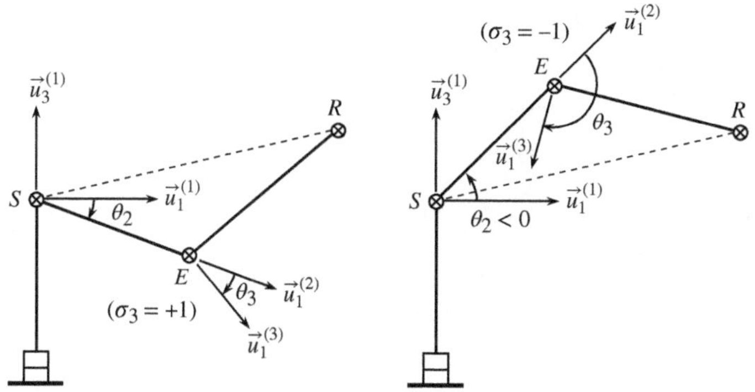

Figure 9.4 Elbow-down and elbow-up poses of a Puma manipulator.

in Figure 9.3 when the manipulator is in one of the left and right shouldered poses. Of course, θ_2 also assumes a value that is compatible with one of these poses as implied by Eqs. (9.50) and (9.51).

(b) Second Kind of Multiplicity

The second kind of multiplicity is associated with the sign variable σ_3 that arises in the process of finding θ_3 by using Eqs. (9.44)–(9.46). The manipulator attains the same location of the wrist point by assuming one of the two alternative poses that correspond to $\sigma_3 = +1$ and $\sigma_3 = -1$. These poses are illustrated in Figure 9.4. The poses corresponding to $\sigma_3 = +1$ and $\sigma_3 = -1$ are designated, respectively, as *elbow-down pose* and *elbow-up pose*. These designations are justified by Eq. (9.46), because it indicates that, for the same value of $\eta_3 = \sin\theta_3$, θ_3 becomes acute if $\sigma_3 = +1$ and θ_3 becomes obtuse if $\sigma_3 = -1$. The corresponding values of θ_2 are determined by Eqs. (9.50) and (9.51).

(c) Third Kind of Multiplicity

The third kind of multiplicity is associated with the sign variable σ_5 that arises in the process of finding the wrist angles $(\theta_5, \theta_4, \theta_6)$ by using Eqs. (9.57)–(9.59). The manipulator attains the same orientation of the end-effector with both $\sigma_5 = +1$ and $\sigma_5 = -1$. As opposed to σ_1 and σ_3, σ_5 does not lead to visually distinct poses. It only leads to a phenomenon called *wrist flip*. This is a phenomenon, in which the end-effector arrives at an orientation that looks like the initial orientation after three successive rotations. This phenomenon is a characteristic feature of a spherical wrist. It can be explained mathematically as in the following.

Suppose that $\sigma_5 = +1$ leads to the angles θ_5, θ_4, and θ_6. If so, Eqs. (9.57)–(9.59) imply that $\sigma_5 = -1$ leads to the angles $\theta_5' = -\theta_5$, $\theta_4' = \theta_4 + \pi$, and $\theta_6' = \theta_6 + \pi$. However, these two sets of angles, lead in turn to the same orientation matrix $\hat{C}^*$, which is expressed by Eq. (9.52). In order to verify this fact, let

$$\hat{C}^{**} = e^{\tilde{u}_3\theta_4'} e^{\tilde{u}_2\theta_5'} e^{\tilde{u}_3\theta_6'} \tag{9.60}$$

After inserting the primed angles with their expressions in terms of the unprimed angles, Eq. (9.60) can be manipulated as shown below by using the *shifting property* of the rotation matrices.

$$\hat{C}^{**} = e^{\tilde{u}_3(\theta_4+\pi)} e^{-\tilde{u}_2\theta_5} e^{\tilde{u}_3(\theta_6+\pi)} = e^{\tilde{u}_3\theta_4} e^{\tilde{u}_3\pi} e^{-\tilde{u}_2\theta_5} e^{\tilde{u}_3\pi} e^{\tilde{u}_3\theta_6} \Rightarrow$$

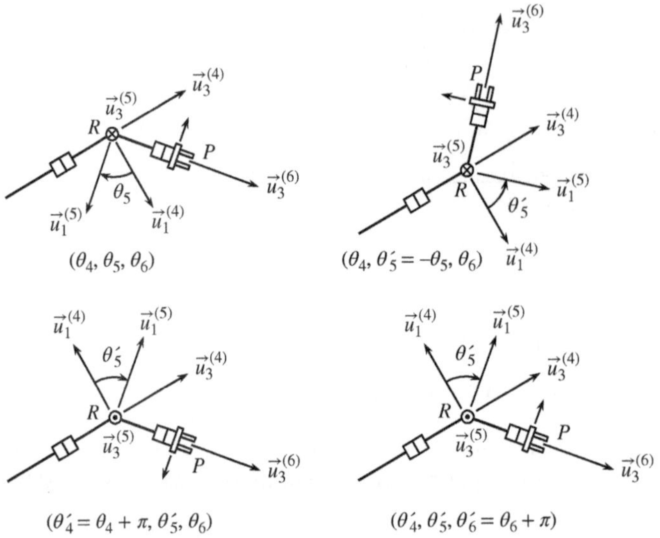

Figure 9.5 The wrist flip phenomenon of the third kind of multiplicity.

$$\hat{C}^{**} = e^{\tilde{u}_3\theta_4}e^{\tilde{u}_2\theta_5}e^{\tilde{u}_3(2\pi)}e^{\tilde{u}_3\theta_6} = e^{\tilde{u}_3\theta_4}e^{\tilde{u}_2\theta_5}e^{\tilde{u}_3(\theta_6+2\pi)} \Rightarrow$$

$$\hat{C}^{**} = e^{\tilde{u}_3\theta_4}e^{\tilde{u}_2\theta_5}e^{\tilde{u}_3\theta_6} = \hat{C}^{*} \tag{9.61}$$

As verified above, $\hat{C}^{**}$ turns out to be the same as $\hat{C}^{*}$. The wrist flip phenomenon can also be verified figuratively as illustrated in Figure 9.5 by means of three successive rotations. Thus, it is concluded that σ_5 does not lead to visually distinct poses. Therefore, it can be taken as $\sigma_5 = +1$ without loss of generality.

9.1.5 Singularity Analysis in the Position Domain

A Puma manipulator may have three distinct kinds of position singularities, which are described and discussed below.

(a) First Kind of Position Singularity

Equation (9.30) implies that the first kind of position singularity occurs if the position of the end-effector is specified so that $r_1 = r_2 = 0$. The same equation also implies that such a specification can be made only if the manipulator happens to be so special that $d_2 = 0$. This singularity is illustrated in Figure 9.6. As a noticeable feature of this singularity, the wrist point is located on the axis of the first joint. If this singularity occurs, θ_1 becomes indefinite and ineffective. In other words, θ_1 can be assigned an arbitrary value, but it cannot cause any change in the position of the end-effector, whatever the assigned value is. Thus, the manipulator gains a positioning freedom in the joint space, but this freedom happens to be useless in the task space as long as the wrist point is kept on the axis of the first joint. On the other hand, if the task to be executed necessitates to keep the wrist point as such, then the freedom in θ_1 may be used to orient the arm of the manipulator conveniently depending on the environmental conditions, e.g. the locations of the obstacles that may exist in the vicinity.

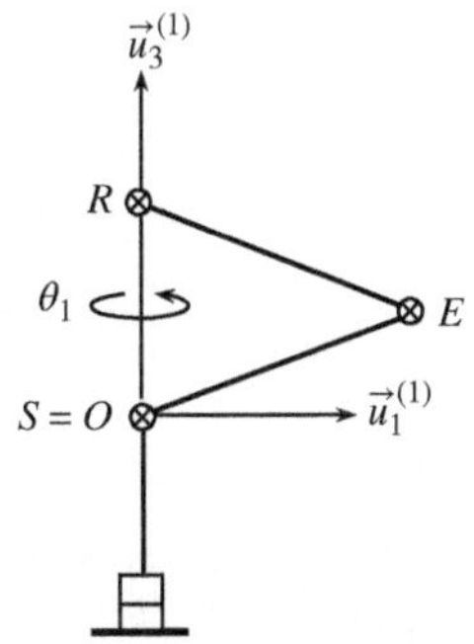

Figure 9.6 First kind of position singularity of a Puma manipulator.

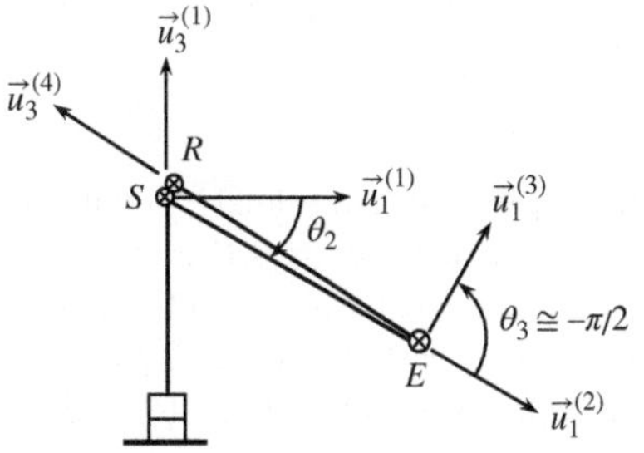

Figure 9.7 Second kind of position singularity of a Puma manipulator.

(b) Second Kind of Position Singularity

Equation (9.48) implies that the second kind of position singularity occurs if $D_2 = 0$, i.e. if the determinant of the coefficient matrix vanishes. According to Eq. (9.49), D_2 can vanish only if the following equations are both satisfied.

$$b_2 = d_4 \tag{9.62}$$

$$s\theta_3 = -1 \text{ or } \theta_3 = -\pi/2 \tag{9.63}$$

Equation (9.48) also implies that this singularity occurs if the position of the end-effector is specified in such a way that the wrist point coincides with the shoulder point. That is,

$$r_{12} = \sqrt{r_1^2 + r_2^2} = d_2 \tag{9.64}$$

$$r_3 = 0 \tag{9.65}$$

The above equations express the noticeable feature of this singularity that the upper and front arms of the manipulator have equal lengths and the front arm is folded completely over the upper arm. However, this singularity cannot occur exactly because of the physical shapes of the relevant links and joints. This singularity is illustrated (approximately) in Figure 9.7.

If this singularity occurs, θ_2 becomes indefinite and ineffective. In other words, θ_2 can be assigned an arbitrary value, but it cannot cause any change in the position of the end-effector, whatever the assigned value is. Thus, the manipulator gains a positioning freedom in the joint space, but this freedom happens to be useless in the task space as

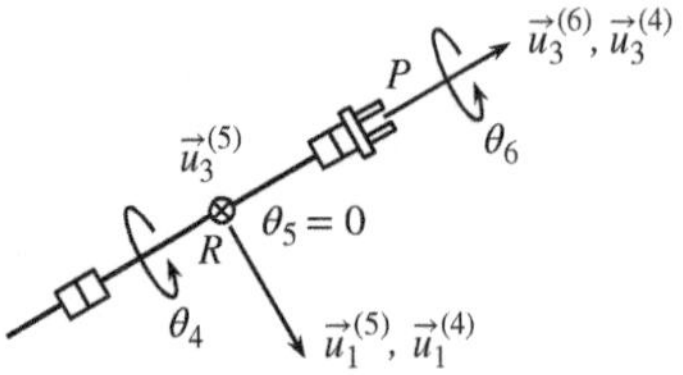

Figure 9.8 Third kind of position singularity of a Puma manipulator.

long as the wrist point is kept coincident (actually almost coincident) with the shoulder point. On the other hand, if the task to be executed necessitates keeping the wrist point as such, then the freedom in θ_2 may be used to orient the folded arm links of the manipulator conveniently depending on the environmental conditions. For example, the coincident upper and front arms may be allowed to get aligned with the gravity direction so that the task can be executed with minimal potential energy.

(c) Third Kind of Position Singularity

Equations (9.58) and (9.59) imply that the third kind of position singularity occurs if $s\theta_5 = 0$, i.e. if $\theta_5 = 0$ or $\theta_5 = \pi$. In other words, this singularity occurs if the orientation of the end-effector is specified so that the approach vector is aligned with the front arm. However, while this singularity can occur easily with $\theta_5 = 0$, it can hardly occur with $\theta_5 = \pi$ due to the physical shapes of the relevant links and joints. Therefore, this singularity is illustrated in Figure 9.8 with $\theta_5 = 0$.

If this singularity occurs, θ_4 and θ_6 become indefinite so that they cannot be found separately. This is because they turn out to be rotation angles about coincident axes, which are the axes of the fourth and sixth joints, and therefore they cannot be distinguished from each other. Nevertheless, their combination, i.e. $\theta_{46} = \theta_4 + \theta_6$, can still be found as explained below.

In the singularity with $\theta_5 = 0$, Eq. (9.52) becomes

$$e^{\tilde{u}_3\theta_4}\hat{I}e^{\tilde{u}_3\theta_6} = e^{\tilde{u}_3\theta_4}e^{\tilde{u}_3\theta_6} = e^{\tilde{u}_3(\theta_4+\theta_6)} = e^{\tilde{u}_3\theta_{46}} = \hat{C}^* \tag{9.66}$$

Equation (9.66) gives θ_{46} as follows:

$$c_{11}^* = \overline{u}_1^t\hat{C}^*\overline{u}_1 = c\theta_{46}, c_{21}^* = \overline{u}_2^t\hat{C}^*\overline{u}_1 = s\theta_{46} \Rightarrow$$

$$\theta_{46} = \text{atan}_2(c_{21}^*, c_{11}^*) \tag{9.67}$$

Since θ_{46} becomes known, only one of θ_4 and θ_6, e.g. θ_6, can be assigned an arbitrary value and the other one, i.e. θ_4, can then be found as $\theta_4 = \theta_{46} - \theta_6$. Thus, the manipulator gains a positioning freedom in the joint space, but this freedom happens to be useless in the task space as long as the approach vector is kept aligned with the front arm.

Note that, in such a pose of the manipulator, it becomes impossible to rotate the end-effector with respect to Link 4 about the coincident unit vectors $\vec{u}_1^{(4)}$ and $\vec{u}_1^{(5)}$.

9.1.6 Forward Kinematics in the Velocity Domain

(a) Angular Velocities of the Links with Respect to the Base Frame

Let $\overline{\omega}_k = \overline{\omega}_{0,k}^{(0)} = \text{colm}[\dot{\hat{C}}_k\hat{C}_k^t]$ for $k = 1, 2, \ldots, 6$. Then, Eqs. (9.7)–(9.13) lead to the following angular velocity expressions.

$$\overline{\omega}_1 = \dot{\theta}_1\overline{u}_3 \tag{9.68}$$

$$\overline{\omega}_2 = \dot{\theta}_1\overline{u}_3 + \dot{\theta}_2 e^{\tilde{u}_3\theta_1}\overline{u}_2 \tag{9.69}$$

$$\overline{\omega}_3 = \dot{\theta}_1\overline{u}_3 + \dot{\theta}_{23}e^{\tilde{u}_3\theta_1}\overline{u}_2 \tag{9.70}$$

$$\overline{\omega}_4 = \dot{\theta}_1\overline{u}_3 + \dot{\theta}_{23}e^{\tilde{u}_3\theta_1}\overline{u}_2 + \dot{\theta}_4 e^{\tilde{u}_3\theta_1}e^{\tilde{u}_2\theta_{23}}\overline{u}_3 \tag{9.71}$$

$$\overline{\omega}_5 = \dot{\theta}_1\overline{u}_3 + \dot{\theta}_{23}e^{\tilde{u}_3\theta_1}\overline{u}_2 + \dot{\theta}_4 e^{\tilde{u}_3\theta_1}e^{\tilde{u}_2\theta_{23}}\overline{u}_3 + \dot{\theta}_5 e^{\tilde{u}_3\theta_1}e^{\tilde{u}_2\theta_{23}}e^{\tilde{u}_3\theta_4}\overline{u}_2 \tag{9.72}$$

$$\begin{aligned}\overline{\omega}_6 = {}& \dot{\theta}_1\overline{u}_3 + \dot{\theta}_{23}e^{\tilde{u}_3\theta_1}\overline{u}_2 + \dot{\theta}_4 e^{\tilde{u}_3\theta_1}e^{\tilde{u}_2\theta_{23}}\overline{u}_3 + \dot{\theta}_5 e^{\tilde{u}_3\theta_1}e^{\tilde{u}_2\theta_{23}}e^{\tilde{u}_3\theta_4}\overline{u}_2 \\ & + \dot{\theta}_6 e^{\tilde{u}_3\theta_1}e^{\tilde{u}_2\theta_{23}}e^{\tilde{u}_3\theta_4}e^{\tilde{u}_2\theta_5}\overline{u}_3\end{aligned} \tag{9.73}$$

For the sake of notational convenience, $\overline{\omega}_6$ is denoted here simply as $\overline{\omega}$. That is, $\overline{\omega}_6 = \overline{\omega}$.

(b) Velocity of the Wrist Point with Respect to the Base Frame

Let $\overline{w} = \overline{w}^{(0)} = \overline{v}_R = \dot{\overline{r}}$. Then, $\overline{w}$ can be obtained from Eq. (9.16) as follows:

$$\begin{aligned}\overline{w} = {}& (\dot{\theta}_1 e^{\tilde{u}_3\theta_1}\tilde{u}_3)\left[\overline{u}_1(d_4 s\theta_{23} + b_2 c\theta_2) + \overline{u}_2 d_2 + \overline{u}_3(d_4 c\theta_{23} - b_2 s\theta_2)\right] \\ & + \left(e^{\tilde{u}_3\theta_1}\right)\left[\overline{u}_1(d_4\dot{\theta}_{23}c\theta_{23} - b_2\dot{\theta}_2 s\theta_2) + \overline{u}_3(-d_4\dot{\theta}_{23}s\theta_{23} - b_2\dot{\theta}_2 c\theta_2)\right] \Rightarrow\end{aligned}$$

$$\begin{aligned}\overline{w} = {}& e^{\tilde{u}_3\theta_1}[\overline{u}_1(d_4\dot{\theta}_{23}c\theta_{23} - b_2\dot{\theta}_2 s\theta_2 - d_2\dot{\theta}_1) + \overline{u}_2(d_4 s\theta_{23} + b_2 c\theta_2)\dot{\theta}_1 \\ & - \overline{u}_3(d_4\dot{\theta}_{23}s\theta_{23} + b_2\dot{\theta}_2 c\theta_2)]\end{aligned} \tag{9.74}$$

With $\overline{w}' = \overline{w}^{(1)}$, Eq. (9.74) can also be written as

$$\overline{w} = e^{\tilde{u}_3\theta_1}\overline{w}' \tag{9.75}$$

(c) Velocity of the Tip Point with Respect to the Base Frame

Let $\overline{v} = \overline{v}^{(0)} = \overline{v}_P = \dot{\overline{p}}$. Then, $\overline{v}$ can be obtained from Eq. (9.25) as follows:

$$\dot{\overline{p}} = \dot{\overline{r}} + d_6\dot{\hat{C}}\overline{u}_3 = \dot{\overline{r}} + d_6(\dot{\hat{C}}\hat{C}^t)\hat{C}\overline{u}_3 \Rightarrow$$

$$\overline{v} = \overline{w} + d_6\tilde{\omega}\hat{C}\overline{u}_3 \tag{9.76}$$

9.1.7 Inverse Kinematics in the Velocity Domain

Since $\overline{p}$ and $\hat{C}$ are specified as functions of time, the velocities $\overline{v}$, $\overline{\omega}$, and $\overline{w}$ are also known. They are obtained as shown below.

$$\overline{v} = \dot{\overline{p}}, \quad \overline{\omega} = \text{colm}(\dot{\hat{C}}\hat{C}^t); \quad \overline{w} = \overline{v} - d_6\tilde{\omega}\hat{C}\overline{u}_3 \tag{9.77}$$

(a) Solution for the Arm Joint Velocities

The solution for the first three joint velocities can be obtained from Eqs. (9.74) and (9.75) by combining them as follows so that the terms containing the unknown joint velocities appear with simple expressions:

$$\overline{w}' = \overline{u}_1 w_1' + \overline{u}_2 w_2' + \overline{u}_3 w_3' = e^{-\tilde{u}_3\theta_1}\overline{w} \tag{9.78}$$

Upon premultiplications by $\overline{u}_1^t$, $\overline{u}_2^t$, and $\overline{u}_3^t$, Eq. (9.78) leads to the following scalar equations.

$$d_4\dot{\theta}_{23}c\theta_{23} - b_2\dot{\theta}_2 s\theta_2 - d_2\dot{\theta}_1 = w_1' = \overline{u}_1^t e^{-\tilde{u}_3\theta_1}\overline{w} = w_1 c\theta_1 + w_2 s\theta_1 \tag{9.79}$$

$$(d_4 s\theta_{23} + b_2 c\theta_2)\dot{\theta}_1 = w_2' = \overline{u}_2^t e^{-\tilde{u}_3\theta_1}\overline{w} = w_2 c\theta_1 - w_1 s\theta_1 \tag{9.80}$$

$$d_4\dot{\theta}_{23}s\theta_{23} + b_2\dot{\theta}_2 c\theta_2 = -w_3' = -\overline{u}_3^t e^{-\tilde{u}_3\theta_1}\overline{w} = -w_3 \tag{9.81}$$

Note that $\dot{\theta}_1$ is the only unknown in Eq. (9.80). Its coefficient is $k_1 = d_4 s\theta_{23} + b_2 c\theta_2$. Hence, if $k_1 \neq 0$, $\dot{\theta}_1$ can be found as

$$\dot{\theta}_1 = (w_2 c\theta_1 - w_1 s\theta_1)/(d_4 s\theta_{23} + b_2 c\theta_2) \tag{9.82}$$

With the availability of $\dot{\theta}_1$, Eqs. (9.79) and (9.81) can be written together as the following matrix equation.

$$\begin{bmatrix} d_4 c\theta_{23} & -b_2 s\theta_2 \\ d_4 s\theta_{23} & b_2 c\theta_2 \end{bmatrix} \begin{bmatrix} \dot{\theta}_{23} \\ \dot{\theta}_2 \end{bmatrix} = \begin{bmatrix} w_1^* \\ -w_3 \end{bmatrix} \tag{9.83}$$

In Eq. (9.83), w_1^* and the determinant D_{23} of the coefficient matrix have the following expressions.

$$w_1^* = w_1 c\theta_1 + w_2 s\theta_1 + d_2 \dot{\theta}_1 \tag{9.84}$$

$$D_{23} = b_2 d_4 (c\theta_{23} c\theta_2 + s\theta_{23} s\theta_2) = b_2 d_4 \cos(\theta_{23} - \theta_2) = b_2 d_4 \cos\theta_3 \tag{9.85}$$

If $D_{23} \neq 0$, i.e. if $\cos\theta_3 \neq 0$, Eq. (9.83) gives $\dot{\theta}_2$ and $\dot{\theta}_3$ as follows:

$$\begin{bmatrix} \dot{\theta}_{23} \\ \dot{\theta}_2 \end{bmatrix} = \frac{1}{b_2 d_4 \cos\theta_3} \begin{bmatrix} b_2 c\theta_2 & b_2 s\theta_2 \\ -d_4 s\theta_{23} & d_4 c\theta_{23} \end{bmatrix} \begin{bmatrix} w_1^* \\ -w_3 \end{bmatrix} \Rightarrow$$

$$\dot{\theta}_{23} = (w_1^* \cos\theta_2 - w_3 \sin\theta_2)/(d_4 \cos\theta_3) \tag{9.86}$$

$$\dot{\theta}_2 = -(w_1^* \sin\theta_{23} + w_3 \cos\theta_{23})/(b_2 \cos\theta_3) \tag{9.87}$$

$$\dot{\theta}_3 = \dot{\theta}_{23} - \dot{\theta}_2 \tag{9.88}$$

(b) Solution for the Wrist Joint Velocities

In order to find the wrist joint velocities $(\dot{\theta}_4, \dot{\theta}_5, \dot{\theta}_6)$ after finding the arm joint velocities, the angular velocity equation of the end-effector, i.e. Eq. (9.73), can be reduced to the following considerably simplified form.

$$\dot{\theta}_4 \overline{u}_3 + \dot{\theta}_5 \overline{u}_2 + \dot{\theta}_6 e^{\tilde{u}_2 \theta_5} \overline{u}_3 = \overline{\omega}^* \Rightarrow$$

$$\overline{u}_1(\dot{\theta}_6 s\theta_5) + \overline{u}_2 \dot{\theta}_5 + \overline{u}_3(\dot{\theta}_4 + \dot{\theta}_6 c\theta_5) = \overline{\omega}^* \tag{9.89}$$

In Eq. (9.89), $\overline{\omega}^*$ is known as expressed below.

$$\overline{\omega}^* = e^{-\tilde{u}_3 \theta_4} e^{-\tilde{u}_2 \theta_{23}} e^{-\tilde{u}_3 \theta_1} (\overline{\omega} - \dot{\theta}_1 \overline{u}_3 - \dot{\theta}_{23} e^{\tilde{u}_3 \theta_1} \overline{u}_2) \Rightarrow$$

$$\overline{\omega}^* = e^{-\tilde{u}_3 \theta_4} e^{-\tilde{u}_2 \theta_{23}} [e^{-\tilde{u}_3 \theta_1} \overline{\omega} - (\dot{\theta}_1 \overline{u}_3 + \dot{\theta}_{23} \overline{u}_2)] \tag{9.90}$$

Incidentally, it happens that $\overline{\omega}^* = \overline{\omega}_{6/3}^{(4)}$. In other words, $\overline{\omega}^*$ represents the relative angular velocity of the end-effector with respect to Link 3 as expressed in the frame of Link 4. Upon premultiplications by $\overline{u}_1^t$, $\overline{u}_2^t$, and $\overline{u}_3^t$, Eq. (9.89) leads to the following scalar equations.

$$\dot{\theta}_6 \sin\theta_5 = \omega_1^* = \overline{u}_1^t \overline{\omega}^* \tag{9.91}$$

$$\dot{\theta}_5 = \omega_2^* = \overline{u}_2^t \overline{\omega}^* \tag{9.92}$$

$$\dot{\theta}_4 + \dot{\theta}_6 \cos\theta_5 = \omega_3^* = \overline{u}_3^t \overline{\omega}^* \tag{9.93}$$

Note that Eq. (9.92) has already given $\dot{\theta}_5$. As for $\dot{\theta}_6$ and $\dot{\theta}_4$, they can be found from Eqs. (9.91) and (9.93) as follows, if $\sin\theta_5 \neq 0$:

$$\dot{\theta}_6 = \omega_1^* / \sin\theta_5 \tag{9.94}$$

$$\dot{\theta}_4 = \omega_3^* - \dot{\theta}_6 \cos\theta_5 = (\omega_3^* \sin\theta_5 - \omega_1^* \cos\theta_5)/\sin\theta_5 \tag{9.95}$$

9.1.8 Singularity Analysis in the Velocity Domain

A Puma manipulator may have three distinct kinds of motion singularities, which are described and discussed below.

(a) First Kind of Motion Singularity

Equation (9.82) implies that the first kind of motion singularity occurs if

$$k_1 = d_4 s\theta_{23} + b_2 c\theta_2 = 0 \tag{9.96}$$

On the other hand, referring to Part (c) of Section 9.1.2, it is seen that

$$d_4 s\theta_{23} + b_2 c\theta_2 = r_1' \tag{9.97}$$

Recall that r_1' is the projection of the vector $\vec{r}_{SR}$ on the first axis of the link frame $\mathcal{F}_1(O)$. Therefore, in this singularity with $r_1' = 0$, $\vec{r}_{SR}$ becomes parallel to the axis of the first joint. In other words, the wrist point R becomes located on a *cylindrical surface*, whose radius is d_2. This surface is a noticeable feature of this singularity. This singularity is illustrated in Figure 9.9.

The singularity surface mentioned above also constitutes a workspace boundary for the wrist point. That is, as also indicated in Part (a) of Section 9.1.3 by Inequality (9.41), the wrist point cannot get into the region bounded by this cylindrical surface. Note that the appearance of this motion singularity becomes the same as the appearance of the first kind of position singularity if $d_2 = 0$.

If the manipulator has to pass through this singularity or keep on moving on the singularity surface as a task requirement, then, according to Eq. (9.82), $\dot{\theta}_1$ can be kept finite by planning the task motion so that it obeys the following *compatibility condition*.

$$w_2 c\theta_1 = w_1 s\theta_1 \tag{9.98}$$

If the task motion obeys the above condition, which is a singularity-induced restriction in the task space, $\dot{\theta}_1$ becomes finite but indefinite. This is because, when Eq. (9.80) is left out, which is trivially satisfied as $0 = 0$, there remains two equations, i.e. Eqs. (9.79) and (9.81), which are to be satisfied by three joint velocities $\dot{\theta}_1$, $\dot{\theta}_2$, and $\dot{\theta}_3$. In other words, as an expected singularity feature, this motion singularity also induces a motion freedom in the joint space. Thus, an arbitrary value can be assigned to $\dot{\theta}_1$ and the others ($\dot{\theta}_2$ and $\dot{\theta}_3$) can then be found as explained in Part (a) of Section 9.1.7.

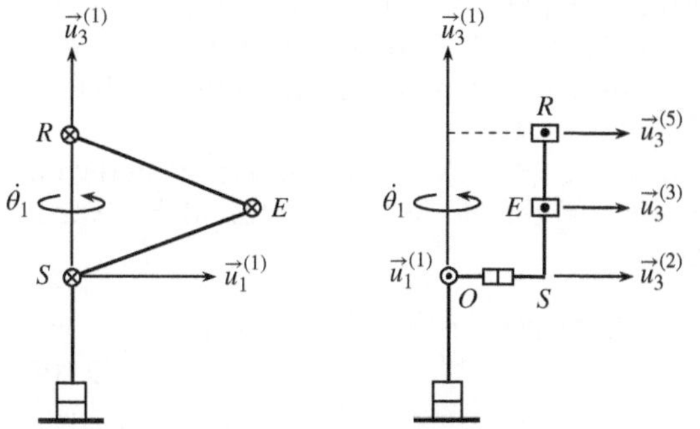

Figure 9.9 First kind of motion singularity of a Puma manipulator.

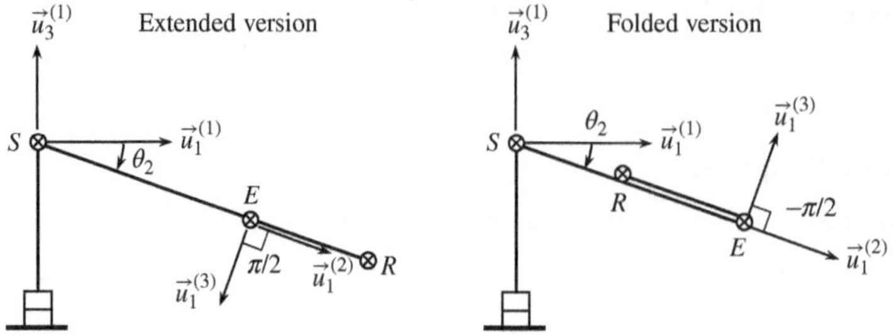

Figure 9.10 Second kind of motion singularity of a Puma manipulator.

(b) Second Kind of Motion Singularity

Equations (9.86) and (9.87) imply that the second kind of motion singularity occurs if

$$\cos\theta_3 = 0 \text{ or } \theta_3 = \sigma_3'\pi/2 \text{ with } \sigma_3' = \pm 1 \tag{9.99}$$

This singularity is illustrated in Figure 9.10 in its *extended version* ($\theta_3 = \pi/2$) and *folded version* ($\theta_3 = -\pi/2$). Note that the appearance of the folded version of this motion singularity becomes the same as the appearance of the second kind of position singularity if $d_4 = b_2$.

In this singularity, with $\theta_{23} = \theta_2 + \sigma_3'\pi/2$, Eqs. (9.79) and (9.81) can be manipulated into the following forms.

$$(b_2\dot{\theta}_2 + \sigma_3' d_4 \dot{\theta}_{23})s\theta_2 = -w_1^* \tag{9.100}$$

$$(b_2\dot{\theta}_2 + \sigma_3' d_4 \dot{\theta}_{23})c\theta_2 = -w_3 \tag{9.101}$$

The above equations become consistent and thus they can give finite values for $\dot{\theta}_2$ and $\dot{\theta}_{23}$ if the following *compatibility condition* is satisfied by the planned task motion.

$$w_1^* c\theta_2 = w_3 s\theta_2 \tag{9.102}$$

The same equations can be combined into the following equation, which relates the finite but indefinite values of $\dot{\theta}_2$ and $\dot{\theta}_3$.

$$b_2\dot{\theta}_2 + \sigma_3' d_4\dot{\theta}_{23} = (b_2 + \sigma_3' d_4)\dot{\theta}_2 + \sigma_3' d_4 \dot{\theta}_3 = -(w_1^* s\theta_2 + w_3 c\theta_2) \tag{9.103}$$

As seen above, like the first kind, this second kind of motion singularity also induces a restriction in the task space as expressed by Eq. (9.102) and it also induces a motion freedom in the joint space as expressed by Eq. (9.103). Thus, as long as the task-space compatibility condition is satisfied, one of $\dot{\theta}_2$ and $\dot{\theta}_3$ can be assigned a finite arbitrary value and the other one can be found from Eq. (9.103).

The task-space compatibility condition, i.e. Eq. (9.102), can also be written in the following detailed form upon inserting the equality that $\theta_{23} = \theta_2 + \sigma_3'\pi/2$ and the expression of w_1^* given by Eqs. (9.84) and (9.82).

$$\begin{aligned} &w_1[(b_2 + \sigma_3' d_4)c\theta_1 c\theta_2 - d_2 s\theta_1] + w_2[(b_2 + \sigma_3' d_4)s\theta_1 c\theta_2 + d_2 c\theta_1] \\ &\quad - w_3(b_2 + \sigma_3' d_4)s\theta_2 = 0 \end{aligned} \tag{9.104}$$

Equation (9.104) implies that the wrist point velocity $\vec{w} = w_1\vec{u}_1^{(0)} + w_2\vec{u}_2^{(0)} + w_3\vec{u}_3^{(0)}$ must be tangent to a singularity surface, whose *gradient vector* ($\vec{\gamma}$) is expressed as follows in

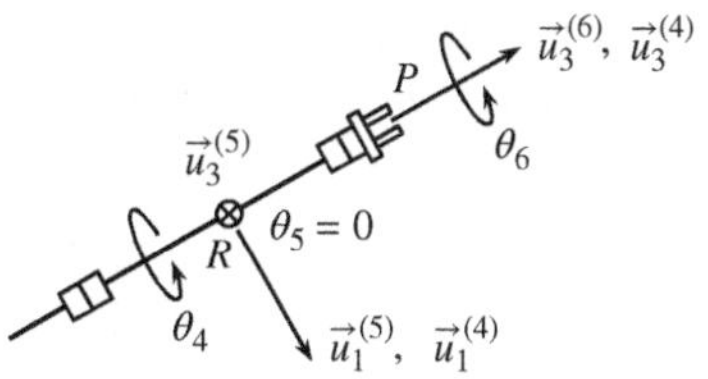

Figure 9.11 Third kind of motion singularity of a Puma manipulator.

the base frame $\mathcal{F}_0(O)$ and in the frame $\mathcal{F}_2(O)$ attached to the upper arm, i.e. the second link:

$$\vec{\gamma} = \gamma_1\vec{u}_1^{(0)} + \gamma_2\vec{u}_2^{(0)} + \gamma_3\vec{u}_3^{(0)} = \gamma_1''\vec{u}_1^{(2)} + \gamma_2''\vec{u}_2^{(2)} + \gamma_3''\vec{u}_3^{(2)} \tag{9.105}$$

According to Eq. (9.104), the components of $\vec{\gamma}$ are defined as shown below in $\mathcal{F}_0(O)$.

$$\left.\begin{aligned} \gamma_1 &= (b_2 + \sigma_3' d_4)c\theta_1 c\theta_2 - d_2 s\theta_1 \\ \gamma_2 &= (b_2 + \sigma_3' d_4)s\theta_1 c\theta_2 + d_2 c\theta_1 \\ \gamma_3 &= -(b_2 + \sigma_3' d_4)s\theta_2 \end{aligned}\right\} \tag{9.106}$$

As for the components of $\vec{\gamma}$ in $\mathcal{F}_2(O)$, it can be shown that

$$\left.\begin{aligned} \gamma_1'' &= b_2 + \sigma_3' d_4 \\ \gamma_2'' &= d_2 \\ \gamma_3'' &= 0 \end{aligned}\right\} \tag{9.107}$$

(c) Third Kind of Motion Singularity

Equations (9.94) and (9.95) imply that the third kind of motion singularity occurs if

$$s\theta_5 = 0, \text{i.e.} \quad \theta_5 = 0 \;\text{ or }\; \theta_5 = \pi \tag{9.108}$$

However, while this singularity can occur easily with $\theta_5 = 0$, it can hardly occur with $\theta_5 = \pi$ due to the physical shapes of the relevant links and joints. Therefore, this singularity is illustrated in Figure 9.11 with $\theta_5 = 0$. Note that the appearance of this motion singularity is the same as the appearance of the third kind of position singularity.

In this singularity with $\theta_5 = 0$, Eqs. (9.91)–(9.93) take the following forms.

$$\omega_1^* = \overline{u}_1^t \overline{\omega}^* = 0 \tag{9.109}$$

$$\dot{\theta}_5 = \omega_2^* = \overline{u}_2^t \overline{\omega}^* \tag{9.110}$$

$$\dot{\theta}_4 + \dot{\theta}_6 = \omega_3^* = \overline{u}_3^t \overline{\omega}^* \tag{9.111}$$

Equation (9.109) expresses the restriction (i.e. the compatibility condition) induced by this singularity in the task space. According to this restriction, the end-effector cannot rotate about the coincident unit vectors $\vec{u}_1^{(4)}$ and $\vec{u}_1^{(5)}$ with respect to Link 3. If the specified motion of the end-effector obeys this condition, then $\dot{\theta}_4$ and $\dot{\theta}_6$ become finite but indefinite. Yet, their indefinite values become related by Eq. (9.111). Thus, this singularity induces a motion freedom in the joint space so that one of $\dot{\theta}_4$ and $\dot{\theta}_6$ can be assigned an arbitrary value.

If the manipulator is to be operated in this singular pose for a while, as in the case of a drilling task, then it will be reasonable to assign the duty evenly to the actuators of the fourth and sixth joints so that

$$\dot{\theta}_4 = \dot{\theta}_6 = \omega_3^*/2 \tag{9.112}$$

9.2 Stanford Manipulator

A Stanford manipulator is shown in Figures 9.12 and 9.13. Its generic name originates from the research manipulator developed by *Victor Scheinman* at *Stanford University*. It comprises one prismatic joint and five revolute joints. Indicating the order of the joints as well, the manipulator is designated symbolically as 2R-P-3R or R^2PR^3. The joint axes are also shown in the figures together with the relevant unit vectors. The prismatic joint $\mathcal{J}_3$ and the succeeding revolute joint $\mathcal{J}_4$ form a *cylindrical* joint arrangement. Moreover, the last three revolute joints ($\mathcal{J}_4$, $\mathcal{J}_5$, $\mathcal{J}_6$) form a *spherical* joint arrangement. The kinematic details (the joint variables and the constant geometric parameters) of the manipulator are shown in the line diagrams in Figure 9.13. The line diagrams comprise the *side view*, the *top view*, and two *auxiliary views* that show the joint variables that are not seen in the side and top views. The significant points of the manipulator are named as follows:

O: Center point or neck point (origin of the base frame)
S: Shoulder point, R: Wrist point, P: Tip point

9.2.1 Kinematic Description According to the D–H Convention

(a) Rotation Angles

$$\theta_1, \theta_2, \theta_3 = \delta_3 = 0, \theta_4, \theta_5, \theta_6$$

Five of the rotation angles are joint variables, which are shown in Figure 9.13. The third one is associated with the cylindrical arrangement of the joints $\mathcal{J}_3$ and $\mathcal{J}_4$. So, it is taken to be zero as discussed in Chapter 7. Thus, $\vec{u}_1^{(3)}$ is arranged to be parallel to $\vec{u}_1^{(2)}$.

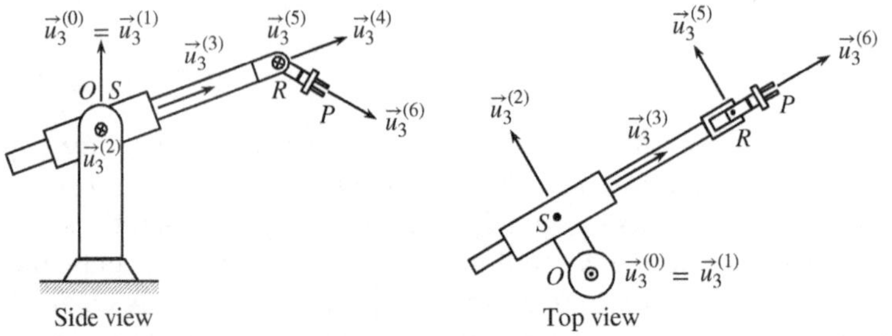

Figure 9.12 A Stanford manipulator in its side and top views.

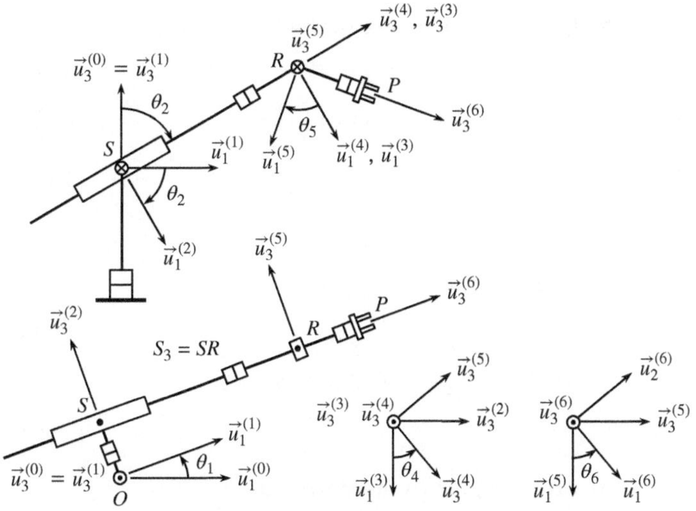

Figure 9.13 Line diagrams that show the kinematic details.

(b) Twist Angles

$$\beta_1 = 0, \beta_2 = -\pi/2, \beta_3 = \pi/2, \beta_4 = 0, \beta_5 = -\pi/2, \beta_6 = \pi/2$$

(c) Offsets

$$d_1 = 0, d_2 = OS, s_3 = SR, d_4 = 0, d_5 = 0, d_6 = RP$$

Five of the offsets are constant. The one associated with the prismatic joint is naturally variable. The variable offset and the nonzero constant offsets are given the following special names.

s_3: Variable length of the boom, i.e. the sliding arm (Links 3 and 4)
d_2: Shoulder offset, d_6: Tip point offset

(d) Effective Link Lengths

$$b_0 = 0, b_1 = 0, b_2 = 0, b_3 = 0, b_4 = 0, b_5 = 0$$

(e) Link Frame Origins

$$O_0 = O, O_1 = O, O_2 = S, O_3 = R, O_4 = R, O_5 = R, O_6 = P$$

9.2.2 Forward Kinematics in the Position Domain

(a) Link-to-Link Orientation Matrices

$$\widehat{C}^{(0,1)} = e^{\tilde{u}_1\beta_1}e^{\tilde{u}_3\theta_1} = e^{\tilde{u}_3\theta_1} \tag{9.113}$$

$$\widehat{C}^{(1,2)} = e^{\tilde{u}_1\beta_2}e^{\tilde{u}_3\theta_2} = e^{-\tilde{u}_1\pi/2}e^{\tilde{u}_3\theta_2} \tag{9.114}$$

$$\widehat{C}^{(2,3)} = e^{\tilde{u}_1\beta_3}e^{\tilde{u}_3\theta_3} = e^{\tilde{u}_1\pi/2}e^{\tilde{u}_3 0} = e^{\tilde{u}_1\pi/2} \tag{9.115}$$

$$\widehat{C}^{(3,4)} = e^{\tilde{u}_1\beta_4}e^{\tilde{u}_3\theta_4} = e^{\tilde{u}_3\theta_4} \tag{9.116}$$

$$\widehat{C}^{(4,5)} = e^{\tilde{u}_1\beta_5}e^{\tilde{u}_3\theta_5} = e^{-\tilde{u}_1\pi/2}e^{\tilde{u}_3\theta_5} \tag{9.117}$$

$$\widehat{C}^{(5,6)} = e^{\tilde{u}_1\beta_6}e^{\tilde{u}_3\theta_6} = e^{\tilde{u}_1\pi/2}e^{\tilde{u}_3\theta_6} \tag{9.118}$$

(b) Orientation Matrices of the Links with Respect to the Base Frame

$$\hat{C}_1 = \hat{C}^{(0,1)} = e^{\tilde{u}_3\theta_1} \tag{9.119}$$

$$\hat{C}_2 = \hat{C}^{(0,2)} = \hat{C}^{(0,1)}\hat{C}^{(1,2)} = e^{\tilde{u}_3\theta_1}e^{-\tilde{u}_1\pi/2}e^{\tilde{u}_3\theta_2} = e^{\tilde{u}_3\theta_1}e^{\tilde{u}_2\theta_2}e^{-\tilde{u}_1\pi/2} \tag{9.120}$$

$$\hat{C}_3 = \hat{C}^{(0,3)} = \hat{C}^{(0,2)}\hat{C}^{(2,3)} = e^{\tilde{u}_3\theta_1}e^{\tilde{u}_2\theta_2}e^{-\tilde{u}_1\pi/2}e^{\tilde{u}_1\pi/2} = e^{\tilde{u}_3\theta_1}e^{\tilde{u}_2\theta_2} \tag{9.121}$$

$$\hat{C}_4 = \hat{C}^{(0,4)} = \hat{C}^{(0,3)}\hat{C}^{(3,4)} = e^{\tilde{u}_3\theta_1}e^{\tilde{u}_2\theta_2}e^{\tilde{u}_3\theta_4} \tag{9.122}$$

$$\hat{C}_5 = \hat{C}^{(0,5)} = \hat{C}^{(0,4)}\hat{C}^{(4,5)} = e^{\tilde{u}_3\theta_1}e^{\tilde{u}_2\theta_2}e^{\tilde{u}_3\theta_4}e^{-\tilde{u}_1\pi/2}e^{\tilde{u}_3\theta_5} \Rightarrow$$

$$\hat{C}_5 = e^{\tilde{u}_3\theta_1}e^{\tilde{u}_2\theta_2}e^{\tilde{u}_3\theta_4}e^{\tilde{u}_2\theta_5}e^{-\tilde{u}_1\pi/2} \tag{9.123}$$

$$\hat{C}_6 = \hat{C}^{(0,6)} = \hat{C}^{(0,5)}\hat{C}^{(5,6)} = e^{\tilde{u}_3\theta_1}e^{\tilde{u}_2\theta_2}e^{\tilde{u}_3\theta_4}e^{\tilde{u}_2\theta_5}e^{-\tilde{u}_1\pi/2}e^{\tilde{u}_1\pi/2}e^{\tilde{u}_3\theta_6} \Rightarrow$$

$$\hat{C}_6 = \hat{C} = e^{\tilde{u}_3\theta_1}e^{\tilde{u}_2\theta_2}e^{\tilde{u}_3\theta_4}e^{\tilde{u}_2\theta_5}e^{\tilde{u}_3\theta_6} \tag{9.124}$$

(c) Location of the Wrist Point with Respect to the Base Frame

$$\vec{r} = \vec{r}_{OR} = \vec{r}_{OS} + \vec{r}_{SR} = d_2\vec{u}_3^{(2)} + s_3\vec{u}_3^{(3)} \tag{9.125}$$

The vector Eq. (9.125) implies the following matrix equation in the base frame.

$$\overline{r} = \overline{r}^{(0)} = d_2\overline{u}_3^{(2/0)} + s_3\overline{u}_3^{(3/0)} \tag{9.126}$$

Equation (9.126) can be manipulated as shown below.

$$\overline{r} = d_2\hat{C}^{(0,2)}\overline{u}_3^{(2/2)} + s_3\hat{C}^{(0,3)}\overline{u}_3^{(3/3)} \Rightarrow$$

$$\overline{r} = d_2e^{\tilde{u}_3\theta_1}e^{\tilde{u}_2\theta_2}e^{-\tilde{u}_1\pi/2}\overline{u}_3 + s_3e^{\tilde{u}_3\theta_1}e^{\tilde{u}_2\theta_2}\overline{u}_3 \Rightarrow$$

$$\overline{r} = d_2e^{\tilde{u}_3\theta_1}\overline{u}_2 + s_3e^{\tilde{u}_3\theta_1}e^{\tilde{u}_2\theta_2}\overline{u}_3 \Rightarrow$$

$$\overline{r} = e^{\tilde{u}_3\theta_1}(d_2\overline{u}_2 + s_3e^{\tilde{u}_2\theta_2}\overline{u}_3) \Rightarrow$$

$$\overline{r} = e^{\tilde{u}_3\theta_1}[d_2\overline{u}_2 + s_3(\overline{u}_3c\theta_2 + \overline{u}_1s\theta_2)] \Rightarrow$$

$$\overline{r} = e^{\tilde{u}_3\theta_1}[\overline{u}_1(s_3s\theta_2) + \overline{u}_2d_2 + \overline{u}_3(s_3c\theta_2)] \tag{9.127}$$

Equation (9.127) can also be written as

$$\overline{r} = e^{\tilde{u}_3\theta_1}\overline{r}' \tag{9.128}$$

In Eq. (9.128),

$$\overline{r}' = \overline{u}_1(s_3s\theta_2) + \overline{u}_2d_2 + \overline{u}_3(s_3c\theta_2) \tag{9.129}$$

Note that $\overline{r}'$ is the matrix representation of the vector $\vec{r}$ in the link frame $\mathcal{F}_1(O)$, i.e. $\overline{r}' = \overline{r}^{(1)}$. In other words, Eq. (9.129) indicates the following coordinates of the wrist point R in the frame $\mathcal{F}_1(O)$, which can also be written down directly by inspection referring to Figure 9.13.

$$\left.\begin{aligned} r_1' &= s_3s\theta_2 \\ r_2' &= d_2 \\ r_3' &= s_3c\theta_2 \end{aligned}\right\} \tag{9.130}$$

On the other hand, the expansion of the expression in Eq. (9.127) leads to the following coordinates of the wrist point R in the base frame $\mathcal{F}_0(O)$.

$$\overline{r}^{(0)} = \overline{r} = \overline{u}_1 r_1 + \overline{u}_2 r_2 + \overline{u}_3 r_3 \tag{9.131}$$

$$\overline{r} = (e^{\tilde{u}_3\theta_1}\overline{u}_1)(s_3 s\theta_2) + (e^{\tilde{u}_3\theta_1}\overline{u}_2)d_2 + (e^{\tilde{u}_3\theta_1}\overline{u}_3)(s_3 c\theta_2) \Rightarrow$$

$$\overline{r} = (\overline{u}_1 c\theta_1 + \overline{u}_2 s\theta_1)(s_3 s\theta_2) + (\overline{u}_2 c\theta_1 - \overline{u}_1 s\theta_1)d_2 + (\overline{u}_3)(s_3 c\theta_2) \Rightarrow$$

$$\overline{r} = \overline{u}_1(s_3 s\theta_2 c\theta_1 - d_2 s\theta_1) + \overline{u}_2(s_3 s\theta_2 s\theta_1 + d_2 c\theta_1) + \overline{u}_3(s_3 c\theta_2) \tag{9.132}$$

Equation (9.132) indicates that

$$\left.\begin{aligned} r_1 &= s_3 s\theta_2 c\theta_1 - d_2 s\theta_1 \\ r_2 &= s_3 s\theta_2 s\theta_1 + d_2 c\theta_1 \\ r_3 &= s_3 c\theta_2 \end{aligned}\right\} \tag{9.133}$$

(d) Location of the Tip Point with Respect to the Base Frame

$$\vec{p} = \vec{r}_{OP} = \vec{r}_{OR} + \vec{r}_{RP} = \vec{r} + d_6 \vec{u}_3^{(6)} \tag{9.134}$$

Equation (9.134) implies the following matrix equation in the base frame.

$$\overline{p} = \overline{p}^{(0)} = \overline{r}^{(0)} + d_6 \overline{u}_3^{(6/0)} = \overline{r}^{(0)} + d_6 \widehat{C}^{(0,6)} \overline{u}_3^{(6/6)} \Rightarrow$$

$$\overline{p} = \overline{r} + d_6 \widehat{C} \overline{u}_3 \tag{9.135}$$

When Eq. (9.124) is substituted, Eq. (9.135) becomes

$$\overline{p} = \overline{r} + d_6 e^{\tilde{u}_3\theta_1} e^{\tilde{u}_2\theta_2} e^{\tilde{u}_3\theta_4} e^{\tilde{u}_2\theta_5} \overline{u}_3 \tag{9.136}$$

9.2.3 Inverse Kinematics in the Position Domain

Here, $\overline{p}$ and $\widehat{C}$ are specified. Hence, $\overline{r}$ also becomes specified according to Eq. (9.135). That is,

$$\overline{r} = \overline{p} - d_6 \widehat{C} \overline{u}_3 \tag{9.137}$$

(a) Solution for the Arm Joint Variables

The wrist of a Stanford manipulator is spherical. Therefore, like the Puma manipulator, the inverse kinematic solution can be started from the wrist location equation, i.e. Eq. (9.127), which can be rearranged as follows in order to distribute the unknowns (θ_1, θ_2, s_3) more or less evenly to both sides of the equation:

$$\overline{u}_1(s_3 s\theta_2) + \overline{u}_2 d_2 + \overline{u}_3(s_3 c\theta_2) = e^{-\tilde{u}_3\theta_1}\overline{r} \tag{9.138}$$

Upon premultiplications by $\overline{u}_1^t$, $\overline{u}_2^t$, and $\overline{u}_3^t$, Eq. (9.138) leads to the following scalar equations.

$$s_3 s\theta_2 = r_1 c\theta_1 + r_2 s\theta_1 \tag{9.139}$$

$$r_2 c\theta_1 - r_1 s\theta_1 = d_2 \tag{9.140}$$

$$s_3 c\theta_2 = r_3 \tag{9.141}$$

Note that Eq. (9.140) contains θ_1 as the only unknown. It is the same *sine–cosine equation*, which was obtained previously for the Puma manipulator. Therefore, with a posture mode sign variable $\sigma_1 = \pm 1$, it can be solved similarly so that

$$\theta_1 = \phi_1 - \psi_1 = \text{atan}_2(r_2, r_1) - \text{atan}_2(d_2, \sigma_1\sqrt{r_1^2 + r_2^2 - d_2^2}) \tag{9.142}$$

Equation (9.142) implies the following workspace limitation.

$$r_1^2 + r_2^2 \geq d_2^2 \tag{9.143}$$

With the availability of θ_1 and noting that $s_3 \geq s_3^\circ > 0$, Eqs. (9.139) and (9.141) give s_3 and θ_2 as shown below.

$$s_3 = \sqrt{(r_1 c\theta_1 + r_2 s\theta_1)^2 + r_3^2} \tag{9.144}$$

$$\theta_2 = \text{atan}_2[(r_1 c\theta_1 + r_2 s\theta_1), r_3] \tag{9.145}$$

On the other hand, as shown before for the Puma manipulator, it happens that

$$r_1 c\theta_1 + r_2 s\theta_1 = \sigma_1\sqrt{r_1^2 + r_2^2 - d_2^2} \tag{9.146}$$

Hence, s_3 and θ_2 can also be expressed as follows:

$$s_3 = \sqrt{r_1^2 + r_2^2 + r_3^2 - d_2^2} \tag{9.147}$$

$$\theta_2 = \text{atan}_2(\sigma_1\sqrt{r_1^2 + r_2^2 - d_2^2}, r_3) = \sigma_1 \text{atan}_2(\sqrt{r_1^2 + r_2^2 - d_2^2}, r_3) \tag{9.148}$$

(b) Solution for the Wrist Joint Variables

In order to find the wrist joint variables $(\theta_4, \theta_5, \theta_6)$ after finding the arm joint variables, the end-effector orientation equation, i.e. Eq. (9.124), can be written as follows:

$$e^{\tilde{u}_3\theta_4} e^{\tilde{u}_2\theta_5} e^{\tilde{u}_3\theta_6} = \hat{C}^* \tag{9.149}$$

In Eq. (9.149), $\hat{C}^*$ is known as

$$\hat{C}^* = e^{-\tilde{u}_2\theta_2} e^{-\tilde{u}_3\theta_1} \hat{C} \tag{9.150}$$

Note that Eq. (9.149) is the same as Eq. (9.52) that was derived for the Puma manipulator. Therefore, it can be solved similarly for the wrist joint variables as explained in Section 9.1.3.

9.2.4 Multiplicity Analysis

(a) First Kind of Multiplicity

The first kind of multiplicity is associated with the sign variable σ_1 that arises in the process of finding θ_1 by using Eq. (9.142). The manipulator attains the same location of the wrist point by assuming one of the two alternative poses that correspond to $\sigma_1 = +1$ and $\sigma_1 = -1$. These poses are very similar to those illustrated in Figure 9.3 for the Puma manipulator. Therefore, they are also designated as *left shouldered* if $\sigma_1 = +1$ and *right shouldered* if $\sigma_1 = -1$. They are illustrated in Figure 9.14.

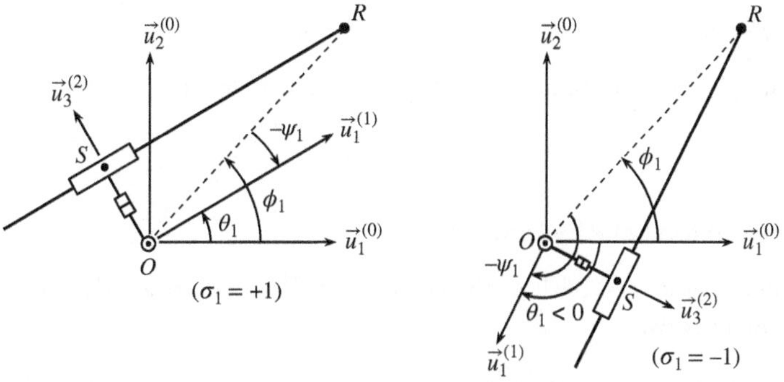

Figure 9.14 Left and right shouldered poses of a Stanford manipulator.

(b) Second Kind of Multiplicity

The second kind of multiplicity is associated with the sign variable σ_5 that arises in the process of finding the wrist joint variables $(\theta_5, \theta_4, \theta_6)$. Therefore, it is the same as the third kind of multiplicity of the Puma manipulator. In other words, as illustrated in Figure 9.5, it occurs similarly as a wrist flip phenomenon without any visual distinction.

9.2.5 Singularity Analysis in the Position Domain

A Stanford manipulator with $d_2 > 0$ may have only one kind of position singularity. Equation (9.149) implies that it is the same as the third kind of position singularity of the Puma manipulator, which is explained in Section 9.1.5 and illustrated in Figure 9.8.

9.2.6 Forward Kinematics in the Velocity Domain

(a) Angular Velocity of the End-Effector with Respect to the Base Frame

Let $\overline{\omega} = \overline{\omega}_6$. Then, Eq. (9.124) leads to the following expression.

$$\overline{\omega} = \dot{\theta}_1\overline{u}_3 + \dot{\theta}_2 e^{\tilde{u}_3\theta_1}\overline{u}_2 + \dot{\theta}_4 e^{\tilde{u}_3\theta_1}e^{\tilde{u}_2\theta_2}\overline{u}_3 + \dot{\theta}_5 e^{\tilde{u}_3\theta_1}e^{\tilde{u}_2\theta_2}e^{\tilde{u}_3\theta_4}\overline{u}_2 + \dot{\theta}_6 e^{\tilde{u}_3\theta_1}e^{\tilde{u}_2\theta_2}e^{\tilde{u}_3\theta_4}e^{\tilde{u}_2\theta_5}\overline{u}_3 \tag{9.151}$$

(b) Velocity of the Wrist Point with Respect to the Base Frame

Let $\overline{w} = \overline{w}^{(0)} = \overline{v}_R = \dot{\overline{r}}$. Then, $\overline{w}$ can be obtained from Eq. (9.127) as follows:

$$\overline{w} = (\dot{\theta}_1 e^{\tilde{u}_3\theta_1}\tilde{u}_3)[\overline{u}_1(s_3 s\theta_2) + \overline{u}_2 d_2 + \overline{u}_3(s_3 c\theta_2)] + (e^{\tilde{u}_3\theta_1})[\overline{u}_1(\dot{s}_3 s\theta_2 + s_3\dot{\theta}_2 c\theta_2) + \overline{u}_3(\dot{s}_3 c\theta_2 - s_3\dot{\theta}_2 s\theta_2)] \Rightarrow$$

$$\overline{w} = e^{\tilde{u}_3\theta_1}[\overline{u}_1(\dot{s}_3 s\theta_2 + s_3\dot{\theta}_2 c\theta_2 - d_2\dot{\theta}_1) + \overline{u}_2(s_3\dot{\theta}_1 s\theta_2] + \overline{u}_3(\dot{s}_3 c\theta_2 - s_3\dot{\theta}_2 s\theta_2)] \tag{9.152}$$

With $\overline{w}' = \overline{w}^{(1)}$, Eq. (9.152) can also be written as

$$\overline{w} = e^{\tilde{u}_3\theta_1}\overline{w}' \tag{9.153}$$

(c) Velocity of the Tip Point with Respect to the Base Frame

Let $\overline{v} = \overline{v}^{(0)} = \overline{v}_P = \dot{\overline{p}}$. Then, $\overline{v}$ can be obtained from Eq. (9.135) as follows:

$$\overline{v} = \overline{w} + d_6\tilde{\omega}\hat{C}\overline{u}_3 \tag{9.154}$$

9.2.7 Inverse Kinematics in the Velocity Domain

Since $\overline{p}$ and $\hat{C}$ are specified as functions of time, the velocities $\overline{v}$, $\overline{\omega}$, and $\overline{w}$ are also known. They are obtained as shown below.

$$\overline{v} = \dot{\overline{p}},\quad \overline{\omega} = \mathrm{colm}(\dot{\hat{C}}\hat{C}^t);\quad \overline{w} = \overline{v} - d_6\tilde{\omega}\hat{C}\overline{u}_3 \tag{9.155}$$

(a) Solution for the Arm Joint Velocities

The solution can be obtained from Eqs. (9.152) and (9.153) by combining them as follows so that the terms containing the unknown joint velocities appear with simple expressions:

$$\overline{w}' = \overline{u}_1 w_1' + \overline{u}_2 w_2' + \overline{u}_3 w_3' = e^{-\tilde{u}_3\theta_1}\overline{w} \tag{9.156}$$

Upon premultiplications by $\overline{u}_1^t$, $\overline{u}_2^t$, and $\overline{u}_3^t$, Eq. (9.156) leads to the following scalar equations.

$$\dot{s}_3 s\theta_2 + s_3\dot{\theta}_2 c\theta_2 - d_2\dot{\theta}_1 = w_1' = \overline{u}_1^t e^{-\tilde{u}_3\theta_1}\overline{w} = w_1 c\theta_1 + w_2 s\theta_1 \tag{9.157}$$

$$(s_3 s\theta_2)\dot{\theta}_1 = w_2' = \overline{u}_2^t e^{-\tilde{u}_3\theta_1}\overline{w} = w_2 c\theta_1 - w_1 s\theta_1 \tag{9.158}$$

$$\dot{s}_3 c\theta_2 - s_3\dot{\theta}_2 s\theta_2 = w_3' = \overline{u}_3^t e^{-\tilde{u}_3\theta_1}\overline{w} = w_3 \tag{9.159}$$

Note that $\dot{\theta}_1$ is the only unknown in Eq. (9.158). Its coefficient is $k_1 = s_3 s\theta_2$. Hence, if $k_1 = s_3 s\theta_2 \neq 0$, $\dot{\theta}_1$ can be found as

$$\dot{\theta}_1 = (w_2 c\theta_1 - w_1 s\theta_1)/(s_3 s\theta_2) \tag{9.160}$$

With the availability of $\dot{\theta}_1$, Eqs. (9.157) and (9.159) can be written together as the following matrix equation.

$$\begin{bmatrix} c\theta_2 & s\theta_2 \\ -s\theta_2 & c\theta_2 \end{bmatrix}\begin{bmatrix} s_3\dot{\theta}_2 \\ \dot{s}_3 \end{bmatrix} = \begin{bmatrix} w_1^* \\ w_3 \end{bmatrix} \tag{9.161}$$

In Eq. (9.161),

$$w_1^* = w_1' + d_2\dot{\theta}_1 = w_1 c\theta_1 + w_2 s\theta_1 + d_2\dot{\theta}_1 \tag{9.162}$$

Note that, in Eq. (9.161), $s_3\dot{\theta}_2$ is deliberately treated as an unknown instead of just $\dot{\theta}_2$. Thus, the coefficient matrix comes out to be simplified to an orthonormal matrix with unity determinant. Hence, recalling that $s_3 \geq s_3^\circ > 0$, Eq. (9.161) can be solved simply as follows:

$$\begin{bmatrix} s_3\dot{\theta}_2 \\ \dot{s}_3 \end{bmatrix} = \begin{bmatrix} c\theta_2 & -s\theta_2 \\ s\theta_2 & c\theta_2 \end{bmatrix}\begin{bmatrix} w_1^* \\ w_3 \end{bmatrix} \Rightarrow$$

$$\dot{\theta}_2 = (w_1^* c\theta_2 - w_3 s\theta_2)/s_3 \tag{9.163}$$

$$\dot{s}_3 = w_1^* s\theta_2 + w_3 c\theta_2 \tag{9.164}$$

(b) Solution for the Wrist Joint Velocities

In order to find the wrist joint velocities $(\dot{\theta}_4, \dot{\theta}_5, \dot{\theta}_6)$ after finding the arm joint velocities, the angular velocity equation of the end-effector, i.e. Eq. (9.151), can be reduced to the following considerably simplified form.

$$\dot{\theta}_4\overline{u}_3 + \dot{\theta}_5\overline{u}_2 + \dot{\theta}_6 e^{\widetilde{u}_2\theta_5}\overline{u}_3 = \overline{\omega}^* \Rightarrow$$

$$\overline{u}_1(\dot{\theta}_6 s\theta_5) + \overline{u}_2\dot{\theta}_5 + \overline{u}_3(\dot{\theta}_4 + \dot{\theta}_6 c\theta_5) = \overline{\omega}^* \quad (9.165)$$

In Eq. (9.165), $\overline{\omega}^*$ is known as expressed below.

$$\overline{\omega}^* = e^{-\widetilde{u}_3\theta_4}e^{-\widetilde{u}_2\theta_2}e^{-\widetilde{u}_3\theta_1}(\overline{\omega} - \dot{\theta}_1\overline{u}_3 - \dot{\theta}_2 e^{\widetilde{u}_3\theta_1}\overline{u}_2) \Rightarrow$$

$$\overline{\omega}^* = e^{-\widetilde{u}_3\theta_4}e^{-\widetilde{u}_2\theta_2}[e^{-\widetilde{u}_3\theta_1}\overline{\omega} - (\dot{\theta}_1\overline{u}_3 + \dot{\theta}_2\overline{u}_2)] \quad (9.166)$$

Note that Eq. (9.165) is the same as Eq. (9.89) obtained for the Puma manipulator. Furthermore, it similarly happens that $\overline{\omega}^* = \overline{\omega}_{6/3}^{(4)}$, which represents the relative angular velocity of the end-effector with respect to Link 3 as expressed in the frame of Link 4. Hence, referring to Section 9.1.5, the solution to Eq. (9.165) can be obtained as follows:

$$\dot{\theta}_5 = \omega_2^* \quad (9.167)$$

If $s\theta_5 \neq 0$,

$$\dot{\theta}_6 = \omega_1^*/s\theta_5 \quad (9.168)$$

$$\dot{\theta}_4 = \omega_3^* - \dot{\theta}_6 c\theta_5 = (\omega_3^* s\theta_5 - \omega_1^* c\theta_5)/s\theta_5 \quad (9.169)$$

9.2.8 Singularity Analysis in the Velocity Domain

A Stanford manipulator may have two distinct kinds of motion singularities, although it can have only one kind of position singularity as mentioned in Section 9.2.5. The motion singularities are described and discussed below.

(a) First Kind of Motion Singularity

As implied by Eq. (9.160) and the fact that $s_3 > 0$, the first kind of motion singularity occurs if

$$s\theta_2 = 0, \text{i.e.} \quad \theta_2 = 0 \ \text{ or } \ \theta_2 = \pi \quad (9.170)$$

On the other hand, referring to Part (c) of Section 9.2.2, it is seen that $s_3 s\theta_2 = r_1'$ and r_1' is the projection of the vector $\vec{r}_{SR}$ on the first axis of the link frame $\mathcal{F}_1(O)$. Therefore, in this singularity, $\vec{r}_{SR}$ becomes parallel to the axis of the first joint. In other words, the wrist point R becomes located on a *cylindrical surface*, whose base radius is d_2. This surface is a noticeable feature of this singularity. The singularity surface mentioned above also constitutes a workspace boundary for the wrist point. That is, as also indicated in Part (a) of Section 9.2.3 by Inequality (9.143), the wrist point cannot get into the region bounded by this cylindrical surface. Two occurrences of this singularity are illustrated in Figure 9.15 with $\theta_2 = 0$ and $\theta_2 = \pi$.

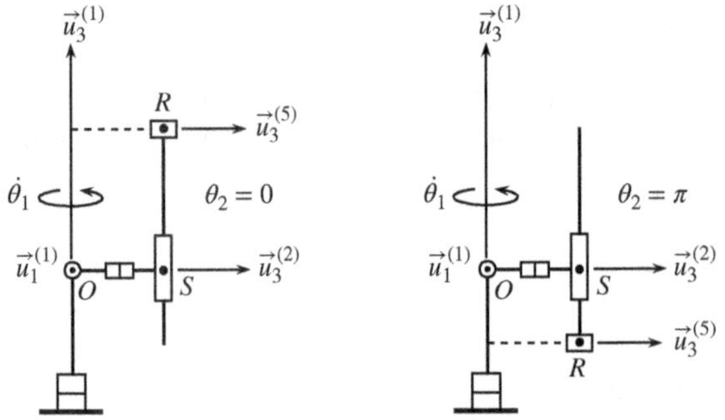

Figure 9.15 First kind of motion singularity of a Stanford manipulator.

If the manipulator has to pass through this singularity or keep on moving on the singularity surface as a task requirement, then, according to Eq. (9.160), $\dot{\theta}_1$ can be kept finite by planning the task motion so that it obeys the following *compatibility condition.*

$$w_2 c\theta_1 = w_1 s\theta_1 \tag{9.171}$$

If the task motion obeys the above condition, which is a singularity-induced restriction in the task space, $\dot{\theta}_1$ becomes finite but indefinite. This is because, when Eq. (9.158) is left out, which is trivially satisfied as $0 = 0$, there remains two equations, i.e. Eqs. (9.157) and (9.159), which are to be satisfied by three joint velocities $\dot{\theta}_1$, $\dot{\theta}_2$, and $\dot{s}_3$. In other words, the singularity also induces a motion freedom in the joint space. Thus, an arbitrary value can be assigned to $\dot{\theta}_1$ and the others ($\dot{\theta}_2$ and $\dot{s}_3$) can then be found as explained in Part (a) of Section 9.2.7.

(b) Second Kind of Motion Singularity

Equations (9.168) and (9.169) imply that the second kind of motion singularity occurs if $s\theta_5 = 0$, i.e. if $\theta_5 = 0$ or $\theta_5 = \pi$. However, while this singularity can occur easily with $\theta_5 = 0$, it can hardly occur with $\theta_5 = \pi$ due to the physical shapes of the relevant links and joints. In other words, this singularity is the same as the third kind of motion singularity of the Puma manipulator, which is explained in Section 9.1.5 and illustrated in Figure 9.11.

9.3 Elbow Manipulator

An elbow manipulator is shown in Figure 9.16. The first example of manipulators with this generic name is the *Cincinnati Milacron T3 Robotic Arm* developed by *Richard Hohn* for the Cincinnati Milacron Corporation.

Actually, the class of elbow manipulators is more inclusive than the group of manipulators like the one depicted in Figure 9.16. It covers all the *articulated* manipulators with *anthropomorphic* arms. An articulated manipulator comprises revolute joints only and an anthropomorphic arm possesses an elbow joint. In this general sense, a Puma manipulator is also an elbow manipulator. However, since the Cincinnati Milacron T3 is the first anthropomorphic robot with an elbow joint, it is considered to be the namer for the class of elbow manipulators.

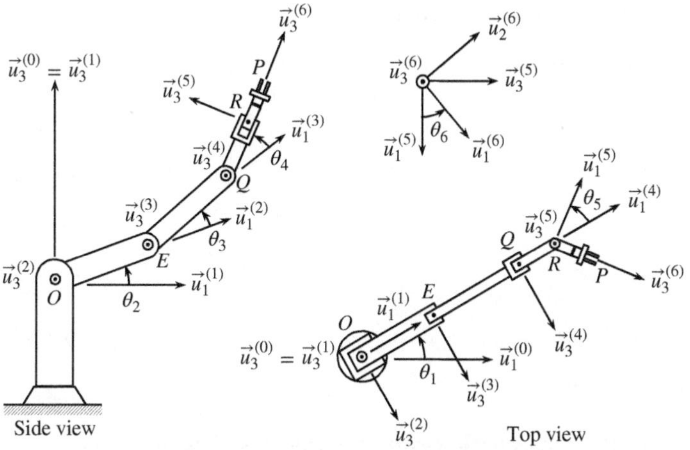

Figure 9.16 An elbow manipulator in its side and top views.

An elbow manipulator comprises six revolute joints. So, it is designated symbolically as 6R or R^6. Its joint axes are also shown in Figure 9.16 together with the relevant unit vectors. The significant points of the manipulator are named as follows:

O: Center point (origin of the base frame)
E: Elbow point, Q: Pre-wrist point, R: Wrist point, P: Tip point

The shoulder point of this manipulator is coincident with its center point, i.e. $S = O$.
Differently from the Puma and Stanford manipulators, an elbow manipulator has a noticeable feature that its wrist is not spherical, i.e. the axes of the last three joints do not intersect at a single point.

9.3.1 Kinematic Description According to the D–H Convention

(a) Joint Variables

$$\theta_1, \theta_2, \theta_3, \theta_4, \theta_5, \theta_6$$

They are the rotation angles about the joint axes. They are shown in Figure 9.16.

(b) Twist Angles

$$\beta_1 = 0, \beta_2 = \pi/2, \beta_3 = 0, \beta_4 = 0, \beta_5 = -\pi/2, \beta_6 = \pi/2$$

(c) Offsets

$$d_1 = 0, d_2 = 0, d_3 = 0, d_4 = 0, d_5 = 0, d_6 = RP \text{ (tip point offset)}$$

(d) Effective Link Lengths

$$b_0 = 0, b_1 = 0, b_2 = OE, b_3 = EQ, b_4 = QR, b_5 = 0$$

(e) Link Frame Origins

$$O_0 = O, O_1 = O, O_2 = O, O_3 = E, O_4 = Q, O_5 = R, O_6 = P$$

9.3.2 Forward Kinematics in the Position Domain

(a) Link-to-Link Orientation Matrices

$$\hat{C}^{(0,1)} = e^{\tilde{u}_1\beta_1}e^{\tilde{u}_3\theta_1} = e^{\tilde{u}_3\theta_1} \tag{9.172}$$

$$\hat{C}^{(1,2)} = e^{\tilde{u}_1\beta_2}e^{\tilde{u}_3\theta_2} = e^{\tilde{u}_1\pi/2}e^{\tilde{u}_3\theta_2} \tag{9.173}$$

$$\hat{C}^{(2,3)} = e^{\tilde{u}_1\beta_3}e^{\tilde{u}_3\theta_3} = e^{\tilde{u}_3\theta_3} \tag{9.174}$$

$$\hat{C}^{(3,4)} = e^{\tilde{u}_1\beta_4}e^{\tilde{u}_3\theta_4} = e^{\tilde{u}_3\theta_4} \tag{9.175}$$

$$\hat{C}^{(4,5)} = e^{\tilde{u}_1\beta_5}e^{\tilde{u}_3\theta_5} = e^{-\tilde{u}_1\pi/2}e^{\tilde{u}_3\theta_5} \tag{9.176}$$

$$\hat{C}^{(5,6)} = e^{\tilde{u}_1\beta_6}e^{\tilde{u}_3\theta_6} = e^{\tilde{u}_1\pi/2}e^{\tilde{u}_3\theta_6} \tag{9.177}$$

(b) Orientation Matrices of the Links with Respect to the Base Frame

$$\hat{C}_1 = \hat{C}^{(0,1)} = e^{\tilde{u}_3\theta_1} \tag{9.178}$$

$$\hat{C}_2 = \hat{C}^{(0,2)} = \hat{C}^{(0,1)}\hat{C}^{(1,2)} = e^{\tilde{u}_3\theta_1}e^{\tilde{u}_1\pi/2}e^{\tilde{u}_3\theta_2} \Rightarrow$$

$$\hat{C}_2 = e^{\tilde{u}_3\theta_1}e^{-\tilde{u}_2\theta_2}e^{\tilde{u}_1\pi/2} \tag{9.179}$$

$$\hat{C}_3 = \hat{C}^{(0,3)} = \hat{C}^{(0,2)}\hat{C}^{(2,3)} = e^{\tilde{u}_3\theta_1}e^{-\tilde{u}_2\theta_2}e^{\tilde{u}_1\pi/2}e^{\tilde{u}_3\theta_3} \Rightarrow$$

$$\hat{C}_3 = e^{\tilde{u}_3\theta_1}e^{-\tilde{u}_2\theta_2}e^{-\tilde{u}_2\theta_3}e^{\tilde{u}_1\pi/2} = e^{\tilde{u}_3\theta_1}e^{-\tilde{u}_2\theta_{23}}e^{\tilde{u}_1\pi/2} \tag{9.180}$$

$$\hat{C}_4 = \hat{C}^{(0,4)} = \hat{C}^{(0,3)}\hat{C}^{(3,4)} = e^{\tilde{u}_3\theta_1}e^{-\tilde{u}_2\theta_{23}}e^{\tilde{u}_1\pi/2}e^{\tilde{u}_3\theta_4} \Rightarrow$$

$$\hat{C}_4 = e^{\tilde{u}_3\theta_1}e^{-\tilde{u}_2\theta_{23}}e^{-\tilde{u}_2\theta_4}e^{\tilde{u}_1\pi/2} = e^{\tilde{u}_3\theta_1}e^{-\tilde{u}_2\theta_{234}}e^{\tilde{u}_1\pi/2} \tag{9.181}$$

$$\hat{C}_5 = \hat{C}^{(0,5)} = \hat{C}^{(0,4)}\hat{C}^{(4,5)} = e^{\tilde{u}_3\theta_1}e^{-\tilde{u}_2\theta_{234}}e^{\tilde{u}_1\pi/2}e^{-\tilde{u}_1\pi/2}e^{\tilde{u}_3\theta_5} \Rightarrow$$

$$\hat{C}_5 = e^{\tilde{u}_3\theta_1}e^{-\tilde{u}_2\theta_{234}}e^{\tilde{u}_3\theta_5} \tag{9.182}$$

$$\hat{C}_6 = \hat{C}^{(0,6)} = \hat{C}^{(0,5)}\hat{C}^{(5,6)} = e^{\tilde{u}_3\theta_1}e^{-\tilde{u}_2\theta_{234}}e^{\tilde{u}_3\theta_5}e^{\tilde{u}_1\pi/2}e^{\tilde{u}_3\theta_6} \Rightarrow$$

$$\hat{C}_6 = \hat{C} = e^{\tilde{u}_3\theta_1}e^{-\tilde{u}_2\theta_{234}}e^{\tilde{u}_3\theta_5}e^{-\tilde{u}_2\theta_6}e^{\tilde{u}_1\pi/2} \tag{9.183}$$

The above equations contain the following *combined joint variables.*

$$\theta_{23} = \theta_2 + \theta_3 \tag{9.184}$$

$$\theta_{234} = \theta_2 + \theta_3 + \theta_4 \tag{9.185}$$

The angles θ_{23} and θ_{234} are also called *absolute joint angles* because they describe the overall orientations of $\vec{u}_1^{(3)}$ and $\vec{u}_1^{(4)}$ with respect to $\vec{u}_1^{(1)}$ of the first link frame. They are so called considering the fact that θ_3 and θ_4 are actually *relative joint angles* because they describe the orientations of $\vec{u}_1^{(3)}$ and $\vec{u}_1^{(4)}$ with respect to $\vec{u}_1^{(2)}$ and $\vec{u}_1^{(3)}$, respectively.

(c) Location of the Wrist Point with Respect to the Base Frame

$$\vec{r} = \vec{r}_{OR} = \vec{r}_{OE} + \vec{r}_{EQ} + \vec{r}_{QR} = b_2\vec{u}_1^{(2)} + b_3\vec{u}_1^{(3)} + b_4\vec{u}_1^{(4)} \tag{9.186}$$

The vector Eq. (9.186) implies the following matrix equation in the base frame.

$$\overline{r} = \overline{r}^{(0)} = b_2\overline{u}_1^{(2/0)} + b_3\overline{u}_1^{(3/0)} + b_4\overline{u}_1^{(4/0)} \tag{9.187}$$

Equation (9.187) can be manipulated as shown below.

$$\overline{r} = b_2\widehat{C}^{(0,2)}\overline{u}_1^{(2/2)} + b_3\widehat{C}^{(0,3)}\overline{u}_1^{(3/3)} + b_4\widehat{C}^{(0,4)}\overline{u}_1^{(4/4)} \Rightarrow$$

$$\overline{r} = b_2 e^{\tilde{u}_3\theta_1} e^{-\tilde{u}_2\theta_2} e^{\tilde{u}_1\pi/2}\overline{u}_1 + b_3 e^{\tilde{u}_3\theta_1} e^{-\tilde{u}_2\theta_{23}} e^{\tilde{u}_1\pi/2}\overline{u}_1 + b_4 e^{\tilde{u}_3\theta_1} e^{-\tilde{u}_2\theta_{234}} e^{\tilde{u}_1\pi/2}\overline{u}_1 \Rightarrow$$

$$\overline{r} = b_2 e^{\tilde{u}_3\theta_1} e^{-\tilde{u}_2\theta_2}\overline{u}_1 + b_3 e^{\tilde{u}_3\theta_1} e^{-\tilde{u}_2\theta_{23}}\overline{u}_1 + b_4 e^{\tilde{u}_3\theta_1} e^{-\tilde{u}_2\theta_{234}}\overline{u}_1 \Rightarrow$$

$$\overline{r} = e^{\tilde{u}_3\theta_1}(b_2 e^{-\tilde{u}_2\theta_2}\overline{u}_1 + b_3 e^{-\tilde{u}_2\theta_{23}}\overline{u}_1 + b_4 e^{-\tilde{u}_2\theta_{234}}\overline{u}_1) \tag{9.188}$$

Equation (9.188) can also be written as

$$\overline{r} = e^{\tilde{u}_3\theta_1}\overline{r}' \tag{9.189}$$

In Eq. (9.189), $\overline{r}' = \overline{r}^{(1)}$. That is,

$$\overline{r}' = \overline{u}_1(b_2c\theta_2 + b_3c\theta_{23} + b_4c\theta_{234}) + \overline{u}_3(b_2s\theta_2 + b_3s\theta_{23} + b_4s\theta_{234}) \tag{9.190}$$

Note that, referring to Figure 9.16, $\overline{r}' = \overline{u}_1r_1' + \overline{u}_2r_2' + \overline{u}_3r_3'$ can also be written down directly by inspection. In other words, Figure 9.16 readily implies the following coordinates of the wrist point R in the frame $\mathcal{F}_1(O)$.

$$\left.\begin{aligned} r_1' &= b_2c\theta_2 + b_3c\theta_{23} + b_4c\theta_{234} \\ r_2' &= 0 \\ r_3' &= b_2s\theta_2 + b_3s\theta_{23} + b_4s\theta_{234} \end{aligned}\right\} \tag{9.191}$$

On the other hand, the expansion of the expression in Eq. (9.188) leads to the coordinates of the wrist point R in the base frame $\mathcal{F}_0(O)$ as follows:

$$\overline{r}^{(0)} = \overline{r} = \overline{u}_1r_1 + \overline{u}_2r_2 + \overline{u}_3r_3 \tag{9.192}$$

$$\overline{r} = (e^{\tilde{u}_3\theta_1}\overline{u}_1)r_1' + (e^{\tilde{u}_3\theta_1}\overline{u}_3)r_3' \Rightarrow$$

$$\overline{r} = (\overline{u}_1c\theta_1 + \overline{u}_2s\theta_1)r_1' + (\overline{u}_3)r_3' \Rightarrow$$

$$\overline{r} = \overline{u}_1(r_1'c\theta_1) + \overline{u}_2(r_1's\theta_1) + \overline{u}_3(r_3') \tag{9.193}$$

Equations (9.193) and (9.191) indicate that

$$\left.\begin{aligned} r_1 &= (b_2c\theta_2 + b_3c\theta_{23} + b_4c\theta_{234})c\theta_1 \\ r_2 &= (b_2c\theta_2 + b_3c\theta_{23} + b_4c\theta_{234})s\theta_1 \\ r_3 &= b_2s\theta_2 + b_3s\theta_{23} + b_4s\theta_{234} \end{aligned}\right\} \tag{9.194}$$

(d) Location of the Tip Point with Respect to the Base Frame

$$\vec{p} = \vec{r}_{OP} = \vec{r}_{OR} + \vec{r}_{RP} = \vec{r} + d_6\vec{u}_3^{(6)} \tag{9.195}$$

Equation (9.195) implies the following matrix equation in the base frame.

$$\overline{p} = \overline{p}^{(0)} = \overline{r}^{(0)} + d_6\overline{u}_3^{(6/0)} = \overline{r}^{(0)} + d_6\widehat{C}^{(0,6)}\overline{u}_3^{(6/6)} \Rightarrow$$

$$\overline{p} = \overline{r} + d_6\widehat{C}\overline{u}_3 \tag{9.196}$$

When Eq. (9.183) is substituted, Eq. (9.196) becomes

$$\begin{aligned}
\overline{p} &= \overline{r} + d_6 e^{\tilde{u}_3\theta_1} e^{-\tilde{u}_2\theta_{234}} e^{\tilde{u}_3\theta_5} e^{-\tilde{u}_2\theta_6} e^{\tilde{u}_1\pi/2} \overline{u}_3 \Rightarrow \\
\overline{p} &= \overline{r} - d_6 e^{\tilde{u}_3\theta_1} e^{-\tilde{u}_2\theta_{234}} e^{\tilde{u}_3\theta_5} e^{-\tilde{u}_2\theta_6} \overline{u}_2 \Rightarrow \\
\overline{p} &= \overline{r} - d_6 e^{\tilde{u}_3\theta_1} e^{-\tilde{u}_2\theta_{234}} e^{\tilde{u}_3\theta_5} \overline{u}_2 \qquad (9.197)
\end{aligned}$$

9.3.3 Inverse Kinematics in the Position Domain

Here, $\overline{p}$ and $\hat{C}$ are specified. Hence, $\overline{r}$ also becomes specified according to Eq. (9.196). That is,

$$\overline{r} = \overline{p} - d_6 \hat{C} \overline{u}_3 \qquad (9.198)$$

Because the wrist of an elbow manipulator is not spherical, the joint variables can no longer be categorized as arm and wrist joint variables. Therefore, they have to be found in a mixed manner without the convenience of this categorization. Another aspect of the nonspherical wrist is that the wrist location equation (9.188) contains more than three unknowns, which are θ_2, θ_{23}, θ_{234}, and θ_1. Therefore, it cannot be solved alone independently of the end-effector orientation equation (9.183). However, owing to the particular configuration of this manipulator, θ_1 can be singled out and thus the solution can still be obtained analytically. To start the solution, Eq. (9.188) can be rearranged as follows:

$$\overline{u}_1(b_2 c\theta_2 + b_3 c\theta_{23} + b_4 c\theta_{234}) + \overline{u}_3(b_2 s\theta_2 + b_3 s\theta_{23} + b_4 s\theta_{234}) = e^{-\tilde{u}_3\theta_1}\overline{r} \qquad (9.199)$$

Upon premultiplications by $\overline{u}_1^t$, $\overline{u}_2^t$, and $\overline{u}_3^t$, Eq. (9.199) leads to the following scalar equations.

$$\begin{aligned}
& b_2 c\theta_2 + b_3 c\theta_{23} + b_4 c\theta_{234} = \overline{u}_1^t e^{-\tilde{u}_3\theta_1}\overline{r} = (\overline{u}_1^t c\theta_1 + \overline{u}_2^t s\theta_1)\overline{r} \Rightarrow \\
& b_2 c\theta_2 + b_3 c\theta_{23} + b_4 c\theta_{234} = r_1 c\theta_1 + r_2 s\theta_1 \qquad (9.200) \\
& 0 = \overline{u}_2^t e^{-\tilde{u}_3\theta_1}\overline{r} = (\overline{u}_2^t c\theta_1 - \overline{u}_1^t s\theta_1)\overline{r} \Rightarrow \\
& r_2 c\theta_1 = r_1 s\theta_1 \qquad (9.201) \\
& b_2 s\theta_2 + b_3 s\theta_{23} + b_4 s\theta_{234} = \overline{u}_3^t e^{-\tilde{u}_3\theta_1}\overline{r} = \overline{u}_3^t \overline{r} \Rightarrow \\
& b_2 s\theta_2 + b_3 s\theta_{23} + b_4 s\theta_{234} = r_3 \qquad (9.202)
\end{aligned}$$

Note that Eq. (9.201) contains θ_1 as the only unknown. Hence, θ_1 can be found by noting the following implication of Eq. (9.201) with a sign variable $\sigma_1 = \pm 1$.

$$\left.\begin{aligned} r_1 &= \sigma_1 r_{12} c\theta_1 \\ r_2 &= \sigma_1 r_{12} s\theta_1 \end{aligned}\right\} \qquad (9.203)$$

In Eq. (9.32),

$$r_{12} = \sqrt{r_1^2 + r_2^2} \qquad (9.204)$$

If $r_{12} \neq 0$, i.e. if r_1 and r_2 are not both zero, Eq. (9.203) gives θ_1 as

$$\theta_1 = \mathrm{atan}_2(\sigma_1 r_2, \sigma_1 r_1) \qquad (9.205)$$

Even though θ_1 becomes available, the remaining two scalar equations (9.200) and (9.202) still contain more than two unknowns, which are θ_2, θ_{23}, and θ_{234}. Therefore, the solution process cannot be continued by using those equations. However, it can be continued by skipping over to the end-effector orientation equation (9.183). For this purpose, with the availability of θ_1, Eq. (9.183) can be written as follows:

$$e^{-\tilde{u}_2\theta_{234}} e^{\tilde{u}_3\theta_5} e^{-\tilde{u}_2\theta_6} = \widehat{C}^* \tag{9.206}$$

In Eq. (9.206), $\widehat{C}^*$ is known as

$$\widehat{C}^* = e^{-\tilde{u}_3\theta_1} \widehat{C} e^{-\tilde{u}_1\pi/2} \tag{9.207}$$

The unknown angles in Eq. (9.206) are arranged like the Euler angles of the 2-3-2 sequence. Therefore, they can be extracted by using a procedure similar to that explained in Chapter 3. This procedure leads to the following scalar equation pairs generated from Eq. (9.206).

$$c_{22}^* = \overline{u}_2^t \widehat{C}^* \overline{u}_2 = c\theta_5 \Rightarrow s\theta_5 = \sigma_5 \sqrt{1-(c_{22}^*)^2};\quad \sigma_5 = \pm 1 \tag{9.208}$$

$$c_{23}^* = \overline{u}_2^t \widehat{C}^* \overline{u}_3 = -s\theta_5 s\theta_6;\quad c_{21}^* = \overline{u}_2^t \widehat{C}^* \overline{u}_1 = s\theta_5 c\theta_6 \tag{9.209}$$

$$c_{32}^* = \overline{u}_3^t \widehat{C}^* \overline{u}_2 = -s\theta_{234} s\theta_5;\quad c_{12}^* = \overline{u}_1^t \widehat{C}^* \overline{u}_2 = -c\theta_{234} s\theta_5 \tag{9.210}$$

Equation Pair (9.208) gives θ_5 as

$$\theta_5 = \mathrm{atan}_2(\sigma_5\sqrt{1-(c_{22}^*)^2}, c_{22}^*) = \sigma_5 \mathrm{atan}_2(\sqrt{1-(c_{22}^*)^2}, c_{22}^*) \tag{9.211}$$

If $s\theta_5 \neq 0$, the other equation pairs give θ_6 and θ_{234} as follows without additional superfluous sign variables:

$$\theta_6 = \mathrm{atan}_2(-\sigma_5 c_{23}^*, \sigma_5 c_{21}^*) \tag{9.212}$$

$$\theta_{234} = \mathrm{atan}_2(-\sigma_5 c_{32}^*, -\sigma_5 c_{12}^*) \tag{9.213}$$

With the availability of θ_1 and θ_{234}, Eqs. (9.200) and (9.202) can be revisited for the final stage of the solution. At this stage, the mentioned equations can be written as follows:

$$b_2 c\theta_2 + b_3 c\theta_{23} = x_1 \tag{9.214}$$

$$b_2 s\theta_2 + b_3 s\theta_{23} = x_3 \tag{9.215}$$

In Eqs. (9.214) and (9.215), x_1 and x_3 are defined as shown below. They happen to be the coordinates of the pre-wrist point Q in the frame $\mathcal{F}_1(O)$ of the first link. They are known at this stage owing to the availability of θ_1 and θ_{234}.

$$x_1 = r_1 c\theta_1 + r_2 s\theta_1 - b_4 c\theta_{234} = \sigma_1\sqrt{r_1^2 + r_2^2} - b_4 c\theta_{234} \tag{9.216}$$

$$x_3 = r_3 - b_4 s\theta_{234} \tag{9.217}$$

In order to find θ_3, the squares of Eqs. (9.214) and (9.215) can be added to obtain the following equation.

$$\begin{aligned} x_1^2 + x_3^2 &= b_2^2 + b_3^2 + 2b_2b_3(c\theta_{23}c\theta_2 + s\theta_{23}s\theta_{23}) = b_2^2 + b_3^2 + 2b_2b_3\cos(\theta_{23} - \theta_2) \\ &= b_2^2 + b_3^2 + 2b_2b_3c\theta_3 \Rightarrow \end{aligned}$$

$$c\theta_3 = \xi_3 = [(x_1^2 + x_3^2) - (b_2^2 + b_3^2)]/(2b_2b_3) \tag{9.218}$$

Equation (9.218) implies that

$$s\theta_3 = \sigma_3\sqrt{1-\xi_3^2};\quad \sigma_3 = \pm 1 \tag{9.219}$$

Hence, θ_3 is found with the following alternative values of opposite signs.

$$\theta_3 = \text{atan}_2(\sigma_3\sqrt{1-\xi_3^2}, \xi_3) = \sigma_3\text{atan}_2(\sqrt{1-\xi_3^2}, \xi_3) \tag{9.220}$$

With known θ_3, Eqs. (9.214) and (9.215) can be written as follows:

$$(b_2 + b_3c\theta_3)c\theta_2 - (b_3s\theta_3)s\theta_2 = x_1 \tag{9.221}$$

$$(b_2 + b_3c\theta_3)s\theta_2 + (b_3s\theta_3)c\theta_2 = x_3 \tag{9.222}$$

Equations (9.221) and (9.222) can be written together as the following matrix equation in order to obtain $s\theta_2$ and $c\theta_2$ separately so that an additional superfluous sign variable is not introduced in finding θ_2.

$$\begin{bmatrix}(b_2 + b_3c\theta_3) & -b_3s\theta_3 \\ b_3s\theta_3 & (b_2 + b_3c\theta_3)\end{bmatrix}\begin{bmatrix}c\theta_2 \\ s\theta_2\end{bmatrix} = \begin{bmatrix}x_1 \\ x_3\end{bmatrix} \tag{9.223}$$

In Eq. (9.223), the determinant of the coefficient matrix is

$$D_2 = (b_2 + b_3c\theta_3)^2 + (b_3s\theta_3)^2 = b_2^2 + b_3^2 + 2b_2b_3c\theta_3 \tag{9.224}$$

If $D_2 \neq 0$, i.e. if $D_2 > 0$, Eq. (9.223) gives $c\theta_2$ and $s\theta_2$ as follows:

$$\begin{bmatrix}c\theta_2 \\ s\theta_2\end{bmatrix} = \frac{1}{D_2}\begin{bmatrix}(b_2 + b_3c\theta_3) & b_3s\theta_3 \\ -b_3s\theta_3 & (b_2 + b_3c\theta_3)\end{bmatrix}\begin{bmatrix}x_1 \\ x_3\end{bmatrix} \Rightarrow$$

$$\left.\begin{aligned} c\theta_2 &= [(b_2 + b_3c\theta_3)x_1 + (b_3s\theta_3)x_3]/D_2 = A_2/D_2 \\ c\theta_2 &= [(b_2 + b_3c\theta_3)x_3 - (b_3s\theta_3)x_1]/D_2 = B_2/D_2 \end{aligned}\right\} \tag{9.225}$$

Hence, θ_2 is found as

$$\theta_2 = \text{atan}_2(B_2, A_2) \tag{9.226}$$

Finally, with the availability of θ_2, θ_3, and θ_{234}, θ_4 also becomes available as

$$\theta_4 = \theta_{234} - (\theta_2 + \theta_3) \tag{9.227}$$

9.3.4 Multiplicity Analysis

(a) First Kind of Multiplicity

The first kind of multiplicity is associated with the sign variable σ_1 that arises in the process of finding θ_1 by using Eq. (9.205). The manipulator attains the same location of the wrist point by assuming one of the two alternative poses that correspond to $\sigma_1 = +1$ and $\sigma_1 = -1$. These poses are illustrated in Figure 9.17. The poses corresponding to $\sigma_1 = +1$ and $\sigma_1 = -1$ are designated, respectively, as *wrist-ahead pose* and *wrist-behind pose*. They are so called because Eq. (9.205) implies that if $\sigma_1 = +1$ leads to θ_1, then $\sigma_1 = -1$ leads to $\theta_1' = \theta_1 + \pi$, which means that $\vec{u}_1^{(1)}$, i.e. the look-ahead direction of the manipulator, is reversed.

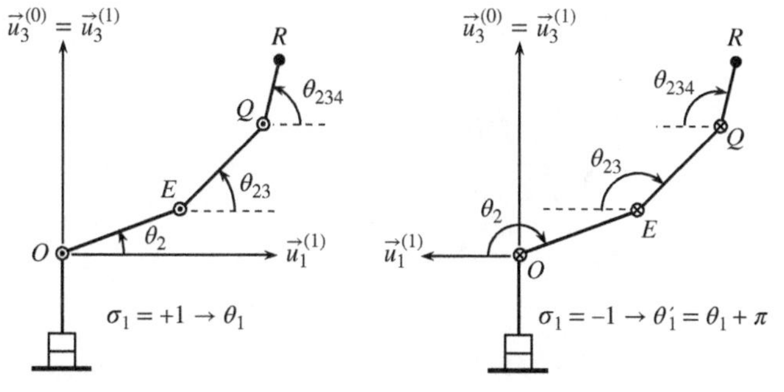

Figure 9.17 Wrist-ahead and wrist-behind poses of an elbow manipulator.

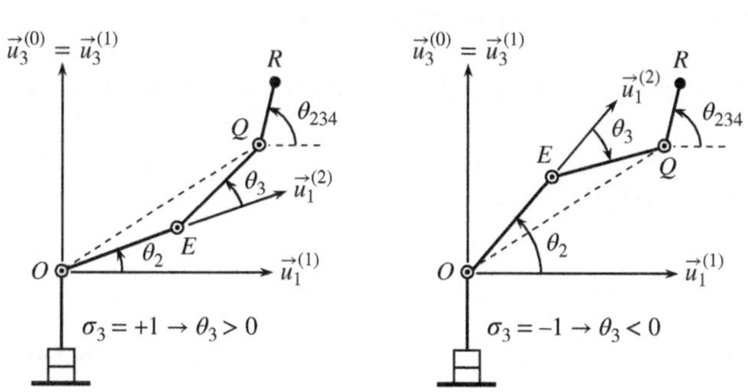

Figure 9.18 Elbow-down and elbow-up poses of an elbow manipulator.

(b) Second Kind of Multiplicity

The second kind of multiplicity is associated with the sign variable σ_3 that arises in the process of finding θ_3 by using Eqs. (9.218)–(9.220). The manipulator attains the same location of the *pre-wrist point* by assuming one of the two alternative poses that correspond to $\sigma_3 = +1$ and $\sigma_3 = -1$. These poses are illustrated in Figure 9.18. The poses corresponding to $\sigma_3 = +1$ and $\sigma_3 = -1$ are designated, respectively, as *elbow-down pose* and *elbow-up pose.* These designations are justified by Eq. (9.220), because it indicates that, for the same value of $\xi_3 = c\theta_3$, θ_3 becomes positive if $\sigma_3 = +1$ and θ_3 becomes negative if $\sigma_3 = -1$. The corresponding values of θ_2 are determined by Eqs. (9.225) and (9.226).

(c) Third Kind of Multiplicity

The third kind of multiplicity is associated with the sign variable σ_5 that arises in the process of finding the angles θ_5, θ_6, and θ_{234} by using Eqs. (9.211)–(9.213). The manipulator attains the same orientation of the end-effector with both $\sigma_5 = +1$ and $\sigma_5 = -1$. However, unlike the Puma and Stanford manipulators, the elbow manipulator assumes distinct poses depending on the value of σ_5 because of its nonspherical wrist. These poses are illustrated in Figure 9.19. The way they come out is explained below.

Suppose that $\sigma_5 = +1$ leads to the angles θ_5, θ_6, and θ_{234}. If so, Eqs. (9.211)–(9.213) imply that $\sigma_5 = -1$ leads to the angles $\theta_5' = -\theta_5$, $\theta_6' = \theta_6 + \pi$, and $\theta_{234}' = \theta_{234} + \pi$. However, the orientation of Link 4 with θ_{234}' happens to be the reverse of its orientation with

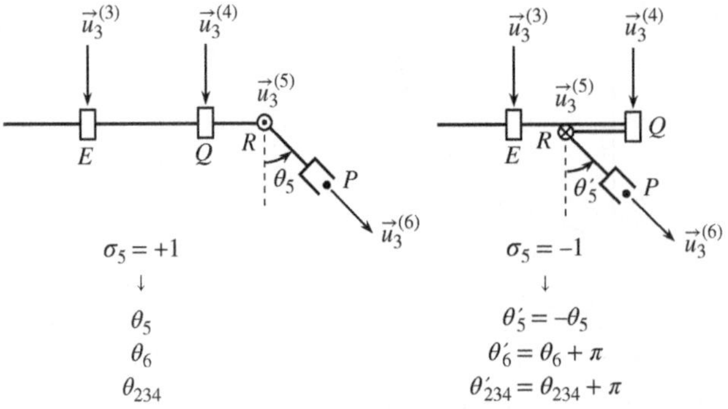

Figure 9.19 Extended gripper and folded gripper poses of an elbow manipulator.

θ_{234}. That is why Link 4 appears to be folded back if σ_5 is switched from $\sigma_5 = +1$ to $\sigma_5 = -1$.

Based on the above explanation, it can be said that $\sigma_5 = +1$ leads to an *extended gripper pose* and $\sigma_5 = -1$ leads to a *folded gripper pose*.

9.3.5 Singularity Analysis in the Position Domain

An elbow manipulator may have three distinct kinds of position singularities, which are described and discussed below.

(a) First Kind of Position Singularity

Equation (9.205) implies that the first kind of position singularity occurs if the position of the end-effector is specified so that $r_1 = r_2 = 0$. This singularity is illustrated in Figure 9.20. As a noticeable feature of this singularity, the wrist point is located on the axis of the first joint. If this singularity occurs, θ_1 becomes indefinite and ineffective. In other words, θ_1 can be assigned an arbitrary value, but it cannot cause any change in the position of the end-effector, whatever the assigned value is. Thus, the manipulator gains a positioning freedom in the joint space, but this freedom happens to be useless in the task space as long as the wrist point is kept on the axis of the first joint. On other

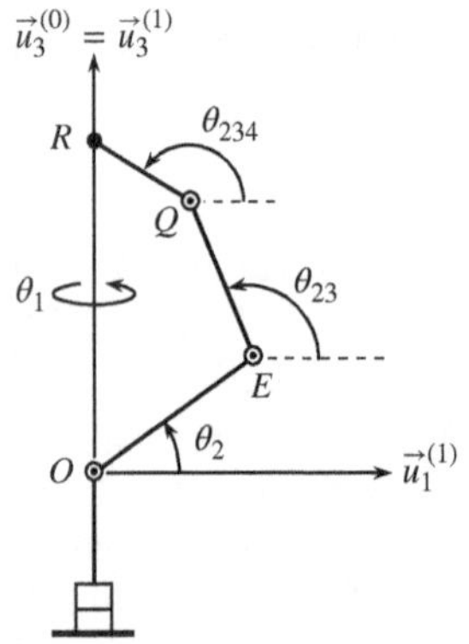

Figure 9.20 First kind of position singularity of an elbow manipulator.

hand, if the task to be executed necessitates keeping the wrist point as such, then the freedom in θ_1 may be used to orient the arm of the manipulator conveniently depending on the environmental conditions, e.g. the locations of the obstacles that may exist in the vicinity.

(b) Second Kind of Position Singularity

Equation (9.225) implies that the second kind of position singularity occurs if $D_2 = 0$, i.e. if the determinant of the coefficient matrix vanishes. According to Eq. (9.224), D_2 can vanish only if the following equations are both satisfied.

$$b_2 = b_3 \tag{9.228}$$

$$c\theta_3 = -1 \text{ or } \theta_3 = \pi \tag{9.229}$$

Equation (9.223) also implies that this singularity occurs if the position of the end-effector is specified in such a way that the pre-wrist point Q coincides with the origin O of the base frame. This can happen if the wrist point coordinates obey the following equations, which are the special forms of Eqs. (9.216) and (9.217).

$$x_1 = 0 \Rightarrow \sqrt{r_1^2 + r_2^2} = \sigma_1 b_4 c\theta_{234} \tag{9.230}$$

$$x_3 = 0 \Rightarrow r_3 = b_4 s\theta_{234} \tag{9.231}$$

Equations (9.228) and (9.229) express the noticeable feature of this singularity that the upper and front arms of the manipulator have equal lengths and the front arm is folded completely over the upper arm. However, this singularity cannot occur exactly because of the physical shapes of the relevant links and joints. This singularity is illustrated (approximately) in Figure 9.21.

If this singularity occurs, θ_2 becomes indefinite and ineffective. In other words, θ_2 can be assigned an arbitrary value, but it cannot cause any change in the position of the end-effector, whatever the assigned value is. Thus, the manipulator gains a positioning freedom in the joint space, but this freedom happens to be useless in the task space as long as the pre-wrist point is kept coincident (actually almost coincident) with the base frame origin. On the other hand, if the task to be executed necessitates keeping the manipulator in such a pose, then the freedom in θ_2 may be used to orient the folded arm links of the manipulator conveniently depending on the environmental conditions. For example, the coincident upper and front arms may be allowed to get aligned with the gravity direction so that the task can be executed with minimal potential energy.

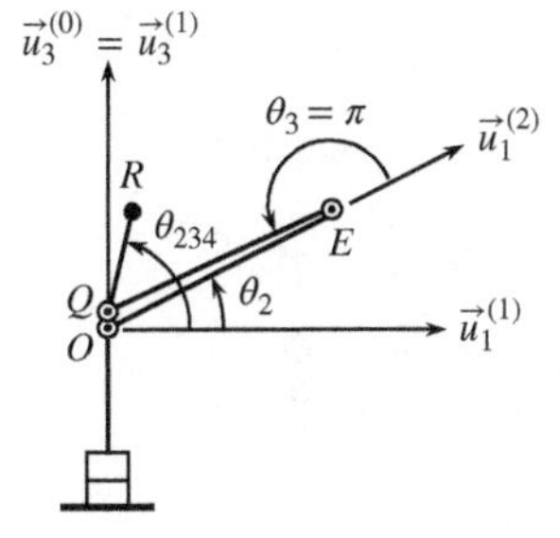

Figure 9.21 Second kind of position singularity of an elbow manipulator.

(c) Third Kind of Position Singularity

Equations (9.212) and (9.213) imply that the third kind of position singularity occurs if $s\theta_5 = 0$, i.e. if $\theta_5 = 0$ or $\theta_5 = \pi$. In other words, as also implied by Figure 9.16, this singularity occurs if the orientation of the end-effector is specified so that the approach vector is perpendicular to the front arm. This singularity is illustrated in Figure 9.22 with both $\theta_5 = 0$ and $\theta_5 = \pi$.

If this singularity occurs, θ_{234} and θ_6 become indefinite so that they cannot be found separately. This is because they turn out to be rotation angles about parallel axes, which are the axis of the sixth joint and the already parallel axes of the fourth, third, and second joints. Therefore, the angles θ_{234} and θ_6 cannot be distinguished from each other. Nevertheless, their combinations, i.e. $\theta_{2346} = \theta_{234} + \theta_6$ or $\theta'_{2346} = \theta_{234} - \theta_6$ can still be found as explained below.

In the singularity with $\theta_5 = 0$, Eq. (9.206) becomes

$$e^{-\tilde{u}_2\theta_{234}}\hat{I}e^{-\tilde{u}_2\theta_6} = e^{-\tilde{u}_2\theta_{234}}e^{-\tilde{u}_2\theta_6} = e^{-\tilde{u}_2(\theta_{234}+\theta_6)} = \hat{C}^* \Rightarrow$$

$$e^{-\tilde{u}_2\theta_{2346}} = \hat{C}^* \tag{9.232}$$

In the singularity with $\theta_5 = \pi$, Eq. (9.206) can be manipulated into the following form by means of the shifting property.

$$e^{-\tilde{u}_2\theta_{234}}e^{\tilde{u}_3\pi}e^{-\tilde{u}_2\theta_6} = e^{-\tilde{u}_2\theta_{234}}e^{\tilde{u}_2\theta_6}e^{\tilde{u}_3\pi} = e^{-\tilde{u}_2(\theta_{234}-\theta_6)}e^{\tilde{u}_3\pi} = \hat{C}^* \Rightarrow$$

$$e^{-\tilde{u}_2\theta'_{2346}} = \hat{C}^*e^{-\tilde{u}_3\pi} \tag{9.233}$$

Equation (9.232) gives the combined angle $\theta_{2346} = \theta_{234} + \theta_6$ as follows:

$$c^*_{11} = \overline{u}_1^t\hat{C}^*\overline{u}_1 = c\theta_{2346},\ c^*_{31} = \overline{u}_3^t\hat{C}^*\overline{u}_1 = s\theta_{2346} \Rightarrow$$

$$\theta_{2346} = \text{atan}_2(c^*_{31}, c^*_{11}) \tag{9.234}$$

Equation (9.233) gives the other combined angle $\theta'_{2346} = \theta_{234} - \theta_6$ as follows:

$$\left.\begin{aligned}\overline{u}_1^t\hat{C}^*e^{-\tilde{u}_3\pi}\overline{u}_1 &= -\overline{u}_1^t\hat{C}^*\overline{u}_1 = -c^*_{11} = c\theta'_{2346}\\ \overline{u}_3^t\hat{C}^*e^{-\tilde{u}_3\pi}\overline{u}_1 &= -\overline{u}_3^t\hat{C}^*\overline{u}_1 = -c^*_{31} = s\theta'_{2346}\end{aligned}\right\} \Rightarrow$$

$$\theta'_{2346} = \text{atan}_2(-c^*_{31}, -c^*_{11}) \tag{9.235}$$

Since either θ_{2346} or θ'_{2346} becomes known, only one of θ_{234} and θ_6, e.g. θ_6, can be assigned an arbitrary value and the other one, i.e. θ_{234}, can then be found either as $\theta_{234} = \theta_{2346} - \theta_6$

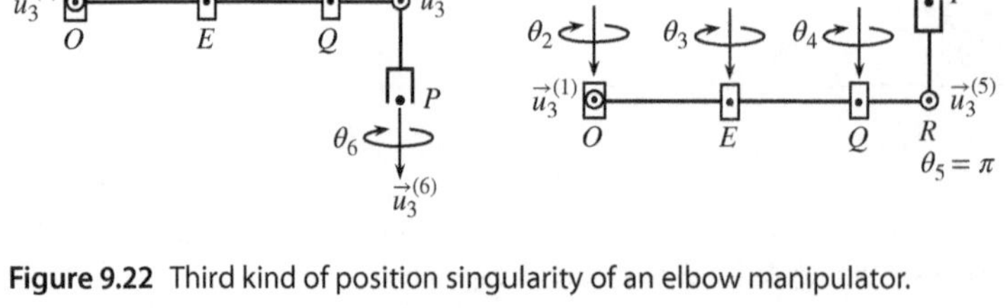

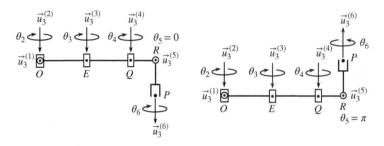

Figure 9.22 Third kind of position singularity of an elbow manipulator.

or as $\theta_{234} = \theta'_{2346} + \theta_6$. Thus, the manipulator gains a positioning freedom in the joint space, but this freedom happens to be useless in the task space as long as the approach vector is kept perpendicular to the front arm.

9.3.6 Forward Kinematics in the Velocity Domain

(a) Angular Velocity of the End-Effector with Respect to the Base Frame

Let $\overline{\omega} = \overline{\omega}_6$. Then, Eq. (9.183) leads to the following expression.

$$\overline{\omega} = \dot{\theta}_1\overline{u}_3 - \dot{\theta}_{234}e^{\tilde{u}_3\theta_1}\overline{u}_2 + \dot{\theta}_5 e^{\tilde{u}_3\theta_1}e^{-\tilde{u}_2\theta_{234}}\overline{u}_3 - \dot{\theta}_6 e^{\tilde{u}_3\theta_1}e^{-\tilde{u}_2\theta_{234}}e^{\tilde{u}_3\theta_5}\overline{u}_2 \quad (9.236)$$

(b) Velocity of the Wrist Point with Respect to the Base Frame

Let $\overline{w} = \overline{w}^{(0)} = \overline{v}_R = \dot{\overline{r}}$. Then, $\overline{w}$ can be obtained from Eqs. (9.189) and (9.190) as follows:

$$\overline{w} = (\dot{\theta}_1 e^{\tilde{u}_3\theta_1}\tilde{u}_3)(\overline{u}_1 r'_1 + \overline{u}_3 r'_3) + (e^{\tilde{u}_3\theta_1})(\overline{u}_1\dot{r}'_1 + \overline{u}_3\dot{r}'_3) \Rightarrow$$

$$\overline{w} = e^{\tilde{u}_3\theta_1}(\overline{u}_1\dot{r}'_1 + \overline{u}_2 r'_1\dot{\theta}_1 + \overline{u}_3\dot{r}'_3) \Rightarrow$$

$$\overline{w} = e^{\tilde{u}_3\theta_1}(\overline{u}_1 w'_1 + \overline{u}_2 w'_2 + \overline{u}_3 w'_3) = e^{\tilde{u}_3\theta_1}\overline{w}' \quad (9.237)$$

In Eq. (9.237),

$$\left.\begin{aligned} w'_1 &= -(b_2\dot{\theta}_2 s\theta_2 + b_3\dot{\theta}_{23}s\theta_{23} + b_4\dot{\theta}_{234}s\theta_{234}) \\ w'_2 &= r'_1\dot{\theta}_1 = (b_2 c\theta_2 + b_3 c\theta_{23} + b_4 c\theta_{234})\dot{\theta}_1 \\ w'_3 &= b_2\dot{\theta}_2 c\theta_2 + b_3\dot{\theta}_{23}c\theta_{23} + b_4\dot{\theta}_{234}c\theta_{234} \end{aligned}\right\} \quad (9.238)$$

(c) Velocity of the Tip Point with Respect to the Base Frame

Let $\overline{v} = \overline{v}^{(0)} = \overline{v}_P = \dot{\overline{p}}$. Then, $\overline{v}$ can be obtained from Eq. (9.186) as follows:

$$\overline{v} = \overline{w} + d_6\tilde{\omega}\hat{C}\overline{u}_3 \quad (9.239)$$

9.3.7 Inverse Kinematics in the Velocity Domain

Since $\overline{p}$ and $\hat{C}$ are specified as functions of time, the velocities $\overline{v}$, $\overline{\omega}$, and $\overline{w}$ are also known. They are obtained as shown below.

$$\overline{v} = \dot{\overline{p}}, \quad \overline{\omega} = \text{colm}(\dot{\hat{C}}\hat{C}^t); \quad \overline{w} = \overline{v} - d_6\tilde{\omega}\hat{C}\overline{u}_3 \quad (9.240)$$

The solution for the joint velocities can be obtained from Eqs. (9.237) and (9.236). For this purpose, Eq. (9.237) can be written as follows:

$$\overline{w}' = \overline{u}_1 w'_1 + \overline{u}_2 w'_2 + \overline{u}_3 w'_3 = e^{-\tilde{u}_3\theta_1}\overline{w} \quad (9.241)$$

Upon pre-multiplications by $\overline{u}_1^t$, $\overline{u}_2^t$, and $\overline{u}_3^t$, Eq. (9.241) leads to the following scalar equations.

$$b_2\dot{\theta}_2 s\theta_2 + b_3\dot{\theta}_{23}s\theta_{23} + b_4\dot{\theta}_{234}s\theta_{234} = -w'_1 = -(w_1 c\theta_1 + w_2 s\theta_1) \quad (9.242)$$

$$(b_2 c\theta_2 + b_3 c\theta_{23} + b_4 c\theta_{234})\dot{\theta}_1 = w'_2 = w_2 c\theta_1 - w_1 s\theta_1 \quad (9.243)$$

$$b_2\dot{\theta}_2 c\theta_2 + b_3\dot{\theta}_{23}c\theta_{23} + b_4\dot{\theta}_{234}c\theta_{234} = w'_3 = w_3 \quad (9.244)$$

In Eq. (9.243), the only unknown is $\dot{\theta}_1$ and its coefficient is $r_1' = b_2c\theta_2 + b_3c\theta_{23} + b_4c\theta_{234}$. If $r_1' \neq 0$, $\dot{\theta}_1$ can be found from Eq. (9.243) as

$$\dot{\theta}_1 = (w_2c\theta_1 - w_1s\theta_1)/(b_2c\theta_2 + b_3c\theta_{23} + b_4c\theta_{234}) \tag{9.245}$$

With the availability of $\dot{\theta}_1$, Eq. (9.236) can be arranged as follows:

$$\begin{aligned} &\dot{\theta}_{234}\overline{u}_2 - \dot{\theta}_5\overline{u}_3 + \dot{\theta}_6 e^{\tilde{u}_3\theta_5}\overline{u}_2 = -\overline{\omega}^* \Rightarrow \\ &\overline{u}_1(\dot{\theta}_6 s\theta_5) - \overline{u}_2(\dot{\theta}_{234} + \dot{\theta}_6 c\theta_5) + \overline{u}_3\dot{\theta}_5 = \overline{\omega}^* \end{aligned} \tag{9.246}$$

In Eq. (9.246), $\overline{\omega}^*$ is known as expressed below.

$$\overline{\omega}^* = e^{-\tilde{u}_2\theta_{234}} e^{-\tilde{u}_3\theta_1}(\overline{\omega} - \dot{\theta}_1\overline{u}_3) = e^{-\tilde{u}_2\theta_{234}}(e^{-\tilde{u}_3\theta_1}\overline{\omega} - \dot{\theta}_1\overline{u}_3) \tag{9.247}$$

At this point, noting that $\overline{\omega} - \dot{\theta}_1\overline{u}_3 = \overline{\omega}_{6/1}^{(0)}$ and recalling that $\widehat{C}^{(0,4)} = e^{\tilde{u}_3\theta_1} e^{-\tilde{u}_2\theta_{234}} e^{\tilde{u}_1\pi/2}$, it can be shown that

$$\overline{\omega}^* = e^{\tilde{u}_1\pi/2}\overline{\omega}_{6/1}^{(4)} \tag{9.248}$$

Hence, it is seen that, with a modification factor $e^{\tilde{u}_1\pi/2}$, $\overline{\omega}^*$ represents the relative angular velocity of the end-effector with respect to Link 1 as expressed in the frame of Link 4.

Upon premultiplications by $\overline{u}_1^t$, $\overline{u}_2^t$, and $\overline{u}_3^t$, Eq. (9.248) leads to the following scalar equations.

$$\dot{\theta}_6 s\theta_5 = \omega_1^* \tag{9.249}$$

$$\dot{\theta}_{234} + \dot{\theta}_6 c\theta_5 = -\omega_2^* \tag{9.250}$$

$$\dot{\theta}_5 = \omega_3^* \tag{9.251}$$

Equation (9.251) has already given $\dot{\theta}_5$. The other two equations give $\dot{\theta}_6$ and $\dot{\theta}_{234}$ if $s\theta_3 \neq 0$. That is,

$$\dot{\theta}_6 = \omega_1^*/s\theta_5 \tag{9.252}$$

$$\dot{\theta}_{234} = -(\omega_2^* + \dot{\theta}_6 c\theta_5) = -(\omega_1^* c\theta_5 + \omega_2^* s\theta_5)/s\theta_5 \tag{9.253}$$

With the availability of $\dot{\theta}_1$ and $\dot{\theta}_{234}$, Eqs. (9.242) and (9.244) can be written together as the following matrix equation.

$$\begin{bmatrix} b_3s\theta_{23} & b_2s\theta_2 \\ b_3c\theta_{23} & b_2c\theta_2 \end{bmatrix} \begin{bmatrix} \dot{\theta}_{23} \\ \dot{\theta}_2 \end{bmatrix} = \begin{bmatrix} -w_1^* \\ w_3^* \end{bmatrix} \tag{9.254}$$

In Eq. (9.254), w_1^*, w_3^*, and the determinant D_{23} of the coefficient matrix have the following expressions.

$$w_1^* = w_1' + b_4\dot{\theta}_{234}s\theta_{234} = w_1c\theta_1 + w_2s\theta_1 + b_4\dot{\theta}_{234}s\theta_{234} \tag{9.255}$$

$$w_3^* = w_3 - b_4\dot{\theta}_{234}c\theta_{234} \tag{9.256}$$

$$D_{23} = b_2b_3(s\theta_{23}c\theta_2 - c\theta_{23}s\theta_2) = b_2b_3\sin(\theta_{23} - \theta_2) = b_2b_3s\theta_3 \tag{9.257}$$

If $D_{23} \neq 0$, i.e. if $s\theta_3 \neq 0$, Eq. (9.254) gives $\dot{\theta}_2$ and $\dot{\theta}_3$ as follows:

$$\begin{aligned} &\begin{bmatrix} \dot{\theta}_{23} \\ \dot{\theta}_2 \end{bmatrix} = \frac{1}{b_2b_3s\theta_3} \begin{bmatrix} b_2c\theta_2 & -b_2s\theta_2 \\ -b_3c\theta_{23} & b_3s\theta_{23} \end{bmatrix} \begin{bmatrix} -w_1^* \\ w_3^* \end{bmatrix} \Rightarrow \\ &\dot{\theta}_{23} = -(w_1^*c\theta_2 + w_3^*s\theta_2)/(b_3s\theta_3) \end{aligned} \tag{9.258}$$

$$\dot{\theta}_2 = (w_1^* c\theta_{23} + w_3^* s\theta_{23})/(b_2 s\theta_3) \tag{9.259}$$

$$\dot{\theta}_3 = \dot{\theta}_{23} - \dot{\theta}_2 \tag{9.260}$$

Finally, $\dot{\theta}_4$ is also found as

$$\dot{\theta}_4 = \dot{\theta}_{234} - \dot{\theta}_{23} \tag{9.261}$$

9.3.8 Singularity Analysis in the Velocity Domain

An elbow manipulator may have three distinct kinds of motion singularities, which are described and discussed below.

(a) First Kind of Motion Singularity

Equation (9.245) implies that the first kind of motion singularity occurs if

$$r_1' = b_2 c\theta_2 + b_3 c\theta_{23} + b_4 c\theta_{234} = 0 \tag{9.262}$$

On the other hand, referring to Part (c) of Section 9.3.2, it is seen that $r_1' = 0$ implies that $r_1 = r_2 = 0$. Therefore, the appearance of this motion singularity is the same as the appearance of the first kind of position singularity illustrated in Figure 9.20.
If the manipulator has to pass through this singularity as a task requirement, then, according to Eq. (9.245), $\dot{\theta}_1$ can be kept finite by planning the task motion so that it obeys the following *compatibility condition.*

$$w_2 c\theta_1 = w_1 s\theta_1 \tag{9.263}$$

If the task motion obeys the above condition, which is a singularity-induced restriction in the task space, $\dot{\theta}_1$ becomes finite but indefinite. This is because, when Eq. (9.243) is left out, which is trivially satisfied as $0 = 0$, there remains five scalar equations to be satisfied by all the six joint velocities $(\dot{\theta}_1, \ldots, \dot{\theta}_6)$. These equations are (9.242), (9.244), and (9.249)–(9.251) noting that $\overline{\omega}^*$ depends on $\dot{\theta}_1$. In other words, the singularity also induces a motion freedom in the joint space. Thus, an arbitrary value can be assigned to $\dot{\theta}_1$ and the others $(\dot{\theta}_2, \ldots, \dot{\theta}_6)$ can then be found as explained above in Section 9.3.7.

(b) Second Kind of Motion Singularity

Equations (9.258) and (9.259) imply that the second kind of motion singularity occurs if $s\theta_3 = 0$, i.e. if

$$\theta_3 = 0 \text{ or } \theta_3 = \pi \tag{9.264}$$

This singularity is illustrated in Figure 9.23 in its *extended* ($\theta_3 = 0$) and *folded* ($\theta_3 = \pi$) versions. Note that the appearance of the folded version of this motion singularity becomes the same as the appearance of the second kind of position singularity if $b_3 = b_2$. In this singularity, with $\sigma_3' = \text{sgn}(c\theta_3)$, Eqs. (9.242) and (9.244) can be manipulated into the following forms.

$$(b_2\dot{\theta}_2 + \sigma_3' b_3 \dot{\theta}_{23}) s\theta_2 = -w_1^* \tag{9.265}$$

$$(b_2\dot{\theta}_2 + \sigma_3' b_3 \dot{\theta}_{23}) c\theta_2 = w_3^* \tag{9.266}$$

Here, $\sigma_3' = +1$ and $\sigma_3' = -1$ represent the *extended* and *folded* versions of this singularity.

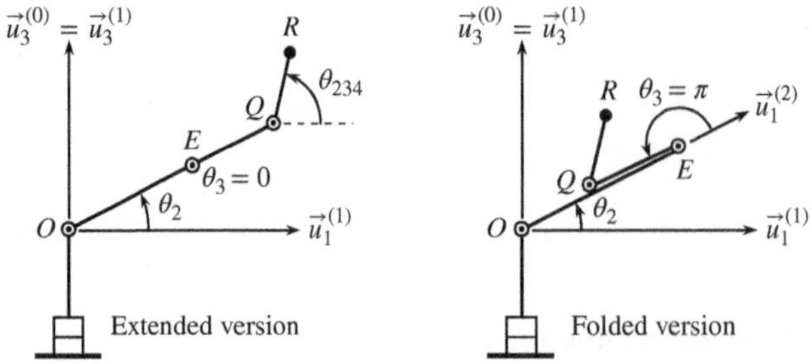

Figure 9.23 Second kind of motion singularity of an elbow manipulator.

Equations (9.265) and (9.266) become consistent and thus they can give finite values for $\dot{\theta}_2$ and $\dot{\theta}_{23}$ if the following *compatibility condition* is satisfied by the planned task motion.

$$w_1^* c\theta_2 + w_3^* s\theta_2 = 0 \tag{9.267}$$

The same equations can be combined into the following equation, which relates the finite but indefinite values of $\dot{\theta}_2$ and $\dot{\theta}_3$.

$$b_2\dot{\theta}_2 + \sigma_3' b_3\dot{\theta}_{23} = (b_2 + \sigma_3' b_3)\dot{\theta}_2 + \sigma_3' b_3\dot{\theta}_3 = w_3^* c\theta_2 - w_1^* s\theta_2 \tag{9.268}$$

As seen above, like the first kind, this second kind of motion singularity also induces a restriction in the task space as expressed by Eq. (9.267) and it also induces a motion freedom in the joint space as expressed by Eq. (9.268). Thus, as long as the task-space compatibility condition is satisfied, one of $\dot{\theta}_2$ and $\dot{\theta}_3$ can be assigned a finite arbitrary value and the other one can be found from Eq. (9.268).

(c) Third Kind of Motion Singularity

Equations (9.252) and (9.253) imply that the third kind of motion singularity occurs if $s\theta_5 = 0$, i.e. if

$$\theta_5 = 0 \text{ or } \theta_5 = \pi \tag{9.269}$$

Note that the appearance of this motion singularity is the same as the appearance of the third kind of position singularity illustrated in Figure 9.22.

In this singularity, with $\sigma_5' = \text{sgn}\left(c\theta_5\right)$, Eqs. (9.249) and (9.250) take the following forms.

$$\omega_1^* = 0 \tag{9.270}$$

$$\dot{\theta}_{234} + \sigma_5' \dot{\theta}_6 = -\omega_2^* \tag{9.271}$$

Equation (9.270) expresses the restriction (i.e. the compatibility condition) induced by this singularity in the task space. According to this restriction, the end-effector cannot rotate about $\vec{u}_1^{(4)}$ with respect to Link 1. If the specified motion of the end-effector obeys this condition, then $\dot{\theta}_{234}$ and $\dot{\theta}_6$ become finite but indefinite. Yet, their indefinite values become related by Eq. (9.271). Thus, this singularity induces a motion freedom in the joint space so that one of $\dot{\theta}_{234}$ and $\dot{\theta}_6$, say $\dot{\theta}_6$, can be assigned an arbitrary value and the other one, i.e. $\dot{\theta}_{234}$, can be found accordingly.

9.4 Scara Manipulator

A Scara manipulator is shown in Figure 9.24. Its generic name comes from the acronym SCARA that stands for "Selective Compliance Assembly Robot Arm." It was developed originally under the guidance of *Professor Hiroshi Makino* at the *University of Yamanashi* in Japan. It was purposefully developed to be somewhat compliant along the first (x) and second (y) axes but definitely stiff along the third (z) axis of the base frame. Thus, it was intended to be used particularly for certain assembly operations such as inserting a pin into a hole without causing excessive lateral contact forces and moments.

Normally, it is a 4-DoF (degree of freedom) manipulator without the fifth and sixth joints and its end-effector is attached to the fourth link so that $\vec{u}_3^{(4)}$ is the approach vector. However, for the sake of generality, it is treated here as a 6-DoF manipulator as shown in Figure 9.24. With this general configuration, its wrist is arranged to be spherical and it is designated symbolically as 2R-P-3R or R^2PR^3. Its joint axes and the special kinematic details (the joint variables and the constant geometric parameters) are also shown in the figure together with the relevant unit vectors. The significant points of the manipulator are named as follows:

O: Center point or shoulder point (origin of the base frame)
E: Elbow point, Q: Sliding reference point, R: Wrist point, P: Tip point

9.4.1 Kinematic Description According to the D–H Convention

(a) Rotation Angles

$$\theta_1, \theta_2, \theta_3 = \delta_3 = 0, \theta_4, \theta_5, \theta_6$$

Five of the rotation angles are joint variables. The third one is associated with the cylindrical arrangement of the joints $\mathcal{J}_3$ and $\mathcal{J}_4$. So, it is taken to be zero as discussed in Chapter 7. Thus, $\vec{u}_1^{(3)}$ is arranged to be parallel to $\vec{u}_1^{(2)}$.

(b) Twist Angles

$$\beta_1 = 0, \beta_2 = 0, \beta_3 = \pi, \beta_4 = 0, \beta_5 = -\pi/2, \beta_6 = \pi/2$$

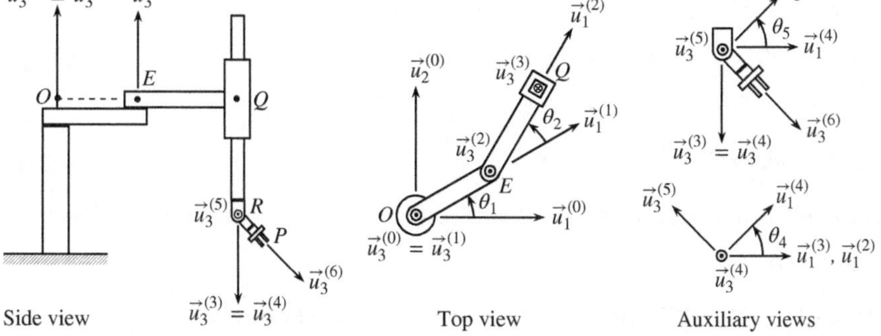

Figure 9.24 A Scara manipulator in its various views.

(c) Offsets

$$d_1 = 0, d_2 = 0, s_3 = QR, d_4 = 0, d_5 = 0, d_6 = RP$$

Five of the offsets are constant. The one associated with the prismatic joint is naturally variable.

(d) Effective Link Lengths

$$b_0 = 0, b_1 = 0, b_2 = OE, b_3 = EQ, b_4 = 0, b_5 = 0$$

(e) Link Frame Origins

$$O_0 = O, O_1 = O, O_2 = E, O_3 = R, O_4 = R, O_5 = R, O_6 = P$$

9.4.2 Forward Kinematics in the Position Domain

(a) Link-to-Link Orientation Matrices

$$\hat{C}^{(0,1)} = e^{\tilde{u}_1\beta_1}e^{\tilde{u}_3\theta_1} = e^{\tilde{u}_3\theta_1} \tag{9.272}$$

$$\hat{C}^{(1,2)} = e^{\tilde{u}_1\beta_2}e^{\tilde{u}_3\theta_2} = e^{\tilde{u}_3\theta_2} \tag{9.273}$$

$$\hat{C}^{(2,3)} = e^{\tilde{u}_1\beta_3}e^{\tilde{u}_3\theta_3} = e^{\tilde{u}_1\pi} \tag{9.274}$$

$$\hat{C}^{(3,4)} = e^{\tilde{u}_1\beta_4}e^{\tilde{u}_3\theta_4} = e^{\tilde{u}_3\theta_4} \tag{9.275}$$

$$\hat{C}^{(4,5)} = e^{\tilde{u}_1\beta_5}e^{\tilde{u}_3\theta_5} = e^{-\tilde{u}_1\pi/2}e^{\tilde{u}_3\theta_5} \tag{9.276}$$

$$\hat{C}^{(5,6)} = e^{\tilde{u}_1\beta_6}e^{\tilde{u}_3\theta_6} = e^{\tilde{u}_1\pi/2}e^{\tilde{u}_3\theta_6} \tag{9.277}$$

(b) Orientation Matrices of the Links with Respect to the Base Frame

$$\hat{C}_1 = \hat{C}^{(0,1)} = e^{\tilde{u}_3\theta_1} \tag{9.278}$$

$$\hat{C}_2 = \hat{C}^{(0,2)} = \hat{C}^{(0,1)}\hat{C}^{(1,2)} = e^{\tilde{u}_3\theta_1}e^{\tilde{u}_3\theta_2} = e^{\tilde{u}_3(\theta_1+\theta_2)} = e^{\tilde{u}_3\theta_{12}} \tag{9.279}$$

$$\hat{C}_3 = \hat{C}^{(0,3)} = \hat{C}^{(0,2)}\hat{C}^{(2,3)} = e^{\tilde{u}_3\theta_{12}}e^{\tilde{u}_1\pi} \tag{9.280}$$

$$\hat{C}_4 = \hat{C}^{(0,4)} = \hat{C}^{(0,3)}\hat{C}^{(3,4)} = e^{\tilde{u}_3\theta_{12}}e^{\tilde{u}_1\pi}e^{\tilde{u}_3\theta_4} = e^{\tilde{u}_3\theta_{12}}e^{-\tilde{u}_3\theta_4}e^{\tilde{u}_1\pi} \Rightarrow$$

$$\hat{C}_4 = e^{\tilde{u}_3(\theta_{12}-\theta_4)}e^{\tilde{u}_1\pi} \tag{9.281}$$

$$\hat{C}_5 = \hat{C}^{(0,5)} = \hat{C}^{(0,4)}\hat{C}^{(4,5)} = e^{\tilde{u}_3(\theta_{12}-\theta_4)}e^{\tilde{u}_1\pi}e^{-\tilde{u}_1\pi/2}e^{\tilde{u}_3\theta_5} \Rightarrow$$

$$\hat{C}_5 = e^{\tilde{u}_3(\theta_{12}-\theta_4)}e^{\tilde{u}_1\pi/2}e^{\tilde{u}_3\theta_5} = e^{\tilde{u}_3(\theta_{12}-\theta_4)}e^{-\tilde{u}_2\theta_5}e^{\tilde{u}_1\pi/2} \tag{9.282}$$

$$\hat{C}_6 = \hat{C}^{(0,6)} = \hat{C}^{(0,5)}\hat{C}^{(5,6)} = e^{\tilde{u}_3(\theta_{12}-\theta_4)}e^{-\tilde{u}_2\theta_5}e^{\tilde{u}_1\pi/2}e^{\tilde{u}_1\pi/2}e^{\tilde{u}_3\theta_6} \Rightarrow$$

$$\hat{C}_6 = \hat{C} = e^{\tilde{u}_3(\theta_{12}-\theta_4)}e^{-\tilde{u}_2\theta_5}e^{-\tilde{u}_3\theta_6}e^{\tilde{u}_1\pi} \tag{9.283}$$

In Eq. (9.279), θ_{12} is introduced as a *combined joint variable*. It is defined as follows.

$$\theta_{12} = \theta_1 + \theta_2 \tag{9.284}$$

(c) Location of the Wrist Point with Respect to the Base Frame

$$\vec{r} = \vec{r}_{OR} = \vec{r}_{OE} + \vec{r}_{EQ} + \vec{r}_{QR} = b_1\vec{u}_1^{(1)} + b_2\vec{u}_1^{(2)} + s_3\vec{u}_3^{(3)} \tag{9.285}$$

The vector Eq. (9.285) implies the following matrix equation in the base frame.

$$\bar{r} = \bar{r}^{(0)} = b_1\bar{u}_1^{(1/0)} + b_2\bar{u}_1^{(2/0)} + s_3\bar{u}_3^{(3/0)} \tag{9.286}$$

Eq. (9.286) can be manipulated as shown below.

$$\bar{r} = b_1\hat{C}^{(0,1)}\bar{u}_1^{(1/1)} + b_2\hat{C}^{(0,2)}\bar{u}_1^{(2/2)} + s_3\hat{C}^{(0,3)}\bar{u}_3^{(3/3)} \Rightarrow$$

$$\bar{r} = b_1 e^{\tilde{u}_3\theta_1}\bar{u}_1 + b_2 e^{\tilde{u}_3\theta_{12}}\bar{u}_1 + s_3 e^{\tilde{u}_3\theta_{12}} e^{\tilde{u}_1\pi}\bar{u}_3 \Rightarrow$$

$$\bar{r} = b_1 e^{\tilde{u}_3\theta_1}\bar{u}_1 + b_2 e^{\tilde{u}_3\theta_{12}}\bar{u}_1 - s_3 e^{\tilde{u}_3\theta_{12}}\bar{u}_3 \Rightarrow$$

$$\bar{r} = b_1 e^{\tilde{u}_3\theta_1}\bar{u}_1 + b_2 e^{\tilde{u}_3\theta_{12}}\bar{u}_1 - s_3\bar{u}_3 \tag{9.287}$$

Equation (9.287) leads to the following equation that shows the coordinates of the wrist point R in the base frame $\mathcal{F}_0(O)$.

$$\bar{r} = b_1(\bar{u}_1 c\theta_1 + \bar{u}_2 s\theta_1) + b_2(\bar{u}_1 c\theta_{12} + \bar{u}_2 s\theta_{12}) - s_3\bar{u}_3 \Rightarrow$$

$$\bar{r} = \bar{u}_1 r_1 + \bar{u}_2 r_2 + \bar{u}_3 r_3 = \bar{u}_1(b_1 c\theta_1 + b_2 c\theta_{12}) + \bar{u}_2(b_1 s\theta_1 + b_2 s\theta_{12}) - \bar{u}_3 s_3 \tag{9.288}$$

(d) Location of the Tip Point with Respect to the Base Frame

$$\vec{p} = \vec{r}_{OP} = \vec{r}_{OR} + \vec{r}_{RP} = \vec{r} + d_6\vec{u}_3^{(6)} \tag{9.289}$$

Equation (9.289) implies the following matrix equation in the base frame.

$$\bar{p} = \bar{p}^{(0)} = \bar{r}^{(0)} + d_6\bar{u}_3^{(6/0)} = \bar{r}^{(0)} + d_6\hat{C}^{(0,6)}\bar{u}_3^{(6/6)} \Rightarrow$$

$$\bar{p} = \bar{r} + d_6\hat{C}\bar{u}_3 \tag{9.290}$$

When Eq. (9.283) is substituted, Eq. (9.290) becomes

$$\bar{p} = \bar{r} - d_6 e^{\tilde{u}_3(\theta_{12}-\theta_4)} e^{-\tilde{u}_2\theta_5}\bar{u}_3 \tag{9.291}$$

9.4.3 Inverse Kinematics in the Position Domain

Here, $\bar{p}$ and $\hat{C}$ are specified. Hence, $\bar{r}$ also becomes specified according to Eq. (9.290). That is,

$$\bar{r} = \bar{p} - d_6\hat{C}\bar{u}_3 \tag{9.292}$$

(a) Solution for the Arm Joint Variables

The wrist of a Scara manipulator is spherical. Therefore, the inverse kinematic solution can be started from the wrist location Eq. (9.288), which implies the following scalar equations.

$$b_1 c\theta_1 + b_2 c\theta_{12} = r_1 \tag{9.293}$$

$$b_1 s\theta_1 + b_2 s\theta_{12} = r_2 \tag{9.294}$$

$$s_3 = -r_3 \tag{9.295}$$

Note that Eq. (9.295) has already given s_3. This equation is consistent with the fact that $s_3 > 0$ because $r_3 < 0$ always for a Scara manipulator.
In order to find θ_2, the squares of Eqs. (9.293) and (9.294) can be added so that

$$b_1^2 + b_2^2 + 2b_1b_2(c\theta_{12}c\theta_1 + s\theta_{12}s\theta_1) = r_1^2 + r_2^2 \Rightarrow$$

$$b_1^2 + b_2^2 + 2b_1b_2\cos(\theta_{12} - \theta_1) = r_1^2 + r_2^2 \Rightarrow$$
$$c\theta_2 = \xi_2 = [(r_1^2 + r_2^2) - (b_1^2 + b_2^2)]/(2b_1b_2) \tag{9.296}$$

Equation (9.296) leads to $s\theta_2$ and then θ_2 as follows:

$$s\theta_2 = \eta_2 = \sigma_2\sqrt{1 - \xi_2^2}; \quad \sigma_2 = \pm 1 \tag{9.297}$$

$$\theta_2 = \text{atan}_2(\sigma_2\sqrt{1 - \xi_2^2}, \xi_2) = \sigma_2\text{atan}_2(\sqrt{1 - \xi_2^2}, \xi_2) \tag{9.298}$$

With the availability of θ_2, Eqs. (9.293) and (9.294) can be written together as the following matrix equation in order to obtain $c\theta_1$ and $s\theta_1$ separately so that an additional superfluous sign variable is avoided in finding θ_1.

$$\begin{bmatrix} b_1 + b_2c\theta_2 & -b_2s\theta_2 \\ b_2s\theta_2 & b_1 + b_2c\theta_2 \end{bmatrix} \begin{bmatrix} c\theta_1 \\ s\theta_1 \end{bmatrix} = \begin{bmatrix} r_1 \\ r_2 \end{bmatrix} \tag{9.299}$$

In Eq. (9.299), the determinant of the coefficient matrix is

$$D_1 = (b_1 + b_2c\theta_2)^2 + (b_2s\theta_2)^2 = b_1^2 + b_2^2 + 2b_1b_2c\theta_2 \tag{9.300}$$

If $D_1 \neq 0$, i.e. if $D_1 > 0$, Eq. (9.299) gives $c\theta_1$ and $s\theta_1$ as follows:

$$\begin{bmatrix} c\theta_1 \\ s\theta_1 \end{bmatrix} = \frac{1}{D_1} \begin{bmatrix} b_1 + b_2c\theta_2 & b_2s\theta_2 \\ -b_2s\theta_2 & b_1 + b_2c\theta_2 \end{bmatrix} \begin{bmatrix} r_1 \\ r_2 \end{bmatrix} \Rightarrow$$
$$\left.\begin{aligned} c\theta_1 &= [(b_1 + b_2c\theta_2)r_1 + (b_2s\theta_2)r_2]/D_1 = A_1/D_1 \\ s\theta_1 &= [(b_1 + b_2c\theta_2)r_2 - (b_2s\theta_2)r_1]/D_1 = B_1/D_1 \end{aligned}\right\} \tag{9.301}$$

Hence, θ_1 is found as

$$\theta_1 = \text{atan}_2(B_1, A_1) \tag{9.302}$$

(b) Solution for the Wrist Joint Variables

In order to find the wrist joint variables (θ_4, θ_5, θ_6) after finding the arm joint variables, the end-effector orientation equation, i.e. Eq. (9.283), can be written as follows:

$$e^{-\tilde{u}_3\theta_4}e^{-\tilde{u}_2\theta_5}e^{-\tilde{u}_3\theta_6} = e^{-\tilde{u}_3\theta_{12}}\hat{C}e^{-\tilde{u}_1\pi} \Rightarrow$$
$$e^{\tilde{u}_3\theta_6}e^{\tilde{u}_2\theta_5}e^{\tilde{u}_3\theta_4} = \hat{C}^* \tag{9.303}$$

In Eq. (9.303), $\hat{C}^*$ is known as

$$\hat{C}^* = e^{\tilde{u}_1\pi}\hat{C}^t e^{\tilde{u}_3\theta_{12}} \tag{9.304}$$

Note that the unknown angles in Eq. (9.303) are arranged like the Euler angles of the 3-2-3 sequence. Therefore, they can be extracted by using the similar procedure explained in Chapter 3. This procedure leads to the following scalar equation pairs generated from Eq. (9.303).

$$c_{33}^* = \overline{u}_3^t\hat{C}^*\overline{u}_3 = c\theta_5 \Rightarrow s\theta_5 = \sigma_5\sqrt{1 - (c_{33}^*)^2}; \quad \sigma_5 = \pm 1 \tag{9.305}$$

$$c_{23}^* = \overline{u}_2^t\hat{C}^*\overline{u}_3 = s\theta_6 s\theta_5; \quad c_{13}^* = \overline{u}_1^t\hat{C}^*\overline{u}_3 = c\theta_6 s\theta_5 \tag{9.306}$$

$$c_{32}^* = \overline{u}_3^t\hat{C}^*\overline{u}_2 = s\theta_5 s\theta_4; \quad c_{31}^* = \overline{u}_3^t\hat{C}^*\overline{u}_1 = -s\theta_5 c\theta_4 \tag{9.307}$$

Equation Pair (9.305) gives θ_5 as

$$\theta_5 = \text{atan}_2(\sigma_5\sqrt{1-(c^*_{33})^2}, c^*_{33}) = \sigma_5\text{atan}_2(\sqrt{1-(c^*_{33})^2}, c^*_{33}) \tag{9.308}$$

If $s\theta_5 \neq 0$, the other equation pairs give θ_6 and θ_4 as follows without additional superfluous sign variables.

$$\theta_6 = \text{atan}_2(\sigma_5 c^*_{23}, \sigma_5 c^*_{13}) \tag{9.309}$$

$$\theta_4 = \text{atan}_2(\sigma_5 c^*_{32}, -\sigma_5 c^*_{31}) \tag{9.310}$$

9.4.4 Multiplicity Analysis

(a) First Kind of Multiplicity

The first kind of multiplicity is associated with the sign variable σ_2 that arises in the process of finding θ_2 by using Eqs. (9.296)–(9.298). The manipulator attains the same location of the wrist point by assuming one of the two alternative poses that correspond to $\sigma_2 = +1$ and $\sigma_2 = -1$. These poses are illustrated in Figure 9.25. The poses corresponding to $\sigma_2 = +1$ and $\sigma_2 = -1$ are designated, respectively, as *right elbowed pose* and *left elbowed pose*. These designations are justified by Eq. (9.298), because it indicates that, for the same value of $c\theta_2 = \xi_2$, θ_2 becomes positive if $\sigma_2 = +1$ and θ_2 becomes negative if $\sigma_2 = -1$. The corresponding values of θ_1 are determined by Eqs. (9.301) and (9.302).

(b) Second Kind of Multiplicity

The second kind of multiplicity is associated with the sign variable σ_5 that arises in the process of finding the wrist joint variables $(\theta_5, \theta_4, \theta_6)$. Therefore, it is the same as the third kind of multiplicity of the Puma manipulator. In other words, as illustrated in Figure 9.5, it occurs similarly as a wrist flip phenomenon without any visual distinction.

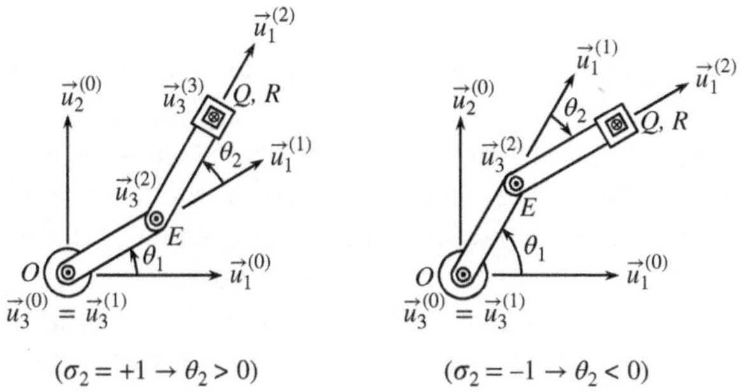

Figure 9.25 Right elbowed and left elbowed poses of a Scara manipulator.

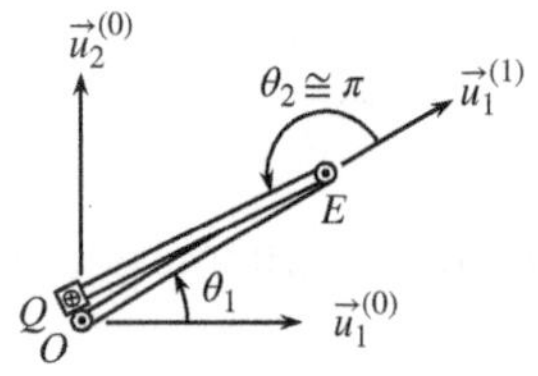

Figure 9.26 First kind of position singularity of a Scara manipulator.

9.4.5 Singularity Analysis in the Position Domain

A Scara manipulator may have two distinct kinds of position singularities, which are described and discussed below.

(a) First Kind of Position Singularity

Equation (9.301) implies that the first kind of position singularity occurs if $D_1 = 0$, i.e. if the determinant of the coefficient matrix vanishes. According to Eq. (9.300), D_1 can vanish only if the following equations are both satisfied.

$$b_1 = b_2 \tag{9.311}$$

$$c\theta_2 = -1 \quad \text{or} \quad \theta_2 = \pi \tag{9.312}$$

Equation (9.299) also implies that this singularity occurs if the position of the end-effector is specified in such a way that the point Q coincides with the base frame origin O. That is,

$$r_1 = r_2 = 0 \tag{9.313}$$

The above equations express the noticeable feature of this singularity that the upper and front arms of the manipulator have equal lengths and the front arm is folded completely over the upper arm. However, this singularity cannot occur exactly because of the physical shapes of the relevant links and joints. This singularity is illustrated (approximately) in Figure 9.26.

If this singularity occurs, θ_1 becomes indefinite and ineffective. In other words, θ_1 can be assigned an arbitrary value, but it cannot cause any change in the position of the end-effector, whatever the assigned value is. Thus, the manipulator gains a positioning freedom in the joint space, but this freedom happens to be useless in the task space as long as the points Q and O are kept coincident (actually almost coincident). On the other hand, if the task to be executed necessitates to keep the manipulator in such a pose, then the freedom in θ_1 may be used to orient the folded arm links of the manipulator conveniently depending on the environmental conditions, e.g. for avoiding possible obstacles.

(b) Second Kind of Position Singularity

The second kind of position singularity of a Scara manipulator is the same as the third kind of position singularity of the Puma manipulator, which is explained in Section 9.1.3 and illustrated in Figure 9.8.

9.4.6 Forward Kinematics in the Velocity Domain

(a) Angular Velocity of the End-Effector with Respect to the Base Frame

Let $\overline{\omega} = \overline{\omega}_6$. Then, Eq. (9.283) leads to the following expression.

$$\overline{\omega} = (\dot{\theta}_{12} - \dot{\theta}_4)\overline{u}_3 - \dot{\theta}_5 e^{\tilde{u}_3(\theta_{12}-\theta_4)}\overline{u}_2 - \dot{\theta}_6 e^{\tilde{u}_3(\theta_{12}-\theta_4)} e^{-\tilde{u}_2\theta_5}\overline{u}_3 \tag{9.314}$$

(b) Velocity of the Wrist Point with Respect to the Base Frame

Let $\overline{w} = \overline{w}^{(0)} = \overline{v}_R = \dot{\overline{r}}$. Then, $\overline{w}$ can be obtained from Eq. (9.288) as follows:

$$\overline{w} = -\overline{u}_1(b_1\dot{\theta}_1 s\theta_1 + b_2\dot{\theta}_{12}s\theta_{12}) + \overline{u}_2(b_1\dot{\theta}_1 c\theta_1 + b_2\dot{\theta}_{12}c\theta_{12}) - \overline{u}_3\dot{s}_3 \tag{9.315}$$

(c) Velocity of the Tip Point with Respect to the Base Frame

Let $\overline{v} = \overline{v}^{(0)} = \overline{v}_P = \dot{\overline{p}}$. Then, $\overline{v}$ can be obtained from Eq. (9.290) as follows:

$$\overline{v} = \overline{w} + d_6\tilde{\omega}\hat{C}\overline{u}_3 \tag{9.316}$$

9.4.7 Inverse Kinematics in the Velocity Domain

Since $\overline{p}$ and $\hat{C}$ are specified as functions of time, the velocities $\overline{v}$, $\overline{\omega}$, and $\overline{w}$ are also known. They are obtained as shown below.

$$\overline{v} = \dot{\overline{p}}, \quad \overline{\omega} = \operatorname{colm}(\dot{\hat{C}}\hat{C}^t); \quad \overline{w} = \overline{v} - d_6\tilde{\omega}\hat{C}\overline{u}_3 \tag{9.317}$$

(a) Solution for the Arm Joint Velocities

Upon premultiplications by $\overline{u}_1^t$, $\overline{u}_2^t$, and $\overline{u}_3^t$, Eq. (9.315) leads to the following scalar equations.

$$b_1\dot{\theta}_1 s\theta_1 + b_2\dot{\theta}_{12}s\theta_{12} = -w_1 \tag{9.318}$$

$$b_1\dot{\theta}_1 c\theta_1 + b_2\dot{\theta}_{12}c\theta_{12} = w_2 \tag{9.319}$$

$$\dot{s}_3 = -w_3 \tag{9.320}$$

Note that Eq. (9.320) has already given $\dot{s}_3$.
In order to find $\dot{\theta}_1$ and $\dot{\theta}_2$, Eqs. (9.318) and (9.319) can be written together as the following matrix equation.

$$\begin{bmatrix} b_2 s\theta_{12} & b_1 s\theta_1 \\ b_2 c\theta_{12} & b_1 c\theta_1 \end{bmatrix} \begin{bmatrix} \dot{\theta}_{12} \\ \dot{\theta}_1 \end{bmatrix} = \begin{bmatrix} -w_1 \\ w_2 \end{bmatrix} \tag{9.321}$$

In Eq. (9.321), the determinant of the coefficient matrix is

$$D_{12} = b_1 b_2(s\theta_{12}c\theta_1 - c\theta_{12}s\theta_1) = b_1 b_2 \sin(\theta_{12} - \theta_1) = b_1 b_2 s\theta_2 \tag{9.322}$$

If $D_{12} \neq 0$, i.e. if $s\theta_2 \neq 0$, Eq. (9.321) gives $\dot{\theta}_{12}$, $\dot{\theta}_1$, and $\dot{\theta}_2$ as follows:

$$\begin{bmatrix} \dot{\theta}_{12} \\ \dot{\theta}_1 \end{bmatrix} = \frac{1}{b_1 b_2 s\theta_2} \begin{bmatrix} b_1 c\theta_1 & -b_1 s\theta_1 \\ -b_2 c\theta_{12} & b_2 s\theta_{12} \end{bmatrix} \begin{bmatrix} -w_1 \\ w_2 \end{bmatrix} \Rightarrow$$

$$\dot{\theta}_{12} = -(w_1 c\theta_1 + w_2 s\theta_1)/(b_2 s\theta_2) \tag{9.323}$$

$$\dot{\theta}_1 = (w_1 c\theta_{12} + w_2 s\theta_{12})/(b_1 s\theta_2) \tag{9.324}$$

$$\dot{\theta}_2 = \dot{\theta}_{12} - \dot{\theta}_1 \tag{9.325}$$

(b) Solution for the Wrist Joint Velocities

In order to find the wrist joint velocities $(\dot{\theta}_4, \dot{\theta}_5, \dot{\theta}_6)$ after finding the arm joint velocities, the angular velocity equation of the end-effector, i.e. Eq. (9.314), can be reduced to the following simplified form.

$$\dot{\theta}_4\overline{u}_3 + \dot{\theta}_5\overline{u}_2 + \dot{\theta}_6 e^{\tilde{u}_2\theta_5}\overline{u}_3 = -\overline{\omega}^* \Rightarrow$$
$$\overline{u}_1(\dot{\theta}_6 s\theta_5) + \overline{u}_2\dot{\theta}_5 + \overline{u}_3(\dot{\theta}_4 + \dot{\theta}_6 c\theta_5) = -\overline{\omega}^* \tag{9.326}$$

In Eq. (9.326), $\overline{\omega}^*$ is known as expressed below.

$$\overline{\omega}^* = e^{-\tilde{u}_3(\theta_{12}-\theta_4)}\overline{\omega} - \overline{u}_3\dot{\theta}_{12} \tag{9.327}$$

Equation (9.326) gives $\dot{\theta}_5$ as follows:

$$\dot{\theta}_5 = -\omega_2^* \tag{9.328}$$

If $s\theta_5 \neq 0$, the same equation gives the following values for $\dot{\theta}_6$ and $\dot{\theta}_4$.

$$\dot{\theta}_6 = -\omega_1^*/s\theta_5 \tag{9.329}$$
$$\dot{\theta}_4 = -\omega_3^* - \dot{\theta}_6 c\theta_5 = (\omega_1^* c\theta_5 - \omega_3^* s\theta_5)/s\theta_5 \tag{9.330}$$

9.4.8 Singularity Analysis in the Velocity Domain

A Scara manipulator may have two distinct kinds of motion singularities, which are described and discussed below.

(a) First Kind of Motion Singularity

Equations (9.323) and (9.324) imply that the first kind of motion singularity occurs if $s\theta_2 = 0$, i.e. if $\theta_2 = 0$ or $\theta_2 = \pi$. This singularity is illustrated in Figure 9.27 in its *extended* $(\theta_2 = 0)$ and *folded* $(\theta_2 = \pi)$ versions. Note that the appearance of the folded version of this motion singularity becomes the same as the appearance of the first kind of position singularity if it happens that $b_2 = b_1$.
In this singularity, i.e. when $\theta_{12} = \theta_1 + (1 - \sigma_2')\pi/2$ with $\sigma_2' = \mathrm{sgn}(c\theta_2)$, Eqs. (9.318) and (9.319) can be manipulated into the following forms.

$$(b_1\dot{\theta}_1 + \sigma_2' b_2\dot{\theta}_{12})s\theta_1 = -w_1 \tag{9.331}$$
$$(b_1\dot{\theta}_1 + \sigma_2' b_2\dot{\theta}_{12})c\theta_1 = w_2 \tag{9.332}$$

The above equations become consistent and thus they can give finite values for $\dot{\theta}_1$ and $\dot{\theta}_2$ if the following *compatibility condition* is satisfied by the planned task motion.

$$w_1 c\theta_1 + w_2 s\theta_1 = 0 \tag{9.333}$$

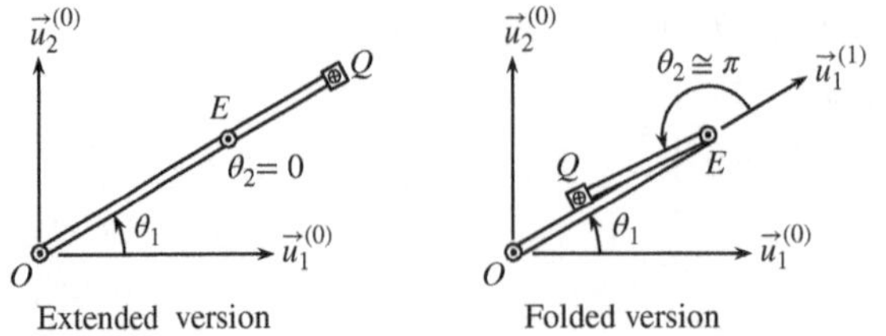

Figure 9.27 First kind of motion singularity of a Scara manipulator.

The same equations can be combined into the following equation, which relates the finite but indefinite values of $\dot{\theta}_1$ and $\dot{\theta}_2$.

$$b_1\dot{\theta}_1 + \sigma_2' b_2\dot{\theta}_{12} = (b_1 + \sigma_2' b_2)\dot{\theta}_1 + \sigma_2' b_2\dot{\theta}_2 = w_2 c\theta_1 - w_1 s\theta_1 \tag{9.334}$$

As seen above, this motion singularity induces a restriction in the task space as expressed by Eq. (9.333) and it also induces a motion freedom in the joint space as expressed by Eq. (9.334). Thus, as long as the task-space compatibility condition is satisfied, one of $\dot{\theta}_1$ and $\dot{\theta}_2$ can be assigned a finite arbitrary value and the other one can be found from Eq. (9.334).

(b) Second Kind of Motion Singularity

Equations (9.329) and (9.330) imply that the second kind of motion singularity occurs if $s\theta_5 = 0$, i.e. if $\theta_5 = 0$ or $\theta_5 = \pi$. However, while this singularity can occur easily with $\theta_5 = 0$, it can hardly occur with $\theta_5 = \pi$ due to the physical shapes of the relevant links and joints. In other words, this singularity is the same as the third kind of motion singularity of the Puma manipulator, which is explained in Section 9.1.5 and illustrated in Figure 9.11.

9.5 An RP²R³ Manipulator without an Analytical Solution

Figure 9.28 shows an RP²R³ manipulator in its various views. As noticed, its wrist is not spherical due to the nonzero *wrist offset* $d_5 = QR$. This is a conceptual manipulator. It is studied here as a typical example of a manipulator that does not have an analytical inverse kinematic solution. Yet, it is shown that the solution can be obtained in a *first-order semi-analytical way,* in which *only one* variable need be found by using a numerical method.

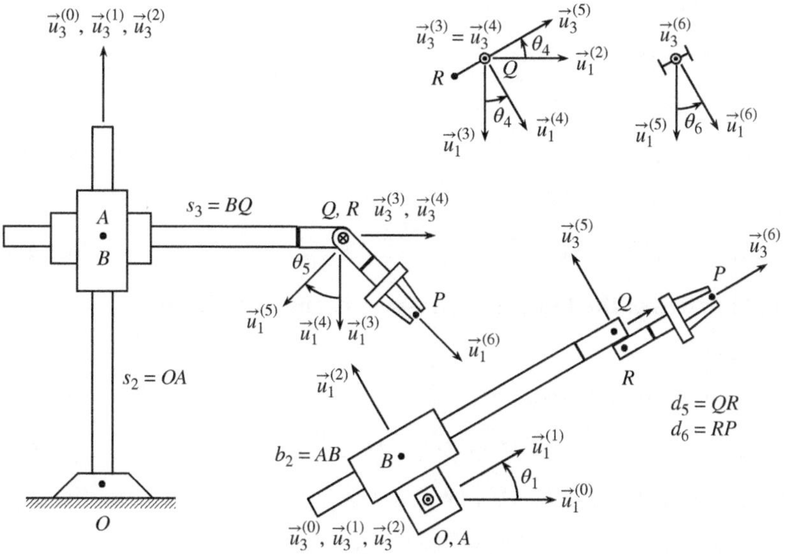

Figure 9.28 An RP²R³ manipulator in its various views.

9.5.1 Kinematic Description According to the D–H Convention

(a) Rotation Angles

$$\theta_1, \theta_2 = \delta_2 = \pi/2, \theta_3 = \delta_3 = -\pi/2, \theta_4, \theta_5, \theta_6$$

Four of the rotation angles (θ_1, θ_4, θ_5, θ_6) are joint variables. The second and third rotation angles are associated with the prismatic joints $\mathcal{J}_2$ and $\mathcal{J}_3$. So, their constant values are identified as observed in Figure 9.28.

(b) Twist Angles

$$\beta_1 = 0, \beta_2 = 0, \beta_3 = \pi/2, \beta_4 = 0, \beta_5 = -\pi/2, \beta_6 = \pi/2$$

(c) Offsets

$$d_1 = 0, s_2 = OA, s_3 = BQ, d_4 = 0, d_5 = QR, d_6 = RP$$

Four of the offsets (d_1, d_4, d_5, d_6) are constant. Two of the offsets associated with the prismatic joints (s_2 and s_3) are naturally variable.

(d) Effective Link Lengths

$$b_0 = 0, b_1 = 0, b_2 = AB, b_3 = 0, b_4 = 0, b_5 = 0$$

(e) Link Frame Origins

$$O_0 = O, O_1 = O, O_2 = A, O_3 = Q, O_4 = Q, O_5 = R, O_6 = P$$

9.5.2 Forward Kinematics in the Position Domain

(a) Link-to-Link Orientation Matrices

$$\hat{C}^{(0,1)} = e^{\tilde{u}_1\beta_1}e^{\tilde{u}_3\theta_1} = e^{\tilde{u}_3\theta_1} \tag{9.335}$$

$$\hat{C}^{(1,2)} = e^{\tilde{u}_1\beta_2}e^{\tilde{u}_3\theta_2} = e^{\tilde{u}_3\pi/2} \tag{9.336}$$

$$\hat{C}^{(2,3)} = e^{\tilde{u}_1\beta_3}e^{\tilde{u}_3\theta_3} = e^{\tilde{u}_1\pi/2}e^{-\tilde{u}_3\pi/2} \tag{9.337}$$

$$\hat{C}^{(3,4)} = e^{\tilde{u}_1\beta_4}e^{\tilde{u}_3\theta_4} = e^{\tilde{u}_3\theta_4} \tag{9.338}$$

$$\hat{C}^{(4,5)} = e^{\tilde{u}_1\beta_5}e^{\tilde{u}_3\theta_5} = e^{-\tilde{u}_1\pi/2}e^{\tilde{u}_3\theta_5} \tag{9.339}$$

$$\hat{C}^{(5,6)} = e^{\tilde{u}_1\beta_6}e^{\tilde{u}_3\theta_6} = e^{\tilde{u}_1\pi/2}e^{\tilde{u}_3\theta_6} \tag{9.340}$$

(b) Orientation Matrices of the Links with Respect to the Base Frame

$$\hat{C}_1 = \hat{C}^{(0,1)} = e^{\tilde{u}_3\theta_1} \tag{9.341}$$

$$\hat{C}_2 = \hat{C}^{(0,2)} = \hat{C}^{(0,1)}\hat{C}^{(1,2)} = e^{\tilde{u}_3\theta_1}e^{\tilde{u}_3\pi/2} \tag{9.342}$$

$$\hat{C}_3 = \hat{C}^{(0,3)} = \hat{C}^{(0,2)}\hat{C}^{(2,3)} = e^{\tilde{u}_3\theta_1}e^{\tilde{u}_3\pi/2}e^{\tilde{u}_1\pi/2}e^{-\tilde{u}_3\pi/2} = e^{\tilde{u}_3\theta_1}e^{\tilde{u}_2\pi/2} \tag{9.343}$$

$$\hat{C}_4 = \hat{C}^{(0,4)} = \hat{C}^{(0,3)}\hat{C}^{(3,4)} = e^{\tilde{u}_3\theta_1}e^{\tilde{u}_2\pi/2}e^{\tilde{u}_3\theta_4} = e^{\tilde{u}_3\theta_1}e^{\tilde{u}_1\theta_4}e^{\tilde{u}_2\pi/2} \tag{9.344}$$

$$\hat{C}_5 = \hat{C}^{(0,5)} = \hat{C}^{(0,4)}\hat{C}^{(4,5)} = e^{\tilde{u}_3\theta_1}e^{\tilde{u}_1\theta_4}e^{\tilde{u}_2\pi/2}e^{-\tilde{u}_1\pi/2}e^{\tilde{u}_3\theta_5} \Rightarrow$$

$$\hat{C}_5 = e^{\tilde{u}_3\theta_1}e^{\tilde{u}_1\theta_4}e^{\tilde{u}_2\pi/2}e^{\tilde{u}_2\theta_5}e^{-\tilde{u}_1\pi/2} = e^{\tilde{u}_3\theta_1}e^{\tilde{u}_1\theta_4}e^{\tilde{u}_2\theta'_5}e^{-\tilde{u}_1\pi/2} \tag{9.345}$$

$$\hat{C}_6 = \hat{C}^{(0,6)} = \hat{C}^{(0,5)}\hat{C}^{(5,6)} = e^{\tilde{u}_3\theta_1}e^{\tilde{u}_1\theta_4}e^{\tilde{u}_2\theta'_5}e^{-\tilde{u}_1\pi/2}e^{\tilde{u}_1\pi/2}e^{\tilde{u}_3\theta_6} \Rightarrow$$

$$\hat{C}_6 = \hat{C} = e^{\tilde{u}_3\theta_1}e^{\tilde{u}_1\theta_4}e^{\tilde{u}_2\theta'_5}e^{\tilde{u}_3\theta_6} \quad (9.346)$$

In Eq. (9.346), θ'_5 is introduced as a *modified joint variable*. It is defined as follows:

$$\theta'_5 = \theta_5 + \pi/2 \quad (9.347)$$

(c) Location of the Wrist Point with Respect to the Base Frame

$$\vec{r} = \vec{r}_{OR} = \vec{r}_{OA} + \vec{r}_{AB} + \vec{r}_{BQ} + \vec{r}_{QR} = s_2\vec{u}_3^{(2)} + b_2\vec{u}_1^{(2)} + s_3\vec{u}_3^{(3)} - d_5\vec{u}_3^{(5)} \quad (9.348)$$

The vector Eq. (9.348) implies the following matrix equation in the base frame.

$$\bar{r} = \bar{r}^{(0)} = s_2\bar{u}_3^{(2/0)} + b_2\bar{u}_1^{(2/0)} + s_3\bar{u}_3^{(3/0)} - d_5\bar{u}_3^{(5/0)} \quad (9.349)$$

Equation (9.349) can be manipulated as shown below.

$$\bar{r} = s_2\hat{C}^{(0,2)}\bar{u}_3^{(2/2)} + b_2\hat{C}^{(0,2)}\bar{u}_1^{(2/2)} + s_3\hat{C}^{(0,3)}\bar{u}_3^{(3/3)} - d_5\hat{C}^{(0,5)}\bar{u}_3^{(5/5)} \Rightarrow$$

$$\bar{r} = s_2e^{\tilde{u}_3\theta_1}e^{\tilde{u}_3\pi/2}\bar{u}_3 + b_2e^{\tilde{u}_3\theta_1}e^{\tilde{u}_3\pi/2}\bar{u}_1 + s_3e^{\tilde{u}_3\theta_1}e^{\tilde{u}_2\pi/2}\bar{u}_3 - d_5e^{\tilde{u}_3\theta_1}e^{\tilde{u}_1\theta_4}e^{\tilde{u}_2\theta'_5}e^{-\tilde{u}_1\pi/2}\bar{u}_3 \Rightarrow$$

$$\bar{r} = s_2e^{\tilde{u}_3\theta_1}\bar{u}_3 + b_2e^{\tilde{u}_3\theta_1}\bar{u}_2 + s_3e^{\tilde{u}_3\theta_1}\bar{u}_1 - d_5e^{\tilde{u}_3\theta_1}e^{\tilde{u}_1\theta_4}\bar{u}_2 \Rightarrow$$

$$\bar{r} = e^{\tilde{u}_3\theta_1}(s_2\bar{u}_3 + b_2\bar{u}_2 + s_3\bar{u}_1 - d_5e^{\tilde{u}_1\theta_4}\bar{u}_2) \Rightarrow$$

$$\bar{r} = e^{\tilde{u}_3\theta_1}\bar{r}' = e^{\tilde{u}_3\theta_1}[\bar{u}_1s_3 + \bar{u}_2(b_2 - d_5c\theta_4) + \bar{u}_3(s_2 - d_5s\theta_4)] \quad (9.350)$$

(d) Location of the Tip Point with Respect to the Base Frame

$$\vec{p} = \vec{r}_{OP} = \vec{r}_{OR} + \vec{r}_{RP} = \vec{r} + d_6\vec{u}_3^{(6)} \quad (9.351)$$

Equation (9.351) implies the following matrix equation in the base frame.

$$\bar{p} = \bar{p}^{(0)} = \bar{r}^{(0)} + d_6\bar{u}_3^{(6/0)} = \bar{r}^{(0)} + d_6\hat{C}^{(0,6)}\bar{u}_3^{(6/6)} \Rightarrow$$

$$\bar{p} = \bar{r} + d_6\hat{C}\bar{u}_3 \quad (9.352)$$

When Eq. (9.346) is substituted, Eq. (9.352) takes either of the following forms.

$$\bar{p} = \bar{r} + d_6e^{\tilde{u}_3\theta_1}e^{\tilde{u}_1\theta_4}e^{\tilde{u}_2\theta'_5}\bar{u}_3 \quad (9.353)$$

$$\bar{p} = \bar{r} + d_6e^{\tilde{u}_3\theta_1}e^{\tilde{u}_1\theta_4}e^{\tilde{u}_2\theta_5}\bar{u}_1 \quad (9.354)$$

9.5.3 Inverse Kinematics in the Position Domain

Here, $\bar{p}$ and $\hat{C}$ are specified. Hence, $\bar{r}$ also becomes specified according to Eq. (9.352). That is,

$$\bar{r} = \bar{p} - d_6\hat{C}\bar{u}_3 \quad (9.355)$$

When Eqs. (9.346) and (9.350) are examined, it is seen that each of these equations contains four unknowns and none of these unknowns can be singled out from any of these equations. Therefore, the present manipulator cannot have an analytical inverse kinematic solution. Nevertheless, a first-order semi-analytical solution can still be obtained. Such a solution necessitates only one equation to be solved numerically as explained below.

The semi-analytical solution process can be started by treating any one of the six joint variables as if it is known temporarily. For the present manipulator, θ_4 can be selected to undertake this role. Then, both of Eqs. (9.350) and (9.346) can be solved analytically for the other joint variables so as to express them in terms of θ_4.

To prepare for the solution, Eq. (9.350) can be written as

$$\overline{u}_1 s_3 + \overline{u}_2(b_2 - d_5 c\theta_4) + \overline{u}_3(s_2 - d_5 s\theta_4) = e^{-\tilde{u}_3\theta_1}\overline{r} \tag{9.356}$$

Upon premultiplications by $\overline{u}_1^t$, $\overline{u}_2^t$, and $\overline{u}_3^t$, Eq. (9.356) leads to the following scalar equations.

$$s_3 = r_1 c\theta_1 + r_2 s\theta_1 \tag{9.357}$$

$$b_2 - d_5 c\theta_4 = r_2 c\theta_1 - r_1 s\theta_1 \tag{9.358}$$

$$s_2 - d_5 s\theta_4 = r_3 \tag{9.359}$$

Note that Eq. (9.359) has already given s_2 as a function of θ_4, i.e.

$$s_2 = f_2(\theta_4) = r_3 + d_5 s\theta_4 \tag{9.360}$$

If $r_1^2 + r_2^2 \neq 0$, Eq. (9.358) gives $c\theta_1$, $s\theta_1$, and θ_1 as the following functions of θ_4.

$$c\theta_1 = (r_2 z_1 + \sigma_1 r_1 z_2)/(r_1^2 + r_2^2) \tag{9.361}$$

$$s\theta_1 = (-r_1 z_1 + \sigma_1 r_2 z_2)/(r_1^2 + r_2^2) \tag{9.362}$$

$$\theta_1 = f_1(\theta_4) = \operatorname{atan}_2[(-r_1 z_1 + \sigma_1 r_2 z_2), (r_2 z_1 + \sigma_1 r_1 z_2)] \tag{9.363}$$

In the above equations, $\sigma_1 = \pm 1$ and the functions z_1 and z_2 are defined as follows:

$$z_1 = z_1(\theta_4) = b_2 - d_5 c\theta_4 \tag{9.364}$$

$$z_2 = z_2(\theta_4) = \sqrt{r_1^2 + r_2^2 - z_1^2(\theta_4)} \tag{9.365}$$

By using Eqs. (9.361) and (9.362), Eq. (9.357) gives s_3 as the following function of θ_4.

$$s_3 = f_3(\theta_4) = \sigma_1 z_2(\theta_4) = \sigma_1\sqrt{r_1^2 + r_2^2 - z_1^2(\theta_4)} \tag{9.366}$$

Equation (9.366) implies that s_3 can be *real* and *positive* (as the only physical possibility) if σ_1 and the wrist point coordinates are such that

$$\sigma_1 = +1 \tag{9.367}$$

$$r_1^2 + r_2^2 \geq (b_2 - d_5 c\theta_4)^2 \tag{9.368}$$

On the other hand, Eq. (9.346) can be prepared for the solution by writing it as

$$e^{\tilde{u}_2\theta_5'}e^{\tilde{u}_3\theta_6} = e^{-\tilde{u}_1\theta_4}e^{-\tilde{u}_3\theta_1}\hat{C} \tag{9.369}$$

Equation (9.369) leads primarily to the following consistency equation.

$$\overline{u}_2^t e^{\tilde{u}_2\theta_5'}e^{\tilde{u}_3\theta_6}\overline{u}_3 = \overline{u}_2^t e^{-\tilde{u}_1\theta_4}e^{-\tilde{u}_3\theta_1}\hat{C}\overline{u}_3 \Rightarrow$$

$$\overline{u}_2^t\overline{u}_3 = 0 = (\overline{u}_2^t c\theta_4 + \overline{u}_3^t s\theta_4)e^{-\tilde{u}_3\theta_1}\hat{C}\overline{u}_3 = (\overline{u}_2^t e^{-\tilde{u}_3\theta_1}c\theta_4 + \overline{u}_3^t s\theta_4)\hat{C}\overline{u}_3 \Rightarrow$$

$$[(\overline{u}_2^t c\theta_1 - \overline{u}_1^t s\theta_1)c\theta_4 + \overline{u}_3^t s\theta_4]\hat{C}\overline{u}_3 = 0 \Rightarrow$$

$$(c_{23}c\theta_1 - c_{13}s\theta_1)c\theta_4 + c_{33}s\theta_4 = 0 \tag{9.370}$$

When Eqs. (9.361) and (9.362) are substituted with $\sigma_1 = +1$, Eq. (9.370) becomes

$$[(r_1c_{13} + r_2c_{23})z_1(\theta_4) + (r_1c_{23} - r_2c_{13})z_2(\theta_4)]c\theta_4 + (r_1^2 + r_2^2)c_{33}s\theta_4 = 0 \tag{9.371}$$

Equation (9.369) also leads to the following scalar equations.

$$\overline{u}_3^t e^{\tilde{u}_2\theta_5'} e^{\tilde{u}_3\theta_6}\overline{u}_3 = \overline{u}_3^t e^{\tilde{u}_2\theta_5'}\overline{u}_3 = \overline{u}_3^t e^{-\tilde{u}_1\theta_4} e^{-\tilde{u}_3\theta_1}\hat{C}\overline{u}_3 = g_{33}(\theta_4, \theta_1) \Rightarrow$$

$$c\theta_5' = g_{33}(\theta_4, f_1(\theta_4)) \tag{9.372}$$

$$\overline{u}_1^t e^{\tilde{u}_2\theta_5'} e^{\tilde{u}_3\theta_6}\overline{u}_3 = \overline{u}_1^t e^{\tilde{u}_2\theta_5'}\overline{u}_3 = \overline{u}_1^t e^{-\tilde{u}_1\theta_4} e^{-\tilde{u}_3\theta_1}\hat{C}\overline{u}_3 = g_{13}(\theta_4, \theta_1) \Rightarrow$$

$$s\theta_5' = g_{13}(\theta_4, f_1(\theta_4)) \tag{9.373}$$

$$\overline{u}_2^t e^{\tilde{u}_2\theta_5'} e^{\tilde{u}_3\theta_6}\overline{u}_1 = \overline{u}_2^t e^{\tilde{u}_3\theta_6}\overline{u}_1 = \overline{u}_2^t e^{-\tilde{u}_1\theta_4} e^{-\tilde{u}_3\theta_1}\hat{C}\overline{u}_1 = g_{21}(\theta_4, \theta_1) \Rightarrow$$

$$s\theta_6 = g_{21}(\theta_4, f_1(\theta_4)) \tag{9.374}$$

$$\overline{u}_2^t e^{\tilde{u}_2\theta_5'} e^{\tilde{u}_3\theta_6}\overline{u}_2 = \overline{u}_2^t e^{\tilde{u}_3\theta_6}\overline{u}_2 = \overline{u}_2^t e^{-\tilde{u}_1\theta_4} e^{-\tilde{u}_3\theta_1}\hat{C}\overline{u}_2 = g_{22}(\theta_4, \theta_1) \Rightarrow$$

$$c\theta_6 = g_{22}(\theta_4, f_1(\theta_4)) \tag{9.375}$$

Equations (9.372)–(9.375) give θ_5' and θ_6 as the following functions of θ_4.

$$\theta_5' = f_5(\theta_4) = \text{atan}_2[g_{13}(\theta_4, f_1(\theta_4)), g_{33}(\theta_4, f_1(\theta_4))] \Rightarrow \theta_5 = \theta_5' - \pi/2 \tag{9.376}$$

$$\theta_6 = f_6(\theta_4) = \text{atan}_2[g_{21}(\theta_4, f_1(\theta_4)), g_{22}(\theta_4, f_1(\theta_4))] \tag{9.377}$$

As for θ_4, it can be found by solving the consistency Eq. (9.371). Clearly, Eq. (9.371) is quite complicated and it cannot be solved analytically for θ_4. However, it can be solved rather easily by using a suitable numerical method, in which θ_4 is assigned successive values differing with a sufficiently small increment (such as $\delta\theta_4 = 0.1°$) so as to sweep the interval $[-\pi, +\pi]$ until Eq. (9.371) is satisfied with a tolerable error.

Afterwards, with the availability of θ_4, the other joint variables can be obtained readily from the preceding relevant equations.

9.5.4 Multiplicity Analysis

The inverse kinematic solution presented above shows that the present manipulator has only one kind of multiplicity, which is associated with Eq. (9.371). This multiplicity occurs because the numerical solution mentioned in Section 9.5.3 gives in general more than one value for θ_4.

The outcome of the numerical solution obtained in a case study is illustrated in Figure 9.29, which shows the graph of the following error function $\varepsilon(\theta_4)$ plotted versus θ_4 and indicates the two solutions, $\theta_4^{(1)}$ and $\theta_4^{(2)}$, which make $\varepsilon(\theta_4) = 0$ and thus satisfy Eq. (9.371).

$$\varepsilon(\theta_4) = [(r_1c_{13} + r_2c_{23})z_1(\theta_4) + (r_1c_{23} - r_2c_{13})z_2(\theta_4)]c\theta_4 + (r_1^2 + r_2^2)c_{33}s\theta_4 = 0 \tag{9.378}$$

The case study is carried out with the following manipulator parameters and the specified position data for the end-effector.

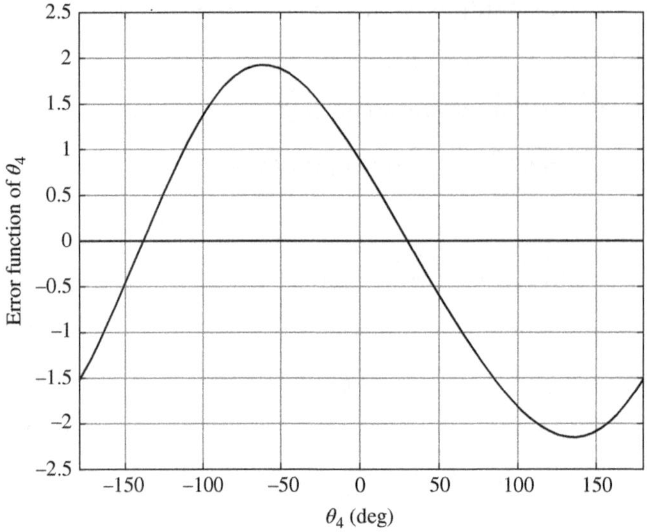

Figure 9.29 Graphical representation of the solution to Eq. (9.371).

$$\text{Fixed parameters: } b_2 = 0.4\,\text{m}, d_5 = 0.2\,\text{m}, d_6 = 0.3\,\text{m}$$

$$\text{Tip point location: } p_1 = 2.0\,\text{m}, p_2 = 1.5\,\text{m}, p_3 = 1.0\,\text{m}$$

$$\text{3-2-3 Euler angles: } \phi_1 = \pi/4, \phi_2 = 3\pi/4, \phi_3 = \pi/4 \Rightarrow$$

$$\hat{C} = e^{\tilde{u}_3\phi_1} e^{\tilde{u}_2\phi_2} e^{\tilde{u}_3\phi_3}$$

As observed in Figure 9.29, the numerical solution of Eq. (9.371) results in the following two values for θ_4, which satisfy the equality $\varepsilon(\theta_4) = 0$.

$$\theta_4^{(1)} = -138.40^\circ \text{ and } \theta_4^{(2)} = 29.90^\circ \tag{9.379}$$

When the other joint variables are computed in correspondence to each of the above values, the following results are obtained.

$$\theta_1^{(1)} = 22.24^\circ \text{ and } \theta_1^{(2)} = 30.44^\circ \tag{9.380}$$

$$s_2^{(1)} = 1.079\,\text{m} \text{ and } s_2^{(2)} = 1.312\,\text{m} \tag{9.381}$$

$$s_3^{(1)} = 2.223\,\text{m} \text{ and } s_3^{(2)} = 2.279\,\text{m} \tag{9.382}$$

$$\theta_5^{(1)} = -32.32^\circ \text{ and } \theta_5^{(2)} = 27.50^\circ \tag{9.383}$$

$$\theta_6^{(1)} = -165.71^\circ \text{ and } \theta_6^{(2)} = 24.84^\circ \tag{9.384}$$

The pose corresponding to the first solution (with a *negative* and *obtuse* θ_4) is such that the wrist point R is on the *upper left-hand* side of the point Q, whereas the pose corresponding to the second solution (with a *positive* and *acute* θ_4) is such that the wrist point R is on the *lower right-hand* side of the point Q (as illustrated in Figure 9.28).

9.5.5 Singularity Analysis in the Position Domain

According to Eq. (9.371), the manipulator gets into a pose of position singularity if $r_1 = r_2 = 0$. However, due to the physical shapes of the prismatic joints, such a pose may never occur. Therefore, the present manipulator does not have any position singularity.

9.5.6 Forward Kinematics in the Velocity Domain

(a) Angular Velocity of the End-Effector with Respect to the Base Frame

Let $\overline{\omega} = \overline{\omega}_6$. Then, Eq. (9.346) leads to the following expression.

$$\overline{\omega} = \dot{\theta}_1 \overline{u}_3 + \dot{\theta}_4 e^{\tilde{u}_3\theta_1}\overline{u}_1 + \dot{\theta}_5 e^{\tilde{u}_3\theta_1} e^{\tilde{u}_1\theta_4}\overline{u}_2 + \dot{\theta}_6 e^{\tilde{u}_3\theta_1} e^{\tilde{u}_1\theta_4} e^{\tilde{u}_2\theta_5'}\overline{u}_3$$

Upon substituting $\theta_5' = \theta_5 + \pi/2$, $\overline{\omega}$ can also be expressed as

$$\overline{\omega} = \dot{\theta}_1 \overline{u}_3 + \dot{\theta}_4 e^{\tilde{u}_3\theta_1}\overline{u}_1 + \dot{\theta}_5 e^{\tilde{u}_3\theta_1} e^{\tilde{u}_1\theta_4}\overline{u}_2 + \dot{\theta}_6 e^{\tilde{u}_3\theta_1} e^{\tilde{u}_1\theta_4} e^{\tilde{u}_2\theta_5}\overline{u}_1 \tag{9.385}$$

(b) Velocity of the Wrist Point with Respect to the Base Frame

Let $\overline{w} = \overline{w}^{(0)} = \overline{v}_R = \dot{\overline{r}}$. Then, $\overline{w}$ can be obtained from Eq. (9.350) as follows:

$$\begin{aligned}\overline{w} &= (\dot{\theta}_1 e^{\tilde{u}_3\theta_1}\tilde{u}_3)[\overline{u}_1 s_3 + \overline{u}_2(b_2 - d_5 c\theta_4) + \overline{u}_3(s_2 - d_5 s\theta_4)]\\ &\quad + (e^{\tilde{u}_3\theta_1})[\overline{u}_1\dot{s}_3 + \overline{u}_2(d_5\dot{\theta}_4 s\theta_4) + \overline{u}_3(\dot{s}_2 - d_5\dot{\theta}_4 c\theta_4)] \Rightarrow\end{aligned}$$

$$\begin{aligned}\overline{w} &= (e^{\tilde{u}_3\theta_1})[\overline{u}_2 s_3\dot{\theta}_1 - \overline{u}_1(b_2 - d_5 c\theta_4)\dot{\theta}_1]\\ &\quad + (e^{\tilde{u}_3\theta_1})[\overline{u}_1\dot{s}_3 + \overline{u}_2(d_5\dot{\theta}_4 s\theta_4) + \overline{u}_3(\dot{s}_2 - d_5\dot{\theta}_4 c\theta_4)] \Rightarrow\end{aligned}$$

$$\overline{w} = e^{\tilde{u}_3\theta_1}(\overline{u}_1 w_1' + \overline{u}_2 w_2' + \overline{u}_3 w_3') = e^{\tilde{u}_3\theta_1}\overline{w}' \tag{9.386}$$

In Eq. (9.386),

$$\left.\begin{aligned} w_1' &= \dot{s}_3 - (b_2 - d_5 c\theta_4)\dot{\theta}_1\\ w_2' &= s_3\dot{\theta}_1 + d_5\dot{\theta}_4 s\theta_4\\ w_3' &= \dot{s}_2 - d_5\dot{\theta}_4 c\theta_4 \end{aligned}\right\} \tag{9.387}$$

(c) Velocity of the Tip Point with Respect to the Base Frame

Let $\overline{v} = \overline{v}^{(0)} = \overline{v}_P = \dot{\overline{p}}$. Then, $\overline{v}$ can be obtained from Eq. (9.352) as follows:

$$\overline{v} = \overline{w} + d_6\tilde{\omega}\hat{C}\overline{u}_3 \tag{9.388}$$

9.5.7 Inverse Kinematics in the Velocity Domain

Since $\overline{p}$ and $\hat{C}$ are specified as functions of time, the velocities $\overline{v}$, $\overline{\omega}$, and $\overline{w}$ are also known. They are obtained as shown below.

$$\overline{v} = \dot{\overline{p}},\quad \overline{\omega} = \mathrm{colm}(\dot{\hat{C}}\hat{C}^t);\quad \overline{w} = \overline{v} - d_6\tilde{\omega}\hat{C}\overline{u}_3 \tag{9.389}$$

(a) Solution for the Joint Velocities

The velocity equations contain the joint velocities linearly. Therefore, the velocity equations can be solved analytically for any manipulator whatsoever, even though the position equations cannot be solved analytically for the present manipulator. The scalar velocity equations required for the analytical solution are obtained as explained below. Equation (9.386) leads to the following scalar equations.

$$\overline{u}_1 w_1' + \overline{u}_2 w_2' + \overline{u}_3 w_3' = \overline{w}' = e^{-\tilde{u}_3\theta_1}\overline{w} \Rightarrow$$

$$\dot{s}_3 - (b_2 - d_5 c\theta_4)\dot{\theta}_1 = \overline{u}_1^t \overline{w}' = w_1' = w_1 c\theta_1 + w_2 s\theta_1 \tag{9.390}$$

$$s_3\dot{\theta}_1 + d_5\dot{\theta}_4 s\theta_4 = \overline{u}_2^t \overline{w}' = w_2' = w_2 c\theta_1 - w_1 s\theta_1 \tag{9.391}$$

$$\dot{s}_2 - d_5\dot{\theta}_4 c\theta_4 = \overline{u}_3^t \overline{w}' = w_3' = w_3 \tag{9.392}$$

Equation (9.385) leads to the following scalar equations.

$$\dot{\theta}_1 e^{-\tilde{u}_1\theta_4}\overline{u}_3 + \dot{\theta}_4\overline{u}_1 + \dot{\theta}_5\overline{u}_2 + \dot{\theta}_6 e^{\tilde{u}_2\theta_5}\overline{u}_1 = \overline{\omega}^* \Rightarrow$$

$$\dot{\theta}_4 + \dot{\theta}_6 c\theta_5 = \overline{u}_1^t \overline{\omega}^* = \omega_1^* \tag{9.393}$$

$$\dot{\theta}_5 + \dot{\theta}_1 s\theta_4 = \overline{u}_2^t \overline{\omega}^* = \omega_2^* \tag{9.394}$$

$$\dot{\theta}_1 c\theta_4 - \dot{\theta}_6 s\theta_5 = \overline{u}_3^t \overline{\omega}^* = \omega_3^* \tag{9.395}$$

In the above angular velocity equations,

$$\overline{\omega}^* = e^{-\tilde{u}_1\theta_4} e^{-\tilde{u}_3\theta_1}\overline{\omega} \tag{9.396}$$

As seen above, each set of scalar equations derived from $\overline{w}$ and $\overline{\omega}$ contains four unknown joint velocities. Therefore, similarly as done in Section 9.5.3, the inverse velocity solution can be obtained as follows by treating one of the joint velocities, say $\dot{\theta}_4$, as if it is known temporarily.

Equations (9.390)–(9.392) give $\dot{\theta}_1$, $\dot{s}_2$, and $\dot{s}_3$ in terms of $\dot{\theta}_4$ as shown below.

$$\dot{s}_2 = w_3 + d_5\dot{\theta}_4 c\theta_4 \tag{9.397}$$

$$\dot{\theta}_1 = (w_2' - d_5\dot{\theta}_4 s\theta_4)/s_3 \tag{9.398}$$

$$\dot{s}_3 = [s_3 w_1' + (b_2 - d_5 c\theta_4)(w_2' - d_5\dot{\theta}_4 s\theta_4)]/s_3 \tag{9.399}$$

Using Eq. (9.398), Eqs. (9.393)–(9.395) can be written again in the following forms.

$$\dot{\theta}_5 = \omega_2^* - \dot{\theta}_1 s\theta_4 = [s_3\omega_2^* - (w_2' - d_5\dot{\theta}_4 s\theta_4)s\theta_4]/s_3 \tag{9.400}$$

$$\dot{\theta}_6 c\theta_5 = \omega_1^* - \dot{\theta}_4 \tag{9.401}$$

$$\dot{\theta}_6 s\theta_5 = \dot{\theta}_1 c\theta_4 - \omega_3^* = [(w_2' - d_5\dot{\theta}_4 s\theta_4)c\theta_4 - s_3\omega_3^*]/s_3 \tag{9.402}$$

Equations (9.401) and (9.402) can be combined into the following equations.

$$\dot{\theta}_6 = (\omega_1^* - \dot{\theta}_4)c\theta_5 + (\dot{\theta}_1 c\theta_4 - \omega_3^*)s\theta_5 \Rightarrow$$

$$\dot{\theta}_6 = \{s_3(\omega_1^* - \dot{\theta}_4)c\theta_5 + [(w_2' - d_5\dot{\theta}_4 s\theta_4)c\theta_4 - s_3\omega_3^*]s\theta_5\}/s_3 \tag{9.403}$$

$$(\omega_1^* - \dot{\theta}_4)s\theta_5 = (\dot{\theta}_1 c\theta_4 - \omega_3^*)c\theta_5 \Rightarrow$$

$$s_3(\omega_1^* - \dot{\theta}_4)s\theta_5 = [(w_2' - d_5\dot{\theta}_4 s\theta_4)c\theta_4 - s_3\omega_3^*]c\theta_5 \tag{9.404}$$

Equations (9.400) and (9.403) have already given $\dot{\theta}_5$ and $\dot{\theta}_6$ in terms of $\dot{\theta}_4$. As for Eq. (9.404), it is the *consistency equation* to be satisfied by $\dot{\theta}_4$. It can be rearranged as follows:

$$(s_3 s\theta_5 - d_5 s\theta_4 c\theta_4 c\theta_5)\dot{\theta}_4 = s_3(\omega_1^* s\theta_5 + \omega_3^* c\theta_5) - w_2' c\theta_4 c\theta_5 \tag{9.405}$$

If the coefficient $k_4 = s_3 s\theta_5 - d_5 s\theta_4 c\theta_4 c\theta_5 \neq 0$, Eq. (9.405) gives $\dot{\theta}_4$ as

$$\dot{\theta}_4 = [s_3(\omega_1^* s\theta_5 + \omega_3^* c\theta_5) - w_2' c\theta_4 c\theta_5]/(s_3 s\theta_5 - d_5 s\theta_4 c\theta_4 c\theta_5) \tag{9.406}$$

With the availability of $\dot{\theta}_4$, the other joint velocities can be obtained readily from the preceding relevant equations.

9.5.8 Singularity Analysis in the Velocity Domain

According to Eq. (9.406), the manipulator gets into a pose of motion singularity if

$$s_3 s\theta_5 - d_5 s\theta_4 c\theta_4 c\theta_5 = 0 \tag{9.407}$$

This singularity can occur in one of the two distinct ways that are described below.

(a) First Kind of Motion Singularity

According to Eq. (9.371), the first kind of motion singularity occurs if

$$s\theta_5 = 0 \text{ (i.e. } \theta_5 = 0) \text{ and } s\theta_4 = 0 \text{ (i.e. } \theta_4 = 0 \text{ or } \theta_4 = \pi) \tag{9.408}$$

Note that $s\theta_5 = 0$ occurs only with $\theta_5 = 0$ due to the physical shapes of the prismatic joints.

In a pose of this singularity, the wrist point R gets located either on the right-hand side (if $\theta_4 = 0$) or on the left-hand side ($\theta_4 = \pi$) of the point Q. Meanwhile, Eqs. (9.390)–(9.395) reduce to the following special equations with $\sigma_4' = \text{sgn}(c\theta_4)$.

$$\dot{\theta}_1 = w_2'/s_3 = \sigma_4' \omega_3^* \tag{9.409}$$

$$\dot{s}_3 = w_1' + (b_2 - \sigma_4' d_5)\dot{\theta}_1 \tag{9.410}$$

$$\dot{\theta}_5 = \omega_2^* \tag{9.411}$$

$$\dot{\theta}_4 + \dot{\theta}_6 = \omega_1^* \tag{9.412}$$

$$\dot{s}_2 = w_3 + \sigma_4' d_5 \dot{\theta}_4 \tag{9.413}$$

The above equations have the following implications in the task and joint spaces.

(i) Task space motion restriction implied by Eq. (9.409):

$$w_2' = \sigma_4' s_3 \omega_3^* \tag{9.414}$$

(ii) Joint space motion freedom implied by Eq. (9.412):

$$\dot{\theta}_4 + \dot{\theta}_6 = \omega_1^* \tag{9.415}$$

Owing to this freedom, one of $\dot{\theta}_4$ and $\dot{\theta}_6$ can be selected arbitrarily. Then, Eq. (9.413) gives $\dot{s}_2$ accordingly.

(b) Second Kind of Motion Singularity

According to Eq. (9.371), the second kind of motion singularity occurs if

$$s\theta_5 = 0 \text{ (i.e. } \theta_5 = 0) \text{ and } c\theta_4 = 0 \text{ (i.e. } \theta_4 = \pm\pi/2) \tag{9.416}$$

Note that $s\theta_5 = 0$ occurs only with $\theta_5 = 0$ due to the physical shapes of the prismatic joints.

In a pose of this singularity, the wrist point R gets located either below (if $\theta_4 = \pi/2$) or above (if $\theta_4 = -\pi/2$) the point Q. Meanwhile, Eqs. (9.390)–(9.395) reduce to the following special equations with $\sigma_4'' = \text{sgn}(s\theta_4)$.

$$\omega_3^* = 0 \tag{9.417}$$

$$\dot{s}_2 = w_3 \tag{9.418}$$

$$\dot{\theta}_4 + \dot{\theta}_6 = \omega_1^* \tag{9.419}$$

$$\dot{\theta}_1 = (w_2' - \sigma_4'' d_5\dot{\theta}_4)/s_3 \tag{9.420}$$

$$\dot{s}_3 = w_1' + b_2\dot{\theta}_1 = [s_3 w_1' + b_2(w_2' - \sigma_4'' d_5\dot{\theta}_4)]/s_3 \tag{9.421}$$

$$\dot{\theta}_5 = \omega_2^* - \sigma_4''\dot{\theta}_1 = (s_3\omega_2^* + d_5\dot{\theta}_4 - \sigma_4'' w_2')/s_3 \tag{9.422}$$

The above equations have the following implications in the task and joint spaces.

(i) Task space motion restriction implied by Eq. (9.417):

$$\omega_3^* = 0 \tag{9.423}$$

(ii) Joint space motion freedom implied by Eq. (9.419):

$$\dot{\theta}_4 + \dot{\theta}_6 = \omega_1^* \tag{9.424}$$

Owing to this freedom, one of $\dot{\theta}_4$ and $\dot{\theta}_6$ can be selected arbitrarily. Then, Eqs. (9.420)–(9.422) give $\dot{\theta}_1$, $\dot{s}_3$, and $\dot{\theta}_5$ accordingly.

9.6 An RPRPR^2 Manipulator with an Uncustomary Analytical Solution

Figure 9.30 shows an RPRPR^2 manipulator in its various views. As noticed, its wrist is not spherical, because the fourth joint is prismatic. It is another conceptual manipulator like the one studied in Section 9.5. It is studied here as a typical example of a manipulator that looks as if it does not have an analytical inverse kinematic solution. Yet, it is shown that the solution can still be obtained analytically in an unusual way, which is similar to the way of obtaining the first-order semi-analytical solution for the manipulator of Section 9.5.

9.6.1 Kinematic Description According to the D–H Convention

(a) Rotation Angles

$$\theta_1, \theta_2 = \delta_2 = 0, \theta_3, \theta_4 = \delta_4 = -\pi/2, \theta_5, \theta_6$$

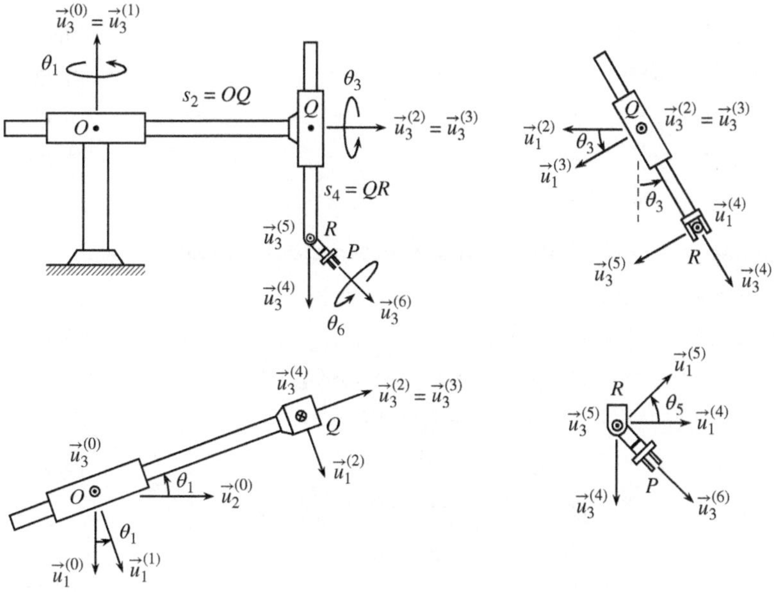

Figure 9.30 An RPRPR^2 manipulator in its various views.

Four of the rotation angles (θ_1, θ_3, θ_5, θ_6) are joint variables. The second rotation angle is associated with the cylindrical arrangement of the joints $\mathcal{J}_2$ and $\mathcal{J}_3$. So, it is taken to be zero as discussed in Chapter 7. Thus, $\vec{u}_1^{(2)}$ is arranged to be parallel to $\vec{u}_1^{(1)}$. On the other hand, the fourth rotation angle is associated with the prismatic joint $\mathcal{J}_4$. As seen in Figure 9.30, the unit vectors $\vec{u}_1^{(3)}$ and $\vec{u}_1^{(4)}$ imply that $\delta_4 = -\pi/2$ about $\vec{u}_3^{(4)}$.

(b) Twist Angles

$$\beta_1 = 0, \beta_2 = -\pi/2, \beta_3 = 0, \beta_4 = -\pi/2, \beta_5 = -\pi/2, \beta_6 = \pi/2$$

(c) Offsets

$$d_1 = 0, s_2 = OQ, d_3 = 0, s_4 = QR, d_5 = 0, d_6 = RP$$

Four of the offsets (d_1, d_3, d_5, d_6) are constant. Two of the offsets associated with the prismatic joints (s_2 and s_4) are joint variables.

(d) Effective Link Lengths

$$b_0 = 0, b_1 = 0, b_2 = 0, b_3 = 0, b_4 = 0, b_5 = 0$$

(e) Link Frame Origins

$$O_0 = O, O_1 = O, O_2 = Q, O_3 = Q, O_4 = R, O_5 = R, O_6 = P$$

9.6.2 Forward Kinematics in the Position Domain

(a) Link-to-Link Orientation Matrices

$$\hat{C}^{(0,1)} = e^{\tilde{u}_1\beta_1}e^{\tilde{u}_3\theta_1} = e^{\tilde{u}_3\theta_1} \tag{9.425}$$

$$\hat{C}^{(1,2)} = e^{\tilde{u}_1\beta_2}e^{\tilde{u}_3\theta_2} = e^{-\tilde{u}_1\pi/2} \tag{9.426}$$

$$\hat{C}^{(2,3)} = e^{\tilde{u}_1\beta_3}e^{\tilde{u}_3\theta_3} = e^{\tilde{u}_3\theta_3} \tag{9.427}$$

$$\hat{C}^{(3,4)} = e^{\tilde{u}_1\beta_4}e^{\tilde{u}_3\theta_4} = e^{-\tilde{u}_1\pi/2}e^{-\tilde{u}_3\pi/2} \tag{9.428}$$

$$\hat{C}^{(4,5)} = e^{\tilde{u}_1\beta_5}e^{\tilde{u}_3\theta_5} = e^{-\tilde{u}_1\pi/2}e^{\tilde{u}_3\theta_5} \tag{9.429}$$

$$\hat{C}^{(5,6)} = e^{\tilde{u}_1\beta_6}e^{\tilde{u}_3\theta_6} = e^{\tilde{u}_1\pi/2}e^{\tilde{u}_3\theta_6} \tag{9.430}$$

(b) Orientation Matrices of the Links with Respect to the Base Frame

$$\hat{C}_1 = \hat{C}^{(0,1)} = e^{\tilde{u}_3\theta_1} \tag{9.431}$$

$$\hat{C}_2 = \hat{C}^{(0,2)} = \hat{C}^{(0,1)}\hat{C}^{(1,2)} = e^{\tilde{u}_3\theta_1}e^{-\tilde{u}_1\pi/2} \tag{9.432}$$

$$\hat{C}_3 = \hat{C}^{(0,3)} = \hat{C}^{(0,2)}\hat{C}^{(2,3)} = e^{\tilde{u}_3\theta_1}e^{-\tilde{u}_1\pi/2}e^{\tilde{u}_3\theta_3} = e^{\tilde{u}_3\theta_1}e^{\tilde{u}_2\theta_3}e^{-\tilde{u}_1\pi/2} \tag{9.433}$$

$$\hat{C}_4 = \hat{C}^{(0,4)} = \hat{C}^{(0,3)}\hat{C}^{(3,4)} = e^{\tilde{u}_3\theta_1}e^{\tilde{u}_2\theta_3}e^{-\tilde{u}_1\pi/2}e^{-\tilde{u}_1\pi/2}e^{-\tilde{u}_3\pi/2} \Rightarrow$$

$$\hat{C}_4 = e^{\tilde{u}_3\theta_1}e^{\tilde{u}_2\theta_3}e^{\tilde{u}_3\pi/2}e^{-\tilde{u}_1\pi} = e^{\tilde{u}_3\theta_1}e^{\tilde{u}_2\theta_3}e^{\tilde{u}_3\pi/2}e^{\tilde{u}_1\pi} \tag{9.434}$$

$$\hat{C}_5 = \hat{C}^{(0,5)} = \hat{C}^{(0,4)}\hat{C}^{(4,5)} = e^{\tilde{u}_3\theta_1}e^{\tilde{u}_2\theta_3}e^{\tilde{u}_3\pi/2}e^{\tilde{u}_1\pi}e^{-\tilde{u}_1\pi/2}e^{\tilde{u}_3\theta_5} \Rightarrow$$

$$\hat{C}_5 = e^{\tilde{u}_3\theta_1}e^{\tilde{u}_2\theta_3}e^{\tilde{u}_3\pi/2}e^{\tilde{u}_1\pi/2}e^{\tilde{u}_3\theta_5} = e^{\tilde{u}_3\theta_1}e^{\tilde{u}_2\theta_3}e^{\tilde{u}_3\pi/2}e^{-\tilde{u}_2\theta_5}e^{\tilde{u}_1\pi/2} \Rightarrow$$

$$\hat{C}_5 = e^{\tilde{u}_3\theta_1}e^{\tilde{u}_2\theta_3}e^{\tilde{u}_1\theta_5}e^{\tilde{u}_3\pi/2}e^{\tilde{u}_1\pi/2} \tag{9.435}$$

$$\hat{C}_6 = \hat{C}^{(0,6)} = \hat{C}^{(0,5)}\hat{C}^{(5,6)} = e^{\tilde{u}_3\theta_1}e^{\tilde{u}_2\theta_3}e^{\tilde{u}_1\theta_5}e^{\tilde{u}_3\pi/2}e^{\tilde{u}_1\pi/2}e^{\tilde{u}_1\pi/2}e^{\tilde{u}_3\theta_6} \Rightarrow$$

$$\hat{C}_6 = e^{\tilde{u}_3\theta_1}e^{\tilde{u}_2\theta_3}e^{\tilde{u}_1\theta_5}e^{\tilde{u}_3\pi/2}e^{\tilde{u}_1\pi}e^{\tilde{u}_3\theta_6} = e^{\tilde{u}_3\theta_1}e^{\tilde{u}_2\theta_3}e^{\tilde{u}_1\theta_5}e^{\tilde{u}_3\pi/2}e^{-\tilde{u}_3\theta_6}e^{\tilde{u}_1\pi} \Rightarrow$$

$$\hat{C}_6 = \hat{C} = e^{\tilde{u}_3\theta_1}e^{\tilde{u}_2\theta_3}e^{\tilde{u}_1\theta_5}e^{-\tilde{u}_3\theta'_6}e^{\tilde{u}_1\pi} \tag{9.436}$$

In Eq. (9.436), θ'_6 is introduced as a *modified joint variable.* It is defined as follows:

$$\theta'_6 = \theta_6 - \pi/2 \tag{9.437}$$

(c) Location of the Wrist Point with Respect to the Base Frame

$$\vec{r} = \vec{r}_{OR} = \vec{r}_{OQ} + \vec{r}_{QR} = s_2\vec{u}_3^{(2)} + s_4\vec{u}_3^{(4)} \tag{9.438}$$

The vector Eq. (9.438) implies the following matrix equation in the base frame.

$$\overline{r} = \overline{r}^{(0)} = s_2\overline{u}_3^{(2/0)} + s_4\overline{u}_3^{(4/0)} \tag{9.439}$$

Eq. (9.439) can be manipulated as shown below.

$$\overline{r} = s_2\hat{C}^{(0,2)}\overline{u}_3^{(2/2)} + s_4\hat{C}^{(0,4)}\overline{u}_3^{(4/4)} \Rightarrow$$

$$\overline{r} = s_2e^{\tilde{u}_3\theta_1}e^{-\tilde{u}_1\pi/2}\overline{u}_3 + s_4e^{\tilde{u}_3\theta_1}e^{\tilde{u}_2\theta_3}e^{\tilde{u}_3\pi/2}e^{\tilde{u}_1\pi}\overline{u}_3 \Rightarrow$$

$$\overline{r} = s_2e^{\tilde{u}_3\theta_1}\overline{u}_2 - s_4e^{\tilde{u}_3\theta_1}e^{\tilde{u}_2\theta_3}e^{\tilde{u}_3\pi/2}\overline{u}_3 = s_2e^{\tilde{u}_3\theta_1}\overline{u}_2 - s_4e^{\tilde{u}_3\theta_1}e^{\tilde{u}_2\theta_3}\overline{u}_3 \Rightarrow$$

$$\overline{r} = e^{\tilde{u}_3\theta_1}(s_2\overline{u}_2 - s_4e^{\tilde{u}_2\theta_3}\overline{u}_3) \Rightarrow$$

$$\overline{r} = e^{\tilde{u}_3\theta_1}[\overline{u}_2s_2 - \overline{u}_3(s_4c\theta_3) - \overline{u}_1(s_4s\theta_3)] \tag{9.440}$$

Equation (9.440) leads to the following equation that shows the coordinates of the wrist point R in the base frame $\mathcal{F}_0(O)$.

$$\overline{r} = \overline{u}_1r_1 + \overline{u}_2r_2 + \overline{u}_3r_3 = (e^{\tilde{u}_3\theta_1}\overline{u}_2)s_2 - (e^{\tilde{u}_3\theta_1}\overline{u}_3)(s_4c\theta_3) - (e^{\tilde{u}_3\theta_1}\overline{u}_1)(s_4s\theta_3) \Rightarrow$$

$$\overline{r} = (\overline{u}_2c\theta_1 - \overline{u}_1s\theta_1)s_2 - (\overline{u}_3)(s_4c\theta_3) - (\overline{u}_1c\theta_1 + \overline{u}_2s\theta_1)(s_4s\theta_3) \Rightarrow$$

$$\overline{r} = -\overline{u}_1(s_2s\theta_1 + s_4s\theta_3c\theta_1) + \overline{u}_2(s_2c\theta_1 - s_4s\theta_3s\theta_1) - \overline{u}_3(s_4c\theta_3) \tag{9.441}$$

(d) Location of the Tip Point with Respect to the Base Frame

$$\vec{p} = \vec{r}_{OP} = \vec{r}_{OR} + \vec{r}_{RP} = \vec{r} + d_6\vec{u}_3^{(6)} \tag{9.442}$$

Equation (9.289) implies the following matrix equation in the base frame.

$$\overline{p} = \overline{p}^{(0)} = \overline{r}^{(0)} + d_6\overline{u}_3^{(6/0)} = \overline{r}^{(0)} + d_6\hat{C}^{(0,6)}\overline{u}_3^{(6/6)} \Rightarrow$$

$$\overline{p} = \overline{r} + d_6\hat{C}\overline{u}_3 \tag{9.443}$$

When Eq. (9.436) is substituted, Eq. (9.443) becomes

$$\overline{p} = \overline{r} + d_6 e^{\tilde{u}_3\theta_1} e^{\tilde{u}_2\theta_3} e^{\tilde{u}_1\theta_5} e^{-\tilde{u}_3\theta_6'} e^{\tilde{u}_1\pi}\overline{u}_3 \Rightarrow$$

$$\overline{p} = \overline{r} - d_6 e^{\tilde{u}_3\theta_1} e^{\tilde{u}_2\theta_3} e^{\tilde{u}_1\theta_5}\overline{u}_3 \tag{9.444}$$

9.6.3 Inverse Kinematics in the Position Domain

Here, $\overline{p}$ and $\hat{C}$ are specified. Hence, $\overline{r}$ also becomes specified according to Eq. (9.443). That is,

$$\overline{r} = \overline{p} - d_6\hat{C}\overline{u}_3 \tag{9.445}$$

(a) Solution for the Joint Variables

When Eqs. (9.436) and (9.440) are examined, it is seen that each of these equations contain four unknowns and none of these unknowns can be singled out from any of these equations. Therefore, the present manipulator does not seem to have an analytical inverse kinematic solution. This suggests to use an approach, which is similar to the approach of obtaining the first-order semi-analytical solution for the manipulator studied in Section 9.5.

∗ General Solution

The general solution process can be started by treating any one of the six joint variables as if it is known temporarily. For the present manipulator, θ_1 can be selected to undertake this role. Then, both of Eqs. (9.440) and (9.436) can be solved rather straightforwardly for the other joint variables.
To prepare for the solution, Eq. (9.440) can be written as

$$\overline{u}_2 s_2 - \overline{u}_3(s_4 c\theta_3) - \overline{u}_1(s_4 s\theta_3) = e^{-\tilde{u}_3\theta_1}\overline{r} \tag{9.446}$$

Upon premultiplications by $\overline{u}_1^t$, $\overline{u}_2^t$, and $\overline{u}_3^t$, Eq. (9.446) leads to the following scalar equations.

$$s_4 s\theta_3 = -(r_1 c\theta_1 + r_2 s\theta_1) \tag{9.447}$$

$$s_2 = r_2 c\theta_1 - r_1 s\theta_1 \tag{9.448}$$

$$s_4 c\theta_3 = -r_3 \tag{9.449}$$

Note that Eq. (9.448) has already given s_2 as a function of θ_1, i.e.

$$s_2 = f_2(\theta_1) = r_2 c\theta_1 - r_1 s\theta_1 \tag{9.450}$$

Note also that $s_4 \geq s_4^\circ > 0$. Therefore, Eqs. (9.447) and (9.449) give θ_3 and s_4 as the following functions of θ_1.

$$\theta_3 = f_3(\theta_1) = \text{atan}_2[-(r_1 c\theta_1 + r_2 s\theta_1), -r_3] \tag{9.451}$$

$$s_4 = f_4(\theta_1) = \sqrt{(r_1 c\theta_1 + r_2 s\theta_1)^2 + r_3^2} \tag{9.452}$$

On the other hand, Eq. (9.436) can be prepared for the solution by writing it as

$$e^{\tilde{u}_2\theta_3} e^{\tilde{u}_1\theta_5} e^{-\tilde{u}_3\theta_6'} = \hat{C}^*(\theta_1) = e^{-\tilde{u}_3\theta_1} \hat{C} e^{-\tilde{u}_1\pi} \tag{9.453}$$

Equation (9.453) contains the three joint angles in an arrangement like the Euler angles of the 2-1-3 sequence. Therefore, they can be extracted by using the similar procedure explained in Chapter 3. This procedure leads to the following scalar equation pairs when it is applied to Eq. (9.453).

$$c_{23}^*(\theta_1) = \overline{u}_2^t \hat{C}^*(\theta_1)\overline{u}_3 = -s\theta_5 \Rightarrow c\theta_5 = \sigma_5\sqrt{1 - [c_{23}^*(\theta_1)]^2}; \quad \sigma_5 = \pm 1 \tag{9.454}$$

$$c_{13}^*(\theta_1) = \overline{u}_1^t \hat{C}^*(\theta_1)\overline{u}_3 = s\theta_3 c\theta_5; \quad c_{33}^*(\theta_1) = \overline{u}_3^t \hat{C}^*(\theta_1)\overline{u}_3 = c\theta_3 c\theta_5 \tag{9.455}$$

$$c_{21}^*(\theta_1) = \overline{u}_2^t \hat{C}^*(\theta_1)\overline{u}_1 = -c\theta_5 s\theta_6'; \quad c_{22}^*(\theta_1) = \overline{u}_2^t \hat{C}^*(\theta_1)\overline{u}_2 = c\theta_5 c\theta_6' \tag{9.456}$$

Equation Pair (9.454) gives θ_5 as the following function of θ_1.

$$\theta_5 = g_5(\theta_1) = \text{atan}_2\{-c_{23}^*(\theta_1), \sigma_5\sqrt{1 - [c_{23}^*(\theta_1)]^2}\} \tag{9.457}$$

If $c\theta_5 \neq 0$, the other equation pairs give θ_3 and θ_6 as the following functions of θ_1 without additional superfluous sign variables.

$$\theta_3 = g_3(\theta_1) = \text{atan}_2[\sigma_5 c_{13}^*(\theta_1), \sigma_5 c_{33}^*(\theta_1)] \tag{9.458}$$

$$\theta_6' = g_6(\theta_1) = \text{atan}_2[-\sigma_5 c_{21}^*(\theta_1), \sigma_5 c_{22}^*(\theta_1)]; \quad \theta_6 = \theta_6' + \pi/2 \tag{9.459}$$

As noticed above, θ_3 has been obtained twice, first from Eq. (9.451) and then from Eq. (9.458). Of course, the two results must be equal to each other. In other words, θ_1 must be determined in such a way that the following *consistency equation* is satisfied.

$$f_3(\theta_1) = g_3(\theta_1) \Rightarrow$$

$$\text{atan}_2[-(r_1 c\theta_1 + r_2 s\theta_1), -r_3] = \text{atan}_2[\sigma_5 c_{13}^*(\theta_1), \sigma_5 c_{33}^*(\theta_1)] \tag{9.460}$$

At first glance, Eq. (9.460) looks as if it can hardly be solved analytically for θ_1. However, actually it can be solved. The analytical solution can be obtained based on the following implication of Eq. (9.460) together with Eqs. (9.451) and (9.458).

$$\tan\theta_3 = [-(r_1 c\theta_1 + r_2 s\theta_1)]/(-r_3) = [\sigma_5 c_{13}^*(\theta_1)]/[\sigma_5 c_{33}^*(\theta_1)] \Rightarrow$$

$$(r_1 c\theta_1 + r_2 s\theta_1)/r_3 = c_{13}^*(\theta_1)/c_{33}^*(\theta_1) \Rightarrow$$

$$(r_1 c\theta_1 + r_2 s\theta_1)c_{33}^*(\theta_1) = r_3 c_{13}^*(\theta_1) \tag{9.461}$$

On the other hand, according to Eq. (9.453),

$$c_{33}^*(\theta_1) = \overline{u}_3^t e^{-\tilde{u}_3\theta_1} \hat{C} e^{-\tilde{u}_1\pi}\overline{u}_3 = -\overline{u}_3^t \hat{C}\overline{u}_3 = -c_{33} \tag{9.462}$$

$$c_{13}^*(\theta_1) = \overline{u}_1^t e^{-\tilde{u}_3\theta_1} \hat{C} e^{-\tilde{u}_1\pi}\overline{u}_3 = -(\overline{u}_1^t c\theta_1 + \overline{u}_2^t s\theta_1)\hat{C}\overline{u}_3 \Rightarrow$$

$$c_{13}^*(\theta_1) = -(c_{13} c\theta_1 + c_{23} s\theta_1) \tag{9.463}$$

Hence, Eq. (9.461) becomes

$$(r_1 c\theta_1 + r_2 s\theta_1)c_{33} = r_3(c_{13}c\theta_1 + c_{23}s\theta_1) \Rightarrow$$

$$(r_1 c_{33} - r_3 c_{13})c\theta_1 + (r_2 c_{33} - r_3 c_{23})s\theta_1 = 0 \quad (9.464)$$

To obtain θ_1 from Eq. (9.464), let

$$q_{12} = \sqrt{(r_1 c_{33} - r_3 c_{13})^2 + (r_2 c_{33} - r_3 c_{23})^2} \quad (9.465)$$

If $q_{12} \neq 0$, Eq. (9.464) gives $s\theta_1$, $c\theta_1$, and θ_1 as follows with a sign variable $\sigma_1 = \pm 1$:

$$s\theta_1 = -\sigma_1(r_1 c_{33} - r_3 c_{13})/q_{12} \quad (9.466)$$

$$c\theta_1 = +\sigma_1(r_2 c_{33} - r_3 c_{23})/q_{12} \quad (9.467)$$

$$\theta_1 = \mathrm{atan}_2[-\sigma_1(r_1 c_{33} - r_3 c_{13}), \sigma_1(r_2 c_{33} - r_3 c_{23})] \quad (9.468)$$

As noticed above, θ_1 requires the third column elements (c_{13}, c_{23}, c_{33}) of the matrix $\hat{C}$. If $\hat{C}$ is specified by means of the Euler angles (ϕ_1, ϕ_2, ϕ_3) of the 3-2-3 sequence, the required elements are expressed as follows:

$$c_{33} = \overline{u}_3^t e^{\tilde{u}_3\phi_1} e^{\tilde{u}_2\phi_2} e^{\tilde{u}_3\phi_3} \overline{u}_3 = \overline{u}_3^t e^{\tilde{u}_2\phi_2} \overline{u}_3 = c\phi_2 \quad (9.469)$$

$$c_{13} = \overline{u}_1^t e^{\tilde{u}_3\phi_1} e^{\tilde{u}_2\phi_2} e^{\tilde{u}_3\phi_3} \overline{u}_3 = \overline{u}_1^t e^{\tilde{u}_3\phi_1} e^{\tilde{u}_2\phi_2} \overline{u}_3 = c\phi_1 s\phi_2 \quad (9.470)$$

$$c_{23} = \overline{u}_2^t e^{\tilde{u}_3\phi_1} e^{\tilde{u}_2\phi_2} e^{\tilde{u}_3\phi_3} \overline{u}_3 = \overline{u}_2^t e^{\tilde{u}_3\phi_1} e^{\tilde{u}_2\phi_2} \overline{u}_3 = s\phi_1 s\phi_2 \quad (9.471)$$

Afterwards, with the availability of θ_1, the other joint variables can readily be obtained from Eqs. (9.450), (9.451), (9.452), (9.457), and (9.459).

* **Special Solution with $c\theta_5 = 0$**

As indicated above concerning Eqs. (9.458) and (9.459), the general solution procedure cannot be used if it turns out that $c\theta_5 = 0$ or $\theta_5 = \sigma_5'\pi/2$ with $\sigma_5' = \pm 1$. Such special poses of the manipulator are illustrated locally in Figure 9.31.

In such a special pose, which resembles the third kind of position singularity of the elbow manipulator, the solution for the other joint variables can be obtained as explained below.

With $\theta_5 = \sigma_5'\pi/2$, Eq. (9.436) can be written as follows:

$$\hat{C} = e^{\tilde{u}_3\theta_1} e^{\tilde{u}_2\theta_3} e^{\tilde{u}_1\sigma_5'\pi/2} e^{-\tilde{u}_3\theta_6'} e^{\tilde{u}_1\pi} = e^{\tilde{u}_3\theta_1} e^{\tilde{u}_2\theta_3} e^{\tilde{u}_2\sigma_5'\theta_6'} e^{\tilde{u}_1\sigma_5'\pi/2} e^{\tilde{u}_1\pi} \Rightarrow$$

$$e^{\tilde{u}_3\theta_1} e^{\tilde{u}_2(\theta_3+\sigma_5'\theta_6')} = \hat{C}e^{-\tilde{u}_1\pi} e^{-\tilde{u}_1\sigma_5'\pi/2} = \hat{C}e^{-\tilde{u}_1\sigma_5'\pi/2} e^{-\tilde{u}_1\pi} \quad (9.472)$$

The variables θ_3 and θ_6' cannot be found separately from Eq. (9.472) because they happen to be rotation angles about parallel axes, which are the axes of the third and sixth joints. Nevertheless, their combination, i.e. $\theta_{36}' = \theta_3 + \sigma_5'\theta_6'$, can still be found from Eq. (9.472), which can be worked out to the following scalar equations.

$$\overline{u}_3^t e^{\tilde{u}_3\theta_1} e^{\tilde{u}_2\theta_{36}'} \overline{u}_2 = \overline{u}_3^t \hat{C} e^{-\tilde{u}_1\sigma_5'\pi/2} e^{-\tilde{u}_1\pi} \overline{u}_2 \Rightarrow$$

$$\overline{u}_3^t \overline{u}_2 = 0 = -\overline{u}_3^t \hat{C} e^{-\tilde{u}_1(\sigma_5'\pi/2)} \overline{u}_2 = \sigma_5' \overline{u}_3^t \hat{C} \overline{u}_3 = \sigma_5' c_{33} \Rightarrow$$

$$c_{33} = 0 \quad (9.473)$$

$$\overline{u}_2^t e^{\tilde{u}_3\theta_1} e^{\tilde{u}_2\theta_{36}} \overline{u}_2 = \overline{u}_2^t \hat{C} e^{-\tilde{u}_1\sigma_5'\pi/2} e^{-\tilde{u}_1\pi} \overline{u}_2 \Rightarrow$$

$$\overline{u}_2^t e^{\tilde{u}_3\theta_1} \overline{u}_2 = -\overline{u}_2^t \hat{C} e^{-\tilde{u}_1(\sigma_5'\pi/2)} \overline{u}_2 = \sigma_5' \overline{u}_2^t \hat{C} \overline{u}_3 \Rightarrow$$

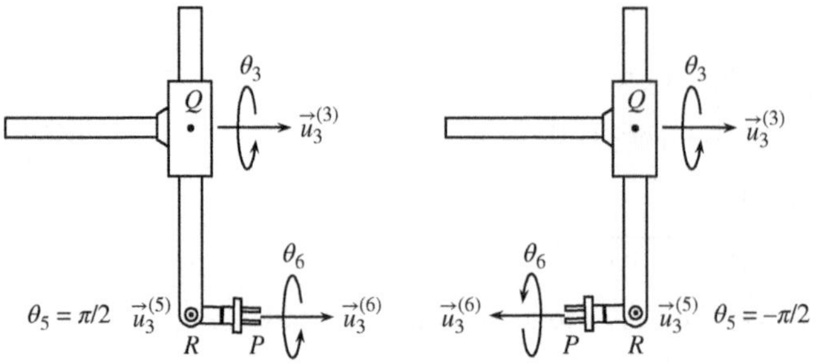

Figure 9.31 Special poses of the manipulator with $c\theta_5 = 0$.

$$c\theta_1 = \sigma'_5 c_{23} \tag{9.474}$$

$$\overline{u}_1^t e^{\tilde{u}_3\theta_1} e^{\tilde{u}_2\theta_{36}} \overline{u}_2 = \overline{u}_1^t \widehat{C} e^{-\tilde{u}_1\sigma'_5\pi/2} e^{-\tilde{u}_1\pi} \overline{u}_2 \Rightarrow$$

$$\overline{u}_1^t e^{\tilde{u}_3\theta_1} \overline{u}_2 = -\overline{u}_1^t \widehat{C} e^{-\tilde{u}_1(\sigma'_5\pi/2)} \overline{u}_2 = \sigma'_5 \overline{u}_1^t \widehat{C} \overline{u}_3 \Rightarrow$$

$$s\theta_1 = -\sigma'_5 c_{13} \tag{9.475}$$

$$\overline{u}_3^t e^{\tilde{u}_3\theta_1} e^{\tilde{u}_2\theta'_{36}} \overline{u}_3 = \overline{u}_3^t \widehat{C} e^{-\tilde{u}_1\sigma'_5\pi/2} e^{-\tilde{u}_1\pi} \overline{u}_3 \Rightarrow$$

$$\overline{u}_3^t e^{\tilde{u}_2\theta'_{36}} \overline{u}_3 = -\overline{u}_3^t \widehat{C} e^{-\tilde{u}_1(\sigma'_5\pi/2)} \overline{u}_3 = -\sigma'_5 \overline{u}_3^t \widehat{C} \overline{u}_2 \Rightarrow$$

$$c\theta'_{36} = -\sigma'_5 c_{32} \tag{9.476}$$

$$\overline{u}_3^t e^{\tilde{u}_3\theta_1} e^{\tilde{u}_2\theta'_{36}} \overline{u}_1 = \overline{u}_3^t \widehat{C} e^{-\tilde{u}_1\sigma'_5\pi/2} e^{-\tilde{u}_1\pi} \overline{u}_1 \Rightarrow$$

$$\overline{u}_3^t e^{\tilde{u}_2\theta'_{36}} \overline{u}_1 = \overline{u}_3^t \widehat{C} \overline{u}_1 = c_{31} \Rightarrow$$

$$s\theta'_{36} = -c_{31} \tag{9.477}$$

Equations (9.474) and (9.475) give θ_1 as

$$\theta_1 = \text{atan}_2(-\sigma'_5 c_{13}, \sigma'_5 c_{23}) \tag{9.478}$$

Equations (9.476) and (9.477) give θ'_{36} as

$$\theta'_{36} = \theta_3 + \sigma'_5\theta'_6 = \text{atan}_2(-c_{31}, -\sigma'_5 c_{32}) \tag{9.479}$$

If $\widehat{C}$ is specified by means of the 3-2-3 Euler angles ϕ_1, ϕ_2, and ϕ_3, the elements of $\widehat{C}$ required in Eqs. (9.473)–(9.479) are expressed as follows:

$$c_{33} = \overline{u}_3^t e^{\tilde{u}_3\phi_1} e^{\tilde{u}_2\phi_2} e^{\tilde{u}_3\phi_3} \overline{u}_3 = \overline{u}_3^t e^{\tilde{u}_2\phi_2} \overline{u}_3 = c\phi_2 \tag{9.480}$$

$$c_{13} = \overline{u}_1^t e^{\tilde{u}_3\phi_1} e^{\tilde{u}_2\phi_2} e^{\tilde{u}_3\phi_3} \overline{u}_3 = \overline{u}_1^t e^{\tilde{u}_3\phi_1} e^{\tilde{u}_2\phi_2} \overline{u}_3 = c\phi_1 s\phi_2 \tag{9.481}$$

$$c_{23} = \overline{u}_2^t e^{\tilde{u}_3\phi_1} e^{\tilde{u}_2\phi_2} e^{\tilde{u}_3\phi_3} \overline{u}_3 = \overline{u}_2^t e^{\tilde{u}_3\phi_1} e^{\tilde{u}_2\phi_2} \overline{u}_3 = s\phi_1 s\phi_2 \tag{9.482}$$

$$c_{31} = \overline{u}_3^t e^{\tilde{u}_3\phi_1} e^{\tilde{u}_2\phi_2} e^{\tilde{u}_3\phi_3} \overline{u}_1 = \overline{u}_3^t e^{\tilde{u}_2\phi_2} e^{\tilde{u}_3\phi_3} \overline{u}_1 = -s\phi_2 c\phi_3 \tag{9.483}$$

$$c_{32} = \overline{u}_3^t e^{\tilde{u}_3\phi_1} e^{\tilde{u}_2\phi_2} e^{\tilde{u}_3\phi_3} \overline{u}_2 = \overline{u}_3^t e^{\tilde{u}_2\phi_2} e^{\tilde{u}_3\phi_3} \overline{u}_2 = s\phi_2 s\phi_3 \tag{9.484}$$

On the other hand, Eqs. (9.473) and (9.480) indicate that ϕ_2 is specified so that

$$c_{33} = c\phi_2 = 0 \Rightarrow \phi_2 = \sigma'_2\pi/2; \sigma'_2 = \pm 1 \tag{9.485}$$

Then, with $\phi_2 = \sigma'_2\pi/2$, Eqs. (9.478) and (9.479) give θ_1 and θ'_{36} as shown below.

$$\theta_1 = \text{atan}_2(-\sigma'_2\sigma'_5 c\phi_1, \sigma'_2\sigma'_5 s\phi_1) \tag{9.486}$$

$$\theta'_{36} = \theta_3 + \sigma'_5\theta'_6 = \text{atan}_2(\sigma'_2 c\phi_3, -\sigma'_2\sigma'_5 s\phi_3) \tag{9.487}$$

Note that, in such a special pose, θ_1 and θ'_{36} become available only in terms of the specified Euler angles ϕ_1 and ϕ_3. This availability is different from the availability of θ_1 by Eq. (9.468) in a general different pose of the manipulator, which requires not only the Euler angles but also the coordinates of the wrist point.

With the availability of θ_1 as expressed in Eq. (9.486), Eqs. (9.450)–(9.452) can be used to express s_2, s_4, and θ_3 as follows in terms of the specified quantities:

$$s_2 = r_2 c\theta_1 - r_1 s\theta_1 \Rightarrow$$

$$s_2 = \sigma'_2\sigma'_5(r_2 s\phi_1 + r_1 c\phi_1) \tag{9.488}$$

$$s_4 = \sqrt{(r_1 c\theta_1 + r_2 s\theta_1)^2 + r_3^2} \Rightarrow$$

$$s_4 = \sqrt{(r_2 c\phi_1 - r_1 s\phi_1)^2 + r_3^2} \tag{9.489}$$

$$\theta_3 = \text{atan}_2[-(r_1 c\theta_1 + r_2 s\theta_1), -r_3] \Rightarrow$$

$$\theta_3 = \text{atan}_2[\sigma'_2\sigma'_5(r_2 c\phi_1 - r_1 s\phi_1), -r_3] \tag{9.490}$$

Hence, θ_6 also becomes known. That is,

$$\theta'_6 = \sigma'_5(\theta'_{36} - \theta_3) \Rightarrow \theta_6 = \theta'_6 + \pi/2 \tag{9.491}$$

9.6.4 Multiplicity Analysis

(a) First Kind of Multiplicity of the General Solution

The first kind of multiplicity is associated with the sign variable σ_1 that arises in the process of finding θ_1 by using Eq. (9.468). The valid value of σ_1 must be such that $s_2 \geq s_2^\circ > 0$, because of the physical shape of the second joint. Equations (9.466), (9.467), and (9.448) give s_2 as follows:

$$s_2 = \sigma_1[(r_1^2 + r_2^2)c_{33} - (r_1 c_{13} + r_2 c_{23})r_3]/q_{12} \tag{9.492}$$

According to Eq. (9.492), the fact that $s_2 > 0$ can be realized if

$$\sigma_1 = \text{sgn}[(r_1^2 + r_2^2)c_{33} - (r_1 c_{13} + r_2 c_{23})r_3] \tag{9.493}$$

(b) Second Kind of Multiplicity of the General Solution

The second kind of multiplicity is associated with the sign variable σ_5 that arises in the process of finding the angles θ_3, θ_5, and θ_6. The sign variable σ_5 has the responsibility of maintaining the consistency of Eq. (9.460). In other words, σ_5 must be such that

$$\text{sgn}(-r_3) = \text{sgn}[\sigma_5 c^*_{33}(\theta_1)] \tag{9.494}$$

On the other hand, $c^*_{33}(\theta_1) = -c_{33}$ according to Eq. (9.462) and $c_{33} \neq 0$ in the general solution. Therefore, Eq. (9.494) implies that

$$\sigma_5 = \text{sgn}(r_3 c_{33}) \tag{9.495}$$

(c) Multiplicity of the Special Solution

The multiplicity of the special solution is associated with the sign variable σ_5' that arises in the process of finding the angles θ_1 and θ_{36}'. Note again that s_2 can only be positive due to the physical shape of the second joint. On the other hand, σ_2' is known owing to the specification of ϕ_2 according to Eq. (9.485). Therefore, Eq. (9.488) implies that

$$\sigma_5' = \sigma_2' \mathrm{sgn}(r_2 s\phi_1 + r_1 c\phi_1) \tag{9.496}$$

9.6.5 Singularity Analysis in the Position Domain

Equations (9.466) and (9.467) imply that a position singularity may occur if $q_{12} = 0$, i.e. if

$$r_1 c_{33} - r_3 c_{13} = r_2 c_{33} - r_3 c_{23} = 0 \tag{9.497}$$

In the general solution described above, $c_{33} \neq 0$. In that case, Eq. (9.497) implies further that

$$r_1 = r_3 c_{13}/c_{33} \text{ and } r_2 = r_3 c_{23}/c_{33} \tag{9.498}$$

However, this singularity is not likely to occur. This is because, referring to Figure 9.30, Eq. Pair (9.498) implies that the points Q and O get coincident, but this coincidence is not possible due to the physical shape of the second joint.

On the other hand, as also described above, the special solution with $c_{33} = 0$ (i.e. with $c\theta_5 = 0$) can be obtained without encountering any singularity.

9.6.6 Forward Kinematics in the Velocity Domain

(a) Angular Velocity of the End-Effector with Respect to the Base Frame

Let $\overline{\omega} = \overline{\omega}_6$. Then, Eq. (9.436) leads to the following expression.

$$\overline{\omega} = \dot{\theta}_1 \overline{u}_3 + \dot{\theta}_3 e^{\tilde{u}_3 \theta_1} \overline{u}_2 + \dot{\theta}_5 e^{\tilde{u}_3 \theta_1} e^{\tilde{u}_2 \theta_3} \overline{u}_1 - \dot{\theta}_6 e^{\tilde{u}_3 \theta_1} e^{\tilde{u}_2 \theta_3} e^{\tilde{u}_1 \theta_5} \overline{u}_3 \tag{9.499}$$

(b) Velocity of the Wrist Point with Respect to the Base Frame

Let $\overline{w} = \overline{w}^{(0)} = \overline{v}_R = \dot{\overline{r}}$. Then, $\overline{w}$ can be obtained from Eq. (9.440) as follows:

$$\begin{aligned} \overline{w} &= (\dot{\theta}_1 e^{\tilde{u}_3 \theta_1} \tilde{u}_3)[\overline{u}_2 s_2 - \overline{u}_3 (s_4 c\theta_3) - \overline{u}_1 (s_4 s\theta_3)] \\ &\quad + (e^{\tilde{u}_3 \theta_1})[\overline{u}_2 \dot{s}_2 - \overline{u}_3 (\dot{s}_4 c\theta_3 - s_4 \dot{\theta}_3 s\theta_3) - \overline{u}_1 (\dot{s}_4 s\theta_3 + s_4 \dot{\theta}_3 c\theta_3)] \Rightarrow \\ \overline{w} &= e^{\tilde{u}_3 \theta_1} \overline{w}' = e^{\tilde{u}_3 \theta_1} (\overline{u}_1 w_1' + \overline{u}_2 w_2' + \overline{u}_3 w_3') \end{aligned} \tag{9.500}$$

In Eq. (9.500),

$$\begin{aligned} \overline{w} &= e^{\tilde{u}_3 \theta_1} [-\overline{u}_1 s_2 \dot{\theta}_1 - \overline{u}_2 (s_4 \dot{\theta}_1 s\theta_3)] \\ &\quad + e^{\tilde{u}_3 \theta_1} [\overline{u}_2 \dot{s}_2 - \overline{u}_3 (\dot{s}_4 c\theta_3 - s_4 \dot{\theta}_3 s\theta_3) - \overline{u}_1 (\dot{s}_4 s\theta_3 + s_4 \dot{\theta}_3 c\theta_3)] \Rightarrow \end{aligned}$$

$$\left.\begin{aligned} w_1' &= -(s_2 \dot{\theta}_1 + s_4 \dot{\theta}_3 c\theta_3 + \dot{s}_4 s\theta_3) \\ w_2' &= \dot{s}_2 - s_4 \dot{\theta}_1 s\theta_3 \\ w_3' &= s_4 \dot{\theta}_3 s\theta_3 - \dot{s}_4 c\theta_3 \end{aligned}\right\} \tag{9.501}$$

(c) Velocity of the Tip Point with Respect to the Base Frame

Let $\overline{v} = \overline{v}^{(0)} = \overline{v}_P = \dot{\overline{p}}$. Then, $\overline{v}$ can be obtained from Eq. (9.443) as follows:

$$\overline{v} = \overline{w} + d_6\tilde{\omega}\hat{C}\overline{u}_3 \tag{9.502}$$

9.6.7 Inverse Kinematics in the Velocity Domain

Since $\overline{p}$ and $\hat{C}$ are specified as functions of time, the velocities $\overline{v}$, $\overline{\omega}$, and $\overline{w}$ are also known. They are obtained as shown below.

$$\overline{v} = \dot{\overline{p}}, \quad \overline{\omega} = \text{colm}(\dot{\hat{C}}\hat{C}^t); \quad \overline{w} = \overline{v} - d_6\tilde{\omega}\hat{C}\overline{u}_3 \tag{9.503}$$

(a) Solution for the Joint Velocities

* **General Solution**

Although each of the equations obtained above for $\overline{w}$ and $\overline{\omega}$ contain four unknown joint velocities, the solution can be obtained quite straightforwardly by treating one of the joint velocities, say $\dot{\theta}_1$, as if it is known temporarily. With this treatment, Eqs. (9.500) and (9.499) can be manipulated into the following scalar equations.
Equation (9.500) can be manipulated as follows:

$$\overline{u}_1 w_1' + \overline{u}_2 w_2' + \overline{u}_3 w_3' = \overline{w}' = e^{-\tilde{u}_3\theta_1}\overline{w} \Rightarrow$$

$$s_2\dot{\theta}_1 + s_4\dot{\theta}_3 c\theta_3 + \dot{s}_4 s\theta_3 = -w_1' = -(w_1 c\theta_1 + w_2 s\theta_1) \tag{9.504}$$

$$\dot{s}_2 - s_4\dot{\theta}_1 s\theta_3 = w_2' = w_2 c\theta_1 - w_1 s\theta_1 \tag{9.505}$$

$$s_4\dot{\theta}_3 s\theta_3 - \dot{s}_4 c\theta_3 = w_3' = w_3 \tag{9.506}$$

Equation (9.499) can be manipulated as follows:

$$\dot{\theta}_3\overline{u}_2 + \dot{\theta}_5\overline{u}_1 - \dot{\theta}_6 e^{\tilde{u}_1\theta_5}\overline{u}_3 = \overline{u}_1\dot{\theta}_5 + \overline{u}_2(\dot{\theta}_3 + \dot{\theta}_6 s\theta_5) - \overline{u}_3(\dot{\theta}_6 c\theta_5)$$
$$= e^{-\tilde{u}_2\theta_3}e^{-\tilde{u}_3\theta_1}(\overline{\omega} - \dot{\theta}_1\overline{u}_3) = \overline{\omega}^* - \dot{\theta}_1 e^{-\tilde{u}_2\theta_3}\overline{u}_3 = \overline{\omega}^* + \dot{\theta}_1(\overline{u}_1 s\theta_3 - \overline{u}_3 c\theta_3) \Rightarrow$$

$$\dot{\theta}_5 = \omega_1^* + \dot{\theta}_1 s\theta_3 \tag{9.507}$$

$$\dot{\theta}_3 + \dot{\theta}_6 s\theta_5 = \omega_2^* \tag{9.508}$$

$$\dot{\theta}_6 c\theta_5 = -\omega_3^* + \dot{\theta}_1 c\theta_3 \tag{9.509}$$

In the above angular velocity equations, $\overline{\omega}^*$ is known as

$$\overline{\omega}^* = e^{-\tilde{u}_2\theta_3}e^{-\tilde{u}_3\theta_1}\overline{\omega} \tag{9.510}$$

Equation (9.505) gives $\dot{s}_2$ as

$$\dot{s}_2 = w_2 c\theta_1 - w_1 s\theta_1 - s_4\dot{\theta}_1 s\theta_3 \tag{9.511}$$

Equations (9.504) and (9.506) can be written together as

$$\begin{bmatrix} c\theta_3 & s\theta_3 \\ -s\theta_3 & c\theta_3 \end{bmatrix}\begin{bmatrix} s_4\dot{\theta}_3 \\ \dot{s}_4 \end{bmatrix} = -\begin{bmatrix} w_1 c\theta_1 + w_2 s\theta_1 + s_2\dot{\theta}_1 \\ w_3 \end{bmatrix} \tag{9.512}$$

Equation (9.512) gives $\dot{\theta}_3$ and $\dot{s}_4$ with the following expressions.

$$\begin{bmatrix} s_4\dot{\theta}_3 \\ \dot{s}_4 \end{bmatrix} = -\begin{bmatrix} c\theta_3 & -s\theta_3 \\ s\theta_3 & c\theta_3 \end{bmatrix}\begin{bmatrix} w_1 c\theta_1 + w_2 s\theta_1 + s_2\dot{\theta}_1 \\ w_3 \end{bmatrix} \Rightarrow$$

$$\dot{\theta}_3 = [w_3 s\theta_3 - (w_1 c\theta_1 + w_2 s\theta_1 + s_2\dot{\theta}_1)c\theta_3]/s_4 \tag{9.513}$$

$$\dot{s}_4 = -[w_3 c\theta_3 + (w_1 c\theta_1 + w_2 s\theta_1 + s_2\dot{\theta}_1)s\theta_3] \tag{9.514}$$

On the other hand, the angular velocity equations give $\dot{\theta}_5$, $\dot{\theta}_6$, and $\dot{\theta}_3$ as follows:

$$\dot{\theta}_5 = \omega_1^* + \dot{\theta}_1 s\theta_3 \tag{9.515}$$

In the general solution with $c\theta_5 \neq 0$,

$$\dot{\theta}_6 = (\dot{\theta}_1 c\theta_3 - \omega_3^*)/c\theta_5 \tag{9.516}$$

$$\dot{\theta}_3 = \omega_2^* - \dot{\theta}_6 s\theta_5 = (\omega_2^* c\theta_5 + \omega_3^* s\theta_5 - \dot{\theta}_1 c\theta_3 s\theta_5)/c\theta_5 \tag{9.517}$$

As noticed above, $\dot{\theta}_3$ is given twice, first by Eq. (9.513) and then by Eq. (9.517). Hence, $\dot{\theta}_1$ can be obtained from the following consistency equation.

$$[w_3 s\theta_3 - (w_1 c\theta_1 + w_2 s\theta_1 + s_2\dot{\theta}_1)c\theta_3]/s_4 = (\omega_2^* c\theta_5 + \omega_3^* s\theta_5 - \dot{\theta}_1 c\theta_3 s\theta_5)/c\theta_5 \tag{9.518}$$

Equation (9.518) can be manipulated so that

$$D_1\dot{\theta}_1 = [(w_1 c\theta_1 + w_2 s\theta_1)c\theta_3 - w_3 s\theta_3]c\theta_5 + s_4(\omega_2^* c\theta_5 + \omega_3^* s\theta_5) \tag{9.519}$$

In Eq. (9.519), the coefficient D_1 is defined as

$$D_1 = (s_4 s\theta_5 - s_2 c\theta_5)c\theta_3 \tag{9.520}$$

If $D_1 = (s_4 s\theta_5 - s_2 c\theta_5)c\theta_3 \neq 0$, Eq. (9.519) gives $\dot{\theta}_1$ as follows:

$$\dot{\theta}_1 = \frac{[(w_1 c\theta_1 + w_2 s\theta_1)c\theta_3 - w_3 s\theta_3]c\theta_5 + s_4(\omega_2^* c\theta_5 + \omega_3^* s\theta_5)}{(s_4 s\theta_5 - s_2 c\theta_5)c\theta_3} \tag{9.521}$$

Once $\dot{\theta}_1$ becomes available as such, the other joint velocities can also be obtained readily from the preceding relevant equations.

* **Special Solution with $c\theta_5 = 0$**

As a special case, if $c\theta_5 = 0$, i.e. if $\theta_5 = \sigma_5'\pi/2$ with $\sigma_5' = \pm 1$, then the joint velocities can be found through a different procedure as explained below. Here, σ_5' is to be taken according to Eq. (9.496) in order to be consistent with the position specifications.

In this special case, Eqs. (9.507)–(9.509) reduce to the following forms.

$$\dot{\theta}_5 = \omega_1^* + \dot{\theta}_1 s\theta_3 \tag{9.522}$$

$$\dot{\theta}_3 + \sigma_5'\dot{\theta}_6 = \omega_2^* \tag{9.523}$$

$$\dot{\theta}_1 c\theta_3 = \omega_3^* \tag{9.524}$$

If $c\theta_3 \neq 0$, the preceding equations give $\dot{\theta}_1$, $\dot{\theta}_5$, and $\dot{\theta}_6$ with the following expressions.

$$\dot{\theta}_1 = \omega_3^*/c\theta_3 \tag{9.525}$$

$$\dot{\theta}_5 = (\omega_1^* c\theta_3 + \omega_3^* s\theta_3)/c\theta_3 \tag{9.526}$$

$$\dot{\theta}_6 = \sigma_5'(\omega_2^* - \dot{\theta}_3) \tag{9.527}$$

Note that $\dot{\theta}_1$ and $\dot{\theta}_5$ have already been found, but $\dot{\theta}_6$ has been expressed in terms of $\dot{\theta}_3$, which is not yet known. In order to find $\dot{\theta}_3$ together with $\dot{s}_2$ and $\dot{s}_4$, Eqs. (9.504)–(9.506) can be written again as follows:

$$s_4\dot{\theta}_3 c\theta_3 + \dot{s}_4 s\theta_3 = -(w_1 c\theta_1 + w_2 s\theta_1 + s_2\dot{\theta}_1) \tag{9.528}$$

$$\dot{s}_2 = w_2 c\theta_1 - w_1 s\theta_1 + s_4\dot{\theta}_1 s\theta_3 \tag{9.529}$$

$$s_4\dot{\theta}_3 s\theta_3 - \dot{s}_4 c\theta_3 = w_3 \tag{9.530}$$

With the availability of $\dot{\theta}_1$ from Eq. (9.525), Eq. (9.529) has already given $\dot{s}_2$. As for $\dot{\theta}_3$ and $\dot{s}_4$, Eqs. (9.528) and (9.530) can be solved similarly as in the general case with the results given by Eqs. (9.513) and (9.514). That is,

$$\dot{\theta}_3 = [w_3 s\theta_3 - (w_1 c\theta_1 + w_2 s\theta_1 + s_2\dot{\theta}_1)c\theta_3]/s_4 \tag{9.531}$$

$$\dot{s}_4 = -[w_3 c\theta_3 + (w_1 c\theta_1 + w_2 s\theta_1 + s_2\dot{\theta}_1)s\theta_3] \tag{9.532}$$

After finding $\dot{\theta}_3$ as above, $\dot{\theta}_6$ can also be found by using Eq. (9.527)

9.6.8 Singularity Analysis in the Velocity Domain

Equations (9.520) and (9.521) imply that this manipulator may have a motion singularity if

$$D_1 = (s_4 s\theta_5 - s_2 c\theta_5)c\theta_3 = 0 \tag{9.533}$$

According to Eq. (9.521), the manipulator can pass through this singularity with a finite (but indefinite) value of $\dot{\theta}_1$ if the following task space compatibility condition is satisfied.

$$[(w_1 c\theta_1 + w_2 s\theta_1)c\theta_3 - w_3 s\theta_3]c\theta_5 + s_4(\omega_2^* c\theta_5 + \omega_3^* s\theta_5) = 0 \tag{9.534}$$

Owing to this condition, the joint space gains a motion freedom. Thus, $\dot{\theta}_1$ can be assigned a finite arbitrary value.

This motion singularity has several versions, which are described and explained below.

(a) First Version of the Motion Singularity

This version occurs if $c\theta_3 = 0$, i.e. if $\theta_3 = \sigma_3'\pi/2$ with $\sigma_3' = \pm 1$, while $s_4 s\theta_5 \neq s_2 c\theta_5$ and $c\theta_5 \neq 0$. When it occurs, the following special forms of Eqs. (9.504)–(9.509) must still be satisfied in order to have finite (but indefinite) values for the joint velocities.

$$s_2\dot{\theta}_1 + \sigma_3'\dot{s}_4 = -(w_1 c\theta_1 + w_2 s\theta_1) \tag{9.535}$$

$$\dot{s}_2 - \sigma_3' s_4\dot{\theta}_1 = w_2 c\theta_1 - w_1 s\theta_1 \tag{9.536}$$

$$\sigma_3' s_4\dot{\theta}_3 = w_3 \tag{9.537}$$

$$\dot{\theta}_5 = \omega_1^* + \sigma_3'\dot{\theta}_1 \tag{9.538}$$

$$\dot{\theta}_3 + \dot{\theta}_6 s\theta_5 = \omega_2^* \tag{9.539}$$

$$\dot{\theta}_6 c\theta_5 = -\omega_3^* \tag{9.540}$$

As the joint space consequence of the singularity, $\dot{\theta}_1$ can be assigned a finite arbitrary value. Then, three of the joint velocities are found as follows from Eqs. (9.535), (9.536), and (9.538) depending on the arbitrarily assigned value of $\dot{\theta}_1$:

$$\dot{s}_4 = -\sigma_3'(w_1 c\theta_1 + w_2 s\theta_1 + s_2\dot{\theta}_1) \tag{9.541}$$

$$\dot{s}_2 = w_2 c\theta_1 - w_1 s\theta_1 + \sigma_3' s_4 \dot{\theta}_1 \tag{9.542}$$

$$\dot{\theta}_5 = \omega_1^* + \sigma_3'\dot{\theta}_1 \tag{9.543}$$

The remaining three equations, i.e. Eqs. (9.537), (9.539), and (9.540), must be satisfied by two joint velocities, which are $\dot{\theta}_3$ and $\dot{\theta}_6$. This requirement can be achieved by the following values of $\dot{\theta}_6$ and $\dot{\theta}_3$, which happen to be independent of $\dot{\theta}_1$. The same equations also lead to the following task space compatibility equation, i.e. Eq. (9.546), which relates w_3 to the angular velocity components ω_2^* and ω_3^*.

$$\dot{\theta}_6 = -\omega_3^*/c\theta_5 \tag{9.544}$$

$$\dot{\theta}_3 = \omega_2^* - \dot{\theta}_6 s\theta_5 = (\omega_2^* c\theta_5 + \omega_3^* s\theta_5)/c\theta_5 \tag{9.545}$$

$$w_3 = \sigma_3' s_4(\omega_2^* c\theta_5 + \omega_3^* s\theta_5)/c\theta_5 \tag{9.546}$$

(b) Second Version of the Motion Singularity

This version occurs if $s_4 s\theta_5 = s_2 c\theta_5$ or $s_2 = s_4 \tan\theta_5$, while $c\theta_3 \neq 0$ and $c\theta_5 \neq 0$. In this version, the tip point P becomes coincident with the base frame origin O. This pose of singularity is illustrated in Figure 9.32 with a somewhat exaggerated appearance. It is exaggerated because actually it is hardly possible to bring P very close to O due to the physical shapes of the relevant links and joints. If the manipulator assumes the pose of this singularity with $s_4 s\theta_5 = s_2 c\theta_5$ and $c\theta_5 \neq 0$, Eq. (9.525) implies that the task-space compatibility condition becomes

$$s_4\omega_2^* + s_2\omega_3^* = w_3 s\theta_3 - (w_1 c\theta_1 + w_2 s\theta_1)c\theta_3 \tag{9.547}$$

(c) Third Version of the Motion Singularity

This version occurs if $s_4 s\theta_5 \neq s_2 c\theta_5$, while $c\theta_3 = c\theta_5 = 0$. This version is a rather special form of the first version, in which Eqs. (9.535)–(9.538) remain the same but Eqs. (9.539) and (9.540) reduce to the following forms.

$$\dot{\theta}_3 + \sigma_5'\dot{\theta}_6 = \omega_2^* \tag{9.548}$$

$$\omega_3^* = 0 \tag{9.549}$$

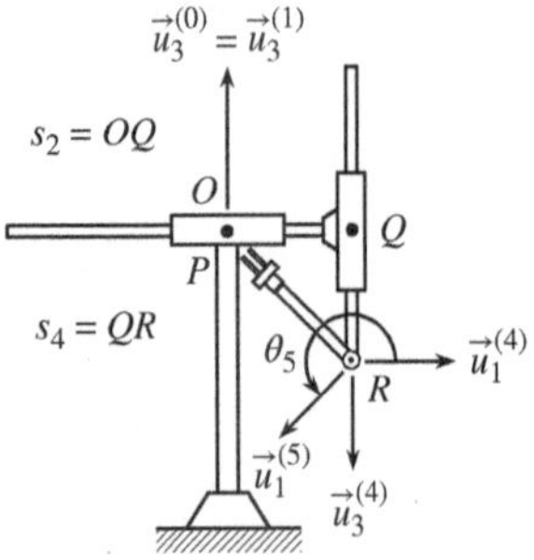

Figure 9.32 Pose of motion singularity with $s_2 = s_4 \tan\theta_5$

Equation (9.549) constitutes the task-space restriction of this version. In this version, the joint velocities $\dot{s}_2$, $\dot{s}_4$, and $\dot{\theta}_5$ are still expressed in the same way as in the first version, but $\dot{\theta}_3$ and $\dot{\theta}_6$ assume the following different expressions.

$$\dot{\theta}_3 = \sigma_3' w_3 / s_4 \tag{9.550}$$

$$\dot{\theta}_6 = \sigma_5'(\omega_2^* - \dot{\theta}_3) = \sigma_5'(s_4\omega_2^* - \sigma_3' w_3)/s_4 \tag{9.551}$$

9.7 A Deficient Puma Manipulator with Five Active Joints

As an example of a deficient manipulator, consider the special Puma manipulator shown in Figure 9.33, which is used deficiently, either deliberately or due to an actuator defect, so that the fourth joint remains fixed with $\theta_4 = 0$. Therefore, instead of studying as a new manipulator, this deficient manipulator is studied here with the necessary relevant modifications made in Section 9.1, where a regular Puma manipulator is studied.

9.7.1 Kinematic Description According to the D–H Convention

(a) Joint Variables

$$\theta_1, \theta_2, \theta_3, \theta_5, \theta_6$$

Note that $\theta_4 = 0$ because the fourth joint is fixed.

(b) Twist Angles

$$\beta_1 = 0, \beta_2 = -\pi/2, \beta_3 = 0, \beta_4 = \pi/2, \beta_5 = -\pi/2, \beta_6 = \pi/2$$

(c) Offsets

$$d_1 = 0, d_2 = OS, d_3 = 0, d_4 = ER, d_5 = 0, d_6 = RP$$

(d) Effective Link Lengths

$$b_0 = 0, b_1 = 0, b_2 = SE, b_3 = 0, b_4 = 0, b_5 = 0$$

(e) Link Frame Origins

$$O_0 = O, O_1 = O, O_2 = S, O_3 = E, O_4 = R, O_5 = R, O_6 = P$$

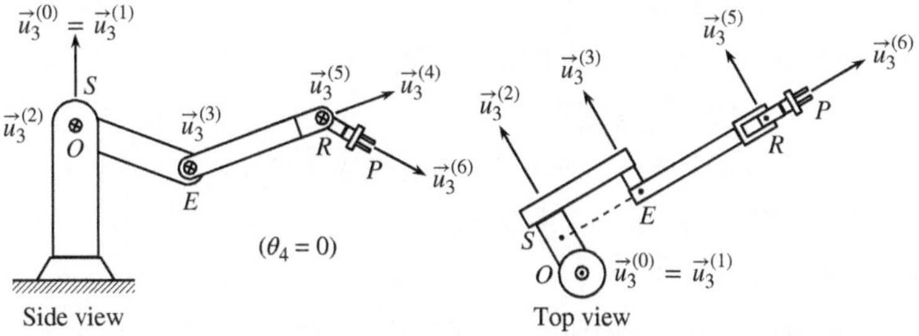

Figure 9.33 A Puma manipulator having the fourth joint fixed.

9.7.2 Forward Kinematics in the Position Domain

(a) Link-to-Link Orientation Matrices

$$\hat{C}^{(0,1)} = e^{\tilde{u}_1\beta_1}e^{\tilde{u}_3\theta_1} = e^{\tilde{u}_3\theta_1} \tag{9.552}$$

$$\hat{C}^{(1,2)} = e^{\tilde{u}_1\beta_2}e^{\tilde{u}_3\theta_2} = e^{-\tilde{u}_1\pi/2}e^{\tilde{u}_3\theta_2} \tag{9.553}$$

$$\hat{C}^{(2,3)} = e^{\tilde{u}_1\beta_3}e^{\tilde{u}_3\theta_3} = e^{\tilde{u}_3\theta_3} \tag{9.554}$$

$$\hat{C}^{(3,4)} = e^{\tilde{u}_1\beta_4}e^{\tilde{u}_3\theta_4} = e^{\tilde{u}_1\pi/2};\quad \theta_4 = 0 \tag{9.555}$$

$$\hat{C}^{(4,5)} = e^{\tilde{u}_1\beta_5}e^{\tilde{u}_3\theta_5} = e^{-\tilde{u}_1\pi/2}e^{\tilde{u}_3\theta_5} \tag{9.556}$$

$$\hat{C}^{(5,6)} = e^{\tilde{u}_1\beta_6}e^{\tilde{u}_3\theta_6} = e^{\tilde{u}_1\pi/2}e^{\tilde{u}_3\theta_6} \tag{9.557}$$

(b) Orientation Matrices of the Links with Respect to the Base Frame

$$\hat{C}_1 = \hat{C}^{(0,1)} = e^{\tilde{u}_3\theta_1} \tag{9.558}$$

$$\hat{C}_2 = \hat{C}^{(0,2)} = \hat{C}^{(0,1)}\hat{C}^{(1,2)} = e^{\tilde{u}_3\theta_1}e^{-\tilde{u}_1\pi/2}e^{\tilde{u}_3\theta_2} \Rightarrow$$

$$\hat{C}_2 = e^{\tilde{u}_3\theta_1}e^{\tilde{u}_2\theta_2}e^{-\tilde{u}_1\pi/2} \tag{9.559}$$

$$\hat{C}_3 = \hat{C}^{(0,3)} = \hat{C}^{(0,2)}\hat{C}^{(2,3)} = e^{\tilde{u}_3\theta_1}e^{\tilde{u}_2\theta_2}e^{-\tilde{u}_1\pi/2}e^{\tilde{u}_3\theta_3} \Rightarrow$$

$$\hat{C}_3 = e^{\tilde{u}_3\theta_1}e^{\tilde{u}_2\theta_2}e^{\tilde{u}_3\theta_3}e^{-\tilde{u}_1\pi/2} = e^{\tilde{u}_3\theta_1}e^{\tilde{u}_2(\theta_2+\theta_3)}e^{-\tilde{u}_1\pi/2} \Rightarrow$$

$$\hat{C}_3 = e^{\tilde{u}_3\theta_1}e^{\tilde{u}_2\theta_{23}}e^{-\tilde{u}_1\pi/2} \tag{9.560}$$

$$\hat{C}_4 = \hat{C}^{(0,4)} = \hat{C}^{(0,3)}\hat{C}^{(3,4)} = e^{\tilde{u}_3\theta_1}e^{\tilde{u}_2\theta_{23}}e^{-\tilde{u}_1\pi/2}e^{\tilde{u}_1\pi/2} \Rightarrow$$

$$\hat{C}_4 = e^{\tilde{u}_3\theta_1}e^{\tilde{u}_2\theta_{23}} \tag{9.561}$$

$$\hat{C}_5 = \hat{C}^{(0,5)} = \hat{C}^{(0,4)}\hat{C}^{(4,5)} = e^{\tilde{u}_3\theta_1}e^{\tilde{u}_2\theta_{23}}e^{-\tilde{u}_1\pi/2}e^{\tilde{u}_3\theta_5} \Rightarrow$$

$$\hat{C}_5 = e^{\tilde{u}_3\theta_1}e^{\tilde{u}_2\theta_{23}}e^{\tilde{u}_2\theta_5}e^{-\tilde{u}_1\pi/2} = e^{\tilde{u}_3\theta_1}e^{\tilde{u}_2\theta_{235}}e^{-\tilde{u}_1\pi/2} \tag{9.562}$$

$$\hat{C}_6 = \hat{C}^{(0,6)} = \hat{C}^{(0,5)}\hat{C}^{(5,6)} = e^{\tilde{u}_3\theta_1}e^{\tilde{u}_2\theta_{235}}e^{-\tilde{u}_1\pi/2}e^{\tilde{u}_1\pi/2}e^{\tilde{u}_3\theta_6} \Rightarrow$$

$$\hat{C}_6 = \hat{C} = e^{\tilde{u}_3\theta_1}e^{\tilde{u}_2\theta_{235}}e^{\tilde{u}_3\theta_6} \tag{9.563}$$

In the above equations, θ_{23} and θ_{235} are introduced as *combined joint variables*. They are defined as follows:

$$\theta_{23} = \theta_2 + \theta_3,\quad \theta_{235} = \theta_2 + \theta_3 + \theta_5 \tag{9.564}$$

(c) Location of the Wrist Point with Respect to the Base Frame

$$\vec{r} = \vec{r}_{OR} = \vec{r}_{OS} + \vec{r}_{SE} + \vec{r}_{ER} = d_2\vec{u}_3^{(2)} + b_2\vec{u}_1^{(2)} + d_4\vec{u}_3^{(4)} \tag{9.565}$$

The vector Eq. (9.565) implies the following matrix equation in the base frame.

$$\overline{r} = \overline{r}^{(0)} = d_2\overline{u}_3^{(2/0)} + b_2\overline{u}_1^{(2/0)} + d_4\overline{u}_3^{(4/0)} \tag{9.566}$$

Equation (9.566) can be manipulated as in Section 9.1. That is,

$$\overline{r} = d_2\hat{C}^{(0,2)}\overline{u}_3^{(2/2)} + b_2\hat{C}^{(0,2)}\overline{u}_1^{(2/2)} + d_4\hat{C}^{(0,4)}\overline{u}_3^{(4/4)} \Rightarrow$$

$$\overline{r} = d_2e^{\tilde{u}_3\theta_1}e^{\tilde{u}_2\theta_2}e^{-\tilde{u}_1\pi/2}\overline{u}_3 + b_2e^{\tilde{u}_3\theta_1}e^{\tilde{u}_2\theta_2}e^{-\tilde{u}_1\pi/2}\overline{u}_1 + d_4e^{\tilde{u}_3\theta_1}e^{\tilde{u}_2\theta_{23}}\overline{u}_3 \Rightarrow$$

$$\overline{r} = d_2e^{\tilde{u}_3\theta_1}\overline{u}_2 + b_2e^{\tilde{u}_3\theta_1}e^{\tilde{u}_2\theta_2}\overline{u}_1 + d_4e^{\tilde{u}_3\theta_1}e^{\tilde{u}_2\theta_{23}}\overline{u}_3 \Rightarrow$$

$$\overline{r} = e^{\tilde{u}_3\theta_1}(d_2\overline{u}_2 + b_2e^{\tilde{u}_2\theta_2}\overline{u}_1 + d_4e^{\tilde{u}_2\theta_{23}}\overline{u}_3) \Rightarrow$$

$$\overline{r} = e^{\tilde{u}_3\theta_1}[d_2\overline{u}_2 + b_2(\overline{u}_1c\theta_2 - \overline{u}_3s\theta_2) + d_4(\overline{u}_3c\theta_{23} + \overline{u}_1s\theta_{23})] \Rightarrow$$

$$\overline{r} = e^{\tilde{u}_3\theta_1}[\overline{u}_1(d_4s\theta_{23} + b_2c\theta_2) + \overline{u}_2d_2 + \overline{u}_3(d_4c\theta_{23} - b_2s\theta_2)] \tag{9.567}$$

(d) Location of the Tip Point with Respect to the Base Frame

$$\vec{p} = \vec{r}_{OP} = \vec{r}_{OR} + \vec{r}_{RP} = \vec{r} + d_6\vec{u}_3^{(6)} \tag{9.568}$$

Equation (9.575) implies the following matrix equation in the base frame.

$$\overline{p} = \overline{p}^{(0)} = \overline{r}^{(0)} + d_6\overline{u}_3^{(6/0)} = \overline{r}^{(0)} + d_6\widehat{C}^{(0,6)}\overline{u}_3^{(6/6)} \Rightarrow$$
$$\overline{p} = \overline{r} + d_6\widehat{C}\overline{u}_3 \tag{9.569}$$

When Eq. (9.563) is substituted, Eq. (9.569) becomes

$$\overline{p} = \overline{r} + d_6 e^{\tilde{u}_3\theta_1} e^{\tilde{u}_2\theta_{235}} e^{\tilde{u}_3\theta_6}\overline{u}_3 \Rightarrow$$
$$\overline{p} = \overline{r} + d_6 e^{\tilde{u}_3\theta_1} e^{\tilde{u}_2\theta_{235}}\overline{u}_3 \tag{9.570}$$

Equation (9.570) can be written in a more detailed way as shown below.

$$\overline{p} = e^{\tilde{u}_3\theta_1}(d_2\overline{u}_2 + b_2 e^{\tilde{u}_2\theta_2}\overline{u}_1 + d_4 e^{\tilde{u}_2\theta_{23}}\overline{u}_3 + d_6 e^{\tilde{u}_2\theta_{235}}\overline{u}_3) = e^{\tilde{u}_3\theta_1}\overline{p}' \tag{9.571}$$

In Eq. (9.571),

$$\overline{p}' = \overline{p}^{(1)} = d_2\overline{u}_2 + b_2 e^{\tilde{u}_2\theta_2}\overline{u}_1 + d_4 e^{\tilde{u}_2\theta_{23}}\overline{u}_3 + d_6 e^{\tilde{u}_2\theta_{235}}\overline{u}_3 \Rightarrow$$
$$\overline{p}' = \begin{bmatrix} p_1' \\ p_2' \\ p_3' \end{bmatrix} = \begin{bmatrix} d_6 s\theta_{235} + d_4 s\theta_{23} + b_2 c\theta_2 \\ d_2 \\ d_6 c\theta_{235} + d_4 c\theta_{23} - b_2 s\theta_2 \end{bmatrix} \tag{9.572}$$

9.7.3 Inverse Kinematics in the Position Domain

For a deficient manipulator such as the present one with DoF = 5, $\overline{p}$ and $\widehat{C}$ cannot both be specified completely. One of them has to be specified with only two parameters. This leads to two different cases of inverse kinematic solution. In one of the cases, $\overline{p}$ is fully specified but $\widehat{C}$ is specified partially with only two of its three independent parameters. In the other case, $\widehat{C}$ is fully specified but $\overline{p}$ is specified partially with only two of its three components.

A general peculiarity of a deficient manipulator is that its inverse kinematic solution cannot be based on the wrist point location matrix $\overline{r}$ because $\overline{r} = \overline{p} - d_6\widehat{C}\overline{u}_3$ and therefore $\overline{r}$ contains unspecified components due to either $\overline{p}$ or $\widehat{C}$.

9.7.3.1 Solution in the Case of Fully Specified Tip Point Location

This case involves operations such as painting or welding, in which the exact orientation of the end-effector (i.e. the painting or welding tool) is not very critical as long as the line segment RP is properly located and directed. In this case, Eq. (9.571) can be written as follows:

$$d_2\overline{u}_2 + b_2 e^{\tilde{u}_2\theta_2}\overline{u}_1 + d_4 e^{\tilde{u}_2\theta_{23}}\overline{u}_3 + d_6 e^{\tilde{u}_2\theta_{235}}\overline{u}_3 = \overline{p}' = e^{-\tilde{u}_3\theta_1}\overline{p} \tag{9.573}$$

Upon premultiplications by $\overline{u}_1^t$, $\overline{u}_2^t$, and $\overline{u}_3^t$, Eq. (9.573) leads to the following scalar equations.

$$d_6 s\theta_{235} + d_4 s\theta_{23} + b_2 c\theta_2 = p_1 c\theta_1 + p_2 s\theta_1 \tag{9.574}$$
$$p_2 c\theta_1 - p_1 s\theta_1 = d_2 \tag{9.575}$$
$$d_6 c\theta_{235} + d_4 c\theta_{23} - b_2 s\theta_2 = p_3 \tag{9.576}$$

Note that Eq. (9.575) contains θ_1 as the only unknown. It is very similar to Eq. (9.30) with the only difference that r_1 and r_2 are replaced with p_1 and p_2. Therefore, θ_1 can be found similarly as explained in Section 9.1 with the same *left shouldered* versus *right shouldered* multiplicity.

On the other hand, the other two Eqs. (9.574) and (9.576) contain more than two unknowns, which are θ_2, θ_{23}, and θ_{235}. Therefore, similarly as done before in Section 9.3 for the Elbow manipulator, the solution process can be continued by using the orientation Eq. (9.563).

Note that Eq. (9.563) contains the relevant three angles (θ_1, θ_{235}, θ_6) in a 3-2-3 arrangement. Therefore, the *task-space representation* of the end-effector orientation can also be expressed conveniently and judiciously by means of the 3-2-3 Euler angles (ϕ_1, ϕ_2, ϕ_3) so that

$$\hat{C} = e^{\tilde{u}_3\phi_1} e^{\tilde{u}_2\phi_2} e^{\tilde{u}_3\phi_3} \tag{9.577}$$

Equations (9.563) and (9.577) can be combined into the following equation.

$$e^{\tilde{u}_3\theta_1} e^{\tilde{u}_2\theta_{235}} e^{\tilde{u}_3\theta_6} = e^{\tilde{u}_3\phi_1} e^{\tilde{u}_2\phi_2} e^{\tilde{u}_3\phi_3} \tag{9.578}$$

Equation (9.578) implies that

$$\theta_1 = \phi_1 \tag{9.579}$$

$$\theta_{235} = \phi_2 \tag{9.580}$$

$$\theta_6 = \phi_3 \tag{9.581}$$

The above equations imply further that ϕ_2 and ϕ_3 can be specified freely as desired. Thus, θ_{235} and θ_6 are found accordingly. However, ϕ_1 cannot be specified freely. It must be equal to θ_1, which has already been found from Eq. (9.575) depending on the specified position of the tip point. In other words, $\hat{C}$ can be specified in a limited way with only two of its three parameters.

Returning back to the tip point position Eqs. (9.574) and (9.576), they can be written as follows with the availability of θ_1 and θ_{235}:

$$d_4 s\theta_{23} + b_2 c\theta_2 = x_1 = p_1 c\theta_1 + p_2 s\theta_1 - d_6 s\theta_{235} \tag{9.582}$$

$$d_4 c\theta_{23} - b_2 s\theta_2 = x_3 = p_3 - d_6 c\theta_{235} \tag{9.583}$$

Here, it is to be noted that $x_1 = r_1'$ and $x_3 = r_3' = r_3$. That is, they are the coordinates of the wrist point R in the frame $\mathcal{F}_1(O)$. Equations (9.582) and (9.583) can be solved for θ_2 and θ_3 as explained below.

The squares of Eqs. (9.582) and (9.583) can be added so that

$$d_4^2 + b_2^2 + 2b_2 d_4 (s\theta_{23} c\theta_2 - c\theta_{23} s\theta_2) = x_1^2 + x_3^2 \Rightarrow$$

$$s\theta_3 = \eta_3 = [(x_1^2 + x_3^2) - (b_2^2 + d_4^2)]/(2b_2 d_4) \tag{9.584}$$

Equation (9.584) gives $c\theta_3$ and θ_3 as follows:

$$c\theta_3 = \sigma_3\sqrt{1 - \eta_3^2}; \quad \sigma_3 = \pm 1 \tag{9.585}$$

$$\theta_3 = \text{atan}_2(\eta_3, \sigma_3\sqrt{1 - \eta_3^2}) \tag{9.586}$$

With the availability of θ_3, Eqs. (9.582) and (9.583) can be written again together as the following matrix equation.

$$\begin{bmatrix} (d_4 c\theta_3) & (b_2 + d_4 s\theta_3) \\ -(b_2 + d_4 s\theta_3) & (d_4 c\theta_3) \end{bmatrix} \begin{bmatrix} s\theta_2 \\ c\theta_2 \end{bmatrix} = \begin{bmatrix} x_1 \\ x_3 \end{bmatrix} \tag{9.587}$$

In Eq. (9.587), the determinant of the coefficient matrix is

$$D_2 = (d_4 c\theta_3)^2 + (b_2 + d_4 s\theta_3)^2 = b_2^2 + d_4^2 + 2b_2 d_4 s\theta_3 \tag{9.588}$$

If $D_2 \neq 0$, i.e. if $D_2 > 0$, Eq. (9.587) gives $s\theta_2$ and $c\theta_2$ as follows:

$$\begin{bmatrix} s\theta_2 \\ c\theta_2 \end{bmatrix} = \frac{1}{D_2} \begin{bmatrix} (d_4 c\theta_3) & -(b_2 + d_4 s\theta_3) \\ (b_2 + d_4 s\theta_3) & (d_4 c\theta_3) \end{bmatrix} \begin{bmatrix} x_1 \\ x_3 \end{bmatrix} \Rightarrow$$

$$\left. \begin{aligned} s\theta_2 &= [(d_4 c\theta_3)x_1 - (b_2 + d_4 s\theta_3)x_3]/D_2 = B_2/D_2 \\ c\theta_2 &= [(d_4 c\theta_3)x_3 + (b_2 + d_4 s\theta_3)x_1]/D_2 = A_2/D_2 \end{aligned} \right\} \tag{9.589}$$

Hence, θ_2 is found as

$$\theta_2 = \operatorname{atan}_2(B_2, A_2) \tag{9.590}$$

9.7.3.2 Solution in the Case of Fully Specified End-Effector Orientation

This case involves operations such as orienting a camera toward a remote object with a characteristic point X. In such an operation, the exact location of the tip point is not very critical as long as the camera is oriented as desired to grasp a proper view of the object. In this case, regarding Eqs. (9.579)–(9.581), all three Euler angles are freely specified as desired. Hence, the angles θ_1, θ_{235}, and θ_6 become readily available as follows:

$$\theta_1 = \phi_1 \tag{9.591}$$

$$\theta_{235} = \phi_2 \tag{9.592}$$

$$\theta_6 = \phi_3 \tag{9.593}$$

With the availability of θ_1 and θ_{235}, Eqs. (9.574) and (9.576) can be used to find θ_3 and θ_2 with the same expressions provided by Eqs. (9.586) and (9.590).

As for Eq. (9.575), it constitutes a constraint on the tip point location so that only one of the coordinates p_1 and p_2, say p_1, can be specified freely as desired. The other one must satisfy the following equation, which represents a straight line whose slope is $\tan\theta_1$.

$$p_2 = p_1 \tan\theta_1 + d_2 \sec\theta_1 \tag{9.594}$$

9.7.4 Multiplicity Analysis in the Position Domain

9.7.4.1 Analysis in the Case of Fully Specified Tip Point Location

(a) First Kind of Multiplicity

The first kind of multiplicity is associated with the sign variable σ_1 that arises in the process of finding θ_1 from Eq. (9.575). The *left shouldered* and *right shouldered* poses of the manipulator that correspond to $\sigma_1 = +1$ and $\sigma_1 = -1$ are shown in Figure 9.34. They are almost the same as those of a regular Puma manipulator with the difference that the wrist point R is replaced with the tip point P.

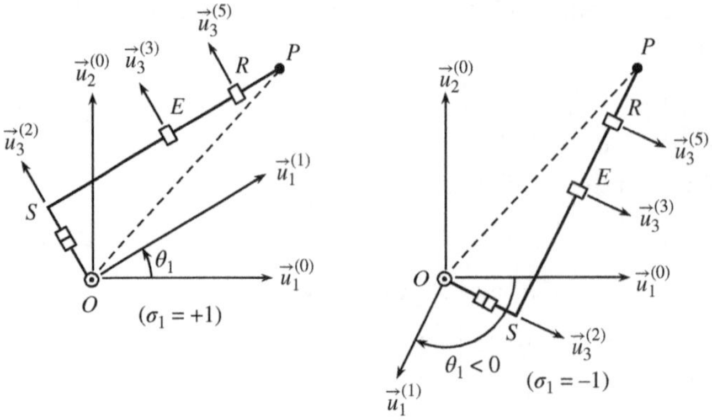

Figure 9.34 Left and right shouldered poses of the manipulator.

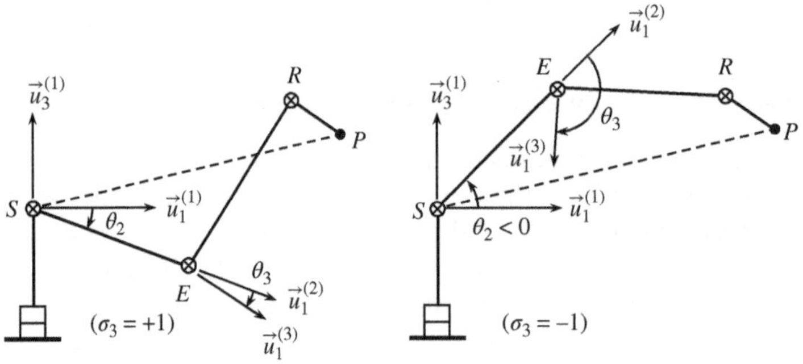

Figure 9.35 Elbow-down and elbow-up poses of the manipulator.

(b) Second Kind of Multiplicity

The second kind of multiplicity is associated with the sign variable σ_3 that arises in the process of finding θ_3 by using Eqs. (9.584)–(9.586). The *elbow-down* and *elbow-up* poses of the manipulator that correspond to $\sigma_3 = +1$ and $\sigma_3 = -1$ are shown in Figure 9.35. They are similar to those of a regular Puma manipulator with the difference that both the wrist and tip points are involved in the formation of these poses.

9.7.4.2 Analysis in the Case of Fully Specified End-Effector Orientation

In this case, the manipulator can have only one kind of multiplicity, which is the same as the second kind of multiplicity mentioned above in Section 9.7.4.1.

9.7.5 Singularity Analysis in the Position Domain

9.7.5.1 Analysis in the Case of Fully Specified Tip Point Location

In this case, the manipulator may have two kinds of position singularities, which are described and discussed below.

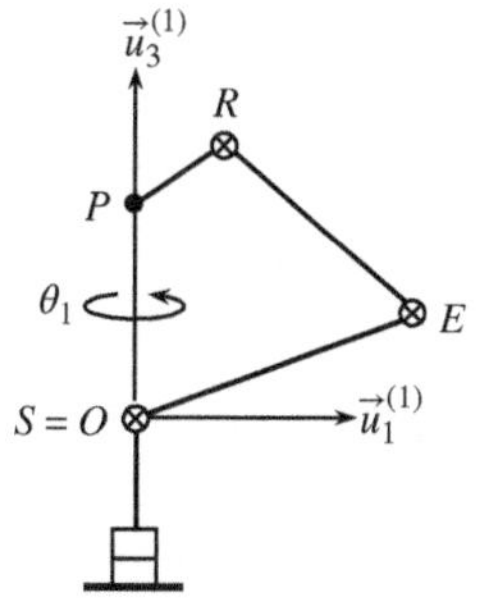

Figure 9.36 First kind of position singularity of the manipulator.

(a) First Kind of Position Singularity

Equation (9.575) implies that the first kind of position singularity occurs if the position of the end-effector is specified so that $p_1 = p_2 = 0$. The same equation also implies that such a specification can be made only if the manipulator happens to be so special that $d_2 = 0$. This singularity is illustrated in Figure 9.36. When this singularity occurs, Eq. (9.575) leaves θ_1 free. However, owing to Eq. (9.591), this freedom allows ϕ_1 to be specified freely as desired. Thus, as a useful feature, this singularity helps the orientation of the end-effector to be fully specified as long as the tip point is kept on the axis of the first joint.

(b) Second Kind of Position Singularity

Equation (9.589) implies that the second kind of position singularity occurs if $D_2 = 0$, i.e. if the determinant of the coefficient matrix vanishes. According to Eq. (9.588), D_2 can vanish only if the following equations are both satisfied.

$$b_2 = d_4 \tag{9.595}$$

$$s\theta_3 = -1 \quad \text{or} \quad \theta_3 = -\pi/2 \tag{9.596}$$

Equation (9.589) also implies that this singularity occurs if the position of the end-effector is specified in such a way that the wrist point coincides with the shoulder point. That is,

$$x_1 = x_3 = 0 \tag{9.597}$$

The above equations express the noticeable feature of this singularity that the upper and front arms of the manipulator have equal lengths and the front arm is folded completely over the upper arm. However, this singularity cannot occur exactly because of the physical shapes of the relevant links and joints. This singularity is illustrated (approximately) in Figure 9.37. Note that it is the same as that of a regular Puma manipulator. Therefore, the two singularities have the same consequences concerning the indefiniteness of θ_2 as discussed in Section 9.1.

9.7.5.2 Analysis in the Case of Fully Specified End-Effector Orientation

In this case, the manipulator can have only one kind of position singularity, which is the same as the second kind of position singularity explained above in Section 9.7.5.1.

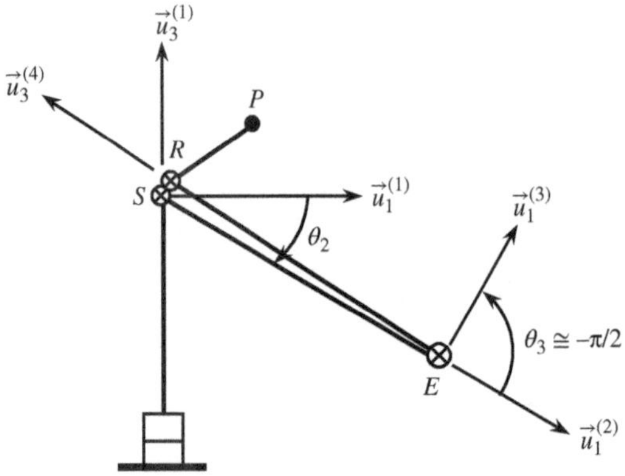

Figure 9.37 Second kind of position singularity of the manipulator.

9.7.6 Forward Kinematics in the Velocity Domain

(a) Angular Velocity of the End-Effector with Respect to the Base Frame

Let $\overline{\omega} = \overline{\omega}_6$. Then, Eq. (9.563) leads to the following expression.

$$\overline{\omega} = \dot{\theta}_1 \overline{u}_3 + \dot{\theta}_{235} e^{\tilde{u}_3\theta_1} \overline{u}_2 + \dot{\theta}_6 e^{\tilde{u}_3\theta_1} e^{\tilde{u}_2\theta_{235}} \overline{u}_3 \tag{9.598}$$

(b) Velocity of the Tip Point with Respect to the Base Frame

Let $\overline{v} = \overline{v}^{(0)} = \overline{v}_P = \dot{\overline{p}}$. Then, $\overline{v}$ can be obtained from Eq. (9.572) as follows:

$$\overline{v} = (\dot{\theta}_1 e^{\tilde{u}_3\theta_1} \tilde{u}_3)\overline{p}' + e^{\tilde{u}_3\theta_1} \dot{\overline{p}}' = e^{\tilde{u}_3\theta_1}(\dot{\overline{p}}' + \tilde{u}_3 \overline{p}' \dot{\theta}_1) = e^{\tilde{u}_3\theta_1} \overline{v}' \tag{9.599}$$

In Eq. (9.599), $\overline{v}'$ comes out with the following components.

$$v_1' = d_6 \dot{\theta}_{235} c\theta_{235} + d_4 \dot{\theta}_{23} c\theta_{23} - b_2 \dot{\theta}_2 s\theta_2 - d_2 \dot{\theta}_1 \tag{9.600}$$

$$v_2' = (d_6 s\theta_{235} + d_4 s\theta_{23} + b_2 c\theta_2)\dot{\theta}_1 \tag{9.601}$$

$$v_3' = -(d_6 \dot{\theta}_{235} s\theta_{235} + d_4 \dot{\theta}_{23} s\theta_{23} + b_2 \dot{\theta}_2 c\theta_2) \tag{9.602}$$

9.7.7 Inverse Kinematics in the Velocity Domain

For a deficient manipulator such as the present one with DoF = 5, $\overline{v}$ and $\overline{\omega}$ cannot both be specified completely. One of them has to be specified with only two parameters. This leads to two different cases of inverse kinematic solution. In one of the cases, $\overline{v}$ is fully specified but $\overline{\omega}$ is specified partially with only two Euler angle rates. In the other case, $\overline{\omega}$ is fully specified but $\overline{v}$ is specified partially with only two components.

9.7.7.1 Solution in the Case of Fully Specified Tip Point Velocity

In this case, Eq. (9.599) can be written as follows:

$$\overline{v}' = e^{-\tilde{u}_3\theta_1} \overline{v} \tag{9.603}$$

Upon premultiplications by $\overline{u}_1^t$, $\overline{u}_2^t$, and $\overline{u}_3^t$, Eq. (9.603) leads to the following scalar equations.

$$d_6 \dot{\theta}_{235} c\theta_{235} + d_4 \dot{\theta}_{23} c\theta_{23} - b_2 \dot{\theta}_2 s\theta_2 - d_2 \dot{\theta}_1 = v_1 c\theta_1 + v_2 s\theta_1 \tag{9.604}$$

$$(d_6 s\theta_{235} + d_4 s\theta_{23} + b_2 c\theta_2)\dot{\theta}_1 = p_1' \dot{\theta}_1 = v_2 c\theta_1 - v_1 s\theta_1 \tag{9.605}$$

$$d_6 \dot{\theta}_{235} s\theta_{235} + d_4 \dot{\theta}_{23} s\theta_{23} + b_2 \dot{\theta}_2 c\theta_2 = -v_3 \tag{9.606}$$

If the coefficient $p_1' = d_6 s\theta_{235} + d_4 s\theta_{23} + b_2 c\theta_2 \neq 0$, Eq. (9.605) gives $\dot{\theta}_1$ as follows:

$$\dot{\theta}_1 = (v_2 c\theta_1 - v_1 s\theta_1)/(d_6 s\theta_{235} + d_4 s\theta_{23} + b_2 c\theta_2) \tag{9.607}$$

On the other hand, the other two Eqs. (9.604) and (9.606) contain more than two unknowns, which are $\dot{\theta}_2$, $\dot{\theta}_{23}$, and $\dot{\theta}_{235}$. Therefore, similarly as done in Section 9.7.3, the solution process can be continued by using the angular velocity equation Eq. (9.598). As for the *task-space representation* of $\overline{\omega}$, it can be obtained as follows by using the 3-2-3 Euler angle expression of $\widehat{C}$ given in Eq. (9.577):

$$\overline{\omega} = \dot{\phi}_1 \overline{u}_3 + \dot{\phi}_2 e^{\tilde{u}_3 \phi_1} \overline{u}_2 + \dot{\phi}_3 e^{\tilde{u}_3 \phi_1} e^{\tilde{u}_2 \phi_2} \overline{u}_3 \tag{9.608}$$

When Eqs. (9.608) and (9.598) are compared by considering Eqs. (9.579)–(9.581), it is seen that

$$\dot{\theta}_1 = \dot{\phi}_1 \tag{9.609}$$

$$\dot{\theta}_{235} = \dot{\phi}_2 \tag{9.610}$$

$$\dot{\theta}_6 = \dot{\phi}_3 \tag{9.611}$$

The above equations imply that $\dot{\phi}_2$ and $\dot{\phi}_3$ can be specified freely as desired. Thus, $\dot{\theta}_{235}$ and $\dot{\theta}_6$ are found accordingly. However, $\dot{\phi}_1$ cannot be specified freely. It must be equal to $\dot{\theta}_1$, which has already been found depending on the specified velocity of the tip point. In other words, $\overline{\omega}$ can be specified in a limited way with only two of the three Euler angle rates.

Returning back to the tip point velocity Eqs. (9.604) and (9.606), they can be written as follows with the availability of $\dot{\theta}_1$ and $\dot{\theta}_{235}$:

$$d_4 \dot{\theta}_{23} c\theta_{23} - b_2 \dot{\theta}_2 s\theta_2 = v_1^* = v_1 c\theta_1 + v_2 s\theta_1 + d_2 \dot{\theta}_1 - d_6 \dot{\theta}_{235} c\theta_{235} \tag{9.612}$$

$$d_4 \dot{\theta}_{23} s\theta_{23} + b_2 \dot{\theta}_2 c\theta_2 = -v_3^* = -(v_3 + d_6 \dot{\theta}_{235} s\theta_{235}) \tag{9.613}$$

If $c\theta_3 \neq 0$, Eqs. (9.612) and (9.613) can be solved for $\dot{\theta}_{23}$ and $\dot{\theta}_2$ as shown below.

$$d_4 \dot{\theta}_{23}(c\theta_{23} c\theta_2 + s\theta_{23} s\theta_2) = v_1^* c\theta_2 - v_3^* s\theta_2 \Rightarrow$$

$$\dot{\theta}_{23} = (v_1^* c\theta_2 - v_3^* s\theta_2)/(d_4 c\theta_3) \tag{9.614}$$

$$b_2 \dot{\theta}_2 (c\theta_{23} c\theta_2 + s\theta_{23} s\theta_2) = -(v_3^* c\theta_{23} + v_1^* s\theta_{23}) \Rightarrow$$

$$\dot{\theta}_2 = -(v_3^* c\theta_{23} + v_1^* s\theta_{23})/(b_2 c\theta_3) \tag{9.615}$$

9.7.7.2 Solution in the Case of Fully Specified End-Effector Angular Velocity

In this case, regarding Eqs. (9.609)–(9.611), all three Euler angles rates are freely specified as desired. Hence, the angles $\dot{\theta}_1$, $\dot{\theta}_{235}$, and $\dot{\theta}_6$ become readily available as follows:

$$\dot{\theta}_1 = \dot{\phi}_1 \tag{9.616}$$

$$\dot{\theta}_{235} = \dot{\phi}_2 \tag{9.617}$$

$$\dot{\theta}_6 = \dot{\phi}_3 \tag{9.618}$$

With the availability of $\dot{\theta}_1$ and $\dot{\theta}_{235}$, Eqs. (9.604) and (9.606) can be used to find $\dot{\theta}_{23}$ and $\dot{\theta}_2$ with the same expressions provided by Eqs. (9.614) and (9.615), if $c\theta_3 \neq 0$.

As for Eq. (9.605), it constitutes a constraint on the tip point velocity so that only one of the components v_1 and v_2, say v_1, can be specified freely as desired. The other one must satisfy the following velocity constraint equation in the task space.

$$v_2 = v_1 \tan\theta_1 + (d_6 s\theta_{235} + d_4 s\theta_{23} + b_2 c\theta_2)\dot{\theta}_1 \sec\theta_1 \tag{9.619}$$

9.7.8 Singularity Analysis in the Velocity Domain

9.7.8.1 Analysis in the Case of Fully Specified Tip Point Velocity

In this case, the manipulator may have two kinds of motion singularities, which are described and discussed below.

(a) First Kind of Motion Singularity

Equation (9.607) implies that the first kind of motion singularity occurs if

$$p_1' = d_6 s\theta_{235} + d_4 s\theta_{23} + b_2 c\theta_2 = 0 \tag{9.620}$$

On the other hand, referring to Figure 9.33 and Part (d) of Section 9.7.2, it is seen that p_1' is the projection of the vector $\vec{r}_{SP}$ on the first axis of the link frame $\mathcal{F}_1(O)$. Therefore, in this singularity, $\vec{r}_{SP}$ becomes parallel to the axis of the first joint. In other words, the tip point P becomes located on a *cylindrical surface,* whose base radius is d_2. This singularity is illustrated in Figure 9.38.

If the manipulator has to pass through this singularity or keep on moving on the singularity surface as a task requirement, then, according to Eq. (9.607), $\dot{\theta}_1$ can be kept finite by planning the task motion so as to obey the following *compatibility condition.*

$$v_2 c\theta_1 = v_1 s\theta_1 \tag{9.621}$$

If the task motion obeys the above velocity restriction, Eq. (9.605) leaves $\dot{\theta}_1$ free. However, owing to Eq. (9.609), this freedom allows $\dot{\phi}_1$ to be specified freely as desired. Thus, as a useful feature, this singularity helps the angular velocity of the end-effector to be fully specified as long as $\vec{r}_{SP}$ is kept parallel to the axis of the first joint.

(b) Second Kind of Motion Singularity

Equations (9.614) and (9.615) imply that the second kind of motion singularity occurs if $\cos\theta_3 = 0$ or $\theta_3 = \sigma_3'\pi/2$ with $\sigma_3' = \pm 1$. This singularity is illustrated in Figure 9.39 in its *extended* ($\theta_3 = \pi/2$) and *folded* ($\theta_3 = -\pi/2$) versions.

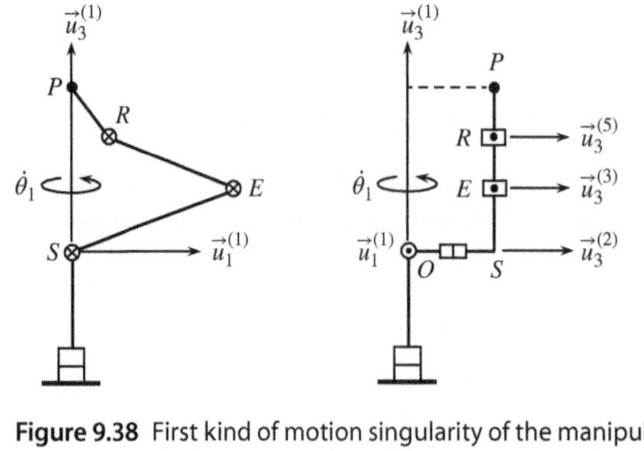

Figure 9.38 First kind of motion singularity of the manipulator.

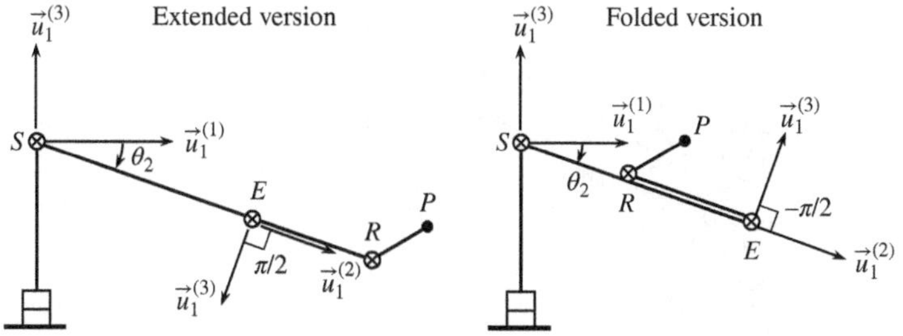

Figure 9.39 Second kind of motion singularity of the manipulator.

In this singularity, with $\theta_{23} = \theta_2 + \sigma_3'\pi/2$, Eqs. (9.612) and (9.613) can be manipulated into the following forms.

$$(b_2\dot{\theta}_2 + \sigma_3' d_4 \dot{\theta}_{23}) s\theta_2 = -v_1^* \tag{9.622}$$

$$(b_2\dot{\theta}_2 + \sigma_3' d_4 \dot{\theta}_{23}) c\theta_2 = -v_3^* \tag{9.623}$$

The above equations become consistent and thus they can give finite values for $\dot{\theta}_2$ and $\dot{\theta}_{23}$ if the following *compatibility condition* is satisfied by the planned task motion.

$$v_1^* c\theta_2 = v_3^* s\theta_2 \tag{9.624}$$

The same equations can be combined into the following equation, which relates the finite but indefinite values of $\dot{\theta}_2$ and $\dot{\theta}_3$.

$$b_2\dot{\theta}_2 + \sigma_3' d_4 \dot{\theta}_{23} = (b_2 + \sigma_3' d_4)\dot{\theta}_2 + \sigma_3' d_4 \dot{\theta}_3 = -(v_1^* s\theta_2 + v_3^* c\theta_2) \tag{9.625}$$

Equation (9.625) expresses the motion freedom induced in the joint space. Thus, as long as the task-space compatibility condition is satisfied, one of $\dot{\theta}_2$ and $\dot{\theta}_3$ can be assigned a finite arbitrary value and the other one can be found from Eq. (9.625).

9.7.8.2 Analysis in the Case of Fully Specified End-Effector Angular Velocity

In this case, the manipulator can have only one kind of motion singularity, which is the same as the second kind of motion singularity explained in Section 9.7.8.1.

9.8 A Redundant Humanoid Manipulator with Eight Joints

A humanoid manipulator is shown in Figure 9.40 with an 8-DoF kinematic model including the rotation of the torso about the vertical axis. As a spatial manipulator, it has two degrees of redundancy. The significant points of the manipulator are as follows:

O: Torso center point (origin of the base frame)
S: Shoulder point, *E*: Elbow point, *R*: Wrist point, *P*: Tip point

9.8.1 Kinematic Description According to the D–H Convention

(a) Joint Variables

$$\theta_1, \theta_2, \theta_3, \theta_4, \theta_5, \theta_6, \theta_7, \theta_8$$

They are the rotation angles about the joint axes. They are all shown in Figure 9.40.

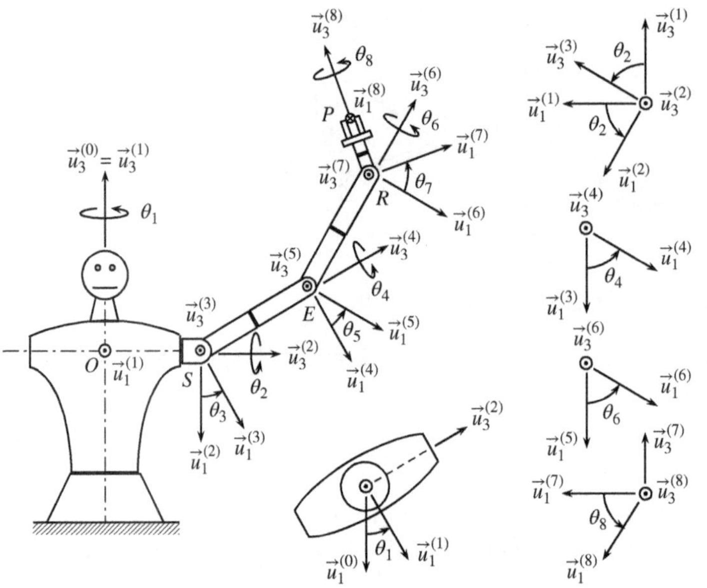

Figure 9.40 A humanoid manipulator in its front and auxiliary views.

(b) Twist Angles

$$\beta_1 = 0, \beta_2 = -\pi/2, \beta_3 = \pi/2, \beta_4 = -\pi/2,$$

$$\beta_5 = \pi/2, \beta_6 = -\pi/2, \beta_7 = \pi/2, \beta_8 = -\pi/2$$

(c) Offsets

$$d_1 = 0, d_2 = OS, d_3 = 0, d_4 = SE, d_5 = 0, d_6 = ER, d_7 = 0, d_8 = RP$$

The nonzero offsets are given the following special names.

d_2: Shoulder length (shoulder offset), d_4: Upper arm length,
d_6: Front arm length, d_8: Hand length (tip point offset)

(d) Effective Link Lengths

$$b_0 = 0, b_1 = 0, b_2 = 0, b_3 = 0, b_4 = 0, b_5 = 0, b_6 = 0, b_7 = 0$$

As noticed, they are all zero. The actual lengths of the members are represented by the offsets.

(e) Link Frame Origins

$$O_0 = O_1 = O, O_2 = O_3 = S, O_4 = O_5 = E, O_6 = O_7 = R, O_8 = P$$

9.8.2 Forward Kinematics in the Position Domain

(a) Link-to-Link Orientation Matrices

$$\hat{C}^{(0,1)} = e^{\tilde{u}_1\beta_1}e^{\tilde{u}_3\theta_1} = e^{\tilde{u}_3\theta_1} \quad (9.626)$$

$$\hat{C}^{(1,2)} = e^{\tilde{u}_1\beta_2}e^{\tilde{u}_3\theta_2} = e^{-\tilde{u}_1\pi/2}e^{\tilde{u}_3\theta_2} \quad (9.627)$$

$$\hat{C}^{(2,3)} = e^{\tilde{u}_1\beta_3}e^{\tilde{u}_3\theta_3} = e^{\tilde{u}_1\pi/2}e^{\tilde{u}_3\theta_3} \quad (9.628)$$

$$\hat{C}^{(3,4)} = e^{\tilde{u}_1\beta_4}e^{\tilde{u}_3\theta_4} = e^{-\tilde{u}_1\pi/2}e^{\tilde{u}_3\theta_4} \quad (9.629)$$

$$\hat{C}^{(4,5)} = e^{\tilde{u}_1\beta_5}e^{\tilde{u}_3\theta_5} = e^{\tilde{u}_1\pi/2}e^{\tilde{u}_3\theta_5} \quad (9.630)$$

$$\hat{C}^{(5,6)} = e^{\tilde{u}_1\beta_6}e^{\tilde{u}_3\theta_6} = e^{-\tilde{u}_1\pi/2}e^{\tilde{u}_3\theta_6} \quad (9.631)$$

$$\hat{C}^{(6,7)} = e^{\tilde{u}_1\beta_6}e^{\tilde{u}_3\theta_7} = e^{\tilde{u}_1\pi/2}e^{\tilde{u}_3\theta_7} \quad (9.632)$$

$$\hat{C}^{(7,8)} = e^{\tilde{u}_1\beta_6}e^{\tilde{u}_3\theta_8} = e^{-\tilde{u}_1\pi/2}e^{\tilde{u}_3\theta_8} \quad (9.633)$$

(b) Orientation Matrices of the Links with Respect to the Base Frame

$$\hat{C}_1 = \hat{C}^{(0,1)} = e^{\tilde{u}_3\theta_1} \quad (9.634)$$

$$\hat{C}_2 = \hat{C}^{(0,2)} = \hat{C}^{(0,1)}\hat{C}^{(1,2)} = e^{\tilde{u}_3\theta_1}e^{-\tilde{u}_1\pi/2}e^{\tilde{u}_3\theta_2} = e^{\tilde{u}_3\theta_1}e^{\tilde{u}_2\theta_2}e^{-\tilde{u}_1\pi/2} \quad (9.635)$$

$$\hat{C}_3 = \hat{C}^{(0,3)} = \hat{C}^{(0,2)}\hat{C}^{(2,3)} = e^{\tilde{u}_3\theta_1}e^{\tilde{u}_2\theta_2}e^{-\tilde{u}_1\pi/2}e^{\tilde{u}_1\pi/2}e^{\tilde{u}_3\theta_3} \Rightarrow$$

$$\hat{C}_3 = e^{\tilde{u}_3\theta_1}e^{\tilde{u}_2\theta_2}e^{\tilde{u}_3\theta_3} \quad (9.636)$$

$$\hat{C}_4 = \hat{C}^{(0,4)} = \hat{C}^{(0,3)}\hat{C}^{(3,4)} = e^{\tilde{u}_3\theta_1}e^{\tilde{u}_2\theta_2}e^{\tilde{u}_3\theta_3}e^{-\tilde{u}_1\pi/2}e^{\tilde{u}_3\theta_4} \Rightarrow$$

$$\hat{C}_4 = e^{\tilde{u}_3\theta_1}e^{\tilde{u}_2\theta_2}e^{\tilde{u}_3\theta_3}e^{\tilde{u}_2\theta_4}e^{-\tilde{u}_1\pi/2} \quad (9.637)$$

$$\hat{C}_5 = \hat{C}^{(0,5)} = \hat{C}^{(0,4)}\hat{C}^{(4,5)} = e^{\tilde{u}_3\theta_1}e^{\tilde{u}_2\theta_2}e^{\tilde{u}_3\theta_3}e^{\tilde{u}_2\theta_4}e^{-\tilde{u}_1\pi/2}e^{\tilde{u}_1\pi/2}e^{\tilde{u}_3\theta_5} \Rightarrow$$

$$\hat{C}_5 = e^{\tilde{u}_3\theta_1}e^{\tilde{u}_2\theta_2}e^{\tilde{u}_3\theta_3}e^{\tilde{u}_2\theta_4}e^{\tilde{u}_3\theta_5} \quad (9.638)$$

$$\hat{C}_6 = \hat{C}^{(0,6)} = \hat{C}^{(0,5)}\hat{C}^{(5,6)} = e^{\tilde{u}_3\theta_1}e^{\tilde{u}_2\theta_2}e^{\tilde{u}_3\theta_3}e^{\tilde{u}_2\theta_4}e^{\tilde{u}_3\theta_5}e^{-\tilde{u}_1\pi/2}e^{\tilde{u}_3\theta_6} \Rightarrow$$

$$\hat{C}_6 = e^{\tilde{u}_3\theta_1}e^{\tilde{u}_2\theta_2}e^{\tilde{u}_3\theta_3}e^{\tilde{u}_2\theta_4}e^{\tilde{u}_3\theta_5}e^{\tilde{u}_2\theta_6}e^{-\tilde{u}_1\pi/2} \quad (9.639)$$

$$\hat{C}_7 = \hat{C}^{(0,7)} = \hat{C}^{(0,6)}\hat{C}^{(6,7)} = e^{\tilde{u}_3\theta_1}e^{\tilde{u}_2\theta_2}e^{\tilde{u}_3\theta_3}e^{\tilde{u}_2\theta_4}e^{\tilde{u}_3\theta_5}e^{\tilde{u}_2\theta_6}e^{-\tilde{u}_1\pi/2}e^{\tilde{u}_1\pi/2}e^{\tilde{u}_3\theta_7} \Rightarrow$$

$$\hat{C}_7 = e^{\tilde{u}_3\theta_1}e^{\tilde{u}_2\theta_2}e^{\tilde{u}_3\theta_3}e^{\tilde{u}_2\theta_4}e^{\tilde{u}_3\theta_5}e^{\tilde{u}_2\theta_6}e^{\tilde{u}_3\theta_7} \quad (9.640)$$

$$\hat{C}_8 = \hat{C}^{(0,8)} = \hat{C}^{(0,7)}\hat{C}^{(7,8)} = e^{\tilde{u}_3\theta_1}e^{\tilde{u}_2\theta_2}e^{\tilde{u}_3\theta_3}e^{\tilde{u}_2\theta_4}e^{\tilde{u}_3\theta_5}e^{\tilde{u}_2\theta_6}e^{\tilde{u}_3\theta_7}e^{-\tilde{u}_1\pi/2}e^{\tilde{u}_3\theta_8} \Rightarrow$$

$$\hat{C}_8 = \hat{C} = e^{\tilde{u}_3\theta_1}e^{\tilde{u}_2\theta_2}e^{\tilde{u}_3\theta_3}e^{\tilde{u}_2\theta_4}e^{\tilde{u}_3\theta_5}e^{\tilde{u}_2\theta_6}e^{\tilde{u}_3\theta_7}e^{\tilde{u}_2\theta_8}e^{-\tilde{u}_1\pi/2} \quad (9.641)$$

(c) Location of the Wrist Point with Respect to the Base Frame

$$\vec{r} = \vec{r}_{OR} = \vec{r}_{OS} + \vec{r}_{SE} + \vec{r}_{ER} = d_2\vec{u}_3^{(2)} + d_4\vec{u}_3^{(4)} + d_6\vec{u}_3^{(6)} \quad (9.642)$$

The vector equation Eq. (9.642) implies the following matrix equation in the base frame.

$$\overline{r} = \overline{r}^{(0)} = d_2\overline{u}_3^{(2/0)} + d_4\overline{u}_3^{(4/0)} + d_6\overline{u}_3^{(6/0)} \quad (9.643)$$

Equation (9.643) can be manipulated as shown below.

$$\overline{r} = d_2\hat{C}^{(0,2)}\overline{u}_3^{(2/2)} + d_4\hat{C}^{(0,4)}\overline{u}_3^{(4/4)} + d_6\hat{C}^{(0,6)}\overline{u}_3^{(6/6)} \Rightarrow$$

$$\overline{r} = d_2e^{\tilde{u}_3\theta_1}e^{\tilde{u}_2\theta_2}e^{-\tilde{u}_1\pi/2}\overline{u}_3 + d_4e^{\tilde{u}_3\theta_1}e^{\tilde{u}_2\theta_2}e^{\tilde{u}_3\theta_3}e^{\tilde{u}_2\theta_4}e^{-\tilde{u}_1\pi/2}\overline{u}_3$$
$$+ d_6e^{\tilde{u}_3\theta_1}e^{\tilde{u}_2\theta_2}e^{\tilde{u}_3\theta_3}e^{\tilde{u}_2\theta_4}e^{\tilde{u}_3\theta_5}e^{\tilde{u}_2\theta_6}e^{-\tilde{u}_1\pi/2}\overline{u}_3 \Rightarrow$$

$$\overline{r} = d_2e^{\tilde{u}_3\theta_1}\overline{u}_2 + d_4e^{\tilde{u}_3\theta_1}e^{\tilde{u}_2\theta_2}e^{\tilde{u}_3\theta_3}\overline{u}_2 + d_6e^{\tilde{u}_3\theta_1}e^{\tilde{u}_2\theta_2}e^{\tilde{u}_3\theta_3}e^{\tilde{u}_2\theta_4}e^{\tilde{u}_3\theta_5}\overline{u}_2 \Rightarrow$$

$$\overline{r} = e^{\tilde{u}_3\theta_1}(d_2\overline{u}_2 + d_4e^{\tilde{u}_2\theta_2}e^{\tilde{u}_3\theta_3}\overline{u}_2 + d_6e^{\tilde{u}_2\theta_2}e^{\tilde{u}_3\theta_3}e^{\tilde{u}_2\theta_4}e^{\tilde{u}_3\theta_5}\overline{u}_2) \quad (9.644)$$

(d) Location of the Tip Point with Respect to the Base Frame

$$\vec{p} = \vec{r}_{OP} = \vec{r}_{OR} + \vec{r}_{RP} = \vec{r} + d_8\vec{u}_3^{(8)} \tag{9.645}$$

Equation (9.645) implies the following matrix equation in the base frame.

$$\overline{p} = \overline{p}^{(0)} = \overline{r}^{(0)} + d_8\overline{u}_3^{(8/0)} = \overline{r}^{(0)} + d_8\widehat{C}^{(0,8)}\overline{u}_3^{(8/8)} \Rightarrow$$

$$\overline{p} = \overline{r} + d_8\widehat{C}\overline{u}_3 \tag{9.646}$$

When Eq. (9.641) is substituted, Eq. (9.646) becomes

$$\overline{p} = \overline{r} + d_8 e^{\tilde{u}_3\theta_1} e^{\tilde{u}_2\theta_2} e^{\tilde{u}_3\theta_3} e^{\tilde{u}_2\theta_4} e^{\tilde{u}_3\theta_5} e^{\tilde{u}_2\theta_6} e^{\tilde{u}_3\theta_7} \overline{u}_2 \tag{9.647}$$

9.8.3 Inverse Kinematics in the Position Domain

Here, $\overline{p}$ and $\widehat{C}$ are specified. Hence, $\overline{r}$ also becomes specified according to Eq. (9.646). That is,

$$\overline{r} = \overline{p} - d_8\widehat{C}\overline{u}_3 \tag{9.648}$$

(a) Solution for the Arm Joint Variables

The wrist of a human being is spherical and so is the wrist of the humanoid manipulator studied here. Therefore, the inverse kinematic solution can be started from the wrist location equation Eq. (9.644) in order to find the arm joint variables (i.e. θ_1, θ_2, ..., θ_5). However, due to the redundancy of the manipulator, the number of the arm joint variables exceeds three by two. Therefore, two more requirements must be posed in addition to achieving the specified location of the wrist point. The additional requirements may be either fixing two of the joints or stipulating an optimization criterion along with fixing one or none of the joints. For example, the arm joint variables are found here in the following two operation modes with different additional requirements.

∗ First Operation Mode with Two Fixed Joints

In this operation mode, it is required to fix the first and fourth joints so that

$$\theta_1 = \theta_4 = 0 \tag{9.649}$$

As for fixing the indicated particular joints, the following arguments can be put forward:

- The torso is kept fixed because an actual human being also does so in many operations.
- One of the last three joints is not fixed in order to keep the spherical nature of the wrist.
- The fourth joint is fixed, rather than one of the second, third, and fifth joints, in order to avoid reduction in the dexterity and the working volume of the manipulator.

With the indicated fixed joints, Eq. (9.644) becomes

$$\overline{r} = d_2\overline{u}_2 + d_4 e^{\tilde{u}_2\theta_2} e^{\tilde{u}_3\theta_3} \overline{u}_2 + d_6 e^{\tilde{u}_2\theta_2} e^{\tilde{u}_3\theta_3} e^{\tilde{u}_3\theta_5} \overline{u}_2 \tag{9.650}$$

Equation (9.650) can be manipulated as follows:

$$e^{\tilde{u}_2\theta_2} e^{\tilde{u}_3\theta_3} (d_4\overline{u}_2 + d_6 e^{\tilde{u}_3\theta_5}\overline{u}_2) = \overline{x} \Rightarrow$$

$$d_4\overline{u}_2 + d_6 e^{\widetilde{u}_3\theta_5}\overline{u}_2 = e^{-\widetilde{u}_3\theta_3}e^{-\widetilde{u}_2\theta_2}\overline{x} \Rightarrow$$
$$\overline{u}_2(d_4 + d_6 c\theta_5) - \overline{u}_1(d_6 s\theta_5) = e^{-\widetilde{u}_3\theta_3}e^{-\widetilde{u}_2\theta_2}\overline{x} \tag{9.651}$$

In Eq. (9.651), $\overline{x} = \overline{r}_{SR}^{(0)}$. It is known as

$$\overline{x} = x_1\overline{u}_1 + x_2\overline{u}_2 + x_3\overline{u}_3 = \overline{r} - d_2\overline{u}_2 = r_1\overline{u}_1 + (r_2 - d_2)\overline{u}_2 + r_3\overline{u}_3 \tag{9.652}$$

In order to eliminate θ_2 and θ_3, both sides of Eq. (9.651) can be premultiplied by their transposes. This manipulation leads to the following equation for θ_5.

$$[\overline{u}_2(d_4 + d_6 c\theta_5) - \overline{u}_1(d_6 s\theta_5)]^t[\overline{u}_2(d_4 + d_6 c\theta_5) - \overline{u}_1(d_6 s\theta_5)]$$
$$= (\overline{x}^t e^{\widetilde{u}_2\theta_2}e^{\widetilde{u}_3\theta_3})(e^{-\widetilde{u}_3\theta_3}e^{-\widetilde{u}_2\theta_2}\overline{x}) = \overline{x}^t\overline{x} = x_1^2 + x_2^2 + x_3^2 \Rightarrow$$
$$(d_4 + d_6 c\theta_5)^2 + (d_6 s\theta_5)^2 = x_1^2 + x_2^2 + x_3^2 \Rightarrow$$
$$c\theta_5 = \xi_5 = [(x_1^2 + x_2^2 + x_3^2) - (d_4^2 + d_6^2)]/(2d_4 d_6) \tag{9.653}$$

Equation (9.653) gives $s\theta_5$ and θ_5 as follows:

$$s\theta_5 = \sigma_5\sqrt{1 - \xi_5^2};\quad \sigma_5 = \pm 1 \tag{9.654}$$
$$\theta_5 = \mathrm{atan}_2(\sigma_5\sqrt{1 - \xi_5^2}, \xi_5) = \sigma_5 \mathrm{atan}_2(\sqrt{1 - \xi_5^2}, \xi_5) \tag{9.655}$$

According to Eqs. (9.653) and (9.654), θ_5 can be found as a real angle if $\xi_5^2 \leq 1$. This condition can be satisfied if the specified position of the hand obeys the following working volume restriction.

$$(d_4 - d_6)^2 \leq x_1^2 + x_2^2 + x_3^2 \leq (d_4 + d_6)^2 \tag{9.656}$$

Note that Inequality (9.656) states that the wrist point is naturally confined into the space between two spherical surfaces of radii $r_{max} = d_4 + d_6$ and $r_{min} = |d_4 - d_6|$.
Note about σ_5:
In an actual human arm, the elbow angle θ_5 is limited so that $0 \leq \theta_5 < \pi$. In other words, $s\theta_5 \geq 0$ always and this fact suggests taking $\sigma_5 = +1$. Although a mechanical humanoid arm need not obey this limitation on θ_5, it can still be driven with $\sigma_5 = +1$ in order to imitate the motion of an actual arm as closely as possible. Based on this argument, $\sigma_5 = +1$ is selected here as the natural option.
With the availability of θ_5, Eq. (9.651) can be rearranged as shown below so that the remaining unknowns (θ_2 and θ_3) are distributed evenly to the two sides of the equation.

$$e^{\widetilde{u}_3\theta_3}[\overline{u}_2(d_4 + d_6 c\theta_5) - \overline{u}_1(d_6 s\theta_5)] = e^{-\widetilde{u}_2\theta_2}\overline{x} \Rightarrow$$
$$(e^{\widetilde{u}_3\theta_3}\overline{u}_2)(d_4 + d_6 c\theta_5) - (e^{\widetilde{u}_3\theta_3}\overline{u}_1)(d_6 s\theta_5) = e^{-\widetilde{u}_2\theta_2}\overline{x} \Rightarrow$$
$$(\overline{u}_2 c\theta_3 - \overline{u}_1 s\theta_3)(d_4 + d_6 c\theta_5) - (\overline{u}_1 c\theta_3 + \overline{u}_2 s\theta_3)(d_6 s\theta_5) = e^{-\widetilde{u}_2\theta_2}\overline{x} \tag{9.657}$$

Upon premultiplications by $\overline{u}_1^t$, $\overline{u}_2^t$, and $\overline{u}_3^t$, Eq. (9.657) leads to the following scalar equations.

$$(d_4 + d_6 c\theta_5)s\theta_3 + (d_6 s\theta_5)c\theta_3 = x_3 s\theta_2 - x_1 c\theta_2 \tag{9.658}$$
$$(d_4 + d_6 c\theta_5)c\theta_3 - (d_6 s\theta_5)s\theta_3 = x_2 \tag{9.659}$$
$$x_3 c\theta_2 + x_1 s\theta_2 = 0 \tag{9.660}$$

If $x_1^2 + x_3^2 \neq 0$, i.e. if the wrist point is not located on the axis of the second joint, Eq. (9.660) gives $s\theta_2$, $c\theta_2$, and θ_2 as follows with $\sigma_2 = \pm 1$.

$$s\theta_2 = \sigma_2 x_3 / \sqrt{x_1^2 + x_3^2} \tag{9.661}$$

$$c\theta_2 = -\sigma_2 x_1 / \sqrt{x_1^2 + x_3^2} \tag{9.662}$$

$$\theta_2 = \mathrm{atan}_2(\sigma_2 x_3, -\sigma_2 x_1) \tag{9.663}$$

With the availability of θ_2, Eqs. (9.658) and (9.659) can be written together as the following matrix equation.

$$\begin{bmatrix} d_4 + d_6 c\theta_5 & d_6 s\theta_5 \\ -d_6 s\theta_5 & d_4 + d_6 c\theta_5 \end{bmatrix} \begin{bmatrix} s\theta_3 \\ c\theta_3 \end{bmatrix} = \begin{bmatrix} \sigma_2 \sqrt{x_1^2 + x_3^2} \\ x_2 \end{bmatrix} \tag{9.664}$$

In Eq. (9.664), the determinant of the coefficient matrix is

$$D_3 = (d_4 + d_6 c\theta_5)^2 + (d_6 s\theta_5)^2 = d_4^2 + d_6^2 + 2 d_4 d_6 c\theta_5 \tag{9.665}$$

Referring to Eq. (9.653), it is seen that D_3 has the following alternative expression.

$$D_3 = \overline{x}^t \overline{x} = x_1^2 + x_2^2 + x_3^2 \tag{9.666}$$

If $D_3 \neq 0$, i.e. if the points R and S are not coincident, Eq. (9.664) gives $s\theta_3$ and $c\theta_3$ as follows:

$$\left.\begin{aligned} &\begin{bmatrix} s\theta_3 \\ c\theta_3 \end{bmatrix} = \frac{1}{D_3} \begin{bmatrix} d_4 + d_6 c\theta_5 & -d_6 s\theta_5 \\ d_6 s\theta_5 & d_4 + d_6 c\theta_5 \end{bmatrix} \begin{bmatrix} \sigma_2 \sqrt{x_1^2 + x_3^2} \\ x_2 \end{bmatrix} \Rightarrow \\ &s\theta_3 = [\sigma_2 (d_4 + d_6 c\theta_5) \sqrt{x_1^2 + x_3^2} - (d_6 s\theta_5) x_2] / D_3 = B_3 / D_3 \\ &c\theta_3 = [\sigma_2 (d_6 s\theta_5) \sqrt{x_1^2 + x_3^2} + (d_4 + d_6 c\theta_5) x_2] / D_3 = A_3 / D_3 \end{aligned}\right\} \tag{9.667}$$

Hence, without an extra superfluous sign variable, θ_3 is found as

$$\theta_3 = \mathrm{atan}_2(B_3, A_3) \tag{9.668}$$

* Second Operation Mode with One Fixed Joint and Minimum Potential Energy

In this operation mode, it is again required to fix the first joint so that

$$\theta_1 = 0 \tag{9.669}$$

Additionally, it is required to minimize the potential energy of the arm. Here, only the potential energy of the upper and front arms is to be taken into account because the potential energy of the hand (i.e. the end-effector) is not changeable due to its specified location and orientation. On the other hand, as far as the minimization of the potential energy is concerned, the mass m_{uf} of the upper and front arms can be assumed to be concentrated at the elbow point E. Thus, the potential energy to be minimized can be expressed as follows:

$$U = m_{uf} g h_E = m_{uf} g [\vec{u}_3^{(0)} \cdot \vec{r}_{OE}] = m_{uf} g [\overline{u}_3^t \overline{r}_{OE}^{(0)}] \tag{9.670}$$

As for the expression of $\overline{r}_{OE}^{(0)}$ with $\theta_1 = 0$, Eqs. (9.643) and (9.644) imply that

$$\overline{r}_{OE}^{(0)} = \overline{r}_{OR}^{(0)}|_{d_6=0} = d_2\overline{u}_2 + d_4 e^{\tilde{u}_2\theta_2} e^{\tilde{u}_3\theta_3}\overline{u}_2 \tag{9.671}$$

Hence,

$$U = m_{uf} g \overline{u}_3^t (d_2\overline{u}_2 + d_4 e^{\tilde{u}_2\theta_2} e^{\tilde{u}_3\theta_3}\overline{u}_2) = m_{uf} g d_4 (\overline{u}_3^t e^{\tilde{u}_2\theta_2} e^{\tilde{u}_3\theta_3}\overline{u}_2) \Rightarrow$$

$$U = m_{uf} g d_4 (\overline{u}_3^t c\theta_2 - \overline{u}_1^t s\theta_2)(\overline{u}_2 c\theta_3 - \overline{u}_1 s\theta_3) \Rightarrow$$

$$U = m_{uf} g d_4 s\theta_2 s\theta_3 \tag{9.672}$$

Noting that the factor $U^* = m_{uf} g d_4$ is constant, the potential energy can also be represented by the following *nondimensional potential energy function.*

$$V = V(\theta_2, \theta_3) = U/U^* = U/(m_{uf} g d_4) = s\theta_2 s\theta_3 \tag{9.673}$$

In this operation mode, the solution process can be started by expressing the joint variables θ_2, θ_3, and θ_5 in terms of θ_4. Afterwards, the interval $[-\pi, \pi]$ can be swept by θ_4 in order to find its optimal value that minimizes the function V.

To start the solution process, Eq. (9.644) can be written as follows by noting that $\theta_1 = 0$ and $e^{\tilde{u}_2\theta_4}\overline{u}_2 = \overline{u}_2$:

$$\overline{r} = d_2\overline{u}_2 + d_4 e^{\tilde{u}_2\theta_2} e^{\tilde{u}_3\theta_3}\overline{u}_2 + d_6 e^{\tilde{u}_2\theta_2} e^{\tilde{u}_3\theta_3} e^{\tilde{u}_2\theta_4} e^{\tilde{u}_3\theta_5}\overline{u}_2 \Rightarrow$$

$$e^{\tilde{u}_2\theta_2} e^{\tilde{u}_3\theta_3} e^{\tilde{u}_2\theta_4}(d_4\overline{u}_2 + d_6 e^{\tilde{u}_3\theta_5}\overline{u}_2) = \overline{r} - d_2\overline{u}_2 = \overline{x} \Rightarrow$$

$$d_4\overline{u}_2 + d_6 e^{\tilde{u}_3\theta_5}\overline{u}_2 = e^{-\tilde{u}_2\theta_4} e^{-\tilde{u}_3\theta_3} e^{-\tilde{u}_2\theta_2}\overline{x} \Rightarrow$$

$$\overline{u}_2(d_4 + d_6 c\theta_5) - \overline{u}_1(d_6 s\theta_5) = e^{-\tilde{u}_2\theta_4} e^{-\tilde{u}_3\theta_3} e^{-\tilde{u}_2\theta_2}\overline{x} \tag{9.674}$$

As done in the first operation mode, both sides of Eq. (9.674) can be premultiplied by their transposes in order to obtain the following equation for θ_5.

$$[\overline{u}_2(d_4 + d_6 c\theta_5) - \overline{u}_1(d_6 s\theta_5)]^t [\overline{u}_2(d_4 + d_6 c\theta_5) - \overline{u}_1(d_6 s\theta_5)]$$

$$= (\overline{x}^t e^{\tilde{u}_2\theta_2} e^{\tilde{u}_3\theta_3} e^{\tilde{u}_2\theta_4})(e^{-\tilde{u}_2\theta_4} e^{-\tilde{u}_3\theta_3} e^{-\tilde{u}_2\theta_2}\overline{x}) = \overline{x}^t\overline{x} \Rightarrow$$

$$(d_4 + d_6 c\theta_5)^2 + (d_6 s\theta_5)^2 = \overline{x}^t\overline{x} = x_1^2 + x_2^2 + x_3^2 \Rightarrow$$

$$c\theta_5 = \xi_5 = [(x_1^2 + x_2^2 + x_3^2) - (d_4^2 + d_6^2)]/(2d_4 d_6) \tag{9.675}$$

Equation (9.675) gives $s\theta_5$ and θ_5 as follows:

$$s\theta_5 = \sigma_5\sqrt{1 - \xi_5^2};\ \sigma_5 = \pm 1 \tag{9.676}$$

$$\theta_5 = \text{atan}_2(\sigma_5\sqrt{1 - \xi_5^2}, \xi_5) \tag{9.677}$$

According to Eqs. (9.675) and (9.676), θ_5 can be found as a real angle if the previously identified working volume restriction, i.e. Inequality (9.656), is again satisfied so that

$$(d_4 - d_6)^2 \le x_1^2 + x_2^2 + x_3^2 \le (d_4 + d_6)^2$$

Notes about σ_5 and θ_5:

As done in the first operation mode, the natural option of $\sigma_5 = +1$ can be selected in the second operation mode, too. As for the elbow angle θ_5, it is noticeable that it has come out to be independent of θ_4. This means that, unless the second requirement about the potential energy is stipulated, θ_4 can be changed arbitrarily without affecting θ_5 and thus the arm can be rotated freely about the line *SR* with the same elbow angle.

With the availability of θ_5, Eq. (9.674) can be rearranged as shown below.

$$e^{\tilde{u}_3\theta_3}e^{\tilde{u}_2\theta_4}[\overline{u}_2(d_4 + d_6c\theta_5) - \overline{u}_1(d_6s\theta_5)] = e^{-\tilde{u}_2\theta_2}\overline{x} \Rightarrow$$

$$(e^{\tilde{u}_3\theta_3}\overline{u}_2)(d_4 + d_6c\theta_5) - (e^{\tilde{u}_3\theta_3}e^{\tilde{u}_2\theta_4}\overline{u}_1)(d_6s\theta_5) = e^{-\tilde{u}_2\theta_2}\overline{x} \Rightarrow$$

$$(\overline{u}_2c\theta_3 - \overline{u}_1s\theta_3)(d_4 + d_6c\theta_5) - e^{\tilde{u}_3\theta_3}(\overline{u}_1c\theta_4 - \overline{u}_3s\theta_4)(d_6s\theta_5) = e^{-\tilde{u}_2\theta_2}\overline{x} \Rightarrow$$

$$(\overline{u}_2c\theta_3 - \overline{u}_1s\theta_3)(d_4 + d_6c\theta_5) - (\overline{u}_1c\theta_3 + \overline{u}_2s\theta_3)\left(d_6c\theta_4s\theta_5\right) + \overline{u}_3(d_6s\theta_4s\theta_5) = e^{-\tilde{u}_2\theta_2}\overline{x} \tag{9.678}$$

Upon premultiplications by $\overline{u}_1^t$, $\overline{u}_2^t$, and $\overline{u}_3^t$, Eq. (9.678) leads to the following scalar equations.

$$(d_4 + d_6c\theta_5)s\theta_3 + (d_6c\theta_4s\theta_5)c\theta_3 = x_3s\theta_2 - x_1c\theta_2 \tag{9.679}$$

$$(d_4 + d_6c\theta_5)c\theta_3 - (d_6c\theta_4s\theta_5)s\theta_3 = x_2 \tag{9.680}$$

$$x_3c\theta_2 + x_1s\theta_2 = d_6s\theta_4s\theta_5 \tag{9.681}$$

If $x_1^2 + x_3^2 \neq 0$, i.e. if the wrist point is not located on the axis of the second joint, Eq. (9.681) can be solved for $s\theta_2$ and $c\theta_2$ as follows with $\sigma_2 = \pm 1$.

$$s\theta_2 = (x_1y_1 + \sigma_2x_3y_2)/(x_1^2 + x_3^2) \tag{9.682}$$

$$c\theta_2 = (x_3y_1 - \sigma_2x_1y_2)/(x_1^2 + x_3^2) \tag{9.683}$$

In the above equations, y_1 and y_2 are defined as the following functions of θ_4.

$$y_1 = y_1(\theta_4) = d_6s\theta_4s\theta_5 \tag{9.684}$$

$$y_2 = y_2(\theta_4) = \sqrt{x_1^2 + x_3^2 - y_1^2} = \sqrt{x_1^2 + x_3^2 - (d_6s\theta_4s\theta_5)^2} \tag{9.685}$$

Hence, θ_2 is found as the following function of θ_4.

$$\theta_2 = f_2(\theta_4) = \mathrm{atan}_2[(x_1y_1 + \sigma_2x_3y_2), (x_3y_1 - \sigma_2x_1y_2)] \tag{9.686}$$

According to Eq. (9.685), θ_2 can be found as a real angle if

$$(d_6s\theta_4s\theta_5)^2 \leq x_1^2 + x_3^2 \tag{9.687}$$

If $s\theta_5 \neq 0$, more specifically, if $s\theta_5 > 0$ with $\sigma_5 = +1$, Inequality (9.687) implies that

$$-s\theta_4^* \leq s\theta_4 \leq s\theta_4^* \tag{9.688}$$

In Inequality (9.688),

$$s\theta_4^* = \begin{cases} \sqrt{x_1^2 + x_3^2}/(d_6s\theta_5), & \text{if } \sqrt{x_1^2 + x_3^2} < d_6s\theta_5 \\ 1, \text{otherwise} \end{cases} \tag{9.689}$$

Thus, whenever $s\theta_4^* < 1$, i.e. $\theta_4^* < \pi/2$, instead of sweeping the whole interval $[-\pi, \pi]$, the optimal solution can be obtained by sweeping an interval ϑ_4^*, which is narrower but split into three parts as shown below.

$$\vartheta_4^* = [-\pi, -(\pi - \theta_4^*)] \cup [-\theta_4^*, \theta_4^*] \cup [(\pi - \theta_4^*), \pi] \tag{9.690}$$

With the availability of θ_2, Eqs. (9.679) and (9.680) can be written together as the following matrix equation by using Eqs. (9.682) and (9.683) as well.

$$\begin{bmatrix} d_4 + d_6 c\theta_5 & d_6 c\theta_4 s\theta_5 \\ -d_6 c\theta_4 s\theta_5 & d_4 + d_6 c\theta_5 \end{bmatrix} \begin{bmatrix} s\theta_3 \\ c\theta_3 \end{bmatrix} = \begin{bmatrix} x_3 s\theta_2 - x_1 c\theta_2 \\ x_2 \end{bmatrix} = \begin{bmatrix} \sigma_2 y_2(\theta_4) \\ x_2 \end{bmatrix} \quad (9.691)$$

In Eq. (9.691), the determinant of the coefficient matrix is

$$D_3 = D_3(\theta_4) = (d_4 + d_6 c\theta_5)^2 + (d_6 c\theta_4 s\theta_5)^2 \quad (9.692)$$

Referring to Eq. (9.675) together with Eqs. (9.684) and (9.685), D_3 can also be expressed as follows:

$$\begin{aligned} &D_3 = [(d_4 + d_6 c\theta_5)^2 + (d_6 s\theta_5)^2] - [(d_6 s\theta_5)^2 - (d_6 c\theta_4 s\theta_5)^2] \Rightarrow \\ &D_3 = D_3(\theta_4) = (x_1^2 + x_2^2 + x_3^2) - (d_6 s\theta_4 s\theta_5)^2 \Rightarrow \\ &D_3 = D_3(\theta_4) = x_1^2 + x_2^2 + x_3^2 - y_1^2(\theta_4) = x_2^2 + y_2^2(\theta_4) \end{aligned} \quad (9.693)$$

If $D_3(\theta_4) \neq 0$, i.e. if x_2 and $y_2(\theta_4)$ are not both zero, Eq. (9.691) gives $s\theta_3$ and $c\theta_3$ as the following functions of θ_4.

$$\begin{bmatrix} s\theta_3 \\ c\theta_3 \end{bmatrix} = \frac{1}{D_3} \begin{bmatrix} d_4 + d_6 c\theta_5 & -d_6 c\theta_4 s\theta_5 \\ d_6 c\theta_4 s\theta_5 & d_4 + d_6 c\theta_5 \end{bmatrix} \begin{bmatrix} \sigma_2 y_2 \\ x_2 \end{bmatrix} \Rightarrow$$

$$\left.\begin{aligned} s\theta_3 &= [\sigma_2 (d_4 + d_6 c\theta_5) y_2 - (d_6 c\theta_4 s\theta_5) x_2]/D_3 = B_3(\theta_4)/D_3(\theta_4) \\ c\theta_3 &= [\sigma_2 (d_6 c\theta_4 s\theta_5) y_2 + (d_4 + d_6 c\theta_5) x_2]/D_3 = A_3(\theta_4)/D_3(\theta_4) \end{aligned}\right\} \quad (9.694)$$

Hence, without an extra superfluous sign variable, θ_3 is also found as a function of θ_4. That is,

$$\theta_3 = f_3(\theta_4) = \text{atan}_2[B_3(\theta_4), A_3(\theta_4)] \quad (9.695)$$

When the expressions of $s\theta_2$ and $s\theta_3$ given by Eqs. (9.682) and (9.694) are inserted into Eq. (9.673), the nondimensional potential energy is obtained as the following function of θ_4.

$$V = V(\theta_4) = [\sin f_2(\theta_4)][\sin f_3(\theta_4)] \quad (9.696)$$

As noticed, the dependence of $V(\theta_4)$ on θ_4 is quite complicated. Therefore, the minimization of $V(\theta_4)$ cannot be achieved analytically. However, without much difficulty, its minimization can be achieved numerically by sweeping either the whole interval $[-\pi, \pi]$ or the split interval ϑ_4^* defined by Eq. (9.690) with a sufficiently small increment $(\delta\theta_4)$ of θ_4 and spotting the minimum value of V together with the corresponding optimal value of θ_4.

After finding θ_4 as explained above, θ_2 and θ_3 can then be found readily by using Eqs. (9.686) and (9.695).

∗ An Example of the Second Operation Mode

In this example, it is assumed that the manipulator has the following geometric parameters.

$$d_2 = 0.2\,\text{m}, d_4 = 0.4\,\text{m}, d_6 = 0.4\,\text{m}$$

This example contains four cases, which demonstrate the effects of the two optional values of σ_2 and the two possible values of θ_4^* (which are $\theta_4^* < \pi/2$ and $\theta_4^* = \pi/2$).

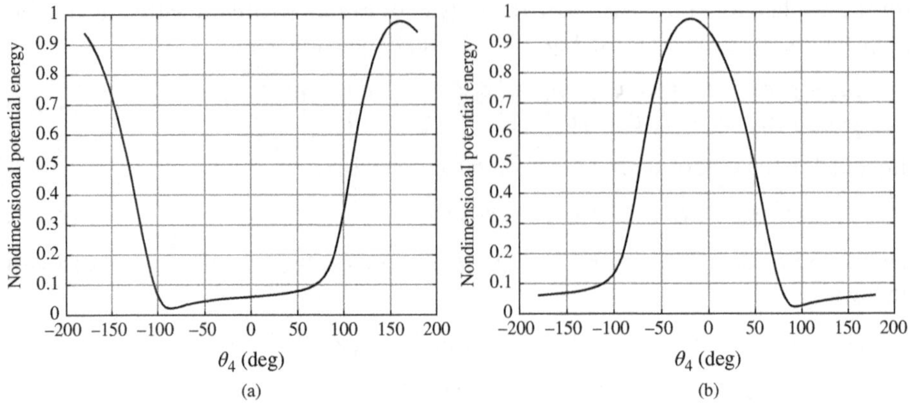

Figure 9.41 $V(\theta_4)$ vs θ_4 in the (a) first and (b) second cases.

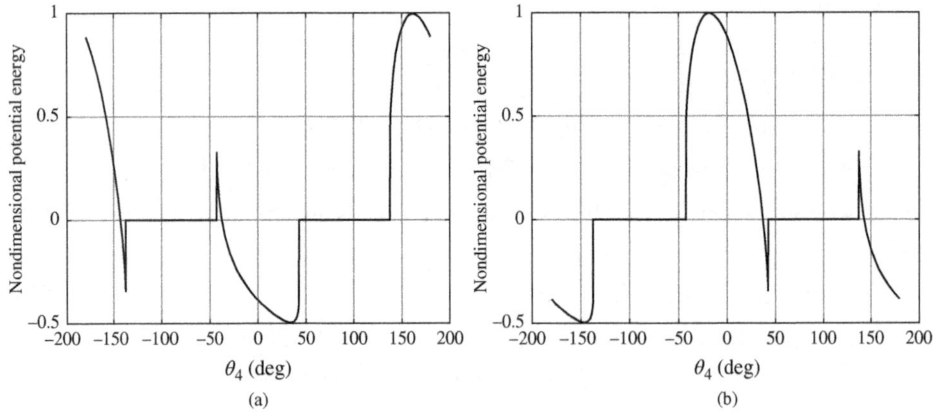

Figure 9.42 $V(\theta_4)$ vs θ_4 in the (a) third and (b) fourth cases.

In each case, the optimization is illustrated by means of a plot that shows $V(\theta_4)$ versus θ_4 as θ_4 is varied with increments of $\delta\theta_4 = 0.1°$ in its proper interval defined by θ_4^*. The optimization results obtained in the four cases are presented below and illustrated in Figure 9.41 for the first and second cases and in Figure 9.42 for the third and fourth cases. The results show that σ_2 does not affect the minimum value of V. It only affects the values of the joint variables. Indeed, it is observed that if $\sigma_2 = +1$ leads to θ_4, θ_2, and θ_3, then $\sigma_2 = -1$ leads to $\theta_4' = \theta_4 - \sigma_4'\pi$, $\theta_2' = \theta_2 - \sigma_2'\pi$, and $\theta_3' = -\theta_3$, where $\sigma_4' = \mathrm{sgn}(\theta_4)$ and $\sigma_2' = \mathrm{sgn}(\theta_2)$.

First and second cases:

$$r_1 = 0.1\text{m}, r_2 = 0.5\text{m}, r_3 = 0.4\text{m} \Rightarrow \theta_5 = 100.81°, \theta_4^* = 90°$$

$$Case\,1: \quad \sigma_2 = +1 \Rightarrow V_{min} = 0.0221 \Rightarrow \theta_4 = -86.20°, \theta_2 = 176.00°, \theta_3 = 18.47°$$

$$Case\,2: \quad \sigma_2 = -1 \Rightarrow V_{min} = 0.0221 \Rightarrow \theta_4 = 93.80°, \theta_2 = -4.00°, \theta_3 = -18.47°$$

Third and fourth cases:

$$r_1 = 0.1\text{m}, r_2 = 0.5\text{m}, r_3 = 0.2\text{m} \Rightarrow \theta_5 = 124.23°, \theta_4^* = 42.54°$$

$$\begin{aligned} &Case\,3: \quad \sigma_2 = +1 \Rightarrow V_{min} = -0.4970 \Rightarrow \theta_4 = 33.80°, \theta_2 = 61.20°, \theta_3 = -34.55° \\ &Case\,4: \quad \sigma_2 = -1 \Rightarrow V_{min} = -0.4970 \Rightarrow \theta_4 = -146.20°, \theta_2 = -118.80°, \\ &\qquad\qquad \theta_3 = 34.55° \end{aligned}$$

In these cases, with $\theta_4^* < \pi/2$, the plots are drawn by replacing the complex values of $V(\theta_4)$ with $V_0 = 0$. The function $V(\theta_4)$ assumes complex values whenever θ_4 violates the constraint defined by Inequality (9.688) with $s\theta_4^* = \sin(42.54°) = 0.676$.

(b) Solution for the Wrist Joint Variables

In order to find the wrist joint variables $(\theta_6, \theta_7, \theta_8)$ after finding the arm joint variables, the end-effector orientation equation, i.e. Eq. (9.641), can be written as follows:

$$e^{\tilde{u}_2\theta_6} e^{\tilde{u}_3\theta_7} e^{\tilde{u}_2\theta_8} = \widehat{C}^* \tag{9.697}$$

In Eq. (9.697), $\widehat{C}^*$ is known as

$$\widehat{C}^* = e^{-\tilde{u}_3\theta_5} e^{-\tilde{u}_2\theta_4} e^{-\tilde{u}_3\theta_3} e^{-\tilde{u}_2\theta_2} e^{-\tilde{u}_3\theta_1} \widehat{C} e^{\tilde{u}_1\pi/2} \tag{9.698}$$

Note that the unknown angles in Eq. (9.697) are arranged like the Euler angles of the 2-3-2 sequence. Therefore, they can be extracted by using the similar procedure explained in Chapter 3. This procedure leads to the following scalar equation pairs generated from Eq. (9.697).

$$c_{22}^* = \overline{u}_2^t \widehat{C}^* \overline{u}_2 = c\theta_7 \Rightarrow s\theta_7 = \sigma_7\sqrt{1-(c_{22}^*)^2}; \quad \sigma_7 = \pm 1 \tag{9.699}$$

$$c_{23}^* = \overline{u}_2^t \widehat{C}^* \overline{u}_3 = s\theta_7 s\theta_8; \quad c_{21}^* = \overline{u}_2^t \widehat{C}^* \overline{u}_1 = s\theta_7 c\theta_8 \tag{9.700}$$

$$c_{32}^* = \overline{u}_3^t \widehat{C}^* \overline{u}_2 = s\theta_6 s\theta_7; \quad c_{12}^* = \overline{u}_1^t \widehat{C}^* \overline{u}_2 = -c\theta_6 s\theta_7 \tag{9.701}$$

Equation Pair (9.699) gives θ_7 as

$$\theta_7 = \text{atan}_2(\sigma_7\sqrt{1-(c_{22}^*)^2}, c_{22}^*) = \sigma_7 \text{atan}_2(\sqrt{1-(c_{22}^*)^2}, c_{22}^*) \tag{9.702}$$

If $s\theta_7 \neq 0$, the other equation pairs give θ_8 and θ_6 as follows without additional superfluous sign variables.

$$\theta_8 = \text{atan}_2(\sigma_7 c_{23}^*, \sigma_7 c_{21}^*) \tag{9.703}$$

$$\theta_6 = \text{atan}_2(\sigma_7 c_{32}^*, -\sigma_7 c_{12}^*) \tag{9.704}$$

9.8.4 Multiplicity Analysis

(a) First Kind of Multiplicity

The first kind of multiplicity is associated with the sign variable σ_5 that arises in the process of finding θ_5. However, as discussed in Part (a) of Section 9.8.3, it is naturally taken as $\sigma_5 = +1$ in order to imitate an actual human arm so that $0 \leq \theta_5 < \pi$.

(b) Second Kind of Multiplicity

The second kind of multiplicity is associated with the sign variable σ_2 that arises in the process of finding θ_2.

As demonstrated in the preceding example, in the *second operation mode* with optimization, σ_2 can be taken as $\sigma_2 = +1$ without any significant loss of generality because both values of σ_2 lead to the same minimum potential energy without a clear visual distinction between the corresponding poses of the arm. The similarity of these poses is a manifestation of the *shoulder flip phenomenon,* which is similar to the *wrist flip phenomenon* of a spherical wrist. It can occur because the shoulder of a humanoid manipulator is also spherical. The shoulder flip phenomenon can be verified as explained below by considering the orientation matrix of the fourth link expressed by Eq. (9.637). With $\theta_1 = 0$, the mentioned matrix becomes

$$\hat{C}_4 = e^{\tilde{u}_2\theta_2} e^{\tilde{u}_3\theta_3} e^{\tilde{u}_2\theta_4} e^{-\tilde{u}_1\pi/2} \tag{9.705}$$

As pointed out in the preceding example, the shoulder flip phenomenon is such that, if $\sigma_2 = +1$ leads to θ_4, θ_2, and θ_3, then $\sigma_2 = -1$ leads to $\theta'_4 = \theta_4 - \sigma'_4\pi$, $\theta'_2 = \theta_2 - \sigma'_2\pi$, and $\theta'_3 = -\theta_3$, where $\sigma'_4 = \text{sgn}(\theta_4)$ and $\sigma'_2 = \text{sgn}(\theta_2)$. In order to verify this phenomenon, Eq. (9.705) is used so that $\hat{C}'_4$ is first expressed in terms of the primed angles $(\theta'_2, \theta'_3, \theta'_4)$ and then the following manipulations are carried out on $\hat{C}'_4$ by using the shifting property of the rotation matrices.

$$\hat{C}'_4 = e^{\tilde{u}_2\theta'_2} e^{\tilde{u}_3\theta'_3} e^{\tilde{u}_2\theta'_4} e^{-\tilde{u}_1\pi/2} \Rightarrow$$
$$\hat{C}'_4 = e^{\tilde{u}_2(\theta_2-\sigma'_2\pi)} e^{\tilde{u}_3(-\theta_3)} e^{\tilde{u}_2(\theta_4-\sigma'_4\pi)} e^{-\tilde{u}_1\pi/2} \Rightarrow$$
$$\hat{C}'_4 = e^{\tilde{u}_2\theta_2} e^{-\tilde{u}_2\sigma'_2\pi} e^{-\tilde{u}_3\theta_3} e^{-\tilde{u}_2\sigma'_4\pi} e^{\tilde{u}_2\theta_4} e^{-\tilde{u}_1\pi/2} \Rightarrow$$
$$\hat{C}'_4 = e^{\tilde{u}_2\theta_2} e^{\tilde{u}_3\theta_3} e^{-\tilde{u}_2(\sigma'_2+\sigma'_4)\pi} e^{\tilde{u}_2\theta_4} e^{-\tilde{u}_1\pi/2}$$

Note that, depending on σ'_2 and σ'_4, $(\sigma'_2 + \sigma'_4)\pi$ turns out to be either 0 or $\pm 2\pi$. In any case, it happens that $e^{-\tilde{u}_2(\sigma'_2+\sigma'_4)\pi} = \hat{I}$ and

$$\hat{C}'_4 = e^{\tilde{u}_2\theta_2} e^{\tilde{u}_3\theta_3} e^{\tilde{u}_2\theta_4} e^{-\tilde{u}_1\pi/2}$$

The above expression of $\hat{C}'_4$ is equal to the expression of $\hat{C}_4$ given by Eq. (9.705). This equality verifies the shoulder flip phenomenon.

On the other hand, in the *first operation mode* with two fixed joints such that $\theta_1 = \theta_4 = 0$, the two values of σ_2 lead to two distinct poses of the arm, which may be called *outward elbow pose* and *inward elbow pose.* The elbow is more distant from the torso center in the former pose but closer to it in the latter pose. In this operation mode, the shoulder flip phenomenon cannot occur because the fourth joint is fixed. The outward and inward elbow poses are defined based on the sense of the angle between the vectors $\vec{x}_E = \vec{r}_{SE}$ and $\vec{x}_R = \vec{x} = \vec{r}_{SR}$. This angle is related to the cross product between the mentioned vectors. In other words, the value of σ_2 that leads to one of the outward and inward elbow poses can be determined referring to the following cross product.

$$\vec{c}_{ER} = \vec{r}_{SE} \times \vec{r}_{SR} = \vec{r}_{SE} \times (\vec{r}_{SE} + \vec{r}_{ER}) = \vec{r}_{SE} \times \vec{r}_{ER} \tag{9.706}$$

In the base frame $\mathcal{F}_0$, the matrix equivalent of Eq. (9.706) can be written as follows:

$$\bar{c}_{ER} = \tilde{r}_{SE}\bar{r}_{SR} = \tilde{r}_{SE}\bar{r}_{ER} \Rightarrow \bar{c}_{ER} = \tilde{x}_E\bar{x} = \tilde{x}_E\bar{x}_{ER} \tag{9.707}$$

With $\theta_1 = \theta_4 = 0$, Eqs. (9.650) and (9.652) imply that

$$\overline{x} = d_4 e^{\widetilde{u}_2\theta_2} e^{\widetilde{u}_3\theta_3} \overline{u}_2 + d_6 e^{\widetilde{u}_2\theta_2} e^{\widetilde{u}_3\theta_3} e^{\widetilde{u}_3\theta_5} \overline{u}_2 \tag{9.708}$$

$$\overline{x}_E = d_4 e^{\widetilde{u}_2\theta_2} e^{\widetilde{u}_3\theta_3} \overline{u}_2 \tag{9.709}$$

$$\overline{x}_{ER} = \overline{x} - \overline{x}_E = d_6 e^{\widetilde{u}_2\theta_2} e^{\widetilde{u}_3\theta_3} e^{\widetilde{u}_3\theta_5} \overline{u}_2 \tag{9.710}$$

Referring to Chapter 1 for the properties of the skew symmetric cross product matrices, it is to be noted that

$$\widetilde{x}_E = \mathrm{ssm}(\overline{x}_E) = d_4 e^{\widetilde{u}_2\theta_2} e^{\widetilde{u}_3\theta_3} \widetilde{u}_2 e^{-\widetilde{u}_3\theta_3} e^{-\widetilde{u}_2\theta_2}$$

By using the above expression of $\widetilde{x}_E$, Eq. (9.707) can be manipulated as follows:

$$\overline{c}_{ER} = [d_4 e^{\widetilde{u}_2\theta_2} e^{\widetilde{u}_3\theta_3} \widetilde{u}_2 e^{-\widetilde{u}_3\theta_3} e^{-\widetilde{u}_2\theta_2}][d_6 e^{\widetilde{u}_2\theta_2} e^{\widetilde{u}_3\theta_3} e^{\widetilde{u}_3\theta_5} \overline{u}_2] \Rightarrow$$

$$\overline{c}_{ER} = d_4 d_6 e^{\widetilde{u}_2\theta_2} e^{\widetilde{u}_3\theta_3} \widetilde{u}_2 e^{\widetilde{u}_3\theta_5} \overline{u}_2 = d_4 d_6 e^{\widetilde{u}_2\theta_2} e^{\widetilde{u}_3\theta_3} \widetilde{u}_2 [\overline{u}_2 c\theta_5 - \overline{u}_1 s\theta_5] \Rightarrow$$

$$\overline{c}_{ER} = d_4 d_6 e^{\widetilde{u}_2\theta_2} e^{\widetilde{u}_3\theta_3} \overline{u}_3 s\theta_5 = d_4 d_6 e^{\widetilde{u}_2\theta_2} \overline{u}_3 s\theta_5 \; \overline{c}_{ER} = d_4 d_6 (\overline{u}_3 c\theta_2 + \overline{u}_1 s\theta_2) s\theta_5 \tag{9.711}$$

Upon substituting Eqs. (9.661) and (9.662), Eq. (9.711) becomes

$$\overline{c}_{ER} = \sigma_2 d_4 d_6 (\overline{u}_1 x_3 - \overline{u}_3 x_1) s\theta_5 / \sqrt{x_1^2 + x_3^2} \tag{9.712}$$

At this point, it is also to be noted that

$$\widetilde{u}_2 \overline{x} = \widetilde{u}_2 (\overline{\mathrm{u}}_1 x_1 + \overline{u}_2 x_2 + \overline{u}_3 x_3) = \overline{u}_1 x_3 - \overline{u}_3 x_1 \tag{9.713}$$

Hence, $\overline{c}_{ER}$ is finally expressed as

$$\overline{c}_{ER} = \widetilde{x}_E \overline{x} = (\sigma_2 \widetilde{u}_2 \overline{x}) d_4 d_6 \sqrt{1 - \xi_5^2} / \sqrt{x_1^2 + x_3^2} \tag{9.714}$$

As for the role of $\overline{c}_{ER}$ in determining the pose of the arm, it is illustrated in Figure 9.43. This figure implies that the arm assumes the *outward elbow pose* if the senses of the cross products $\overline{c}_{ER} = \widetilde{x}_E \overline{x}$ and $\overline{c}_{2x} = \widetilde{u}_2 \overline{x}$ are equal. According to Eq. (9.714), this equality can be realized if $\sigma_2 = +1$. Otherwise, i.e. if $\sigma_2 = -1$, the arm assumes the *inward elbow pose.*

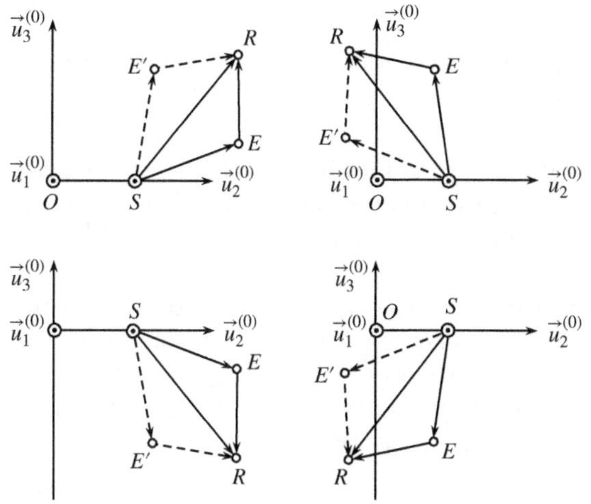

Figure 9.43 Outward elbow (E) and inward elbow (E') poses of the manipulator.

(c) Third Kind of Multiplicity

The third kind of multiplicity is associated with the sign variable σ_7 that arises in the process of finding θ_7. However, the wrist of the humanoid manipulator is spherical. Therefore, like the typical feature of any manipulator with a spherical wrist, the two values of σ_7 do not lead to visually distinct hand poses. They lead only to the wrist flip phenomenon. Therefore, σ_7 is normally taken as $\sigma_7 = +1$.

9.8.5 Singularity Analysis in the Position Domain

The humanoid manipulator considered here may have three distinct kinds of position singularities, which are described and discussed below.

(a) First Kind of Position Singularity

With $\theta_1 = 0$, Eqs. (9.660) and (9.681) imply that the first kind of position singularity occurs if

$$x_1^2 + x_3^2 = r_1^2 + r_3^2 = 0 \quad \text{or} \quad r_1 = r_3 = 0 \tag{9.715}$$

If $\theta_1 \neq 0$, Eq. (9.715) can be generalized to

$$r_1' = r_3' = 0 \tag{9.716}$$

In Eq. (9.716), r_k' denotes the kth component of the wrist point position vector $\vec{r} = \vec{r}_{OR}$ in the torso frame $\mathcal{F}_1(O)$.That is,

$$r_k' = r_k^{(1)} \quad \text{for} \quad k = 1, 2, 3 \tag{9.717}$$

Equations (9.715) or (9.716) implies that this singularity occurs if the position of the end-effector is specified in such a way that the wrist point R is located on the axis of the second joint along $\vec{u}_3^{(2)}$.

According to Eq. (9.681), another feature of this singularity is that

$$d_6 s\theta_4 s\theta_5 = 0 \tag{9.718}$$

Unless $s\theta_5 = 0$, i.e. unless the arm is fully extended or folded, Eq. (9.718) implies that

$$s\theta_4 = 0 \Rightarrow \theta_4 = 0 \quad \text{or} \quad \theta_4 = \pi \tag{9.719}$$

Thus, Eqs. (9.660) and (9.681) become the same and they both imply that θ_2 becomes arbitrary. Consequently, the arm can rotate freely about the line SR due to the arbitrariness of θ_2 while θ_4 can no longer be freely chosen due to its singularity value, which has become definite as either $\theta_4 = 0$ or $\theta_4 = \pi$.

On the other hand, the arbitrariness of θ_2 can be resolved optimally so as to minimize the potential energy of the arm. Equation (9.673) implies that this optimal resolution can be achieved if $\theta_2 = \sigma_2'\pi/2$ with $\sigma_2' = \pm 1$, while θ_3 assumes a value such that $\text{sgn}(\theta_3) = -\sigma_2'$. This value of θ_3 can be found from the following versions of Eqs. (9.679) and (9.680), which are written with $\theta_1 = 0$, $x_1 = x_3 = 0$, and $\sigma_4' = \text{sgn}(c\theta_4)$.

$$(d_4 + d_6 c\theta_5)s\theta_3 + (\sigma_4' d_6 s\theta_5)c\theta_3 = 0 \tag{9.720}$$

$$(d_4 + d_6 c\theta_5)c\theta_3 - (\sigma_4' d_6 s\theta_5)s\theta_3 = x_2 \tag{9.721}$$

If $D_3 = (d_4 + d_6 c\theta_5)^2 + (\sigma_4' d_6 s\theta_5)^2 \neq 0$, the above equations give θ_3 as

$$\theta_3 = -\text{atan}_2[(\sigma_4' d_6 s\theta_5)x_2, (d_4 + d_6 c\theta_5)x_2] \tag{9.722}$$

Recalling that $s\theta_5 > 0$, Eq. (9.722) leads to the following equation in order to achieve the optimality condition that $\text{sgn}(\theta_3) = -\sigma_2'$.

$$\sigma_2' = \sigma_4' \text{sgn}(x_2) \tag{9.723}$$

If the manipulator is operated in the first mode so that $\theta_4 = \theta_1 = 0$, i.e. $\sigma_4' = +1$, then Eq. (9.723) dictates that $\sigma_2' = \text{sgn}(x_2)$. On the other hand, if the manipulator is operated in the second mode so that $\theta_1 = 0$ but θ_4 is free, then $\sigma_4' = +1$ does not violate generality and therefore $\sigma_2' = \text{sgn}(x_2)$ can be used again.
As the extreme version of this singularity, if $x_2 = 0$, too, i.e. if $x_1 = x_2 = x_3 = 0$, then Eq. (9.722) implies that θ_3 also becomes arbitrary along with θ_2. In such a case, the optimal (minimum potential energy) resolution of this double-fold arbitrariness can be achieved with $\theta_2 = \pi/2$ and $\theta_3 = -\pi/2$ or vice versa.

(b) Second Kind of Position Singularity

With $\theta_1 = 0$ and an arbitrary value of θ_4, Eq. (9.692) implies that the second kind of position singularity occurs if

$$D_3 = D_3(\theta_4) = (d_4 + d_6 c\theta_5)^2 + (d_6 c\theta_4 s\theta_5)^2 = 0 \tag{9.724}$$

Equation (9.724) can be satisfied only if the following special features occur simultaneously.

$$d_4 = d_6 \tag{9.725}$$

$$c\theta_5 = -1, \text{i.e. } \theta_5 = \pi \tag{9.726}$$

When this singularity occurs, Eqs. (9.725) and (9.726) indicate that the wrist point R coincides with the shoulder point S. In other words,

$$x_1^2 + x_2^2 + x_3^2 = 0 \Rightarrow x_1 = x_2 = x_3 = 0 \Rightarrow r_1 = r_3 = 0, r_2 = d_2 \tag{9.727}$$

Moreover, Eqs. (9.679) and (9.680) reduce to $0 = 0$ and therefore θ_3 becomes indefinite. Thus, as long as R is kept coincident with S, it is possible to assign an arbitrary value to θ_3. Note that this singularity is nothing but the extreme version of the first kind of singularity considered in Part (a) of this section. Therefore, the arbitrariness of θ_3 can be resolved similarly.

(c) Third Kind of Position Singularity

Equations (9.703) and (9.704) imply that the third kind of position singularity occurs if $s\theta_7 = 0$. Except the differences in the subscripts of the relevant angles, this singularity has the same appearance as the third kind of position singularity of the Puma manipulator, which is explained in Section 9.1.3 and illustrated in Figure 9.8.

9.8.6 Forward Kinematics in the Velocity Domain

(a) Angular Velocity of the End-Effector with Respect to the Base Frame

Let $\overline{\omega} = \overline{\omega}_6$. Then, Eq. (9.641) leads to the following expression.

$$\begin{aligned}\overline{\omega} &= \dot{\theta}_1 \overline{u}_3 + \dot{\theta}_2 e^{\tilde{u}_3\theta_1}\overline{u}_2 + \dot{\theta}_3 e^{\tilde{u}_3\theta_1}e^{\tilde{u}_2\theta_2}\overline{u}_3 + \dot{\theta}_4 e^{\tilde{u}_3\theta_1}e^{\tilde{u}_2\theta_2}e^{\tilde{u}_3\theta_3}\overline{u}_2 + \dot{\theta}_5 e^{\tilde{u}_3\theta_1}e^{\tilde{u}_2\theta_2}e^{\tilde{u}_3\theta_3}e^{\tilde{u}_2\theta_4}\overline{u}_3\\ &+ \dot{\theta}_6 e^{\tilde{u}_3\theta_1}e^{\tilde{u}_2\theta_2}e^{\tilde{u}_3\theta_3}e^{\tilde{u}_2\theta_4}e^{\tilde{u}_3\theta_5}\overline{u}_2 + \dot{\theta}_7 e^{\tilde{u}_3\theta_1}e^{\tilde{u}_2\theta_2}e^{\tilde{u}_3\theta_3}e^{\tilde{u}_2\theta_4}e^{\tilde{u}_3\theta_5}e^{\tilde{u}_2\theta_6}\overline{u}_3\\ &+ \dot{\theta}_8 e^{\tilde{u}_3\theta_1}e^{\tilde{u}_2\theta_2}e^{\tilde{u}_3\theta_3}e^{\tilde{u}_2\theta_4}e^{\tilde{u}_3\theta_5}e^{\tilde{u}_2\theta_6}e^{\tilde{u}_3\theta_7}\overline{u}_2\end{aligned} \tag{9.728}$$

(b) Velocity of the Wrist Point with Respect to the Base Frame

Let $\overline{w} = \overline{w}^{(0)} = \overline{v}_R = \dot{\overline{r}}$. Then, $\overline{w}$ can be obtained from Eq. (9.644) as follows:

$$\begin{aligned}\overline{w} &= \dot{\theta}_1 e^{\tilde{u}_3\theta_1}\tilde{u}_3(d_2\overline{u}_2 + d_4 e^{\tilde{u}_2\theta_2}e^{\tilde{u}_3\theta_3}\overline{u}_2 + d_6 e^{\tilde{u}_2\theta_2}e^{\tilde{u}_3\theta_3}e^{\tilde{u}_2\theta_4}e^{\tilde{u}_3\theta_5}\overline{u}_2)\\ &+e^{\tilde{u}_3\theta_1}(d_4\dot{\theta}_2 e^{\tilde{u}_2\theta_2}\tilde{u}_2 e^{\tilde{u}_3\theta_3}\overline{u}_2 + d_6\dot{\theta}_2 e^{\tilde{u}_2\theta_2}\tilde{u}_2 e^{\tilde{u}_3\theta_3}e^{\tilde{u}_2\theta_4}e^{\tilde{u}_3\theta_5}\overline{u}_2)\\ &+e^{\tilde{u}_3\theta_1}(d_4\dot{\theta}_3 e^{\tilde{u}_2\theta_2}e^{\tilde{u}_3\theta_3}\tilde{u}_3\overline{u}_2 + d_6\dot{\theta}_3 e^{\tilde{u}_2\theta_2}e^{\tilde{u}_3\theta_3}\tilde{u}_3 e^{\tilde{u}_2\theta_4}e^{\tilde{u}_3\theta_5}\overline{u}_2)\\ &+e^{\tilde{u}_3\theta_1}(d_6\dot{\theta}_4 e^{\tilde{u}_2\theta_2}e^{\tilde{u}_3\theta_3}e^{\tilde{u}_2\theta_4}\tilde{u}_2 e^{\tilde{u}_3\theta_5}\overline{u}_2 + d_6\dot{\theta}_5 e^{\tilde{u}_2\theta_2}e^{\tilde{u}_3\theta_3}e^{\tilde{u}_2\theta_4}e^{\tilde{u}_3\theta_5}\tilde{u}_3\overline{u}_2) \Rightarrow\\ \overline{w} &= e^{\tilde{u}_3\theta_1}(-d_2\overline{u}_1 + d_4\tilde{u}_3 e^{\tilde{u}_2\theta_2}e^{\tilde{u}_3\theta_3}\overline{u}_2 + d_6\tilde{u}_3 e^{\tilde{u}_2\theta_2}e^{\tilde{u}_3\theta_3}e^{\tilde{u}_2\theta_4}e^{\tilde{u}_3\theta_5}\overline{u}_2)\dot{\theta}_1\\ &+e^{\tilde{u}_3\theta_1}(d_4 e^{\tilde{u}_2\theta_2}\tilde{u}_2 e^{\tilde{u}_3\theta_3}\overline{u}_2 + d_6 e^{\tilde{u}_2\theta_2}\tilde{u}_2 e^{\tilde{u}_3\theta_3}e^{\tilde{u}_2\theta_4}e^{\tilde{u}_3\theta_5}\overline{u}_2)\dot{\theta}_2\\ &+e^{\tilde{u}_3\theta_1}(-d_4 e^{\tilde{u}_2\theta_2}e^{\tilde{u}_3\theta_3}\overline{u}_1 + d_6 e^{\tilde{u}_2\theta_2}e^{\tilde{u}_3\theta_3}\tilde{u}_3 e^{\tilde{u}_2\theta_4}e^{\tilde{u}_3\theta_5}\overline{u}_2)\dot{\theta}_3\\ &+e^{\tilde{u}_3\theta_1}(d_6\dot{\theta}_4 e^{\tilde{u}_2\theta_2}e^{\tilde{u}_3\theta_3}e^{\tilde{u}_2\theta_4}\tilde{u}_2 e^{\tilde{u}_3\theta_5}\overline{u}_2 - d_6\dot{\theta}_5 e^{\tilde{u}_2\theta_2}e^{\tilde{u}_3\theta_3}e^{\tilde{u}_2\theta_4}e^{\tilde{u}_3\theta_5}\overline{u}_1) \Rightarrow\\ \overline{w} &= e^{\tilde{u}_3\theta_1}[\tilde{u}_3 e^{\tilde{u}_2\theta_2}e^{\tilde{u}_3\theta_3}e^{\tilde{u}_2\theta_4}(d_4\overline{u}_2 + d_6 e^{\tilde{u}_3\theta_5}\overline{u}_2) - d_2\overline{u}_1]\dot{\theta}_1\\ &+e^{\tilde{u}_3\theta_1}e^{\tilde{u}_2\theta_2}\tilde{u}_2 e^{\tilde{u}_3\theta_3}e^{\tilde{u}_2\theta_4}(d_4\overline{u}_2 + d_6 e^{\tilde{u}_3\theta_5}\overline{u}_2)\dot{\theta}_2\\ &+e^{\tilde{u}_3\theta_1}e^{\tilde{u}_2\theta_2}e^{\tilde{u}_3\theta_3}(d_6\tilde{u}_3 e^{\tilde{u}_2\theta_4}e^{\tilde{u}_3\theta_5}\overline{u}_2 - d_4\overline{u}_1)\dot{\theta}_3\\ &+(e^{\tilde{u}_3\theta_1}e^{\tilde{u}_2\theta_2}e^{\tilde{u}_3\theta_3}e^{\tilde{u}_2\theta_4}\tilde{u}_2 e^{\tilde{u}_3\theta_5}\overline{u}_2)d_6\dot{\theta}_4\\ &-(e^{\tilde{u}_3\theta_1}e^{\tilde{u}_2\theta_2}e^{\tilde{u}_3\theta_3}e^{\tilde{u}_2\theta_4}e^{\tilde{u}_3\theta_5}\overline{u}_1)d_6\dot{\theta}_5\end{aligned} \tag{9.729}$$

(c) Velocity of the Tip Point with Respect to the Base Frame

Let $\overline{v} = \overline{v}^{(0)} = \overline{v}_P = \dot{\overline{p}}$. Then, $\overline{v}$ can be obtained from Eq. (9.646) as follows:

$$\overline{v} = \overline{w} + d_8\tilde{\omega}\hat{C}\overline{u}_3 \tag{9.730}$$

9.8.7 Inverse Kinematics in the Velocity Domain

Since $\overline{p}$ and $\hat{C}$ are specified as functions of time, the velocities $\overline{v}$, $\overline{\omega}$, and $\overline{w}$ are also known. They are obtained as shown below.

$$\overline{v} = \dot{\overline{p}}, \quad \overline{\omega} = \text{colm}(\dot{\hat{C}}\hat{C}^t); \quad \overline{w} = \overline{v} - d_8\tilde{\omega}\hat{C}\overline{u}_3 \tag{9.731}$$

(a) Solution for the Arm Joint Velocities

The wrist of a human being is spherical and so is the wrist of the humanoid manipulator studied here. Therefore, the inverse velocity solution can be started from the wrist point velocity equation Eq. (9.729) in order to find the arm joint velocities (i.e. $\dot{\theta}_1, \dot{\theta}_2, \ldots, \dot{\theta}_5$). However, due to the redundancy of the manipulator, the number of the arm joint velocities exceeds three by two. Therefore, two more requirements must be posed in addition to achieving the specified velocity of the wrist point. The additional requirements may be either fixing two of the joints or stipulating an optimization criterion along with fixing one or none of the joints. For example, the arm joint velocities are found here in the following two operation modes with different additional requirements.

∗ **First Operation Mode with Two Fixed Joints**

In this operation mode, it is required to fix the first and fourth joints so that

$$\theta_1 = \theta_4 = 0 \text{ and } \dot{\theta}_1 = \dot{\theta}_4 = 0 \tag{9.732}$$

With this requirement, Eq. (9.729) can be written in its reduced form and then manipulated as shown below.

$$\begin{aligned}
\overline{w} &= e^{\tilde{u}_2\theta_2}\tilde{u}_2 e^{\tilde{u}_3\theta_3}(d_4\overline{u}_2 + d_6 e^{\tilde{u}_3\theta_5}\overline{u}_2)\dot{\theta}_2 \\
&\quad + e^{\tilde{u}_2\theta_2}e^{\tilde{u}_3\theta_3}(d_6\tilde{u}_3 e^{\tilde{u}_3\theta_5}\overline{u}_2 - d_4\overline{u}_1)\dot{\theta}_3 - (e^{\tilde{u}_2\theta_2}e^{\tilde{u}_3\theta_3}e^{\tilde{u}_3\theta_5}\overline{u}_1)d_6\dot{\theta}_5 \Rightarrow \\
\overline{w} &= e^{\tilde{u}_2\theta_2}\tilde{u}_2 e^{\tilde{u}_3\theta_3}[\overline{u}_2(d_4 + d_6 c\theta_5) - \overline{u}_1(d_6 s\theta_5)]\dot{\theta}_2 \\
&\quad + e^{\tilde{u}_2\theta_2}e^{\tilde{u}_3\theta_3}[(d_6 e^{\tilde{u}_3\theta_5}\tilde{u}_3\overline{u}_2 - d_4\overline{u}_1)\dot{\theta}_3 - (e^{\tilde{u}_3\theta_5}\overline{u}_1)d_6\dot{\theta}_5] \Rightarrow \\
\overline{w} &= e^{\tilde{u}_2\theta_2}\tilde{u}_2 e^{\tilde{u}_3\theta_3}[\overline{u}_2(d_4 + d_6 c\theta_5) - \overline{u}_1(d_6 s\theta_5)]\dot{\theta}_2 \\
&\quad - e^{\tilde{u}_2\theta_2}e^{\tilde{u}_3\theta_3}[\overline{u}_1(d_4\dot{\theta}_3) + (e^{\tilde{u}_3\theta_5}\overline{u}_1)d_6(\dot{\theta}_3 + \dot{\theta}_5)] \Rightarrow \\
\overline{w} &= e^{\tilde{u}_2\theta_2}(e^{\tilde{u}_3\theta_3}e^{-\tilde{u}_3\theta_3})\tilde{u}_2 e^{\tilde{u}_3\theta_3}[\overline{u}_2(d_4 + d_6 c\theta_5) - \overline{u}_1(d_6 s\theta_5)]\dot{\theta}_2 \\
&\quad - e^{\tilde{u}_2\theta_2}e^{\tilde{u}_3\theta_3}[\overline{u}_1(d_4\dot{\theta}_3 + d_6\dot{\theta}_{35}c\theta_5) + \overline{u}_2(d_6\dot{\theta}_{35}s\theta_5)] \Rightarrow \\
\overline{w} &= e^{\tilde{u}_2\theta_2}e^{\tilde{u}_3\theta_3}(e^{-\tilde{u}_3\theta_3}\tilde{u}_2 e^{\tilde{u}_3\theta_3})[\overline{u}_2(d_4 + d_6 c\theta_5) - \overline{u}_1(d_6 s\theta_5)]\dot{\theta}_2 \\
&\quad - e^{\tilde{u}_2\theta_2}e^{\tilde{u}_3\theta_3}[\overline{u}_1(d_4\dot{\theta}_3 + d_6\dot{\theta}_{35}c\theta_5) + \overline{u}_2(d_6\dot{\theta}_{35}s\theta_5)]
\end{aligned}$$

Note that

$$e^{-\tilde{u}_3\theta_3}\tilde{u}_2 e^{\tilde{u}_3\theta_3} = \text{ssm}(e^{-\tilde{u}_3\theta_3}\overline{u}_2) = \text{ssm}(\overline{u}_2 c\theta_3 + \overline{u}_1 s\theta_3) = \tilde{u}_2 c\theta_3 + \tilde{u}_1 s\theta_3$$

Hence, the expression of $\overline{w}$ can be manipulated further so that

$$\begin{aligned}
\overline{w} &= e^{\tilde{u}_2\theta_2}e^{\tilde{u}_3\theta_3}(\tilde{u}_2 c\theta_3 + \tilde{u}_1 s\theta_3)[\overline{u}_2(d_4 + d_6 c\theta_5) - \overline{u}_1(d_6 s\theta_5)]\dot{\theta}_2 \\
&\quad - e^{\tilde{u}_2\theta_2}e^{\tilde{u}_3\theta_3}[\overline{u}_1(d_4\dot{\theta}_3 + d_6\dot{\theta}_{35}c\theta_5) + \overline{u}_2(d_6\dot{\theta}_{35}s\theta_5)] \Rightarrow \\
\overline{w} &= e^{\tilde{u}_2\theta_2}e^{\tilde{u}_3\theta_3}\overline{u}_3[(d_4 + d_6 c\theta_5)s\theta_3 + (d_6 s\theta_5 c\theta_3)]\dot{\theta}_2 \\
&\quad - e^{\tilde{u}_2\theta_2}e^{\tilde{u}_3\theta_3}[\overline{u}_1(d_4\dot{\theta}_3 + d_6\dot{\theta}_{35}c\theta_5) + \overline{u}_2(d_6\dot{\theta}_{35}s\theta_5)] \Rightarrow \\
\overline{w} &= e^{\tilde{u}_2\theta_2}e^{\tilde{u}_3\theta_3}[\overline{u}_3(d_4 s\theta_3 + d_6 s\theta_{35})\dot{\theta}_2 \\
&\quad - \overline{u}_1(d_4\dot{\theta}_3 + d_6\dot{\theta}_{35}c\theta_5) - \overline{u}_2(d_6\dot{\theta}_{35}s\theta_5)]
\end{aligned} \tag{9.733}$$

In Eq. (9.733),

$$\theta_{35} = \theta_3 + \theta_5 \text{ and } \dot{\theta}_{35} = \dot{\theta}_3 + \dot{\theta}_5 \tag{9.734}$$

Equation (9.733) can also be written as

$$\overline{u}_1(d_4\dot{\theta}_3 + d_6\dot{\theta}_{35}c\theta_5) + \overline{u}_2(d_6\dot{\theta}_{35}s\theta_5) - \overline{u}_3(d_4s\theta_3 + d_6s\theta_{35})\dot{\theta}_2 = -\overline{w}^* \quad (9.735)$$

In Eq. (9.735), $\overline{w}^*$ is known as

$$\overline{w}^* = e^{-\widetilde{u}_3\theta_3}e^{-\widetilde{u}_2\theta_2}\overline{w} \quad (9.736)$$

Upon premultiplications by $\overline{u}_1^t$, $\overline{u}_2^t$, and $\overline{u}_3^t$, Eq. (9.735) leads to the following scalar equations.

$$d_4\dot{\theta}_3 + (d_6c\theta_5)\dot{\theta}_{35} = -w_1^* \quad (9.737)$$

$$(d_6s\theta_5)\dot{\theta}_{35} = -w_2^* \quad (9.738)$$

$$(d_4s\theta_3 + d_6s\theta_{35})\dot{\theta}_2 = w_3^* \quad (9.739)$$

If $(d_4s\theta_3 + d_6s\theta_{35}) \neq 0$, Eq. (9.739) gives $\dot{\theta}_2$ as

$$\dot{\theta}_2 = w_3^*/(d_4s\theta_3 + d_6s\theta_{35}) \quad (9.740)$$

If $s\theta_5 \neq 0$, Eq. (9.738) gives $\dot{\theta}_{35}$ as

$$\dot{\theta}_{35} = -w_2^*/(d_6s\theta_5) \quad (9.741)$$

Then, Eqs. (9.737) and (9.741) give $\dot{\theta}_3$ as

$$\dot{\theta}_3 = -[w_1^* + (d_6c\theta_5)\dot{\theta}_{35}]/d_4 \Rightarrow$$

$$\dot{\theta}_3 = (w_2^*c\theta_5 - w_1^*s\theta_5)/(d_4s\theta_5) \quad (9.742)$$

Finally, $\dot{\theta}_5$ is obtained as

$$\dot{\theta}_5 = \dot{\theta}_{35} - \dot{\theta}_3 \quad (9.743)$$

∗ Second Operation Mode with One Fixed Joint and Minimum Joint Velocities

In this operation mode, it is again required to fix the first joint so that

$$\theta_1 = 0 \text{ and } \dot{\theta}_1 = 0 \quad (9.744)$$

Additionally, it is required to minimize a positive definite function of the joint velocities. Such a function can typically be expressed as

$$F = k_2\dot{\theta}_2^2 + k_3\dot{\theta}_3^2 + k_4\dot{\theta}_4^2 + k_5\dot{\theta}_5^2 \quad (9.745)$$

In Eq. (9.745), k_2 to k_5 are positive weighting coefficients. If $\dot{\theta}_i$ is desired to have a smaller magnitude than the other joint velocities, then k_i must be selected to be larger than the other weighting coefficients.

In this operation mode, Eq. (9.729) is first written in its reduced form with $\theta_1 = 0$ and $\dot{\theta}_1 = 0$. Then, it can be manipulated as shown below.

$$\overline{w} = e^{\widetilde{u}_2\theta_2}\widetilde{u}_2e^{\widetilde{u}_3\theta_3}e^{\widetilde{u}_2\theta_4}(d_4\overline{u}_2 + d_6e^{\widetilde{u}_3\theta_5}\overline{u}_2)\dot{\theta}_2 + e^{\widetilde{u}_2\theta_2}e^{\widetilde{u}_3\theta_3}(d_6\widetilde{u}_3e^{\widetilde{u}_2\theta_4}e^{\widetilde{u}_3\theta_5}\overline{u}_2 - d_4\overline{u}_1)\dot{\theta}_3$$
$$+ (e^{\widetilde{u}_2\theta_2}e^{\widetilde{u}_3\theta_3}e^{\widetilde{u}_2\theta_4}\widetilde{u}_2e^{\widetilde{u}_3\theta_5}\overline{u}_2)d_6\dot{\theta}_4 - (e^{\widetilde{u}_2\theta_2}e^{\widetilde{u}_3\theta_3}e^{\widetilde{u}_2\theta_4}e^{\widetilde{u}_3\theta_5}\overline{u}_1)d_6\dot{\theta}_5 \Rightarrow$$

$$\overline{w} = (e^{\widetilde{u}_2\theta_2}e^{\widetilde{u}_3\theta_3})(e^{-\widetilde{u}_3\theta_3}\widetilde{u}_2e^{\widetilde{u}_3\theta_3})(d_4\overline{u}_2 + d_6e^{\widetilde{u}_2\theta_4}e^{\widetilde{u}_3\theta_5}\overline{u}_2)\dot{\theta}_2$$
$$+ (e^{\widetilde{u}_2\theta_2}e^{\widetilde{u}_3\theta_3})(d_6\widetilde{u}_3e^{\widetilde{u}_2\theta_4}e^{\widetilde{u}_3\theta_5}\overline{u}_2 - d_4\overline{u}_1)\dot{\theta}_3$$
$$+ (e^{\widetilde{u}_2\theta_2}e^{\widetilde{u}_3\theta_3})[(e^{\widetilde{u}_2\theta_4}\widetilde{u}_2e^{\widetilde{u}_3\theta_5}\overline{u}_2)d_6\dot{\theta}_4 - (e^{\widetilde{u}_2\theta_4}e^{\widetilde{u}_3\theta_5}\overline{u}_1)d_6\dot{\theta}_5] \Rightarrow$$

$$\overline{w} = (e^{\tilde{u}_2\theta_2}e^{\tilde{u}_3\theta_3})(\tilde{u}_2c\theta_3 + \tilde{u}_1s\theta_3)(d_4\overline{u}_2 + d_6e^{\tilde{u}_2\theta_4}e^{\tilde{u}_3\theta_5}\overline{u}_2)\dot{\theta}_2 + (e^{\tilde{u}_2\theta_2}e^{\tilde{u}_3\theta_3})(d_6\tilde{u}_3e^{\tilde{u}_2\theta_4}e^{\tilde{u}_3\theta_5}\overline{u}_2 - d_4\overline{u}_1)\dot{\theta}_3 + (e^{\tilde{u}_2\theta_2}e^{\tilde{u}_3\theta_3})[(\tilde{u}_2e^{\tilde{u}_2\theta_4}e^{\tilde{u}_3\theta_5}\overline{u}_2)d_6\dot{\theta}_4 - (e^{\tilde{u}_2\theta_4}e^{\tilde{u}_3\theta_5}\overline{u}_1)d_6\dot{\theta}_5] \quad (9.746)$$

The optimal solution can be obtained by solving Eq. (9.746) in order to express the first three joint velocities ($\dot{\theta}_2$, $\dot{\theta}_3$, and $\dot{\theta}_5$) in terms of $\dot{\theta}_4$. Afterwards, $\dot{\theta}_4$ can be found so as to minimize F.

To start the solution process, Eq. (9.746) can be written again as follows by using Eq. (9.736), i.e. $\overline{w}^* = e^{-\tilde{u}_3\theta_3}e^{-\tilde{u}_2\theta_2}\overline{w}$:

$$\overline{w}^* = (\tilde{u}_2c\theta_3 + \tilde{u}_1s\theta_3)(d_4\overline{u}_2 + d_6e^{\tilde{u}_2\theta_4}e^{\tilde{u}_3\theta_5}\overline{u}_2)\dot{\theta}_2 + (d_6\tilde{u}_3e^{\tilde{u}_2\theta_4}e^{\tilde{u}_3\theta_5}\overline{u}_2 - d_4\overline{u}_1)\dot{\theta}_3 + [(\tilde{u}_2e^{\tilde{u}_2\theta_4}e^{\tilde{u}_3\theta_5}\overline{u}_2)d_6\dot{\theta}_4 - (e^{\tilde{u}_2\theta_4}e^{\tilde{u}_3\theta_5}\overline{u}_1)d_6\dot{\theta}_5] \quad (9.747)$$

Upon premultiplications by $\overline{u}_1^t$, $\overline{u}_2^t$, and $\overline{u}_3^t$, Eq. (9.747) leads to the following scalar equations.

$$w_1^* = (d_6c\theta_3s\theta_4s\theta_5)\dot{\theta}_2 - (d_4 + d_6c\theta_5)\dot{\theta}_3 + (d_6s\theta_4s\theta_5)\dot{\theta}_4 - (d_6c\theta_4c\theta_5)\dot{\theta}_5 \quad (9.748)$$

$$w_2^* = -(d_6s\theta_3s\theta_4s\theta_5)\dot{\theta}_2 - (d_6c\theta_4s\theta_5)\dot{\theta}_3 - (d_6s\theta_5)\dot{\theta}_5 \quad (9.749)$$

$$w_3^* = [(d_4 + d_6c\theta_5)s\theta_3 + (d_6c\theta_3c\theta_4s\theta_5)]\dot{\theta}_2 + (d_6c\theta_4s\theta_5)\dot{\theta}_4 + (d_6s\theta_4c\theta_5)\dot{\theta}_5 \quad (9.750)$$

On the other hand, like θ_5, $\dot{\theta}_5$ can also be found independently of $\dot{\theta}_4$. This can be done by taking the derivative of Eq. (9.675) as shown below.

$$-\dot{\theta}_5s\theta_5 = (x_1\dot{x}_1 + x_2\dot{x}_2 + x_3\dot{x}_3)/(d_4d_6) \quad (9.751)$$

Note that

$$\dot{x}_k = \dot{r}_k = w_k \text{ for } k = 1, 2, 3 \quad (9.752)$$

Hence, if $s\theta_5 \neq 0$, $\dot{\theta}_5$ is obtained as

$$\dot{\theta}_5 = -\frac{x_1w_1 + x_2w_2 + x_3w_3}{d_4d_6s\theta_5} = -\frac{\overline{x}^t\overline{w}}{d_4d_6s\theta_5} \quad (9.753)$$

As seen above, $\dot{\theta}_5$ has come out to be independent of $\dot{\theta}_4$. Therefore, $\dot{\theta}_5$ can be dropped from the expression of the function F by taking $k_5 = 0$ in Eq. (9.745).

Meanwhile, Eqs. (9.748)–(9.750) can be rearranged as follows:

$$(d_6s\theta_3s\theta_4s\theta_5)\dot{\theta}_2 + (d_6c\theta_4s\theta_5)\dot{\theta}_3 + (d_6s\theta_5)\dot{\theta}_5 = -w_2^* \quad (9.754)$$

$$[(d_4 + d_6c\theta_5)s\theta_3c\theta_4 + d_6c\theta_3s\theta_5]\dot{\theta}_2 - [(d_4 + d_6c\theta_5)s\theta_4]\dot{\theta}_3 + (d_6s\theta_5)\dot{\theta}_4 = w_1^*s\theta_4 + w_3^*c\theta_4 \quad (9.755)$$

$$[(d_4 + d_6c\theta_5)s\theta_3s\theta_4]\dot{\theta}_2 + [(d_4 + d_6c\theta_5)c\theta_4]\dot{\theta}_3 + (d_6c\theta_5)\dot{\theta}_5 = w_3^*s\theta_4 - w_1^*c\theta_4 \quad (9.756)$$

As for Eqs. (9.754) and (9.756), they can be manipulated further as follows in order to retain and eliminate $\dot{\theta}_5$:

$$[(d_6 + d_4c\theta_5)s\theta_3s\theta_4]\dot{\theta}_2 + [(d_6 + d_4c\theta_5)c\theta_4]\dot{\theta}_3 + d_6\dot{\theta}_5 = (w_3^*s\theta_4 - w_1^*c\theta_4)c\theta_5 - w_2^*s\theta_5 \quad (9.757)$$

$$(d_4s\theta_3s\theta_4s\theta_5)\dot{\theta}_2 + (d_4c\theta_4s\theta_5)\dot{\theta}_3 = (w_3^*s\theta_4 - w_1^*c\theta_4)s\theta_5 + w_2^*c\theta_5 \quad (9.758)$$

However, after replacing $\dot{\theta}_5$ with its expression given by Eq. (9.753), it can be shown that Eqs. (9.757) and (9.758) are identical. Therefore, Eqs. (9.755) and (9.758) can be used as two independent equations in order to find $\dot{\theta}_2$ and $\dot{\theta}_3$ in terms of $\dot{\theta}_4$. For this purpose, the two equations can be written together as the following matrix equation.

$$\begin{bmatrix} (d_4 + d_6c\theta_5)s\theta_3c\theta_4 + d_6c\theta_3s\theta_5 & -(d_4 + d_6c\theta_5)s\theta_4 \\ d_4s\theta_3s\theta_4s\theta_5 & d_4c\theta_4s\theta_5 \end{bmatrix} \begin{bmatrix} \dot{\theta}_2 \\ \dot{\theta}_3 \end{bmatrix} = \begin{bmatrix} w_1^*s\theta_4 + w_3^*c\theta_4 - (d_6s\theta_5)\dot{\theta}_4 \\ (w_3^*s\theta_4 - w_1^*c\theta_4)s\theta_5 + w_2^*c\theta_5 \end{bmatrix} \tag{9.759}$$

In Eq. (9.759), the determinant of the coefficient matrix is

$$D_{23} = [(d_4 + d_6c\theta_5)s\theta_3c\theta_4 + d_6c\theta_3s\theta_5](d_4c\theta_4s\theta_5) + [(d_4 + d_6c\theta_5)s\theta_4](d_4s\theta_3s\theta_4s\theta_5) \Rightarrow$$

$$D_{23} = (d_4s\theta_5)[d_4s\theta_3 + d_6(s\theta_3c\theta_5 + c\theta_3c\theta_4s\theta_5)] \tag{9.760}$$

Referring to Eq. (9.679), D_{23} can also be expressed as

$$D_{23} = (d_4s\theta_5)(x_3s\theta_2 - x_1c\theta_2) \tag{9.761}$$

If $D_{23} \neq 0$, then Eq. (9.759) can be solved so that

$$\begin{bmatrix} \dot{\theta}_2 \\ \dot{\theta}_3 \end{bmatrix} = \frac{1}{D_{23}} \begin{bmatrix} d_4c\theta_4s\theta_5 & (d_4 + d_6c\theta_5)s\theta_4 \\ -d_4s\theta_3s\theta_4s\theta_5 & (d_4 + d_6c\theta_5)s\theta_3c\theta_4 + d_6c\theta_3s\theta_5 \end{bmatrix} \times \begin{bmatrix} w_1^*s\theta_4 + w_3^*c\theta_4 - (d_6s\theta_5)\dot{\theta}_4 \\ (w_3^*s\theta_4 - w_1^*c\theta_4)s\theta_5 + w_2^*c\theta_5 \end{bmatrix} \tag{9.762}$$

After the necessary manipulations, $\dot{\theta}_2$ and $\dot{\theta}_3$ are obtained with the following expressions.

$$\dot{\theta}_2 = \frac{d_4(w_2^*s\theta_4c\theta_5 + w_3^*s\theta_5) + (d_6s\theta_4c\theta_5)(w_2^*c\theta_5 + w_3^*s\theta_4s\theta_5 - w_1^*c\theta_4s\theta_5)}{(d_4s\theta_5)(x_3s\theta_2 - x_1c\theta_2)} - \frac{d_6c\theta_4s\theta_5}{x_3s\theta_2 - x_1c\theta_2}\dot{\theta}_4 \tag{9.763}$$

$$\dot{\theta}_3 = \frac{(d_4s\theta_3)(w_2^*c\theta_4c\theta_5 - w_1^*s\theta_5) + d_6(s\theta_3c\theta_4c\theta_5 + c\theta_3s\theta_5)(w_2^*c\theta_5 + w_3^*s\theta_4s\theta_5 - w_1^*c\theta_4s\theta_5)}{(d_4s\theta_5)(x_3s\theta_2 - x_1c\theta_2)} + \frac{d_6s\theta_3s\theta_4s\theta_5}{x_3s\theta_2 - x_1c\theta_2}\dot{\theta}_4 \tag{9.764}$$

The joint velocities obtained above can also be expressed briefly as follows:

$$\dot{\theta}_2 = g_{24}\dot{\theta}_4 + h_2 \tag{9.765}$$

$$\dot{\theta}_3 = g_{34}\dot{\theta}_4 + h_3 \tag{9.766}$$

Note that Eqs. (9.763) and (9.764) already indicate the definitions of the relevant *velocity influence coefficients* (g_{24} and g_{34}) together with the *bias terms* (h_2 and h_3).

In the optimization stage of the solution process, the function F can be written as follows by omitting $\dot{\theta}_5$ (due to its independence of $\dot{\theta}_4$) and by inserting the expressions of $\dot{\theta}_2$ and $\dot{\theta}_3$ given by Eqs. (9.765) and (9.766):

$$F = k_2(g_{24}\dot{\theta}_4 + h_2)^2 + k_3(g_{34}\dot{\theta}_4 + h_3)^2 + k_4\dot{\theta}_4^2 \tag{9.767}$$

The minimization of F necessitates that

$$dF/d\dot{\theta}_4 = 2k_2(g_{24}\dot{\theta}_4 + h_2)(g_{24}) + 2k_3(g_{34}\dot{\theta}_4 + h_3)(g_{34}) + 2k_4\dot{\theta}_4 = 0$$

Hence, the optimal value of $\dot{\theta}_4$ is found as

$$\dot{\theta}_4 = -(k_2 g_{24} h_2 + k_3 g_{34} h_3)/(k_2 g_{24}^2 + k_3 g_{34}^2 + k_4) \tag{9.768}$$

The corresponding values of $\dot{\theta}_2$ and $\dot{\theta}_3$ are then obtained by using Eqs. (9.765) and (9.766).

(b) Solution for the Wrist Joint Velocities

In order to find the wrist joint velocities $(\dot{\theta}_6, \dot{\theta}_7, \dot{\theta}_8)$ after finding the arm joint velocities, the angular velocity equation of the end-effector, i.e. Eq. (9.728), can be reduced to the following simplified form.

$$\dot{\theta}_6\overline{u}_2 + \dot{\theta}_7\overline{u}_3 + \dot{\theta}_8 e^{\tilde{u}_3\theta_7}\overline{u}_2 = \overline{\omega}^* \Rightarrow$$

$$\overline{u}_3\dot{\theta}_7 + \overline{u}_2(\dot{\theta}_6 + \dot{\theta}_8 c\theta_7) - \overline{u}_1(\dot{\theta}_8 s\theta_7) = \overline{\omega}^* \tag{9.769}$$

In Eq. (9.769), $\overline{\omega}^*$ is known as expressed by the following two consecutive equations.

$$\overline{\omega}^* = e^{-\tilde{u}_2\theta_6} e^{-\tilde{u}_3\theta_5} e^{-\tilde{u}_2\theta_4} e^{-\tilde{u}_3\theta_3} e^{-\tilde{u}_2\theta_2} e^{-\tilde{u}_3\theta_1} \overline{\omega}' \tag{9.770}$$

$$\overline{\omega}' = \overline{\omega} - \dot{\theta}_1\overline{u}_3 - \dot{\theta}_2 e^{\tilde{u}_3\theta_1}\overline{u}_2 - \dot{\theta}_3 e^{\tilde{u}_3\theta_1} e^{\tilde{u}_2\theta_2}\overline{u}_3 - \dot{\theta}_4 e^{\tilde{u}_3\theta_1} e^{\tilde{u}_2\theta_2} e^{\tilde{u}_3\theta_3}\overline{u}_2$$
$$- \dot{\theta}_5 e^{\tilde{u}_3\theta_1} e^{\tilde{u}_2\theta_2} e^{\tilde{u}_3\theta_3} e^{\tilde{u}_2\theta_4}\overline{u}_3 \tag{9.771}$$

The solution to Eq. (9.769) can be obtained as follows:

$$\dot{\theta}_7 = \omega_3^* \tag{9.772}$$

If $s\theta_7 \neq 0$,

$$\dot{\theta}_8 = -\omega_1^*/s\theta_7 \tag{9.773}$$

$$\dot{\theta}_6 = \omega_2^* - \dot{\theta}_8 c\theta_7 = (\omega_2^* s\theta_7 + \omega_1^* c\theta_7)/s\theta_7 \tag{9.774}$$

9.8.8 Singularity Analysis in the Velocity Domain

The humanoid manipulator considered here may have three distinct kinds of motion singularities, which are described and discussed below for the somewhat special case with $\theta_1 = 0$ and $\dot{\theta}_1 = 0$.

(a) First Kind of Motion Singularity

Equation (9.753) implies that the first kind of motion singularity occurs if

$$s\theta_5 = 0 \tag{9.775}$$

This singularity may occur either in the *extended version* with $\theta_5 = 0$ or in the *folded version* with $\theta_5 = \pi$. The folded version may occur only approximately though due to the physical shapes of the relevant links and joints.
Equation (9.753) also implies that $\dot{\theta}_5$ becomes indefinite in this singularity. However, it can be kept finite if the specified motion of the hand obeys the following task space compatibility condition.

$$\overline{x}^t\overline{w} = 0 \tag{9.776}$$

When this singularity occurs, Eqs. (9.754)–(9.756) simplify to the following equations with $\sigma_5' = \text{sgn}(c\theta_5)$.

$$w_2^* = 0 \tag{9.777}$$

$$(s\theta_3 c\theta_4)\dot{\theta}_2 - (s\theta_4)\dot{\theta}_3 = (w_1^* s\theta_4 + w_3^* c\theta_4)/(d_4 + \sigma_5' d_6) \tag{9.778}$$

$$(s\theta_3 s\theta_4)\dot{\theta}_2 + (c\theta_4)\dot{\theta}_3 = (w_3^* s\theta_4 - w_1^* c\theta_4 - \sigma_5' d_6 \dot{\theta}_5)/(d_4 + \sigma_5' d_6) \tag{9.779}$$

Note that $\overline{x}$ assumes the following expression in this singularity.

$$\overline{x} = \overline{r} - d_2 \overline{u}_2 = (d_4 + \sigma_5' d_6) e^{\tilde{u}_2 \theta_2} e^{\tilde{u}_3 \theta_3} \overline{u}_2 \tag{9.780}$$

Hence, the equivalence of Eqs. (9.776) and (9.777) can be shown as follows:

$$\overline{x}^t \overline{w} = (d_4 + \sigma_5' d_6)(e^{\tilde{u}_2 \theta_2} e^{\tilde{u}_3 \theta_3} \overline{u}_2)^t \overline{w} = (d_4 + \sigma_5' d_6) \overline{u}_2^t e^{-\tilde{u}_3 \theta_3} e^{-\tilde{u}_2 \theta_2} \overline{w} = 0 \Rightarrow$$

$$\overline{x}^t \overline{w} = (d_4 + \sigma_5' d_6) \overline{u}_2^t \overline{w}^* = (d_4 + \sigma_5' d_6) w_2^* = 0 \Rightarrow w_2^* = 0$$

Equation (9.776) or (9.777) indicates the task space restriction imposed by this singularity on the motion of the wrist point.

On the other hand, Eqs. (9.778) and (9.779) indicate the joint space consequence of this singularity that $\dot{\theta}_2$ and $\dot{\theta}_3$ become dependent on $\dot{\theta}_5$ instead of $\dot{\theta}_4$. However, $\dot{\theta}_5$ also becomes arbitrary in addition to $\dot{\theta}_4$.

Thus, as long as $\overline{w}$ obeys the above-mentioned task space restriction, the manipulator can be driven through this singularity by assigning some suitable values to $\dot{\theta}_4$ and $\dot{\theta}_5$, which can be their values immediately before the instant of singularity.

(b) Second Kind of Motion Singularity

Equations (9.763) and (9.764) imply that the second kind of motion singularity occurs if

$$x_1 c\theta_2 - x_3 s\theta_2 = 0 \tag{9.781}$$

According to Eqs. (9.760) and (9.761), Eq. (9.781) is equivalent to

$$d_4 s\theta_3 + d_6(s\theta_3 c\theta_5 + c\theta_3 c\theta_4 s\theta_5) = 0 \tag{9.782}$$

Equation (9.782) can be used to eliminate d_4 in Eq. (9.755). Then, the resulting equation can be written as follows along with Eq. (9.758):

$$(d_6 s\theta_3 c\theta_3 s\theta_4 s\theta_4 s\theta_5)\dot{\theta}_2 + (d_6 c\theta_3 s\theta_4 c\theta_4 s\theta_5)\dot{\theta}_3$$
$$= (w_1^* s\theta_4 + w_3^* c\theta_4) s\theta_3 - (d_6 s\theta_3 s\theta_5)\dot{\theta}_4 \tag{9.783}$$

$$(d_4 s\theta_3 s\theta_4 s\theta_5)\dot{\theta}_2 + (d_4 c\theta_4 s\theta_5)\dot{\theta}_3 = (w_3^* s\theta_4 - w_1^* c\theta_4) s\theta_5 + w_2^* c\theta_5 \tag{9.784}$$

The present singularity makes the above equations dependent. Consequently, $\dot{\theta}_2$ and $\dot{\theta}_3$ become related through a generic equation such as

$$\dot{\theta}_3 = \gamma_{32}\dot{\theta}_2 + \gamma_{30} \tag{9.785}$$

After substituting Eq. (9.785), Eqs. (9.783) and (9.784) can be manipulated into the following forms.

$$(c\theta_3 s\theta_4 s\theta_5)(s\theta_3 s\theta_4 + \gamma_{32} c\theta_4)\dot{\theta}_2$$
$$= [(w_3^* s\theta_4 - w_1^* c\theta_4) s\theta_5 + w_2^* c\theta_5] c\theta_3 s\theta_4 / d_4 - \gamma_{30} c\theta_3 s\theta_4 c\theta_4 s\theta_5 \tag{9.786}$$

$$(c\theta_3 s\theta_4 s\theta_5)(s\theta_3 s\theta_4 + \gamma_{32} c\theta_4)\dot{\theta}_2$$
$$= [(w_1^* s\theta_4 + w_3^* c\theta_4) s\theta_3 - (d_6 s\theta_3 s\theta_5)\dot{\theta}_4]/d_6 - \gamma_{30} c\theta_3 s\theta_4 c\theta_4 s\theta_5 \quad (9.787)$$

Note that the left-hand sides of the above equations are equal. Therefore, their right-hand sides must also be equal. Due to this equality, $\dot{\theta}_4$ becomes determined as expressed below.

$$\dot{\theta}_4 = \frac{(w_2^* c\theta_3 - w_1^* s\theta_3) s\theta_4 c\theta_5 - w_3^*(s\theta_3 c\theta_4 c\theta_5 + c\theta_3 s\theta_5)}{d_4 s\theta_3 s\theta_5} \quad (9.788)$$

On the other hand, Eqs. (9.786) and (9.787) must both be satisfied for any arbitrary $\dot{\theta}_2$. This condition leads to the following expressions for γ_{32} and γ_{30}.

$$\gamma_{32} = -s\theta_3 s\theta_4 / c\theta_4 \quad (9.789)$$

$$\gamma_{30} = [(w_3^* s\theta_4 - w_1^* c\theta_4) s\theta_5 + w_2^* c\theta_5]/(d_4 c\theta_4 s\theta_5) \quad (9.790)$$

Hence, Eq. (9.785) can be expressed in the following detailed form.

$$\dot{\theta}_3 c\theta_4 + \dot{\theta}_2 s\theta_3 s\theta_4 = [(w_3^* s\theta_4 - w_1^* c\theta_4) s\theta_5 + w_2^* c\theta_5]/(d_4 s\theta_5) \quad (9.791)$$

Equations (9.788) and (9.791) show that the second type of motion singularity shifts the arbitrariness from $\dot{\theta}_4$ to $\dot{\theta}_2$ and $\dot{\theta}_3$. Thus, the manipulator can be driven through this singularity by assigning a suitable value to one of $\dot{\theta}_2$ and $\dot{\theta}_3$ and then finding the other from Eq. (9.791).

(c) Third Kind of Motion Singularity

Equations (9.768) and (9.769) imply that the third kind of motion singularity occurs if $s\theta_7 = 0$. Except the different subscripts of the relevant angles, this singularity has the same appearance as the third kind of motion singularity of the Puma manipulator, which is explained in Section 9.1.5 and illustrated in Figure 9.11.

9.8.9 Consistency of the Inverse Kinematics in the Position and Velocity Domains

As seen in Sections 9.8.3 and 9.8.7, the optimal inverse kinematic solutions are obtained in the position and velocity domains by, respectively, stipulating a position domain criterion and a velocity domain criterion. The solutions are described in a self-contained manner without considering the consistency relationships between them. Actually, of course, the optimal inverse kinematic solutions obtained in the position and velocity domains must be consistent with each other. In other words, if the inverse kinematic solution is obtained in the position domain by stipulating a position domain criterion, then the inverse kinematic solution in the velocity domain must be obtained via differentiation in order to be consistent with that position domain criterion instead of stipulating an independent velocity domain criterion. Conversely, if the inverse kinematic solution is obtained in the velocity domain by stipulating a velocity domain criterion, then the inverse kinematic solution in the position domain must be obtained via integration in order to be consistent with that velocity domain criterion instead of stipulating an independent position domain criterion.

The consistency of the optimal inverse kinematic solutions in the position and velocity domains is discussed below.

(a) Consistency of the Velocity Domain Solution with the Position Domain Criterion

Regarding the velocity domain solution in the second operation mode described in Part (a) of Section 9.8.7, $\dot{\theta}_5$ was obtained irrespective of the other joint velocities, but $\dot{\theta}_2$ and $\dot{\theta}_3$ were obtained as follows in terms of $\dot{\theta}_4$.

$$\dot{\theta}_2 = g_{24}\dot{\theta}_4 + h_2 \tag{9.792}$$

$$\dot{\theta}_3 = g_{34}\dot{\theta}_4 + h_3 \tag{9.793}$$

On the other hand, regarding the position domain solution in the second operation mode described in Part (a) of Section 9.8.3, θ_5 was also obtained irrespective of the other joint variables, but the interrelated joint variables θ_2, θ_3, and θ_4 were determined so as to minimize the following nondimensional potential energy function.

$$V = V(\theta_2, \theta_3) = s\theta_2 s\theta_3 \tag{9.794}$$

If the velocity domain solution is to be consistent with the position domain solution, it must also be obtained according to a criterion derived from the function $V = V(\theta_2, \theta_3)$, instead of a different criterion such as the minimization of the function F expressed by Eq. (9.745).

At this point, referring to Eqs. (9.686) and (9.695), note that both θ_2 and θ_3 are expressed in terms of θ_4 and the specified location of the wrist point. That is,

$$\theta_2 = f_2(\theta_4, \bar{r}) \text{ and } \theta_3 = f_3(\theta_4, \bar{r}) \tag{9.795}$$

Therefore, as explained in Section 9.8.3, the minimization of V is achieved numerically by changing θ_4 in an appropriate interval and thus finding its optimal value.

On the other hand, in order to set up the consistency of the solutions in the position and velocity domains, the optimality of the above-mentioned solution can also be expressed by the following equation, even though it was not used directly in Section 9.8.3:

$$\frac{\partial V}{\partial \theta_4} = \frac{\partial V}{\partial \theta_2} \cdot \frac{\partial \theta_2}{\partial \theta_4} + \frac{\partial V}{\partial \theta_3} \cdot \frac{\partial \theta_2}{\partial \theta_4} = 0 \tag{9.796}$$

Meanwhile, referring to Eqs. (9.792) and (9.793), note that

$$\frac{\partial \theta_2}{\partial \theta_4} = \frac{\partial \dot{\theta}_2}{\partial \dot{\theta}_4} = g_{24} \text{ and } \frac{\partial \theta_3}{\partial \theta_4} = \frac{\partial \dot{\theta}_3}{\partial \dot{\theta}_4} = g_{34} \tag{9.797}$$

Equations (9.796) and (9.797) lead to the following optimality condition:

$$(c\theta_2 s\theta_3)(g_{24}) + (s\theta_2 c\theta_3)(g_{34}) = 0 \Rightarrow$$

$$g_{24} c\theta_2 s\theta_3 + g_{34} s\theta_2 c\theta_3 = 0 \tag{9.798}$$

As for g_{24} and g_{34}, Eqs. (9.763) and (9.764) imply that

$$g_{24} = g_{24}(\theta_4, \bar{r}) \text{ and } g_{34} = g_{34}(\theta_4, \bar{r}) \tag{9.799}$$

With the presence of Eq. Pairs (9.795) and (9.799), Eq. (9.798) can be solved numerically by following a similar procedure described in Section 9.8.3 in order to find θ_4 optimally as the following function of the specified wrist point location, which must be matching the result obtained in Section 9.8.3.

$$\theta_4 = f_4(\bar{r}) \tag{9.800}$$

Moreover, the position domain optimality condition, i.e. Eq. (9.798), must actually be satisfied continuously at every instant of time during the operation of the manipulator. In other words, the derivative of Eq. (9.798) must also be satisfied so that

$$(g_{34}c\theta_2 c\theta_3 - g_{24}s\theta_2 s\theta_3)\dot{\theta}_2 + (g_{24}c\theta_2 c\theta_3 - g_{34}s\theta_2 s\theta_3)\dot{\theta}_3 + \dot{g}_{24}c\theta_2 s\theta_3 + \dot{g}_{34}s\theta_2 c\theta_3 = 0 \tag{9.801}$$

The derivatives $\dot{g}_{24}$ and $\dot{g}_{34}$ can be obtained as follows from Eq. Pair (9.799).

$$\left.\begin{aligned} \dot{g}_{24} &= \frac{\partial g_{24}}{\partial \theta_4}\dot{\theta}_4 + \sum_{k=1}^{3}\frac{\partial g_{24}}{\partial r_k}\dot{r}_k = k_{24}\dot{\theta}_4 + l_{24} \\ \dot{g}_{34} &= \frac{\partial g_{34}}{\partial \theta_4}\dot{\theta}_4 + \sum_{k=1}^{3}\frac{\partial g_{34}}{\partial r_k}\dot{r}_k = k_{34}\dot{\theta}_4 + l_{34} \end{aligned}\right\} \tag{9.802}$$

After the necessary substitutions, Eq. (9.801) becomes

$$\begin{aligned} &(g_{34}c\theta_2 c\theta_3 - g_{24}s\theta_2 s\theta_3)(g_{24}\dot{\theta}_4 + h_2) \\ &\quad + (g_{24}c\theta_2 c\theta_3 - g_{34}s\theta_2 s\theta_3)(g_{34}\dot{\theta}_4 + h_3) \\ &\quad + (k_{24}\dot{\theta}_4 + l_{24})c\theta_2 s\theta_3 + (k_{34}\dot{\theta}_4 + l_{34})s\theta_2 c\theta_3 = 0 \Rightarrow \\ &m_4\dot{\theta}_4 + n_4 = 0 \end{aligned} \tag{9.803}$$

In Eq. (9.803),

$$m_4 = k_{24}c\theta_2 s\theta_3 + k_{34}c\theta_3 s\theta_2 + 2g_{24}g_{34}c\theta_2 c\theta_3 - (g_{24}^2 + g_{34}^2)s\theta_2 s\theta_3 \tag{9.804}$$

$$n_4 = (g_{34}c\theta_2 c\theta_3 - g_{24}s\theta_2 s\theta_3)h_2 + (g_{24}c\theta_2 c\theta_3 - g_{34}s\theta_2 s\theta_3)h_3 + l_{24}c\theta_2 s\theta_3 + l_{34}c\theta_3 s\theta_2 \tag{9.805}$$

Hence, provided that $m_4 \neq 0$, $\dot{\theta}_4$ is found as follows as the optimal joint velocity that is consistent with position domain optimization.

$$\dot{\theta}_4 = -n_4/m_4 \tag{9.806}$$

Afterwards, Eqs. (9.792) and (9.793) provide the corresponding optimal values of $\dot{\theta}_2$ and $\dot{\theta}_3$.

(b) Consistency of the Position Domain Solution with the Velocity Domain Criterion

If the redundancy is desired to be resolved optimally in the velocity domain by stipulating the minimization of the function F expressed by Eq. (9.745), then the resultant expressions of $\dot{\theta}_2$, $\dot{\theta}_3$, and $\dot{\theta}_4$ that are given by Eqs. (9.765)–(9.767) constitute three differential equations, which are written briefly as follows:

$$\dot{\theta}_2 = f_2(\theta_2, \theta_3, \theta_4; w_1^*, w_2^*, w_3^*) \tag{9.807}$$

$$\dot{\theta}_3 = f_3(\theta_2, \theta_3, \theta_4; w_1^*, w_2^*, w_3^*) \tag{9.808}$$

$$\dot{\theta}_4 = f_4(\theta_2, \theta_3, \theta_4; w_1^*, w_2^*, w_3^*) \tag{9.809}$$

In this case, the position domain does not need an optimization criterion, because θ_2, θ_3, and θ_4 will be determined as functions of time by integrating the preceding differential equations. However, the initial values θ_2°, θ_3°, and θ_4° that are required to start the integration necessitate a once-to-use optimization criterion so that they can be determined from the specified initial position of the end-effector.

Note that θ_5 and the wrist joint variables (θ_6, θ_7, θ_8) remain out of the integration process, because they can be determined at any instant directly from the specified location and orientation of the end-effector.

The numerical integration of the preceding differential equations can be achieved conveniently by using the *prediction-correction method*. The prediction and correction stages of this method are explained below.

Prediction stage:

At this stage, the values of the joint variables at a certain time instant t_k can be predicted simply as follows by using their values at the previous time instant $t_{k-1} = t_k - \tau_k$.

$$\theta_i'(t_k) = \theta_i(t_{k-1}) + f_i(t_{k-1})\tau_k; \quad i = 2, 3, 4 \tag{9.810}$$

Here, τ_k is a sufficiently small time increment, $f_i(t_{k-1})$ is the value of the ith function at t_{k-1}, k is an integer such that $1 \le k \le n$, and $t_n = t_f$ is the completion time of the task to be performed.

Correction stage:

At this stage, the rough prediction $\theta_i'(t_k)$ is refined (i.e. corrected) as shown below.

$$\theta_i(t_k) = \theta_i'(t_k) + \delta\theta_i(t_k) \tag{9.811}$$

Here, the corrective increments denoted by $\delta\theta_i(t_k)$ for $i = 2, 3, 4$ are to be determined in such a way that the position equations, i.e. Eqs. (9.679)–(9.681), are satisfied. These equations are written again below for the sake of convenience.

$$(d_4 + d_6 c\theta_5)s\theta_3 + (d_6 s\theta_5)c\theta_3 c\theta_4 = x_3 s\theta_2 - x_1 c\theta_2 \tag{9.812}$$

$$(d_4 + d_6 c\theta_5)c\theta_3 - (d_6 s\theta_5)s\theta_3 c\theta_4 = x_2 \tag{9.813}$$

$$x_3 c\theta_2 + x_1 s\theta_2 = d_6 s\theta_4 s\theta_5 \tag{9.814}$$

However, as seen before, Eqs. (9.812) and (9.813) are not independent of each other, because $s\theta_3$ and $c\theta_3$ were determined, as expressed by Eq. (9.694), so as to satisfy both of them by obeying the consistency relationship that $(s\theta_3)^2 + (c\theta_3)^2 = 1$. Since they are dependent, Eqs. (9.812) and (9.813) can be combined effectively into the following simpler equivalent equation, in which θ_4 is eliminated.

$$(x_3 s\theta_2 - x_1 c\theta_2)s\theta_3 + x_2 c\theta_3 = d_4 + d_6 c\theta_5 \tag{9.815}$$

Now, assuming that all the corrective increments are small owing to the suitably selected sufficiently small value of τ_k, the incremental forms of the independent Eqs. (9.814) and (9.815) can be written at the instant t_k as the following equations that contain the corrective increments linearly.

$$(x_1 c\theta_2' - x_3 s\theta_2')\delta\theta_2 - (d_6 c\theta_4' s\theta_5)\delta\theta_4 = d_6 s\theta_4' s\theta_5 - (x_3 c\theta_2' + x_1 s\theta_2') \tag{9.816}$$

$$\begin{aligned}&[(x_3 c\theta_2' + x_1 s\theta_2')s\theta_3']\delta\theta_2 + [(x_3 s\theta_2' - x_1 c\theta_2')c\theta_3' - x_2 s\theta_3']\delta\theta_3 \\ &= (d_4 + d_6 c\theta_5) - (x_3 s\theta_2' - x_1 c\theta_2')s\theta_3' - x_2 c\theta_3'\end{aligned} \tag{9.817}$$

Equations (9.816) and (9.817) can be written compactly as follows:

$$\gamma_{12}'\delta\theta_2 - \gamma_{14}'\delta\theta_4 = \varepsilon_1' \tag{9.818}$$

$$\gamma_{22}'\delta\theta_2 + \gamma_{23}'\delta\theta_3 = \varepsilon_2' \tag{9.819}$$

In Eqs. (9.818) and (9.819), the coefficients ($\gamma'_{12}, \ldots, \gamma'_{23}$) and the error terms ($\varepsilon'_1$ and ε'_2) are defined according to Eqs. (9.816) and (9.817).

Equations (9.818) and (9.819) contain three corrective increments ($\delta\theta_2$, $\delta\theta_3$, and $\delta\theta_4$) to be determined. Therefore, two of them, e.g. $\delta\theta_2$ and $\delta\theta_3$, can be obtained in terms of the third one, i.e. $\delta\theta_4$, as follows, if $\gamma'_{12} \neq 0$ and $\gamma'_{23} \neq 0$.

$$\delta\theta_2 = (\varepsilon'_1 + \gamma'_{14}\delta\theta_4)/\gamma'_{12} = (\varepsilon'_1/\gamma'_{12}) + (\gamma'_{14}/\gamma'_{12})\delta\theta_4 \Rightarrow$$

$$\delta\theta_2 = h'_2 + g'_{24}\delta\theta_4 \tag{9.820}$$

$$\delta\theta_3 = (\varepsilon'_2 - \gamma'_{22}\delta\theta_2)/\gamma'_{23} = [\varepsilon'_2 - \gamma'_{22}(h'_2 + g'_{24}\delta\theta_4)]/\gamma'_{23} \Rightarrow$$

$$\delta\theta_3 = [\varepsilon'_2 - (\gamma'_{22}/\gamma'_{12})\varepsilon'_1]/\gamma'_{23} - [(\gamma'_{22}\gamma'_{14})/(\gamma'_{12}\gamma'_{23})]\delta\theta_4 \Rightarrow$$

$$\delta\theta_3 = h'_3 + g'_{34}\delta\theta_4 \tag{9.821}$$

If it happens that $\gamma'_{12} = 0$ and/or $\gamma'_{23} = 0$, at a certain instant t_k, then t_k can be skipped by replacing it with $t_{k+1} = t_{k-1} + 2\tau_k$. In other words, the singularities can be avoided by taking a larger time step during the integration process.

On the other hand, since both $\delta\theta_2$ and $\delta\theta_3$ are obtained in terms of $\delta\theta_4$, an additional criterion is needed in order to determine $\delta\theta_4$. Although it is not necessary to do so, the previous criterion function F, which was introduced by Eq. (9.745) for the joint velocities, can be used for the corrective increments as well by expressing it in the following form, again with $k_5 = 0$.

$$F = k_2(\delta\theta_2)^2 + k_3(\delta\theta_3)^2 + k_4(\delta\theta_4)^2 \tag{9.822}$$

Hence, $\delta\theta_4$ is determined as follows so that F is minimized.

$$\delta\theta_4 = -(k_2 g'_{24} h'_2 + k_3 g'_{34} h'_3)/[k_2(g'_{24})^2 + k_3(g'_{34})^2 + k_4] \tag{9.823}$$

Subsequently, $\delta\theta_2$ and $\delta\theta_3$ are also determined by using Eqs. (9.820) and (9.821).

10

Position and Velocity Analyses of Parallel Manipulators

Synopsis

This chapter covers the position and velocity analyses of parallel manipulators both in the forward and inverse senses. The sections of this chapter are arranged with the contents summarized below. The content of each section is explained and discussed with the help of suitably selected illustrative and instructive examples.

Section 10.1 describes the general kinematic features of the parallel manipulators and introduces the relevant terminology with the necessary explanations and illustrations.

Section 10.2 describes the general format of writing the necessary equations in order to express the kinematic relationships among the position parameters of the end-effector, the primary variables, and the secondary variables. The primary and secondary variables are generally taken as the variables of the active (actuated) and passive (unactuated) joints, respectively. In some cases, the secondary variables may also be taken as certain combinations of the passive joint variables. As for the position parameters of the end-effector, they consist of the coordinates that describe the location of its tip point and the Euler angles of a suitable sequence that describe its orientation with respect to the base frame.

Section 10.3 is about the *forward kinematics* in the *position domain*. The main concern of this section is to obtain the position parameters of the end-effector and the secondary variables as functions of the specified primary variables. This section is also concerned with the *posture multiplicities of forward kinematics* (PMFKs) (i.e. the different posture modes that the manipulator can assume for the same set of specified primary variables) and the *position singularities of forward kinematics* (PSFKs) (i.e. the *uncontrollable configurations* that the manipulator assumes due to some *critically specified* primary variables). These singularities must of course be avoided in order to keep the manipulator controllable.

Section 10.4 is about the *inverse kinematics* in the *position domain*. The main concern of this section is to obtain both the primary and secondary variables as functions of the specified position parameters of the end-effector. This section is also concerned with the *posture multiplicities of inverse kinematics* (PMIKs) (i.e. the different posture modes that the manipulator can assume for the same specified position of the end-effector) and the *position singularities of inverse kinematics* (PSIKs) (i.e. the configurations such that some

Kinematics of General Spatial Mechanical Systems, First Edition. M. Kemal Ozgoren.

Companion Website: www.wiley.com/go/ozgoren/spatialmechanicalsystems

of the actuators become *ineffective* as caused by certain *critically specified positions* of the end-effector). However, even though some of the actuators become ineffective, these singularities need not be avoided because the manipulator can still be controlled by the other actuators if these critical positions of the end-effector happen to be acceptable (or even desirable) for some particular tasks.

Section 10.5 describes how the velocity equations can be obtained through the differentiation of the position equations. These equations are needed in order to express the relationships among the velocity parameters of the end-effector, the rates of the primary variables, and the rates of the secondary variables. The velocity parameters of the end-effector consist of the components of its tip point velocity and its angular velocity with respect to the base frame. The velocity equations turn out to be linear in the velocity parameters of the end-effector and the rates of the primary and secondary variables. However, the relevant coefficient matrices involved in the velocity equations turn out to be functions of position, which are expressed in terms of both primary and secondary variables. Therefore, in working with the velocity equations, it is assumed that the primary and secondary variables are already available from the previously carried out kinematic analyses in the position domain.

Section 10.6 is about the *forward kinematics* in the *velocity domain*. The main concern of this section is to obtain the velocity parameters of the end-effector and the rates of the secondary variables as functions of the specified rates of the primary variables. Since the velocity equations are linear in the rates of the variables, the velocity domain does not contain multiplicities but it still contains singularities. Therefore, this section is also concerned with the singularities that are called *motion singularities of forward kinematics* (MSFKs) (i.e. the *uncontrollable configurations* that the manipulator assumes due to some *critically specified* rates of the primary variables). These singularities must of course be avoided in order to keep the motion of the manipulator under control at all instants during the execution of a task.

Section 10.7 is about the *inverse kinematics* in the *velocity domain*. The main concern of this section is to obtain the rates of both the primary and secondary variables as functions of the specified velocity parameters of the end-effector. As mentioned before, the velocity domain does not contain multiplicities but it contains singularities. Therefore, this section is also concerned with the singularities that are called *motion singularities of inverse kinematics* (MSIKs) (i.e. the configurations such that some of the actuators become *ineffective* as caused by certain *critically specified velocity parameters* of the end-effector). However, even though some of the actuators become ineffective, these singularities need not be avoided because the motion of the manipulator can still be controlled by the other actuators if the motion of the end-effector with these critically specified velocity parameters happen to be acceptable (or even desirable) for some particular tasks.

Sections 10.8 and 10.9 are allocated for comprehensive position and velocity analyses of two popular spatial parallel manipulator configurations. They are the configurations with six and three degrees of freedom, which are typically used in *Stewart–Gough platforms* and *delta robots*. These two sections are especially included, because the examples of the preceding sections are based on a planar parallel manipulator for the sake of clarity and neatness in demonstrating and discussing the essential aspects of the position and velocity analyses together with the associated multiplicities and singularities.

10.1 General Kinematic Features of Parallel Manipulators

A parallel manipulator is a multi-DoF (degree of freedom) mechanical system that has at least two separate kinematic chains between the end-effector and the base. The link carrying the end-effector and the link attached to the base are conceived as *platforms*. So, they are called *terminal platform* and *fixed platform*, respectively. The terminal platform is also called a *moving platform* if the parallel manipulator has only two platforms. In general, however, a parallel manipulator may also contain *intermediate platforms*, which are inserted in order to increase the mobility of the manipulator and the reachability range of the end-effector. As for the word *parallel*, it implies the presence of at least two kinematic chains that are parallel, not in the geometrical sense, but in the functional sense. In other words, the functionally parallel kinematic chains share the duty of supporting and positioning the end-effector with respect to the base. These kinematic chains are usually visualized as legs and therefore they are called *legs*, but occasionally some of them may also be called *arms* due to the visualization of the parallel manipulator formed by the arms of a human being by holding the hands together. Incidentally, a leg or an arm is also called a *limb* for the sake of having a unified term. On the other hand, similarly to the just mentioned biomechanical example, a parallel manipulator may also be formed as a combination of at least two collaborating serial manipulators that firmly hold a manipulated object, which becomes the terminal platform. The platforms of a parallel manipulator together with the limbs (the legs and/or the arms) form at least one kinematic loop. However, in general, a parallel manipulator happens to be a multi-loop mechanical system with several kinematic loops.

As compared with a serial manipulator, the multi-leg (or multi-arm) feature makes a parallel manipulator sturdier, increases its load-carrying capacity, and provides a possibility of more robust and precise position control for the end-effector. On the other hand, the presence of the loop or loops formed by the legs (or the arms) reduces the size of the working volume and the reachability range of the end-effector.

The DoF or the *mobility* (μ) of a *spatial parallel manipulator* can be determined by using the following *Kutzbach–Gruebler formula*.

$$\mu = 6n_m - (5j_1 + 4j_2 + 3j_3) \tag{10.1}$$

Equation (10.1) is written with $\lambda = 6$, which is the DoF of the *three-dimensional working space*. In the same equation, n_m is the number of the *moving links* and j_k is the number of the joints with a *relative mobility* of k. Here, $k = 1, 2, 3$ because a practicable parallel manipulator may contain only revolute ($k = 1$), prismatic ($k = 1$), universal ($k = 2$), cylindrical ($k = 2$), and spherical joints ($k = 3$).

A parallel manipulator contains n_{ikl} number of *independent kinematic loops* (IKLs). It happens that

$$n_{ikl} = j_1 + j_2 + j_3 - n_m \tag{10.2}$$

The positions (i.e. locations and orientations) of the moving links of a parallel manipulator can be described completely by n_v variables. These variables are divided into two groups so that

$$n_v = n_{pv} + n_{sv} \tag{10.3}$$

In Eq. (10.3), n_{pv} is the number of the *primary variables* that are *necessary* and *sufficient* to describe the pose of a parallel manipulator with μ degrees of freedom and n_{sv} is the number of the *secondary variables* that are *dependent* on the primary variables through the *loop equations*. Usually, but not necessarily always, the primary variables can be taken as the joint variables of the active (i.e. actuated) joints. If so, they are also called *active joint variables*. Then, the other variables that belong to the passive (i.e. unactuated) joints are called *passive joint variables*. According to the preceding explanations, the numbers n_{pv} and n_{sv} are given by the following formulas.

$$n_{pv} = \mu \tag{10.4}$$

$$n_{sv} = 6n_{ikl} \tag{10.5}$$

Equation (10.5) implies that, in a specified pose of the parallel manipulator, the secondary variables must assume such values that they satisfy the $6n_{ikl}$ scalar equations imposed by the n_{ikl} independent kinematic loops. The set of six scalar equations that belong to each kinematic loop consists of three independent scalar *orientation equations* and three independent scalar *characteristic point location equations*. The orientation equations relate the relative orientations of the consecutive links that form the loop, whereas the characteristic point location equations relate the relative positions of the consecutive characteristic points located around the loop. The characteristic points comprise the *joint centers* and the *link centers*.

When a *planar parallel manipulator* is of concern, the preceding equations reduce to the following forms.

$$\mu = 3n_m - 2j_1 \tag{10.6}$$

$$n_{ikl} = j_1 - n_m \tag{10.7}$$

$$n_{pv} = \mu \tag{10.8}$$

$$n_{sv} = 3n_{ikl} \tag{10.9}$$

Equation (10.6) is written with $\lambda = 3$, which is the DoF of the *two-dimensional working space*. Besides, in Eqs. (10.6) and (10.7), only j_1 occurs because a planar parallel manipulator can have only revolute and/or prismatic joints.

Three examples of parallel manipulators are given below. They are deliberately selected to be planar for the sake of simplicity and clarity.

Example 10.1 ***3RPR Planar Parallel Manipulator***

The parallel manipulator shown in Figure 10.1 consists of a fixed platform (link $\mathcal{L}_0$), a moving platform (link $\mathcal{L}_7$) that carries the end-effector (QP), and three legs (B_1A_1, B_2A_2, B_3A_3). The legs consist of the link pairs $\{\mathcal{L}_1,\mathcal{L}_4\}$, $\{\mathcal{L}_2,\mathcal{L}_5\}$, and $\{\mathcal{L}_3,\mathcal{L}_6\}$. Each leg is connected to the platforms with a pair of revolute joints and the two links of each leg are connected to each other with a prismatic joint. Thus, according to the popular *designation convention*, this manipulator is designated as a 3RPR planar parallel manipulator. This designation implies that the manipulator has three legs and each leg comprises one prismatic and two revolute joints that are arranged in the indicated order.

For this manipulator, $n_m = 7$ and $j_1 = 9$. Therefore, according to Eqs. (10.6)–(10.9),

$$\mu = 3, n_{ikl} = 2, n_{pv} = 3, n_{sv} = 6$$

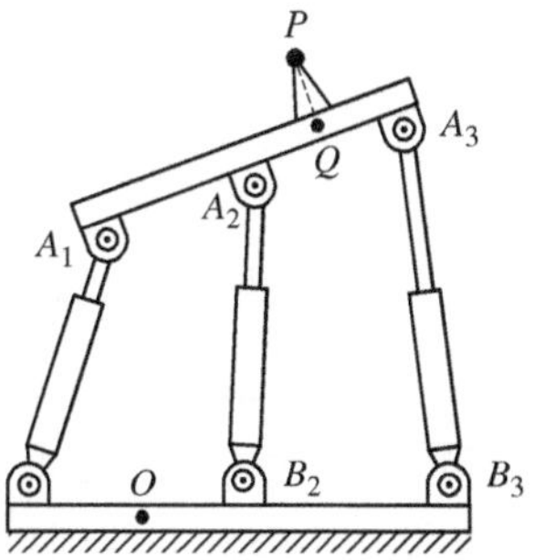

Figure 10.1 A 3RPR planar parallel manipulator with three legs.

The active joints are the prismatic joints, which are controlled by means of suitable linear actuators. Therefore, the primary variables are taken conveniently and appropriately as the active joint variables, which are the lengths of the legs. That is,

$$s_1 = s_{14} = B_1A_1, s_2 = s_{25} = B_2A_2, s_3 = s_{36} = B_3A_3$$

The passive joints are the remaining revolute joints. Therefore, the secondary variables happen to be the passive joint variables, which are the angles between the legs and the platforms. That is,

$$\theta_{01} = \sphericalangle(\mathcal{L}_0 \to \mathcal{L}_1), \theta_{02} = \sphericalangle(\mathcal{L}_0 \to \mathcal{L}_2), \theta_{03} = \sphericalangle(\mathcal{L}_0 \to \mathcal{L}_3)$$

$$\theta_{47} = \sphericalangle(\mathcal{L}_4 \to \mathcal{L}_7), \theta_{57} = \sphericalangle(\mathcal{L}_5 \to \mathcal{L}_7), \theta_{67} = \sphericalangle(\mathcal{L}_6 \to \mathcal{L}_7)$$

The two IKLs (independent kinematic loops) can be taken as IKL-1 $= B_1B_2A_2A_1B_1$ and IKL-2 $= B_2B_3A_3A_2B_2$. In that case, the third kinematic loop KL-3 $= B_1B_3A_3A_1B_1$ becomes dependent on IKL-1 and IKL-2 because it is equal to their *union*, i.e. KL-3 = (IKL-1) ∪ (IKL-2). Alternatively, the two independent loops could be taken as IKL$'$-1 $= B_1B_2A_2A_1B_1$ and IKL$'$-2 $= B_1B_3A_3A_1B_1$. In that case, the third loop KL$'$-3 $= B_2B_3A_3A_2B_2$ would again be dependent on IKL$'$-1 and IKL$'$-2 because these three loops would still be related to each other so that IKL$'$-2 = (IKL$'$-1) ∪ (KL$'$ − 3).

Example 10.2 ***3PRR + 3RPR Planar Parallel Manipulator***

The parallel manipulator shown in Figure 10.2 consists of three platforms. The fixed platform is the base link $\mathcal{L}_0$ that accommodates the guideways for the three sliders, i.e. the links $\mathcal{L}_1$, $\mathcal{L}_2$, and $\mathcal{L}_3$, which function as the *feet* of the manipulator. Their pivot centers are D_1, D_2, and D_3. The terminal platform (link $\mathcal{L}_{14}$) carries the end-effector (QP). The intermediate platform (link $\mathcal{L}_7$) accommodates six bearings. The centers of the upper bearings are B_1, B_2, and B_3. They belong to the legs B_1A_1, B_2A_2, and B_3A_3 that consist of the link pairs $\{\mathcal{L}_8,\mathcal{L}_{11}\}$, $\{\mathcal{L}_9,\mathcal{L}_{12}\}$, and $\{\mathcal{L}_{10},\mathcal{L}_{13}\}$. The centers of the lower bearings are C_1, C_2, C_3. They belong to the legs D_1C_1, D_2C_2, and D_3C_3 that consist of the link pairs $\{\mathcal{L}_1,\mathcal{L}_4\}$, $\{\mathcal{L}_2,\mathcal{L}_5\}$, and $\{\mathcal{L}_3,\mathcal{L}_6\}$. According to the configuration described above, this manipulator is designated as a 3PRR + 3RPR planar parallel manipulator.

For this manipulator, $n_m = 14$ and $j_1 = 18$. Therefore, according to Eqs. (10.6)–(10.9),

$$\mu = 6, n_{ikl} = 4, n_{pv} = 6, n_{sv} = 12$$

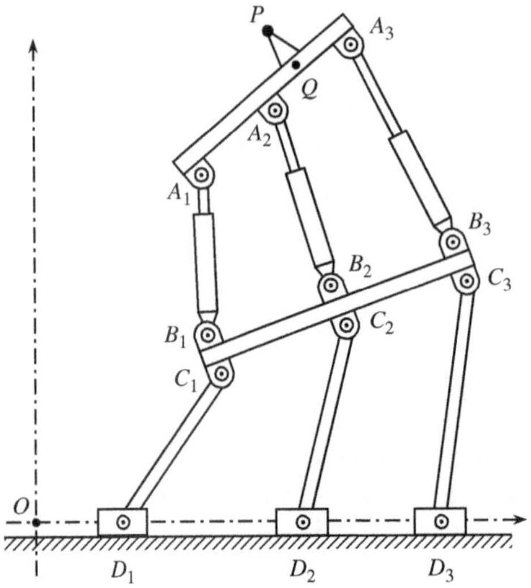

Figure 10.2 A 3PRR + 3RPR planar parallel manipulator with an intermediate platform.

As noticed above, the mobility of this manipulator happens to be $\mu = 6$. Therefore, it requires six active joints in order to control the position of its end-effector. In other words, this manipulator has three degrees of *kinematic redundancy* in its planar working space with $\lambda = 3$. This redundancy can be resolved optimally by minimizing a performance function that is defined appropriately for the task to be executed.

At this point, it is worth noting that, as compared with a serial manipulator, the kinematic redundancy may be more desirable for a parallel manipulator, because it can compensate for the range limitation caused by the kinematic loop or loops.

The active joints of this manipulator are the six prismatic joints. Three of them are the prismatic joints between the sliding feet and the fixed platform. The other three are the prismatic joints between the links of the upper legs. These active joints suggest the following six active joint variables to be taken as the primary variables.

$$s_1 = s_{01} = OD_1, s_2 = s_{02} = OD_2, s_3 = s_{03} = OD_3$$

$$s_4 = s_{11/8} = B_1A_1, s_5 = s_{12/9} = B_2A_2, s_6 = s_{13/10} = B_3A_3$$

Then, the secondary variables happen to be the passive joint variables associated with the 12 revolute joints. They are the angles defined similarly as in Example 10.1.

The four IKLs of this parallel manipulator can be taken as follows:

$$\text{IKL-1} = D_1D_2C_2C_1D_1, \text{IKL-2} = D_2D_3C_3C_2D_2$$

$$\text{IKL-3} = B_1B_2A_2A_1B_1, \text{IKL-4} = B_2B_3A_3A_2B_2$$

Example 10.3 ***Planar Parallel Manipulator Formed by Two Planar Serial Manipulators***

The planar parallel manipulator shown in Figure 10.3 is formed as a combination of two coplanar serial manipulators that are collaborating in order to manipulate a rigid bar that they tightly grip. Owing to the tight grip, the rigid bar and the two grippers may be

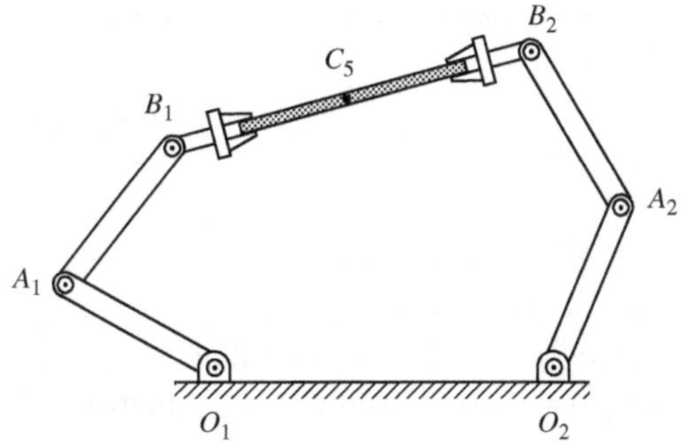

Figure 10.3 A planar parallel manipulator formed by two planar serial manipulators.

assumed to be integrated into a single link, which becomes the moving platform of the manipulator. Thus, $n_m = 5$ and $j_1 = 6$. Therefore, according to Eqs. (10.6)–(10.9),

$$\mu = 3, n_{ikl} = 1, n_{pv} = 3, n_{sv} = 3$$

Note that this parallel manipulator has only one kinematic loop, which is $O_1O_2A_2B_2B_1A_1O_1$, and its mobility is $\mu = 3$. Since $\mu = 3$, only three actuators are sufficient to control the motion of this manipulator as desired. However, since each of the serial manipulators comes in with three actuators, the combined system contains six actuators. If this system is driven by using all the six actuators, then it will have an *actuator redundancy* with three excess actuators. This actuator redundancy can be resolved in various ways. For example, the actuators of the joints at B_1 and B_2 can be left free (i.e. unenergized), while the actuators of the other joints (i.e. the joints at O_1, O_2, A_1, A_2) can be used optimally so as to minimize the torques generated by them.

As for the joint variables, the number $n_{pv} = \mu = 3$ obtained from Eq. (10.8) necessitates that only three of the joint variables be taken as the primary variables, even though more than three actuators may be employed for an optimization purpose. In other words, if a parallel manipulator, such as this one, has actuator redundancy, then the number of the active joint variables may exceed the number of the primary variables.

For this parallel manipulator, the links can be numbered so that $\mathcal{L}_0 = O_1O_2$ (fixed platform), $\mathcal{L}_1 = O_1A_1$, $\mathcal{L}_2 = O_2A_2$, $\mathcal{L}_3 = A_1B_1$, $\mathcal{L}_4 = A_2B_2$, $\mathcal{L}_5 = B_1B_2$ (moving platform). The angles $\theta_1 = \theta_{01}$, $\theta_2 = \theta_{02}$, and $\theta_3 = \theta_{13}$ of the joints at O_1, O_2, and A_1 can be taken as the primary variables. Then, the other angles θ_{24}, θ_{35}, and θ_{45} of the joints at A_2, B_1, and B_2 become the secondary variables. On the other hand, since the four actuators, which are located at O_1, O_2, A_1, and A_2, are used redundantly as mentioned before, the active joint variables are θ_{01}, θ_{02}, θ_{13} and θ_{24} and the passive joint variables are θ_{35} and θ_{45}.

Note that, as it happens in this example, the active and passive joint variables may constitute different sets as compared with the sets of the primary and secondary variables.

Note also that the active joint variables belong necessarily to the active joints, whereas some of the primary variables (or even all of them in general) do not have to belong to the joints. For example, the set of primary variables may include the orientation angles of the links and/or the coordinates of a characteristic point, such as the tip point.

10.2 Position Equations of a Parallel Manipulator

For a parallel manipulator, the following complementary equations can be written in order to express the position of the end-effector in terms of the relevant primary and secondary variables and to set up all the necessary relationships among the primary and secondary variables.

$$\overline{\rho} = \overline{f}(\overline{x}, \overline{y}) \tag{10.10}$$

$$\overline{g}(\overline{x}, \overline{y}) = \overline{0} \tag{10.11}$$

The symbols in the above equations have the following meanings.

$\overline{\rho} \in \mathcal{R}^{\lambda}$ is a column matrix that represents the position (i.e. the orientation and the tip point location) of the end-effector in the *working space*, which is also called *task space*. Note that $\lambda = 6$ for the three-dimensional task space and $\lambda = 3$ for the two-dimensional (planar) task space.
$\overline{x} \in \mathcal{R}^m$ is a column matrix that consists of the *primary variables*. Note that $m = \mu = n_{pv}$.
$\overline{y} \in \mathcal{R}^n$ is a column matrix that consists of the *secondary variables*. Note that $n = n_{sv}$.
$\overline{f}$ is a continuous and differentiable function such that $\overline{f} : \mathcal{R}^{m+n} \to \mathcal{R}^{\lambda}$.
$\overline{g}$ is a continuous and differentiable function such that $\overline{g} : \mathcal{R}^{m+n} \to \mathcal{R}^{n}$.

The elements of $\overline{g}$ are the scalar functions that are obtained from the closure equations written for the IKLs (independent kinematic loops) of the manipulator.

The elements of $\overline{f}$ are the scalar functions that are obtained from the position equation written for the end-effector along a *selected kinematic branch* that connects the terminal platform to the fixed platform. Obviously, there is not only one of such branches. Indeed, even a simple parallel manipulator with a single loop has two such branches. In general, a parallel manipulator may have n_{br} such branches. Naturally, the branch selected to generate the function $\overline{f}$ is supposed to be as short as possible for the sake of kinematic simplicity.

If the working space of the parallel manipulator is three dimensional, then $\overline{\rho}$ is partitioned as

$$\overline{\rho} = \begin{bmatrix} \overline{p} \\ \overline{\phi} \end{bmatrix} \tag{10.12}$$

In Eq. (10.12), $\overline{p}$ consists of the coordinates of the tip point of the end-effector with respect to the reference frame attached to the fixed platform and $\overline{\phi}$ consists of the Euler angles of a suitable sequence (e.g. RFB-123 or RFB-323) that describe the orientation of the terminal platform with respect to the fixed platform. In more detail,

$$\overline{p} = \begin{bmatrix} p_1 \\ p_2 \\ p_3 \end{bmatrix}, \quad \overline{\phi} = \begin{bmatrix} \phi_1 \\ \phi_2 \\ \phi_3 \end{bmatrix} \tag{10.13}$$

If the working space of the parallel manipulator is two dimensional, then $\overline{\rho}$ consists of the two coordinates of the tip point and the orientation angle of the end-effector. That is,

$$\overline{\rho} = \begin{bmatrix} p_1 \\ p_2 \\ \phi \end{bmatrix} \tag{10.14}$$

A parallel manipulator is called *regular* if it does not have any kinematic or actuator redundancy or deficiency. For a regular manipulator, it happens that $m = \mu = \lambda$.

Example 10.4 ***Position Equations of a 3RRR Planar Parallel Manipulator***

Consider the 3RRR planar parallel manipulator shown in Figure 10.4. The joint centers A_1 and A_2 are coincident. The origin O of the base frame is located at the joint center

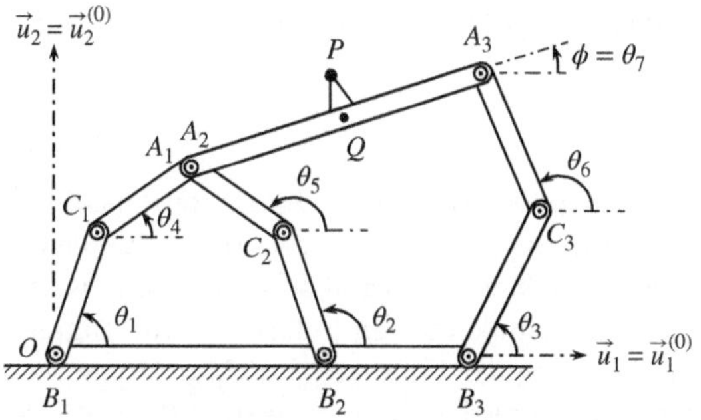

Figure 10.4 A 3RRR planar parallel manipulator.

B_1. The links of the manipulator are numbered so that $\mathcal{L}_0$ is the fixed platform ($B_1B_2B_3$), $\mathcal{L}_1$ and $\mathcal{L}_4$ form the first leg L_1, $\mathcal{L}_2$ and $\mathcal{L}_5$ form the second leg L_2, $\mathcal{L}_3$ and $\mathcal{L}_6$ form the third leg L_3, and $\mathcal{L}_7$ is the moving platform ($A_1A_2A_3$). The end-effector is a tool shown as QP. The active joints of the manipulator are the three revolute joints located at the points B_1, B_2, and B_3.

The constant geometric parameters of the manipulator are listed below.

$$b_2 = B_1B_2, b_3 = B_1B_3, b_3' = B_2B_3$$

$$d_1 = A_1Q, d_3 = A_3Q, d_7 = A_1A_3, h_7 = QP$$

$$r_1 = B_1C_1, r_2 = B_2C_2, r_3 = B_3C_3$$

$$r_4 = C_1A_1, r_5 = C_2A_2, r_6 = C_3A_3$$

According to Eqs. (10.6)–(10.9),

$$\mu = 3n_m - 2j_1 = 3 \times 7 - 2 \times 9 = 3, n_{ikl} = j_1 - n_m = 9 - 7 = 2$$

$$n_{pv} = \mu = 3, n_{sv} = 3n_{ikl} = 6$$

The three primary variables are taken as the following active joint variables.

$$\theta_1 = \theta_{01}, \theta_2 = \theta_{02}, \theta_3 = \theta_{03}$$

Then, the passive joint variables become the six secondary variables. They are expressed as follows by using the orientation angles of the links that are shown in Figure 10.4 and by noting that $\phi = \theta_7$:

$$\theta_{14} = \theta_4 - \theta_1, \theta_{25} = \theta_5 - \theta_2, \theta_{36} = \theta_6 - \theta_3$$

$$\theta_{47} = \phi - \theta_4, \theta_{57} = \phi - \theta_5, \theta_{67} = \phi - \theta_6$$

The two IKLs can be taken as IKL-1 = $B_1B_2C_2A_2A_1C_1B_1$ and IKL-2 = $B_1B_3C_3A_3A_2A_1C_1B_1$. For these loops, the loop equations can be written as explained below.

∗ Equation for the link orientations around IKL-1:

$$\theta_{01} + \theta_{14} + \theta_{47} = \theta_{02} + \theta_{25} + \theta_{57} \Rightarrow$$

$$\theta_1 + (\theta_4 - \theta_1) + (\phi - \theta_4) = \theta_2 + (\theta_5 - \theta_2) + (\phi - \theta_5) \Rightarrow 0 = 0 \tag{10.15}$$

∗ Equation for the link orientations around IKL-2:

$$\theta_{01} + \theta_{14} + \theta_{47} = \theta_{03} + \theta_{36} + \theta_{67} \Rightarrow$$

$$\theta_1 + (\theta_4 - \theta_1) + (\phi - \theta_4) = \theta_3 + (\theta_6 - \theta_3) + (\phi - \theta_6) \Rightarrow 0 = 0 \tag{10.16}$$

* End-effector orientation through the kinematic branch $OC_1A_1A_2QP$:

$$\theta_{07} = \theta_{01} + \theta_{14} + \theta_{47} \Rightarrow$$

$$\phi = \theta_1 + (\theta_4 - \theta_1) + (\phi - \theta_4) \Rightarrow 0 = 0 \tag{10.17}$$

As noticed above, the loop and branch equations written for the orientations of the relevant links are identically satisfied. Actually, this is a general property of the planar mechanisms. Therefore, for the planar mechanisms, the loop and branch equations can be simplified by skipping the orientation relationships written in terms of the angular joint variables. This can be done because the angular joint variable θ_{ij} of a revolute joint between two adjacent links $\mathcal{L}_i$ and $\mathcal{L}_j$ can simply be obtained as $\theta_{ij} = \theta_{j/i} = \theta_j - \theta_i$ without requiring the solution of any equation. Here, $\theta_i = \theta_{0i}$ and $\theta_j = \theta_{0j}$ are the orientation angles of the links $\mathcal{L}_i$ and $\mathcal{L}_j$ with respect to the base link $\mathcal{L}_0$.

* Equations for the joint center locations around IKL-1:

$$\overrightarrow{B_1C_1} + \overrightarrow{C_1A_1} = \overrightarrow{B_1B_2} + \overrightarrow{B_2C_2} + \overrightarrow{C_2A_2} + \overrightarrow{A_2A_1} \Rightarrow$$

$$r_1\vec{u}(\theta_1) + r_4\vec{u}(\theta_4) = b_2\vec{u}(0) + r_2\vec{u}(\theta_2) + r_5\vec{u}(\theta_5) + \vec{0} \Rightarrow$$

$$r_1c\theta_1 + r_4c\theta_4 = b_2 + r_2c\theta_2 + r_5c\theta_5 \tag{10.18}$$

$$r_1s\theta_1 + r_4s\theta_4 = r_2s\theta_2 + r_5s\theta_5 \tag{10.19}$$

* Equations for the joint center locations around IKL-2:

$$\overrightarrow{B_1C_1} + \overrightarrow{C_1A_1} = \overrightarrow{B_1B_3} + \overrightarrow{B_3C_3} + \overrightarrow{C_3A_3} + \overrightarrow{A_3A_1} \Rightarrow$$

$$r_1\vec{u}(\theta_1) + r_4\vec{u}(\theta_4) = b_3\vec{u}(0) + r_3\vec{u}(\theta_3) + r_6\vec{u}(\theta_6) - d_7\vec{u}(\theta_7) \Rightarrow$$

$$r_1c\theta_1 + r_4c\theta_4 = b_3 + r_3c\theta_3 + r_6c\theta_6 - d_7c\theta_7 \tag{10.20}$$

$$r_1s\theta_1 + r_4s\theta_4 = r_3s\theta_3 + r_6s\theta_6 - d_7s\theta_7 \tag{10.21}$$

* End-effector location through the kinematic branch $OC_1A_1A_2QP$:

$$\overrightarrow{OP} = \overrightarrow{B_1P} = \overrightarrow{B_1C_1} + \overrightarrow{C_1A_1} + \overrightarrow{A_1Q} + \overrightarrow{QP} \Rightarrow$$

$$p_1\vec{u}_1 + p_2\vec{u}_2 = r_1\vec{u}(\theta_1) + r_4\vec{u}(\theta_4) + d_1\vec{u}(\theta_7) + h_7\vec{u}(\theta_7 + \pi/2) \Rightarrow$$

$$p_1 = r_1c\theta_1 + r_4c\theta_4 + d_1c\theta_7 - h_7s\theta_7 \tag{10.22}$$

$$p_2 = r_1s\theta_1 + r_4s\theta_4 + d_1s\theta_7 + h_7c\theta_7 \tag{10.23}$$

In the above vector equations, $\vec{u}(\theta)$ is the *directional unit vector*. Referring to the base frame $\mathcal{F}_0(O)$, it is defined as follows by using the abbreviation that $\vec{u}_k = \vec{u}_k^{(0)}$ for $k = 1, 2, 3$:

$$\vec{u}(\theta) = \vec{u}_1c\theta + \vec{u}_2s\theta \tag{10.24}$$

Finally, Eqs. (10.10) and (10.11) can be written as follows by using the preceding scalar equations, in which the secondary variables are taken as the link angles θ_4, θ_5, θ_6, and θ_7:

$$\bar{p} = \bar{f}(\bar{x}, \bar{y}) \Rightarrow \begin{bmatrix} p_1 \\ p_2 \\ \phi \end{bmatrix} = \begin{bmatrix} r_1c\theta_1 + r_4c\theta_4 + d_1c\theta_7 - h_7s\theta_7 \\ r_1s\theta_1 + r_4s\theta_4 + d_1s\theta_7 + h_7c\theta_7 \\ \theta_7 \end{bmatrix} \tag{10.25}$$

$$\bar{g}(\bar{x}, \bar{y}) = \bar{0} \Rightarrow \begin{bmatrix} 0 \\ 0 \\ b_2 + r_2c\theta_2 + r_5c\theta_5 - r_1c\theta_1 - r_4c\theta_4 \\ r_2s\theta_2 + r_5s\theta_5 - r_1s\theta_1 - r_4s\theta_4 \\ b_3 + r_3c\theta_3 + r_6c\theta_6 - d_7c\theta_7 - r_1c\theta_1 - r_4c\theta_4 \\ r_3s\theta_3 + r_6s\theta_6 - d_7s\theta_7 - r_1s\theta_1 - r_4s\theta_4 \end{bmatrix} = \begin{bmatrix} 0 \\ 0 \\ 0 \\ 0 \\ 0 \\ 0 \end{bmatrix} \tag{10.26}$$

However, as seen above, the orientation equations written for a planar parallel manipulator are identically satisfied when the orientation angles of the links are used instead of the angular joint variables. Owing to this fact, Eqs. (10.10) and (10.11) can be modified to the following dimensionally reduced equations, in which the angle ϕ of the end-effector (i.e. the terminal platform) is treated, not as an element of $\overline{\rho}$, but as an element of $\overline{y}$.

$$\overline{\rho} \;\Rightarrow\; \overline{p} = \overline{f}(\overline{x}, \overline{y}) \;\Rightarrow\; \begin{bmatrix} p_1 \\ p_2 \end{bmatrix} = \begin{bmatrix} r_1 c\theta_1 + r_4 c\theta_4 + d_1 c\phi - h_7 s\phi \\ r_1 s\theta_1 + r_4 s\theta_4 + d_1 s\phi + h_7 c\phi \end{bmatrix} \tag{10.27}$$

$$\overline{g}(\overline{x}, \overline{y}) = \overline{0} \;\Rightarrow\; \begin{bmatrix} b_2 + r_2 c\theta_2 + r_5 c\theta_5 - r_1 c\theta_1 - r_4 c\theta_4 \\ r_2 s\theta_2 + r_5 s\theta_5 - r_1 s\theta_1 - r_4 s\theta_4 \\ b_3 + r_3 c\theta_3 + r_6 c\theta_6 - d_7 c\phi - r_1 c\theta_1 - r_4 c\theta_4 \\ r_3 s\theta_3 + r_6 s\theta_6 - d_7 s\phi - r_1 s\theta_1 - r_4 s\theta_4 \end{bmatrix} = \begin{bmatrix} 0 \\ 0 \\ 0 \\ 0 \end{bmatrix} \tag{10.28}$$

10.3 Forward Kinematics in the Position Domain

The forward kinematics of a parallel manipulator in the position domain involves the *forward kinematic solution* together with the identification and analysis of the associated multiplicities and singularities. More specifically, the multiplicities are called *posture multiplicities of forward kinematics* (PMFKs) and the singularities are called *position singularities of forward kinematics* (PSFKs).

The forward kinematic solution is needed in order to determine the position of the end-effector (i.e. $\overline{p}$) and the secondary variables (i.e. $\overline{y}$) that are induced by the specified values of the primary variables (i.e. $\overline{x}$). The solution can be obtained in two stages.

In the first stage of the forward kinematic solution, Eq. (10.11) is solved in order to obtain $\overline{y}$ as a function of $\overline{x}$. In general, this stage does not allow an analytical solution, that is, the solution can only be obtained either in a completely numerical way or in a semi-analytical way. In either case, $\overline{y}$ becomes available discretely as a *tabulated function* of discretely specified $\overline{x}$.

If the manipulator is *not* in a pose of PSFK, then Eq. (10.11) can be solved so that

$$\overline{y} = \overline{y}^{(k)}(\overline{x}); \quad k = 1, 2, \ldots, n_{pm} \tag{10.29}$$

In Eq. (10.29), n_{pm} is the number of the different *posture modes*, any of which may be assumed by the manipulator for the same specified values of the primary variables (i.e. $\overline{x}$). In the same equation, $\overline{y}^{(k)}(\overline{x})$ is a function that represents the kth posture mode denoted as PM-k.

For a manipulator, an appropriate posture mode is actually selected while it is assembled. Therefore, a posture mode may also be called an *assembly mode*. The selection of an appropriate posture mode depends on the environment and the operating conditions of the manipulator. Once the manipulator is assembled in a certain posture mode, then it keeps operating in the same posture mode, unless it gets into a *posture mode changing pose* (PMCP-*ij*), which is the *borderline* between two posture modes PM-*i* and PM-*j*. If it gets into a PMCP-*ij*, then the current posture mode PM-*i* may be changed into the other posture mode PM-*j* easily by a small external intervention (intentional or unintentional) without the necessity of disassembling and reassembling the manipulator.

In the second stage of the forward kinematic solution, the position of the end-effector (i.e. $\overline{p}$) is determined simply by substituting Eq. (10.29) into Eq. (10.10). That is,

$$\overline{\rho} = \overline{p}^{(k)}(\overline{x}) = \overline{f}(\overline{x}, \overline{y}^{(k)}(\overline{x})) \tag{10.30}$$

Equation (10.30) indicates that $\overline{\rho}$ is determined not only by the specified $\overline{x}$ but also by the selected posture mode PM-k.

If the manipulator happens to be in a pose of PSFK, then that pose of the manipulator and the corresponding position of the end-effector become indefinite and therefore uncontrollable. This phenomenon is explained below.

A pose of PSFK occurs if $\overline{x}$ is specified in such a special way that a certain partition of it, e.g. $\overline{x}_b \in \mathcal{R}^{m_b}$, is tied up to its complementary partition $\overline{x}_a \in \mathcal{R}^{m_a}$, where $m_a + m_b = m$. In other words, $\overline{x}_a$ is still specified arbitrarily but $\overline{x}_b$ is specified so that it depends on $\overline{x}_a$ through a *singularity relationship* such as

$$\overline{x}_b = \overline{\xi}_{ba}(\overline{x}_a) \tag{10.31}$$

When $\overline{x}$ is inserted with its above-mentioned partitions, Eq. (10.11) can be separated into the following consistently partitioned equations, in which $\overline{y}$ is also similarly partitioned.

$$\overline{g}_a(\overline{x}_a, \overline{\xi}_{ba}(\overline{x}_a), \overline{y}_a, \overline{y}_b) = \overline{0} \tag{10.32}$$

$$\overline{g}_b(\overline{x}_a, \overline{\xi}_{ba}(\overline{x}_a), \overline{y}_a, \overline{y}_b) = \overline{0} \tag{10.33}$$

In the above equations,

$$\overline{y}_b \in \mathcal{R}^{m_b}, \ \ \overline{g}_b \in \mathcal{R}^{m_b}; \ \ \overline{y}_a \in \mathcal{R}^{n_a}, \ \ \overline{g}_a \in \mathcal{R}^{n_a}; \ \ n_a = n - m_b \tag{10.34}$$

The characterizing feature of a PSFK is that Eq. (10.33) turns into an identity and therefore it is satisfied no matter what $\overline{y}_b$ is. Thus, $\overline{y}_b$ becomes indefinite. Meanwhile, Eqs. (10.32) and (10.10) give $\overline{y}_a$ and $\overline{\rho}$ (again with multiplicities) as the following functions of the freely specified partition $\overline{x}_a$ and the indefinite partition $\overline{y}_b$.

$$\overline{y}_a = \overline{y}_a^{(k)}(\overline{x}_a, \overline{y}_b); \ \ k = 1, 2, \ldots, n'_{pm} \tag{10.35}$$

$$\overline{\rho} = \overline{\rho}^{(k)}(\overline{x}_a, \overline{y}_b) = \overline{f}(\overline{x}_a, \overline{\xi}_{ba}(\overline{x}_a), \overline{y}_a^{(k)}(\overline{x}_a, \overline{y}_b), \overline{y}_b) \tag{10.36}$$

Consequently, $\overline{y}_a$ and $\overline{\rho}$ also become indefinite due to their dependence on $\overline{y}_b$. This means that the manipulator becomes uncontrollable in a pose of PSFK. Therefore, it must be prevented from getting into a pose of PSFK. This can be achieved in two ways. One way is to design the manipulator so that its geometric parameters do not allow any PSFK to occur. The other way is to use the manipulator carefully so that it does not get into a pose of PSFK. In other words, the planned task must be such that it does not give any chance for the primary variables to become interrelated as in Eq. (10.31).

Example 10.5 ***Forward Kinematics of the 3RRR Planar Parallel Manipulator***

For the 3RRR parallel manipulator shown in Figure 10.4, the scalar loop equations, i.e. Eqs. (10.18)–(10.21), can be written again as shown below.

$$r_5 c\theta_5 = r_4 c\theta_4 - (b_2 + r_2 c\theta_2 - r_1 c\theta_1) \tag{10.37}$$

$$r_5 s\theta_5 = r_4 s\theta_4 - (r_2 s\theta_2 - r_1 s\theta_1) \tag{10.38}$$

$$d_7 c\phi = r_6 c\theta_6 - (r_1 c\theta_1 + r_4 c\theta_4 - r_3 c\theta_3 - b_3) \tag{10.39}$$

$$d_7 s\phi = r_6 s\theta_6 - (r_1 s\theta_1 + r_4 s\theta_4 - r_3 s\theta_3) \tag{10.40}$$

(a) Forward Kinematic Solution

Here, the primary variables (i.e. the angles θ_1, θ_2, and θ_3) are specified and the consequent secondary variables together with the tip point coordinates are found as explained below. Fortunately, it has been possible to find them in a completely analytical way for this particular manipulator.
With known θ_1 and θ_2, Eqs. (10.37) and (10.38) of the first kinematic loop IKL-1 can be written briefly as follows:

$$r_5 c\theta_5 = r_4 c\theta_4 - x_{12} \tag{10.41}$$

$$r_5 s\theta_5 = r_4 s\theta_4 - y_{12} \tag{10.42}$$

In Eqs. (10.41) and (10.42), x_{12} and y_{12} are the known components of the relative position vector $\overrightarrow{C_1C_2}$. That is,

$$x_{12} = b_2 + r_2 c\theta_2 - r_1 c\theta_1 \tag{10.43}$$

$$y_{12} = r_2 s\theta_2 - r_1 s\theta_1 \tag{10.44}$$

The squares of Eqs. (10.41) and (10.42) can be added up side by side so that

$$x_{12} c\theta_4 + y_{12} s\theta_4 = f_{12} \tag{10.45}$$

In Eq. (10.45), f_{12} is known as

$$f_{12} = (x_{12}^2 + y_{12}^2 + r_4^2 - r_5^2)/(2r_4) \tag{10.46}$$

If x_{12} and y_{12} are not both zero, then Eq. (10.45) gives θ_4 as described below.

$$\theta_4 = \psi_{12} + \sigma_1 \gamma_{12} \tag{10.47}$$

$$\psi_{12} = \text{atan}_2(y_{12}, x_{12}) \tag{10.48}$$

$$\gamma_{12} = \text{atan}_2(g_{12}, f_{12}) \tag{10.49}$$

$$g_{12} = \sqrt{x_{12}^2 + y_{12}^2 - f_{12}^2} \tag{10.50}$$

$$\sigma_1 = \pm 1 \tag{10.51}$$

In Eq. (10.47), $\psi_{12} = \sphericalangle(\overrightarrow{C_1C_2})$ and σ_1 is the *posture mode sign variable* of the first kinematic loop IKL-1. This sign variable is to be selected according to the desired posture mode. The way it is selected is explained in the following Part (b) about the posture modes.
As a special case, if $x_{12} = y_{12} = 0$, then the manipulator will be in a pose of position singularity, which will be discussed later in Part (c).
With the availability of θ_4, Eqs. (10.41) and (10.42) give θ_5 as follows without any additional sign variable:

$$\theta_5 = \text{atan}_2[(r_4 s\theta_4 - y_{12}), (r_4 c\theta_4 - x_{12})] \tag{10.52}$$

As for Eqs. (10.39) and (10.40) of the second kinematic loop IKL-2, they can be written briefly as follows by using the angles available at this stage, i.e. θ_1, θ_2, θ_3, and θ_4:

$$d_7 c\phi = r_6 c\theta_6 - x_{67} \tag{10.53}$$

$$d_7 s\phi = r_6 s\theta_6 - y_{67} \tag{10.54}$$

In Eqs. (10.53) and (10.54), x_{67} and y_{67} are the known components of the relative position vector $\overrightarrow{C_3A_1} = \overrightarrow{C_3A_2}$. That is,

$$x_{67} = r_1 c\theta_1 + r_4 c\theta_4 - r_3 c\theta_3 - b_3 \tag{10.55}$$

$$y_{67} = r_1 s\theta_1 + r_4 s\theta_4 - r_3 s\theta_3 \tag{10.56}$$

If x_{67} and y_{67} are not both zero, then Eqs. (10.53) and (10.54) can be solved similarly to Eqs. (10.41) and (10.42) in order to find θ_6 and ϕ. The solution is summarized below.

$$\theta_6 = \psi_{67} + \sigma_2 \gamma_{67} \tag{10.57}$$

$$\psi_{67} = \text{atan}_2(y_{67}, x_{67}) \tag{10.58}$$

$$\gamma_{67} = \text{atan}_2(g_{67}, f_{67}) \tag{10.59}$$

$$f_{67} = (x_{67}^2 + y_{67}^2 + r_6^2 - d_7^2)/(2r_6) \tag{10.60}$$

$$g_{67} = \sqrt{x_{67}^2 + y_{67}^2 - f_{67}^2} \tag{10.61}$$

$$\sigma_2 = \pm 1 \tag{10.62}$$

$$\phi = \text{atan}_2[(r_6 s\theta_6 - y_{67}), (r_6 c\theta_6 - x_{67})] \tag{10.63}$$

In Eq. (10.57), $\psi_{67} = \sphericalangle(\overrightarrow{C_3A_1})$ and σ_2 is the *posture mode sign variable* of the second kinematic loop IKL-2. Like the previous σ_1, it is to be selected according to the desired posture mode. The way it is selected is also explained in the following Part (b) about the posture modes.

As a special case, if $x_{67} = y_{67} = 0$, then the manipulator will be in another pose of position singularity, which will be discussed later in Part (c).

After finding ϕ and all the secondary variables as explained above, the coordinates of the tip point (p_1 and p_2) can be obtained readily from Eq. (10.27). That is,

$$p_1 = r_1 c\theta_1 + r_4 c\theta_4 + d_1 c\phi - h_7 s\phi \tag{10.64}$$

$$p_2 = r_1 s\theta_1 + r_4 s\theta_4 + d_1 s\phi + h_7 c\phi \tag{10.65}$$

Note that θ_4 is affected by σ_1 whereas ϕ is affected by both σ_1 and σ_2. Therefore, for the same specifed values of the primary angles θ_1, θ_2, and θ_3, the end-effector may assume four different positions depending on the selected values of σ_1 and σ_2. These positions are illustrated in Figure 10.5 associated with Part (b) about the posture modes.

(b) Posture Modes of Forward Kinematics and Posture Mode Changing Poses

As seen in Part (a), when actuated by means of the primary angles θ_1, θ_2, and θ_3, the manipulator may be operated in one of the four different posture modes, which are represented by the sign variables σ_1 and σ_2. The relationships between the posture modes and the sign variables σ_1 and σ_2 are explained below.

Considering σ_1 with its role in Eq. (10.47) and referring to Figure 10.4 by recalling that $\psi_{12} = \sphericalangle(\overrightarrow{C_1C_2})$ and noting that $\gamma_{12} \geq 0$ (because $g_{12} \geq 0$), σ_1 leads to the following posture modes. They depend on x_{12} because $x_{12} > 0$ makes ψ_{12} an acute angle, whereas $x_{12} < 0$ makes ψ_{12} an obtuse angle. They are illustrated schematically in Figure 10.5.

$$\text{PM-1 } (A_1 \text{ above } C_1C_2) \Rightarrow \sigma_1 = +\text{sgn}(x_{12}) \tag{10.66}$$

$$\text{PM-2 } (A_1 \text{ below } C_1C_2) \Rightarrow \sigma_1 = -\text{sgn}(x_{12}) \tag{10.67}$$

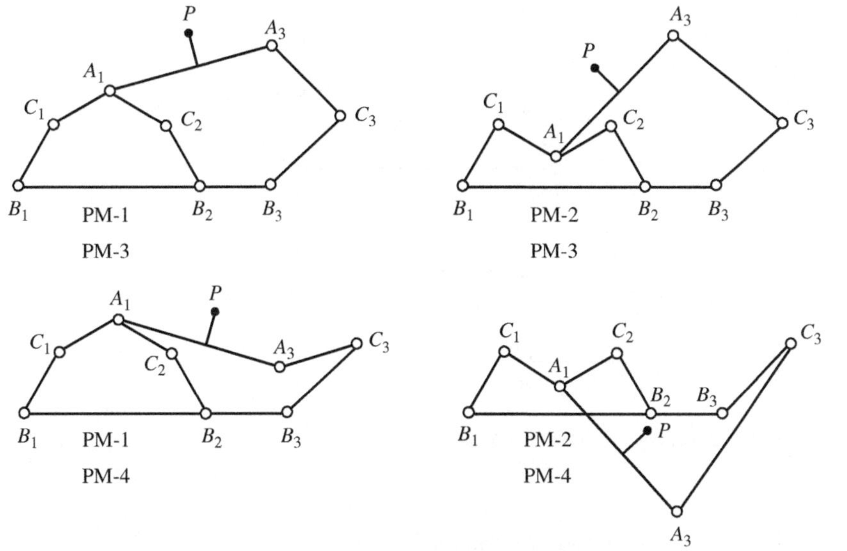

Figure 10.5 Posture modes of the 3RRR planar parallel manipulator.

The borderline between the posture modes considered above is the PMCP (PMCP-12) between them. Such a pose occurs if $\gamma_{12} = 0$ so that

$$\theta_4 = \psi_{12} \tag{10.68}$$

Equation (10.68) implies that the lines C_1A_1 and C_1C_2 are aligned. In other words, in PMCP-12, the coincident points A_1 and A_2 become located right on the line between the points C_1 and C_2. Note that, in such a critical pose, the manipulator can easily be perturbed to make it continue its subsequent operation in any one of PM-1 and PM-2.

Considering σ_2 with its role in Eq. (10.57) and referring to Figure 10.4 by recalling that $\psi_{67} = \sphericalangle(\overrightarrow{C_3A_1})$ and noting that $\gamma_{67} \geq 0$ (because $g_{67} \geq 0$), σ_2 leads to the following posture modes. They depend on x_{67} because $x_{67} > 0$ makes ψ_{67} an acute angle, whereas $x_{67} < 0$ makes ψ_{67} an obtuse angle. They are illustrated schematically in Figure 10.5.

$$\text{PM-3 } (A_3 \text{ above } C_3A_1) \Rightarrow \sigma_2 = -\text{sgn}(x_{67}) \tag{10.69}$$

$$\text{PM-4 } (A_3 \text{ below } C_3A_1) \Rightarrow \sigma_2 = +\text{sgn}(x_{67}) \tag{10.70}$$

The borderline between the posture modes considered above is the PMCP (PMCP-34) between them. Such a pose occurs if $\gamma_{67} = 0$ so that

$$\theta_6 = \psi_{67} \tag{10.71}$$

Equation (10.71) implies that the lines C_3A_2 and C_3A_3 are aligned. In other words, in PMCP-34, the point A_3 becomes located right on the line between the points C_3 and A_2. Note that this is another critical pose, in which the manipulator can easily be perturbed to make it continue its subsequent operation in any one of PM-3 and PM-4.

The two PMCPs described above and their combination as the third PMCP are illustrated schematically in Figure 10.6.

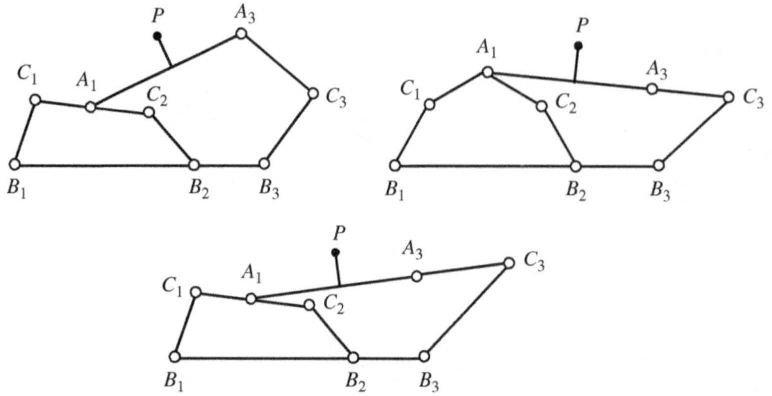

Figure 10.6 PMCPs of the 3RRR planar parallel manipulator.

(c) Position Singularities of Forward Kinematics

As mentioned in Part (a), there are two distinct versions of PSFKs. They are denoted as PSFK-1 and PSFK-2. They may occur separately at different times or simultaneously at the same time. They are described and discussed below.

The first kind of position singularity (PSFK-1) occurs if $x_{12} = y_{12} = 0$ in Eqs. (10.41) and (10.42). When it occurs, Eqs. (10.41) and (10.42) reduce to the following equations.

$$r_5 c\theta_5 = r_4 c\theta_4 \tag{10.72}$$

$$r_5 s\theta_5 = r_4 s\theta_4 \tag{10.73}$$

Equations (10.72) and (10.73) can be satisfied nontrivially (i.e. with nonzero link lengths) only if the link lengths and the angles are equal to each other. That is,

$$r_5 = r_4 \tag{10.74}$$

$$\theta_5 = \theta_4 \tag{10.75}$$

Equation (10.74) implies that PSFK-1 can never occur if $r_5 \neq r_4$. In other words, if the manipulator is designed so that $r_5 \neq r_4$, then PSFK-1 is prevented for sure.

On the other hand, if PSFK-1 occurs due to having $r_5 = r_4$, then Eq. (10.75) implies that θ_4 and θ_5 cannot be found even though they turn out to be equal. In other words, θ_4 and θ_5 become equal but indefinite. Consequently, the links $\mathcal{L}_4$ and $\mathcal{L}_5$ become coincident and they gain a DoF to rotate arbitrarily in an uncontrollable way about the points C_1 and C_2 that are now coincident. Since the rotation of the coincident links $\mathcal{L}_4$ and $\mathcal{L}_5$ becomes uncontrollable, the positions of the moving platform and the end-effector also become uncontrollable. This situation is illustrated in Figure 10.7.

As for the active joint variables θ_1 and θ_2, Eqs. (10.43) and (10.44) imply that they must satisfy the following equations due to having $x_{12} = y_{12} = 0$ in a pose of PSFK-1.

$$x_{12} = 0 \Rightarrow r_2 c\theta_2 = r_1 c\theta_1 - b_2 \tag{10.76}$$

$$y_{12} = 0 \Rightarrow r_2 s\theta_2 = r_1 s\theta_1 \tag{10.77}$$

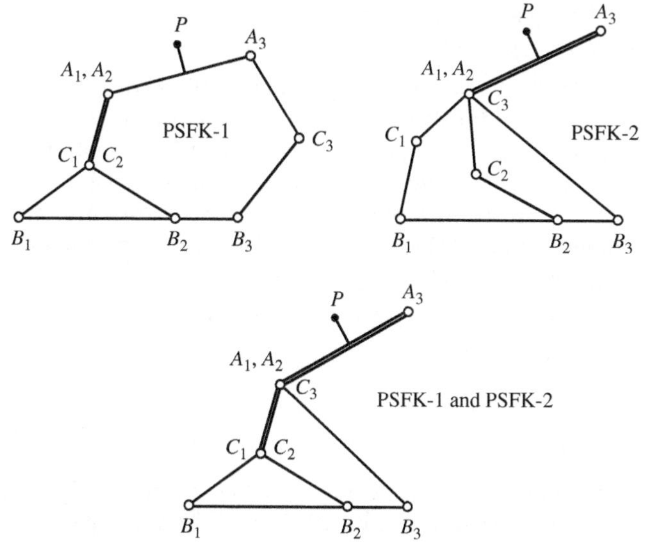

Figure 10.7 PSFK poses of the 3RRR planar parallel manipulator.

Equations (10.76) and (10.77) imply that PSFK-1 occurs if θ_1 and θ_2 are specified with the following *singularity relationship* in between for a manipulator with $r_5 = r_4$.

$$\theta_2 = \xi_{21}(\theta_1) = \text{atan}_2[(r_1 s\theta_1), (r_1 c\theta_1 - b_2)] \tag{10.78}$$

Equations (10.76) and (10.77) also imply that θ_1 and θ_2 lead to PSFK-1 if they are specified with the following special values.

$$c\theta_1 = \xi_1 = (r_1^2 - r_2^2 + b_2^2)/(2b_2r_1) \Rightarrow s\theta_1 = \eta_1 = \sigma_1'\sqrt{1-\xi_1^2} \Rightarrow$$
$$\theta_1 = \sigma_1'\text{atan}_2\left(\sqrt{1-\xi_1^2}, \xi_1\right) \tag{10.79}$$

$$c\theta_2 = \xi_2 = (r_1^2 - r_2^2 - b_2^2)/(2b_2r_2) \Rightarrow s\theta_2 = \eta_2 = \sigma_1'\sqrt{1-\xi_2^2} \Rightarrow$$
$$\theta_2 = \sigma_1'\text{atan}_2\left(\sqrt{1-\xi_2^2}, \xi_2\right) \tag{10.80}$$

In Eqs. (10.79) and (10.80), $\sigma_1' = \pm 1$ and this sign variable is shared by θ_1 and θ_2 due to the presence of Eq. (10.77). Note that $\sigma_1' = +1$ corresponds to a pose of PSFK-1 such that the points C_1 and C_2 become coincident above the line B_1B_2 and $\sigma_1' = -1$ corresponds to the antisymmetric pose such that the points C_1 and C_2 become coincident below the line B_1B_2.

The second kind of position singularity (PSFK-2) occurs if $x_{67} = y_{67} = 0$ in Eqs. (10.53) and (10.54). When it occurs, Eqs. (10.53) and (10.54) reduce to the following equations.

$$d_7 c\phi = r_6 c\theta_6 \tag{10.81}$$
$$d_7 s\phi = r_6 s\theta_6 \tag{10.82}$$

Equations (10.81) and (10.82) can be satisfied nontrivially (i.e. with nonzero link lengths) only if the link lengths and the angles are equal to each other. That is,

$$d_7 = r_6 \tag{10.83}$$
$$\phi = \theta_6 \tag{10.84}$$

Equation (10.83) implies that PSFK-2 can never occur if $d_7 \neq r_6$. In other words, if the manipulator is designed so that $d_7 \neq r_6$, then PSFK-2 is prevented for sure.

On the other hand, if PSFK-2 occurs due to having $d_7 = r_6$, then Eq. (10.84) implies that θ_6 and ϕ cannot be found even though they turn out to be equal. In other words, θ_6 and ϕ become equal but indefinite. Consequently, the links $\mathcal{L}_6$ and $\mathcal{L}_7$ (the moving platform) become coincident and they gain a DoF to rotate arbitrarily in an uncontrollable way about the points A_1, A_2, and C_3 that are all coincident now. Since the rotation of the coincident links $\mathcal{L}_6$ and $\mathcal{L}_7$ becomes uncontrollable, the position of the end-effector also becomes uncontrollable. This situation is also illustrated in Figure 10.7.

As for the active joint variables θ_1, θ_2, and θ_3, Eqs. (10.55) and (10.56) imply that they must satisfy the following equations due to having $x_{67} = y_{67} = 0$ in a pose of PSFK-2.

$$x_{67} = 0 \Rightarrow r_3 c\theta_3 = r_1 c\theta_1 + r_4 c\theta_4 - b_3 \tag{10.85}$$

$$y_{67} = 0 \Rightarrow r_3 s\theta_3 = r_1 s\theta_1 + r_4 s\theta_4 \tag{10.86}$$

Note that, if PSFK-1 is *not* concurrent, then θ_4 is already available in terms of θ_1 and θ_2 within a selected posture mode as expressed by Eq. (10.47) and the accompanying equations. Otherwise, i.e. if PSFK-1 *is* concurrent, then as discussed before, θ_4 becomes unrelated to θ_1 and θ_2.

Hence, in the absence of PSFK-1, Eqs. (10.85) and (10.86) imply that PSFK-2 occurs if θ_1, θ_2, and θ_3 are specified with the following *singularity relationship* among them.

$$\theta_3 = \xi_{312}(\theta_1, \theta_2) = \operatorname{atan}_2[(r_1 s\theta_1 + r_4 s\theta_4), (r_1 c\theta_1 + r_4 c\theta_4 - b_3)] \tag{10.87}$$

In Eq. (10.87),

$$\theta_4 = f_4(\theta_1, \theta_2, \sigma_1) \tag{10.88}$$

On the other hand, Eqs. (10.85) and (10.86) also imply that θ_3 and θ_4 assume special values induced by the specified value of θ_1. In order to find these special values, Eqs. (10.85) and (10.86) can be rearranged as follows:

$$r_4 c\theta_4 = r_3 c\theta_3 - (r_1 c\theta_1 - b_3) \tag{10.89}$$

$$r_4 s\theta_4 = r_3 s\theta_3 - r_1 s\theta_1 \tag{10.90}$$

Upon the elimination of θ_4, Eqs. (10.89) and (10.90) lead to the following equation.

$$x_{31} c\theta_3 + y_{31} s\theta_3 = f_{31} \tag{10.91}$$

In Eq. (10.91),

$$x_{31} = r_1 c\theta_1 - b_3 = (\vec{r}_{B_3C_1})_x, \quad y_{31} = r_1 s\theta_1 = (\vec{r}_{B_3C_1})_y \tag{10.92}$$

$$f_{31} = \frac{x_{31}^2 + y_{31}^2 + r_3^2 - r_4^2}{2r_3} \tag{10.93}$$

Equation (10.91) gives θ_3 by means of the following set of equations.

$$\theta_3 = \psi_{31} + \sigma_3' \gamma_{31} \tag{10.94}$$

$$\psi_{31} = \operatorname{atan}_2(y_{31}, x_{31}) = \sphericalangle(\vec{r}_{B_3C_1}) \tag{10.95}$$

$$\gamma_{31} = \operatorname{atan}_2(g_{31}, f_{31}) \tag{10.96}$$

$$g_{31} = \sqrt{x_{31}^2 + y_{31}^2 - f_{31}^2} \tag{10.97}$$

In Eq. (10.94), the sign variable σ_3' is such that the point C_3 is *above* the line segment B_3C_1 (as seen in Figure 10.7) if $\sigma_3' = -1$ and vice versa if $\sigma_3' = +1$.
After finding θ_3, Eqs. (10.89) and (10.90) can be used to express θ_4 in terms of θ_3. That is,

$$\theta_4 = \text{atan}_2[(r_3c\theta_3 - x_{31}), (r_3s\theta_3 - y_{31})] \tag{10.98}$$

As seen above, Eqs. (10.94) and (10.98) constitute the following singularity relationships that make θ_3 and θ_4 dependent on θ_1.

$$\theta_3 = f_3'(\theta_1, \sigma_3') \tag{10.99}$$

$$\theta_4 = f_4'(\theta_1, \sigma_3') \tag{10.100}$$

Note that Eq. (10.100) provides another expression for θ_4 in addition to Eq. (10.88). Therefore, the consistency of Eqs. (10.88) and Eq. (10.100) necessitates that

$$f_4(\theta_1, \theta_2, \sigma_1) = f_4'(\theta_1, \sigma_3') \tag{10.101}$$

Equations (10.99) and (10.101) indicate that, in a pose of PSFK-2, the primary variables θ_2 and θ_3 can no longer be specified arbitrarily as desired, because they both become dependent on the other primary variable θ_1. In other words, leaving aside the uncontrollable mobility caused by the indefiniteness of $\phi = \theta_6$, the controllable mobility of the manipulator reduces to $\mu' = 1$. This mobility reduction caused by PSFK-2 can be seen in Figure 10.7.

Furthermore, if PSFK-1 and PSFK-2 happen to be concurrent, then the controllable mobility of the manipulator reduces further down to $\mu'' = 0$, because PSFK-1 eliminates the freedom left in θ_1, too, according to Eq. (10.79).

10.4 Inverse Kinematics in the Position Domain

The inverse kinematics of a parallel manipulator in the position domain involves the *inverse kinematic solution* together with the identification and analysis of the associated multiplicities and singularities. More specifically, the multiplicities are called *posture multiplicities of inverse kinematics* (PMIKs) and the singularities are called *position singularities of inverse kinematics* (PSIKs).

The inverse kinematic solution is needed in order to determine all the primary and secondary variables (i.e. both $\overline{x}$ and $\overline{y}$) that correspond to a specified position (i.e. $\overline{p}$) of the end-effector. However, the main purpose of the inverse kinematics is to determine the *active joint variables*, which are directly or indirectly related to the primary variables (i.e. $\overline{x}$). Thus, it becomes possible to command the actuators in order to achieve the specified desired position of the end-effector.

In this section, the inverse kinematics will be considered for a *regular parallel manipulator*, which does not have any kinematic or actuator redundancy or deficiency. The mobility of such a manipulator is equal to the DoF of the working space. That is, $\mu = \lambda$.

The inverse kinematic solution is normally obtained *leg by leg*. This is because, almost without exception, a practicable parallel manipulator consists of several legs that support only one moving platform and the legs are equally actuated, that is, each leg has the same number of active joints. Thus, the particular solution obtained for one of the legs provides the relevant active joint variables irrespective of the other legs. Besides, the kinematic equations written for each leg happen to be considerably simpler than the loop equations written for all the IKLs (independent kinematic loops) of the manipulator. Therefore, the process of obtaining the inverse kinematic solution separately for each leg turns out to be much simpler than the process of obtaining the forward kinematic solution for the whole manipulator.

For a parallel manipulator with a single moving platform and n_L number of legs, the following set of equations can be written in order to express the position of the end-effector by going through each of the legs.

$$\overline{p} = \overline{f}_k(\overline{z}_k); \quad k = 1, 2, \ldots, n_L \tag{10.102}$$

In Eq. (10.102), $\overline{z}_k$ is the *leg-specific partition* of $\overline{z}$ that belongs to the leg L_k and $\overline{z} \in \mathcal{R}^{m+n}$ is the column matrix that contains all the primary and secondary variables. That is,

$$\overline{z} = \begin{bmatrix} \overline{x} \\ \overline{y} \end{bmatrix} \tag{10.103}$$

The partition $\overline{z}_k$ is also similarly partitioned so that

$$\overline{z}_k = \begin{bmatrix} \overline{x}_k \\ \overline{y}_k \end{bmatrix} \tag{10.104}$$

In Eq. (10.104), the subpartition $\overline{x}_k$ consists of the primary variables that belong to the leg L_k. The number of these variables is equal to the number of the actuators that drive L_k. In fact, they are normally taken as the associated active joint variables.

Another feature of a practicable regular parallel manipulator is that each of its legs consists of the same number of links that are connected in series. In other words, each leg can be envisioned as a serial manipulator. On the other hand, a regular parallel manipulator (with $\mu = \lambda$) that has n_L legs possesses $n_{ikl} = n_L - 1$ independent kinematic loops, which are formed by taking the legs pair by pair. Therefore, the total number of the primary and secondary variables can be expressed as

$$n_v = n_{pv} + n_{sv} = \mu + \lambda n_{ikl} = \lambda + \lambda(n_L - 1) = \lambda n_L \tag{10.105}$$

Equation (10.105) shows that each leg comes in with $n_v/n_L = \lambda$ variables. This, in turn, shows that $\overline{z}_k$ is co-dimensional with $\overline{x}$ and $\overline{p}$. That is, $\dim(\overline{z}_k) = \dim(\overline{x}) = \dim(\overline{p}) = m = \mu = \lambda$. Owing to this fact, $\overline{z}_k$ can be extracted from Eq. (10.102) for L_k if L_k is *not* in an associated pose of PSIK, which is denoted as PSIK-k. Note that the PSIKs are associated specifically with the legs, but not with the whole manipulator. The solution to Eq. (10.102) can be expressed as follows for each leg, i.e. for $k = 1, 2, \ldots, n_L$.

$$\overline{z}_k = \overline{z}_k^{(i)}(\overline{p}); \quad i = 1, 2, \ldots, n_{kpm} \tag{10.106}$$

In Eq. (10.106), n_{kpm} is the number of the different posture modes, any of which may be asssumed by the leg L_k for the same specified $\overline{p}$. In the same equation, $\overline{z}_k^{(i)}(\overline{p})$ is a function that represents the ith *posture mode* of L_k, which is denoted as PML-ki.

For the legs of a parallel manipulator, the appropriate posture modes are actually selected while the manipulator is assembled. Therefore, the posture modes of the legs may also be called *assembly modes of the legs.* The selection of appropriate posture modes for the legs depends on the environment and the operating conditions of the manipulator. Once a leg, e.g. L_k, is assembled in a certain posture mode, then it keeps operating in the same posture mode, unless it gets into a PMCPL (posture mode changing pose of a leg) (i.e. PMCPL-kij). If it gets into an PMCPL-kij, then the current posture mode of L_k (e.g. PML-*ki*) may be changed into another posture mode (e.g. PML-*kj*) easily by a small external intervention (intentional or unintentional) without the necessity of disassembling and reassembling L_k.

If L_k happens to be in a pose of PSIK (i.e. PSIK-*k*), then $\overline{z}_k$ cannot be extracted as a definite solution from Eq. (10.102). A pose of PSIK-*k* occurs if $\overline{\rho}$ (i.e. the position of the end-effector) is specified in such a special way that a certain partition of it, e.g. $\overline{\rho}_{kd} \in \mathcal{R}^{m_{kd}}$, is tied up to its complementary partition $\overline{\rho}_{kc} \in \mathcal{R}^{m_{kc}}$, where $m_{kc} + m_{kd} = m$. In other words, $\overline{\rho}_{kc}$ is still specified arbitrarily but $\overline{\rho}_{kd}$ is specified in a special way so that it depends on $\overline{\rho}_{kc}$ through a *singularity relationship*. That is,

$$\overline{\rho}_{kd} = \overline{\xi}_{kdc}(\overline{\rho}_{kc}) \tag{10.107}$$

When $\overline{\rho}$ is inserted with its above-mentioned partitions, Eq. (10.102) can be separated into the following consistently partitioned equations, in which $\overline{z}_k$ is also similarly partitioned.

$$\overline{\rho}_{kc} = \overline{f}_{kc}(\overline{z}_{kc}, \overline{z}_{kd}) \tag{10.108}$$

$$\overline{\rho}_{kd} = \overline{f}_{kd}(\overline{z}_{kc}, \overline{z}_{kd}) \tag{10.109}$$

When Eqs. (10.107) and (10.108) are substituted, Eq. (10.109) becomes

$$\overline{\xi}_{kdc}(\overline{f}_{kc}(\overline{z}_{kc}, \overline{z}_{kd})) = \overline{f}_{kd}(\overline{z}_{kc}, \overline{z}_{kd}) \tag{10.110}$$

The characterizing feature of PSIK-*k* is that Eq. (10.110) turns into an identity and therefore it is satisfied no matter what $\overline{z}_{kd}$ is. Thus, $\overline{z}_{kd}$ becomes indefinite. As for the other partition $\overline{z}_{kc}$, it can be obtained from Eq. (10.108) depending on the specified partition $\overline{\rho}_{kc}$ and the indefinite partition $\overline{z}_{kd}$. The solution can be expressed as follows:

$$\overline{z}_{kc} = \overline{z}_{kc}^{(i)}(\overline{\rho}_{kc}, \overline{z}_{kd}); \quad i = 1, 2, \ldots, n'_{kpm} \tag{10.111}$$

Consequently, $\overline{z}_{kc}$ also becomes indefinite due to its dependence on $\overline{z}_{kd}$. This means that $\overline{z}_k$ becomes indefinite as a whole and thus the active joint variable or variables contained in $\overline{z}_k$ can be changed arbitrarily without creating any effect on the position of the end-effector which is specified according to the singularity relationship expressed by Eq. (10.107). In other words, as long as the end-effector is kept in this special singularity position, possibly due to a specific task requirement, L_k loses its effect on the end-effector completely no matter how arbitrarily it is actuated.

However, unlike the PSFK poses of the manipulator, the PSIK poses of the legs need not be avoided. Let alone avoiding, they must sometimes be deliberately allowed in order to execute some special tasks, which could not be executed otherwise. In other words, the reachable range of the end-effector and the dexterity of the manipulator can be increased by allowing the PSIKs of the legs to occur. Moreover, when they are allowed to occur due to such tasks, they may be used even in some advantageous ways. For example, if PSIK-*k* occurs, then L_k becomes useless. So, instead of spending power in order to keep it in an arbitrarily selected pose, it may be allowed to assume a pose caused by its own weight so that no power is spent against gravity.

Example 10.6 ***Inverse Kinematics of the 3RRR Planar Parallel Manipulator***

For the parallel manipulator shown in Figure 10.4, the following equations can be written in order to express the position of the end-effector by going through each of the three legs.

* Vector and scalar equations written for the leg L_1:

$$\overrightarrow{OP} = \overrightarrow{B_1P} = \overrightarrow{B_1C_1} + \overrightarrow{C_1A_1} + \overrightarrow{A_1Q} + \overrightarrow{QP} \Rightarrow$$

$$\vec{p} = p_1\vec{u}_1 + p_2\vec{u}_2 = r_1\vec{u}(\theta_1) + r_4\vec{u}(\theta_4) + d_1\vec{u}(\phi) + h_7\vec{u}\left(\phi + \frac{\pi}{2}\right) \tag{10.112}$$

$$p_1 = r_1c\theta_1 + r_4c\theta_4 + d_1c\phi - h_7s\phi \tag{10.113}$$

$$p_2 = r_1s\theta_1 + r_4s\theta_4 + d_1s\phi + h_7c\phi \tag{10.114}$$

* Vector and scalar equations written for the leg L_2:

$$\overrightarrow{OP} = \overrightarrow{B_1P} = \overrightarrow{B_1B_2} + \overrightarrow{B_2C_2} + \overrightarrow{C_2A_2} + \overrightarrow{A_2Q} + \overrightarrow{QP} \Rightarrow$$

$$\vec{p} = p_1\vec{u}_1 + p_2\vec{u}_2 = b_2\vec{u}_1 + r_2\vec{u}(\theta_2) + r_5\vec{u}(\theta_5) + d_1\vec{u}(\phi) + h_7\vec{u}\left(\phi + \frac{\pi}{2}\right) \tag{10.115}$$

$$p_1 = b_2 + r_2c\theta_2 + r_5c\theta_5 + d_1c\phi - h_7s\phi \tag{10.116}$$

$$p_2 = r_2s\theta_2 + r_5s\theta_5 + d_1s\phi + h_7c\phi \tag{10.117}$$

* Vector and scalar equations written for the leg L_3:

$$\overrightarrow{OP} = \overrightarrow{B_1P} = \overrightarrow{B_1B_3} + \overrightarrow{B_3C_3} + \overrightarrow{C_3A_3} + \overrightarrow{A_3Q} + \overrightarrow{QP} \Rightarrow$$

$$\vec{p} = p_1\vec{u}_1 + p_2\vec{u}_2 = b_3\vec{u}_1 + r_3\vec{u}(\theta_3) + r_6\vec{u}(\theta_6) - d_3\vec{u}(\phi) + h_7\vec{u}\left(\phi + \frac{\pi}{2}\right) \tag{10.118}$$

$$p_1 = b_3 + r_3c\theta_3 + r_6c\theta_6 - d_3c\phi - h_7s\phi \tag{10.119}$$

$$p_2 = r_3s\theta_3 + r_6s\theta_6 - d_3s\phi + h_7c\phi \tag{10.120}$$

(a) Inverse Kinematic Solutions for the Legs

Here, the position of the end-effector is specified. That is, the tip point coordinates (p_1, p_2) and the orientation angle (ϕ) are known. The corresponding primary and secondary variables are found as explained below. The primary variables are the angles $(\theta_1, \theta_2, \theta_3)$ of the active joints and the secondary variables are the orientation angles $(\theta_4, \theta_5, \theta_6)$ of the upper links of the legs.

* Inverse kinematic solution for the leg L_1:

Equations (10.113) and (10.114) can be written as follows:

$$r_4c\theta_4 = x_1 - r_1c\theta_1 \tag{10.121}$$

$$r_4s\theta_4 = y_1 - r_1s\theta_1 \tag{10.122}$$

In Eqs. (10.121) and (10.122), x_1 and y_1 are the known components of the relative position vector $\overrightarrow{B_1A_1}$. That is,

$$x_1 = p_1 + h_7s\phi - d_1c\phi \tag{10.123}$$

$$y_1 = p_2 - h_7c\phi - d_1s\phi \tag{10.124}$$

If x_1 and y_1 are not both zero, then Eqs. (10.121) and (10.122) can be solved in order to find θ_1 and θ_4. The solution can be obtained similarly as in Section 10.3. It is summarized below.

$$\theta_1 = \psi_1 + \sigma_1\gamma_1 \tag{10.125}$$

$$\psi_1 = \text{atan}_2(y_1, x_1) \tag{10.126}$$

$$\gamma_1 = \text{atan}_2(g_1, f_1) \tag{10.127}$$

$$f_1 = (x_1^2 + y_1^2 + r_1^2 - r_4^2)/(2r_1) \tag{10.128}$$

$$g_1 = \sqrt{x_1^2 + y_1^2 - f_1^2} \tag{10.129}$$

$$\sigma_1 = \pm 1 \tag{10.130}$$

$$\theta_4 = \text{atan}_2[(y_1 - r_1 s\theta_1), (x_1 - r_1 c\theta_1)] \tag{10.131}$$

* Inverse kinematic solution for the leg L_2:

Equations (10.116) and (10.117) can be written as follows:

$$r_5 c\theta_5 = x_2 - r_2 c\theta_2 \tag{10.132}$$

$$r_5 s\theta_5 = y_2 - r_2 s\theta_2 \tag{10.133}$$

In Eqs. (10.132) and (10.133), x_2 and y_2 are the known components of the relative position vector $\overrightarrow{B_2A_2}$. That is,

$$x_2 = p_1 + h_7 s\phi - d_1 c\phi - b_2 \tag{10.134}$$

$$y_2 = p_2 - h_7 c\phi - d_1 s\phi \tag{10.135}$$

If x_2 and y_2 are not both zero, then Eqs. (10.132) and (10.133) can be solved in order to find θ_2 and θ_5. The solution can be obtained similarly as described for L_1. It is summarized below.

$$\theta_2 = \psi_2 + \sigma_2 \gamma_2 \tag{10.136}$$

$$\psi_2 = \text{atan}_2(y_2, x_2) \tag{10.137}$$

$$\gamma_2 = \text{atan}_2(g_2, f_2) \tag{10.138}$$

$$f_2 = (x_2^2 + y_2^2 + r_2^2 - r_5^2)/(2r_2) \tag{10.139}$$

$$g_2 = \sqrt{x_2^2 + y_2^2 - f_2^2} \tag{10.140}$$

$$\sigma_2 = \pm 1 \tag{10.141}$$

$$\theta_5 = \text{atan}_2[(y_2 - r_2 s\theta_2), (x_2 - r_2 c\theta_2)] \tag{10.142}$$

* Inverse kinematic solution for the leg L_3:

Equations (10.119) and (10.120) can be written as follows:

$$r_6 c\theta_6 = x_3 - r_3 c\theta_3 \tag{10.143}$$

$$r_6 s\theta_6 = y_3 - r_3 s\theta_3 \tag{10.144}$$

In Eqs. (10.143) and (10.144), x_3 and y_3 are the known components of the relative position vector $\overrightarrow{B_3A_3}$. That is,

$$x_3 = p_1 + h_7 s\phi + d_3 c\phi - b_3 \tag{10.145}$$

$$y_3 = p_2 - h_7 c\phi + d_3 s\phi \tag{10.146}$$

If x_3 and y_3 are not both zero, then Eqs. (10.143) and (10.144) can be solved in order to find θ_3 and θ_6. The solution can be obtained similarly as described for the other legs. It is summarized below.

$$\theta_3 = \psi_3 + \sigma_3\gamma_3 \tag{10.147}$$

$$\psi_3 = \text{atan}_2(y_3, x_3) \tag{10.148}$$

$$\gamma_3 = \text{atan}_2(g_3, f_3) \tag{10.149}$$

$$f_3 = (x_3^2 + y_3^2 + r_3^2 - r_6^2)/(2r_3) \tag{10.150}$$

$$g_3 = \sqrt{x_3^2 + y_3^2 - f_3^2} \tag{10.151}$$

$$\sigma_3 = \pm 1 \tag{10.152}$$

$$\theta_6 = \text{atan}_2[(y_3 - r_3 s\theta_3), (x_3 - r_3 c\theta_3)] \tag{10.153}$$

(b) Posture Modes of Inverse Kinematics and Posture Mode Changing Poses

∗ PML-11, PML-12, and PMCPL-1 of the leg L_1:

Equation (10.126) implies that

$$\psi_1 = \sphericalangle(\overrightarrow{B_1A_1}) \tag{10.154}$$

Based on ψ_1, Eq. (10.125) and Figure 10.4 lead to the following conclusions.

(i) PML-11 occurs if $\theta_1 > \psi_1 > 0$ or $\theta_1 < \psi_1 < 0$. This PML may be called *knee-behind*. For this PML, $\sigma_1 = +\text{sgn}(\psi_1)$.
(ii) PML-12 occurs if $\theta_1 < \psi_1 > 0$ or $\theta_1 > \psi_1 < 0$. This PML may be called *knee-ahead*. For this PML, $\sigma_1 = -\text{sgn}(\psi_1)$.
(iii) PMCPL-1 occurs if $\theta_1 = \psi_1$ or $\theta_1 = \psi_1 \pm \pi$.

∗ PML-21, PML-22, and PMCPL-2 of the leg L_2:

Equation (10.137) implies that

$$\psi_2 = \sphericalangle(\overrightarrow{B_2A_2}) \tag{10.155}$$

Based on ψ_2, Eq. (10.136) and Figure 10.4 lead to the following conclusions.

(i) PML-21 occurs if $\theta_2 > \psi_2 > 0$ or $\theta_2 < \psi_2 < 0$. This PML may be called *knee-behind*. For this PML, $\sigma_2 = +\text{sgn}(\psi_2)$.
(ii) PML-22 occurs if $\theta_2 < \psi_2 > 0$ or $\theta_2 > \psi_2 < 0$. This PML may be called *knee-ahead*. For this PML, $\sigma_2 = -\text{sgn}(\psi_2)$.
(iii) PMCPL-2 occurs if $\theta_2 = \psi_2$ or $\theta_2 = \psi_2 \pm \pi$.

∗ PML-31, PML-32, and PMCPL-3 of the leg L_3:

Equation (10.148) implies that

$$\psi_3 = \sphericalangle(\overrightarrow{B_3A_3}) \tag{10.156}$$

Based on ψ_3, Eq. (10.147) and Figure 10.4 lead to the following conclusions.

(i) PML-31 occurs if $\theta_3 > \psi_3 > 0$ or $\theta_3 < \psi_3 < 0$. This PML may be called *knee-behind*. For this PML, $\sigma_3 = +\text{sgn}(\psi_3)$.
(ii) PML-32 occurs if $\theta_3 < \psi_3 > 0$ or $\theta_3 > \psi_3 < 0$. This PML may be called *knee-ahead*. For this PML, $\sigma_3 = -\text{sgn}(\psi_3)$.
(iii) PMCPL-3 occurs if $\theta_3 = \psi_3$ or $\theta_3 = \psi_3 \pm \pi$.
The posture modes and the PMCPs of the legs are illustrated in Figures 10.8 and 10.9.

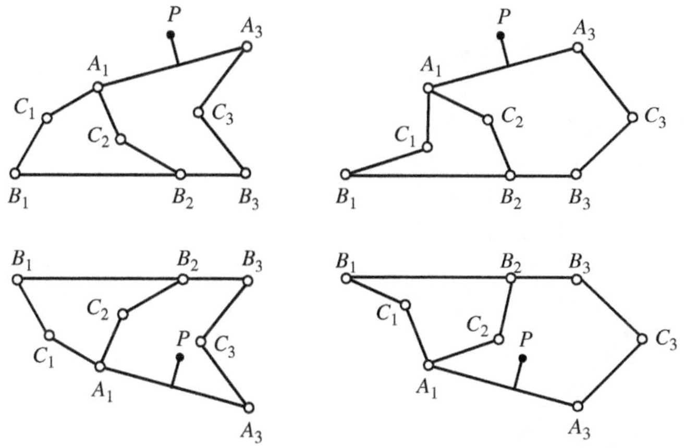

Knee-behind posture modes Knee-ahead posture modes

Figure 10.8 Four of the PMLs of the 3RRR planar parallel manipulator.

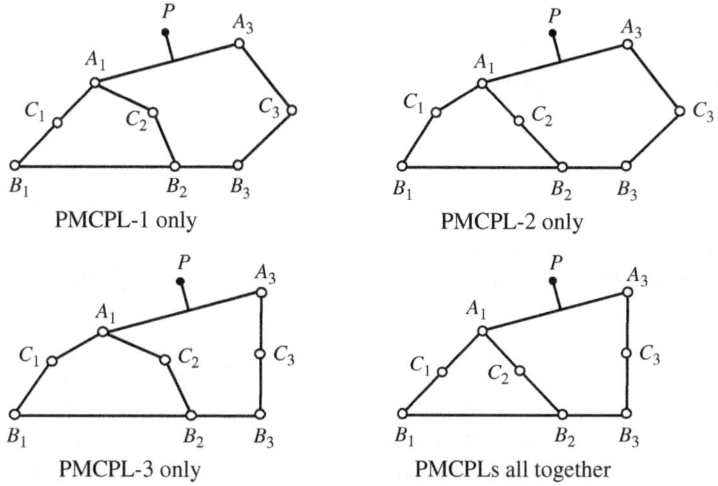

Figure 10.9 PMCPLs of the 3RRR planar parallel manipulator.

(c) Position Singularities of Inverse Kinematics

As mentioned before, each leg may have a PSIK of its own. The PSIK of L_k is denoted here as PSIK-k. The PSIKs of the legs may occur separately at different times or simultaneously at the same time. They are described and discussed below.

* PSIK-1 of the leg L_1:

The leg L_1 gets into a pose of PSIK-1, if $x_1 = y_1 = 0$ in Eqs. (10.121) and (10.122). In such a pose, Eqs. (10.121) and (10.122) reduce to the following equations.

$$r_4 c\theta_4 = -r_1 c\theta_1 \tag{10.157}$$

$$r_4 s\theta_4 = -r_1 s\theta_1 \tag{10.158}$$

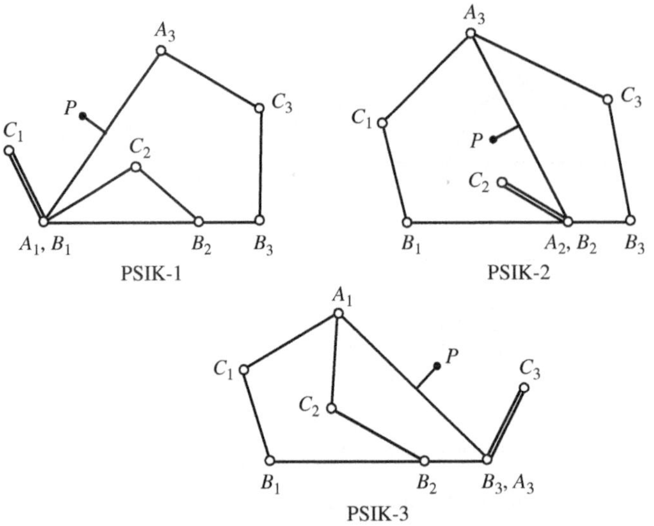

Figure 10.10 PSIK poses of the 3RRR planar parallel manipulator.

Equations (10.157) and (10.158) can be satisfied nontrivially (i.e. with nonzero link lengths) only if the link lengths are equal and the angles complement each other to $\pm\pi$. That is,

$$r_4 = r_1 \tag{10.159}$$

$$\theta_4 = \theta_1 \pm \pi \tag{10.160}$$

Equation (10.159) implies that PSIK-1 can never occur if $r_4 \neq r_1$. In other words, if the links of L_1 have different lengths, i.e. if $r_4 \neq r_1$, then PSIK-1 will be geometrically prevented.

On the other hand, if PSIK-1 occurs due to having $r_4 = r_1$, then Eq. (10.160) implies that θ_1 and θ_4 cannot be found even though they turn out to be complementary angles. In other words, θ_1 and θ_4 become indefinite. Consequently, the links $\mathcal{L}_1$ and $\mathcal{L}_4$ of the leg L_1 become coincident (i.e. folded over) and they gain a DoF to rotate together arbitrarily about the center at the coincident points B_1 and A_1. However, since θ_1 is an active joint variable, this freedom is still controllable even though it has no effect on the particularly specified position of the end-effector so that $x_1 = y_1 = 0$. This situation is illustrated in Figure 10.10. Such a particular position of the end-effector occurs, if p_1, p_2, and ϕ are specified so as to obey the following *singularity relationships*.

$$x_1 = 0 \;\Rightarrow\; p_1 = d_1 c\phi - h_7 s\phi \tag{10.161}$$

$$y_1 = 0 \;\Rightarrow\; p_2 = d_1 s\phi + h_7 c\phi \tag{10.162}$$

Equations (10.161) and (10.162) imply that

$$p_1^2 + p_2^2 = r_{17}^2 = d_1^2 + h_7^2 \tag{10.163}$$

Equation (10.163) implies further that ϕ can still be specified arbitrarily but the location of the tip point P must be specified so that P lies on a circle with radius $r_{17} = \sqrt{d_1^2 + h_7^2}$ and center B_1. The end-effector may deliberately be positioned as such due to a special task requirement. Therefore, instead of avoiding, PSIK-1 must be allowed to occur in order to execute this special task. However, L_1 will be ineffective in the execution of this task. Yet, the other legs will be sufficient to execute this task as seen in Figure 10.10. In such a case, L_1 can be commanded to assume an arbitrarily desired position. For example, it can be let free to align itself with the direction of gravity so that no actuation power is spent.

* PSIK-2 of the leg L_2:

The leg L_2 gets into a pose of PSIK-2, if $x_2 = y_2 = 0$ in Eqs. (10.132) and (10.133). In such a pose, Eqs. (10.132) and (10.133) reduce to the following equations.

$$r_5 c\theta_5 = -r_2 c\theta_2 \tag{10.164}$$

$$r_5 s\theta_5 = -r_2 s\theta_2 \tag{10.165}$$

Equations (10.164) and (10.165) can be satisfied nontrivially (i.e. with nonzero link lengths) only if the link lengths are equal and the angles complement each other to $\pm\pi$. That is,

$$r_5 = r_2 \tag{10.166}$$

$$\theta_5 = \theta_2 \pm \pi \tag{10.167}$$

Equation (10.166) implies that PSIK-2 can never occur if $r_5 \neq r_2$. In other words, if the links of L_2 have different lengths, i.e. if $r_5 \neq r_2$, then PSIK-2 will be geometrically prevented.

On the other hand, if PSIK-2 occurs due to having $r_5 = r_2$, then Eq. (10.167) implies that θ_2 and θ_5 cannot be found even though they turn out to be complementary angles. In other words, θ_2 and θ_5 become indefinite. Consequently, the links $\mathcal{L}_2$ and $\mathcal{L}_5$ of the leg L_2 become coincident (i.e. folded over) and they gain a DoF to rotate together arbitrarily about the center at the coincident points B_2 and A_2. However, since θ_2 is an active joint variable, this freedom is still controllable even though it has no effect on the particularly specified position of the end-effector so that $x_2 = y_2 = 0$. This situation is illustrated in Figure 10.10. Such a particular position of the end-effector occurs, if p_1, p_2, and ϕ are specified so as to obey the following *singularity relationships.*

$$x_2 = 0 \;\Rightarrow\; p_1 = b_2 + d_1 c\phi - h_7 s\phi \tag{10.168}$$

$$y_2 = 0 \;\Rightarrow\; p_2 = d_1 s\phi + h_7 c\phi \tag{10.169}$$

Equations (10.168) and (10.169) imply that

$$(p_1 - b_2)^2 + p_2^2 = r_{17}^2 = d_1^2 + h_7^2 \tag{10.170}$$

Equation (10.170) implies further that ϕ can still be specified arbitrarily but the location of the tip point P must be specified so that P lies on a circle with radius $r_{17} = \sqrt{d_1^2 + h_7^2}$ and center B_2. The end-effector may deliberately be positioned as such due to a special task requirement. Therefore, instead of avoiding, PSIK-2 must be allowed to occur in order to execute this special task. However, L_2 will be ineffective in the execution of this task. Yet, the other legs will be sufficient to execute this task as seen in Figure 10.10. In such a case, L_2 can be commanded to assume an arbitrarily desired position. For example, it can be let free to align itself with the direction of gravity so that no actuation power is spent.

∗ PSIK-3 of the leg L_3:

The leg L_3 gets into a pose of PSIK-3, if $x_3 = y_3 = 0$ in Eqs. (10.143) and (10.144). In such a pose, Eqs. (10.143) and (10.144) reduce to the following equations.

$$r_6 c\theta_6 = -r_3 c\theta_3 \tag{10.171}$$

$$r_6 s\theta_6 = -r_3 s\theta_3 \tag{10.172}$$

Equations (10.171) and (10.172) can be satisfied nontrivially (i.e. with nonzero link lengths) only if the link lengths are equal and the angles complement each other to $\pm\pi$. That is,

$$r_6 = r_3 \tag{10.173}$$

$$\theta_6 = \theta_3 \pm \pi \tag{10.174}$$

Equation (10.173) implies that PSIK-3 can never occur if $r_6 \neq r_3$. In other words, if the links of L_3 have different lengths, i.e. if $r_6 \neq r_3$, then PSIK-3 will be geometrically prevented.

On the other hand, if PSIK-3 occurs due to having $r_6 = r_3$, then Eq. (10.174) implies that θ_3 and θ_6 cannot be found even though they turn out to be complementary angles. In other words, θ_3 and θ_6 become indefinite. Consequently, the links $\mathcal{L}_3$ and $\mathcal{L}_6$ of the leg L_3 become coincident (i.e. folded over) and they gain a DoF to rotate together arbitrarily about the center at the coincident points B_3 and A_3. However, since θ_3 is an active joint variable, this freedom is still controllable even though it has no effect on the particularly specified position of the end-effector so that $x_3 = y_3 = 0$. This situation is illustrated in Figure 10.10. Such a particular position of the end-effector occurs, if p_1, p_2, and ϕ are specified so as to obey the following *singularity relationships.*

$$x_1 = 0 \Rightarrow p_1 = b_3 - (h_7 s\phi + d_3 c\phi) \tag{10.175}$$

$$y_1 = 0 \Rightarrow p_2 = h_7 c\phi - d_3 s\phi \tag{10.176}$$

Equations (10.175) and (10.176) imply that

$$(p_1 - b_3)^2 + p_2^2 = r_{37}^2 = d_3^2 + h_7^2 \tag{10.177}$$

Equation (10.177) implies further that ϕ can still be specified arbitrarily but the location of the tip point P must be specified so that P lies on a circle with radius $r_{37} = \sqrt{d_3^2 + h_7^2}$ and center B_3. The end-effector may deliberately be positioned as such due to a special task requirement. Therefore, instead of avoiding, PSIK-3 must be allowed to occur in order to execute this special task. However, L_3 will be ineffective in the execution of this task. Yet, the other legs will be sufficient to execute this task as seen in Figure 10.10. In such a case, L_3 can be commanded to assume an arbitrarily desired position. For example, it can be let free to align itself with the direction of gravity so that no actuation power is spent.

10.5 Velocity Equations of a Parallel Manipulator

For a parallel manipulator, the following complementary equations can be written in order to express the *velocity state* of the end-effector in terms of the rates of the relevant variables and to set up the relationships among the rates of the primary and secondary variables.

$$\overline{\eta} = \widehat{A}(\overline{x}, \overline{y})\dot{\overline{x}} + \widehat{B}(\overline{x}, \overline{y})\dot{\overline{y}} \tag{10.178}$$

$$\widehat{M}(\overline{x}, \overline{y})\dot{\overline{y}} = \widehat{N}(\overline{x}, \overline{y})\dot{\overline{x}} \tag{10.179}$$

In Eqs. (10.178) and (10.179), $\overline{x}$ and $\overline{y}$ are already defined in Section 10.2. Their definitions are repeated below for the sake of convenience.

$\overline{x} \in \mathcal{R}^m$ is a column matrix that consists of the *primary variables.* Note that $m = \mu = n_{pv}$.

$\overline{y} \in \mathcal{R}^n$ is a column matrix that consists of the *secondary variables.* Note that $n = n_{sv}$.

The new symbols in Eqs. (10.178) and (10.179) have the following meanings.

$\overline{\eta} \in \mathcal{R}^\lambda$ is a column matrix that represents the *velocity state* (i.e. the tip point velocity and the angular velocity) of the end-effector in the *working space,* which is also called the *task space.* Note that $\lambda = 6$ for the three-dimensional task space and $\lambda = 3$ for the two-dimensional (planar) task space.

If the working space of the parallel manipulator is three dimensional, then $\overline{\eta}$ is partitioned as

$$\overline{\eta} = \begin{bmatrix} \overline{v} \\ \overline{\omega} \end{bmatrix} \tag{10.180}$$

In Eq. (10.180), $\overline{v}$ and $\overline{\omega}$, respectively, consist of the *tip point velocity* components and the *angular velocity* components of the end-effector with respect to the reference frame attached to the fixed platform. In more detail,

$$\overline{v} = \begin{bmatrix} v_1 \\ v_2 \\ v_3 \end{bmatrix}, \quad \overline{\omega} = \begin{bmatrix} \omega_1 \\ \omega_2 \\ \omega_3 \end{bmatrix} \tag{10.181}$$

If the working space of the parallel manipulator is two dimensional, then $\overline{\eta}$ consists of the two velocity components of the tip point and the scalar angular velocity of the end-effector. That is,

$$\overline{\eta} = \begin{bmatrix} v_1 \\ v_2 \\ \omega \end{bmatrix} \tag{10.182}$$

In Eq. (10.182), which is written for the *planar manipulators,* $\overline{\eta}$ is related to $\overline{\rho}$ simply as follows:

$$\overline{\eta} = \dot{\overline{\rho}} \;\Rightarrow\; \begin{bmatrix} v_1 \\ v_2 \\ \omega \end{bmatrix} = \begin{bmatrix} \dot{p}_1 \\ \dot{p}_2 \\ \dot{\phi} \end{bmatrix} \tag{10.183}$$

However, in Eq. (10.180), which is written for the *spatial manipulators,* $\overline{\eta} \neq \dot{\overline{\rho}}$ because, although $\overline{v} = \dot{\overline{p}}$, $\overline{\omega} \neq \dot{\overline{\phi}}$. In fact, it happens that

$$\overline{\omega} = \widehat{E}(\overline{\phi})\dot{\overline{\phi}} \tag{10.184}$$

For example, let $\overline{\phi}$ consist of the Euler angles of an $i-j-k$ sequence, i.e. let the orientation matrix of the end-effector be expressed as $\widehat{C} = e^{\tilde{u}_i\phi_1}e^{\tilde{u}_j\phi_2}e^{\tilde{u}_k\phi_3}$. Then, by referring to Chapter 4, it can be shown that

$$\overline{\omega} = \dot{\phi}_1 \overline{u}_i + \dot{\phi}_2 e^{\tilde{u}_i \phi_1} \overline{u}_j + \dot{\phi}_3 e^{\tilde{u}_i \phi_1} e^{\tilde{u}_j \phi_2} \overline{u}_k \Rightarrow$$
$$\hat{E}(\overline{\phi}) = [\overline{u}_i \;\; e^{\tilde{u}_i \phi_1} \overline{u}_j \;\; e^{\tilde{u}_i \phi_1} e^{\tilde{u}_j \phi_2} \overline{u}_k] \quad (10.185)$$

Hence, the relationship between $\overline{\eta}$ and $\overline{\rho}$ can be written as follows:

$$\overline{\eta} = \hat{D}(\overline{\phi})\dot{\overline{\rho}} \quad (10.186)$$

In Eq. (10.186),

$$\hat{D}(\overline{\phi}) = \begin{bmatrix} \hat{I} & \hat{0} \\ \hat{0} & \hat{E}(\overline{\phi}) \end{bmatrix} \quad (10.187)$$

By using Eqs. (10.10) and (10.186), Eq. (10.178) can be derived as follows:

$$\dot{\overline{\rho}} = \hat{F}_x(\overline{x}, \overline{y})\dot{\overline{x}} + \hat{F}_y(\overline{x}, \overline{y})\dot{\overline{y}} \quad (10.188)$$

In Eq. (10.188), $\hat{F}_x(\overline{x}, \overline{y})$ and $\hat{F}_y(\overline{x}, \overline{y})$ are defined so that

$$[\hat{F}_x(\overline{x}, \overline{y})]_{ij} = F_{xij}(\overline{x}, \overline{y}) = \partial f_i(\overline{x}, \overline{y})/\partial x_j \quad (10.189)$$
$$[\hat{F}_y(\overline{x}, \overline{y})]_{ij} = F_{yij}(\overline{x}, \overline{y}) = \partial f_i(\overline{x}, \overline{y})/\partial y_j \quad (10.190)$$

Hence,

$$\overline{\eta} = \hat{D}(\overline{\phi})[\hat{F}_x(\overline{x}, \overline{y})\dot{\overline{x}} + \hat{F}_y(\overline{x}, \overline{y})\dot{\overline{y}}] \quad (10.191)$$

On the other hand, Eq. (10.10) implies that

$$\overline{\phi} = \overline{\phi}(\overline{x}, \overline{y}) \quad (10.192)$$

Thus, the coefficient matrices in Eq. (10.178) are obtained from Eq. (10.191) as shown below.

$$\hat{A}(\overline{x}, \overline{y}) = \hat{D}(\overline{\phi}(\overline{x}, \overline{y}))\hat{F}_x(\overline{x}, \overline{y}) \quad (10.193)$$
$$\hat{B}(\overline{x}, \overline{y}) = \hat{D}(\overline{\phi}(\overline{x}, \overline{y}))\hat{F}_y(\overline{x}, \overline{y}) \quad (10.194)$$

As for the coefficient matrices in Eq. (10.179), they are obtained as follows by differentiating Eq. (10.11)

$$\hat{G}_x(\overline{x}, \overline{y})\dot{\overline{x}} + \hat{G}_y(\overline{x}, \overline{y})\dot{\overline{y}} = \overline{0} \quad (10.195)$$

In Eq. (10.195), $\hat{G}_x(\overline{x}, \overline{y})$ and $\hat{G}_y(\overline{x}, \overline{y})$ are defined so that

$$[\hat{G}_x(\overline{x}, \overline{y})]_{ij} = G_{xij}(\overline{x}, \overline{y}) = \partial g_i(\overline{x}, \overline{y})/\partial x_j \quad (10.196)$$
$$[\hat{G}_y(\overline{x}, \overline{y})]_{ij} = G_{yij}(\overline{x}, \overline{y}) = \partial g_i(\overline{x}, \overline{y})/\partial y_j \quad (10.197)$$

Upon comparing Eqs. (10.195) and (10.179), it is seen that

$$\hat{M}(\overline{x}, \overline{y}) = \hat{G}_y(\overline{x}, \overline{y}) \quad (10.198)$$
$$\hat{N}(\overline{x}, \overline{y}) = -\hat{G}_x(\overline{x}, \overline{y}) \quad (10.199)$$

Example 10.7 ***Velocity Equations of the 3RRR Planar Parallel Manipulator***

Consider the manipulator shown in Figure 10.4. For this manipulator, the position relationships have already been expressed by Eqs. (10.27) and (10.28). These equations lead to the following velocity equations upon differentiation.

Differentiation of Eq. (10.27):

$$\begin{bmatrix} v_1 \\ v_2 \end{bmatrix} = \begin{bmatrix} \dot{p}_1 \\ \dot{p}_2 \end{bmatrix} = \begin{bmatrix} -(r_1 s\theta_1)\dot{\theta}_1 - (r_4 s\theta_4)\dot{\theta}_4 - (d_1 s\phi + h_7 c\phi)\dot{\phi} \\ +(r_1 c\theta_1)\dot{\theta}_1 + (r_4 c\theta_4)\dot{\theta}_4 + (d_1 c\phi - h_7 s\phi)\dot{\phi} \end{bmatrix}$$

Velocity equation for the tip point:

$$\begin{bmatrix} v_1 \\ v_2 \end{bmatrix} = \begin{bmatrix} -r_1 s\theta_1 \\ +r_1 c\theta_1 \end{bmatrix} \dot{\theta}_1 + \begin{bmatrix} -r_4 s\theta_4 \\ +r_4 c\theta_4 \end{bmatrix} \dot{\theta}_4 - \begin{bmatrix} h_7 c\phi + d_1 s\phi \\ h_7 s\phi - d_1 c\phi \end{bmatrix} \dot{\phi} \tag{10.200}$$

Differentiation of Eq. (10.28):

$$\begin{bmatrix} -(r_2 s\theta_2)\dot{\theta}_2 - (r_5 s\theta_5)\dot{\theta}_5 + (r_1 s\theta_1)\dot{\theta}_1 + (r_4 s\theta_4)\dot{\theta}_4 \\ +(r_2 c\theta_2)\dot{\theta}_2 + (r_5 c\theta_5)\dot{\theta}_5 - (r_1 c\theta_1)\dot{\theta}_1 - (r_4 c\theta_4)\dot{\theta}_4 \\ -(r_3 s\theta_3)\dot{\theta}_3 - (r_6 s\theta_6)\dot{\theta}_6 + (d_7 s\phi)\dot{\phi} + (r_1 s\theta_1)\dot{\theta}_1 + (r_4 s\theta_4)\dot{\theta}_4 \\ +(r_3 c\theta_3)\dot{\theta}_3 + (r_6 c\theta_6)\dot{\theta}_6 - (d_7 c\phi)\dot{\phi} - (r_1 c\theta_1)\dot{\theta}_1 - (r_4 c\theta_4)\dot{\theta}_4 \end{bmatrix} = \begin{bmatrix} 0 \\ 0 \\ 0 \\ 0 \end{bmatrix}$$

Velocity equation for the first independent kinematic loop (IKL-1):

$$\begin{bmatrix} -r_4 s\theta_4 & r_5 s\theta_5 \\ -r_4 c\theta_4 & r_5 c\theta_5 \end{bmatrix} \begin{bmatrix} \dot{\theta}_4 \\ \dot{\theta}_5 \end{bmatrix} = \begin{bmatrix} r_1 s\theta_1 & -r_2 s\theta_2 \\ r_1 c\theta_1 & -r_2 c\theta_2 \end{bmatrix} \begin{bmatrix} \dot{\theta}_1 \\ \dot{\theta}_2 \end{bmatrix} \tag{10.201}$$

Velocity equation for the second independent kinematic loop (IKL-2):

$$\begin{bmatrix} r_6 s\theta_6 & -d_7 s\phi \\ r_6 c\theta_6 & -d_7 c\phi \end{bmatrix} \begin{bmatrix} \dot{\theta}_6 \\ \dot{\phi} \end{bmatrix} - \begin{bmatrix} r_4 s\theta_4 \\ r_4 c\theta_4 \end{bmatrix} \dot{\theta}_4 = \begin{bmatrix} r_1 s\theta_1 & -r_3 s\theta_3 \\ r_1 c\theta_1 & -r_3 c\theta_3 \end{bmatrix} \begin{bmatrix} \dot{\theta}_1 \\ \dot{\theta}_3 \end{bmatrix} \tag{10.202}$$

10.6 Forward Kinematics in the Velocity Domain

The forward kinematics of a parallel manipulator in the velocity domain involves the *forward velocity solution* together with the identification and analysis of the associated singularities, which are more specifically designated as MSFKs. As seen previously in Section 10.5, the velocity equations turn out to be linear in the rates of the primary and secondary variables. Therefore, the forward velocity solution does not involve multiplicities.

The forward velocity solution is needed in order to determine the velocity state of the end-effector (i.e. $\overline{\eta}$) and the rates of the secondary variables (i.e. $\dot{\overline{y}}$) that are induced by the specified rates of the primary variables (i.e. $\dot{\overline{x}}$) at a currently known pose of the manipulator. Here, it is assumed that the current pose of the manipulator (described by $\overline{x}$ and $\overline{y}$) is known quantitatively as the outcome of the forward kinematic solution in the position domain with the appropriately selected posture modes. The forward velocity solution can be obtained in two stages.

In the first stage of the solution, Eq. (10.179) is solved in order to obtain $\dot{\overline{y}}$ as a function of $\dot{\overline{x}}$. If the manipulator is *not* in a pose of MSFK, i.e. if the matrix $\widehat{M}(\overline{x},\overline{y})$ is not singular, then Eq. (10.179) can be solved so that

$$\dot{\overline{y}} = \widehat{K}(\overline{x},\overline{y})\dot{\overline{x}} \tag{10.203}$$

In Eq. (10.203),

$$\widehat{K}(\overline{x},\overline{y}) = \widehat{M}^{-1}(\overline{x},\overline{y})\widehat{N}(\overline{x},\overline{y}) \tag{10.204}$$

In the second stage of the solution, the velocity state of the end-effector (i.e. $\overline{\eta}$) is determined simply by substituting Eq. (10.203) into Eq. (10.178). That is,

$$\overline{\eta} = \hat{J}(\overline{x}, \overline{y})\dot{\overline{x}} \tag{10.205}$$

In Eq. (10.205), $\hat{J}(\overline{x}, \overline{y})$ is designated specifically as the *Jacobian matrix* of the manipulator. It is expressed as follows:

$$\hat{J}(\overline{x}, \overline{y}) = \hat{A}(\overline{x}, \overline{y}) + \hat{B}(\overline{x}, \overline{y})\hat{K}(\overline{x}, \overline{y}) \tag{10.206}$$

If the manipulator happens to be in a pose of MSFK, then the velocity state of the manipulator and the end-effector become indefinite and therefore uncontrollable. This phenomenon is explained below.

A pose of MSFK occurs if the matrix $\hat{M}(\overline{x}, \overline{y})$ in Eq. (10.179) becomes singular. In such a pose, the rank of $\hat{M}(\overline{x}, \overline{y})$ reduces from n to $n_a = n - m_b$. Consequently, $\dot{\overline{y}}$ becomes indefinite and $\dot{\overline{x}}$ cannot be specified arbitrarily as desired. In other words, only a certain partition $\dot{\overline{x}}_a \in \mathcal{R}^{m_a}$ of $\dot{\overline{x}}$ can be specified arbitrarily but its complementary partition $\dot{\overline{x}}_b \in \mathcal{R}^{m_b}$ (where $m_b = m - m_a$) becomes enslaved to $\dot{\overline{x}}_a$ through a *singularity relationship* such as

$$\dot{\overline{x}}_b = \hat{S}_{ba}(\overline{x}, \overline{y})\dot{\overline{x}}_a \tag{10.207}$$

In order to verify the above-mentioned consequences, $\dot{\overline{y}}$ and the matrices $\hat{M}(\overline{x}, \overline{y})$ and $\hat{N}(\overline{x}, \overline{y})$ can also be partitioned accordingly. Thus, Eq. (10.179) can be separated into the following two equations.

$$\hat{M}_{aa}(\overline{x}, \overline{y})\dot{\overline{y}}_a + \hat{M}_{ab}(\overline{x}, \overline{y})\dot{\overline{y}}_b = \hat{N}_{aa}(\overline{x}, \overline{y})\dot{\overline{x}}_a + \hat{N}_{ab}(\overline{x}, \overline{y})\dot{\overline{x}}_b \tag{10.208}$$

$$\hat{M}_{ba}(\overline{x}, \overline{y})\dot{\overline{y}}_a + \hat{M}_{bb}(\overline{x}, \overline{y})\dot{\overline{y}}_b = \hat{N}_{ba}(\overline{x}, \overline{y})\dot{\overline{x}}_a + \hat{N}_{bb}(\overline{x}, \overline{y})\dot{\overline{x}}_b \tag{10.209}$$

In the above equations, $\dot{\overline{y}}_a \in \mathcal{R}^{n_a}$, $\dot{\overline{y}}_b \in \mathcal{R}^{m_b}$, and $\hat{M}_{aa}(\overline{x}, \overline{y})$ is a full-rank matrix with rank n_a. Therefore, $\dot{\overline{y}}_a$ can be extracted from Eq. (10.208) as

$$\dot{\overline{y}}_a = \hat{M}_{aa}^{-1}(\overline{x}, \overline{y})[\hat{N}_{aa}(\overline{x}, \overline{y})\dot{\overline{x}}_a + \hat{N}_{ab}(\overline{x}, \overline{y})\dot{\overline{x}}_b - \hat{M}_{ab}(\overline{x}, \overline{y})\dot{\overline{y}}_b] \tag{10.210}$$

When Eq. (10.210) is substituted, Eq. (10.209) becomes

$$\begin{aligned}
&[\hat{M}_{bb}(\overline{x}, \overline{y}) - \hat{M}_{ba}(\overline{x}, \overline{y})\hat{M}_{aa}^{-1}(\overline{x}, \overline{y})\hat{M}_{ab}(\overline{x}, \overline{y})]\dot{\overline{y}}_b \\
&= [\hat{N}_{ba}(\overline{x}, \overline{y}) - \hat{M}_{ba}(\overline{x}, \overline{y})\hat{M}_{aa}^{-1}(\overline{x}, \overline{y})\hat{N}_{aa}(\overline{x}, \overline{y})]\dot{\overline{x}}_a \\
&+ [\hat{N}_{bb}(\overline{x}, \overline{y}) - \hat{M}_{ba}(\overline{x}, \overline{y})\hat{M}_{aa}^{-1}(\overline{x}, \overline{y})\hat{N}_{ab}(\overline{x}, \overline{y})]\dot{\overline{x}}_b
\end{aligned} \tag{10.211}$$

The singularity of $\hat{M}(\overline{x}, \overline{y})$ implies that

$$\hat{M}_{bb}(\overline{x}, \overline{y}) = \hat{M}_{ba}(\overline{x}, \overline{y})\hat{M}_{aa}^{-1}(\overline{x}, \overline{y})\hat{M}_{ab}(\overline{x}, \overline{y}) \tag{10.212}$$

Due to Eq. (10.212), (10.211) can only be satisfied if $\dot{\overline{x}}_b$ is enslaved to $\dot{\overline{x}}_a$ as mentioned before. Hence, the singularity matrix $\hat{S}_{ba}(\overline{x}, \overline{y})$ in Eq. (10.207) comes out as

$$\begin{aligned}
\hat{S}_{ba}(\overline{x}, \overline{y}) &= [\hat{N}_{bb}(\overline{x}, \overline{y}) - \hat{M}_{ba}(\overline{x}, \overline{y})\hat{M}_{aa}^{-1}(\overline{x}, \overline{y})\hat{N}_{ab}(\overline{x}, \overline{y})]^{-1} \\
&\times [\hat{M}_{ba}(\overline{x}, \overline{y})\hat{M}_{aa}^{-1}(\overline{x}, \overline{y})\hat{N}_{aa}(\overline{x}, \overline{y}) - \hat{N}_{ba}(\overline{x}, \overline{y})]
\end{aligned} \tag{10.213}$$

Moreover, due to Eqs. (10.212) and (10.213), Eq. (10.211) reduces to an equation trivially satisfied as $\overline{0} = \overline{0}$ no matter what $\dot{\overline{y}}_b$ is. In other words, $\dot{\overline{y}}_b$ becomes indefinite but it can retain a finite magnitude owing to Eq. (10.207). Therefore, $\dot{\overline{y}}_a$ also becomes indefinite due to its dependence on $\dot{\bar{y}}_b$ according to Eq. (10.210). Consequently, $\dot{\bar{y}}$ becomes indefinite as a whole.

As for the velocity state of the end-effector in this motion singularity, Eq. (10.178) can be written as follows with the relevant partitions:

$$\bar{\eta} = \hat{A}_a(\bar{x},\bar{y})\dot{\bar{x}}_a + \hat{A}_b(\bar{x},\bar{y})\dot{\bar{x}}_b + \hat{B}(\bar{x},\bar{y})\dot{\bar{y}} \Rightarrow$$

$$\bar{\eta} = [\hat{A}_a(\bar{x},\bar{y}) + \hat{A}_b(\bar{x},\bar{y})\hat{S}_{ba}(\bar{x},\bar{y})]\dot{\bar{x}}_a + \hat{B}(\bar{x},\bar{y})\dot{\bar{y}} \tag{10.214}$$

Hence, it is seen that $\bar{\eta}$ also becomes indefinite due to the indefiniteness of $\dot{\bar{y}}$. This means that the manipulator becomes uncontrollable in a pose of MSFK. Therefore, it must be prevented from getting into a pose of MSFK. However, unlike the PSFKs of the position domain, the MSFKs do not depend, in general, on the geometric parameters of the links. This fact is demonstrated in Example 10.8 of this section. Therefore, the only safe way to prevent MSFKs from occurring is to plan the task motion carefully so that the matrix $\hat{M}(\bar{x},\bar{y})$ never becomes singular.

Incidentally, the MSFKs also happen to be the PMCPs of the manipulator. This fact is illustrated in Figures 10.11 and 10.12.

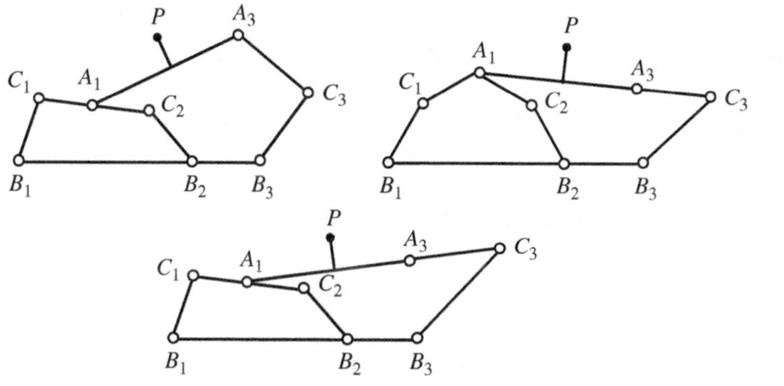

Figure 10.11 Extended MSFK poses of the 3RRR planar parallel manipulator.

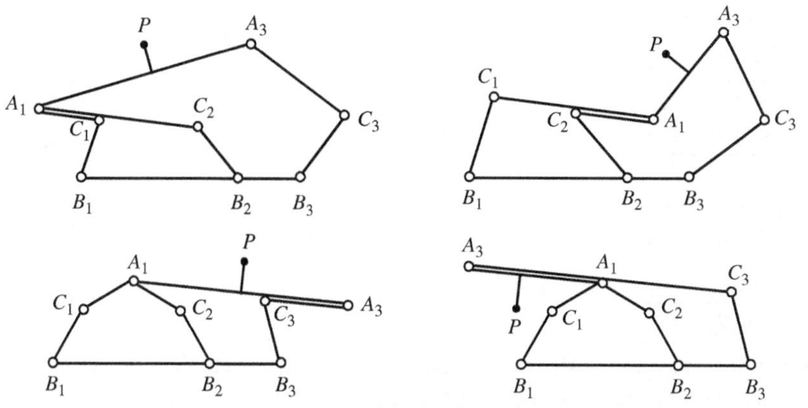

Figure 10.12 Folded MSFK poses of the 3RRR planar parallel manipulator.

Example 10.8 ***Forward Velocity Analysis of the 3RRR Planar Parallel Manipulator***

The following velocity equations have already been obtained in Example 10.7. However, they are repeated here for the sake of convenience.

* Velocity equation for the first kinematic loop (IKL-1):

$$\begin{bmatrix} -r_4 s\theta_4 & r_5 s\theta_5 \\ -r_4 c\theta_4 & r_5 c\theta_5 \end{bmatrix} \begin{bmatrix} \dot{\theta}_4 \\ \dot{\theta}_5 \end{bmatrix} = \begin{bmatrix} r_1 s\theta_1 & -r_2 s\theta_2 \\ r_1 c\theta_1 & -r_2 c\theta_2 \end{bmatrix} \begin{bmatrix} \dot{\theta}_1 \\ \dot{\theta}_2 \end{bmatrix} \tag{10.215}$$

* Velocity equation for the second kinematic loop (IKL-2):

$$\begin{bmatrix} r_6 s\theta_6 & -d_7 s\phi \\ r_6 c\theta_6 & -d_7 c\phi \end{bmatrix} \begin{bmatrix} \dot{\theta}_6 \\ \dot{\phi} \end{bmatrix} - \begin{bmatrix} r_4 s\theta_4 \\ r_4 c\theta_4 \end{bmatrix} \dot{\theta}_4 = \begin{bmatrix} r_1 s\theta_1 & -r_3 s\theta_3 \\ r_1 c\theta_1 & -r_3 c\theta_3 \end{bmatrix} \begin{bmatrix} \dot{\theta}_1 \\ \dot{\theta}_3 \end{bmatrix} \tag{10.216}$$

* Velocity equation for the tip point:

$$\begin{bmatrix} v_1 \\ v_2 \end{bmatrix} = \begin{bmatrix} -r_1 s\theta_1 \\ +r_1 c\theta_1 \end{bmatrix} \dot{\theta}_1 + \begin{bmatrix} -r_4 s\theta_4 \\ +r_4 c\theta_4 \end{bmatrix} \dot{\theta}_4 - \begin{bmatrix} h_7 c\phi + d_1 s\phi \\ h_7 s\phi - d_1 c\phi \end{bmatrix} \dot{\phi} \tag{10.217}$$

(a) Forward Velocity Solution

Here, the rates of the primary variables (i.e. $\dot{\theta}_1$, $\dot{\theta}_2$, and $\dot{\theta}_3$) are specified and the consequent rates of the secondary variables together with the tip point velocity components are found as explained below.

The solution can be started from Eq. (10.215) by inverting the coefficient matrix on the left-hand side. The determinant of this coefficient matrix is

$$D_{45} = r_4 r_5 (s\theta_5 c\theta_4 - c\theta_5 s\theta_4) = r_4 r_5 \sin(\theta_5 - \theta_4) \tag{10.218}$$

If $D_{45} \neq 0$, i.e. if $\sin(\theta_5 - \theta_4) \neq 0$, then Eq. (10.215) gives $\dot{\theta}_4$ and $\dot{\theta}_5$ as follows:

$$\begin{bmatrix} \dot{\theta}_4 \\ \dot{\theta}_5 \end{bmatrix} = \frac{1}{D_{45}} \begin{bmatrix} r_5 c\theta_5 & -r_5 s\theta_5 \\ r_4 c\theta_4 & -r_4 s\theta_4 \end{bmatrix} \begin{bmatrix} r_1 s\theta_1 & -r_2 s\theta_2 \\ r_1 c\theta_1 & -r_2 c\theta_2 \end{bmatrix} \begin{bmatrix} \dot{\theta}_1 \\ \dot{\theta}_2 \end{bmatrix} \Rightarrow$$

$$\dot{\theta}_4 = G_{41}\dot{\theta}_1 + G_{42}\dot{\theta}_2 \tag{10.219}$$

$$\dot{\theta}_5 = G_{51}\dot{\theta}_1 + G_{52}\dot{\theta}_2 \tag{10.220}$$

In Eqs. (10.219) and (10.220),

$$G_{41} = \frac{r_1 \sin(\theta_1 - \theta_5)}{r_4 \sin(\theta_5 - \theta_4)} \tag{10.221}$$

$$G_{42} = \frac{r_2 \sin(\theta_5 - \theta_2)}{r_4 \sin(\theta_5 - \theta_4)} \tag{10.222}$$

$$G_{51} = \frac{r_1 \sin(\theta_1 - \theta_4)}{r_5 \sin(\theta_5 - \theta_4)} \tag{10.223}$$

$$G_{52} = \frac{r_2 \sin(\theta_4 - \theta_2)}{r_5 \sin(\theta_5 - \theta_4)} \tag{10.224}$$

The solution can be continued with Eq. (10.216) by inverting the coefficient matrix on the left-hand side. The determinant of this coefficient matrix is

$$D_{67} = r_6 d_7 (-s\theta_6 c\phi + c\theta_6 s\phi) = -r_6 d_7 \sin(\theta_6 - \phi) \tag{10.225}$$

If $D_{67} \neq 0$, i.e. if $\sin(\theta_6 - \phi) \neq 0$, then Eq. (10.216) gives $\dot{\theta}_6$ and $\dot{\phi}$ as follows:

$$\begin{bmatrix} \dot{\theta}_6 \\ \dot{\phi} \end{bmatrix} = \frac{1}{D_{67}} \begin{bmatrix} -d_7 c\phi & d_7 s\phi \\ -r_6 c\theta_6 & r_6 s\theta_6 \end{bmatrix} \begin{bmatrix} r_1 s\theta_1 & -r_3 s\theta_3 \\ r_1 c\theta_1 & -r_3 c\theta_3 \end{bmatrix} \begin{bmatrix} \dot{\theta}_1 \\ \dot{\theta}_3 \end{bmatrix} + \begin{bmatrix} r_4 s\theta_4 \\ r_4 c\theta_4 \end{bmatrix} (G_{41}\dot{\theta}_1 + G_{42}\dot{\theta}_2) \Rightarrow$$

$$\dot{\theta}_6 = G_{61}\dot{\theta}_1 + G_{62}\dot{\theta}_2 + G_{63}\dot{\theta}_3 \tag{10.226}$$

$$\dot{\phi} = G_{71}\dot{\theta}_1 + G_{72}\dot{\theta}_2 + G_{73}\dot{\theta}_3 \tag{10.227}$$

In Eqs. (10.226) and (10.227),

$$G_{61} = \frac{r_1 \sin(\theta_1 - \phi)}{r_6 \sin(\theta_6 - \phi)} + \frac{r_4 \sin(\theta_4 - \phi)}{r_6 \sin(\theta_6 - \phi)} G_{41} \tag{10.228}$$

$$G_{62} = \frac{r_4 \sin(\theta_4 - \phi)}{r_6 \sin(\theta_6 - \phi)} G_{42} \tag{10.229}$$

$$G_{63} = \frac{r_3 \sin(\phi - \theta_3)}{r_6 \sin(\theta_6 - \phi)} \tag{10.230}$$

$$G_{71} = \frac{r_1 \sin(\theta_1 - \theta_6)}{d_7 \sin(\theta_6 - \phi)} + \frac{r_4 \sin(\theta_4 - \theta_6)}{d_7 \sin(\theta_6 - \phi)} G_{41} \tag{10.231}$$

$$G_{72} = \frac{r_4 \sin(\theta_4 - \theta_6)}{d_7 \sin(\theta_6 - \phi)} G_{42} \tag{10.232}$$

$$G_{73} = \frac{r_3 \sin(\theta_6 - \theta_3)}{d_7 \sin(\theta_6 - \phi)} \tag{10.233}$$

In all the above equations, G_{ij} is defined as the *velocity influence coefficient* that shows the influence of the rate $\dot{\theta}_j$ on the rate $\dot{\theta}_i$. Note that G_{ij} is a function of position only.

As the final step of the solution, the velocity components of the tip point is obtained by inserting the rates obtained above into Eq. (10.217). That is,

$$\begin{bmatrix} v_1 \\ v_2 \end{bmatrix} = \begin{bmatrix} -r_1 s\theta_1 \\ +r_1 c\theta_1 \end{bmatrix} \dot{\theta}_1 + \begin{bmatrix} -r_4 s\theta_4 \\ +r_4 c\theta_4 \end{bmatrix} (G_{41}\dot{\theta}_1 + G_{42}\dot{\theta}_2). - \begin{bmatrix} h_7 c\phi + s\phi \\ h_7 s\phi - d_1 c\phi \end{bmatrix} (G_{71}\dot{\theta}_1 + G_{72}\dot{\theta}_2 + G_{73}\dot{\theta}_3) \tag{10.234}$$

(b) Motion Singularities of Forward Kinematics

The determinants D_{45} and D_{67} expressed in Part (a) imply that there are two distinct versions of MSFKs. They are denoted as MSFK-1 and MSFK-2. They may occur separately at different times or simultaneously at the same time. They are described and discussed below.

The first kind of motion singularity (MSFK-1) occurs if $D_{45} = 0$, i.e. if $\sin(\theta_5 - \theta_4) = 0$. This condition implies that MSFK-1 has two versions, which are characterized as follows:

$$\theta_5 = \theta_4 \text{ or } \theta_5 = \theta_4 \pm \pi \tag{10.235}$$

The version with $\theta_5 = \theta_4$ is called *folded version* and the other version with $\theta_5 = \theta_4 \pm \pi$ is called *extended version*. In either of the versions, Eq. (10.215) can be written as the following two equations.

$$(r_4\dot{\theta}_4 - \sigma_1' r_5\dot{\theta}_5)s\theta_4 = (r_2 s\theta_2)\dot{\theta}_2 - (r_1 s\theta_1)\dot{\theta}_1 \tag{10.236}$$

$$(r_4\dot{\theta}_4 - \sigma_1' r_5\dot{\theta}_5)c\theta_4 = (r_2 c\theta_2)\dot{\theta}_2 - (r_1 c\theta_1)\dot{\theta}_1 \tag{10.237}$$

In the above equations,

$$\sigma_1' = \begin{cases} +1 \text{ if } \theta_5 = \theta_4 \\ -1 \text{ if } \theta_5 = \theta_4 \pm \pi \end{cases} \tag{10.238}$$

Equations (10.236) and (10.237) can be manipulated further into the following equations.

$$r_2\dot{\theta}_2 \sin(\theta_2 - \theta_4) = r_1\dot{\theta}_1 \sin(\theta_1 - \theta_4) \tag{10.239}$$

$$r_4\dot{\theta}_4 - \sigma_1' r_5\dot{\theta}_5 = r_2\dot{\theta}_2 \cos(\theta_2 - \theta_4) - r_1\dot{\theta}_1 \cos(\theta_1 - \theta_4) \tag{10.240}$$

Equation (10.239) is the *singularity relationship* of MSFK-1. It shows how the primary rates $\dot{\theta}_1$ and $\dot{\theta}_2$ become dependent on each other.

Equation (10.240) shows that the secondary rates $\dot{\theta}_4$ and $\dot{\theta}_5$ become indefinite, because, they can assume arbitrary values as long as they satisfy this particular equation.

The second kind of motion singularity (MSFK-2) occurs if $D_{67} = 0$, i.e. if $\sin(\theta_6 - \phi) = 0$. This condition implies that MSFK-2 has two versions, too, which are characterized as follows:

$$\theta_6 = \phi \text{ or } \theta_6 = \phi \pm \pi \tag{10.241}$$

The version with $\theta_6 = \phi$ is called *folded version* and the other version with $\theta_6 = \phi \pm \pi$ is called *extended version*. In either of the versions, Eq. (10.216) can be written as the following two equations.

$$(\sigma_2' r_6\dot{\theta}_6 - d_7\dot{\phi})s\phi = (r_1 s\theta_1)\dot{\theta}_1 - (r_3 s\theta_3)\dot{\theta}_3 + (r_4 s\theta_4)\dot{\theta}_4 \tag{10.242}$$

$$(\sigma_2' r_6\dot{\theta}_6 - d_7\dot{\phi})c\phi = (r_1 c\theta_1)\dot{\theta}_1 - (r_3 c\theta_3)\dot{\theta}_3 + (r_4 c\theta_4)\dot{\theta}_4 \tag{10.243}$$

In the above equations,

$$\sigma_2' = \begin{cases} +1 \text{ if } \theta_6 = \phi \\ -1 \text{ if } \theta_6 = \phi \pm \pi \end{cases} \tag{10.244}$$

Equations (10.242) and (10.243) can be manipulated further into the following equations.

$$r_1\dot{\theta}_1 \sin(\theta_1 - \phi) - r_3\dot{\theta}_3 \sin(\theta_3 - \phi) + r_4\dot{\theta}_4 \sin(\theta_4 - \phi) = 0 \tag{10.245}$$

$$\sigma_2' r_6\dot{\theta}_6 - d_7\dot{\phi} = r_1\dot{\theta}_1 \cos(\theta_1 - \phi) - r_3\dot{\theta}_3 \cos(\theta_3 - \phi) + r_4\dot{\theta}_4 \cos(\theta_4 - \phi) \tag{10.246}$$

If MSFK-2 and MSFK-1 are *not* concurrent, then, noting that $\dot{\theta}_4$ is given definitely by Eqs. (10.219), (10.245), and (10.246) can also be written as follows:

$$\begin{aligned} &[r_1 \sin(\theta_1 - \phi) + r_4 G_{41} \sin(\theta_4 - \phi)]\dot{\theta}_1 \\ &+ r_4 G_{42}\dot{\theta}_2 \sin(\theta_4 - \phi) - r_3\dot{\theta}_3 \sin(\theta_3 - \phi) = 0 \end{aligned} \tag{10.247}$$

$$\begin{aligned} &\sigma_2' r_6\dot{\theta}_6 - d_7\dot{\phi} = [r_1 \cos(\theta_1 - \phi) + r_4 G_{41} \cos(\theta_4 - \phi)]\dot{\theta}_1 \\ &+ r_4 G_{42}\dot{\theta}_2 \cos(\theta_4 - \phi) - r_3\dot{\theta}_3 \cos(\theta_3 - \phi) \end{aligned} \tag{10.248}$$

Equation (10.247) is the *singularity relationship* of MSFK-2. It shows how the primary rates $\dot{\theta}_1$, $\dot{\theta}_2$, and $\dot{\theta}_3$ become interrelated with each other.

Equation (10.248) shows that the rates $\dot{\theta}_6$ and $\dot{\phi}$ become indefinite, because, they can assume arbitrary values as long as they satisfy this particular equation.

If MSFK-2 and MSFK-1 *are* concurrent, then $\dot{\theta}_4$ and $\dot{\theta}_5$ lose their dependence on $\dot{\theta}_1$ and $\dot{\theta}_2$ due to MSFK-1. Consequently, Eq. (10.247) can not be established and therefore $\dot{\theta}_3$ also loses its dependence on $\dot{\theta}_1$ and $\dot{\theta}_2$. In other words, $\dot{\theta}_1$ and $\dot{\theta}_3$ can be specified as desired but $\dot{\theta}_2$ remains tied up to $\dot{\theta}_1$ due to MSFK-1. On the other hand, since Eq. (10.245) must be satisfied due to MSFK-2, $\dot{\theta}_4$ becomes a definite rate that is determined in terms of the specified values of $\dot{\theta}_1$ and $\dot{\theta}_3$. Then, according to Eq. (10.240), $\dot{\theta}_5$ gets determined, too. However, the rates $\dot{\theta}_6$ and $\dot{\phi}$ still remain indefinite but related to each other according to Eq. (10.246).

The extended and folded poses of the manipulator in MSFK-1 and MSFK-2 are illustrated in Figures 10.11 and 10.12.

(c) Comparison of the Position and Motion Singularities

At this point, it is worth comparing Figures 10.11, 10.12, and 10.7 in order to see the differences between the PSFKs and MSFKs. Note that the mathematical sources of PSFKs and MSFKs are different and independent. The PSFKs arise completely from the position equations without any connection with the velocity equations. Conversely, the MSFKs arise completely from the velocity equations without any connection with the position equations. Yet, in the mentioned figures, it is seen that the PSFK poses appear as special versions of the MSFK poses. In other words, the MSFK poses occur simply when certain adjacent links get aligned either in a folded way or in an extended way, whereas the PSFK poses occur more specifically when those adjacent links have equal lengths and get aligned only in a folded way.

10.7 Inverse Kinematics in the Velocity Domain

The inverse kinematics of a parallel manipulator in the velocity domain involves the *inverse velocity solution* together with the identification and analysis of the associated singularities, which are more specifically designated as MSIKs. As seen before, the velocity equations turn out to be linear in the rates of the primary and secondary variables. Therefore, like the forward velocity solution, the inverse velocity solution does not involve multiplicities, either.

The inverse velocity solution is needed in order to determine the rates of all the primary and secondary variables (i.e. both $\dot{\bar{x}}$ and $\dot{\bar{y}}$) that correspond to a specified velocity state (i.e. $\bar{\eta}$) of the end-effector at a currently known pose of the manipulator described quantitatively by $\bar{x}$ and $\bar{y}$. However, the main purpose of the inverse velocity solution is to determine the *rates of the active joint variables*, which are directly or indirectly related to the rates of the primary variables (i.e. $\dot{\bar{x}}$). Thus, it becomes possible to command the actuators in order to achieve the specified desired motion of the end-effector.

In this section, the inverse velocity solution will also be obtained for a *regular parallel manipulator* as done in Section 10.4.

Like the inverse kinematic solution in the position domain, the inverse velocity solution is also obtained *leg by leg*.

For a parallel manipulator with a single moving platform and n_L number of legs, the following set of equations can be written in order to express the velocity state of the end-effector by going through each of the legs.

$$\overline{\eta} = \widehat{L}_k(\overline{z})\dot{\overline{z}}_k; \quad k = 1, 2, \dots, n_L \tag{10.249}$$

In Eq. (10.249), $\dot{\overline{z}}_k$ is the leg-specific partition of $\dot{\overline{z}}$ that belongs to the leg L_k and $\dot{\overline{z}} \in \mathcal{R}^{m+n}$ is the column matrix that contains the rates of all the primary and secondary variables. That is,

$$\dot{\overline{z}} = \begin{bmatrix} \dot{\overline{x}} \\ \dot{\overline{y}} \end{bmatrix} \tag{10.250}$$

The partition $\dot{\overline{z}}_k$ is also similarly partitioned so that

$$\dot{\overline{z}}_k = \begin{bmatrix} \dot{\overline{x}}_k \\ \dot{\overline{y}}_k \end{bmatrix} \tag{10.251}$$

In Eq. (10.251), the subpartition $\dot{\overline{x}}_k$ consists of the rates of the primary variables that belong to L_k. The number of these variables is equal to the number of the actuators that drive L_k. In fact, they are normally taken as the rates of the associated active joint variables.

As for the matrix $\widehat{L}_k(\overline{z})$ in Eq. (10.249), it is obtained from Eq. (10.102) similarly as the matrices $\widehat{A}(\overline{x}, \overline{y})$ and $\widehat{B}(\overline{x}, \overline{y})$ are obtained as explained in Section 10.5. The derivation is shown below.

$$\dot{\overline{p}} = \widehat{F}_k(\overline{z}_k)\dot{\overline{z}}_k; \quad k = 1, 2, \dots, n_L \tag{10.252}$$

In Eq. (10.252),

$$[\widehat{F}_k(\overline{z}_k)]_{ij} = F_{kij}(\overline{z}_k) = \partial f_{ki}(\overline{z}_k)/\partial z_j \tag{10.253}$$

On the other hand, according to Eqs. (10.190) and (10.191) of Section 10.5,

$$\overline{\eta} = \widehat{D}(\overline{\phi})\dot{\overline{p}} = \widehat{D}(\overline{\phi})\widehat{F}_k(\overline{z}_k)\dot{\overline{z}}_k \tag{10.254}$$

$$\overline{\phi} = \overline{\phi}(\overline{x}, \overline{y}) = \overline{\phi}(\overline{z}) \tag{10.255}$$

Hence, it is seen that

$$\widehat{L}_k(\overline{z}) = \widehat{D}(\overline{\phi}(\overline{z}))\widehat{F}_k(\overline{z}_k) \tag{10.256}$$

Suppose the leg L_k is *not* in a pose of MSIK-k, i.e. MSIK associated with L_k. Then, $\det[\widehat{L}_k(\overline{z})] \neq 0$ and therefore Eq. (10.249) can be solved simply as follows for each leg.

$$\dot{\overline{z}}_k = \widehat{L}_k^{-1}(\overline{z})\overline{\eta}; \quad k = 1, 2, \dots, n_L \tag{10.257}$$

A pose of MSIK-k occurs if the matrix $\widehat{L}_k(\overline{z})$ in Eq. (10.249) becomes singular. In such a pose of L_k, the rank of $\widehat{L}_k(\overline{z})$ reduces from m to $m_{kc} = m - m_{kd}$. Consequently, $\dot{\overline{z}}_k$ becomes indefinite and $\overline{\eta}$ cannot be specified arbitrarily as desired. In other words, only a certain partition $\overline{\eta}_{kc} \in \mathcal{R}^{m_{kc}}$ of $\overline{\eta}$ can be specified arbitrarily but its complementary partition $\overline{\eta}_{kd} \in \mathcal{R}^{m_{kd}}$ becomes enslaved to $\overline{\eta}_{kc}$ through a *singularity relationship* such as

$$\overline{\eta}_{kd} = \widehat{S}_{kdc}(\overline{z})\overline{\eta}_{kc} \tag{10.258}$$

In order to verify the above-mentioned consequences, $\dot{\bar{z}}_k$ and $\hat{L}_k(\bar{z})$ can also be partitioned accordingly. Thus, Eq. (10.249) can be separated into the following two equations.

$$\hat{L}_{kcc}(\bar{z})\dot{\bar{z}}_{kc} + \hat{L}_{kcd}(\bar{z})\dot{\bar{z}}_{kd} = \bar{\eta}_{kc} \tag{10.259}$$

$$\hat{L}_{kdc}(\bar{z})\dot{\bar{z}}_{kc} + \hat{L}_{kdd}(\bar{z})\dot{\bar{z}}_{kd} = \bar{\eta}_{kd} \tag{10.260}$$

In the above equations, $\dot{\bar{z}}_{kc} \in \mathcal{R}^{m_{kc}}$, $\dot{\bar{z}}_{kd} \in \mathcal{R}^{m_{kd}}$, and $\hat{L}_{kcc}(\bar{z})$ is a full-rank matrix with rank m_{kc}. Therefore, $\dot{\bar{z}}_{kc}$ can be extracted from Eq. (10.259) as

$$\dot{\bar{z}}_{kc} = \hat{L}_{kcc}^{-1}(\bar{z})[\bar{\eta}_{kc} - \hat{L}_{kcd}(\bar{z})\dot{\bar{z}}_{kd}] \tag{10.261}$$

When Eq. (10.261) is substituted, Eq. (10.260) becomes

$$[\hat{L}_{kdd}(\bar{z}) - \hat{L}_{kdc}(\bar{z})\hat{L}_{kcc}^{-1}(\bar{z})\hat{L}_{kcd}(\bar{z})]\dot{\bar{z}}_{kd} = \bar{\eta}_{kd} - \hat{L}_{kdc}(\bar{z})\hat{L}_{kcc}^{-1}(\bar{z})\bar{\eta}_{kc} \tag{10.262}$$

The singularity of $\hat{L}_k(\bar{z})$ implies that

$$\hat{L}_{kdd}(\bar{z}) = \hat{L}_{kdc}(\bar{z})\hat{L}_{kcc}^{-1}(\bar{z})\hat{L}_{kcd}(\bar{z}) \tag{10.263}$$

Due to Eq. (10.263), (10.262) can only be satisfied if $\bar{\eta}_{kd}$ is enslaved to $\bar{\eta}_{kc}$ as mentioned before. Hence, the singularity matrix $\hat{S}_{kdc}(\bar{z})$ in Eq. (10.258) comes out as

$$\hat{S}_{kdc}(\bar{z}) = \hat{L}_{kdc}(\bar{z})\hat{L}_{kcc}^{-1}(\bar{z}) \tag{10.264}$$

Moreover, due to Eqs. (10.263) and (10.264), Eq. (10.262) reduces to an equation trivially satisfied as $\bar{0} = \bar{0}$ no matter what $\dot{\bar{z}}_{kd}$ is. In other words, $\dot{\bar{z}}_{kd}$ becomes indefinite but it can retain a finite magnitude owing to Eq. (10.258). Therefore, $\dot{\bar{z}}_{kc}$ also becomes indefinite due to its dependence on $\dot{\bar{z}}_{kd}$ according to Eq. (10.261). Consequently, $\dot{\bar{z}}_k$ becomes indefinite as a whole. This means that the rates of the active joint variable or variables contained in $\dot{\bar{z}}_k$ can be changed arbitrarily without creating any effect on the velocity state of the end-effector which obeys the singularity relationship expressed by Eq. (10.258). In other words, as long as the end-effector is kept moving according to this special singularity constraint, possibly due to a specific task requirement, L_k loses its effect on the end-effector completely no matter how arbitrarily it is actuated.

However, unlike the MSFK poses of the manipulator, the MSIK poses of the legs need not be avoided. Let alone avoiding, they must sometimes be deliberately allowed in order to execute some special tasks, which could not be executed otherwise. In other words, the reachable range of the end-effector and the dexterity of the manipulator can be increased by allowing the MSIKs of the legs to occur. Moreover, when they are allowed to occur due to such tasks, they may be used even in an advantageous way. For example, if MSIK-k occurs, then L_k becomes ineffective on the consistently specified motion of the end-effector. However, its supporting action still continues. Besides, at the instant of MSIK-k, the actuator or actuators of L_k can be used in order to select the *posture mode* (PML) of L_k after the MSIK-k. This is because the pose of MSIK-k happens to be the same as the PMCPL of L_k, i.e. PMCPL-k. Thus, with a little actuation effort, the current PML of L_k can either be maintained or it can be changed into another PML for the post-singularity motion of the manipulator.

Example 10.9 ***Inverse Velocity Analysis of the 3RRR Planar Parallel Manipulator***

For the parallel manipulator shown in Figure 10.4, the following velocity equations can be obtained by differentiating the position equations written for each leg in Example 10.6.

* Velocity equations for L_1:

$$v_1 = -(r_1 s\theta_1)\dot{\theta}_1 - (r_4 s\theta_4)\dot{\theta}_4 - (d_1 s\phi + h_7 c\phi)\dot{\phi} \tag{10.265}$$

$$v_2 = +(r_1 c\theta_1)\dot{\theta}_1 + (r_4 c\theta_4)\dot{\theta}_4 + (d_1 c\phi - h_7 s\phi)\dot{\phi} \tag{10.266}$$

* Velocity equations for L_2:

$$v_1 = -(r_2 s\theta_2)\dot{\theta}_2 - (r_5 s\theta_5)\dot{\theta}_5 - (d_1 s\phi + h_7 c\phi)\dot{\phi} \tag{10.267}$$

$$v_2 = +(r_2 c\theta_2)\dot{\theta}_2 + (r_5 c\theta_5)\dot{\theta}_5 + (d_1 c\phi - h_7 s\phi)\dot{\phi} \tag{10.268}$$

* Velocity equations for L_3:

$$v_1 = -(r_3 s\theta_3)\dot{\theta}_3 - (r_6 s\theta_6)\dot{\theta}_6 + (d_3 s\phi - h_7 c\phi)\dot{\phi} \tag{10.269}$$

$$v_2 = +(r_3 c\theta_3)\dot{\theta}_3 + (r_6 c\theta_6)\dot{\theta}_6 - (d_3 c\phi + h_7 s\phi)\dot{\phi} \tag{10.270}$$

(a) Inverse Velocity Solutions for the Legs

Here, the velocity state of the end-effector is specified. That is, the tip point velocity components (v_1, v_2) and the angular velocity of the end-effector $(\omega = \dot{\phi})$ are known. The corresponding rates of the primary and secondary variables are found as explained below.

* Inverse velocity solution for L_1:

Equations (10.265) and (10.266) can be written as the following matrix equation.

$$\begin{bmatrix} -r_1 s\theta_1 & -r_4 s\theta_4 \\ +r_1 c\theta_1 & +r_4 c\theta_4 \end{bmatrix} \begin{bmatrix} \dot{\theta}_1 \\ \dot{\theta}_4 \end{bmatrix} = \begin{bmatrix} w_{11} \\ w_{12} \end{bmatrix} = \begin{bmatrix} v_1 + (d_1 s\phi + h_7 c\phi)\dot{\phi} \\ v_2 - (d_1 c\phi - h_7 s\phi)\dot{\phi} \end{bmatrix} \tag{10.271}$$

In Eq. (10.271), w_{11} and w_{12} are the known components of the velocity $\vec{w}_1$ of the point A_1. On the left-hand side of the same equation, the coefficient matrix has the following determinant.

$$D_{14} = r_1 r_4 \sin(\theta_4 - \theta_1) \tag{10.272}$$

If $D_{14} \neq 0$, i.e. if $\sin(\theta_4 - \theta_1) \neq 0$, then Eq. (10.271) can be solved as follows in order to find the rates $\dot{\theta}_1$ and $\dot{\theta}_4$:

$$\begin{bmatrix} \dot{\theta}_1 \\ \dot{\theta}_4 \end{bmatrix} = \frac{1}{D_{14}} \begin{bmatrix} +r_4 c\theta_4 & +r_4 s\theta_4 \\ -r_1 c\theta_1 & -r_1 s\theta_1 \end{bmatrix} \begin{bmatrix} w_{11} \\ w_{12} \end{bmatrix} \Rightarrow$$

$$\dot{\theta}_1 = (w_{11} c\theta_4 + w_{12} s\theta_4)/[r_1 \sin(\theta_4 - \theta_1)] \tag{10.273}$$

$$\dot{\theta}_4 = -(w_{11} c\theta_1 + w_{12} s\theta_1)/[r_4 \sin(\theta_4 - \theta_1)] \tag{10.274}$$

* Inverse velocity solution for L_2:

Equations (10.267) and (10.268) can be written as the following matrix equation.

$$\begin{bmatrix} -r_2 s\theta_2 & -r_5 s\theta_5 \\ +r_2 c\theta_2 & +r_5 c\theta_5 \end{bmatrix} \begin{bmatrix} \dot{\theta}_2 \\ \dot{\theta}_5 \end{bmatrix} = \begin{bmatrix} w_{21} \\ w_{22} \end{bmatrix} = \begin{bmatrix} v_1 + (d_1 s\phi + h_7 c\phi)\dot{\phi} \\ v_2 - (d_1 c\phi - h_7 s\phi)\dot{\phi} \end{bmatrix} \tag{10.275}$$

In Eq. (10.275), w_{21} and w_{22} are the known components of the velocity $\vec{w}_2$ of the point A_2. Note that $\vec{w}_2 = \vec{w}_1$ because the points A_2 and A_1 are coincident. On the left-hand side of the same equation, the coefficient matrix has the following determinant.

$$D_{25} = r_2 r_5 \sin(\theta_5 - \theta_2) \tag{10.276}$$

If $D_{25} \neq 0$, i.e. if $\sin(\theta_5 - \theta_2) \neq 0$, then Eq. (10.275) can be solved as follows in order to find the rates $\dot{\theta}_2$ and $\dot{\theta}_5$:

$$\begin{bmatrix} \dot{\theta}_2 \\ \dot{\theta}_5 \end{bmatrix} = \frac{1}{D_{25}} \begin{bmatrix} +r_5 c\theta_5 & +r_5 s\theta_5 \\ -r_2 c\theta_2 & -r_2 s\theta_2 \end{bmatrix} \begin{bmatrix} w_{21} \\ w_{22} \end{bmatrix} \Rightarrow$$

$$\dot{\theta}_2 = (w_{21} c\theta_5 + w_{22} s\theta_5)/[r_2 \sin(\theta_5 - \theta_2)] \tag{10.277}$$

$$\dot{\theta}_5 = -(w_{21} c\theta_2 + w_{22} s\theta_2)/[r_5 \sin(\theta_5 - \theta_2)] \tag{10.278}$$

* Inverse velocity solution for L_3:

Equations (10.269) and (10.270) can be written as the following matrix equation.

$$\begin{bmatrix} -r_3 s\theta_3 & -r_6 s\theta_6 \\ +r_3 c\theta_3 & +r_6 c\theta_6 \end{bmatrix} \begin{bmatrix} \dot{\theta}_3 \\ \dot{\theta}_6 \end{bmatrix} = \begin{bmatrix} w_{31} \\ w_{32} \end{bmatrix} = \begin{bmatrix} v_1 - (d_3 s\phi - h_7 c\phi)\dot{\phi} \\ v_2 + (d_3 c\phi + h_7 s\phi)\dot{\phi} \end{bmatrix} \tag{10.279}$$

In Eq. (10.279), w_{31} and w_{32} are the known components of the velocity $\vec{w}_3$ of the point A_3. On the left-hand side of the same equation, the coefficient matrix has the following determinant.

$$D_{36} = r_3 r_6 \sin(\theta_6 - \theta_3) \tag{10.280}$$

If $D_{36} \neq 0$, i.e. if $\sin(\theta_6 - \theta_3) \neq 0$, then Eq. (10.279) can be solved as follows in order to find the rates $\dot{\theta}_3$ and $\dot{\theta}_6$:

$$\begin{bmatrix} \dot{\theta}_3 \\ \dot{\theta}_6 \end{bmatrix} = \frac{1}{D_{36}} \begin{bmatrix} +r_6 c\theta_6 & +r_6 s\theta_6 \\ -r_3 c\theta_3 & -r_3 s\theta_3 \end{bmatrix} \begin{bmatrix} w_{31} \\ w_{32} \end{bmatrix} \Rightarrow$$

$$\dot{\theta}_3 = (w_{31} c\theta_6 + w_{32} s\theta_6)/[r_3 \sin(\theta_6 - \theta_3)] \tag{10.281}$$

$$\dot{\theta}_6 = -(w_{31} c\theta_3 + w_{32} s\theta_3)/[r_6 \sin(\theta_6 - \theta_3)] \tag{10.282}$$

(b) Motion Singularities of Inverse Kinematics

As mentioned before, each leg may have a MSIK of its own. The MSIK of L_k is denoted here as MSIK-k. The MSIKs of the legs may occur separately at different times or simultaneously at the same time. They are described and discussed below.

* MSIK-1 of L_1:

As seen in Part (a), L_1 gets into a pose of MSIK-1, if $\sin(\theta_4 - \theta_1) = 0$, i.e. if $\theta_4 = \theta_1$ or $\theta_4 = \theta_1 \pm \pi$. In a pose of MSIK-1, Eq. (10.271) leads to the following scalar equations.

$$(r_1 \dot{\theta}_1 + \sigma_1' r_4 \dot{\theta}_4) s\theta_1 = -w_{11} \tag{10.283}$$

$$(r_1 \dot{\theta}_1 + \sigma_1' r_4 \dot{\theta}_4) c\theta_1 = +w_{12} \tag{10.284}$$

In the above equations,

$$\sigma_1' = \begin{cases} +1 & \text{if } \theta_4 = \theta_1 \\ -1 & \text{if } \theta_4 = \theta_1 \pm \pi \end{cases} \tag{10.285}$$

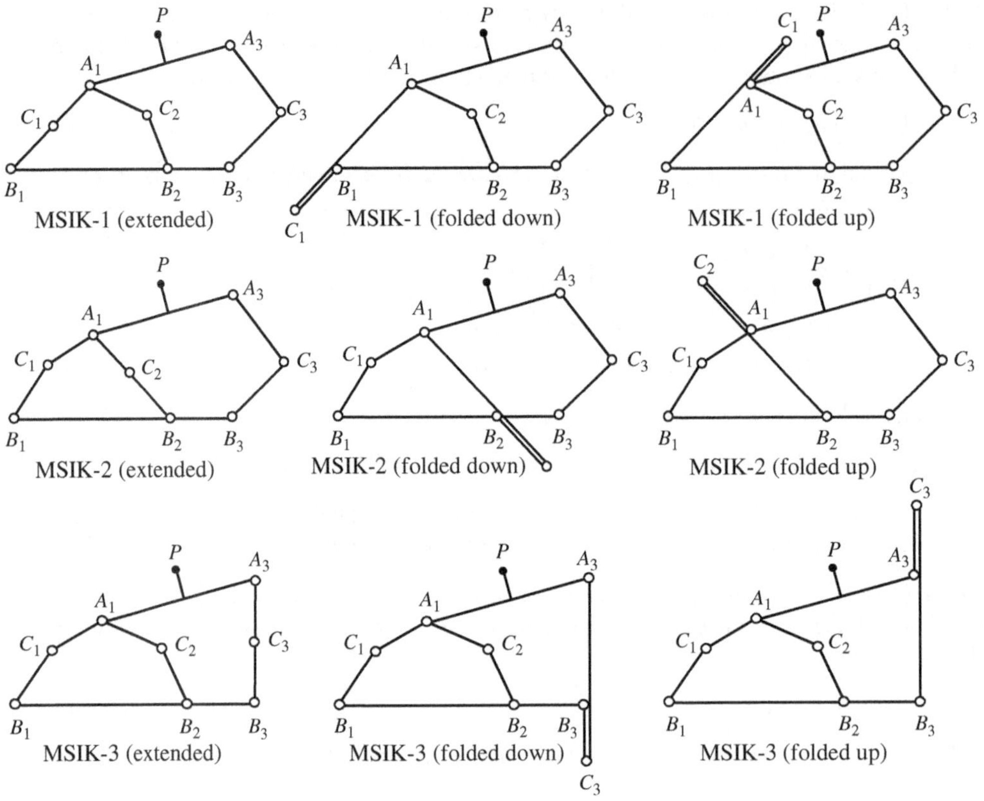

Figure 10.13 MSIK poses of the 3RRR planar parallel manipulator.

Equations (10.283) and (10.284) can be manipulated into the following equations.

$$w_{11}c\theta_1 + w_{12}s\theta_1 = 0 \tag{10.286}$$

$$r_1\dot{\theta}_1 + \sigma_1' r_4\dot{\theta}_4 = w_{12}c\theta_1 - w_{11}s\theta_1 \tag{10.287}$$

Equation (10.286) is the *singularity relationship* of MSIK-1. It can be written in more detail as follows:

$$v_1 c\theta_1 + v_2 s\theta_1 = [d_1 \sin(\theta_1 - \phi) - h_7 \cos(\theta_1 - \phi)]\dot{\phi} \tag{10.288}$$

Equation (10.288) expresses the velocity constraint that must be obeyed by the end-effector in order to keep $\dot{\theta}_1$ and $\dot{\theta}_4$ finite as the manipulator passes through MSIK-1. This constraint is obeyed if the motion of the end-effector is planned so that the velocity of the point A_1 is perpendicular to the line segment B_1A_1. The necessity of this special motion is also evident in Figure 10.13.

As for Eq. (10.287), it shows that the rates $\dot{\theta}_1$ and $\dot{\theta}_4$ become indefinite because they can assume arbitrary values as long as they satisfy this particular equation. However, $\dot{\theta}_1$ can be commanded to have an appropriate desired value because θ_1 is an active joint variable. Then, Eq. (10.287) gives the corresponding value of $\dot{\theta}_4$. Thus, the indefiniteness caused by MSIK-1 can be eliminated.

∗ MSIK-2 of L_2:

As seen in Part (a), L_2 gets into a pose of MSIK-2, if $\sin(\theta_5 - \theta_2) = 0$, i.e. if $\theta_5 = \theta_2$ or $\theta_5 = \theta_2 \pm \pi$. In a pose of MSIK-2, Eq. (10.275) leads to the following scalar equations.

$$(r_2\dot{\theta}_2 + \sigma_2' r_5\dot{\theta}_5)s\theta_2 = -w_{21} \tag{10.289}$$

$$(r_1\dot{\theta}_1 + \sigma_1' r_4\dot{\theta}_4)c\theta_1 = +w_{22} \tag{10.290}$$

In the above equations,

$$\sigma_2' = \begin{cases} +1 & \text{if } \theta_5 = \theta_2 \\ -1 & \text{if } \theta_5 = \theta_2 \pm \pi \end{cases} \tag{10.291}$$

Equations (10.289) and (10.290) can be manipulated into the following equations.

$$w_{21}c\theta_2 + w_{22}s\theta_2 = 0 \tag{10.292}$$

$$r_2\dot{\theta}_2 + \sigma_2' r_5\dot{\theta}_5 = w_{22}c\theta_2 - w_{21}s\theta_2 \tag{10.293}$$

Equation (10.292) is the *singularity relationship* of MSIK-2. It can be written in more detail as follows:

$$v_1c\theta_2 + v_2s\theta_2 = [d_1\sin(\theta_2 - \phi) - h_7\cos(\theta_2 - \phi)]\dot{\phi} \tag{10.294}$$

Equation (10.294) expresses the velocity constraint that must be obeyed by the end-effector in order to keep $\dot{\theta}_2$ and $\dot{\theta}_5$ finite as the manipulator passes through MSIK-2. This constraint is obeyed if the motion of the end-effector is planned so that the velocity of the coincident points A_1 and A_2 is perpendicular to the line segment $B_2A_2 = B_2A_1$. The necessity of this special motion is also evident in Figure 10.13.

As for Eq. (10.293), it shows that the rates $\dot{\theta}_2$ and $\dot{\theta}_5$ become indefinite because they can assume arbitrary values as long as they satisfy this particular equation. However, $\dot{\theta}_2$ can be commanded to have an appropriate desired value because θ_2 is an active joint variable. Then, Eq. (10.293) gives the corresponding value of $\dot{\theta}_5$. Thus, the indefiniteness caused by MSIK-2 can also be eliminated just like MSIK-1.

∗ MSIK-3 of L_3:

As seen in Part (a), L_3 gets into a pose of MSIK-3, if $\sin(\theta_6 - \theta_3) = 0$, i.e. if $\theta_6 = \theta_3$ or $\theta_6 = \theta_3 \pm \pi$. In a pose of MSIK-3, Eq. (10.279) leads to the following scalar equations.

$$(r_3\dot{\theta}_3 + \sigma_3' r_6\dot{\theta}_6)s\theta_3 = -w_{31} \tag{10.295}$$

$$(r_3\dot{\theta}_3 + \sigma_3' r_6\dot{\theta}_6)c\theta_3 = w_{32} \tag{10.296}$$

In the above equations,

$$\sigma_3' = \begin{cases} +1 & \text{if } \theta_6 = \theta_3 \\ -1 & \text{if } \theta_6 = \theta_3 \pm \pi \end{cases} \tag{10.297}$$

Equations (10.295) and (10.296) can be manipulated into the following equations.

$$w_{31}c\theta_3 + w_{32}s\theta_3 = 0 \tag{10.298}$$

$$r_3\dot{\theta}_3 + \sigma_3' r_6\dot{\theta}_6 = w_{32}c\theta_1 - w_{31}s\theta_1 \tag{10.299}$$

Equation (10.298) is the *singularity relationship* of MSIK-3. It can be written in more detail as follows:

$$v_1c\theta_3 + v_2s\theta_3 = -[d_3\sin(\theta_3 - \phi) + h_7\cos(\theta_3 - \phi)]\dot{\phi} = 0 \tag{10.300}$$

Equation (10.300) expresses the velocity constraint that must be obeyed by the end-effector in order to keep $\dot{\theta}_3$ and $\dot{\theta}_6$ finite as the manipulator passes through MSIK-3. This constraint is obeyed if the motion of the end-effector is planned so that the velocity of the point A_3 is perpendicular to the line segment B_3A_3. The necessity of this special motion is also evident in Figure 10.13.

As for Eq. (10.299), it shows that the rates $\dot{\theta}_3$ and $\dot{\theta}_6$ become indefinite, because, they can assume arbitrary values as long as they satisfy this particular equation. However, $\dot{\theta}_3$ can be commanded to have an appropriate desired value because θ_3 is an active joint variable. Then, Eq. (10.299) gives the corresponding value of $\dot{\theta}_6$. Thus, the indefiniteness caused by MSIK-3 can also be eliminated just like MSIK-1 and MSIK-2.

The MSIK poses of the manipulator are illustrated in Figure 10.13. Note that the MSIK poses of the legs are the same as the PMCPLs (posture mode changing poses of the legs).

10.8 Stewart–Gough Platform as a 6UPS Spatial Parallel Manipulator

10.8.1 Kinematic Description

The most typical and classical parallel manipulator configuration is shown in Figure 10.14. A manipulator of this kind is known with the generic name *Stewart–Gough platform*. The configuration of such a manipulator is denoted as 6UPS, which means that the manipulator has six legs and each leg comprises one *universal joint* with the fixed platform, one *prismatic joint* between the lower and upper links of the leg, and one *spherical joint* with the moving platform. The manipulator is actuated through the prismatic joints. If hydraulic actuators are used, then the prismatic joints are replaced with *cylindrical joints* and the configuration of the manipulator is denoted as 6UCS.

For both configurations, the number of the moving links is $n_m = 13$. As for the joint numbers, $j_1 = j_2 = j_3 = 6$ for the 6UPS configuration, whereas $j_1 = 0$, $j_2 = 12$, and $j_3 = 6$ for the 6UCS configuration. Therefore, the Kutzbach–Gruebler formula, i.e. Eq. (10.1), gives the following mobility values.

$$6\text{UPS} \Rightarrow \mu_{ups} = 6; \quad 6\text{UCS} \Rightarrow \mu_{ucs} = 12$$

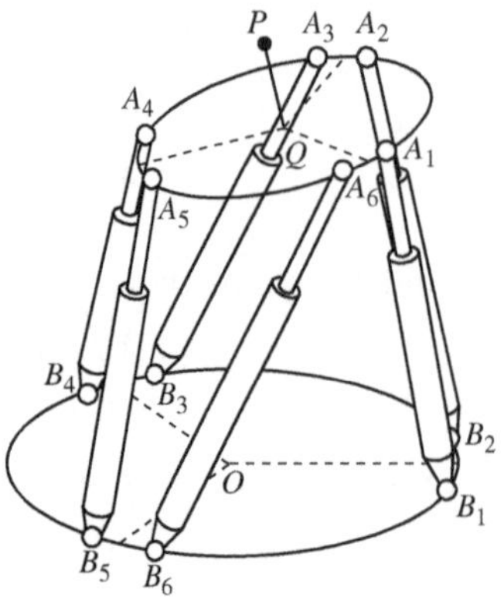

Figure 10.14 Stewart–Gough platform.

However, the 6UCS configuration contains six *insignificant* mobilities, which are the arbitrary relative spinning rotations between the pistons and the cylinders. Therefore, the significant mobility of each configuration is $\mu = 6$.

The Stewart–Gough platforms used in practice have symmetric configurations. Therefore, the constant geometric parameters, which are illustrated in Figures 10.14–10.16, have the following particular values.

$$OB_k = b_0,\ \ QA_k = d_0\ \ \forall\ k = 1, 2, \ldots, 6;\ \ QP = h_0$$

$$\sphericalangle(\overrightarrow{OB_1}) = \beta_1 = -\beta_0,\ \ \sphericalangle(\overrightarrow{OB_2}) = \beta_2 = +\beta_0$$

$$\sphericalangle(\overrightarrow{OB_3}) = \beta_3 = 2\pi/3 - \beta_0,\ \ \sphericalangle(\overrightarrow{OB_4}) = \beta_4 = 2\pi/3 + \beta_0$$

$$\sphericalangle(\overrightarrow{OB_5}) = \beta_5 = -2\pi/3 - \beta_0,\ \ \sphericalangle(\overrightarrow{OB_6}) = \beta_6 = -2\pi/3 + \beta_0$$

$$\sphericalangle(\overrightarrow{QA_1}) = \gamma_1 = -\pi/3 + \gamma_0,\ \ \sphericalangle(\overrightarrow{QA_2}) = \gamma_2 = \pi/3 - \gamma_0$$

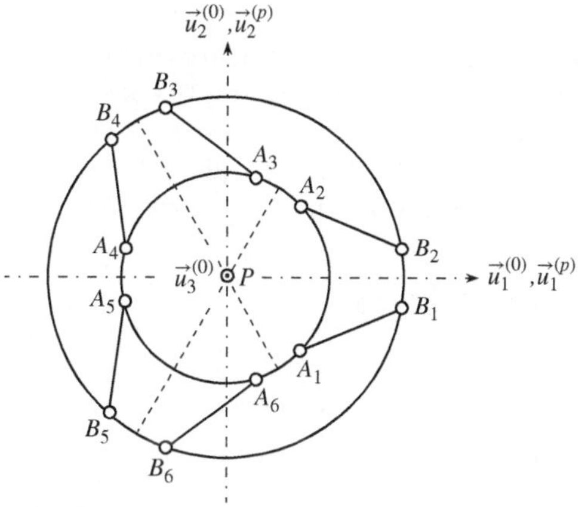

Figure 10.15 Top view of the Stewart-Gough platform in the parking position.

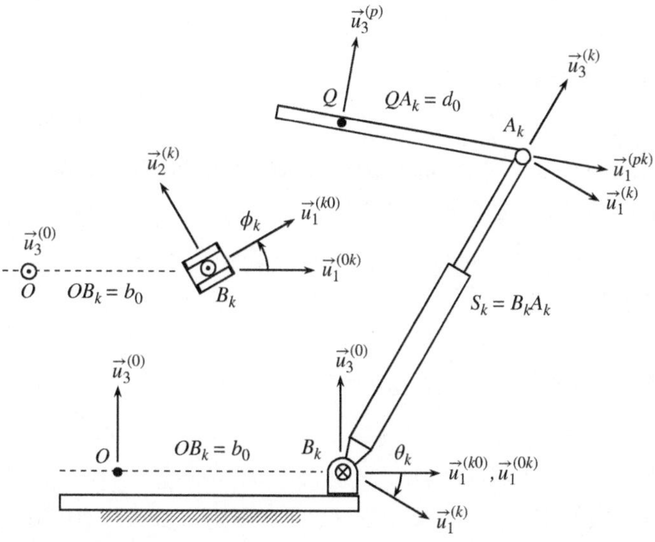

Figure 10.16 Leg details of the Stewart-Gough platform.

$$\measuredangle(\overrightarrow{QA_3}) = \gamma_3 = \pi/3 + \gamma_0, \quad \measuredangle(\overrightarrow{QA_4}) = \gamma_4 = \pi - \gamma_0$$
$$\measuredangle(\overrightarrow{QA_5}) = \gamma_5 = -\pi + \gamma_0, \quad \measuredangle(\overrightarrow{QA_6}) = \gamma_6 = -\pi/3 - \gamma_0$$

The primary variables are taken as the leg lengths, which are the variables of the active prismatic joints. That is, for $k = 1, 2, \ldots, 6$,

$$s_k = B_k A_k$$

The secondary variables are then taken as the angles of the passive universal and spherical joints. They are indicated below for $k = 1, 2, \ldots, 6$.

Universal joint variables according to the 3-2 sequence: ϕ_k, θ_k

Spherical joint variables according to the 3-2-3 sequence: ϕ'_k, θ'_k, ψ'_k

Here, the 3-2-3 sequence is selected for the spherical joints so that they resemble the universal joints apart from the third rotations.

10.8.2 Position Equations

Referring to Figures 10.15 and 10.16 and using the geometrical parameters indicated in Section 10.8.1, the following equations can be written in order to express the position of the end-effector, i.e. the orientation of the moving platform ($\mathcal{L}_p = \mathcal{L}_{13}$) and the location of the tip point (P) with respect to the base frame $\mathcal{F}_0(O)$, by going through each leg. Note that the lower and upper links of the leg L_k are denoted as $\mathcal{L}_k$ and $\mathcal{L}_{k+6}$, respectively, for $k = 1, 2, \ldots, 6$.

(a) End-Effector Orientation Equations Through the Legs

For $k = 1, 2, \ldots, 6$, and $p = 13$,

$$\hat{C} = \hat{C}^{(0,p)} = \hat{C}^{(0,0k)}\hat{C}^{(0k,k)}\hat{C}^{(k,pk)}\hat{C}^{(pk,p)}; \quad \hat{C}^{(pk,p)} = \hat{C}^{(p,pk)t} \Rightarrow$$
$$\hat{C} = [e^{\tilde{u}_3\beta_k}][e^{\tilde{u}_3\phi_k}e^{\tilde{u}_2\theta_k}][e^{\tilde{u}_3\phi'_k}e^{\tilde{u}_2\theta'_k}e^{\tilde{u}_3\psi'_k}][e^{\tilde{u}_3\gamma_k}]^t \Rightarrow$$
$$\hat{C} = e^{\tilde{u}_3(\phi_k+\beta_k)}e^{\tilde{u}_2\theta_k}e^{\tilde{u}_3\phi'_k}e^{\tilde{u}_2\theta'_k}e^{\tilde{u}_3(\psi'_k-\gamma_k)} \tag{10.301}$$

(b) Tip Point Location Equations Through the Legs

For $k = 1, 2, \ldots, 6$, and $p = 13$,

$$\vec{p} = \overrightarrow{OP} = \overrightarrow{OB_k} + \overrightarrow{B_kA_k} + \overrightarrow{A_kQ} + \overrightarrow{QP} \Rightarrow$$
$$\vec{p} = b_0\vec{u}_1^{(0k)} + s_k\vec{u}_3^{(k)} - d_0\vec{u}_1^{(pk)} + h_0\vec{u}_3^{(p)} \tag{10.302}$$

Equation (10.302) can be written as the following matrix equation in the base frame.

$$\overline{p} = \overline{p}^{(0)} = b_0\overline{u}_1^{(0k/0)} + s_k\overline{u}_3^{(k/0)} - d_0\overline{u}_1^{(pk/0)} + h_0\overline{u}_3^{(p/0)} \Rightarrow$$
$$\overline{p} = b_0\hat{C}^{(0,0k)}\overline{u}_1^{(0k/0k)} + s_k\hat{C}^{(0,k)}\overline{u}_3^{(k/k)} - d_0\hat{C}^{(0,pk)}\overline{u}_1^{(pk/pk)} + h_0\hat{C}^{(0,p)}\overline{u}_3^{(p/p)} \Rightarrow$$
$$\begin{aligned}\overline{p} = {} & b_0 e^{\tilde{u}_3\beta_k}\overline{u}_1 + s_k e^{\tilde{u}_3\beta_k}e^{\tilde{u}_3\phi_k}e^{\tilde{u}_2\theta_k}\overline{u}_3 \\ & - d_0 e^{\tilde{u}_3\beta_k}e^{\tilde{u}_3\phi_k}e^{\tilde{u}_2\theta_k}e^{\tilde{u}_3\phi'_k}e^{\tilde{u}_2\theta'_k}e^{\tilde{u}_3\psi'_k}\overline{u}_1 \\ & + h_0 e^{\tilde{u}_3\beta_k}e^{\tilde{u}_3\phi_k}e^{\tilde{u}_2\theta_k}e^{\tilde{u}_3\phi'_k}e^{\tilde{u}_2\theta'_k}e^{\tilde{u}_3\psi'_k}e^{-\tilde{u}_3\gamma_k}\overline{u}_3 \Rightarrow\end{aligned}$$
$$\begin{aligned}\overline{p} = {} & b_0 e^{\tilde{u}_3\beta_k}\overline{u}_1 + s_k e^{\tilde{u}_3(\phi_k+\beta_k)}e^{\tilde{u}_2\theta_k}\overline{u}_3 \\ & - d_0 e^{\tilde{u}_3(\phi_k+\beta_k)}e^{\tilde{u}_2\theta_k}e^{\tilde{u}_3\phi'_k}e^{\tilde{u}_2\theta'_k}e^{\tilde{u}_3\psi'_k}\overline{u}_1 \\ & + h_0 e^{\tilde{u}_3(\phi_k+\beta_k)}e^{\tilde{u}_2\theta_k}e^{\tilde{u}_3\phi'_k}e^{\tilde{u}_2\theta'_k}\overline{u}_3\end{aligned} \tag{10.303}$$

Equation (10.303) can also be written more compactly as follows upon inserting $\widehat{C}$ given by Eq. (10.301).

$$\overline{p} = b_0 e^{\tilde{u}_3\beta_k}\overline{u}_1 + s_k e^{\tilde{u}_3(\phi_k+\beta_k)} e^{\tilde{u}_2\theta_k}\overline{u}_3 + \widehat{C}(h_0\overline{u}_3 - d_0 e^{\tilde{u}_3\gamma_k}\overline{u}_1) \tag{10.304}$$

10.8.3 Inverse Kinematics in the Position Domain

(a) Inverse Kinematic Solution

For a specified position of the end-effector, i.e. for specified matrices $\overline{p}$ and $\widehat{C}$, the corresponding active and passive joint variables can be found leg by leg as explained below. For the leg L_k, Eq. (10.304) can be written as follows:

$$s_k e^{\tilde{u}_3\phi_k} e^{\tilde{u}_2\theta_k}\overline{u}_3 = \overline{r}_k \tag{10.305}$$

In Eq. (10.305), the column matrix $\overline{r}_k$ is known with the following expression.

$$\overline{r}_k = e^{-\tilde{u}_3\beta_k}(\overline{p} - h_0\widehat{C}\overline{u}_3 + d_0\widehat{C}e^{\tilde{u}_3\gamma_k}\overline{u}_1) - b_0\overline{u}_1 \tag{10.306}$$

Note that $\overline{r}_k$ represents the location of the joint center A_k in the *leg-specific base frame* $\mathcal{F}_{0k}(B_k)$ formed by the unit vectors $\vec{u}_1^{(0k)}$ and $\vec{u}_3^{(0k)} = \vec{u}_3^{(0)}$ as illustrated in Figures 10.15 and 10.16.

Note also that $s_k > 0$ always. Therefore, s_k is obtained as follows upon premultiplying both sides of Eq. (10.305) by their transposes.

$$s_k^2(\overline{u}_3^t e^{-\tilde{u}_2\theta_k} e^{-\tilde{u}_3\phi_k})(e^{\tilde{u}_3\phi_k} e^{\tilde{u}_2\theta_k}\overline{u}_3) = s_k^2(\overline{u}_3^t\overline{u}_3) = s_k^2 = \overline{r}_k^t\overline{r}_k \Rightarrow$$

$$s_k = \sqrt{\overline{r}_k^t\overline{r}_k} \tag{10.307}$$

With the availability of s_k, Eq. (10.305) can be rearranged as

$$s_k e^{\tilde{u}_2\theta_k}\overline{u}_3 = e^{-\tilde{u}_3\phi_k}\overline{r}_k \tag{10.308}$$

Upon premultiplications by $\overline{u}_1^t$, $\overline{u}_2^t$, and $\overline{u}_3^t$, Eq. (10.308) leads to the following three scalar equations, in which $r_{ki} = \overline{u}_i^t\overline{r}_k$ for $i = 1, 2, 3$.

$$s_k\overline{u}_1^t e^{\tilde{u}_2\theta_k}\overline{u}_3 = \overline{u}_1^t e^{-\tilde{u}_3\phi_k}\overline{r}_k \;\Rightarrow\; s_k s\theta_k = r_{k1}c\phi_k + r_{k2}s\phi_k \tag{10.309}$$

$$s_k\overline{u}_2^t e^{\tilde{u}_2\theta_k}\overline{u}_3 = \overline{u}_2^t e^{-\tilde{u}_3\phi_k}\overline{r}_k \;\Rightarrow\; 0 = r_{k2}c\phi_k - r_{k1}s\phi_k \tag{10.310}$$

$$s_k\overline{u}_3^t e^{\tilde{u}_2\theta_k}\overline{u}_3 = \overline{u}_3^t e^{-\tilde{u}_3\phi_k}\overline{r}_k \;\Rightarrow\; s_k c\theta_k = r_{k3} \tag{10.311}$$

Equation (10.311) gives θ_k as follows with a sign variable $\sigma_k = \pm 1$.

$$c\theta_k = r_{k3}/s_k \Rightarrow s\theta_k = \sigma_k\sqrt{s_3^2 - r_{k3}^2}/s_k \Rightarrow$$

$$\theta_k = \sigma_k \operatorname{atan}_2(\sqrt{s_3^2 - r_{k3}^2}, r_{k3}) \tag{10.312}$$

With the availability of θ_k as well, Eqs. (10.309) and (10.310) can be manipulated into the following two equations.

$$(r_{k1}^2 + r_{k2}^2)s\phi_k = r_{k2}s_k s\theta_k = \sigma_k r_{k2}\sqrt{s_3^2 - r_{k3}^2} \tag{10.313}$$

$$(r_{k1}^2 + r_{k2}^2)c\phi_k = r_{k1}s_k s\theta_k = \sigma_k r_{k1}\sqrt{s_3^2 - r_{k3}^2} \tag{10.314}$$

If $r_{k1}^2 + r_{k2}^2 \neq 0$, then Eqs. (10.313) and (10.314) give ϕ_k as follows without any additional sign variable.

$$\phi_k = \text{atan}_2(\sigma_k r_{k2}, \sigma_k r_{k1}) \tag{10.315}$$

According to the preceding equations, if $\sigma_k = +1$ leads to $\theta_k > 0$ and ϕ_k, then $\sigma_k = -1$ leads to $\theta'_k = -\theta_k$ with $\phi'_k = \phi_k \pm \pi$. However, the pairs $\{\theta_k, \phi_k\}$ and $\{\theta'_k, \phi'_k\}$ are not visually distinguishable. Therefore, the value of σ_k is not very significant. Nevertheless, ϕ_k is preferred to be an acute angle so that $c\phi_k > 0$ in order to prevent the pedestal of the universal joint, which is shown in Figure 10.16, from unnecessarily excessive rotations. In other words, it is preferred to have $|\phi_k| < \pi/2$ instead of $|\phi_k| < \pi$. Equation (10.314) implies that this preference can be realized with $\sigma_k = \text{sgn}(r_{k1})$.
As for the angles of the spherical joint of the leg L_k, they can be found from Eq. (10.301) by rearranging it as shown below.

$$e^{\tilde{u}_3\phi'_k} e^{\tilde{u}_2\theta'_k} e^{\tilde{u}_3\psi'_k} = \hat{C}_k^* \tag{10.316}$$

In Eq. (10.316), the matrix $\hat{C}_k^*$ is known with the following expression.

$$\hat{C}_k^* = e^{-\tilde{u}_2\theta_k} e^{-\tilde{u}_3(\phi_k+\beta_k)} \hat{C} e^{\tilde{u}_3\gamma_k} \tag{10.317}$$

Equation (10.316) leads to the following five scalar equations.

$$\overline{u}_3^t e^{\tilde{u}_2\theta'_k} \overline{u}_3 = \overline{u}_3^t \hat{C}_k^* \overline{u}_3 \Rightarrow c\theta'_k = c_{k33}^* \tag{10.318}$$

$$\overline{u}_1^t e^{\tilde{u}_3\phi'_k} e^{\tilde{u}_2\theta'_k} \overline{u}_3 = \overline{u}_1^t \hat{C}_k^* \overline{u}_3 \Rightarrow c\phi'_k s\theta'_k = c_{k13}^* \tag{10.319}$$

$$\overline{u}_2^t e^{\tilde{u}_3\phi'_k} e^{\tilde{u}_2\theta'_k} \overline{u}_3 = \overline{u}_2^t \hat{C}_k^* \overline{u}_3 \Rightarrow s\phi'_k s\theta'_k = c_{k23}^* \tag{10.320}$$

$$\overline{u}_3^t e^{\tilde{u}_2\theta'_k} e^{\tilde{u}_3\psi'_k} \overline{u}_1 = \overline{u}_3^t \hat{C}_k^* \overline{u}_1 \Rightarrow -s\theta'_k c\psi'_k = c_{k31}^* \tag{10.321}$$

$$\overline{u}_3^t e^{\tilde{u}_2\theta'_k} e^{\tilde{u}_3\psi'_k} \overline{u}_2 = \overline{u}_3^t \hat{C}_k^* \overline{u}_2 \Rightarrow s\theta'_k s\psi'_k = c_{k32}^* \tag{10.322}$$

Equation (10.318) gives θ'_k as follows:

$$c\theta'_k = c_{k33}^* \Rightarrow s\theta'_k = \sigma'_k \sqrt{1 - (c_{k33}^*)^2}; \quad \sigma'_k = \pm 1 \Rightarrow$$

$$\theta'_k = \sigma'_k \text{atan}_2(\sqrt{1 - (c_{k33}^*)^2}, c_{k33}^*) \tag{10.323}$$

If $s\theta'_k \neq 0$, then Eqs. (10.319)–(10.322) give ϕ'_k and ψ'_k as shown below.

$$\phi'_k = \text{atan}_2(\sigma'_k c_{k23}^*, \sigma'_k c_{k13}^*) \tag{10.324}$$

$$\psi'_k = \text{atan}_2(\sigma'_k c_{k32}^*, -\sigma'_k c_{k31}^*) \tag{10.325}$$

In a spherical joint, the opposite values of σ'_k do not lead to visually distinct solutions. Therefore, σ'_k is normally taken as $\sigma'_k = +1$ without loss of generality.

(b) Position Singularities of Inverse Kinematics

∗ PSIK-1(k) (first kind of PSIK for L_k):

Equation (10.315) implies that the leg L_k gets into a pose of PSIK-1(k) if $r_{k1}^2 + r_{k2}^2 = 0$. In this singularity, L_k becomes perpendicular to the base with $\theta_k = 0$ and ϕ_k becomes indefinite. Note that this singularity is quite likely to occur. However, it is not necessary to avoid this singularity because the active joint variable s_k can still be found with a definite value no matter what ϕ_k is. In other words, this singularity does not affect the controllability of the manipulator.

* PSIK-2(k) (second kind of PSIK for L_k):

Equations (10.319)–(10.322) imply that the leg L_k gets into a pose of PSIK-2(k) if $s\theta'_k = 0$, i.e. if $\theta'_k = 0$ or $\theta'_k = \pm\pi$. In this singularity, the unit vectors $\vec{u}_3^{(k)}$ and $\vec{u}_3^{(p)}$ become codirectional if $\theta'_k = 0$ and opposite if $\theta'_k = \pm\pi$. Thus, L_k becomes perpendicular to the plane of the moving platform $\mathcal{L}_p$. However, the case with $\theta'_k = \pm\pi$ is not likely to occur due to the hindrance of the other legs, whereas the case with $\theta'_k = 0$ is quite likely to occur. Consequently, Eq. (10.316) takes the following form with $\theta'_k = 0$.

$$e^{\tilde{u}_3\phi'_k}e^{\tilde{u}_2 0}e^{\tilde{u}_3\psi'_k} = e^{\tilde{u}_3\phi'_k}e^{\tilde{u}_3\psi'_k} = e^{\tilde{u}_3(\phi'_k+\psi'_k)} = \hat{C}^*_k \tag{10.326}$$

Equation (10.326) implies that ϕ'_k and ψ'_k cannot be found separately and therefore they become indefinite. Nevertheless, their addition can still be found with the following definite value.

$$\phi'_k + \psi'_k = \phi^*_k = \text{atan}_2(c^*_{k21}, c^*_{k11}) \tag{10.327}$$

This singularity does not affect the controllability of the manipulator, either, because the active joint variable s_k can still be found with a definite value no matter what ϕ'_k and ψ'_k are as long as they add up to ϕ^*_k. Therefore, it is not necessary to avoid this singularity, either.

10.8.4 Forward Kinematics in the Position Domain

For specified values of the leg lengths (i.e. s_k for all $k = 1, 2, \ldots, 6$), the passive joint variables and the position of the end-effector (i.e. $\bar{p}$ and $\hat{C}$) can be obtained from the IKLs (independent kinematic loops) of the manipulator. The formation of the IKLs and the process of the forward kinematic solution based on them are explained below.

(a) Independent Kinematic Loops

The manipulator has $n_m = 13$ moving links and its joints are such that $j_1 = 6$ (prismatic joints), $j_2 = 6$ (universal joints), and $j_3 = 6$ (spherical joints). Therefore, Eq. (10.2) gives the number of the IKLs as

$$n_{ikl} = j_1 + j_2 + j_3 - n_m = 5 \tag{10.328}$$

The IKLs can be formed by using five appropriately chosen leg pairs such as $\{L_1, L_2\}$, $\{L_1, L_3\}$, $\{L_1, L_4\}$, $\{L_1, L_5\}$, and $\{L_1, L_6\}$. Then, the *loop equations* can be generated for these loops by equating the position expressions written for the end-effector along the legs of the loops. Such expressions have already been written in Eqs. (10.301) and (10.304) of Section 10.8.2. The loop equations involve both the orientations of the relevant links and the locations of the relevant characteristic points, i.e. the joint centers and the link centers.

The IKLs mentioned above and the associated loop equations are provided below. The loop equations are generated by equating the right-hand sides of Eqs. (10.301) and (10.304) for the index pairs $\{1, 2\}$, $\{1, 3\}$, …, $\{1, 6\}$.

(b) Loop Equations

The loop equations are generated as follows for each of the IKLs, i.e. for IKL-$k = B_1A_1A_k B_kB_1$, where $k = 2, 3, \ldots, 6$:

∗ Orientation equation for IKL-k:

$$e^{\tilde{u}_3(\phi_k+\beta_k)}e^{\tilde{u}_2\theta_k}e^{\tilde{u}_3\phi'_k}e^{\tilde{u}_2\theta'_k}e^{\tilde{u}_3(\psi'_k-\gamma_k)}$$
$$= e^{\tilde{u}_3(\phi_1+\beta_1)}e^{\tilde{u}_2\theta_1}e^{\tilde{u}_3\phi'_1}e^{\tilde{u}_2\theta'_1}e^{\tilde{u}_3(\psi'_1-\gamma_1)} \tag{10.329}$$

∗ Characteristic point location equation for IKL-k:

$$b_0 e^{\tilde{u}_3\beta_k}\overline{u}_1 + s_k e^{\tilde{u}_3(\phi_k+\beta_k)}e^{\tilde{u}_2\theta_k}\overline{u}_3 - d_0\widehat{C}e^{\tilde{u}_3\gamma_k}\overline{u}_1$$
$$= b_0 e^{\tilde{u}_3\beta_1}\overline{u}_1 + s_1 e^{\tilde{u}_3(\phi_1+\beta_1)}e^{\tilde{u}_2\theta_1}\overline{u}_3 - d_0\widehat{C}e^{\tilde{u}_3\gamma_1}\overline{u}_1 \tag{10.330}$$

(c) Solution of the Loop Equations and Finding the Position of the End-Effector

The preceding loop equations are equivalent to 30 independent scalar equations, because each of the equations written for the 3 × 3 orientation matrices and the 3 × 1 point location matrices is equivalent to three independent scalar equations. On the other hand, since the leg lengths are specified, each leg (e.g. L_k) contributes five unknowns, which are the two angles (ϕ_k, θ_k) of its universal joint and the three angles (ϕ'_k, θ'_k, ψ'_k) of its spherical joint. Therefore, there are altogether 30 unknowns to be found. In other words, the number of unknowns and the number of independent scalar equations are balanced.

The solution, of course, cannot be obtained in a completely analytical way. However, it can be obtained in a *fourth-order semi-analytical way*. In order to obtain the solution this way, four of the 30 variables are treated as if they are temporarily known. These four variables are called here *major unknowns*. Then, the other 26 variables are called *minor unknowns*. In the present problem, it happens to be convenient to select the major unknowns as θ'_1, ϕ'_1, ψ'_1, and θ_1, which belong to L_1. The convenience is due to the fact that L_1 is taken as the *invariant leg* of all the five loops. After selecting the major unknowns, the loop equations can be solved analytically for the 26 minor unknowns so that they are obtained as functions of the major unknowns. This solution process also generates four *consistency equations*, which are to be solved to find the major unknowns. However, the consistency equations turn out to be rather complicated and therefore they can only be solved by using a suitable numerical method. After the numerical values of the major unknowns are thus found, the numerical values of the 26 minor unknowns can also be found by using their previously obtained expressions in terms of the major unknowns.

The procedure explained above can be initiated by inserting the following expression of the matrix $\widehat{C}$ into the preceding characteristic point location equations.

$$\widehat{C} = e^{\tilde{u}_3(\phi_1+\beta_1)}e^{\tilde{u}_2\theta_1}e^{\tilde{u}_3\phi'_1}e^{\tilde{u}_2\theta'_1}e^{\tilde{u}_3(\psi'_1-\gamma_1)} \tag{10.331}$$

Note that Eq. (10.331) is the special form of Eq. (10.301) that is written specifically for L_1 because the selected major unknowns belong to L_1.

After the substitution of Eq. (10.331), each of the characteristic point location equations can be arranged as shown below for $k = 2, 3, \ldots, 6$.

$$s_k e^{\tilde{u}_3\phi_k}e^{\tilde{u}_2\theta_k}\overline{u}_3 = e^{\tilde{u}_3\phi_1}\overline{h}_k - \overline{b}_k \tag{10.332}$$

In Eq. (10.332),

$$\overline{b}_k = b_0(\overline{u}_1 - e^{-\tilde{u}_3\beta'_k}\overline{u}_1) = b_0[\overline{u}_1(1 - c\beta'_k) + \overline{u}_2 s\beta'_k] \tag{10.333}$$

$$\overline{h}_k = e^{-\widetilde{u}_3\beta'_k} e^{\widetilde{u}_2\theta_1} [s_1\overline{u}_3 - d_0 e^{\widetilde{u}_3\phi'_1} e^{\widetilde{u}_2\theta'_1} e^{\widetilde{u}_3\psi'_1} (\overline{u}_1 - e^{\widetilde{u}_3\gamma'_k}\overline{u}_1)] \tag{10.334}$$

$$\beta'_k = \beta_k - \beta_1 \tag{10.335}$$

$$\gamma'_k = \gamma_k - \gamma_1 \tag{10.336}$$

Note that $\overline{b}_k$ is constant but $\overline{h}_k$ is a function of the major unknowns. That is,

$$\overline{h}_k = \overline{h}_k(\phi'_1, \theta'_1, \psi'_1, \theta_1) \tag{10.337}$$

The angles ϕ_1, ϕ_k, and θ_k can be extracted from Eq. (10.332) in three steps as explained below.

In the first step, θ_k can be extracted by premultiplying both sides of Eq. (10.332) by $\overline{u}_3^t$. This operation is carried out as follows:

$$s_k \overline{u}_3^t e^{\widetilde{u}_3\phi_k} e^{\widetilde{u}_2\theta_k} \overline{u}_3 = \overline{u}_3^t e^{\widetilde{u}_3\phi_1} \overline{h}_k - \overline{u}_3^t \overline{b}_k \Rightarrow$$

$$s_k \overline{u}_3^t e^{\widetilde{u}_2\theta_k} \overline{u}_3 = \overline{u}_3^t \overline{h}_k - 0;\ \overline{u}_3^t \overline{h}_k = h_{k3} \Rightarrow$$

$$s_k c\theta_k = h_{k3} \tag{10.338}$$

Equation (10.338) implies the following equation with a sign variable $\sigma_k = \pm 1$.

$$s_k s\theta_k = \sigma_k \sqrt{s_k^2 - h_{k3}^2} \tag{10.339}$$

Hence, θ_k is found as

$$\theta_k = \sigma_k \mathrm{atan}_2(\sqrt{s_k^2 - h_{k3}^2}, h_{k3}) \tag{10.340}$$

In the second step, ϕ_1 can be extracted by premultiplying both sides of Eq. (10.332) by their transposes. This operation is carried out as follows by denoting the components of $\overline{h}_k$ as h_{ki}, where $h_{ki} = \overline{u}_i^t \overline{h}_k$ for $i = 1, 2, 3$.

$$(s_k \overline{u}_3^t e^{-\widetilde{u}_2\theta_k} e^{-\widetilde{u}_3\phi_k})(s_k e^{\widetilde{u}_3\phi_k} e^{\widetilde{u}_2\theta_k} \overline{u}_3) = (\overline{h}_k^t e^{-\widetilde{u}_3\phi_1} - \overline{b}_k^t)(e^{\widetilde{u}_3\phi_1} \overline{h}_k - \overline{b}_k) \Rightarrow$$

$$s_k^2 = \overline{h}_k^t \overline{h}_k + \overline{b}_k^t \overline{b}_k - 2\overline{b}_k^t e^{\widetilde{u}_3\phi_1} \overline{h}_k \Rightarrow$$

$$s_k^2 = \overline{h}_k^t \overline{h}_k + 2b_0^2(1 - c\beta'_k) - 2b_0[\overline{u}_1^t(1 - c\beta'_k) + \overline{u}_2^t s\beta'_k] e^{\widetilde{u}_3\phi_1} \overline{h}_k \Rightarrow$$

$$[(\overline{u}_1^t c\phi_1 - \overline{u}_2^t s\phi_1)(1 - c\beta'_k) + (\overline{u}_2^t c\phi_1 + \overline{u}_1^t s\phi_1) s\beta'_k] \overline{h}_k$$

$$= [\overline{h}_k^t \overline{h}_k + 2b_0^2(1 - c\beta'_k) - s_k^2]/(2b_0) \Rightarrow$$

$$(h_{k1} c\phi_1 - h_{k2} s\phi_1)(1 - c\beta'_k) + (h_{k2} c\phi_1 + h_{k1} s\phi_1) s\beta'_k$$

$$= [\overline{h}_k^t \overline{h}_k + 2b_0^2(1 - c\beta'_k) - s_k^2]/(2b_0) \Rightarrow$$

$$m_{1k} c\phi_1 + n_{1k} s\phi_1 = g_{1k} \tag{10.341}$$

In Eq. (10.341),

$$m_{1k} = h_{k1}(1 - c\beta'_k) + h_{k2} s\beta'_k \tag{10.342}$$

$$n_{1k} = h_{k1} s\beta'_k - h_{k2}(1 - c\beta'_k) \tag{10.343}$$

$$g_{1k} = [\overline{h}_k^t \overline{h}_k + 2b_0^2(1 - c\beta'_k) - s_k^2]/(2b_0) \tag{10.344}$$

If m_{1k} and n_{1k} are not both zero, then the following equations can be derived from Eq. (10.341) with a second sign variable $\sigma_{1k} = \pm 1$.

$$c\phi_1 = (m_{1k}g_{1k} + \sigma_{1k}n_{1k}g_{1k}^*)/(m_{1k}^2 + n_{1k}^2) \tag{10.345}$$

$$s\phi_1 = (n_{1k}g_{1k} - \sigma_{1k}m_{1k}g_{1k}^*)/(m_{1k}^2 + n_{1k}^2) \tag{10.346}$$

In Eqs. (10.345) and (10.346),

$$g_{1k}^* = \sqrt{m_{1k}^2 + n_{1k}^2 - g_{1k}^2} \tag{10.347}$$

Equations (10.345) and (10.346) give ϕ_1 as follows:

$$\phi_1 = \mathrm{atan}_2[(n_{1k}g_{1k} - \sigma_{1k}m_{1k}g_{1k}^*), (m_{1k}g_{k1} + \sigma_{1k}n_{1k}g_{1k}^*)] \tag{10.348}$$

Note that, due to its dependence on $\overline{h}_k = \overline{h}_k(\phi_1', \theta_1', \psi_1', \theta_1)$, ϕ_1 has come out as a function of the major unknowns through the loop IKL-k. That is,

$$\phi_1 = f_{1k}(\phi_1', \theta_1', \psi_1', \theta_1, \sigma_{1k}) \tag{10.349}$$

In the third step, with the availability of θ_k from the first step, ϕ_k can be extracted by rearranging Eq. (10.332) as shown below.

$$s_k e^{\tilde{u}_3\phi_k} e^{\tilde{u}_2\theta_k} \overline{u}_3 - \overline{b}_k = e^{\tilde{u}_3\phi_1} \overline{h}_k \tag{10.350}$$

The angle ϕ_1 in Eq. (10.350) can be eliminated by premultiplying both sides of the equation by their transposes. This operation is carried out as follows:

$$\begin{aligned}
&(s_k\overline{u}_3^t e^{-\tilde{u}_2\theta_k} e^{-\tilde{u}_3\phi_k} - \overline{b}_k^t)(s_k e^{\tilde{u}_3\phi_k} e^{\tilde{u}_2\theta_k}\overline{u}_3 - \overline{b}_k) = (\overline{h}_k^t e^{-\tilde{u}_3\phi_1})(e^{\tilde{u}_3\phi_1}\overline{h}_k) \Rightarrow \\
&2s_k\overline{u}_3^t e^{-\tilde{u}_2\theta_k} e^{-\tilde{u}_3\phi_k}\overline{b}_k = s_k^2 + \overline{b}_k^t\overline{b}_k - \overline{h}_k^t\overline{h}_k \Rightarrow \\
&(\overline{u}_3^t c\theta_k + \overline{u}_1^t s\theta_k)e^{-\tilde{u}_3\phi_k}[\overline{u}_1(1 - c\beta_k') + \overline{u}_2 s\beta_k'] \\
&\quad = [s_k^2 + 2b_0(1 - c\beta_k') - \overline{h}_k^t\overline{h}_k]/(2b_0 s_k) \Rightarrow \\
&(\overline{u}_3^t c\theta_k + \overline{u}_1^t c\phi_k s\theta_k + \overline{u}_2^t s\phi_k s\theta_k)[\overline{u}_1(1 - c\beta_k') + \overline{u}_2 s\beta_k'] \\
&\quad = [s_k^2 + 2b_0(1 - c\beta_k') - \overline{h}_k^t\overline{h}_k]/(2b_0 s_k) \Rightarrow \\
&[(1 - c\beta_k')s\theta_k]c\phi_k + (s\beta_k' s\theta_k)s\phi_k = [s_k^2 + 2b_0(1 - c\beta_k') - \overline{h}_k^t\overline{h}_k]/(2b_0 s_k)
\end{aligned} \tag{10.351}$$

If $s\theta_k \neq 0$, Eq. (10.351) can be written further as follows:

$$(1 - c\beta_k')c\phi_k + s\beta_k' s\phi_k = \lambda_k \tag{10.352}$$

In Eq. (10.352),

$$\lambda_k = [s_k^2 + 2b_0(1 - c\beta_k') - \overline{h}_k^t\overline{h}_k]/(2b_0 s_k s\theta_k) \tag{10.353}$$

Since the joint centers are not coincident, $\beta_k \neq \beta_1$, i.e. $\beta_k' \neq 0$. Owing to this fact, the following equations can be derived from Eq. (10.352) with a third sign variable $\sigma_k' = \pm 1$.

$$c\phi_k = \frac{1}{2}[(1 - c\beta_k')\lambda_k + \sigma_k' s\beta_k' \mu_k]/(1 - c\beta_k') \tag{10.354}$$

$$s\phi_k = \frac{1}{2}[s\beta_k'\lambda_k + \sigma_k'(1 - c\beta_k')\mu_k]/(1 - c\beta_k') \tag{10.355}$$

In Eqs. (10.354) and (10.355),

$$\mu_k = \sqrt{2(1 - c\beta_k') - \lambda_k^2} \tag{10.356}$$

Equations (10.354) and (10.355) give ϕ_k as follows:

$$\phi_k = \text{atan}_2\{[s\beta_k'\lambda_k + \sigma_k'(1 - c\beta_k')\mu_k], [(1 - c\beta_k')\lambda_k + \sigma_k' s\beta_k'\mu_k]\} \tag{10.357}$$

As for the multiplicities encountered above, the sign variables σ_k, σ_k', and σ_{1k} are to be selected so that ϕ_1 and ϕ_k come out preferably as acute angles with positive cosines according to the criterion mentioned before in Section 10.8.3. This criterion can be satisfied either completely or at least to a large extent by the following values of the three sign variables.

According to Eq. (10.345), depending on the term $m_{1k}g_{1k}$, $c\phi_1$ will be either positive or it will be negative but with a smaller magnitude if

$$\sigma_{1k} = \text{sgn}(n_{1k}) \tag{10.358}$$

According to Eq. (10.354), $c\phi_k$ will be positive if

$$\lambda_k > 0 \text{ and } \sigma_k' = +1 \tag{10.359}$$

On the other hand, according to Eqs. (10.353) and (10.339), λ_k will be positive if

$$\sigma_k = \text{sgn}[s_k^2 + 2b_0(1 - c\beta_k') - \bar{h}_k^t \bar{h}_k] \tag{10.360}$$

The solution described above is carried out for $k = 2, 3, \ldots, 6$ and thus the angles of all the six universal joints (except θ_1) are obtained as functions of the major unknowns ϕ_1', θ_1', ψ_1', and θ_1. However, one of these angles, which is ϕ_1, is obtained five times; each time with a different expression as seen in Eq. (10.349). Of course, these five expressions must be equal to each other. Hence, the following four *consistency equations* are obtained.

$$f_{13}(\phi_1', \theta_1', \psi_1', \theta_1, \sigma_{13}) = f_{12}(\phi_1', \theta_1', \psi_1', \theta_1, \sigma_{12}) \tag{10.361}$$

$$f_{14}(\phi_1', \theta_1', \psi_1', \theta_1, \sigma_{14}) = f_{12}(\phi_1', \theta_1', \psi_1', \theta_1, \sigma_{12}) \tag{10.362}$$

$$f_{15}(\phi_1', \theta_1', \psi_1', \theta_1, \sigma_{15}) = f_{12}(\phi_1', \theta_1', \psi_1', \theta_1, \sigma_{12}) \tag{10.363}$$

$$f_{16}(\phi_1', \theta_1', \psi_1', \theta_1, \sigma_{16}) = f_{12}(\phi_1', \theta_1', \psi_1', \theta_1, \sigma_{12}) \tag{10.364}$$

The above consistency equations can be solved by means of a suitable numerical method in order to find the numerical values of ϕ_1', θ_1', ψ_1', and θ_1. Afterwards, the numerical values of the other universal joint angles can be found by using Eqs. (10.340), (10.348), and (10.357).

In the next stage of the forward kinematic solution, the angles of the spherical joints are found for $k = 2, 3, \ldots, 6$ in addition to the angles of the first spherical joint that have been found in the previous stage. For this purpose, Eq. (10.329) can be rearranged as follows:

$$e^{\tilde{u}_3\phi_k'} e^{\tilde{u}_2\theta_k'} e^{\tilde{u}_3\psi_k'} = \hat{C}_k^* \tag{10.365}$$

In Eq. (10.365), the matrix $\hat{C}_k^*$ is known with the following expression.

$$\hat{C}_k^* = e^{-\tilde{u}_2\theta_k} e^{\tilde{u}_3(\phi_1 - \phi_k - \beta_k')} e^{\tilde{u}_2\theta_1} e^{\tilde{u}_3\phi_1'} e^{\tilde{u}_2\theta_1'} e^{\tilde{u}_3(\psi_1' + \gamma_k')} \tag{10.366}$$

Note that the left-hand sides of Eqs. (10.316) and (10.365) are the same. Therefore, the angles θ'_k, ϕ'_k, and ψ'_k can be found by solving Eq. (10.365) for them in the same way as Eq. (10.316) was solved before, provided that $\theta'_k \neq 0$.

In the final stage of the forward kinematic solution, the matrices $\widehat{C}$ and $\overline{p}$ are determined by using Eqs. (10.301) and (10.304) for one of the selected legs, say L_1, by inserting the values of the relevant joint variables (s_1, ϕ_1, θ_1, ϕ'_1, θ'_1, ψ'_1) that are either specified or found in the previous stages.

(d) Position Singularities of Forward Kinematics

The forward kinematic solution presented in Part (c) stipulates three conditions in order to produce definite values for the azimuth angles (ϕ_1, …, ϕ_6) of the universal joints and the angle triads of the spherical joints. If these conditions are not satisfied, then three distinct kinds of PSFKs occur. They are described and discussed below.

∗ PSFK-1(k) (first kind of PSFK associated with IKL-k):

This is the singularity that occurs if $\theta_k = 0$ as implied by Eqs. (10.350) and (10.351). In this singularity, the leg L_k becomes perpendicular to the base and the angle ϕ_k becomes indefinite. As noticed, this is a local singularity, in which only the spinning motion of L_k about its centerline becomes uncontrollable. Therefore, this singularity does not cause any loss of controllability in the position of the end-effector, because, even though L_k acquires freedom to spin around, it still performs its main duty of keeping the joint center A_k in a definite position defined by $\theta_k = 0$ and the specified value of s_k. So, this singularity is ignorable and it is not necessary to avoid it.

∗ PSFK-2(k) (second kind of PSFK associated with IKL-k):

This is the singularity that occurs if $\theta'_k = 0$ as implied by Eq. (10.365). In this singularity, the leg L_k becomes perpendicular to the moving platform and Eq. (10.365) reduces to

$$e^{\tilde{u}_3(\phi'_k+\psi'_k)} = e^{\tilde{u}_3\phi^*_k} = \widehat{C}^*_k \tag{10.367}$$

Equation (10.367) implies that ϕ'_k and ψ'_k become indefinite even though their addition ϕ^*_k can still be determined definitely. Note that this is also a local singularity hidden within the encasement of the spherical joint of L_k. In other words, it is not even discernible owing to the definite value of ϕ^*_k, which is sufficient, together with $\theta'_k = 0$, to describe the relative orientation of the moving platform with respect to L_k. Therefore, this singularity does not cause any loss of controllability in the position of the end-effector. So, it is also ignorable and it is not necessary to avoid it.

∗ PSFK-3 (third kind of PSFK associated with the whole manipulator):

This is the singularity that occurs if $m_{1k} = n_{1k} = 0$ as implied by Eq. (10.341). In this singularity, the angle ϕ_1 becomes indefinite. However, this singularity cannot be local, because, as discussed before, ϕ_1 is the common solution obtained from all five of the IKLs. Therefore, when this singularity occurs, its defining condition must hold for all the five loops so that

$$m_{1k} = n_{1k} = 0 \;\; \forall \;\; k = 2, 3, \ldots, 6 \tag{10.368}$$

Due to Eq. (10.368), (10.342)–(10.344) turn into the following singularity relationships.

$$h_{k1}(1 - c\beta'_k) + h_{k2}s\beta'_k = 0 \quad (10.369)$$

$$h_{k1}s\beta'_k - h_{k2}(1 - c\beta'_k) = 0 \quad (10.370)$$

$$\overline{h}_k^t \overline{h}_k + 2b_0^2(1 - c\beta'_k) - s_k^2 = 0 \quad (10.371)$$

Since the joint centers are not coincident, $\beta_k \neq \beta_1$, i.e. $\beta'_k \neq 0$. Therefore, the preceding singularity relationships can be expressed more plainly as follows:

$$h_{k1} = h_{k2} = 0 \quad (10.372)$$

$$s_k^2 = h_{k3}^2 + 2b_0^2(1 - c\beta'_k) \quad (10.373)$$

On the other hand, according to Eq. (10.334),

$$h_{k1} = \overline{u}_1^t e^{-\tilde{u}_3\beta'_k} e^{\tilde{u}_2\theta_1}[s_1\overline{u}_3 - d_0 e^{\tilde{u}_3\phi'_1} e^{\tilde{u}_2\theta'_1} e^{\tilde{u}_3\psi'_1}(\overline{u}_1 - e^{\tilde{u}_3\gamma'_k}\overline{u}_1)] \quad (10.374)$$

$$h_{k2} = \overline{u}_2^t e^{-\tilde{u}_3\beta'_k} e^{\tilde{u}_2\theta_1}[s_1\overline{u}_3 - d_0 e^{\tilde{u}_3\phi'_1} e^{\tilde{u}_2\theta'_1} e^{\tilde{u}_3\psi'_1}(\overline{u}_1 - e^{\tilde{u}_3\gamma'_k}\overline{u}_1)] \quad (10.375)$$

$$h_{k3} = \overline{u}_3^t e^{\tilde{u}_2\theta_1}[s_1\overline{u}_3 - d_0 e^{\tilde{u}_3\phi'_1} e^{\tilde{u}_2\theta'_1} e^{\tilde{u}_3\psi'_1}(\overline{u}_1 - e^{\tilde{u}_3\gamma'_k}\overline{u}_1)] \quad (10.376)$$

Here, it will be shown that this singularity may occur, i.e. Eqs. (10.372) and (10.373) can be satisfied for all $k = 2, 3, \ldots, 6$, if the moving platform happens to be parallel to the base. In other words, this singularity may occur if the moving platform is oriented only by a rotation of an arbitrary angle ψ_0 about $\vec{u}_3^{(0)}$ so that

$$\widehat{C} = e^{\tilde{u}_3\psi_0} = e^{\tilde{u}_3(\phi_1+\beta_1)} e^{\tilde{u}_2\theta_1} e^{\tilde{u}_3\phi'_1} e^{\tilde{u}_2\theta'_1} e^{\tilde{u}_3(\psi'_1-\gamma_1)} \quad (10.377)$$

Equation (10.377) leads to the result that

$$e^{\tilde{u}_3\phi'_1} e^{\tilde{u}_2\theta'_1} e^{\tilde{u}_3\psi'_1} = e^{-\tilde{u}_2\theta_1} e^{\tilde{u}_3(\psi_0-\phi_1-\beta_1+\gamma_1)} = e^{-\tilde{u}_2\theta_1} e^{\tilde{u}_3\psi_0^*} \quad (10.378)$$

Note that both ψ_0 and ϕ_1 are arbitrary angles in this singularity. Consequently, the combined angle $\psi_0^* = \psi_0 - \phi_1 - \beta_1 + \gamma_1$ is also arbitrary. Due to this arbitrariness, the position of the end-effector becomes uncontrollable. Therefore, PSFK-3 must be avoided.

When Eq. (10.378) is substituted, Eqs. (10.374)–(10.376) take the following forms.

$$h_{k1} = \overline{u}_1^t e^{-\tilde{u}_3\beta'_k} e^{\tilde{u}_2\theta_1}[s_1\overline{u}_3 - d_0 e^{-\tilde{u}_2\theta_1} e^{\tilde{u}_3\psi_0^*}(\overline{u}_1 - e^{\tilde{u}_3\gamma'_k}\overline{u}_1)] \quad (10.379)$$

$$h_{k2} = \overline{u}_2^t e^{-\tilde{u}_3\beta'_k} e^{\tilde{u}_2\theta_1}[s_1\overline{u}_3 - d_0 e^{-\tilde{u}_2\theta_1} e^{\tilde{u}_3\psi_0^*}(\overline{u}_1 - e^{\tilde{u}_3\gamma'_k}\overline{u}_1)] \quad (10.380)$$

$$h_{k3} = \overline{u}_3^t e^{\tilde{u}_2\theta_1}[s_1\overline{u}_3 - d_0 e^{-\tilde{u}_2\theta_1} e^{\tilde{u}_3\psi_0^*}(\overline{u}_1 - e^{\tilde{u}_3\gamma'_k}\overline{u}_1)] \quad (10.381)$$

The above equations can be worked into the following simpler forms, in which $\psi_k^* = \psi_0^* - \beta'_k$.

$$h_{k1} = s_1 s\theta_1 c\beta'_k - d_0[(1 - c\gamma'_k)c\psi_k^* + s\gamma'_k s\psi_k^*] \quad (10.382)$$

$$h_{k2} = -s_1 s\theta_1 s\beta'_k - d_0[(1 - c\gamma'_k)s\psi_k^* - s\gamma'_k c\psi_k^*] \quad (10.383)$$

$$h_{k3} = s_1 c\theta_1 \quad (10.384)$$

When the preceding equations are substituted, the previous singularity relationships, i.e. Eqs. (10.372) and (10.373), become more explicit in terms of the manipulator parameters. That is,

$$h_{k1} = 0 \Rightarrow s_1 s\theta_1 c\beta'_k = d_0[(1 - c\gamma'_k)c\psi_k^* + s\gamma'_k s\psi_k^*] \quad (10.385)$$

$$h_{k2} = 0 \Rightarrow s_1 s\theta_1 s\beta'_k = -d_0[(1 - c\gamma'_k)s\psi_k^* - s\gamma'_k c\psi_k^*] \quad (10.386)$$

$$s_k^2 = h_{k3}^2 + 2b_0^2(1 - c\beta'_k) \Rightarrow (s_1 c\theta_1)^2 = s_k^2 - 2b_0^2(1 - c\beta'_k) \quad (10.387)$$

Equations (10.385) and (10.386) can be combined so that

$$(s_1 s\theta_1)^2 = d_0^2[(1 - c\gamma_k')^2 + (s\gamma_k')^2] = 2d_0^2(1 - c\gamma_k') \tag{10.388}$$

Afterwards, Eqs. (10.387) and (10.388) can further be combined so that

$$s_1^2 = (s_1 c\theta_1)^2 + (s_1 s\theta_1)^2 = [s_k^2 - 2b_0^2(1 - c\beta_k')] + [2d_0^2(1 - c\gamma_k')] \Rightarrow$$

$$s_k^2 = s_1^2 + 2[b_0^2(1 - c\beta_k') - d_0^2(1 - c\gamma_k')] \tag{10.389}$$

For $k = 2, 3, \ldots, 6$, Eq. (10.389) expresses the relationships among the active joint variables (i.e. the leg lengths), which cause PSFK-3 to occur. As mentioned before, in such a pose of the manipulator, the moving platform can twist uncontrollably by an arbitrary angle ψ_0 about the third axis of the base frame. Therefore, the leg lengths must never be specified according to Eq. (10.389) in order to avoid PSFK-3.

A special but very typical version of PSFK-3 occurs if all the leg lengths are specified to be equal for a particularly configured manipulator such that $d_0 = b_0$ and $\beta_k = \gamma_k$ for all $k = 1, 2, \ldots, 6$. In such a special version of PSFK-3, the manipulator may either twist arbitrarily about the third axis of the base frame or it may behave like a kind of parallelogram mechanism that is left free. In either case, the end-effector becomes uncontrollable. Therefore, for a manipulator with this particular configuration, the leg lengths must never be specified to be equal in order to keep the end-effector under control.

10.8.5 Velocity Equations

For the Stewart–Gough platform described in Section 10.8.1, the following velocity equations can be obtained by applying the velocity formulations presented in Chapter 4 to the position equations written in Section 10.8.2.

(a) Angular Velocity Equations for the End-Effector Through the Legs

Let $\overline{\omega} = \overline{\omega}_{p/0}^{(0)}$. Then, noting that $\overline{\omega}_{p/0}^{(0)} = \mathrm{colm}[\dot{\hat{C}}^{(0,p)} \hat{C}^{(p,0)}]$ or simply $\overline{\omega} = \mathrm{colm}(\dot{\hat{C}}\hat{C}^t)$, $\overline{\omega}$ can be obtained from Eq. (10.301) as follows for $k = 1, 2, \ldots, 6$.

$$\overline{\omega} = \dot{\phi}_k \overline{u}_3 + \dot{\theta}_k e^{\widetilde{u}_3(\phi_k+\beta_k)} \overline{u}_2 + \dot{\phi}_k' e^{\widetilde{u}_3(\phi_k+\beta_k)} e^{\widetilde{u}_2\theta_k} \overline{u}_3$$
$$+ \dot{\theta}_k' e^{\widetilde{u}_3(\phi_k+\beta_k)} e^{\widetilde{u}_2\theta_k} e^{\widetilde{u}_3\phi_k'} \overline{u}_2 + \dot{\psi}_k' e^{\widetilde{u}_3(\phi_k+\beta_k)} e^{\widetilde{u}_2\theta_k} e^{\widetilde{u}_3\phi_k'} e^{\widetilde{u}_2\theta_k'} \overline{u}_3 \tag{10.390}$$

(b) Tip Point Velocity Equations Through the Legs

Let $\overline{v} = \overline{v}_{P/O/\mathcal{F}_0}^{(0)}$. Then, noting that $\overline{v}_{P/O/\mathcal{F}_0}^{(0)} = \dot{\overline{r}}_{P/O}^{(0)}$ or simply $\overline{v} = \dot{\overline{p}}$, $\overline{v}$ can be obtained from Eq. (10.304) as follows for $k = 1, 2, \ldots, 6$:

$$\overline{v} = \dot{s}_k e^{\widetilde{u}_3(\phi_k+\beta_k)} e^{\widetilde{u}_2\theta_k} \overline{u}_3 + s_k \dot{\phi}_k e^{\widetilde{u}_3(\phi_k+\beta_k)} \widetilde{u}_3 e^{\widetilde{u}_2\theta_k} \overline{u}_3$$
$$+ s_k \dot{\theta}_k e^{\widetilde{u}_3(\phi_k+\beta_k)} e^{\widetilde{u}_2\theta_k} \widetilde{u}_2 \overline{u}_3 + \dot{\hat{C}}(h_0 \overline{u}_3 - d_0 e^{\widetilde{u}_3\gamma_k} \overline{u}_1)$$

Note that

$$\widetilde{u}_2 \overline{u}_3 = \overline{u}_1,\ \widetilde{u}_3 e^{\widetilde{u}_2\theta_k} \overline{u}_3 = \widetilde{u}_3(\overline{u}_3 c\theta_k + \overline{u}_1 s\theta_k) = \overline{u}_2 s\theta_k, \text{ and } \dot{\hat{C}} = \dot{\hat{C}}\hat{C}^t\hat{C} = \widetilde{\omega}\hat{C}$$

Hence, the expression of $\overline{v}$ becomes

$$\overline{v} = \dot{s}_k e^{\widetilde{u}_3(\phi_k+\beta_k)} e^{\widetilde{u}_2\theta_k} \overline{u}_3 + (s_k \dot{\phi}_k s\theta_k) e^{\widetilde{u}_3(\phi_k+\beta_k)} \overline{u}_2$$
$$+ s_k \dot{\theta}_k e^{\widetilde{u}_3(\phi_k+\beta_k)} e^{\widetilde{u}_2\theta_k} \overline{u}_1 + \widetilde{\omega}\hat{C}(h_0 \overline{u}_3 - d_0 e^{\widetilde{u}_3\gamma_k} \overline{u}_1) \tag{10.391}$$

10.8.6 Inverse Kinematics in the Velocity Domain

(a) Inverse Velocity Solution

For a specified velocity state of the end-effector at a currently known pose of the manipulator, i.e. for known matrices $\overline{v}$, $\overline{\omega}$, and $\widehat{C}$, the corresponding rates of the active and passive joint variables can be found leg by leg as explained below.

For the leg L_k, Eq. (10.391) can be written as follows:

$$\overline{u}_1(s_k\dot{\theta}_k) + \overline{u}_2(s_k\dot{\phi}_k s\theta_k) + \overline{u}_3\dot{s}_k = \overline{w}_k \tag{10.392}$$

In Eq. (10.392), the column matrix $\overline{w}_k$ is known with the following expression.

$$\overline{w}_k = e^{-\tilde{u}_2\theta_k}e^{-\tilde{u}_3(\phi_k+\beta_k)}[\overline{v} - \tilde{\omega}\widehat{C}(h_0\overline{u}_3 - d_0 e^{\tilde{u}_3\gamma_k}\overline{u}_1)] \tag{10.393}$$

Note that $\overline{w}_k$ represents the velocity of the joint center A_k in the leg frame $\mathcal{F}_k(B_k)$ attached to L_k. This frame is formed by the basis vectors $\vec{u}_1^{(k)}$, $\vec{u}_2^{(k)}$, and $\vec{u}_3^{(k)}$ as illustrated in Figure 10.16.

Note also that $s_k > 0$ always. Therefore, if $s\theta_k \neq 0$, Eq. (10.392) leads to the following solution, in which $w_{ki} = \overline{u}_i^t\overline{w}_k$ for $i = 1, 2, 3$.

$$\dot{\theta}_k = w_{k1}/s_k \tag{10.394}$$

$$\dot{\phi}_k = w_{k2}/(s_k s\theta_k) \tag{10.395}$$

$$\dot{s}_k = w_{k3} \tag{10.396}$$

As for the rates of the angles of the spherical joint of L_k, they can be found from Eq. (10.390) after rearranging it as shown below.

$$\dot{\phi}'_k\overline{u}_3 + \dot{\theta}'_k\overline{u}_2 + \dot{\psi}'_k e^{\tilde{u}_2\theta'_k}\overline{u}_3 = \overline{\omega}_k^* \Rightarrow$$

$$\overline{u}_1(\dot{\psi}'_k s\theta'_k) + \overline{u}_2\dot{\theta}'_k + \overline{u}_3(\dot{\phi}'_k + \dot{\psi}'_k c\theta'_k) = \overline{\omega}_k^* \tag{10.397}$$

In Eq. (10.397), $\overline{\omega}_k^*$ is known with the following expression.

$$\overline{\omega}_k^* = e^{-\tilde{u}_3\phi'_k}e^{-\tilde{u}_2\theta_k}e^{-\tilde{u}_3(\phi_k+\beta_k)}[\overline{\omega} - \dot{\phi}_k\overline{u}_3 - \dot{\theta}_k e^{\tilde{u}_3(\phi_k+\beta_k)}\overline{u}_2] \tag{10.398}$$

Note that $\overline{\omega}_k^*$ represents the relative angular velocity of the moving platform ($\mathcal{L}_p$) with respect to the leg L_k in the frame $\mathcal{F}_k^*(B_k)$, which is obtained by rotating the leg frame $\mathcal{F}_k(B_k)$ about $\vec{u}_3^{(k)}$ by the angle ϕ'_k. The special reason of using this particular frame is that it provides the simplest expression for $\overline{\omega}_k^*$ in terms of the rates to be found as seen on the left-hand side of Eq. (10.397).

If $s\theta'_k \neq 0$, Eq. (10.397) leads to the following solution, in which $\omega_{ki}^* = \overline{u}_i^t\overline{\omega}_k^*$ for $i = 1, 2, 3$.

$$\dot{\theta}'_k = \omega_{k2}^* \tag{10.399}$$

$$\dot{\psi}'_k = \omega_{k1}^*/s\theta'_k \tag{10.400}$$

$$\dot{\phi}'_k = \omega_{k3}^* - \dot{\psi}'_k c\theta'_k = (\omega_{k3}^* s\theta'_k - \omega_{k1}^* c\theta'_k)/s\theta'_k \tag{10.401}$$

(b) Motion Singularities of Inverse Kinematics

$*$ MSIK-1(k) (first kind of MSIK for L_k):

Equation (10.395) implies that the leg L_k gets into a pose of MSIK-1(k) if $s\theta_k = 0$, i.e. if $\theta_k = 0$ as the physically possible case. Note that the poses of MSIK-1(k) and

PSIK-1(k) happen to be the same for this manipulator. In this singularity, L_k becomes perpendicular to the base with $\theta_k = 0$ and $\dot{\phi}_k$ becomes indefinite. This singularity occurs with a finite value of $\dot{\phi}_k$ (even though it becomes indefinite) if the motion of the end-effector is specified in such a special way that $w_{k2} = 0$, which implies that the joint center A_k does not have any velocity component along $\vec{u}_2^{(k)}$. However, such a motion specification does not cause any problem in finding definite values for $\dot{s}_k$ and $\dot{\theta}_k$. Therefore, if it is acceptable to have $w_{k2} = 0$ for the task to be executed, this singularity need not be avoided.

* MSIK-2(k) (second kind of MSIK for L_k):

Equations (10.400) implies that the leg L_k gets into a pose of MSIK-2(k) if $s\theta'_k = 0$, i.e. if $\theta'_k = 0$ as the physically possible case. Note that the poses of MSIK-2(k) and PSIK-2(k) also happen to be the same for this manipulator. In this singularity with $\theta'_k = 0$, the unit vectors $\vec{u}_3^{(k)}$ and $\vec{u}_3^{(p)}$ become codirectional and the leg L_k becomes perpendicular to the plane of the moving platform. This singularity occurs with a finite value of $\dot{\psi}'_k$ (even though it becomes indefinite) if the motion of the end-effector is specified in such a special way that $\omega^*_{k1} = 0$. Moreover, $\dot{\phi}'_k$ becomes indistinguishable from $\dot{\psi}'_k$ because they represent rotations about two axes that have become parallel. However, $\dot{\phi}'_k$ and $\dot{\psi}'_k$ also become related as $\dot{\phi}'_k + \dot{\psi}'_k = \omega^*_{k3}$ according to Eq. (10.401). In any case, a motion specification with $\omega^*_{k1} = 0$ does not cause any problem in finding definite values for $\dot{s}_k$ and the angle rates other than $\dot{\phi}'_k$ and $\dot{\psi}'_k$. Therefore, if it is acceptable to have $\omega^*_{k1} = 0$ for the task to be executed, this singularity need not be avoided, either.

10.8.7 Forward Kinematics in the Velocity Domain

For specified rates of change of the leg lengths (i.e. $\dot{s}_1, \dot{s}_2, \ldots, \dot{s}_6$), the rates of the passive joint variables and the velocity state of the end-effector (i.e. $\overline{v}$ and $\overline{\omega}$) can be obtained from the IKLs (independent kinematic loops) of the manipulator. The formation of the IKLs has already been explained in the previous sections. The associated velocity equations and their solutions are explained below.

(a) Loop Equations in the Velocity Domain

The loop equations in the velocity domain are generated by writing Eqs. (10.390) and (10.391) for the legs of each IKL and equating their right-hand sides to each other. This action results in the following equations for the kinematic loop IKL-$k = B_1A_1A_kB_kB_1$, where $k = 2, 3, \ldots, 6$.

* Angular velocity equation for IKL-k:

$$\begin{aligned}
&\dot{\phi}_k\overline{u}_3 + \dot{\theta}_k e^{\tilde{u}_3(\phi_k+\beta_k)}\overline{u}_2 + \dot{\phi}'_k e^{\tilde{u}_3(\phi_k+\beta_k)} e^{\tilde{u}_2\theta_k}\overline{u}_3 \\
&\quad + \dot{\theta}'_k e^{\tilde{u}_3(\phi_k+\beta_k)} e^{\tilde{u}_2\theta_k} e^{\tilde{u}_3\phi'_k}\overline{u}_2 + \dot{\psi}'_k e^{\tilde{u}_3(\phi_k+\beta_k)} e^{\tilde{u}_2\theta_k} e^{\tilde{u}_3\phi'_k} e^{\tilde{u}_2\theta'_k}\overline{u}_3 \\
&= \dot{\phi}_1\overline{u}_3 + \dot{\theta}_1 e^{\tilde{u}_3(\phi_1+\beta_1)}\overline{u}_2 + \dot{\phi}'_1 e^{\tilde{u}_3(\phi_1+\beta_1)} e^{\tilde{u}_2\theta_1}\overline{u}_3 \\
&\quad + \dot{\theta}'_1 e^{\tilde{u}_3(\phi_1+\beta_1)} e^{\tilde{u}_2\theta_1} e^{\tilde{u}_3\phi'_1}\overline{u}_2 + \dot{\psi}'_1 e^{\tilde{u}_3(\phi_1+\beta_1)} e^{\tilde{u}_2\theta_1} e^{\tilde{u}_3\phi'_1} e^{\tilde{u}_2\theta'_1}\overline{u}_3.
\end{aligned} \tag{10.402}$$

* Characteristic point velocity equation for IKL-k:

$$\begin{aligned}
&\dot{s}_k e^{\tilde{u}_3(\phi_k+\beta_k)} e^{\tilde{u}_2\theta_k}\overline{u}_3 + (s_k\dot{\phi}_k s\theta_k) e^{\tilde{u}_3(\phi_k+\beta_k)}\overline{u}_2 \\
&\quad + s_k\dot{\theta}_k e^{\tilde{u}_3(\phi_k+\beta_k)} e^{\tilde{u}_2\theta_k}\overline{u}_1 - d_0\tilde{\omega}\widehat{C} e^{\tilde{u}_3\gamma_k}\overline{u}_1 \\
&= \dot{s}_1 e^{\tilde{u}_3(\phi_1+\beta_1)} e^{\tilde{u}_2\theta_1}\overline{u}_3 + (s_1\dot{\phi}_1 s\theta_1) e^{\tilde{u}_3(\phi_1+\beta_1)}\overline{u}_2 \\
&\quad + s_1\dot{\theta}_1 e^{\tilde{u}_3(\phi_1+\beta_1)} e^{\tilde{u}_2\theta_1}\overline{u}_1 - d_0\tilde{\omega}\widehat{C} e^{\tilde{u}_3\gamma_1}\overline{u}_1
\end{aligned} \tag{10.403}$$

(b) Forward Velocity Solution

The solution can be obtained by using a fourth-order semi-analytical method that is similar to the method used in the position domain. In order to obtain the solution this way, four of the 30 velocity parameters to be found are treated as if they are temporarily known. The velocity parameters comprise the rates of the joint angles and the components of the relevant velocities. The mentioned four velocity parameters are called *major unknowns*. Then, the other 26 velocity parameters are called *minor unknowns*. In the present problem, it happens to be convenient to select the major unknowns as $\dot{\theta}_1$ and the three components of $\overline{\omega}$, i.e. ω_1, ω_2, and ω_3. After selecting the major unknowns, the loop equations can be solved analytically for the 26 minor unknowns so that they are obtained as functions of the major unknowns. This solution procedure also generates four *consistency equations*, which are to be solved to find the major unknowns. However, the consistency equations turn out to be rather complicated, but still linear though, owing to the fact that the velocity equations happen to be linear in the velocity parameters. Therefore, in the numerical phase of the procedure, the solution for the major unknowns can be obtained by inverting the relevant 4×4 coefficient matrix by using a suitable numerical method.

The first stage of the procedure explained above can be started by rearranging Eq. (10.403) as follows for $k = 2, 3, \ldots, 6$:

$$s_k\dot{\theta}_k\overline{u}_1 + (s_k\dot{\phi}_k s\theta_k)\overline{u}_2 - (s_1\dot{\phi}_1 s\theta_1)e^{-\widetilde{u}_2\theta_k}e^{-\widetilde{u}_3\phi_k^*}\overline{u}_2 = \overline{v}_k^* \tag{10.404}$$

In Eq. (10.404),

$$\phi_k^* = (\phi_k - \phi_1) + (\beta_k - \beta_1) \tag{10.405}$$

$$\begin{aligned}\overline{v}_k^* &= (\dot{s}_1 e^{-\widetilde{u}_2\theta_k}e^{-\widetilde{u}_3\phi_k^*}e^{\widetilde{u}_2\theta_1}\overline{u}_3 - \dot{s}_k\overline{u}_3) + s_1\dot{\theta}_1 e^{-\widetilde{u}_2\theta_k}e^{-\widetilde{u}_3\phi_k^*}e^{\widetilde{u}_2\theta_1}\overline{u}_1 \\ &\quad + d_0 e^{-\widetilde{u}_2\theta_k}e^{-\widetilde{u}_3(\phi_k+\beta_k)}(\omega_1\widetilde{u}_1 + \omega_2\widetilde{u}_2 + \omega_3\widetilde{u}_3)\widehat{C}(e^{\widetilde{u}_3\gamma_k}\overline{u}_1 - e^{\widetilde{u}_3\gamma_1}\overline{u}_1)\end{aligned} \tag{10.406}$$

Note that

$$\overline{v}_k^* = \overline{v}_k^*(\dot{\theta}_1, \omega_1, \omega_2, \omega_3) \tag{10.407}$$

Upon premultiplications by $\overline{u}_1^t$, $\overline{u}_2^t$, and $\overline{u}_3^t$, Eq. (10.404) leads to the following three scalar equations, in which $v_{ki}^* = \overline{u}_i^t\overline{v}_k^*$ for $i = 1, 2, 3$.

$$s_k\dot{\theta}_k - (s_1 s\theta_1 c\theta_k s\phi_k^*)\dot{\phi}_1 = v_{k1}^* \tag{10.408}$$

$$(s_k s\theta_k)\dot{\phi}_k - (s_1 s\theta_1 c\phi_k^*)\dot{\phi}_1 = v_{k2}^* \tag{10.409}$$

$$(s_1 s\theta_1 s\theta_k s\phi_k^*)\dot{\phi}_1 = -v_{k3}^* \tag{10.410}$$

If $s\theta_1 s\theta_k s\phi_k^* \neq 0$, then Eq. (10.410) gives $\dot{\phi}_1$ as

$$\dot{\phi}_1 = -v_{k3}^*/(s_1 s\theta_1 s\theta_k s\phi_k^*) \tag{10.411}$$

After finding $\dot{\phi}_1$, Eq. (10.409) gives $\dot{\phi}_k$, if $s\theta_k \neq 0$, and Eq. (10.408) gives $\dot{\theta}_k$. That is,

$$\dot{\phi}_k = [v_{k2}^* + (s_1 s\theta_1 c\phi_k^*)\dot{\phi}_1]/(s_k s\theta_k) \tag{10.412}$$

$$\dot{\theta}_k = [v_{k1}^* + (s_1 s\theta_1 c\theta_k s\phi_k^*)\dot{\phi}_1]/s_k \tag{10.413}$$

The above equations can also be written as follows by inserting the expression of $\dot{\phi}_1$.

$$\dot{\phi}_k = (v^*_{k2} s\theta_k s\phi^*_k - v^*_{k3} c\phi^*_k)/(s_k s^2\theta_k s\phi^*_k) \tag{10.414}$$

$$\dot{\theta}_k = (v^*_{k1} s\theta_k - v^*_{k3} c\theta_k)/(s_k s\theta_k) \tag{10.415}$$

The solution described above is carried out for $k = 2, 3, \ldots, 6$ and thus the rates of the six universal joint angles (except $\dot{\theta}_1$) are obtained as functions of the major unknowns ω_1, ω_2, ω_3, and $\dot{\theta}_1$. However, one of these rates, which is $\dot{\phi}_1$, is obtained five times; each time with a different expression as seen in Eq. (10.411). Of course, these five expressions must be equal to each other. Hence, the following four *consistency equations* are obtained.

$$v^*_{33} s\theta_2 s\phi^*_2 = v^*_{23} s\theta_3 s\phi^*_3 \tag{10.416}$$

$$v^*_{43} s\theta_2 s\phi^*_2 = v^*_{23} s\theta_4 s\phi^*_4 \tag{10.417}$$

$$v^*_{53} s\theta_2 s\phi^*_2 = v^*_{23} s\theta_5 s\phi^*_5 \tag{10.418}$$

$$v^*_{63} s\theta_2 s\phi^*_2 = v^*_{23} s\theta_6 s\phi^*_6 \tag{10.419}$$

On the other hand, according to Eq. (10.406),

$$\begin{aligned} v^*_{k3} &= (\dot{s}_1 \overline{u}_3^t e^{-\tilde{u}_2\theta_k} e^{-\tilde{u}_3\phi^*_k} e^{\tilde{u}_2\theta_1} \overline{u}_3 - \dot{s}_k) + s_1 \dot{\theta}_1 \overline{u}_3^t e^{-\tilde{u}_2\theta_k} e^{-\tilde{u}_3\phi^*_k} e^{\tilde{u}_2\theta_1} \overline{u}_1 \\ &\quad + d_0 \overline{u}_3^t e^{-\tilde{u}_2\theta_k} e^{-\tilde{u}_3(\phi_k+\beta_k)} (\omega_1 \tilde{u}_1 + \omega_2 \tilde{u}_2 + \omega_3 \tilde{u}_3) \hat{C} (e^{\tilde{u}_3\gamma_k} \overline{u}_1 - e^{\tilde{u}_3\gamma_1} \overline{u}_1) \end{aligned} \tag{10.420}$$

The above equation can be written compactly as

$$v^*_{k3} = \Theta_{k1} \dot{\theta}_1 + \Omega_{k1}\omega_1 + \Omega_{k2}\omega_2 + \Omega_{k3}\omega_3 - v^\circ_k \tag{10.421}$$

In Eq. (10.421),

$$v^\circ_k = \dot{s}_k - \dot{s}_1 \overline{u}_3^t e^{-\tilde{u}_2\theta_k} e^{-\tilde{u}_3\phi^*_k} e^{\tilde{u}_2\theta_1} \overline{u}_3 \tag{10.422}$$

$$\Theta_{k1} = s_1 \overline{u}_3^t e^{-\tilde{u}_2\theta_k} e^{-\tilde{u}_3\phi^*_k} e^{\tilde{u}_2\theta_1} \overline{u}_1 \tag{10.423}$$

$$\Omega_{k1} = d_0 \overline{u}_3^t e^{-\tilde{u}_2\theta_k} e^{-\tilde{u}_3(\phi_k+\beta_k)} \tilde{u}_1 \hat{C} (e^{\tilde{u}_3\gamma_k} \overline{u}_1 - e^{\tilde{u}_3\gamma_1} \overline{u}_1) \tag{10.424}$$

$$\Omega_{k2} = d_0 \overline{u}_3^t e^{-\tilde{u}_2\theta_k} e^{-\tilde{u}_3(\phi_k+\beta_k)} \tilde{u}_2 \hat{C} (e^{\tilde{u}_3\gamma_k} \overline{u}_1 - e^{\tilde{u}_3\gamma_1} \overline{u}_1) \tag{10.425}$$

$$\Omega_{k3} = d_0 \overline{u}_3^t e^{-\tilde{u}_2\theta_k} e^{-\tilde{u}_3(\phi_k+\beta_k)} \tilde{u}_3 \hat{C} (e^{\tilde{u}_3\gamma_k} \overline{u}_1 - e^{\tilde{u}_3\gamma_1} \overline{u}_1) \tag{10.426}$$

After substituting Eq. (10.420), Eqs. (10.416)–(10.419) can be written compactly as the following matrix equation.

$$\begin{bmatrix} A_{31} & A_{32} & A_{33} & A_{34} \\ A_{41} & A_{42} & A_{43} & A_{44} \\ A_{51} & A_{52} & A_{53} & A_{54} \\ A_{61} & A_{62} & A_{63} & A_{64} \end{bmatrix} \begin{bmatrix} \omega_1 \\ \omega_2 \\ \omega_3 \\ \dot{\theta}_1 \end{bmatrix} = \begin{bmatrix} v'_3 \\ v'_4 \\ v'_5 \\ v'_6 \end{bmatrix} \tag{10.427}$$

In Eq. (10.427), for $k = 3, 4, 5, 6$ and $j = 1, 2, 3$,

$$A_{kj} = \Omega_{kj} s\theta_2 s\phi^*_2 - \Omega_{2j} s\theta_k s\phi^*_k \tag{10.428}$$

$$A_{k4} = \Theta_{k1} s\theta_2 s\phi^*_2 - \Theta_{21} s\theta_k s\phi^*_k \tag{10.429}$$

$$v'_k = v^\circ_k s\theta_2 s\phi^*_2 - v^\circ_2 s\theta_k s\phi^*_k \tag{10.430}$$

Hence, provided that the 4×4 coefficient matrix $\hat{A}$ in Eq. (10.427) is not singular, the major unknowns ω_1, ω_2, ω_3, and $\dot{\theta}_1$ can be found by using a suitable numerical method.

For example, they can be found directly as follows by using *Cramer's rule.*

$$\omega_j = D_j/D_0 \quad \text{for } j = 1, 2, 3 \tag{10.431}$$

$$\dot{\theta}_1 = D_4/D_0 \tag{10.432}$$

In Eqs. (10.431) and (10.432), $D_0 = \det(\hat{A})$ and $D_j = \det(\hat{B}_j)$ for $j = 1, 2, 3, 4$. As for the matrix $\hat{B}_j$, it is obtained from $\hat{A}$ by replacing the jth column of $\hat{A}$ with the column matrix on the right-hand side of Eq. (10.427).

In the next stage of the forward velocity solution, the rates of the spherical joint angles are to be found for $k = 2, 3, \ldots, 6$. Here, the other rates and $\overline{\omega} = \omega_1\overline{u}_1 + \omega_2\overline{u}_2 + \omega_3\overline{u}_3$ are already available from the previous stage. Therefore, the required rates can be found in exactly the same way as they were found in Section 10.8.6 by rearranging Eq. (10.390) as Eq. (10.397) and using the same $\overline{\omega}_k^*$ defined by Eq. (10.398). If $s\theta_k' \neq 0$, then the solution will be the same as expressed by Eqs. (10.399)–(10.401).

In the final stage of the forward velocity solution, $\overline{v}$ is also found by using Eq. (10.391), in addition to $\overline{\omega}$ that was found in the first stage.

(c) Motion Singularities of Forward Kinematics

The forward velocity solution presented above stipulates certain conditions in order to produce definite values for the rates of the azimuth angles of the universal joints ($\phi_1, \ldots, \phi_6$), the rates of the angle triads of the spherical joints, and the angular velocity components of the moving platform. If these conditions are not satisfied, then three distinct kinds of MSFKs occur. They are described and discussed below.

* MSFK-1(k) (first kind of MSFK associated with IKL-k):

This is the singularity that occurs if $\theta_k = 0$ as implied by Eq. (10.412). In this singularity, the leg L_k becomes perpendicular to the base and the rate $\dot{\phi}_k$ becomes indefinite. As noticed, this is a local singularity that has the same appearance as that of PSFK-1(k). In this singularity, only the spinning motion of L_k about its centerline becomes uncontrollable. Therefore, this singularity does not cause any loss of controllability in the motion of the end-effector, because, even though L_k acquires freedom to spin around, it still performs its main duty of contributing the specified value of $\dot{s}_k$ to the velocity of the end-effector. So, this singularity is ignorable and it is not necessary to avoid it.

* MSFK-2(k) (second kind of MSFK associated with IKL-k):

As mentioned before, the rates of the spherical joint angles can be found in exactly the same way within the scopes both inverse and forward kinematics. Therefore, the poses of MSIK-2(k) and MSFK-2(k) are the same with the characteristic feature that $\theta_k' = 0$. As discussed before, in both of these motion singularities, it happens that $\omega_{k1}^* = 0$ while $\dot{\phi}_k'$ and $\dot{\psi}_k'$ become indefinite but related so that $\dot{\phi}_k' + \dot{\psi}_k' = \omega_{k3}^*$ according to Eq. (10.401). On the other hand, like MSFK-1(k), MSFK-2(k) does not cause any loss of controllability either in the motion of the end-effector, because, whatever happens within the encasement of the spherical joint, the leg L_k still performs its main duty of contributing the specified value of $\dot{s}_k$ to the velocity of the end-effector. So, this singularity is also ignorable and it is not necessary to avoid it.

∗ MSFK-3 (third kind of MSFK associated with the whole manipulator):

This is the singularity that occurs if $s\theta_1 s\theta_k s\phi_k^* = 0$ as implied by Eqs. (10.410) and (10.411). In this singularity, $\dot{\phi}_1$ becomes indefinite but remains finite if

$$v_{k3}^* = 0 \ \forall \ k = 2, 3, \ldots, 6 \tag{10.433}$$

The condition of this singularity can typically be satisfied if

$$\theta_k = 0 \ \forall \ k = 1, 2, 3, \ldots, 6 \tag{10.434}$$

In the presence of Eqs. (10.433) and (10.434), Eq. (10.420) becomes

$$(\dot{s}_1 - \dot{s}_k) + d_0 \overline{u}_3^t (\omega_1 \widetilde{u}_1 + \omega_2 \widetilde{u}_2 + \omega_3 \widetilde{u}_3) \widehat{C} (e^{\widetilde{u}_3 \gamma_k} \overline{u}_1 - e^{\widetilde{u}_3 \gamma_1} \overline{u}_1) = 0 \tag{10.435}$$

On the other hand, like PSFK-3 of the position domain, MSFK-3 may also occur if the moving platform happens to be moving parallel to the base. In other words, this singularity may occur if the moving platform is oriented only by a rotation of an arbitrary angle ψ_0 about $\vec{u}_3^{(0)}$ so that

$$\widehat{C} = e^{\widetilde{u}_3 \psi_0} \tag{10.436}$$

Equation (10.436) leads to the following equalities for $\overline{\omega}$ and its components.

$$\overline{\omega} = \text{colm}(\dot{\widehat{C}} \widehat{C}^t) = \dot{\psi}_0 \overline{u}_3 \Rightarrow$$

$$\omega_1 = \omega_2 = 0, \quad \omega_3 = \dot{\psi}_0 \tag{10.437}$$

Due to the above equalities, Eq. (10.436) reduces to the following *singularity relationship* among the rates of the active joint variables.

$$\dot{s}_k = \dot{s}_1 \ \forall \ k = 2, 3, \ldots, 6 \tag{10.438}$$

In such an occurrence of MSFK-3, Eqs. (10.434), (10.437), and (10.438) indicate that the legs are perpendicular to the base, they are parallel to each other, they have equal lengths, and their lengths are specified to be changing at the same rate.

Meanwhile, since $v_{k3}^* = 0$ for all $k = 2, 3, \ldots, 6$, the coefficient matrix $\widehat{A}$ in Eq. (10.427) becomes singular. Therefore, although $\omega_1 = \omega_2 = 0$, ω_3 and $\dot{\theta}_1$ also become indefinite, i.e. arbitrary, in addition to $\dot{\phi}_1$. This conclusion about ω_3 is implied by Eq. (10.437), too, because ψ_0 and $\dot{\psi}_0$ are arbitrary.

As consequences of the pose of MSFK-3 described above, the manipulator may either have an arbitrary twisting motion about the third axis of the base frame or it may behave like a kind of parallelogram mechanism that is left free. In either case, the end-effector becomes uncontrollable. Therefore, MSFK-3 must be avoided. For this purpose, the leg lengths must never be specified to be equal and to be changing with equal rates.

10.9 Delta Robot: A 3RS^{2}S^2 Spatial Parallel Manipulator

10.9.1 Kinematic Description

Another popular parallel manipulator configuration is illustrated in Figures 10.17 and 10.18 with its characteristic features. A robotic manipulator of this kind is known with the generic name *delta robot*. The configuration of such a manipulator is denoted as

$3RS^2S^2$, which means that the manipulator has three legs and each leg comprises one *revolute joint* with the fixed platform, a pair of *spherical joints* between the upper and lower links of the leg, and another pair of *spherical joints* with the moving platform. The lower part of each leg consists of two parallel links. The manipulator is actuated through the revolute joints.

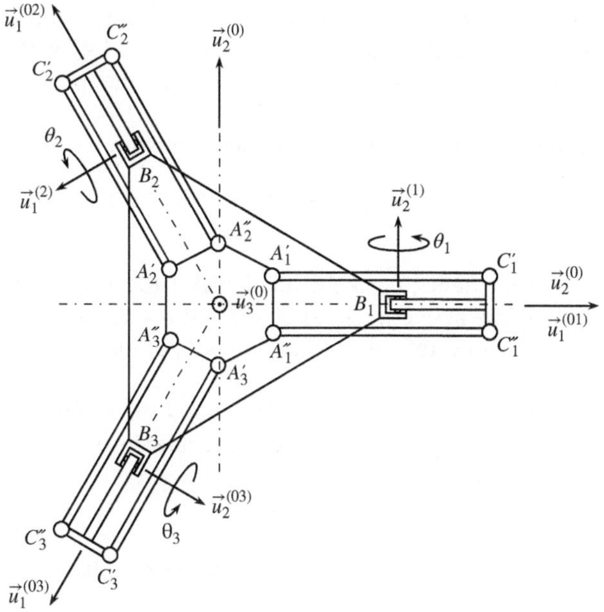

Figure 10.17 Delta robot: a $3RS^2S^2$ parallel manipulator (top view).

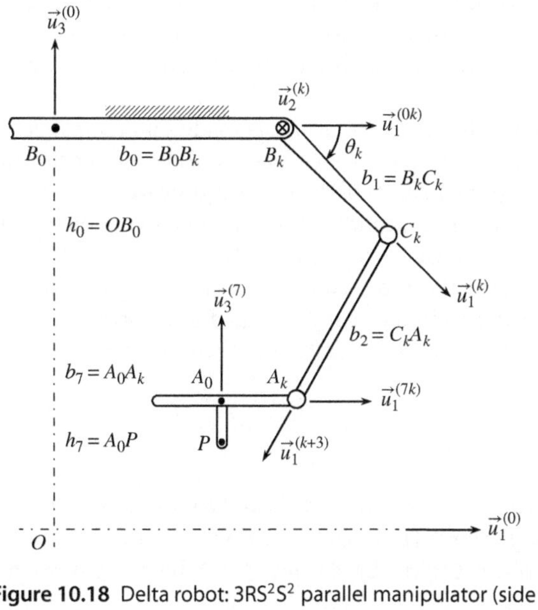

Figure 10.18 Delta robot: $3RS^2S^2$ parallel manipulator (side view of leg L_k).

Figure 10.17 illustrates the top view of the manipulator in a pose such that the center of the moving platform is vertically below the center of the fixed platform. As seen in the figure, the manipulator has a symmetric configuration such that the fixed platform ($\mathcal{L}_0$) can be modeled geometrically as an equilateral triangle. The centers of the revolute joints (B_1, B_2, B_3) are located at the corners of this triangle. Considering one of the three legs, say L_k, its upper link is $\mathcal{L}_k$ and its lower parallel links are $\mathcal{L}'_{k+3}$ and $\mathcal{L}''_{k+3}$. On these parallel links, the centers of the upper spherical joints are C'_k and C''_k with the mid point C_k between them. Similarly, the centers of the lower spherical joints are A'_k and A''_k with the mid point A_k between them. The lower spherical joints are shared by the moving platform $\mathcal{L}_7$.

The geometric parameters associated with Figure 10.17 are indicated below.

$$\beta_1 = \sphericalangle[\overrightarrow{B_0B_1}] = 0, \beta_2 = \sphericalangle[\overrightarrow{B_0B_2}] = \beta_0, \beta_3 = \sphericalangle[\overrightarrow{B_0B_3}] = -\beta_0; \beta_0 = 2\pi/3$$

$$C'_kC''_k = A'_kA''_k = 2d_0 \ \forall \ k = 1, 2, 3$$

Figure 10.18 illustrates the side view of the leg L_k of the manipulator together with the relevant geometric parameters and the active joint variable θ_k. The side views of the other two legs are of course similar due to the symmetric configuration of the manipulator.

This manipulator is such that the number of the moving links is $n_m = 10$. As for the joint numbers, $j_1 = 3, j_2 = 0$, and $j_3 = 12$. Therefore, the Kutzbach–Gruebler formula, i.e. Eq. (10.1), gives the mobility of the manipulator as

$$\mu = 6 \times 10 - (5 \times 3 + 3 \times 12) = 9 \tag{10.439}$$

However, for this manipulator, μ comprises *significant* and *insignificant* mobilities. In other words,

$$\mu = \mu^* + \mu' = 3 + 6 \tag{10.440}$$

In Eq. (10.440), $\mu' = 6$ is the *insignificant mobility* associated with the spinning freedom of the lower links of the legs between the spherical joints and $\mu^* = 3$ is the *significant mobility* of the moving platform that is controlled by the rotary actuators of the three revolute joints.

The preceding mobility analysis shows that the delta robot is made up of a *deficient manipulator* in the three-dimensional working space. Indeed, as it will also be verified in the forthcoming kinematic analysis, the parallel links of the lower legs prevent the moving platform from making any rotational motion. In other words, the moving platform can only move translationally with respect to the base frame $\mathcal{F}_0(O)$.

Incidentally, Figure 10.18 also illustrates the most common scenario, in which the delta robot is used. In this scenario, the fixed platform is situated at a certain height above the horizontal plane of the base frame and the delta robot works downward on the objects it handles.

10.9.2 Position Equations

Referring to Figures 10.17 and 10.18, the following equations can be written in order to express the orientation of the moving platform ($\mathcal{L}_7$) together with the end-effector and the location of the tip point (P) with respect to the base frame $\mathcal{F}_0(O)$ by going through each leg.

(a) End-Effector Orientation Equations Through the Legs

For $k = 1, 2, 3, 7$ and the *"lefty" (left-hand side) legs,*

$$\hat{C} = \hat{C}^{(0,7)} = \hat{C}^{(0,0k)}\hat{C}^{(0k,k)}\hat{C}^{(k,k+3)'}\hat{C}^{(k+3,7k)'}\hat{C}^{(7k,7)'} \Rightarrow$$

$$\hat{C} = [e^{\tilde{u}_3\beta_k}][e^{\tilde{u}_2\theta_k}][e^{\tilde{u}_2\theta'_k}e^{\tilde{u}_3\phi'_k}e^{\tilde{u}_1\psi'_k}][e^{\tilde{u}_1\psi'_{k+3}}e^{\tilde{u}_3\phi'_{k+3}}e^{\tilde{u}_2\theta'_{k+3}}][e^{\tilde{u}_3\beta_k}]^t \Rightarrow$$

$$\hat{C} = e^{\tilde{u}_3\beta_k}e^{\tilde{u}_2(\theta_k+\theta'_k)}e^{\tilde{u}_3\phi'_k}e^{\tilde{u}_1(\psi'_k+\psi'_{k+3})}e^{\tilde{u}_3\phi'_{k+3}}e^{\tilde{u}_2\theta'_{k+3}}e^{-\tilde{u}_3\beta_k} \quad (10.441)$$

For $k = 1, 2, 3, 7$ and the *"righty" (right-hand side) legs,*

$$\hat{C} = \hat{C}^{(0,7)} = \hat{C}^{(0,0k)}\hat{C}^{(0k,k)}\hat{C}^{(k,k+3)''}\hat{C}^{(k+3,7k)''}\hat{C}^{(7k,7)''} \Rightarrow$$

$$\hat{C} = [e^{\tilde{u}_3\beta_k}][e^{\tilde{u}_2\theta_k}][e^{\tilde{u}_2\theta''_k}e^{\tilde{u}_3\phi''_k}e^{\tilde{u}_1\psi''_k}][e^{\tilde{u}_1\psi''_{k+3}}e^{\tilde{u}_3\phi''_{k+3}}e^{\tilde{u}_2\theta''_{k+3}}][e^{\tilde{u}_3\beta_k}]^t \Rightarrow$$

$$\hat{C} = e^{\tilde{u}_3\beta_k}e^{\tilde{u}_2(\theta_k+\theta''_k)}e^{\tilde{u}_3\phi''_k}e^{\tilde{u}_1(\psi''_k+\psi''_{k+3})}e^{\tilde{u}_3\phi''_{k+3}}e^{\tilde{u}_2\theta''_{k+3}}e^{-\tilde{u}_3\beta_k} \quad (10.442)$$

Note that the Euler angle sequences associated with the upper and lower spherical joints are selected differently (i.e. respectively, as 2-3-1 and 1-3-2) for the sake of kinematic convenience as it will be evidenced in the sequel. Here, it can be said that the major convenience of these sequences is that the indefinite angle pairs $\{\psi'_k, \psi'_{k+3}\}$ and $\{\psi''_k, \psi''_{k+3}\}$ that describe the insignificant spinning motions of the lower legs get combined into single equivalent spin angles.

(b) Tip Point Location Equations Through the Legs

$*$ Tip point location equations through the *lefty legs* for $k = 1, 2, 3$:

$$\vec{p} = \overrightarrow{OP} = \overrightarrow{OB_0} + \overrightarrow{B_0B_k} + \overrightarrow{B_kC_k} + \overrightarrow{C_kC'_k} + \overrightarrow{C'_kA'_k} + \overrightarrow{A'_kA_k} + \overrightarrow{A_kA_0} + \overrightarrow{A_0P} \Rightarrow$$

$$\vec{p} = h_0\vec{u}_3^{(0)} + b_0\vec{u}_1^{(0k)} + b_1\vec{u}_1^{(k)} + d_0\vec{u}_2^{(k)} + b_2\vec{u}_1^{(k+3)'} - d_0\vec{u}_2^{(7k)'} - b_7\vec{u}_1^{(7k)'} - h_7\vec{u}_3^{(7)} \quad (10.443)$$

Equation (10.443) can be written as the following matrix equation in the base frame.

$$\bar{p} = \bar{p}^{(0)} = h_0\bar{u}_3^{(0/0)} + b_0\bar{u}_1^{(0k/0)} + b_1\bar{u}_1^{(k/0)} + d_0\bar{u}_2^{(k/0)} + b_2\bar{u}_1^{(k+3/0)'} - d_0\bar{u}_2^{(7k/0)'} - b_7\bar{u}_1^{(7k/0)'} - h_7\bar{u}_3^{(7/0)} \Rightarrow$$

$$\bar{p} = h_0\bar{u}_3 + b_0\hat{C}^{(0,0k)}\bar{u}_1 + b_1\hat{C}^{(0,k)}\bar{u}_1 + d_0\hat{C}^{(0,k)}\bar{u}_2 + b_2\hat{C}^{(0,k+3)'}\bar{u}_1 - d_0\hat{C}^{(0,7k)'}\bar{u}_2 - b_7\hat{C}^{(0,7k)'}\bar{u}_1 - h_7\hat{C}^{(0,7)}\bar{u}_3 \Rightarrow$$

$$\begin{aligned}\bar{p} = {} & h_0\bar{u}_3 + b_0e^{\tilde{u}_3\beta_k}\bar{u}_1 + b_1e^{\tilde{u}_3\beta_k}e^{\tilde{u}_2\theta_k}\bar{u}_1 + d_0e^{\tilde{u}_3\beta_k}e^{\tilde{u}_2\theta_k}\bar{u}_2 \\ & + b_2e^{\tilde{u}_3\beta_k}e^{\tilde{u}_2(\theta_k+\theta'_k)}e^{\tilde{u}_3\phi'_k}e^{\tilde{u}_1\psi'_k}\bar{u}_1 \\ & - d_0e^{\tilde{u}_3\beta_k}e^{\tilde{u}_2(\theta_k+\theta'_k)}e^{\tilde{u}_3\phi'_k}e^{\tilde{u}_1(\psi'_k+\psi'_{k+3})}e^{\tilde{u}_3\phi'_{k+3}}e^{\tilde{u}_2\theta'_{k+3}}\bar{u}_2 \\ & - b_7e^{\tilde{u}_3\beta_k}e^{\tilde{u}_2(\theta_k+\theta'_k)}e^{\tilde{u}_3\phi'_k}e^{\tilde{u}_1(\psi'_k+\psi'_{k+3})}e^{\tilde{u}_3\phi'_{k+3}}e^{\tilde{u}_2\theta'_{k+3}}\bar{u}_1 \\ & - h_7e^{\tilde{u}_3\beta_k}e^{\tilde{u}_2(\theta_k+\theta'_k)}e^{\tilde{u}_3\phi'_k}e^{\tilde{u}_1(\psi'_k+\psi'_{k+3})}e^{\tilde{u}_3\phi'_{k+3}}e^{\tilde{u}_2\theta'_{k+3}}e^{-\tilde{u}_3\beta_k}\bar{u}_3 \Rightarrow\end{aligned}$$

$$\begin{aligned}\bar{p} = {} & h_0\bar{u}_3 + b_0e^{\tilde{u}_3\beta_k}\bar{u}_1 + b_1e^{\tilde{u}_3\beta_k}e^{\tilde{u}_2\theta_k}\bar{u}_1 + d_0e^{\tilde{u}_3\beta_k}\bar{u}_2 \\ & + b_2e^{\tilde{u}_3\beta_k}e^{\tilde{u}_2(\theta_k+\theta'_k)}e^{\tilde{u}_3\phi'_k}\bar{u}_1 \\ & - d_0e^{\tilde{u}_3\beta_k}e^{\tilde{u}_2(\theta_k+\theta'_k)}e^{\tilde{u}_3\phi'_k}e^{\tilde{u}_1(\psi'_k+\psi'_{k+3})}e^{\tilde{u}_3\phi'_{k+3}}\bar{u}_2 \\ & - b_7e^{\tilde{u}_3\beta_k}e^{\tilde{u}_2(\theta_k+\theta'_k)}e^{\tilde{u}_3\phi'_k}e^{\tilde{u}_1(\psi'_k+\psi'_{k+3})}e^{\tilde{u}_3\phi'_{k+3}}e^{\tilde{u}_2\theta'_{k+3}}\bar{u}_1 \\ & - h_7e^{\tilde{u}_3\beta_k}e^{\tilde{u}_2(\theta_k+\theta'_k)}e^{\tilde{u}_3\phi'_k}e^{\tilde{u}_1(\psi'_k+\psi'_{k+3})}e^{\tilde{u}_3\phi'_{k+3}}e^{\tilde{u}_2\theta'_{k+3}}\bar{u}_3 \Rightarrow\end{aligned}$$

$$\begin{aligned}
\overline{p} &= h_0\overline{u}_3 + e^{\tilde{u}_3\beta_k}[\overline{u}_1(b_0 + b_1 c\theta_k) + \overline{u}_2 d_0 - \overline{u}_3 b_1 s\theta_k] \\
&+ b_2 e^{\tilde{u}_3\beta_k} e^{\tilde{u}_2(\theta_k+\theta'_k)} e^{\tilde{u}_3\phi'_k}\overline{u}_1 \\
&- e^{\tilde{u}_3\beta_k} e^{\tilde{u}_2(\theta_k+\theta'_k)} e^{\tilde{u}_3\phi'_k} e^{\tilde{u}_1(\psi'_k+\psi'_{k+3})} e^{\tilde{u}_3\phi'_{k+3}} e^{\tilde{u}_2\theta'_{k+3}}(\overline{u}_1 b_7 + \overline{u}_2 d_0 + \overline{u}_3 h_7)
\end{aligned} \tag{10.444}$$

Equation (10.444) can also be written more compactly as follows upon inserting the matrix $\widehat{C}$ given by Eq. (10.441).

$$\begin{aligned}
\overline{p} &= h_0\overline{u}_3 + e^{\tilde{u}_3\beta_k}[\overline{u}_1(b_0 + b_1 c\theta_k) + \overline{u}_2 d_0 - \overline{u}_3 b_1 s\theta_k] \\
&+ b_2 e^{\tilde{u}_3\beta_k} e^{\tilde{u}_2(\theta_k+\theta'_k)} e^{\tilde{u}_3\phi'_k}\overline{u}_1 - \widehat{C} e^{\tilde{u}_3\beta_k}(\overline{u}_1 b_7 + \overline{u}_2 d_0 + \overline{u}_3 h_7)
\end{aligned} \tag{10.445}$$

* Tip point location equations through the righty legs for $k = 1, 2, 3$:

$$\begin{aligned}
\vec{p} &= \overrightarrow{OP} = \overrightarrow{OB_0} + \overrightarrow{B_0B_k} + \overrightarrow{B_kC_k} + \overrightarrow{C_kC''_k} + \overrightarrow{C''_kA''_k} + \overrightarrow{A''_kA_k} + \overrightarrow{A_kA_0} + \overrightarrow{A_0P} \Rightarrow \\
\vec{p} &= h_0\vec{u}_3^{(0)} + b_0\vec{u}_1^{(0k)} + b_1\vec{u}_1^{(k)} - d_0\vec{u}_2^{(k)} + b_2\vec{u}_1^{(k+3)''} \\
&+ d_0\vec{u}_2^{(7k)''} - b_7\vec{u}_1^{(7k)''} - h_7\vec{u}_3^{(7)}
\end{aligned} \tag{10.446}$$

Equation (10.446) can be written as the following matrix equation in the base frame.

$$\begin{aligned}
\overline{p} &= \overline{p}^{(0)} = h_0\overline{u}_3^{(0/0)} + b_0\overline{u}_1^{(0k/0)} + b_1\overline{u}_1^{(k/0)} - d_0\overline{u}_2^{(k/0)} + b_2\overline{u}_1^{(k+3/0)''} \\
&+ d_0\overline{u}_2^{(7k/0)''} - b_7\overline{u}_1^{(7k/0)''} - h_7\overline{u}_3^{(7/0)} \Rightarrow \\
\overline{p} &= h_0\overline{u}_3 + b_0\widehat{C}^{(0,0k)}\overline{u}_1 + b_1\widehat{C}^{(0,k)}\overline{u}_1 - d_0\widehat{C}^{(0,k)}\overline{u}_2 \\
&+ b_2\widehat{C}^{(0,k+3)''}\overline{u}_1 + d_0\widehat{C}^{(0,7k)''}\overline{u}_2 - b_7\widehat{C}^{(0,7k)''}\overline{u}_1 - h_7\widehat{C}^{(0,7)}\overline{u}_3 \Rightarrow \\
\overline{p} &= h_0\overline{u}_3 + b_0 e^{\tilde{u}_3\beta_k}\overline{u}_1 + b_1 e^{\tilde{u}_3\beta_k} e^{\tilde{u}_2\theta_k}\overline{u}_1 - d_0 e^{\tilde{u}_3\beta_k} e^{\tilde{u}_2\theta_k}\overline{u}_2 \\
&+ b_2 e^{\tilde{u}_3\beta_k} e^{\tilde{u}_2(\theta_k+\theta''_k)} e^{\tilde{u}_3\phi''_k} e^{\tilde{u}_1\psi''_k}\overline{u}_1 \\
&+ d_0 e^{\tilde{u}_3\beta_k} e^{\tilde{u}_2(\theta_k+\theta''_k)} e^{\tilde{u}_3\phi''_k} e^{\tilde{u}_1(\psi''_k+\psi''_{k+3})} e^{\tilde{u}_3\phi''_{k+3}} e^{\tilde{u}_2\theta''_{k+3}}\overline{u}_2 \\
&- b_7 e^{\tilde{u}_3\beta_k} e^{\tilde{u}_2(\theta_k+\theta''_k)} e^{\tilde{u}_3\phi''_k} e^{\tilde{u}_1(\psi''_k+\psi''_{k+3})} e^{\tilde{u}_3\phi''_{k+3}} e^{\tilde{u}_2\theta''_{k+3}}\overline{u}_1 \\
&- h_7 e^{\tilde{u}_3\beta_k} e^{\tilde{u}_2(\theta_k+\theta''_k)} e^{\tilde{u}_3\phi''_k} e^{\tilde{u}_1(\psi''_k+\psi''_{k+3})} e^{\tilde{u}_3\phi''_{k+3}} e^{\tilde{u}_2\theta''_{k+3}} e^{-\tilde{u}_3\beta_k}\overline{u}_3 \Rightarrow \\
\overline{p} &= h_0\overline{u}_3 + b_0 e^{\tilde{u}_3\beta_k}\overline{u}_1 + b_1 e^{\tilde{u}_3\beta_k} e^{\tilde{u}_2\theta_k}\overline{u}_1 - d_0 e^{\tilde{u}_3\beta_k}\overline{u}_2 \\
&+ b_2 e^{\tilde{u}_3\beta_k} e^{\tilde{u}_2(\theta_k+\theta''_k)} e^{\tilde{u}_3\phi''_k}\overline{u}_1 \\
&+ d_0 e^{\tilde{u}_3\beta_k} e^{\tilde{u}_2(\theta_k+\theta''_k)} e^{\tilde{u}_3\phi''_k} e^{\tilde{u}_1(\psi''_k+\psi''_{k+3})} e^{\tilde{u}_3\phi''_{k+3}}\overline{u}_2 \\
&- b_7 e^{\tilde{u}_3\beta_k} e^{\tilde{u}_2(\theta_k+\theta''_k)} e^{\tilde{u}_3\phi''_k} e^{\tilde{u}_1(\psi''_k+\psi''_{k+3})} e^{\tilde{u}_3\phi''_{k+3}} e^{\tilde{u}_2\theta''_{k+3}}\overline{u}_1 \\
&- h_7 e^{\tilde{u}_3\beta_k} e^{\tilde{u}_2(\theta_k+\theta''_k)} e^{\tilde{u}_3\phi''_k} e^{\tilde{u}_1(\psi''_k+\psi''_{k+3})} e^{\tilde{u}_3\phi''_{k+3}} e^{\tilde{u}_2\theta''_{k+3}}\overline{u}_3 \Rightarrow \\
\overline{p} &= h_0\overline{u}_3 + e^{\tilde{u}_3\beta_k}[\overline{u}_1(b_0 + b_1 c\theta_k) - \overline{u}_2 d_0 - \overline{u}_3 b_1 s\theta_k] \\
&+ b_2 e^{\tilde{u}_3\beta_k} e^{\tilde{u}_2(\theta_k+\theta''_k)} e^{\tilde{u}_3\phi''_k}\overline{u}_1 \\
&- e^{\tilde{u}_3\beta_k} e^{\tilde{u}_2(\theta_k+\theta''_k)} e^{\tilde{u}_3\phi''_k} e^{\tilde{u}_1(\psi''_k+\psi''_{k+3})} e^{\tilde{u}_3\phi''_{k+3}} e^{\tilde{u}_2\theta''_{k+3}}(\overline{u}_1 b_7 - \overline{u}_2 d_0 + \overline{u}_3 h_7)
\end{aligned} \tag{10.447}$$

Equation (10.447) can also be written more compactly as follows upon inserting the matrix $\widehat{C}$ given by Eq. (10.442).

$$\begin{aligned}
\overline{p} &= h_0\overline{u}_3 + e^{\tilde{u}_3\beta_k}[\overline{u}_1(b_0 + b_1 c\theta_k) - \overline{u}_2 d_0 - \overline{u}_3 b_1 s\theta_k] \\
&+ b_2 e^{\tilde{u}_3\beta_k} e^{\tilde{u}_2(\theta_k+\theta''_k)} e^{\tilde{u}_3\phi''_k}\overline{u}_1 - \widehat{C} e^{\tilde{u}_3\beta_k}(\overline{u}_1 b_7 - \overline{u}_2 d_0 + \overline{u}_3 h_7)
\end{aligned} \tag{10.448}$$

10.9.3 Independent Kinematic Loops and the Associated Equations

(a) Independent Kinematic Loops

According to Eq. (10.2), the number of the IKLs is

$$n_{ikl} = j_1 + j_2 + j_3 - n_m = 3 + 0 + 12 - 10 = 5 \tag{10.449}$$

The five IKLs can be denoted and formed as indicated below.

* IKL-1 $= B_0B_1C_1C_1'A_1'A_0A_2'C_2'C_2B_2B_0$: Formed basically by the first and second lefty legs.
* IKL-2 $= B_0B_1C_1C_1'A_1'A_0A_3'C_3'C_3B_3B_0$: Formed basically by the first and third lefty legs.
* IKL-3 $= C_1'A_1'A_1''C_1''C_1'$: Formed basically by the pair of first lefty and righty legs.
* IKL-4 $= C_2'A_2'A_2''C_2''C_2'$: Formed basically by the pair of second lefty and righty legs.
* IKL-5 $= C_3'A_3'A_3''C_3''C_3'$: Formed basically by the pair of third lefty and righty legs.

(b) Orientation Equations for the IKLs

These loop equations are generated as shown below by equating the right-hand side expressions that are obtained by writing Eqs. (10.441) and (10.442) for the legs that form the loops.

* Orientation equation for IKL-1 ($k = 1, 2$ with the lefty legs):

$$\begin{aligned} &e^{\tilde{u}_3\beta_2} e^{\tilde{u}_2(\theta_2+\theta_2')} e^{\tilde{u}_3\phi_2'} e^{\tilde{u}_1(\psi_2'+\psi_5')} e^{\tilde{u}_3\phi_5'} e^{\tilde{u}_2\theta_5'} e^{-\tilde{u}_3\beta_2} \\ &= e^{\tilde{u}_3\beta_1} e^{\tilde{u}_2(\theta_1+\theta_1')} e^{\tilde{u}_3\phi_1'} e^{\tilde{u}_1(\psi_1'+\psi_4')} e^{\tilde{u}_3\phi_4'} e^{\tilde{u}_2\theta_4'} e^{-\tilde{u}_3\beta_1} \end{aligned} \tag{10.450}$$

* Orientation equation for IKL-2 ($k = 1, 3$ with the lefty legs):

$$\begin{aligned} &e^{\tilde{u}_3\beta_3} e^{\tilde{u}_2(\theta_3+\theta_3')} e^{\tilde{u}_3\phi_3'} e^{\tilde{u}_1(\psi_3'+\psi_6')} e^{\tilde{u}_3\phi_6'} e^{\tilde{u}_2\theta_6'} e^{-\tilde{u}_3\beta_3} \\ &= e^{\tilde{u}_3\beta_1} e^{\tilde{u}_2(\theta_1+\theta_1')} e^{\tilde{u}_3\phi_1'} e^{\tilde{u}_1(\psi_1'+\psi_4')} e^{\tilde{u}_3\phi_4'} e^{\tilde{u}_2\theta_4'} e^{-\tilde{u}_3\beta_1} \end{aligned} \tag{10.451}$$

* Orientation equation for IKL-3 ($k = 1$ with the lefty and righty legs):

$$\begin{aligned} &e^{\tilde{u}_2\theta_1''} e^{\tilde{u}_3\phi_1''} e^{\tilde{u}_1(\psi_1''+\psi_4'')} e^{\tilde{u}_3\phi_4''} e^{\tilde{u}_2\theta_4''} \\ &= e^{\tilde{u}_2\theta_1'} e^{\tilde{u}_3\phi_1'} e^{\tilde{u}_1(\psi_1'+\psi_4')} e^{\tilde{u}_3\phi_4'} e^{\tilde{u}_2\theta_4'} \end{aligned} \tag{10.452}$$

* Orientation equation for IKL-4 ($k = 2$ with the lefty and righty legs):

$$\begin{aligned} &e^{\tilde{u}_2\theta_2''} e^{\tilde{u}_3\phi_2''} e^{\tilde{u}_1(\psi_2''+\psi_5'')} e^{\tilde{u}_3\phi_5''} e^{\tilde{u}_2\theta_5''} \\ &= e^{\tilde{u}_2\theta_2'} e^{\tilde{u}_3\phi_2'} e^{\tilde{u}_1(\psi_2'+\psi_5')} e^{\tilde{u}_3\phi_5'} e^{\tilde{u}_2\theta_5'} \end{aligned} \tag{10.453}$$

* Orientation equation for IKL-5 ($k = 3$ with the lefty and righty legs):

$$\begin{aligned} &e^{\tilde{u}_2\theta_3''} e^{\tilde{u}_3\phi_3''} e^{\tilde{u}_1(\psi_3''+\psi_6'')} e^{\tilde{u}_3\phi_6''} e^{\tilde{u}_2\theta_6''} \\ &= e^{\tilde{u}_2\theta_3'} e^{\tilde{u}_3\phi_3'} e^{\tilde{u}_1(\psi_3'+\psi_6')} e^{\tilde{u}_3\phi_6'} e^{\tilde{u}_2\theta_6'} \end{aligned} \tag{10.454}$$

(c) Characteristic Point Location Equations for the IKLs

These loop equations are generated as shown below by equating the right-hand side expressions that are obtained by writing Eqs. (10.445) and (10.448) for the basic legs that form the loops.

∗ Characteristic point location equation for IKL-1 ($k = 1, 2$ with the lefty legs):

$$\begin{aligned} & e^{\tilde{u}_3\beta_2}[\overline{u}_1(b_0 + b_1 c\theta_2) + \overline{u}_2 d_0 - \overline{u}_3 b_1 s\theta_2] \\ & + b_2 e^{\tilde{u}_3\beta_2} e^{\tilde{u}_2(\theta_2+\theta'_2)} e^{\tilde{u}_3\phi'_2}\overline{u}_1 - \widehat{C} e^{\tilde{u}_3\beta_2}(\overline{u}_1 b_7 + \overline{u}_2 d_0 + \overline{u}_3 h_7) \\ & = e^{\tilde{u}_3\beta_1}[\overline{u}_1(b_0 + b_1 c\theta_1) + \overline{u}_2 d_0 - \overline{u}_3 b_1 s\theta_1] \\ & + b_2 e^{\tilde{u}_3\beta_1} e^{\tilde{u}_2(\theta_1+\theta'_1)} e^{\tilde{u}_3\phi'_1}\overline{u}_1 - \widehat{C} e^{\tilde{u}_3\beta_1}(\overline{u}_1 b_7 + \overline{u}_2 d_0 + \overline{u}_3 h_7) \end{aligned} \tag{10.455}$$

∗ Characteristic point location equation for IKL-2 ($k = 1, 3$ with the lefty legs):

$$\begin{aligned} & e^{\tilde{u}_3\beta_3}[\overline{u}_1(b_0 + b_1 c\theta_3) + \overline{u}_2 d_0 - \overline{u}_3 b_1 s\theta_3] \\ & + b_2 e^{\tilde{u}_3\beta_3} e^{\tilde{u}_2(\theta_3+\theta'_3)} e^{\tilde{u}_3\phi'_3}\overline{u}_1 - \widehat{C} e^{\tilde{u}_3\beta_3}(\overline{u}_1 b_7 + \overline{u}_2 d_0 + \overline{u}_3 h_7) \\ & = e^{\tilde{u}_3\beta_1}[\overline{u}_1(b_0 + b_1 c\theta_1) + \overline{u}_2 d_0 - \overline{u}_3 b_1 s\theta_1] \\ & + b_2 e^{\tilde{u}_3\beta_1} e^{\tilde{u}_2(\theta_1+\theta'_1)} e^{\tilde{u}_3\phi'_1}\overline{u}_1 - \widehat{C} e^{\tilde{u}_3\beta_1}(\overline{u}_1 b_7 + \overline{u}_2 d_0 + \overline{u}_3 h_7) \end{aligned} \tag{10.456}$$

∗ Characteristic point location equations for IKL-3, IKL-4, and IKL-5:

These loop equations are generated by using Eqs. (10.445) and (10.448) for each pair of the lefty and righty legs, i.e. for $k = 1, 2, 3$. The mentioned equations lead to the following loop equation for the kth leg pair upon carrying out the necessary simplifications.

$$b_2 e^{\tilde{u}_2\theta_k}[e^{\tilde{u}_2\theta''_k} e^{\tilde{u}_3\phi''_k} - e^{\tilde{u}_2\theta'_k} e^{\tilde{u}_3\phi'_k}]\overline{u}_1 = 2d_0[\hat{I} - e^{-\tilde{u}_3\beta_k}\widehat{C} e^{\tilde{u}_3\beta_k}]\overline{u}_2 \tag{10.457}$$

When Eq. (10.441) of $\widehat{C}$ is substituted and the necessary simplifications are carried out, Eq. (10.457) becomes

$$\begin{aligned} & b_2[e^{\tilde{u}_2\theta''_k} e^{\tilde{u}_3\phi''_k} - e^{\tilde{u}_2\theta'_k} e^{\tilde{u}_3\phi'_k}]\overline{u}_1 \\ & = 2d_0 e^{\tilde{u}_2\theta'_k}[\hat{I} - e^{\tilde{u}_3\phi'_k} e^{\tilde{u}_1(\psi'_k+\psi'_{k+3})} e^{\tilde{u}_3\phi'_{k+3}}]\overline{u}_2 \end{aligned} \tag{10.458}$$

Equation (10.458) must be satisfied for all possible values of θ'_k. This necessity leads to the following equalities.

$$e^{\tilde{u}_2\theta''_k} e^{\tilde{u}_3\phi''_k} = e^{\tilde{u}_2\theta'_k} e^{\tilde{u}_3\phi'_k} \tag{10.459}$$

$$e^{\tilde{u}_3\phi'_k} e^{\tilde{u}_1(\psi'_k+\psi'_{k+3})} e^{\tilde{u}_3\phi'_{k+3}} = \hat{I} \tag{10.460}$$

The above equalities can in turn be satisfied if

$$\theta''_k = \theta'_k \tag{10.461}$$

$$\phi''_k = \phi'_k \tag{10.462}$$

$$\psi'_k + \psi'_{k+3} = 0 \tag{10.463}$$

$$\phi'_{k+3} = -\phi'_k \tag{10.464}$$

On the other hand, if Eq. (10.442) of $\widehat{C}$ were substituted into Eq. (10.457), the following companion of Eq. (10.460) would be obtained together with its similar implications.

$$e^{\tilde{u}_3\phi''_k} e^{\tilde{u}_1(\psi''_k+\psi''_{k+3})} e^{\tilde{u}_3\phi''_{k+3}} = \hat{I} \tag{10.465}$$

$$\psi''_k + \psi''_{k+3} = 0 \tag{10.466}$$

$$\phi''_{k+3} = -\phi''_k \tag{10.467}$$

Note that ψ'_k and ψ'_{k+3} are the spherical joint angles that correspond to the insignificant free spinning rotation of the lower leg $C'_kA'_k$ about its centerline. Similarly, ψ''_k and ψ''_{k+3} are the spherical joint angles that correspond to the insignificant free spinning rotation of the lower leg $C''_kA''_k$ about its centerline. Naturally, these spin angles may assume arbitrary values but they must add up to zero according to Eqs. (10.463) and (10.466). In other words, the net rotation of the moving platform with respect to the upper link of the leg L_k must be zero about C_kA_k.

Moreover, all the above equations imply that the leg segments $C'_kA'_k$ and $C''_kA''_k$ remain parallel to each other together with the leg chords $C'_kC''_k$ and $A'_kA''_k$.

Another consequence of Eqs. (10.460) and (10.465) is that Eqs. (10.441) and (10.442) of the orientation matrix $\hat{C}$ of the moving platform become simplified as seen below.

$$\hat{C} = e^{\tilde{u}_3\beta_k}e^{\tilde{u}_2(\theta_k+\theta'_k+\theta'_{k+3})}e^{-\tilde{u}_3\beta_k} \tag{10.468}$$

$$\hat{C} = e^{\tilde{u}_3\beta_k}e^{\tilde{u}_2(\theta_k+\theta''_k+\theta''_{k+3})}e^{-\tilde{u}_3\beta_k} \tag{10.469}$$

Equations (10.468) and (10.469) must both give the same $\hat{C}$ for all of the relevant joint angles and for all $k = 1, 2, 3$. This necessity can be satisfied if

$$\theta'_{k+3} = -(\theta_k + \theta'_k),\ \ \theta''_{k+3} = -(\theta_k + \theta''_k) \tag{10.470}$$

Moreover, since the lower leg segments are always parallel to each other,

$$\theta''_k = \theta'_k,\ \ \theta''_{k+3} = \theta'_{k+3} \tag{10.471}$$

Eq. Pairs (10.470) and (10.471) lead to the conclusion that

$$\hat{C} = \hat{I} \tag{10.472}$$

Equation (10.472) verifies that the moving platform of the delta robot cannot have any angular motion; it can only move translationally. Therefore, it is sufficient to express the location of the tip point in order to describe the position of the end-effector. For this purpose, Eq. (10.445) can be written as follows for the three lefty legs (i.e. for $k = 1, 2, 3$) with $\hat{C} = \hat{I}$:

$$\overline{p} = e^{\tilde{u}_3\beta_k}[\overline{u}_1(b_0 - b_7 + b_1c\theta_k) + \overline{u}_3(h_0 - h_7 - b_1s\theta_k) + b_2e^{\tilde{u}_2\theta^*_k}e^{\tilde{u}_3\phi'_k}\overline{u}_1] \tag{10.473}$$

In Eq. (10.473),

$$\theta^*_k = \theta_k + \theta'_k \tag{10.474}$$

Note that it is sufficient to express $\overline{p}$ for the lefty legs as done above, because its expression for the righty legs would be exactly the same owing to Eqs. (10.461) and (10.462)

As for the nontrivial loop equations, there remains only two of them, which are written below as the simplified forms of Eqs. (10.455) and (10.456) with $\hat{C} = \hat{I}$ and $\beta_1 = 0$.

* Characteristic point location equation for IKL-1 ($k = 1, 2$ with the lefty legs):

$$\begin{aligned} &e^{\tilde{u}_3\beta_2}[\overline{u}_1(b_0 - b_7 + b_1c\theta_2) - \overline{u}_3(b_1s\theta_2) + b_2e^{\tilde{u}_2\theta^*_2}e^{\tilde{u}_3\phi'_2}\overline{u}_1] \\ &= \overline{u}_1(b_0 - b_7 + b_1c\theta_1) - \overline{u}_3(b_1s\theta_1) + b_2e^{\tilde{u}_2\theta^*_1}e^{\tilde{u}_3\phi'_1}\overline{u}_1 \end{aligned} \tag{10.475}$$

* Characteristic point location equation for IKL-2 ($k = 1, 3$ with the lefty legs):

$$\begin{aligned} &e^{\tilde{u}_3\beta_3}[\overline{u}_1(b_0 - b_7 + b_1c\theta_3) - \overline{u}_3b_1s\theta_3 + b_2e^{\tilde{u}_2\theta^*_3}e^{\tilde{u}_3\phi'_3}\overline{u}_1]. \\ &= \overline{u}_1(b_0 - b_7 + b_1c\theta_1) - \overline{u}_3b_1s\theta_1 + b_2e^{\tilde{u}_2\theta^*_1}e^{\tilde{u}_3\phi'_1}\overline{u}_1 \end{aligned} \tag{10.476}$$

10.9.4 Inverse Kinematics in the Position Domain

(a) Inverse Kinematic Solution

For a specified position of the end-effector represented by $\overline{p}$, the corresponding active and passive joint variables $(\theta_k, \theta'_k, \phi'_k)$ can be found leg by leg (i.e. for $k = 1, 2, 3$) as explained below.

For the leg L_k, Eq. (10.473) can be written as follows:

$$b_1(\overline{u}_1 c\theta_k - \overline{u}_3 s\theta_k) + b_2 e^{\tilde{u}_2\theta^*_k} e^{\tilde{u}_3\phi'_k}\overline{u}_1 = \overline{r}_k \tag{10.477}$$

In Eq. (10.477), the column matrix $\overline{r}_k$ is known with the following expression.

$$\overline{r}_k = e^{-\tilde{u}_3\beta_k}\overline{p} - \overline{u}_1(b_0 - b_7) - \overline{u}_3(h_0 - h_7) \tag{10.478}$$

Note that $\overline{r}_k$ represents the position vector $\vec{r}_k = \overrightarrow{B_k A_k}$ in the *leg-specific base frame* $\mathcal{F}_{0k}(B_k)$ formed by the unit vectors $\vec{u}_1^{(0k)}$ and $\vec{u}_3^{(0k)} = \vec{u}_3^{(0)}$ as illustrated in Figures 10.17 and 10.18.

Upon premultiplications by $\overline{u}_1^t, \overline{u}_2^t$, and $\overline{u}_3^t$, Eq. (10.477) leads to the following three scalar equations, in which $r_{ki} = \overline{u}_i^t \overline{r}_k$ for $i = 1, 2, 3$.

$$b_1 c\theta_k + b_2 c\theta^*_k c\phi'_k = r_{k1} \tag{10.479}$$

$$b_2 s\phi'_k = r_{k2} \tag{10.480}$$

$$b_1 s\theta_k + b_2 s\theta^*_k c\phi'_k = -r_{k3} \tag{10.481}$$

Equation (10.480) gives ϕ'_k as follows with a sign variable $\sigma'_k = \pm 1$.

$$s\phi'_k = r_{k2}/b_2 \Rightarrow c\phi'_k = \sigma'_k\sqrt{b_2^2 - r_{k2}^2}/b_2 \Rightarrow$$

$$\phi'_k = \text{atan}_2(r_{k2}, \sigma'_k\sqrt{b_2^2 - r_{k2}^2}) \tag{10.482}$$

After finding ϕ'_k, Eqs. (10.479) and (10.481) can be written as follows:

$$b_2 c\theta^*_k c\phi'_k = r_{k1} - b_1 c\theta_k \tag{10.483}$$

$$b_2 s\theta^*_k c\phi'_k = -r_{k3} - b_1 s\theta_k \tag{10.484}$$

Squares of Eqs. (10.483) and (10.484) can be added side by side in order to obtain the following equation.

$$b_2^2(c\phi'_k)^2 = b_2^2 - r_{k2}^2 = r_{k1}^2 + r_{k3}^2 + b_1^2 - b_1 r_{k1} c\theta_k + b_1 r_{k3} s\theta_k \Rightarrow$$

$$r_{k1} c\theta_k - r_{k3} s\theta_k = f_k \tag{10.485}$$

In Eq. (10.485),

$$f_k = (r_{k1}^2 + r_{k2}^2 + r_{k3}^2 + b_1^2 - b_2^2)/b_1 \tag{10.486}$$

If r_{k1} and r_{k3} are not both zero, then Eq. (10.485) gives θ_k as follows with a second sign variable $\sigma_k = \pm 1$.

$$\theta_k = \theta^\circ_k + \sigma_k \gamma_k \tag{10.487}$$

In Eq. (10.487),

$$\theta^\circ_k = -\text{atan}_2\left(r_{k3}, r_{k1}\right) \tag{10.488}$$

$$\gamma_k = \text{atan}_2(\sqrt{r_{k1}^2 + r_{k3}^2 - f_k^2}, f_k) \tag{10.489}$$

After finding θ_k too, Eqs. (10.483) and (10.484) give θ_k^* as follows without a new sign variable but on condition that $c\phi_k' \neq 0$.

$$\theta_k^* = -\text{atan}_2[\sigma_k'(r_{k3} + b_1 s\theta_k), \sigma_k'(r_{k1} - b_1 c\theta_k)] \tag{10.490}$$

Finally, θ_k' is obtained from Eq. (10.474) as

$$\theta_k' = \theta_k^* - \theta_k \tag{10.491}$$

(b) Multiplicity Analysis of Inverse Kinematics

The inverse kinematic solution presented in Part (a) involves two sign variables σ_k and σ_k'. Their effects are discussed below.

The sign variable σ_k in Eq. (10.487) leads to two visually distinct posture modes for the leg L_k. These posture modes can be visualized by referring to Figure 10.19 and noting that θ_k is the angle of $\overrightarrow{B_kC_k}$ and θ_k° is the angle of the projection of $\vec{r}_k = \overrightarrow{B_kA_k}$ on the 1–3 plane of the frame $\mathcal{F}_{0k}(B_k)$. Depending on the two values of σ_k, let $\theta_k^+ = \theta_k^\circ + \gamma_k$ and $\theta_k^- = \theta_k^\circ - \gamma_k$. Then, the angles θ_k^+ and θ_k^- are ordered so that

$$\theta_k^- < \theta_k^\circ < \theta_k^+ \tag{10.492}$$

The posture modes with θ_k^+ and θ_k^- may, respectively, be designated as *inward knee posture mode* and *outward knee posture mode.* In the usual operations of the delta robot, though, the outward knee posture mode is preferred almost unexceptionally.

As for the sign variable σ_k' in Eqs. (10.482) and (10.490), it affects the angles ϕ_k' and θ_k^* as described below.

$$\sigma_k' = +1 \;\Rightarrow\; \phi_k' \text{ and } \theta_k^* \text{ but } \sigma_k' = -1 \;\Rightarrow\; \phi_k'' \text{ and } \theta_k^{**} \tag{10.493}$$

Here, ϕ_k' is an acute angle with positive cosine and ϕ_k'' is the complementary obtuse angle with negative cosine. That is,

$$\phi_k'' = \pi\,\text{sgn}(\phi_k') - \phi_k' \tag{10.494}$$

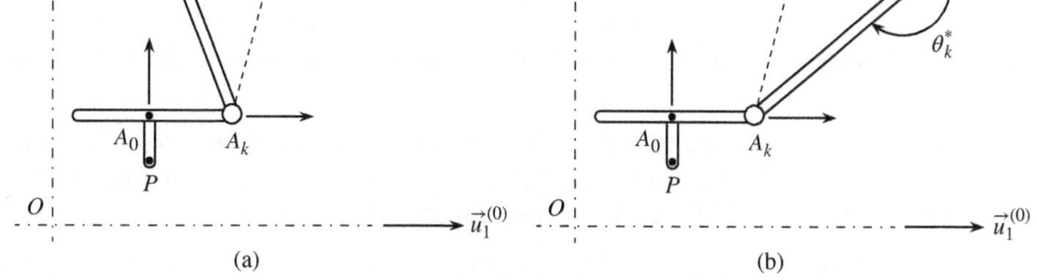

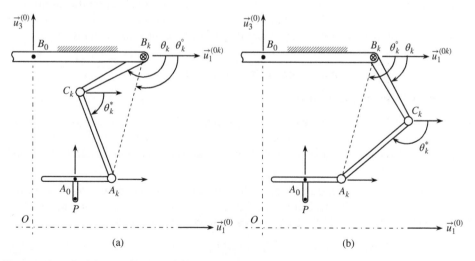

Figure 10.19 The (a) inward knee and (b) outward knee posture modes of the leg L_k.

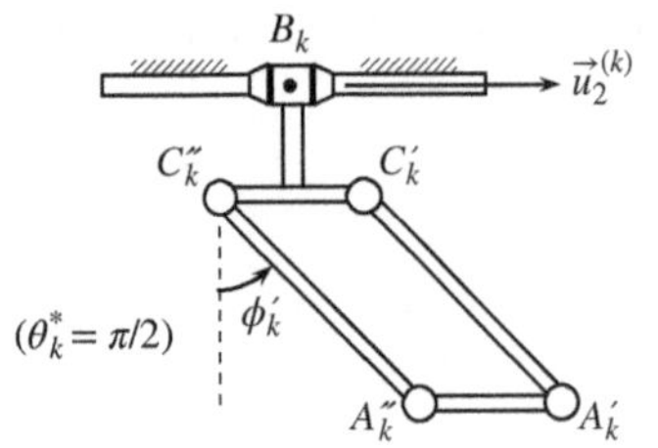

Figure 10.20 Illustration of the angle ϕ_k' when $\theta_k^* = \pi/2$.

On the other hand,

$$\theta_k^{**} = \theta_k^* + \pi \tag{10.495}$$

However, the pair $\{\phi_k', \theta_k^*\}$ led by $\sigma_k' = +1$ is not visually distinct from the pair $\{\phi_k'', \theta_k^{**}\}$ led by $\sigma_k' = -1$. Therefore, it is possible to take $\sigma_k' = +1$ without loss of generality.

The lateral appearance of the leg L_k viewed in the direction of $\vec{u}_1^{(0k)}$, when $\theta_k^* = \pi/2$, with a nonzero acute value of ϕ_k' is illustrated in Figure 10.20.

(c) Position Singularities of Inverse Kinematics

Equation (10.485) implies that the leg L_k may get into a pose of PSIK-1(k) if $r_{k1}^2 + r_{k3}^2 = 0$. In this singularity, the points A_k and B_k become two points located on the axis of the kth revolute (active) joint. Therefore, the active joint variable θ_k becomes indefinite and as such it can be assigned an arbitrary value without affecting the singularity position of the moving platform. In this position, the moving platform happens to be in planar contact with the fixed platform. However, in a normal usage, the delta robot is never allowed to assume such a pose.

Additionally, Eqs. (10.483) and (10.484) imply that L_k may also get into a pose of PSIK-2(k) if $c\phi_k' = 0$, i.e. if $\phi_k' = \pm\pi/2$. In this singularity, the lower links $C_k'A_k'$ and $C_k''A_k''$ get aligned with themselves and the chord $C_k'C_k''$ of L_k. Therefore, the angle θ_k^* becomes indefinite and thus it can assume an arbitrary value that has no effect on the mentioned pose of L_k. However, in a normal usage, the delta robot is never allowed to assume such a pose either in order to prevent the links of L_k from clashing with each other.

10.9.5 Forward Kinematics in the Position Domain

(a) Forward Kinematic Solution

For specified values of the active joint variables (i.e. $\theta_1, \theta_2, \theta_3$), the passive joint variables and the position of the end-effector (i.e. $\overline{p}$) can be found as explained below. Incidentally, it is fortunate that the forward kinematic solution for this manipulator can be obtained analytically.

In the first stage of the solution process, the passive joint variables are found from the equations of the IKLs. For this purpose, Eqs. (10.475) and (10.476) of the independent kinematic loops IKL-1 and IKL-2 can be rearranged as shown below.

$$b_2 e^{\tilde{u}_3\beta_2} e^{\tilde{u}_2\theta_2^*} e^{\tilde{u}_3\phi_2'} \overline{u}_1 = b_2 e^{\tilde{u}_2\theta_1^*} e^{\tilde{u}_3\phi_1'} \overline{u}_1 + \overline{f}_{12}(\theta_1, \theta_2) \tag{10.496}$$

$$b_2 e^{\tilde{u}_3\beta_3} e^{\tilde{u}_2\theta_3^*} e^{\tilde{u}_3\phi_3'} \overline{u}_1 = b_2 e^{\tilde{u}_2\theta_1^*} e^{\tilde{u}_3\phi_1'} \overline{u}_1 + \overline{f}_{13}(\theta_1, \theta_3) \tag{10.497}$$

In Eqs. (10.496) and (10.497),

$$\begin{aligned}\bar{f}_{12}(\theta_1,\theta_2) &= \bar{u}_1(b_0-b_7+b_1c\theta_1)-\bar{u}_3(b_1s\theta_1)\\ &\quad - e^{\tilde{u}_3\beta_2}[\bar{u}_1(b_0-b_7+b_1c\theta_2)-\bar{u}_3(b_1s\theta_2)] \Rightarrow\\ \bar{f}_{12}(\theta_1,\theta_2) &= \bar{u}_1[(b_0-b_7)(1-c\beta_2)+b_1(c\theta_1-c\beta_2c\theta_2)]\\ &\quad -\bar{u}_2(b_0-b_7+b_1c\theta_1)s\beta_2+\bar{u}_3b_1(s\theta_2-s\theta_1)\end{aligned} \tag{10.498}$$

$$\begin{aligned}\bar{f}_{13}(\theta_1,\theta_3) &= \bar{u}_1(b_0-b_7+b_1c\theta_1)-\bar{u}_3(b_1s\theta_1)\\ &\quad - e^{\tilde{u}_3\beta_3}[\bar{u}_1(b_0-b_7+b_1c\theta_3)-\bar{u}_3(b_1s\theta_3)] \Rightarrow\\ \bar{f}_{13}(\theta_1,\theta_3) &= \bar{u}_1[(b_0-b_7)(1-c\beta_3)+b_1(c\theta_1-c\beta_3c\theta_3)]\\ &\quad -\bar{u}_2(b_0-b_7+b_1c\theta_1)s\beta_3+\bar{u}_3b_1(s\theta_3-s\theta_1)\end{aligned} \tag{10.499}$$

In the same equations, $\theta_k^* = \theta_k + \theta_k'$ for $k = 1, 2, 3$ as defined before by Eq. (10.474).

Equations (10.496) and (10.497) lead to the following two scalar equations when they are premultiplied by their transposes and the necessary manipulations are carried out afterward.

$$\bar{f}_{12}^{t}e^{\tilde{u}_2\theta_1^*}e^{\tilde{u}_3\phi_1'}\bar{u}_1 = -(\bar{f}_{12}^{t}\bar{f}_{12})/(2b_2) = -h_{12} \tag{10.500}$$

$$\bar{f}_{13}^{t}e^{\tilde{u}_2\theta_1^*}e^{\tilde{u}_3\phi_1'}\bar{u}_1 = -(\bar{f}_{13}^{t}\bar{f}_{13})/(2b_2) = -h_{13} \tag{10.501}$$

The left-hand side of Eq. (10.500) can be expanded and manipulated as shown below.

$$\begin{aligned}&\bar{f}_{12}^{t}e^{\tilde{u}_2\theta_1^*}(\bar{u}_1c\phi_1'+\bar{u}_2s\phi_1') = -h_{12} \Rightarrow\\ &\bar{f}_{12}^{t}(\bar{u}_1c\theta_1^*c\phi_1'+\bar{u}_2s\phi_1'-\bar{u}_3s\theta_1^*c\phi_1') = -h_{12} \Rightarrow\\ &(f_{121}c\theta_1^*-f_{123}s\theta_1^*)c\phi_1'+f_{122}s\phi_1' = -h_{12}\end{aligned} \tag{10.502}$$

The left-hand side of Eq. (10.501) can also be expanded and manipulated similarly so that

$$(f_{131}c\theta_1^*-f_{133}s\theta_1^*)c\phi_1'+f_{132}s\phi_1' = -h_{13} \tag{10.503}$$

Equations (10.502) and (10.503) can be written together as the following matrix equation.

$$\begin{bmatrix}(f_{121}c\theta_1^*-f_{123}s\theta_1^*) & f_{122}\\ (f_{131}c\theta_1^*-f_{133}s\theta_1^*) & f_{132}\end{bmatrix}\begin{bmatrix}c\phi_1'\\ s\phi_1'\end{bmatrix} = -\begin{bmatrix}h_{12}\\ h_{13}\end{bmatrix} \tag{10.504}$$

In Eq. (10.504), the determinant of the coefficient matrix is

$$\begin{aligned}D &= D(\theta_1^*) = (f_{121}c\theta_1^*-f_{123}s\theta_1^*)f_{132}-(f_{131}c\theta_1^*-f_{133}s\theta_1^*)f_{122} \Rightarrow\\ D &= D(\theta_1^*) = (f_{121}f_{132}-f_{122}f_{131})c\theta_1^*+(f_{122}f_{133}-f_{123}f_{132})s\theta_1^*\end{aligned} \tag{10.505}$$

If $D \neq 0$, Eq. (10.504) can be solved as follows so as to obtain $c\phi_1'$ and $s\phi_1'$ in terms of θ_1^* :

$$\begin{bmatrix}c\phi_1'\\ s\phi_1'\end{bmatrix} = -\frac{1}{D}\begin{bmatrix}f_{132} & -f_{122}\\ -(f_{131}c\theta_1^*-f_{133}s\theta_1^*) & (f_{121}c\theta_1^*-f_{123}s\theta_1^*)\end{bmatrix}\begin{bmatrix}h_{12}\\ h_{13}\end{bmatrix} \Rightarrow$$

$$c\phi_1' = C_1/D = (f_{122}h_{13}-f_{132}h_{12})/D \tag{10.506}$$

$$\begin{aligned}s\phi_1' &= S_1/D = [(f_{131}c\theta_1^*-f_{133}s\theta_1^*)h_{12}-(f_{121}c\theta_1^*-f_{123}s\theta_1^*)h_{13}]/D \Rightarrow\\ s\phi_1' &= S_1/D = [(f_{131}h_{12}-f_{121}h_{13})c\theta_1^*+(f_{123}h_{13}-f_{133}h_{12})s\theta_1^*]/D\end{aligned} \tag{10.507}$$

Hence, ϕ_1' is found as a function of θ_1^*. That is,

$$\phi_1' = \phi_1'(\theta_1^*) = \text{atan}_2[(S_1/D),(C_1/D)] \tag{10.508}$$

The solution obtained above must satisfy the following *compatibility condition.*

$$(c\phi_1')^2 + (s\phi_1')^2 = 1 \;\Rightarrow\; C_1^2 + S_1^2 = D^2 \tag{10.509}$$

Upon the necessary substitutions, Eq. (10.509) can be written as shown below.

$$\begin{aligned}&(f_{122}h_{13} - f_{132}h_{12})^2 + [(f_{131}h_{12} - f_{121}h_{13})c\theta_1^* + (f_{123}h_{13} - f_{133}h_{12})s\theta_1^*]^2\\&= [(f_{121}f_{132} - f_{122}f_{131})c\theta_1^* + (f_{122}f_{133} - f_{123}f_{132})s\theta_1^*]^2\end{aligned} \tag{10.510}$$

Equation (10.510) can be manipulated into the following equation.

$$M_1(c\theta_1^*)^2 + N_1(s\theta_1^*)^2 + L_1(2s\theta_1^*c\theta_1^*) = F_1 \tag{10.511}$$

In Eq. (10.511),

$$F_1 = (f_{122}h_{13} - f_{132}h_{12})^2 \tag{10.512}$$

$$M_1 = (f_{121}f_{132} - f_{122}f_{131})^2 - (f_{131}h_{12} - f_{121}h_{13})^2 \tag{10.513}$$

$$N_1 = (f_{122}f_{133} - f_{123}f_{132})^2 - (f_{123}h_{13} - f_{133}h_{12})^2 \tag{10.514}$$

$$\begin{aligned}L_1 &= (f_{121}f_{132} - f_{122}f_{131})(f_{122}f_{133} - f_{123}f_{132})\\&\quad - (f_{131}h_{12} - f_{121}h_{13})(f_{123}h_{13} - f_{133}h_{12})\end{aligned} \tag{10.515}$$

Equation (10.511) can be manipulated further by using the following identities.

$$2s\theta_1^*c\theta_1^* \equiv \sin(2\theta_1^*) \tag{10.516}$$

$$(c\theta_1^*)^2 \equiv [1 + \cos(2\theta_1^*)]/2 \;\text{ and }\; (s\theta_1^*)^2 \equiv [1 - \cos(2\theta_1^*)]/2 \tag{10.517}$$

By means of the above identities, Eq. (10.511) becomes

$$K_1\cos(2\theta_1^*) + L_1\sin(2\theta_1^*) = G_1 \tag{10.518}$$

In Eq. (10.518),

$$K_1 = (M_1 - N_1)/2 \tag{10.519}$$

$$G_1 = F_1 - (M_1 + N_1)/2 \tag{10.520}$$

If K_1 and L_1 are not both zero, Eq. (10.518) gives θ_1^* as follows with a sign variable $\sigma_1 = \pm 1$.

$$\left.\begin{aligned}\sin(2\theta_1^*) &= (L_1G_1 + \sigma_1K_1H_1)/(K_1^2 + L_1^2)\\\cos(2\theta_1^*) &= (K_1G_1 - \sigma_1L_1H_1)/(K_1^2 + L_1^2)\end{aligned}\right\} \Rightarrow$$

$$\theta_1^* = [(L_1G_1 + \sigma_1K_1H_1),(K_1G_1 - \sigma_1L_1H_1)]/2 \tag{10.521}$$

In Eq. (10.521),

$$H_1 = \sqrt{K_1^2 + L_1^2 - G_1^2} \tag{10.522}$$

Then, going back to Eq. (10.508), ϕ_1' can also be obtained without an additional sign variable.

Afterwards, going further back to Eqs. (10.496) and (10.497) with θ_1^* and ϕ_1' at hand, they can be rearranged as follows in order to find θ_k^* and ϕ_k' for $k = 2$ and 3:

$$e^{\tilde{u}_3\phi_k'}\overline{u}_1 = e^{-\tilde{u}_2\theta_k^*}\overline{n}_k \tag{10.523}$$

In Eq. (10.523), $\overline{n}_k$ is known with the following expression.

$$\overline{n}_k = e^{-\tilde{u}_3\beta_k}(e^{\tilde{u}_2\theta_1^*}e^{\tilde{u}_3\phi_1'}\overline{u}_1 + \overline{f}_{1k}/b_2) \tag{10.524}$$

Upon premultiplications by $\overline{u}_1^t$, $\overline{u}_2^t$, and $\overline{u}_3^t$, Eq. (10.523) leads to the following three scalar equations, in which $n_{ki} = \overline{u}_i^t\overline{n}_k$ for $i = 1, 2, 3$.

$$c\phi_k' = n_{k1}c\theta_k^* - n_{k3}s\theta_k^* \tag{10.525}$$

$$s\phi_k' = n_{k2} \tag{10.526}$$

$$0 = n_{k3}c\theta_k^* + n_{k1}s\theta_k^* \tag{10.527}$$

If n_{k1} and n_{k3} are not both zero, Eq. (10.527) gives θ_k^* as follows with a sign variable $\sigma_k = \pm 1$.

$$\theta_k^* = \text{atan}_2(\sigma_k n_{k3}, -\sigma_k n_{k1}) \tag{10.528}$$

Then, with the availability of θ_k^*, Eqs. (10.525) and (10.525) give ϕ_k' without a new sign variable. That is,

$$\phi_k' = \text{atan}_2[n_{k2}, (n_{k1}c\theta_k^* - n_{k3}s\theta_k^*)] \tag{10.529}$$

Finally, referring to Eq. (10.491), θ_k' is also found as follows for $k = 1, 2, 3$:

$$\theta_k' = \theta_k^* - \theta_k \tag{10.530}$$

In the second stage of the forward kinematic solution, $\overline{p}$ can be obtained going through one of the three legs. For this manipulator, the first leg with $\beta_1 = 0$ is preferable for the sake of simplicity. Thus, the location of the tip point is determined as follows by using Eq. (10.473)

$$\overline{p} = \overline{u}_1(b_0 - b_7 + b_1c\theta_1) + \overline{u}_3(h_0 - h_7 - b_1s\theta_1) + b_2e^{\tilde{u}_2\theta_1^*}e^{\tilde{u}_3\phi_1'}\overline{u}_1 \tag{10.531}$$

(b) Multiplicity Analysis of Forward Kinematics

The forward kinematic solution presented in Part (a) involves three sign variables σ_1, σ_2, and σ_3. They directly affect the angles θ_1^*, θ_2^*, and θ_3^*, respectively. On the other hand, as implied by Figure 10.19, the sines of these angles (i.e. $s\theta_1^*$, $s\theta_2^*$, and $s\theta_3^*$) indicate whether the moving platform is located below the knee points (i.e. C_1, C_2, and C_3) or not. In other words, the moving platform will be below the knee point C_k if $s\theta_k^* > 0$. According to Eq. (10.528), such a pose can occur for the legs L_2 and L_3, i.e. for $k = 2$ and 3, if

$$\sigma_k = \text{sgn}(n_{k3}) \tag{10.532}$$

As for the first leg, Eq. (10.521) implies that such a pose can occur if

$$\sigma_1 = \text{sgn}(K_1) \tag{10.533}$$

However, as verified in Section 10.8.7, the moving platform remains always parallel to the fixed platform. Therefore, the *lower-tip-point* and *higher-tip-point* posture modes of the manipulator must be realized consistently by all the three legs. These two posture modes are illustrated in Figure 10.21, in which only the first and third legs of the

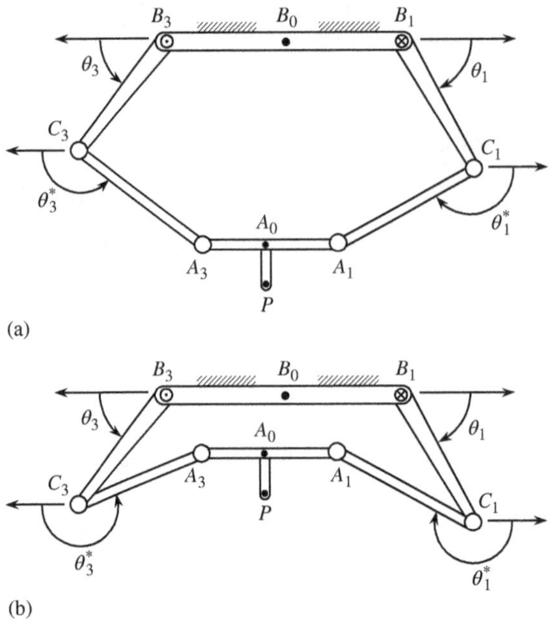

Figure 10.21 The (a) lower-tip-point and (b) higher-tip-point posture modes of the manipulator.

manipulator are taken into the view as if they are coplanar (i.e. as if $\beta_3 = -\pi$) for the sake of neatness. Here, it may be needless to say that the delta robot is of course preferred to be in the lower-tip-point posture mode in its customary operations.

(c) Position Singularities of Forward Kinematics

The forward kinematic solution presented in Part (a) stipulates three conditions in order to obtain a definite pose of the manipulator corresponding to a specified set of active joint variables. If these conditions are not satisfied, then the manipulator gets into a pose of PSFK. In such a pose, the passive joint variables become indefinite and therefore the manipulator becomes uncontrollable. The PSFK conditions are indicated below.

(i) According to Eq. (10.508), ϕ_1' becomes indefinite if

$$D = (f_{121}f_{132} - f_{122}f_{131})c\theta_1^* + (f_{122}f_{133} - f_{123}f_{132})s\theta_1^* = 0 \tag{10.534}$$

(ii) According to Eq. (10.518), θ_1^* becomes indefinite if

$$K_1 = L_1 = 0 \tag{10.535}$$

(iii) According to Eq. (10.528), for $k = 2$ and 3, θ_k^* becomes indefinite if

$$n_{k1} = n_{k3} = 0 \tag{10.536}$$

Actually, like D in Eq. (10.534), the coefficient pairs $\{K_1, L_1\}$ and $\{n_{k1}, n_{k3}\}$ are also expressed in terms of the components of $\overline{f}_{12} = \overline{f}_{12}(\theta_1, \theta_2)$ and $\overline{f}_{13} = \overline{f}_{13}(\theta_1, \theta_3)$ as seen in Part (a). Moreover, the relevant equations show that D and the mention coefficient pairs become all zero if $\overline{f}_{12} = \overline{f}_{13} = \overline{0}$.

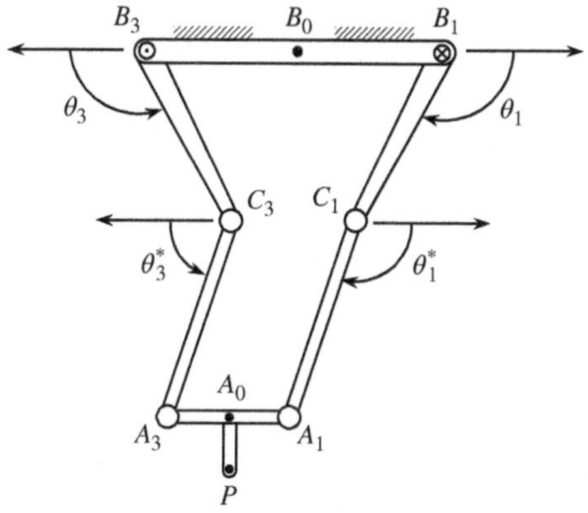

Figure 10.22 The PSFK of the manipulator.

On the other hand, Eqs. (10.498) and (10.499) show that $\overline{f}_{12}$ and $\overline{f}_{13}$ become zero if the following equalities are both satisfied.

$$\theta_3 = \theta_2 = \theta_1 \tag{10.537}$$

$$c\theta_1 = -(b_0 - b_7)/b_1 \tag{10.538}$$

Equation (10.537) indicates the PSFK relationships that arise among the active joint variables. As for Eq. (10.538), it gives the common PSFK value of the active joint variables.

On the other hand, as indicated before, when this PSFK occurs, the angles ϕ_1' and θ_k^* (for $k = 1, 2, 3$) become indefinite according to Eqs. (10.508), (10.518), and (10.528). Consequently, as also implied by the appearance of Figures 10.17 and 10.18 in accordance with Eqs. (10.537) and (10.538), all the six links of the lower legs become parallel to each other and therefore they can swing out of control with arbitrary values of the passive joint variables. Because of this uncontrollability, the mentioned PSFK must of course be avoided. This PSFK of the manipulator is illustrated in Figure 10.22, in which only the first and third legs of the manipulator are shown (again as if $\beta_3 = -\pi$) for the sake of neatness.

10.9.6 Velocity Equations

As verified in the previous position-domain analysis, the moving platform of the delta robot always remains parallel to the fixed platform. Therefore, $\hat{C} = \hat{C}^{(0,7)} = \hat{I}$ and $\overline{\omega} = \overline{\omega}_{7/0}^{(0)} = \overline{0}$. Consequently, Eq. (10.473) for the location of the tip point (i.e. $\overline{p}$) and Eqs. (10.475) and (10.476) for the major independent kinematic loops IKL-1 and IKL-2 happen to be sufficient to describe the kinematic behavior of the manipulator of this robot. Upon differentiation, the mentioned equations lead to the following velocity equations, in which

$$\theta_k^* = \theta_k + \theta_k' \text{ and } \dot{\theta}_k^* = \dot{\theta}_k + \dot{\theta}_k' \tag{10.539}$$

* Tip point velocity equation associated with the leg L_k for $k = 1, 2, 3$:

$$\begin{aligned}\overline{v} = \dot{\overline{p}} &= -b_1 e^{\tilde{u}_3\beta_k}[\overline{u}_1(\dot{\theta}_k s\theta_k) + \overline{u}_3(\dot{\theta}_k c\theta_k)] \\ &+ b_2 e^{\tilde{u}_3\beta_k}[\dot{\theta}_k^* e^{\tilde{u}_2\theta_k^*}\tilde{u}_2 e^{\tilde{u}_3\phi_k'}\overline{u}_1 + \dot{\phi}_k' e^{\tilde{u}_2\theta_k^*} e^{\tilde{u}_3\phi_k'}\tilde{u}_3\overline{u}_1] \Rightarrow \\ \overline{v} &= -b_1\dot{\theta}_k e^{\tilde{u}_3\beta_k}(\overline{u}_3 c\theta_k + \overline{u}_1 s\theta_k) \\ &+ b_2 e^{\tilde{u}_3\beta_k} e^{\tilde{u}_2\theta_k^*} e^{\tilde{u}_3\phi_k'}[\dot{\theta}_k^*(e^{-\tilde{u}_3\phi_k'}\tilde{u}_2 e^{\tilde{u}_3\phi_k'})\overline{u}_1 + \dot{\phi}_k'\overline{u}_2]\end{aligned}$$

Note that

$$\overline{u}_3 c\theta_k + \overline{u}_1 s\theta_k = e^{\tilde{u}_2\theta_k}\overline{u}_3$$

Note also that

$$e^{-\tilde{u}_3\phi_k'}\tilde{u}_2 e^{\tilde{u}_3\phi_k'} = \text{ssm}[e^{-\tilde{u}_3\phi_k'}\overline{u}_2] = \text{ssm}[\overline{u}_2 c\phi_k' + \overline{u}_1 s\phi_k'] = \tilde{u}_2 c\phi_k' + \tilde{u}_1 s\phi_k'$$

Hence, $\overline{v}$ can be expressed as

$$\overline{v} = -b_1\dot{\theta}_k e^{\tilde{u}_3\beta_k} e^{\tilde{u}_2\theta_k}\overline{u}_3 + b_2 e^{\tilde{u}_3\beta_k} e^{\tilde{u}_2\theta_k^*} e^{\tilde{u}_3\phi_k'}(\overline{u}_2\dot{\phi}_k' - \overline{u}_3\dot{\theta}_k^* c\phi_k') \tag{10.540}$$

Equation (10.540) can be used to generate the following velocity equations for the independent kinematic loops IKL-1 and IKL-2.

* Characteristic point velocity equation for IKL-1 ($k = 1, 2$ with the lefty legs):

$$\begin{aligned}&-b_1\dot{\theta}_2 e^{\tilde{u}_3\beta_2} e^{\tilde{u}_2\theta_2}\overline{u}_3 + b_2 e^{\tilde{u}_3\beta_2} e^{\tilde{u}_2\theta_2^*} e^{\tilde{u}_3\phi_2'}(\overline{u}_2\dot{\phi}_2' - \overline{u}_3\dot{\theta}_2^* c\phi_2') \\ &= -b_1\dot{\theta}_1 e^{\tilde{u}_2\theta_1}\overline{u}_3 + b_2 e^{\tilde{u}_2\theta_1^*} e^{\tilde{u}_3\phi_1'}(\overline{u}_2\dot{\phi}_1' - \overline{u}_3\dot{\theta}_1^* c\phi_1')\end{aligned} \tag{10.541}$$

* Characteristic point velocity equation for IKL-2 ($k = 1, 3$ with the lefty legs):

$$\begin{aligned}&-b_1\dot{\theta}_3 e^{\tilde{u}_3\beta_3} e^{\tilde{u}_2\theta_3}\overline{u}_3 + b_2 e^{\tilde{u}_3\beta_3} e^{\tilde{u}_2\theta_3^*} e^{\tilde{u}_3\phi_3'}(\overline{u}_2\dot{\phi}_3' - \overline{u}_3\dot{\theta}_3^* c\phi_3') \\ &= -b_1\dot{\theta}_1 e^{\tilde{u}_2\theta_1}\overline{u}_3 + b_2 e^{\tilde{u}_2\theta_1^*} e^{\tilde{u}_3\phi_1'}(\overline{u}_2\dot{\phi}_1' - \overline{u}_3\dot{\theta}_1^* c\phi_1')\end{aligned} \tag{10.542}$$

10.9.7 Inverse Kinematics in the Velocity Domain

(a) Inverse Velocity Solution

For a specified velocity $\overline{v}$ of the tip point at a currently known pose of the manipulator, the corresponding rates of the active and passive joint variables $(\dot{\theta}_k, \dot{\theta}_k', \dot{\phi}_k')$ can be found leg by leg (i.e. for $k = 1, 2, 3$) as explained below.

For the leg L_k, Eq. (10.540) can be written as follows by recalling that $\theta_k' = \theta_k^* - \theta_k$:

$$b_2 e^{\tilde{u}_3\phi_k'}(\overline{u}_2\dot{\phi}_k' - \overline{u}_3\dot{\theta}_k^* c\phi_k') - b_1\dot{\theta}_k e^{-\tilde{u}_2\theta_k'}\overline{u}_3 = \overline{v}^* \tag{10.543}$$

In Eq. (10.543),

$$\overline{v}^* = e^{-\tilde{u}_2\theta_k^*} e^{-\tilde{u}_3\beta_k}\overline{v} \tag{10.544}$$

Upon premultiplications by $\overline{u}_1^t$, $\overline{u}_2^t$, and $\overline{u}_3^t$, Eq. (10.476) leads to the following three scalar equations, in which $v_i^* = \overline{u}_i^t\overline{v}^*$ for $i = 1, 2, 3$.

$$b_1\dot{\theta}_k s\theta_k' - b_2\dot{\phi}_k' s\phi_k' = v_1^* \tag{10.545}$$

$$b_2\dot{\phi}_k' c\phi_k' = v_2^* \tag{10.546}$$

$$b_1\dot{\theta}_k c\theta_k' + b_2\dot{\theta}_k^* c\phi_k' = -v_3^* \tag{10.547}$$

Equation (10.546) gives $\dot{\phi}'_k$ as follows, if $c\phi'_k \neq 0$:

$$\dot{\phi}'_k = v_2^*/(b_2 c\phi'_k) \tag{10.548}$$

After finding $\dot{\phi}'_k$, Eq. (10.545) gives $\dot{\theta}_k$ as follows, if $s\theta'_k \neq 0$:

$$\dot{\theta}_k = (v_1^* + b_2\dot{\phi}'_k s\phi'_k)/(b_1 s\theta'_k) \tag{10.549}$$

After finding $\dot{\theta}_k$, Eq. (10.547) gives $\dot{\theta}^*_k$ as follows, if again $c\phi'_k \neq 0$:

$$\dot{\theta}^*_k = -(v_3^* + b_1\dot{\theta}_k c\theta'_k)/(b_2 c\phi'_k) \tag{10.550}$$

Finally, $\dot{\theta}'_k$ is also found as

$$\dot{\theta}'_k = \dot{\theta}^*_k - \dot{\theta}_k \tag{10.551}$$

(b) Motion Singularities of Inverse Kinematics

Equation (10.549) implies that the leg L_k gets into a pose of MSIK-1(k) if $s\theta'_k = 0$. In this singularity, the lower part of L_k gets aligned with its upper part, in an extended way if $\theta'_k = 0$ and in a folded way if $\theta'_k = \pm\pi$. However, the folded version is not likely to occur due to the possibility of clashing links. At the instant of this singularity $\dot{\theta}_k$ becomes indefinite. However, since it belongs to an active joint, a suitable finite value can still be assigned to it, provided that the velocity of the end-effector (i.e. $\overline{v}$) be specified so as to obey the following *consistency condition* that is implied by Eqs. (10.545) and (10.546).

$$v_1^* c\phi'_k + v_2^* s\phi'_k = 0 \tag{10.552}$$

In other words, if the motion of the end-effector is planned consistently with Eq. (10.552), then it is not necessary to avoid MSIK-1(k) of L_k. This singularity, which is quite likely to occur with $\theta'_k = 0$, is illustrated in Figure 10.23 by showing the first and third legs (again as if $\beta_3 = -\pi$) when the first leg is in singularity.

Incidentally, when Eqs. (10.445) and (10.441) are compared, it is seen that MSIK-1(k) also happens to be the PMCPL-k, i.e. the PMCPL of the leg L_k.

On the other hand, Eqs. (10.548) and (10.550) imply that the leg L_k may get into a pose of MSIK-2(k) if $c\phi'_k = 0$, i.e. if $\phi'_k = \pm\pi/2$. In this singularity, which looks the same as the PSIK-2(k) of L_k, the lower links $C'_kA'_k$ and $C''_kA''_k$ get aligned with themselves and the chord $C'_kC''_k$ of L_k. Therefore, the rates $\dot{\phi}'_k$ and $\dot{\theta}^*_k$ become indefinite. However, in a normal usage, the delta robot is never allowed to assume such a pose in order to prevent the links of L_k from clashing with each other.

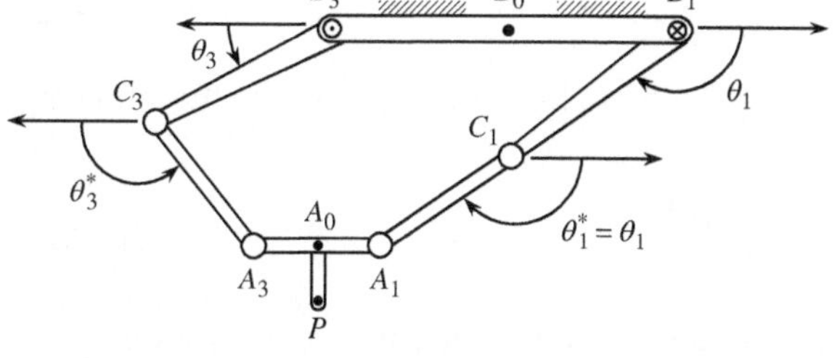

Figure 10.23 MSIK-1 of L_1 viewed together with L_3.

10.9.8 Forward Kinematics in the Velocity Domain

(a) Forward Velocity Solution

For specified rates of the active joint variables (i.e. $\dot{\theta}_1, \dot{\theta}_2, \dot{\theta}_3$), the rates of the passive joint variables and the velocity of the end-effector (i.e. $\overline{v}$) can be found as explained below.

In the first stage of the solution process, the rates of the passive joint variables are found from the velocity equations of the IKLs. For this purpose, Eqs. (10.541) and (10.542) of the IKLs IKL-1 and IKL-2 can be rearranged as shown below.

$$\begin{aligned}&\overline{u}_2\dot{\phi}_2' - \overline{u}_3\dot{\theta}_2^* c\phi_2' \\&= e^{-\tilde{u}_3\phi_2'} e^{-\tilde{u}_2\theta_2^*} e^{-\tilde{u}_3\beta_2} e^{\tilde{u}_2\theta_1^*} e^{\tilde{u}_3\phi_1'} (\overline{u}_2\dot{\phi}_1' - \overline{u}_3\dot{\theta}_1^* c\phi_1') \\&+\rho_{12} e^{-\tilde{u}_3\phi_2'} e^{-\tilde{u}_2\theta_2'} (\dot{\theta}_2\overline{u}_3 - \dot{\theta}_1 e^{-\tilde{u}_2\theta_2} e^{-\tilde{u}_3\beta_2} e^{\tilde{u}_2\theta_1} \overline{u}_3)\end{aligned} \tag{10.553}$$

$$\begin{aligned}&\overline{u}_2\dot{\phi}_3' - \overline{u}_3\dot{\theta}_3^* c\phi_3' \\&= e^{-\tilde{u}_3\phi_3'} e^{-\tilde{u}_2\theta_3^*} e^{-\tilde{u}_3\beta_3} e^{\tilde{u}_2\theta_1^*} e^{\tilde{u}_3\phi_1'} (\overline{u}_2\dot{\phi}_1' - \overline{u}_3\dot{\theta}_1^* c\phi_1') \\&+\rho_{12} e^{-\tilde{u}_3\phi_3'} e^{-\tilde{u}_2\theta_3'} (\dot{\theta}_3\overline{u}_3 - \dot{\theta}_1 e^{-\tilde{u}_2\theta_3} e^{-\tilde{u}_3\beta_3} e^{\tilde{u}_2\theta_1} \overline{u}_3)\end{aligned} \tag{10.554}$$

In Eqs. (10.553) and (10.554),

$$\rho_{12} = b_1/b_2 \tag{10.555}$$

Upon premultiplications by $\overline{u}_1^t$, $\overline{u}_2^t$, and $\overline{u}_3^t$, Eqs. (10.553) and (10.553) lead to the following scalar equations.

$$0 = A_{12}\dot{\phi}_1' - (B_{12}c\phi_1')\dot{\theta}_1^* + E_{12}\dot{\theta}_2 - F_{12}\dot{\theta}_1 \tag{10.556}$$

$$\dot{\phi}_2' = A_{22}\dot{\phi}_1' - (B_{22}c\phi_1')\dot{\theta}_1^* + E_{22}\dot{\theta}_2 - F_{22}\dot{\theta}_1 \tag{10.557}$$

$$-\dot{\theta}_2^* c\phi_2' = A_{32}\dot{\phi}_1' - (B_{32}c\phi_1')\dot{\theta}_1^* + E_{32}\dot{\theta}_2 - F_{32}\dot{\theta}_1 \tag{10.558}$$

$$0 = A_{13}\dot{\phi}_1' - (B_{13}c\phi_1')\dot{\theta}_1^* + E_{13}\dot{\theta}_3 - F_{13}\dot{\theta}_1 \tag{10.559}$$

$$\dot{\phi}_3' = A_{23}\dot{\phi}_1' - (B_{23}c\phi_1')\dot{\theta}_1^* + E_{23}\dot{\theta}_3 - F_{23}\dot{\theta}_1 \tag{10.560}$$

$$-\dot{\theta}_3^* c\phi_3' = A_{33}\dot{\phi}_1' - (B_{33}c\phi_1')\dot{\theta}_1^* + E_{33}\dot{\theta}_3 - F_{33}\dot{\theta}_1 \tag{10.561}$$

In Eqs. (10.556)–(10.561), for the indices $j = 1, 2, 3$ and $k = 2, 3$,

$$A_{jk} = \overline{u}_j^t e^{-\tilde{u}_3\phi_k'} e^{-\tilde{u}_2\theta_k^*} e^{-\tilde{u}_3\beta_k} e^{\tilde{u}_2\theta_1^*} e^{\tilde{u}_3\phi_1'} \overline{u}_2 \tag{10.562}$$

$$B_{jk} = \overline{u}_j^t e^{-\tilde{u}_3\phi_k'} e^{-\tilde{u}_2\theta_k^*} e^{-\tilde{u}_3\beta_k} e^{\tilde{u}_2\theta_1^*} \overline{u}_3 \tag{10.563}$$

$$E_{jk} = \rho_{12}\overline{u}_j^t e^{-\tilde{u}_3\phi_k'} e^{-\tilde{u}_2\theta_k'} \overline{u}_3 \tag{10.564}$$

$$F_{jk} = \rho_{12}\overline{u}_j^t e^{-\tilde{u}_3\phi_k'} e^{-\tilde{u}_2\theta_k^*} e^{-\tilde{u}_3\beta_k} e^{\tilde{u}_2\theta_1} \overline{u}_3 \tag{10.565}$$

To start the solution, Eqs. (10.556) and (10.559) can be written together as the following matrix equation.

$$\begin{bmatrix} A_{12} & -B_{12}c\phi_1' \\ A_{13} & -B_{13}c\phi_1' \end{bmatrix} \begin{bmatrix} \dot{\phi}_1' \\ \dot{\theta}_1^* \end{bmatrix} = \begin{bmatrix} F_{12}\dot{\theta}_1 - E_{12}\dot{\theta}_2 \\ F_{13}\dot{\theta}_1 - E_{13}\dot{\theta}_3 \end{bmatrix} \tag{10.566}$$

On the left-hand side of Eq. (10.566), the determinant of the coefficient matrix is

$$D_1 = (A_{13}B_{12} - A_{12}B_{13})c\phi_1' = D_0 c\phi_1' \tag{10.567}$$

If $D_1 = D_0 c\phi_1' \neq 0$, then the rates $\dot{\phi}_1'$ and $\dot{\theta}_1^*$ can be found from Eq. (10.566) as follows:

$$\dot{\phi}_1' = [(B_{12}F_{13} - B_{13}F_{12})\dot{\theta}_1 + B_{13}E_{12}\dot{\theta}_2 - B_{12}E_{13}\dot{\theta}_3]/D_0 \tag{10.568}$$

$$\dot{\theta}_1^* = [(A_{12}F_{13} - A_{13}F_{12})\dot{\theta}_1 + A_{13}E_{12}\dot{\theta}_2 - A_{12}E_{13}\dot{\theta}_3]/(D_0 c\phi_1') \tag{10.569}$$

With the availability of $\dot{\phi}_1'$ and $\dot{\theta}_1^*$, Eqs. (10.557) and (10.560) give $\dot{\phi}_2'$ and $\dot{\phi}_3'$ directly but Eqs. (10.558) and (10.561) give $\dot{\theta}_2^*$ and $\dot{\theta}_3^*$ if $c\phi_2' \neq 0$ and $c\phi_3' \neq 0$. That is,

$$\dot{\phi}_2' = A_{22}\dot{\phi}_1' - (B_{22}c\phi_1')\dot{\theta}_1^* + E_{22}\dot{\theta}_2 - F_{22}\dot{\theta}_1 \tag{10.570}$$

$$\dot{\phi}_3' = A_{23}\dot{\phi}_1' - (B_{23}c\phi_1')\dot{\theta}_1^* + E_{23}\dot{\theta}_3 - F_{23}\dot{\theta}_1 \tag{10.571}$$

$$\dot{\theta}_2^* = [(B_{32}c\phi_1')\dot{\theta}_1^* - A_{32}\dot{\phi}_1' + F_{32}\dot{\theta}_1 - E_{32}\dot{\theta}_2]/c\phi_2' \tag{10.572}$$

$$\dot{\theta}_3^* = [(B_{33}c\phi_1')\dot{\theta}_1^* - A_{33}\dot{\phi}_1' + F_{33}\dot{\theta}_1 - E_{33}\dot{\theta}_3]/c\phi_3' \tag{10.573}$$

In the second stage of the forward velocity solution, $\overline{v}$ can be obtained going through one of the three legs. For this manipulator, the first leg with $\beta_1 = 0$ is preferable for the sake of simplicity. Thus, the velocity of the tip point is determined as follows by using Eq. (10.540)

$$\overline{v} = -b_1\dot{\theta}_1 e^{\widetilde{u}_2\theta_1}\overline{u}_3 + b_2 e^{\widetilde{u}_2\theta_1^*} e^{\widetilde{u}_3\phi_1'}(\overline{u}_2\dot{\phi}_1' - \overline{u}_3\dot{\theta}_1^* c\phi_1') \tag{10.574}$$

(b) Motion Singularities of Forward Kinematics

The forward velocity solution presented in Part (a) indicates that MSFKs occur if the manipulator assumes a pose such that $D_0 = A_{13}B_{12} - A_{12}B_{13} = 0$ and/or $c\phi_k' = 0$ for one or more values of the index k.

The first kind of MSFK, i.e. MSFK-1, occurs if $D_0 = 0$ according to Eq. (10.568). In other words, MSFK-1 occurs if $\theta_1^* = \theta_2^* = \theta_3^* = 0$ or $\theta_1^* = \theta_2^* = \theta_3^* = \pi$, because Eq. (10.563) implies that B_{1k} vanishes in such a case for $k = 2, 3$. This is verified as shown below.

Case 1 with $\theta_1^* = \theta_2^* = \theta_3^* = 0$,

$$B_{1k} = \overline{u}_1^t e^{-\widetilde{u}_3\phi_k'}\widehat{I} e^{-\widetilde{u}_3\beta_k}\widehat{I}\overline{u}_3 = \overline{u}_1^t e^{-\widetilde{u}_3(\phi_k'+\beta_k)}\overline{u}_3 = \overline{u}_1^t\overline{u}_3 = 0 \tag{10.575}$$

Case 2 with $\theta_1^* = \theta_2^* = \theta_3^* = \pi$,

$$B_{1k} = \overline{u}_1^t e^{-\widetilde{u}_3\phi_k'} e^{-\widetilde{u}_2\pi} e^{-\widetilde{u}_3\beta_k} e^{\widetilde{u}_2\pi}\overline{u}_3 = \overline{u}_1^t e^{\widetilde{u}_3(\beta_k-\phi_k')}\overline{u}_3 = \overline{u}_1^t\overline{u}_3 = 0 \tag{10.576}$$

However, while Case 2 is quite likely to occur, Case 1 is not likely to occur due to the clashing possibility of the relevant links.

When MSFK-1 occurs with $\theta_1^* = \theta_2^* = \theta_3^* = \pi$, Eqs. (10.568)–(10.573) show that the rates of all the passive joint variables become indefinite and consequently the end-effector gets out of control. Therefore, it is necessary to avoid this singularity. The pose of the manipulator in this singularity is illustrated in Figure 10.24 by showing only the first and third legs (again as if $\beta_3 = -\pi$).

Incidentally, when Figures (10.24) and (10.21) are compared, it is seen that MSFK-1 also happens to be the PMCP (posture mode changing pose) of the manipulator.

On the other hand, Eqs. (10.572), (10.573), and (10.569) imply that the manipulator may get into a pose of MSFK-2 if $c\phi_k' = 0$, i.e. if $\phi_k' = \pm\pi/2$ for one or more values of the index k. In any of these singularities, which looks the same as the MSIK-2 and PSIK-2

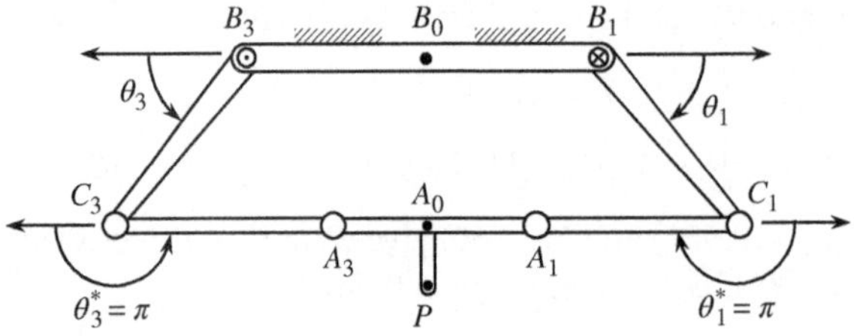

Figure 10.24 MSFK-1 of the manipulator illustrated by means of L_1 and L_3.

of L_k, the lower links $C_k'A_k'$ and $C_k''A_k''$ get aligned with themselves and the chord $C_k'C_k''$ of L_k. Therefore, the rates $\dot{\phi}_k'$ and $\dot{\theta}_k^*$ become indefinite. However, in a normal usage, the delta robot is never allowed to assume such a pose in order to prevent the links of L_k from clashing with each other.

Bibliography

There are a large number of publications about the kinematics of spatial mechanical systems. Therefore, this two-part bibliography list is necessarily limited in content. The first part of the list contains the basic publications that have so far inspired the author in many ways. The second part of the list contains the relevant publications of the author, which constitute the basis of this book.

Basic Publications

Angeles, J. (1982). *Spatial Kinematic Chains*. Springer-Verlag.

Angeles, J. (2014). *Fundamentals of Robotic Mechanical Systems*. Springer-Verlag.

Bahar, L.Y. (1970). Direct determination of finite rotation operators. *Journal of The Franklin Institute* 298 (5): 401–404.

Bottema, O. and Roth, B. (1979). *Theoretical Kinematics*. Amsterdam, Netherlands: North Holland.

Brockett, R.W. (1984). Robotic manipulators and the product of exponential formula. In: *Proceedings of the International Symposium on Mathematical Theory of Networks and Systems*, 120–127. Berlin, Germany.

Coriolis, G.G. (1835). Mémoire sur les équations du mouvement relatif des systèmes de corps. *Journal de l'école Polytechnique* 15: 142–154.

Craig, J.J. (1986). *Introduction to Robotics, Mechanics and Controls*. Reading, MA, USA: Addison-Wesley.

Denavit, J. and Hartenberg, R.S. (1955). A kinematic notation for lower-pair mechanisms based on matrices. *Journal of Applied Mechanics*: 215–221.

Euler, L. (1776). Formulae generales pro translatione quacunque corporum rigidorum. *Novi Commentarii academiae scientiarum Petropolitanae* 20: 189–207.

Hervé, J.M. (1999). The lie group of rigid body displacements, a fundamental tool for mechanism design. *Mechanism and Machine Theory* 34: 719–730.

Maxwell, E.A. (1951). *General Homogeneous Coordinates in Space of Three Dimensions*. Cambridge, UK: Cambridge University Press.

McCarthy, J.M. (1990). *Introduction to Theoretical Kinematics*. Cambridge, MA, USA: MIT Press.

Merlet, J.-P. (2006). *Parallel Robots*, 2e. Dordrecht, Netherlands: Springer.

Murray, R.N., Li, Z., and Sastry, S.S. (1994). *A Mathematical Introduction to Robotic Manipulation*. London, UK, Ch. 2: CRC Press.

Kinematics of General Spatial Mechanical Systems, First Edition. M. Kemal Ozgoren.

Companion Website: www.wiley.com/go/ozgoren/spatialmechanicalsystems

Paul, R.P. (1981). *Robot Manipulators, Mathematics, Programming and Control.* Cambridge, MA, USA: MIT Press.
Rodrigues, O. (1840). Des lois géometriques qui regissent les déplacements d'un systeme solide dans l'espace, et de la variation des coordonnées provenant de ces déplacement considerées indépendant des cause qui peuvent les produire. *Journal de Mathématiques Pures et Appliquées* 5: 380–440.
Tsai, L.W. (1999). *Robot Analysis: The Mechanics of Serial and Parallel Manipulators.* Wiley.

Related Publications of the Author

Ozgoren, M.K. (1994). Some remarks on rotation sequences and associated angular velocities. *Journal of Mechanism and Machine Theory* 29 (7): 933–940.
Ozgoren, M.K. (1999). Kinematic analysis of a manipulator with its position and velocity related singular configurations. *Journal of Mechanism and Machine Theory* 34: 1075–1101.
Ozgoren, M.K. (2002). Topological analysis of six-joint serial manipulators and their inverse kinematic solutions. *Journal of Mechanism and Machine Theory* 37: 511–547.
Ozgoren, M.K. (2007). Kinematic analysis of spatial mechanical systems using exponential rotation matrices. *Journal of Mechanical Design, ASME* 129: 1144–1152.
Ozgoren, M.K. (2013). Optimal inverse kinematic solutions for redundant manipulators by using analytical methods to minimize position and velocity measures. *Journal of Mechanisms and Robotics, ASME* 5: 031009:1–031009:16.
Ozgoren, M.K. (2014). Kinematic analysis of spatial mechanical systems using a new systematic formulation method for lower and higher kinematic pairs. *Journal of Mechanisms and Robotics, ASME* 6: 041003:1–041003:17.
Ozgoren, M.K. (2019). Kinematic and kinetostatic analysis of parallel manipulators with emphasis on position, motion, and actuation singularities. *Robotica* 37 (4): 599–625.

Index

Kinematics of General Spatial Mechanical Systems, First Edition. M. Kemal Ozgoren.

Companion Website: www.wiley.com/go/ozgoren/spatialmechanicalsystems

d

e

f

g

h

i

r